Peter Jung

Advanced Mobile Communications

Also of Interest

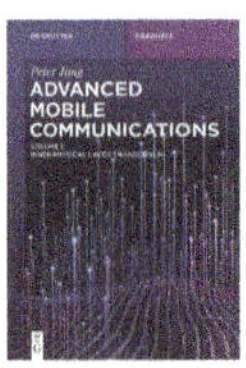

Advanced Mobile Communications
Volume 1: Inner Physical Layer Transceiver
Peter Jung, 2024
ISBN 978-3-11-123909-5, e-ISBN (PDF) 978-3-11-123967-5

5G
The 5th Generation Mobile Networks
Ulrich Trick, 2023
ISBN 978-3-11-118648-1, e-ISBN (PDF) 978-3-11-118661-0

Multiple Access Technologies for 5G
New Approaches and Insight
Edited by: Jie Zeng, Xin Su, Bin Ren and Lin Liang, 2021
ISBN 978-3-11-066581-9, e-ISBN (PDF) 978-3-11-066636-6

Artificial Intelligence for Signal Processing and Wireless Communication
Edited by: Abhinav Sharma, Arpit Jain, Ashwini Kumar Arya and Mangey Ram, 2022
ISBN 978-3-11-073882-7, e-ISBN (PDF) 978-3-11-073465-2

Cloud Security
Techniques and Applications
Edited by: Sirisha Potluri, Katta Subba Rao and Sachi Nandan Mohanty, 2021
ISBN 978-3-11-073750-9, e-ISBN (PDF) 978-3-11-073257-3

Peter Jung

Advanced Mobile Communications

Volume 2: Sophisticated Channel Codes

DE GRUYTER

Author
Prof. Dr.-Ing. habil. Peter Jung
Universität Duisburg-Essen
Fak. IW / Abt. EIT / Kommunikationstechnik
Oststr. 99
47057 Duisburg
Germany

Peter Jung und Guido Bruck
Institut für Kommunikationstechnik (IKT) GbR
Poststraße 87
46562 Voerde
Germany
peter.jung@kommunikationstechnik.org

ISBN 978-3-11-123915-6
e-ISBN (PDF) 978-3-11-123965-1
e-ISBN (EPUB) 978-3-11-123991-0
Set-ISBN 978-3-11-138523-5

Library of Congress Control Number: 2024930004

Bibliographic information published by the Deutsche Nationalbibliothek
The Deutsche Nationalbibliothek lists this publication in the Deutsche Nationalbibliografie; detailed bibliographic data are available on the Internet at http://dnb.dnb.de.

Cover image: Nobi_Prizue / iStock / Getty Images Plus
Typesetting: VTeX UAB, Lithuania
Printing and binding: CPI books GmbH, Leck

www.degruyter.com

Sine Anne et Christa non est vita

"Philosophy (*i. e., natural philosophy*) is written in this grand book—I mean the Universe —which stands continually open to our gaze, but it cannot be understood unless one first learns to comprehend the language and interpret the characters, in which it is written.

It is written in the language of mathematics, and its characters are triangles, circles and other geometrical figures, without which it is humanly impossible to understand a single word of it; without these, one is wandering around in a dark labyrinth."

Galileo Galilei: Il Saggiatore (The Assayer), October 1623.

Preface

"*Take Me Home, Country Roads*" (John Denver, 1971) to this present textbook

> *"Advanced Mobile Communications — Sophisticated Channel Codes,"*

which is the continuation of the first— "*Surprise, Surprise!*" (Mezzoforte, 1982)—volume entitled

> *"Advanced Mobile Communications—Inner Physical Layer Transceiver."*

This second volume is about some basic notions of linear block codes and convolutional codes, discussing fundamentals of finite fields and number theory and polar codes, which extend the well-known Reed–Muller (RM) codes, and low density parity check (LDPC) codes and convolutional turbo codes, which inherit the structure of regular convolutional codes and "that's it, that's it, that's it, baby, what's your mama gonna say when she finds that you party like this girl, sorry, I mean reader"? (Al Jarreau in "*Roof Garden*," 1981, in my opinion the best of all times jazz singer, and George Duke's playing of the Dynomy Rhodes electric piano is breathtakingly fantastic in this song). All the mentioned aspects and, especially, codes have been included in advanced mobile communications systems, and hence deserve a concise discussion.

In what follows and until the end of this preface, let me quote some parts of the preface of the first volume [1] and add some additional remarks.

The field of mobile communications has developed at an increasingly fast pace over the past decades with new and advanced topics arising in various fields, especially regarding the transmission of physical signals, i. e., on the physical layer. This speeding development requires understanding of the foundations, on the one hand, as well as elaborating on new aspects of the fifth, the sixth and following generations of mobile communications, on the other hand, paving the way forward by *"Rockin' Down The Highway" (The Doobie Brothers, 1972)* of mobile communications' history sketched in the cartoon shown in Figure 1.

Like the first volume [1], the present textbook has emerged from the topics that the author has taught in advanced courses at the University of Duisburg-Essen as the Chaired Professor of Communications Technologies. It has been created since 2010, culminating in the author's endeavors to complete the manuscript over the past 3 years. Who would have believed this if I hadn't told it? The author intends this textbook to be an accompanying piece of information for those who are willing to elaborate on the said topics.

In the German edition of his book entitled "A Brief History of Time," Stephen Hawking wrote

> "I've been told that every equation in the book cuts sales in half."
> (in German: "Man hat mir gesagt, dass jede Gleichung im Buch die Verkaufszahlen halbiert.")

https://doi.org/10.1515/9783111239651-201

Figure 1: The author's view on *"Rockin' Down The Highway"* (The Doobie Brothers, 1972) of mobile communications' history.

[2, p. 7]. Well, this textbook mainly consists of equations. Still, I hope that there will be readers who find it useful for their professional work.

Finally, since they are so important to me, let me repeat the two personal remarks that I already included in the preface of the first volume. We all live in a crazy time when plagiarism is sought in all places. Although not every author intentionally steals the intellectual property of others, too many, sometimes perhaps somewhat naive people have had to suffer the most unpleasant consequences of this strange "hunt." I believe I have found a viable defense for myself, in that I include citations in every sentence and in every figure, in which I could discover the thoughts of other people. The good thing about this is that at the same time I am properly honoring the creators of other publications.

I quote those publications, which I have personally used and consulted while writing this textbook, and I am as precise as possible by noting page numbers and even the numbers of equations, tables or figures of the quoted publications in order to enable the reader to quickly identify earlier results. Strangely enough, I have already been severely criticized by anonymous reviewers for this way of quoting. Nevertheless, I do not let them talk me out of it. I would rather bear the criticism of an anonymous reviewer than the public accusations that I would be a "copyist." Perhaps I can even be a role model for future authors.

By the way, I am not claiming to have a complete bibliography, though. In particular, I do not mention or even quote such publications, which I personally evaluate as erroneous or inferior. Since I do not want to hurt the feelings of other authors, I refrain from criticizing these works explicitly. However, I do apologize to the reader already now for possibly ignoring texts, which he or she finds important.

And here comes the second remark. The preface of the autobiography of the Nobel laureate of 1932, Werner *Heisenberg*, entitled "Der Teil und das Ganze" (in English "the part and the whole"), which is obviously derived from the Latin "pars pro toto," begins with the statement [3, p. 7]

> "Wissenschaft wird von Menschen gemacht. Dieser an sich selbstverständliche Sachverhalt gerät leicht in Vergessenheit, [...]"
> (in English: "Science is made by people. This fact, which is self-evident, is easily forgotten, [...]")

When writing this textbook, I had "nice little associations" that human beings tend to have at several occasions, and in some cases I wrote them down. Most of the time these associations are related to songs that all of a sudden appeared in my mind. Usually, only the title of the cited song is of concern to me and reasons the citation. I do not want to create the impression that I would endorse the lyrics or the stories told by the songs, either in part or as a whole. Nevertheless, I hope that these tiny distractions will show to the reader that science must be fun, and without fun there is no science. Remember the German proverb

> "wo man singt, da lass dich ruhig nieder, böse Menschen haben keine Lieder,"

which means as much as

> "settle where they love to sing, singing is no villains' thing."

The reason for my personal selection is given in what follows.

I have been especially fascinated by music of every kind during my whole life. Especially, all kinds of rock'n'roll, jazz, gospel, soul and funk of the 1960s until today, but not the German "Schlager" (sorry, folks, this is just not my taste), have sounded like a dream come true to me.

As if it was today, I remember two occasions most remarkable at least to me. The first one occurred in the morning of a nice day in August 1970. My parents and I were just getting ready for the day, and while preparing, were listening to the morning show on the radio when the 1967 recording of "*A Whiter Shade Of Pale*" by Procul Harum was on. Then I did not know much about Bach and his great piece of music called "*Air*" of his Suite No. 3 in D major but I did hear these wonderful harmonies and this mind blowing sound. Asking my mom what it was, she replied that it was a Hammond organ. I was immediately convinced that I had to learn how to play this instrument. Many years later, I found out that it had been Matthew Fisher playing the Hammond C3 organ on Procul Harum's record. I am glad to say that my parents bought me an organ and allowed me to take lessons.

The second occasion happened in 1972. I remember sitting before the TV set and watching the Temptations perform their smash hit "*Papa Was A Rolling Stone.*" All five musicians were wearing pink suits, the lead singer standing on the right side at the front of the stage, singing into a microphone on a microphone stand and the other four musicians were dancing to the music behind the lead singer. Although I could not understand the lyrics then because I had not yet been taught any English at the age of 8, I was again assured that music makes the difference!

I cannot help it—I still get emotional when thinking of these occasions. Now, you, the reader, may know why some references to well-known pieces of music are scattered around in this textbook. Maybe you might even think that I could be a human, too, and that I might be a scientist, who knows? Anyway, these notions I will not deny.

Vince Ebert, a German physicist and comedian, was recently featured in a local newspaper including a photo of his waving a small paper flag with a slogan printed on it, namely

MAKE SCIENCE GREAT AGAIN,

"*Please, Please, Please*!" (The "Godfather of Soul," James Brown, 1956 — he played this signature song live during his marvelous performance at the "Teenage Awards Music International (TAMI)" show in the most important year of mankind's history, 1964).

One almost last thing that I would like to mention is the following. This monograph may contain mistakes. If you find one, you may keep it, of course. However, I would be glad if you let me know what went wrong in your opinion.

And another almost last thing, I cannot resist to mention is that being a Palatine born in Kaiserslautern and raised in the County of Kaiserslautern, I must proudly second the statement made by my late fellow Palatine, the renowned poet Eugen Damm, who also came "out of Kaiserslautern," that it was the Lord himself who gave us Palatines the everlasting task to lubricate the World Axis. Have you ever heard it squeak? See, how well we Palatines do our job. Now, it is my sincere hope that you all will also be excelling, in this case as communications engineers, after having worked your way through my two volumes.

So, dear reader, let us get the "*Garden Party*" (Mezzoforte, 1982, the author's personal heroes in contemporary jazz and fusion) started. I know that "*We're In This Love Together*" (Al Jarreau, 1981), this everlasting love for science. So, "people all over the world, start a *Love Train*" for science and Nature, "get all on board and keep riding on through. If you miss it, I feel sorry, sorry for you" (O'Jays, 1972)! Let us make science great again, "*All Together Now*" (The Beatles, 1969)!

Duisburg, 15 March 2024

Acknowledgment

I hereby gratefully acknowledge the help of all those people who supported me while I had been working on this textbook. I will not mention many to make sure that most of them can stay incognito and ask for forgiveness by all those who are not named but would have liked to read their names here. Some shall, however, be mentioned explicitly.

First of all, I would like to thank Ms. Anne Jung for her great support during the proofreading process. Also, I would like to thank Dr. Guido Horst Bruck, Dr. Lukas Vincent Grinewitschus, Faris Abdel Rehim, Hamza Almujahed and Kushtrim Dini for countless discussions on the topics and scientific details and for helping during the proofreading process. In addition, I would like to thank Faris Abdel Rehim and Kushtrim Dini for translating my initial lectures manuscript written in Word into LaTeX, so I had a good starting point for writing the two volumes on advanced mobile communications.

Last but not least, I would like to thank all at de Gruyter, especially, Dr. Damiano Sacco, Jessika Kischke and Elena Leuze, for giving me the opportunity to publish this textbook as a title offered by this fine publishing house.

https://doi.org/10.1515/9783111239651-202

Notation

In this textbook, the following notation are used. Matrices are denoted by uppercase in bold italics, e. g., $\boldsymbol{M}$ and vectors are lowercase in bold italics, e. g., $\boldsymbol{v}$. The notation $\boldsymbol{M}^{\mathrm{T}}$ and $\boldsymbol{v}^{\mathrm{T}}$ refer to the transposes of the matrix $\boldsymbol{M}$ and of the vector $\boldsymbol{v}$ while $\boldsymbol{M}^{\mathrm{H}}$ and $\boldsymbol{v}^{\mathrm{H}}$ mean the Hermitian transposes of the matrix $\boldsymbol{M}$ and of the vector $\boldsymbol{v}$, i. e., the complex-conjugate transposes $\boldsymbol{M}^{*\mathrm{T}}$ and $\boldsymbol{v}^{*\mathrm{T}}$. The expression $[\cdot]^*$ denotes the complex conjugation. Complex values are underlined, e. g., $\underline{\boldsymbol{M}}$, $\underline{\boldsymbol{v}}$, $\underline{x}$. The imaginary unit is denoted by

$$\mathrm{j} = \sqrt{-1}.$$

In addition, $\otimes$ is used to indicate Kronecker products such as $\boldsymbol{K} \otimes \boldsymbol{M}$ and Kronecker powers such as $\boldsymbol{K}^{\otimes M}$.

The empty set will be denoted by

$$\emptyset = \{\}.$$

The set of nonnegative integers is denoted by

$$\mathbb{N} = \{0, 1, 2, 3 \cdots\}$$

and

$$\mathbb{N}^* = \{1, 2, 3 \cdots\}$$

is the set of positive integers. The set of all integers is denoted by $\mathbb{Z}$. The set of the real numbers is $\mathbb{R}$ and $\mathbb{R}_0^+$ is the set of nonnegative real numbers. The set of positive real numbers is denoted by $\mathbb{R}^+$. $\mathbb{R}^-$ denotes the set of negative real numbers. The set of the complex numbers is $\mathbb{C}$. Let $a, b \in \mathbb{R}$ with $a \leq b$. Then closed intervals are denoted by $[a, b]$, indicating that both end points a, b belong to the closed interval. Half-open intervals are denoted by $[a, b[$, indicating that only a belongs to the half-open interval but not b, or by $]a, b]$, indicating that only b belongs to the half-open interval but not a. Open intervals are denoted by $]a, b[$, indicating that neither of the end points a, b belongs to the open interval. Let

$$\mathbb{C}^N = \underbrace{\mathbb{C} \times \mathbb{C} \times \cdots \times \mathbb{C}}_{N \text{ terms}}, \quad N \in \mathbb{N}^*, \tag{1}$$

be the N-fold Cartesian product of $\mathbb{C}$ and let

$$\mathbb{R}^N = \underbrace{\mathbb{R} \times \mathbb{R} \times \cdots \times \mathbb{R}}_{N \text{ terms}}, \quad N \in \mathbb{N}^*, \tag{2}$$

be the N-fold Cartesian product of $\mathbb{R}$.

https://doi.org/10.1515/9783111239651-203

Convolution is denoted by $*$. $\mathrm{Re}\{\cdot\}$ denotes the real part of $\cdot$ and $\mathrm{Im}\{\cdot\}$ denotes the imaginary part of $\cdot$.

Channel coding for advanced mobile communications is based on message symbols and code symbols both taken from $\mathbb{F}_2$ equal to $\{0, 1\}$. $\mathbb{F}_2$ is termed the *Galois field* with the two elements 0 and 1. $\mathbb{F}_2$ is a finite field. In order to distinguish linear operations in $\mathbb{F}_2$ from the "regular" addition "+" and the "regular" multiplication "$\cdot$," the additive operation on any two elements of $\mathbb{F}_2$ is denoted by "$\oplus$" and the multiplicative operation on any two elements of $\mathbb{F}_2$ is denoted by "$\odot$." The same notation "$\oplus$" and "$\odot$" are used to indicate operations in any finite field, e. g., in $\mathbb{F}_{2^n}$, $n \in \{2, 3 \cdots\}$. Furthermore, $\mathbb{F}_2[x]$ is the set of polynomials over $\mathbb{F}_2$ and $\mathbb{F}_2[x]_n$ is the set of polynomials over $\mathbb{F}_2$ with a degree not greater than n.

Of Theorems, Lemmata and Corollaries

In this textbook, we will use theorems, lemmata and corollaries in appropriate places. It is therefore imperative to know what these expressions mean. We will give the basic concepts of these in a nutshell without reference as these are considered common knowledge.

A *theorem* is a not self-evident proposition or statement, which has been proven to be true.

A *lemma* is a minor proven proposition or statement, which is usually used as a "stepping stone" toward a larger result, usually a theorem. Lemmata are therefore called "helping theorems" or "auxiliary theorems."

A *corollary* is a proposition or statement, which can be readily deduced from a previous, more notable statement, usually a theorem or a lemma. A corollary can thus be considered as a more or less trivial result.

https://doi.org/10.1515/9783111239651-204

Contents

1 Introduction — Sophisticated Channel Codes for Advanced Mobile Communications

1.1 Oh my! Oh my! Yet Another Book About… Channel Coding

I have never thought that I would ever write a textbook about channel codes. But, as we say in German,

> "unverhofft kommt oft,"

which means as much as "unexpected often comes." So, "*Here We Go Again*" (The Weeknd, 2022), another volume on *channel codes*. Why does the world need this?

First of all, why not? Just remember the great Bavarian comedian Karl *Valentin*, who once realized that "*everything has already been said, just not yet by everyone.*" Of course, he said that in German with a Bavarian touch in his pronunciation. Although this saying includes quite a bit of sense of humor, not necessarily mine but Valentin's, of course, it sounds a bit too sarcastic—doesn't it? Indeed, as far as I am concerned, the true reason is my personal dissatisfaction with the available collection of textbooks about the topic.

In my opinion, it is important to understand that channel coding for advanced mobile communications deserves a look through the "eyes" of advanced mobile communications. Having said this, I have to admit that, of course, the reader will find loads of textbooks that purport to "sail the mobile radio wave" but in the end they are just conventional textbook in disguise. One clear point that is usually not made and followed distinctly is the fact that in advanced mobile communications, only finite length messages, i. e., blocks of message symbols, are transmitted, all considered channel codes are essentially block codes. So, you need to look at, e. g., convolutional codes as being a block code. It is just futile to present theories, which rely on infinite length messages. Period.

But let us look at positive aspects, too. Admittedly, there is a wide variety of textbooks on channel coding. And yes, there are certainly at least some, which are really great. I personally get quite enthusiastic about the texts by Florence Jessie *MacWilliams* and Neil James Alexander *Sloane* [4] as well as by Bernd *Friedrichs* [5].

Nevertheless, in my opinion, although noteworthy textbooks on channel coding are often good in number theory, alas, foundations and reasoning, laid out by information theory [6, 7], are not made clear enough, and hence not put in an understandable and illustrative relation to the why and how of channel coding. To put it in my personal nutshell, usually I am missing clear and easily comprehensible statements about the true and deep reasoning, i. e., I just cannot get satisfactory answers to "why is all that jazz necessary and how does it relate with each other and the rest of the, or rather my, world"?

Moreover, most good textbooks do not cover recent advances.

https://doi.org/10.1515/9783111239651-001

And I am missing precise references to the early publications and, preferably, the origins, which would enable the reader to develop his or her own solid evaluation of the matter.

Remark 1.1 (On the Importance of References). Allow me to give some more details about my complaint with respect to the lack of precise citations. I will refer to the most drastic example that I personally know although it does not lie in the field of channel coding. It concerns the theory of special relativity. As you certainly know, the theory of special relativity is often attributed to the German physicist and Nobel Laureate Albert *Einstein*, although there are certainly more contributors to this theory [8, p. 15f.].

A certainly most remarkable textbook on the theory of special relativity was published in 1919 by the German physicist and Nobel Laureate Max *von Laue*; cf., e. g., [8, p. 226]. Although this textbook is quite clear, a "newbie" in the field, who has only limited command of the German language, might find it hard to understand and translate. Due to its importance, von Laue's textbook served as a starting point for many English essays, some of which contain at least partial translations of the German original. Regrettably, many English authors made translation errors well worth seeing, that caused scientific "nonsense" [8, p. 227].

In information theory speak, considering the translation from German to English a transmission channel, the "average source rate," i. e., the source entropy $\mathrm{H}\{X\}$ was obviously considerably larger than the maximum average mutual information $\max I\{X;Y\}$, i. e., the channel capacity, however, without having a code, which could overcome the equivocation $\mathrm{H}\{X \mid Y\}$.

The said nonsense confused many, especially in the USA, but also elsewhere, for years if not decades [8, p. 227]. For instance, there are translations of the faulty English textbooks back into German, of course, keeping, if not even augmenting, the said nonsense.

Again, using information theory speak, there are cascaded transmission channels each causing a distinct equivocation, and hence suffering from "average source rates" at their specific inputs, which are mismatched to their specific channel capacities at their specific inputs. So, we are all struck by the *"chain rule of information."* A quite entertaining illustration of what cascaded error prone transmission channels can do to us can, e. g., be found at the beginning of the second story "Isnoguds Schüler" (in English "Isnogud's student") of [9], created by the great French cartoonists René *Goscinny* and Jean *Tabary*.

I am sure that this unsatisfactory, probably even detrimental situation could have been avoided by using an "appropriate channel code," in my opinion represented by precise citations.

So, what can we learn from this anecdote? There are at least three lessons to be learned.

- Avoid textbooks, which do not include clear and precise citations including, e. g., page references.
- Do not just believe, what the author tries to tell you.
- Become a critical scientist, check the available references and make up your mind.

Oh, by the way, if you would like to read what the "master mind" Albert Einstein said about the theory of relativity, you may, e. g., want to check the merely 112 pages long monograph [10].

So, again, why not... try to improve the situation?

Being a physicist, I may have a different point of view than someone who has been an engineer for all his professional life and, therefore, I might consider things under a different aspect ratio. Maybe, some of my views might be enlightening.

Figure 1.1 illustrates the answer to the above question by showing some technical details. Of course, these technical details deserve decent definitions, which will be given in the course of this textbook. For the moment, let us just pretend that we knew these definitions and try visualize the relationships among the said technical details, ok?

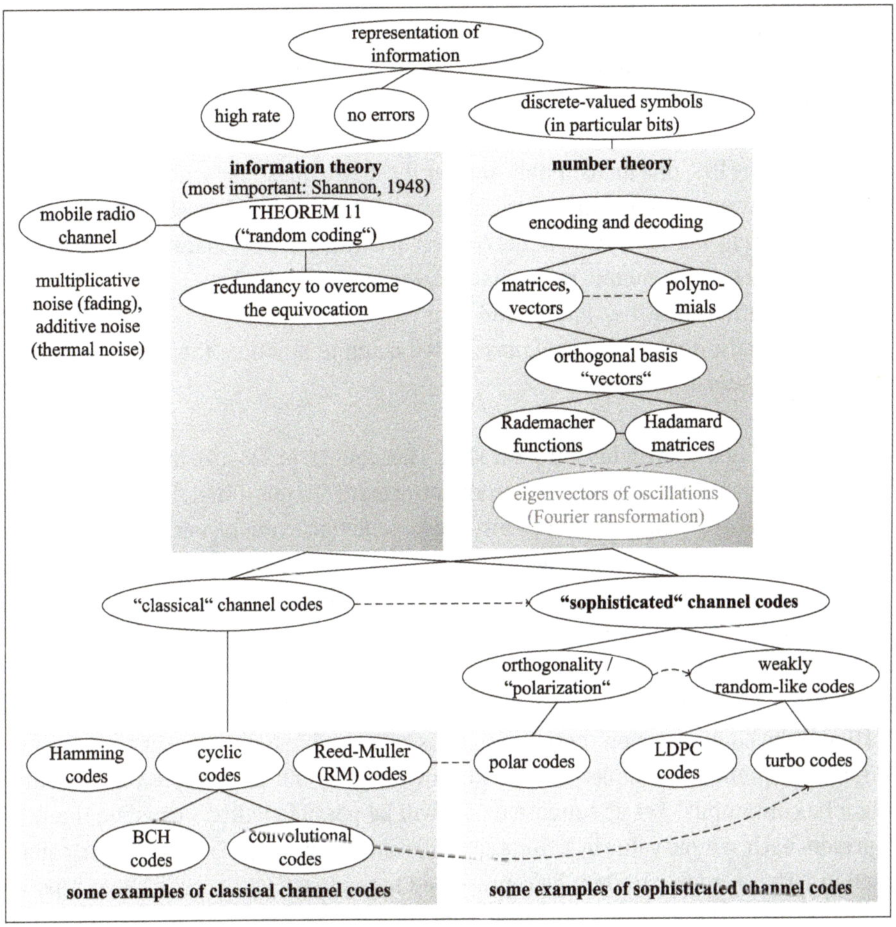

Figure 1.1: The answer to “why does the world need this textbook”?

Also, please understand that for the illustration purpose, I will only give a very limited number of references in the following few paragraphs of this Section 1.1. I promise that detailed citations for further reading will be found in all following sections and chapters.

Channel *coding* is about the transmission of information, which requires an appropriate *representation*. Coding provides this representation. When we think about transmission, we immediately think of an immediate, i. e., an instantaneous, reception, which requires a high *rate*. In this context, rate means *transmission rate* or transmission speed, which is the amount of information, transferred between a transmitter and a receiver in a unit time interval, e. g., one second. Of course, we require the transmission of information to be free of any errors. That’s something!

Therefore, at least two questions need to be answered:

- Is this possible at all?
- If yes, is there a maximum transmission rate for a transmission that is free of any errors?

It seems that the first one to justifiably answer these two questions with a definite "yes," was Claude Elwood *Shannon* in his seminal paper [6], published in 1948, which incidentally founded *information theory*. In the context of a transmission channel, which suffers from noise, such as the mobile radio channel, incorporating both
- multiplicative noise, i. e., *fading*, and
- additive noise such as *thermal noise*, if we deign to attribute thermal noise to the channel, as well,

Shannon provided us with his *Theorem 11* [6, Theorem 11, pp. 22–24], that has also been named the "*random coding theorem*" in the aftermath. To put it in a nutshell, [6, Theorem 11, pp. 22–24] shows that there exist codes, which provide us with redundancy to overcome the equivocation imposed by the channel.

The representation of information could be done in a continuous-time manner by continuous-time functions, usually called signals. However, mathematics teaches us that signals can be "approximated" by a sequence of samples, provided that these signals are bandlimited. Details on this concept known as the "*sampling theorem*" can, e. g., be found in [1] and the citations given therein. So, let us assume that a discrete-time representation by a sequence of samples is at hand. Now, what about the sample values, which might be continuous? Let us suppose that it will be possible with a low error, if any, to represent each sample value by a finite set of distinct levels, which is often called "quantization." These levels can then be represented by a binary vector, which is composed of bits. All in all, let us assume that the information is represented by discrete-valued symbols, and thus by a discrete-time and discrete-valued signal, which is made up of binary symbols, i. e., bits.

This discrete-time and discrete-valued signal is then encoded by the chosen channel code. The encoded signal is then transmitted via the transmission channel and received by the receiver. The receiver has the task to decode the received signal and to form an "estimate," i. e., a detected version of the discrete-time and discrete-valued signal. So, we need some means to mathematically describe the encoding and the decoding. Consequently, we need a mathematical means to represent the code in the first place.

And there we are amidst *number theory*.

To make life easy and convenient, let us assume that the encoding is based on a linear algebra. So, we will only focus on what is usually called *linear codes*. In this case, the encoding and the decoding as well as the code itself can be represented by matrices and vectors. And since there is a one to one correspondence between vectors and polynomials, we may use polynomials instead for the establishment of channel codes, the encoding and the decoding.

Wow! So, would you believe it, we just entered the wide universe of *vector spaces*. Sounds a bit like the mission to explore codes, even those that no one has seen before. Can you hear the Star Trek theme playing in your head, also? Oh, I'm loving it...

Still information hungry?

So, let us continue "*Star Trekkin'* across the Universe" (The Firm, 1988). In vector spaces of the kind that we want to consider, there are orthogonal, if not orthonormal, bases. Since we decided to use binary symbols, i. e., bits, only above already—didn't we?—the basis vectors can be found in, e. g., Hadamard matrices and Rademacher functions, both being closely related to the eigenvectors of oscillations, and hence of many physical systems. So, we cannot avoid to run into relatives of the Fourier transform and the Fourier series.

Needless to mention that eigenvectors represent eigenstates of physical systems and are hence *mutually orthogonal* if they belong to distinct eigenvalues [11, p. 288]. Otherwise they may be chosen mutually orthogonal, e. g., by exploiting the Gram–Schmidt orthogonalization scheme [11, p. 288]. This means that all eigenvectors are linearly independent, in the end [11, p. 288]. Expressing these facts a bit more sloppily, we could get carried away saying that the eigenstates are "*uncorrelated*," meaning that they are not related with each other. Illustratively speaking, each eigenvector adds maximum "novelty" because this "novelty" cannot be anticipated by just looking at all the other eigenvectors. Thus, if we used only eigenvectors as codewords, the corresponding channel code should be a member of the "channel codes premier league," at least regarding its error performance. Moreover, if we used codewords which are linear combinations of the said eigenvectors, we still should have a remarkably well-performing channel code. And would you believe it, we just found a possible design paradigm for, e. g., first-order Reed–Muller (RM) codes, which are closely related with simplex codes and polar codes.

Wow, that was quick, wasn't it?

There also exists a concept in number theory, which is about avoiding relationships between the elements of sets, namely, "*primes*" or, at least, "*irreducibles*." Sounds like a similar idea compared with orthogonality. Indeed, using prime or irreducible polynomials leads to impressively performing weakly random-like codes called turbo codes. And avoiding common ones between any two columns of the parity check matrix being sparse yields another kind of impressively performing weakly random-like codes called low density parity check (LDPC) codes.

Terrific! "*Life Is A Rollercoaster*" (Ronan Keating, 2000), isn't it?

Now, having identified the two supporting pillars of channel coding, namely information theory and number theory, we can move on to the channel codes. Those channel codes, which were proposed in the first about 40 years after 1948, are considered to be "*classical*" *channel codes*, whereas the recent developments since about 1990 shall belong to the "*sophisticated*" *channel codes*. Of course, the sophisticated channel codes rely on the classical channel codes.

Among the classical channel codes we find, e. g., the Hamming codes and cyclic codes such as the Bose–Chaudhuri–Hocquenghem (BCH) codes. Just for the records, yes, you are right, Hamming codes can also be cyclic, but they do not necessarily have to be. Also, we will show that convolutional codes for advanced mobile communications can be regarded as cyclic codes with diversity branches. In addition, the well-known Reed–Muller (RM) codes, which can, e. g., be based on Hadamard matrices and Rademacher functions, respectively, and follow the orthogonality principle, are classical channel codes.

Among the sophisticated channel codes, we find the polar codes, which have received a lot of attention in the past decade and which must be regarded as an evolution of the Reed–Muller (RM) codes, i. e., following the orthogonality principle. The polar codes exploit the chain rule of information, which allows to realize the "channel polarization." However, still the best performing codes are those, which best comply with Shannon's theorem 11 [6, Theorem 11, pp. 22–24], i. e., weakly random-like codes. The low density parity check (LDPC) codes and the turbo codes belong to this class of channel codes.

Let us now illustrate the structure of this textbook. Following this first Section 1.1 of Chapter 1, we will define what we mean by sophisticated channel codes in Section 1.2. The general concept of channel codes is discussed in Section 1.3. In order to provide insight in the mathematical structure of linear channel codes (see Section 1.5), Section 1.4 will illustrate the concept of vector spaces. Sections 1.6, 1.7 and 1.8 will give a brief illustration of error detection and decoding. A selected set of examples of linear channel codes is given in Section 1.9. Section 1.10 discusses six well-known strategies to construct new channel codes from old. Section 1.12 contains a primer on number theory, which is helpful to understand cyclic channel codes; see Sections 1.11 and 1.13. Section 1.14 contains a selected set of examples of cyclic channel codes. After illustrating permutation matrices and the Kronecker product in Sections 1.15 and 1.16, we will illustrate Reed–Muller (RM) codes in Section 1.17. Finally, Section 1.18 contains a brief discussion of convolutional codes.

Indeed, Chapter 1 is quite long, and the reader may think that the topic of the textbook has been missed. Allow me to disagree to this notion. As already pointed out above, this first chapter contains all the basic principles of channel codes. As we already know, these basic principles do not only form the foundation of the classical channel codes but also of the sophisticated channel codes. Consequently, the Chapters 2, 3 and 4, which will discuss the three sophisticated channel codes:

- convolutional turbo codes,
- low density parity check (LDPC) codes and
- polar codes

used in advanced mobile communications, can renounce on a repetition of the said basic principles and can thus be short.

Chapter 2 illustrates convolutional turbo codes with an introductory discussion contained in Section 2.1. Section 2.2 discusses the parallel concatenation of recursive systematic convolutional (RSC) codes, which forms the basis of convolutional turbo codes. Section 2.3 considers the decoding of convolutional turbo codes.

In Chapter 3, low density parity check (LDPC) codes are discussed, beginning with their definition in Section 3.1. Section 3.2 illustrates the decoding of low density parity check (LDPC) codes.

Chapter 4 is devoted to the polar codes, setting out from the channel polarization; cf. Sect. 4.1. The recursive definition of generator matrices is the subject of Section 4.2. The channel combining and the channel splitting are illustrated in Sections 4.3 and 4.4. Section 4.5 discusses the performance of polar codes.

1.2 Of Sophisticated Channel Codes

"Here We Go Again" (Demi Lovato, 2009), in the second volume on *advanced mobile communications*, a term that we already defined in detail in the first volume [1, p. 1]. The said first volume [1] concentrates on the *conversion of digital data into analog signals at the transmitters and back at the receivers*. This is the task accomplished by the *inner physical layer transceiver*.

Hence, in advanced mobile communications systems, the transmitters set out from binary digits, i. e., *bits*, which are grouped in *blocks* or *bursts*, of finite length, conveniently denoted by column vectors $\boldsymbol{b}$. Using a *modulation mapper* [12, Figure 5.3-1, p. 25; Figure 6.3-1, p. 107; pp. 189–193], the bits in the column vectors $\boldsymbol{b}$ are mapped onto *complex data symbols*, grouped in *complex data vectors* $\underline{\boldsymbol{d}}$, also considered to be column vectors. These complex data symbols in the complex data vectors $\underline{\boldsymbol{d}}$ are fed into a digital modulator, which forms the *continuous-time complex baseband signal* $\underline{s}(t)$; cf. the generic block diagram of the transmitter depicted in Figure 1.2.

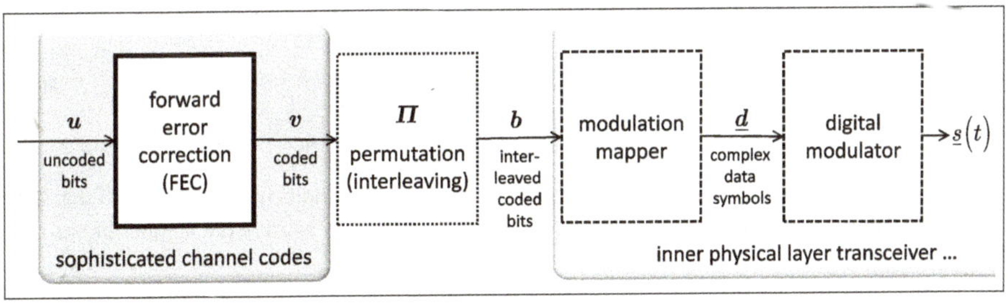

Figure 1.2: Generic block diagram of the physical layer transmitter of an advanced mobile communications system, which is an extended version of, e. g., [1, Figure 3.47, p. 422].

In the present second volume, we will stay solely in the digital domain focusing on *sophisticated channel codes*. What sort of thing is this? Sounds like work is coming up! But let us not be *"Lazy"* (Deep Purple, 1970; there is a terrific cover version of this song by Jimmy Barnes, Joe Bonamassa and others) and shed some light into the *"Black Night"* (Deep Purple, 1970). So, come on all, let us chase away the *"Smoke On The Water"*

(Deep Purple, 1972) and "*Sail Away*" (Deep Purple, 1974) on "*The Wide Ocean*" (Tommy Emmanuel, 2020) of sophisticated channel codes.

Let us start with the expression "codes."

Definition 1.1 (Code). A set of rules used to convert data from one form of representation to another is termed a *code* [13, p. 202].

Obviously, code simply means "representation." Well, that was "*Easy*" (The Commodores, 1977).

And what in the world are "channel codes"? Let us see. In advanced mobile communications systems, the transmission channel is "*noisy*" [6, p. 19], i. e., the transmit signal is perturbed by noise, at least during the transmission due to the time-varying multipath channel, and at the receivers due to thermal noise [1, Section 3.11.2]. [1, Section 3.11.3] briefly discusses the concept of the *channel capacity* C with special focus on noisy transmission channels [6, p. 19]. It is illustrated in [1, Section 3.11.3] that the channel capacity C represents the *maximum average actual transmission rate at which we may transmit with as small a frequency of errors or equivocation as desired*, provided that the transmitted message was appropriately represented by proper coding [6, p. 22]. The said coding is termed *channel coding* [14, p. 3], [1, Section 3.11.3]. [6, Theorem 11, p. 22] claims that "*if* $H\{X\} \leq C$*, there exists a coding system such that the output of the source can be transmitted over the channel with an arbitrarily small frequency of errors (or an arbitrarily small equivocation)*" [1, Section 3.11.3]. Let us have a closer look.

Theorem 1.1 (Channel Capacity is the Maximum Rate for Transmissions Free of Any Errors). *Let a discrete channel have the channel capacity* C *and a discrete source the entropy per second* $H\{X\}$ [6, Theorem 11, p. 22].

1. *If* $H\{X\} \leq C$*, there exists a coding system such that the output of the source can be transmitted over the channel with an arbitrarily small frequency of errors (or an arbitrarily small equivocation)* [6, Theorem 11, p. 22].
2. *If* $H\{X\} > C$*, it is possible to encode the source so that the equivocation is less than* $H\{X\} - C + \epsilon$ *where* ϵ *is arbitrarily small* [6, Theorem 11, p. 22].
3. *There is no method of encoding, which gives an equivocation less than* $H\{X\} - C$ [6, Theorem 11, p. 22].

Proof. We will follow the proof in [6, Theorem 11, pp. 22–24]. We will take into consideration that the entropies defined in [1] make use of the binary logarithm, and hence measure the number of bits per channel access.

Proof of 1

The channel capacity of the noisy transmission channel is given by [1, Section 3.11.3], [6, p. 23]

$$C = \max\left\{\mathrm{H}\{X\} - \mathrm{H}\{X \mid Y\}\right\}, \tag{1.1}$$

$\mathrm{H}\{X \mid Y\}$ being the *equivocation*, also called the *loss*, cf. [1, Section 3.11.2], [6, p. 20]. Y represents the *ensemble* [1, Section 3.11.1], [7, p. 13] of the sink, i. e., of the output of the transmission channel. The maximization in (1.1) is over all the sources, which might be used as input to the transmission channel [6, p. 23].

Let us assume that we have identified that particular source S_0, which achieves the channel capacity [6, p. 23]. We find the following statements to be true for this source S_0 [6, p. 23]:

1. Since the entropy $\mathrm{H}\{X\}$ is the rate of generating bits by the source, the number of distinct binary sequences that are available for transmission and that have a long duration T_S, is approximately $2^{\{T_S \cdot \mathrm{H}\{X\}\}}$ [6, p. 23].
2. With the entropy $\mathrm{H}\{Y\}$ of the output, the number of possible distinct receive sequences during T_S is approximately $2^{\{T_S \cdot \mathrm{H}\{Y\}\}}$.
3. Owing to the average rate of ambiguity $\mathrm{H}\{X \mid Y\}$, i. e., the equivocation, each possible distinct receive sequence could have been generated by approximately $2^{\{T_S \cdot \mathrm{H}\{X \mid Y\}\}}$ distinct input sequences during T_S.

If T_S approaches infinity, the expression "approximately" becomes "exactly" [6, p. 23]. Figure 1.3 illustrates the relations between channel inputs and channel outputs; cf., e. g., [6, Figure 10, p. 23].

Next, suppose that there is another source, S_1 say, which only achieves the transmission rate R lower than C, i. e., $R < C$ [6, p. 23]. Therefore, there are $2^{\{T_S \cdot R\}}$ receive sequences that could have been caused by the maximum approximate number of transmit sequences $2^{\{T_S \cdot \mathrm{H}\{X\}\}}$, being established by the source S_0 [6, p. 24].

The probability that one of these $2^{\{T_S \cdot \mathrm{H}\{X\}\}}$ potential transmit sequences is a receive sequence is hence given by [6, p. 24]

$$\frac{2^{\{T_S \cdot R\}}}{2^{\{T_S \cdot \mathrm{H}\{X\}\}}} = 2^{\{T_S \cdot [R - \mathrm{H}\{X\}]\}}. \tag{1.2}$$

Therefore, the probability that none of these $2^{\{T_S \cdot \mathrm{H}\{X\}\}}$ potential transmit sequences is a receive sequence, i. e., the *error probability* is [6, p. 24]

$$\left\langle 1 - \frac{2^{\{T_S \cdot R\}}}{2^{\{T_S \cdot \mathrm{H}\{X\}\}}} \right\rangle^{2^{\{T_S \cdot \mathrm{H}\{X \mid Y\}\}}} = \left\langle 1 - 2^{\{T_S \cdot [R - \mathrm{H}\{X\}]\}} \right\rangle^{2^{\{T_S \cdot \mathrm{H}\{X \mid Y\}\}}} \tag{1.3}$$

because there is ambiguity measured by the equivocation $\mathrm{H}\{X \mid Y\}$ [6, p. 23f.].

With [6, p. 24]

$$R < \mathrm{H}\{X\} - \mathrm{H}\{X \mid Y\} \quad \Leftrightarrow \quad R - \mathrm{H}\{X\} = -\mathrm{H}\{X \mid Y\} - \eta, \quad \eta \in \mathbb{R}^+, \tag{1.4}$$

(1.3) yields [6, p. 24]

$$\left\langle 1 - 2^{\{T_S \cdot [-\mathrm{H}\{X \mid Y\} - \eta]\}} \right\rangle^{2^{\{T_S \cdot \mathrm{H}\{X \mid Y\}\}}} = \left\langle 1 - 2^{\{-T_S \cdot \mathrm{H}\{X \mid Y\} - T_S \cdot \eta\}} \right\rangle^{2^{\{T_S \cdot \mathrm{H}\{X \mid Y\}\}}}. \tag{1.5}$$

Now, letting T_S approach infinity, the error probability $\langle 1 - 2^{\{-T_S \cdot \mathrm{H}\{X \mid Y\} - T_S \cdot \eta\}} \rangle^{2^{\{T_S \cdot \mathrm{H}\{X \mid Y\}\}}}$ approaches 0.

Proof of 2

In the case of the best possible source, S_0 say, the maximum transmission rate C is possible. Data generated at a rate $\mathrm{H}\{X\}$ beyond this maximum transmission rate C, i. e., $\mathrm{H}\{X\} > C$, is neglected. This neglected part is represented by the equivocation $\mathrm{H}\{X \mid Y\}$ equal to $(\mathrm{H}\{X\} - C)$ at the receiver [6, p. 24]. Adding ϵ to the transmitted part yields $(\mathrm{H}\{X\} + \epsilon)$, and thus the claim 2 [6, p. 24].

Proof of 3

Let us assume that we can encode a source with $\mathrm{H}\{X\}$ equal to $(C + a)$ and obtain an equivocation $\mathrm{H}\{X \mid Y\}$ equal to $(a - \epsilon)$, $\epsilon \in \mathbb{R}^+$, by means of this encoding [6, p. 24]. Then the transmission rate R is equal to $(C + a)$ and we yield

$$\mathrm{H}\{X\} - \mathrm{H}\{X \mid Y\} = C + a - (a - \epsilon) = C + \epsilon, \quad \epsilon \in \mathbb{R}^+. \tag{1.6}$$

This contradicts the definition of C as the maximum of $(\mathrm{H}\{X\} - \mathrm{H}\{X \mid Y\})$, and hence yields the claim 3 [6, p. 24]. □

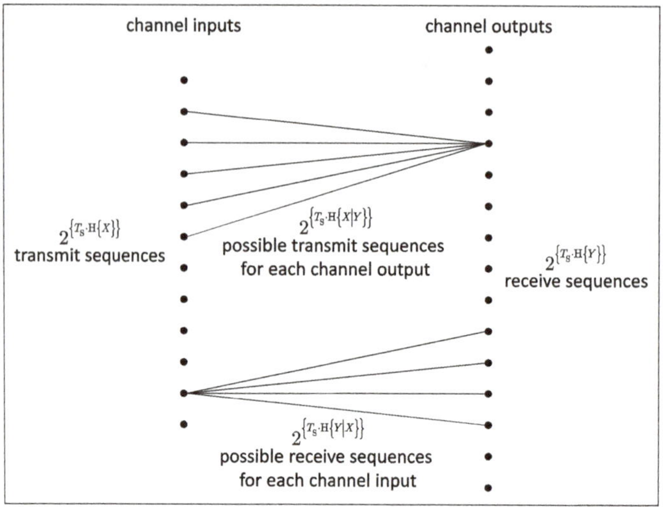

Figure 1.3: Relations between channel inputs and channel outputs (adapted according to [6, Figure 10, p. 23]).

Regrettably, the proof of Theorem 1.1 (cf., e. g., [6, Theorem 11, pp. 22–24]) does not provide us with a procedure for the design of any codes [1, Section 3.11.3], [6, p. 24]. Nevertheless, we have a much clearer picture of what this textbook is all about now: We are striving for the *ideal transmission*.

Definition 1.2 (Ideal Transmission). The transmission of data with the source entropy $\mathrm{H}\{X\}$ via a transmission channel with the channel capacity C is called an *ideal transmission* in the case of

$$\mathrm{H}\{X\} = C, \tag{1.7}$$

consequently yielding the equivocation $\mathrm{H}\{X \mid Y\}$ equal to 0, and hence achieving a reception, which is free of any errors.

And we may now state what channel codes are.

Definition 1.3 (Channel Code). A *channel code* is a means to represent, i. e., to encode, a source S, which has the entropy $\mathrm{H}\{X\}$, in order to approximate the *ideal transmission*.

In other words, channel coding aims at finding codes that facilitate transmitting at the greatest possible transmission rate, trying to avoid any transmission errors at the same time [14, p. 3]. The expression "approximate" in Definition 1.3 as well as the "trying to avoid" indicate that finding a code, which enables the ideal transmission, might be extremely difficult or even impossible. Fortunately, since the foundation of modern communications theory by Claude E. *Shannon* in his seminal paper [6] of 1948, a multitude of channel codes, which approximate the ideal transmission reasonably well, have appeared, many of which have been included in advanced mobile communications systems [1, Section 3.11.3].

In the context of his "Theorem 11" [6, Theorem 11, pp. 22–24] Shannon states the following [6, p. 24]:

> *"An attempt to obtain a good approximation to ideal coding by following the method of the proof (remark by the author: the proof of Theorem 11) is generally impractical. In fact, apart from some rather trivial cases and certain limiting situations, no explicit description of a series of approximation to the ideal has been found. Probably this is no accident but is related to the difficulty of giving an explicit construction for a good approximation to a random sequence."*

Therefore, [6, Theorem 11, pp. 22–24] is often called the *"random coding argument."*

Although it looks somewhat artificial, we will distinguish between *"classical" channel codes*, which were proposed until approximately the end of the 1960s, i. e., in the "early days" of modern communications theory, and *"sophisticated" channel codes*, which were proposed and studied from approximately 1960 until today.

Definition 1.4 (Sophisticated Channel Code). A channel code, which provides an improved approximation of the ideal transmission compared with classical channel codes, will be termed a *"sophisticated" channel code.*

We will consider
- Hamming codes [15, pp. 100–102],
- Hsiao's codes [15, pp. 102–105],
- Reed–Muller codes [15, pp. 105–119],
- Golay codes [15, pp. 125–128],
- Bose–Chaudhuri–Hocquenghem (BCH) codes including Reed–Solomon (RS) codes [15, pp. 194–240] and
- convolutional codes [15, pp. 453–510]

to be classical channel codes. Presently, the group of sophisticated channel codes consists of
- low density parity check (LDPC) codes [15, pp. 851–945],
- turbo codes [15, pp. 766–844] and
- polar codes [16].

In this textbook, we will focus on these three sophisticated channel codes.

Since the modulation mapper expects bits at its input, we will merely consider *binary* channel codes, i. e., such channel codes that generate encoded bits at their outputs. It hence seems reasonable to also restrict the input symbols of the binary sophisticated channel codes to bits. So, this is exactly what we will do.

Restricting us to binary input binary output ("BIBO") channel codes requires a bit of additional mathematical understanding. Therefore, paving our way toward sophisticated channel codes will lead us directly to mathematical discipline of *finite fields* [4, pp. 81, 93], [17, p. 1]. Discussing finite fields seems a bit easier when done in the context of classical channel codes. Hence, Chapter 1 will illustrate some of their important concepts before we "*Get Down On It*" (Kool and the Gang, 1981), i. e., before partying with sophisticated channel codes.

This has got "*Soul With A Capital S*" (Tower of Power, 1993)—

let there be codes!

And there were codes ("*Let There Be Rock*," AC/DC, 1977; sorry, could not resist).

1.3 General Concept of Channel Codes

Let the output of an information source be a sequence $\{u_i\}$ of bits

$$u_i \in \mathbb{F}_2 = \{0,1\}, \quad i \in \mathbb{Z}. \tag{1.8}$$

The set $\mathbb{F}_2$ is of course *countable* and *finite*, i. e., it has the cardinal number $|\mathbb{F}_2|$ equal to 2 and the Lebesgue measure 0. The set $\mathbb{F}_2$ is termed the *Galois field* with two elements [14, p. 407], [18, Definition 1.41, p. 15]. Évariste *Galois* was a French mathematician, who lived in the early nineteenth century and died tragically in a duel at the age of only 20. If this story was not so sad, we might be misled to the conclusion that mathematics is dangerous. Anyway, the author is glad that we are all engineers, aren't we?

It seems quite surprising at first glance to find that a finite set is a field [14, pp. 407–409]. Let us have a closer look in what follows. But before coming to the *finite field*, another expression one has to enter the stage, first, namely "*operation*" [11, p. 308].

Definition 1.5 (An *m*-ary Operation). An *m*-ary operation

$$\varphi : \underbrace{\mathbb{A} \times \mathbb{A} \times \mathbb{A} \times \cdots \times \mathbb{A}}_{m \text{ terms, abbreviated by } \mathbb{A}^m} \mapsto \mathbb{A} \quad \Leftrightarrow \quad \varphi : \mathbb{A}^m \mapsto \mathbb{A} \tag{1.9}$$

on the set $\mathbb{A}$ is a function, which associates a value taken from $\mathbb{A}$ to each *m*-tuple of values taken from $\mathbb{A}$ [11, p. 308], [19, p. 675].

A special case of Definition 1.5 is the *binary operation*.

Definition 1.6 (Binary Operation). A binary operation

$$\circ : \mathbb{A} \times \mathbb{A} \mapsto \mathbb{A}, \tag{1.10}$$

on the set $\mathbb{A}$ is a function, which associates a value taken from $\mathbb{A}$ to each 2-tuple of values taken from $\mathbb{A}$ [11, p. 308].

Remark 1.2 (On Binary Operations). Typical binary operations are the addition of two elements taken from the set $\mathbb{A}$ or the multiplication of two elements taken from the set $\mathbb{A}$ [11, p. 308].

Now, let us first define what we mean by the expressions *field* and *finite field*.

Definition 1.7 (Field). A nonempty set $\mathbb{F}$ is a *field* with two operations "addition" $\oplus$ and "multiplication" $\odot$ if
- $(\mathbb{F}, \oplus)$ is an Abelian group with the neutral or identity element 0,
- $(\mathbb{F}\backslash\{0\}, \odot)$ is an Abelian group with the neutral or identity element 1 and
- the "multiplication" $\odot$ distributes over the "addition" $\oplus$

$$\alpha \odot (\beta \oplus \gamma) = \alpha \odot \beta \;\oplus\; \alpha \odot \gamma, \quad \alpha, \beta, \gamma \in \mathbb{F}, \tag{1.11}$$

[14, p. 407], [18, Definition 1.29, p. 11f.].

Drum roll! Here it comes, the next definition, but not before the "*Fanfare For The Common Man*" (Aaron Copland, 1942, maybe the most well-known cover version was played by Emerson, Lake and Palmer in 1977)!

Definition 1.8 (Finite Field). A field with a finite number of elements is called a *finite field* [14, p. 407], [18, Definition 1.41, p. 15], [4, pp. 81, 93], [17, p. 1].

The finite number of elements of a finite field, i. e., the cardinal number of the set is the *order* of the finite field [4, p. 81].

Now, let us return to $\mathbb{F}_2$.

Definition 1.9 ("Addition" operation $\oplus$ and "multiplication" operation $\odot$ in $\mathbb{F}_2$). In $\mathbb{F}_2$, the "addition" operation $\oplus$ and the "multiplication" operation $\odot$ shall yield the following results [14, Table B.2-1, p. 407]:

$\oplus$	**0**	**1**
0	0	1
1	1	0

$\odot$	**0**	**1**
0	0	0
1	0	1

Obviously, the "addition" operation $\oplus$ operates like the logical *exclusive or* (EXOR, XOR) and the "multiplication" operation $\odot$ operates like the logical *and* [20, Abb. 6.12, p. 626].

Now, let us evaluate whether $\mathbb{F}_2$ is a field. Since we will use the "addition" operation $\oplus$ and the "multiplication" operation $\odot$, we will also denote this finite field by $(\mathbb{F}_2, \oplus, \odot)$.

Theorem 1.2 ($\mathbb{F}_2$ is a Finite Field). *$\mathbb{F}_2$, respectively $(\mathbb{F}_2, \oplus, \odot)$, is a finite field.*

Proof. First of all,

$$|\mathbb{F}_2| = 2 \tag{1.12}$$

and, therefore, $\mathbb{F}_2$ is finite: $\checkmark$.

Now, we have to show that $(\mathbb{F}_2, \oplus)$ is an Abelian group with the neutral or identity element 0.

1. *Closure and uniqueness*: √

For $\alpha, \beta \in \mathbb{F}_2$, it follows from Definition 1.9 that $\alpha \oplus \beta \in \mathbb{F}_2$, and hence the closure is valid. Furthermore, it follows from Definition 1.9 that $\alpha \oplus \beta \in \mathbb{F}_2$ is unique.

2. *Neutral (identity) element*: √

It follows from Definition 1.9 that 0 is the neutral (identity) element.

3. *Inverse element*: √

For every $\alpha \in \mathbb{F}_2$, there is an inverse element $\alpha^{-1} \in \mathbb{F}_2$ with

$$\alpha \oplus \alpha^{-1} = \alpha^{-1} \oplus \alpha = 0. \tag{1.13}$$

It follows from Definition 1.9 that 0 is the inverse element of 0 and that 1 is the inverse element of 1. We yield to the following table:

α	α^{-1}	$\alpha \oplus \alpha^{-1}$	$\alpha^{-1} \oplus \alpha$
0	0	0	0
1	1	0	0

We thus find

$$\alpha^{-1} = \alpha, \quad \alpha \oplus \alpha^{-1} = \alpha^{-1} \oplus \alpha = \alpha \oplus \alpha = 0, \quad \alpha \in \mathbb{F}_2. \tag{1.14}$$

4. *Associative law*: √

For $\alpha, \beta, \gamma \in \mathbb{F}_2$, it follows from Definition 1.9 that

$$\alpha \oplus (\beta \oplus \gamma) = (\alpha \oplus \beta) \oplus \gamma, \quad \alpha, \beta, \gamma \in \mathbb{F}_2. \tag{1.15}$$

We find the following table:

α	β	γ	$\beta \oplus \gamma$	$\alpha \oplus \beta$	$\alpha \oplus (\beta \oplus \gamma)$	$(\alpha \oplus \beta) \oplus \gamma$
0	0	0	0	0	0	0
0	0	1	1	0	1	1
0	1	0	1	1	1	1
0	1	1	0	1	0	0
1	0	0	0	1	1	1
1	0	1	1	1	0	0
1	1	0	1	0	0	0
1	1	1	0	0	1	1

5. *Commutative law*: √

It follows from Definition 1.9 that

$$\alpha \oplus \beta = \beta \oplus \alpha \quad \alpha, \beta \in \mathbb{F}_2, \tag{1.16}$$

holds. We obtain the following table:

α	β	$\alpha \oplus \beta$	$\beta \oplus \alpha$
0	0	0	0
0	1	1	1
1	0	1	1
1	1	0	0

Clearly, $(\mathbb{F}_2, \oplus)$ is an Abelian group with the neutral or identity element 0: ✓.

Next, we have to show that $(\mathbb{F}_2\backslash\{0\}, \odot)$ is an Abelian group with the neutral or identity element 1.

1. *Closure and uniqueness*: ✓

For $\alpha, \beta \in \mathbb{F}_2\backslash\{0\}$, i. e.,

$$\alpha = \beta = 1, \tag{1.17}$$

it follows from Definition 1.9 that

$$\alpha \odot \beta = 1 \odot 1 = 1 \in \mathbb{F}_2\backslash\{0\}, \tag{1.18}$$

hence the closure and the uniqueness are valid.

2. *Neutral (identity) element*: ✓

It follows from Definition 1.9 that 1 is the neutral (identity) element. In fact, it is the only element in $\mathbb{F}_2\backslash\{0\}$.

3. *Inverse element*: ✓

For every $\alpha \in \mathbb{F}_2\backslash\{0\}$, i. e., for α equal to 1, there is an inverse element $\alpha^{-1} \in \mathbb{F}_2\backslash\{0\}$ with

$$\alpha \odot \alpha^{-1} = \alpha^{-1} \odot \alpha = 1 \odot 1 = 1 \tag{1.19}$$

The inverse element of α equal to 1 is α^{-1} equal to 1.

4. *Associative law*: ✓

For $\alpha, \beta, \gamma \in \mathbb{F}_2\backslash\{0\}$, i. e., for

$$\alpha = \beta = \gamma = 1, \tag{1.20}$$

it follows from Definition 1.9 that

$$\begin{gathered} \alpha \odot (\beta \odot \gamma) = 1 \odot (1 \odot 1) = (\alpha \odot \beta) \odot \gamma = (1 \odot 1) \odot 1 = 1, \\ \alpha, \beta, \gamma \in \mathbb{F}_2\backslash\{0\} = \{1\}. \end{gathered} \tag{1.21}$$

5. *Commutative law*: ✓

It follows from Definition 1.9 that

$$\alpha \odot \beta = 1 \odot 1 = \beta \odot \alpha = 1 \odot 1 = 1, \quad \alpha, \beta \in \mathbb{F}_2\backslash\{0\} = \{1\}, \tag{1.22}$$

holds.

Clearly, $(\mathbb{F}_2\backslash\{0\}, \odot)$ is an Abelian group with the neutral or identity element 1: ✓.

Finally, we have to prove the distributive law. We yield to the following table:

α	β	γ	$\beta \oplus \gamma$	$\alpha \odot (\beta \oplus \gamma)$	$\alpha \odot \beta \oplus \alpha \odot \gamma$
0	0	0	0	0	$0 \oplus 0 = 0$
0	0	1	1	0	$0 \oplus 0 = 0$
0	1	0	1	0	$0 \oplus 0 = 0$
0	1	1	0	0	$0 \oplus 0 = 0$
1	0	0	0	0	$0 \oplus 0 = 0$
1	0	1	1	1	$0 \oplus 1 = 1$
1	1	0	1	1	$1 \oplus 0 = 1$
1	1	1	0	0	$1 \oplus 1 = 0$

Obviously, the distributive law holds: ✓. □

Remark 1.3 (No Field With Just a Single Element). There is no finite field with just one element. All fields in abstract algebra must contain at least two distinct elements, namely the additive identity, zero and the multiplicative identity, one. This is not the case in a set with only a single element.

The rather philosophical problem of "the field with just one element" was first addressed by the Belgian mathematician Jacques Tits in 1956 and has attracted some research in mathematics since then.

Now, we can continue with the binary input binary output (BIBO) channel coding. Let the sequence $\{u_i\}$, consisting of the bits $u_i \in \mathbb{F}_2, i \in \mathbb{Z}$, of (1.8), be segmented into message blocks [15, p. 3], each of which we will denote by a row vector [15, p. 3; equation (3.3), p. 67]

$$\boldsymbol{u} = (u_0, u_1 \cdots u_{k-1}), \quad u_i \in \mathbb{F}_2, \quad i \in \{0, 1 \cdots (k-1)\}, \quad k \in \mathbb{N}^*, \tag{1.23}$$

consisting of k message bits u_i, $i \in \{0, 1 \cdots (k-1)\}$, $k \in \mathbb{N}^*$, also termed message bits, message symbols or message digits; see, e. g., [15, p. 3]. We will term this row vector the *message* [15, p. 3], respectively, the *message vector*. Obviously, there are 2^k, $k \in \mathbb{N}^*$, distinct messages [15, p. 3].

Let there be an encoder, which transforms $\boldsymbol{u}$ of (1.23) into a binary n-tuple [15, p. 3; equation (3.5), p. 69], denoted by the row vector

$$\boldsymbol{v} = (v_0, v_1 \cdots v_{n-1}), \quad v_j \in \mathbb{F}_2, \quad j \in \{0, 1 \cdots (n-1)\}, \quad n \geq k, \quad k, n \in \mathbb{N}^*. \tag{1.24}$$

We refer to $\boldsymbol{v}$ of (1.24) to as the *codeword* [15, p. 3] or the *code vector*. The code bits v_j, $j \in \{0, 1 \cdots (n-1)\}$, $n \in \mathbb{N}^*$ are also called code symbols or code digits; see, e. g., [15, p. 3]. Corresponding to the 2^k, $k \in \mathbb{N}^*$, possible messages, there are 2^k, $k \in \mathbb{N}^*$ codewords [15, p. 3]. This set of 2^k codewords is called an (n, k) *block code* [15, p. 3].

The ratio

$$R = \frac{k}{n}, \quad k, n \in \mathbb{N}^*, \tag{1.25}$$

of k, $k \in \mathbb{N}^*$ and n, $n \in \mathbb{N}^*$, is termed the *code rate* [15, p. 3]. For finite $k, n \in \mathbb{N}^*$, we find

$$0 < R \leq 1. \tag{1.26}$$

For a block code to be useful, the 2^k codewords must be distinct [15, p. 66]. Therefore, there a one-to-one correspondence between a particular message $\boldsymbol{u}$ and its codeword $\boldsymbol{v}$ is required [15, p. 66].

1.4 Vector Spaces Over $\mathbb{F}_2$

Just having defined what a finite field is, we have immediately introduced the vectors $\boldsymbol{u}$ of (1.23) and $\boldsymbol{v}$ of (1.24), which consist of components $u_i \in \mathbb{F}_2$, $i \in \{0, 1 \cdots (k-1)\}$, $k \in \mathbb{N}^*$ and $v_j \in \mathbb{F}_2, j \in \{0, 1 \cdots (n-1)\}$, $n \in \mathbb{N}^*$, which are taken from the finite field $\mathbb{F}_2$. Therefore, we say that $\boldsymbol{u}$ is an element of the *k-fold Cartesian product* $\mathbb{F}_2^k$ of $\mathbb{F}_2$ with itself

$$\mathbb{F}_2^k = \underbrace{\mathbb{F}_2 \times \mathbb{F}_2 \times \cdots \times \mathbb{F}_2}_{k \text{ factors}}, \quad k \in \mathbb{N}^*, \tag{1.27}$$

and $\boldsymbol{v}$ is an element of the *n-fold Cartesian product* $\mathbb{F}_2^n$ of $\mathbb{F}_2$ with itself

$$\mathbb{F}_2^n = \underbrace{\mathbb{F}_2 \times \mathbb{F}_2 \times \cdots \times \mathbb{F}_2}_{n \text{ factors}}, \quad n \in \mathbb{N}^*, \tag{1.28}$$

and we write

$$\boldsymbol{u} \in \mathbb{F}_2^k, \quad \boldsymbol{v} \in \mathbb{F}_2^n, \quad k, n \in \mathbb{N}^*. \tag{1.29}$$

Without loss of generality, let us use the vectors taken from $\mathbb{F}_2^k$, $k \in \mathbb{N}^*$. Of course, all aspects discussed in what follows are also valid for the vectors taken from $\mathbb{F}_2^n$, $n \in \mathbb{N}^*$.

Let us introduce the *addition* $\oplus$ *of any two vectors*

$$\boldsymbol{u}^{(1)} = (u_0^{(1)}, u_1^{(1)} \cdots u_{k-1}^{(1)}) \in \mathbb{F}_2^k, \quad \boldsymbol{u}^{(2)} = (u_0^{(2)}, u_1^{(2)} \cdots u_{k-1}^{(2)}) \in \mathbb{F}_2^k, \quad k \in \mathbb{N}^*, \tag{1.30}$$

in the following way:

$$\begin{aligned} \boldsymbol{u}^{(1)} \oplus \boldsymbol{u}^{(2)} &= (u_0^{(1)} \oplus u_0^{(2)}, \ u_1^{(1)} \oplus u_1^{(2)} \ \cdots \ u_{k-1}^{(1)} \oplus u_{k-1}^{(2)}) \in \mathbb{F}_2^k, \\ & u_i^{(1)}, u_i^{(2)} \in \mathbb{F}_2, \quad i \in \{0, 1 \cdots (k-1)\}, \quad k \in \mathbb{N}^*. \end{aligned} \tag{1.31}$$

The addition of any two vectors $\boldsymbol{u}^{(1)} \in \mathbb{F}_2^k, k \in \mathbb{N}^*$ and $\boldsymbol{u}^{(2)} \in \mathbb{F}_2^k, k \in \mathbb{N}^*$, yields the vector $(\boldsymbol{u}^{(1)} \oplus \boldsymbol{u}^{(2)})$, which has the components $(u_i^{(1)} \oplus u_i^{(2)}) \in \mathbb{F}_2, i \in \{0, 1 \cdots (k-1)\}, k \in \mathbb{N}^*$, that are the addition of the respective components $u_i^{(1)} \in \mathbb{F}_2, i \in \{0, 1 \cdots (k-1)\}, k \in \mathbb{N}^*$, of

$\boldsymbol{u}^{(1)}$ and $u_i^{(2)} \in \mathbb{F}_2, i \in \{0,1\cdots(k-1)\}, k \in \mathbb{N}^*$, of $\boldsymbol{u}^{(2)}$. Hence, the vector $(\boldsymbol{u}^{(1)} \oplus \boldsymbol{u}^{(2)})$ is also an element of $\mathbb{F}_2^k$.

Owing to (1.31), $(\mathbb{F}_2^k, \oplus)$ is closed, and the addition declared by (1.31) is unique. Moreover, since $(\mathbb{F}_2, \oplus)$ is an Abelian group (cf. Theorem 1.2), also $(\mathbb{F}_2^k, \oplus)$ *is an Abelian group.*

Now, let us define the *outer multiplication* $\odot$ *of an element* $a \in \mathbb{F}_2$ *and a vector* $\boldsymbol{u} \in \mathbb{F}_2^k$ in the following way:

$$a \odot \boldsymbol{u} = (a \odot u_0,\ a \odot u_1\ \cdots\ a \odot u_{k-1}) \in \mathbb{F}_2^k, \quad a, u_i \in \mathbb{F}_2, \quad i \in \{0,1\cdots(k-1)\}, \quad k \in \mathbb{N}^*. \tag{1.32}$$

Since $u_i \in \mathbb{F}_2, i \in \{0,1\cdots(k-1)\}, k \in \mathbb{N}^*$ and $a \in \mathbb{F}_2$, also $(a \odot u_i) \in \mathbb{F}_2, i \in \{0,1\cdots(k-1)\}$, $k \in \mathbb{N}^*$; cf. Theorem 1.2. Therefore, $(a \odot \boldsymbol{u})$ is an element of $\mathbb{F}_2^k$.

Definition 1.10 (Vector Space Over $\mathbb{F}_2$). The finite field $\mathbb{F}_2$, respectively $(\mathbb{F}_2, \oplus, \odot)$ (cf. Theorem 1.2) and the Abelian group $(\mathbb{F}_2^k, \oplus)$ of the vectors $\boldsymbol{u}$ complying with the outer multiplication $\mathbb{F}_2 \times \mathbb{F}_2^k \to \mathbb{F}_2^k$ introduced in (1.32), which assigns the vector $(a \odot \boldsymbol{u}) \in \mathbb{F}_2^k$ to the ordered tuple $(a, \boldsymbol{u})$, is termed a *vector space over* $\mathbb{F}_2$ [11, p. 327].

The following rules result immediately [11, p. 327]:

(V1) Associative law of addition

$$\left(\boldsymbol{u}^{(1)} \oplus \boldsymbol{u}^{(2)}\right) \oplus \boldsymbol{u}^{(3)} = \boldsymbol{u}^{(1)} \oplus \left(\boldsymbol{u}^{(2)} \oplus \boldsymbol{u}^{(3)}\right) \in \mathbb{F}_2^k, \quad \boldsymbol{u}^{(1)}, \boldsymbol{u}^{(2)}, \boldsymbol{u}^{(3)} \in \mathbb{F}_2^k, \quad k \in \mathbb{N}^*. \tag{1.33}$$

(V2) Neutral (identity) element of addition

There is a vector $\boldsymbol{0}_k \in \mathbb{F}_2^k, k \in \mathbb{N}^*$, termed the *zero vector* with

$$\boldsymbol{u} \oplus \boldsymbol{0}_k = \boldsymbol{u}, \quad \boldsymbol{0}_k, \boldsymbol{u} \in \mathbb{F}_2^k, \quad k \in \mathbb{N}^*. \tag{1.34}$$

This vector $\boldsymbol{0}_k \in \mathbb{F}_2^k$ takes the form

$$\boldsymbol{0}_k = \underbrace{(0, 0\cdots 0)}_{k\text{ components}} \in \mathbb{F}_2^k, \quad 0 \in \mathbb{F}_2, \quad k \in \mathbb{N}^*. \tag{1.35}$$

(V3) Inverse element of addition

There is a vector $(-\boldsymbol{u}) \in \mathbb{F}_2^k, k \in \mathbb{N}^*$, which fulfills

$$\boldsymbol{u} \oplus (-\boldsymbol{u}) = \boldsymbol{0}_k, \quad \boldsymbol{0}_k, \boldsymbol{u}, (-\boldsymbol{u}) \in \mathbb{F}_2^k, \quad k \in \mathbb{N}^*. \tag{1.36}$$

Equation (1.14) of Theorem 1.2 and (1.31) leads us to the conclusion that

$$\begin{aligned}
\boldsymbol{0}_k &= \boldsymbol{u} \oplus (-\boldsymbol{u}) \\
&= \Big(\underbrace{u_0 \oplus (-u_0)}_{=0},\ \underbrace{u_1 \oplus (-u_1)}_{=0}\ \cdots\ \underbrace{u_{k-1} \oplus (-u_{k-1})}_{=0}\Big) \\
&= \Big(\underbrace{u_0 \oplus u_0^{-1}}_{=0},\ \underbrace{u_1 \oplus u_1^{-1}}_{=0}\ \cdots\ \underbrace{u_{k-1} \oplus u_{k-1}^{-1}}_{=0}\Big) \\
&= \Big(\underbrace{u_0 \oplus u_0}_{=0},\ \underbrace{u_1 \oplus u_1}_{=0}\ \cdots\ \underbrace{u_{k-1} \oplus u_{k-1}}_{=0}\Big) \\
&= \boldsymbol{u} \oplus \boldsymbol{u};
\end{aligned} \tag{1.37}$$

thus we obtain

$$(-\boldsymbol{u}) = \boldsymbol{u} \in \mathbb{F}_2. \tag{1.38}$$

(V4) Commutative law of addition

$$\boldsymbol{u}^{(1)} \oplus \boldsymbol{u}^{(2)} = \boldsymbol{u}^{(2)} \oplus \boldsymbol{u}^{(1)} \in \mathbb{F}_2^k, \quad \boldsymbol{u}^{(1)}, \boldsymbol{u}^{(2)} \in \mathbb{F}_2^k, \quad k \in \mathbb{N}^*. \tag{1.39}$$

(V5) Neutral (identity) element of outer multiplication

With $1 \in \mathbb{F}_2$ and $\boldsymbol{u} \in \mathbb{F}_2^k$, $k \in \mathbb{N}^*$, we yield

$$1 \odot \boldsymbol{u} = \boldsymbol{u}, \quad 1 \in \mathbb{F}_2, \quad \boldsymbol{u} \in \mathbb{F}_2^k, \quad k \in \mathbb{N}^*. \tag{1.40}$$

(V6) Associative law of outer multiplication

With $\alpha, \beta \in \mathbb{F}_2$ and $\boldsymbol{u} \in \mathbb{F}_2^k$, $k \in \mathbb{N}^*$, we yield

$$\alpha \odot (\beta \odot \boldsymbol{u}) = (\alpha \odot \beta) \odot \boldsymbol{u}, \quad \alpha, \beta \in \mathbb{F}_2, \quad \boldsymbol{u} \in \mathbb{F}_2^k, \quad k \in \mathbb{N}^*. \tag{1.41}$$

(V7) Distributive law 1

With $\alpha, \beta \in \mathbb{F}_2$ and $\boldsymbol{u} \in \mathbb{F}_2^k$, $k \in \mathbb{N}^*$, we yield

$$(\alpha \oplus \beta) \odot \boldsymbol{u} = \alpha \odot \boldsymbol{u} \oplus \beta \odot \boldsymbol{u}, \quad \alpha, \beta \in \mathbb{F}_2, \quad \boldsymbol{u} \in \mathbb{F}_2^k, \quad k \in \mathbb{N}^*. \tag{1.42}$$

(V8) Distributive law 2

With $\alpha \in \mathbb{F}_2$ and $\boldsymbol{u}^{(1)}, \boldsymbol{u}^{(2)} \in \mathbb{F}_2^k$, $k \in \mathbb{N}^*$, we yield

$$\alpha \odot \left(\boldsymbol{u}^{(1)} \oplus \boldsymbol{u}^{(2)}\right) = \alpha \odot \boldsymbol{u}^{(1)} \oplus \alpha \odot \boldsymbol{u}^{(2)}, \quad \alpha \in \mathbb{F}_2, \quad \boldsymbol{u}^{(1)}, \boldsymbol{u}^{(2)} \in \mathbb{F}_2^k, \quad k \in \mathbb{N}^*. \tag{1.43}$$

Motivated by (1.37), we will define what we mean by two vectors $\boldsymbol{u}^{(1)} \in \mathbb{F}_2^k$, $k \in \mathbb{N}^*$ and $\boldsymbol{u}^{(2)} \in \mathbb{F}_2^k$, $k \in \mathbb{N}^*$, being identical.

Definition 1.11 (Identical Vectors of $\mathbb{F}_2^k$). Two vectors $\boldsymbol{u}^{(1)} \in \mathbb{F}_2^k$, $k \in \mathbb{N}^*$ and $\boldsymbol{u}^{(2)} \in \mathbb{F}_2^k$, $k \in \mathbb{N}^*$, are said to be *identical* if and only if for their components $u_i^{(1)}, u_i^{(2)} \in \mathbb{F}_2$, $i \in \{0, 1 \cdots (k-1)\}$, $k \in \mathbb{N}^*$, the following condition holds:

$$u_i^{(1)} = u_i^{(2)}, \quad u_i^{(1)}, u_i^{(2)} \in \mathbb{F}_2, \quad i \in \left\{0, 1 \cdots (k-1)\right\}, \quad k \in \mathbb{N}^*. \tag{1.44}$$

We then write [11, p. 10f.]

$$\boldsymbol{u}^{(1)} = \boldsymbol{u}^{(2)} \tag{1.45}$$

or

$$\boldsymbol{u}^{(1)} \equiv \boldsymbol{u}^{(2)}. \tag{1.46}$$

Remark 1.4 (On Identical Vectors of $\mathbb{F}_2^k$). Using (1.37), we obtain

$$\boldsymbol{u}^{(1)} \oplus \boldsymbol{u}^{(2)} = \boldsymbol{u}^{(2)} \oplus \boldsymbol{u}^{(1)} = \boldsymbol{0}_k, \quad k \in \mathbb{N}^*, \tag{1.47}$$

if the two vectors $\boldsymbol{u}^{(1)} \in \mathbb{F}_2^k$, $k \in \mathbb{N}^*$ and $\boldsymbol{u}^{(2)} \in \mathbb{F}_2^k$, $k \in \mathbb{N}^*$ are identical.

How can we quantify the case of two vectors $\boldsymbol{u}^{(1)} \in \mathbb{F}_2^k$, $k \in \mathbb{N}^*$ and $\boldsymbol{u}^{(2)} \in \mathbb{F}_2^k$, $k \in \mathbb{N}^*$, that are not identical? Well, obviously, we need some sort of *distance measure*, which operates on two vectors taken from $\mathbb{F}_2^k$, $k \in \mathbb{N}^*$. Let us call this distance measure a *metric* or a *distance* [21, p. 3].

This metric or distance is related to the *"length" of the difference vector* $\boldsymbol{u}^{(1)} \oplus \boldsymbol{u}^{(2)}$ of the two vectors $\boldsymbol{u}^{(1)} \in \mathbb{F}_2^k$, $k \in \mathbb{N}^*$ and $\boldsymbol{u}^{(2)} \in \mathbb{F}_2^k$, $k \in \mathbb{N}^*$ [15, pp. 76–78]. So, let us first think about how we could define "length."

Since the components of any vector $\boldsymbol{u} \in \mathbb{F}_2^k$, $k \in \mathbb{N}^*$, are taken from $\mathbb{F}_2$, the "length" of $\boldsymbol{u} \in \mathbb{F}_2^k$, $k \in \mathbb{N}^*$ could, e. g., be based on the counting of

- those components of $\boldsymbol{u}$, which are all equal to 0, or
- those components of $\boldsymbol{u}$, which are all equal to 1.

Now, if we used the counting of those components of $\boldsymbol{u}$, which are all equal to 0, the zero vector $\boldsymbol{0}_k \in \mathbb{F}_2^k$, $k \in \mathbb{N}^*$, would be the "longest" vector of all vectors of $\mathbb{F}_2^k$, $k \in \mathbb{N}^*$, having the length k. This does sound strange. So, let us not pursue this option.

If we used the counting of those components of $\boldsymbol{u}$, which are all equal to 1, the zero vector $\boldsymbol{0}_k \in \mathbb{F}_2^k$, $k \in \mathbb{N}^*$, would be the "shortest" vector of all vectors of $\mathbb{F}_2^k$, $k \in \mathbb{N}^*$, having the length 0. This sounds quite appealing because it is the same thing in such vector spaces that we might have encountered in the past, e. g., $\mathbb{R}^k$, $k \in \mathbb{N}^*$ or $\mathbb{C}^k$, $k \in \mathbb{N}^*$. So, let us choose this option and define the length of $\boldsymbol{u} \in \mathbb{F}_2^k$, $k \in \mathbb{N}^*$. We will term this length the *Hamming weight* [15, p. 76].

Definition 1.12 (Hamming Weight). Let $\boldsymbol{u}$ equal to $(u_0, u_1 \cdots u_{k-1})$ be a binary k-tuple, i. e., a vector taken from $\mathbb{F}_2^k$, $k \in \mathbb{N}^*$. The *Hamming weight* of $\boldsymbol{u}$, denoted by $w_{\mathrm{H}}\{\boldsymbol{u}\}$, is defined as the number of nonzero components of $\boldsymbol{u}$ [15, p. 76]. We yield

$$w_{\mathrm{H}}\{\boldsymbol{u}\} = \sum_{i=0}^{k-1} u_i, \quad k \in \mathbb{N}^*. \tag{1.48}$$

Let us look at the following example.

Example 1.1. The Hamming weight of

$$\boldsymbol{u} = (1, 0, 0, 1, 0, 1, 1) \tag{1.49}$$

is given by [15, p. 76]

$$w_{\mathrm{H}}\{\boldsymbol{u}\} = w_{\mathrm{H}}\{(1, 0, 0, 1, 0, 1, 1)\} = 4. \tag{1.50}$$

Setting out from the Hamming weight defined in Definition 1.12, we define the metric or distance, which we will call the *Hamming distance* [15, p. 76] in the following way.

Definition 1.13 (Hamming Distance). Let $\boldsymbol{u}^{(1)}$, having the components $u_i^{(1)} \in \mathbb{F}_2$, $i \in \{0, 1 \cdots (k-1)\}$, $k \in \mathbb{N}^*$ and $\boldsymbol{u}^{(2)}$, having the components $u_i^{(2)} \in \mathbb{F}_2$, $i \in \{0, 1 \cdots (k-1)\}$, $k \in \mathbb{N}^*$, be two arbitrary vectors of $\mathbb{F}_2^k$, $k \in \mathbb{N}^*$. The *Hamming distance* between $\boldsymbol{u}^{(1)}$ and $\boldsymbol{u}^{(2)}$ is denoted by $d_{\mathrm{H}}\{\boldsymbol{u}^{(1)}, \boldsymbol{u}^{(2)}\}$ and is defined as the number of components in which $\boldsymbol{u}^{(1)}$ and $\boldsymbol{u}^{(2)}$ differ [15, p. 76].

Remark 1.5 (On the Hamming Distance). The number of the components, in which $\boldsymbol{u}^{(1)}$ and $\boldsymbol{u}^{(2)}$ taken from $\mathbb{F}_2^k$, $k \in \mathbb{N}^*$ differ, is exactly equal to the number of components of the difference vector $\boldsymbol{u}^{(1)} \oplus \boldsymbol{u}^{(2)}$, which are equal to 1, because setting out from the left table of Definition 1.9 and using (1.31),

- all those components of $\boldsymbol{u}^{(1)} \oplus \boldsymbol{u}^{(2)}$, in which $\boldsymbol{u}^{(1)}$ and $\boldsymbol{u}^{(2)}$ are identical, are equal to 0 and
- all those components of $\boldsymbol{u}^{(1)} \oplus \boldsymbol{u}^{(2)}$, in which $\boldsymbol{u}^{(1)}$ and $\boldsymbol{u}^{(2)}$ differ, are equal to 1.

We immediately find

$$d_{\mathrm{H}}\{u_i^{(1)}, u_i^{(2)}\} = u_i^{(1)} \oplus u_i^{(2)} = \begin{cases} 1 & \text{if } u_i^{(1)} \neq u_i^{(2)}, \\ 0 & \text{if } u_i^{(1)} = u_i^{(2)}, \end{cases} \qquad i \in \{0, 1 \cdots (k-1)\}, \quad k \in \mathbb{N}^*. \tag{1.51}$$

Theorem 1.3 (Hamming Distance). *Let $\boldsymbol{u}^{(1)}$, having the components $u_i^{(1)} \in \mathbb{F}_2$, $i \in \{0, 1 \cdots (k-1)\}$, $k \in \mathbb{N}^*$ and $\boldsymbol{u}^{(2)}$, having the components $u_i^{(2)} \in \mathbb{F}_2$, $i \in \{0, 1 \cdots (k-1)\}$, $k \in \mathbb{N}^*$, be two arbitrary vectors of $\mathbb{F}_2^k$, $k \in \mathbb{N}^*$. Then the Hamming distance $d_H\{\boldsymbol{u}^{(1)}, \boldsymbol{u}^{(2)}\}$ is given by*

$$d_H\{\boldsymbol{u}^{(1)}, \boldsymbol{u}^{(2)}\} = \sum_{i=0}^{k-1} d_H\{u_i^{(1)}, u_i^{(2)}\} = \sum_{i=0}^{k-1} u_i^{(1)} \oplus u_i^{(2)}, \quad k \in \mathbb{N}^*. \tag{1.52}$$

Proof. The proof is given in Remark 1.5 and (1.51). □

Remark 1.6 (On the Hamming Distance Again). Let $\boldsymbol{u}^{(1)}$, having the components $u_i^{(1)} \in \mathbb{F}_2$, $i \in \{0, 1 \cdots (k-1)\}$, $k \in \mathbb{N}^*$ and $\boldsymbol{u}^{(2)}$, having the components $u_i^{(2)} \in \mathbb{F}_2$, $i \in \{0, 1 \cdots (k-1)\}$, $k \in \mathbb{N}^*$ be two arbitrary vectors of $\mathbb{F}_2^k$, $k \in \mathbb{N}^*$. Comparing (1.48) and (1.52), we immediately find [15, equation (3.15), p. 76]

$$d_{\mathrm{H}}\{\boldsymbol{u}^{(1)}, \boldsymbol{u}^{(2)}\} = w_{\mathrm{H}}\{\boldsymbol{u}^{(1)} \oplus \boldsymbol{u}^{(2)}\}. \tag{1.53}$$

In particular, if $\boldsymbol{u}^{(2)}$ is equal to the zero vector $\mathbf{0}_k$, $k \in \mathbb{N}^*$, (1.53) yields

$$d_{\mathrm{H}}\{\boldsymbol{u}^{(1)}, \mathbf{0}_k\} = w_{\mathrm{H}}\{\boldsymbol{u}^{(1)} \oplus \mathbf{0}_k\} = w_{\mathrm{H}}\{\boldsymbol{u}^{(1)}\}. \tag{1.54}$$

Again, let us look at an example.

Example 1.2. Let

$$\boldsymbol{u}^{(1)} = (1, 0, 0, 1, 0, 1, 1) \quad \boldsymbol{u}^{(2)} = (0, 1, 0, 0, 0, 1, 1). \tag{1.55}$$

We find

$$d_{\mathrm{H}}\{\boldsymbol{u}^{(1)}, \boldsymbol{u}^{(2)}\} = d_{\mathrm{H}}\{(1, 0, 0, 1, 0, 1, 1), (0, 1, 0, 0, 0, 1, 1)\} = 3 \tag{1.56}$$

because $\boldsymbol{u}^{(1)}$ and $\boldsymbol{u}^{(2)}$ differ in the zeroth, the first and the third places [15, p. 76].

Now, let us look at the properties of the Hamming distance $d_{\mathrm{H}}\{\cdot, \cdot\}$.

Corollary 1.1 (Hamming Distance Equal to 0 Only for Identical Vectors). *Let $\boldsymbol{u}^{(1)}$, having the components $u_i^{(1)} \in \mathbb{F}_2$, $i \in \{0, 1 \cdots (k-1)\}$, $k \in \mathbb{N}^*$ and $\boldsymbol{u}^{(2)}$, having the components $u_i^{(2)} \in \mathbb{F}_2$, $i \in \{0, 1 \cdots (k-1)\}$, $k \in \mathbb{N}^*$, be two arbitrary vectors of $\mathbb{F}_2^k$, $k \in \mathbb{N}^*$. The relation*

$$d_H\left\{\boldsymbol{u}^{(1)}, \boldsymbol{u}^{(2)}\right\} = 0 \tag{1.57}$$

only holds if $\boldsymbol{u}^{(1)}$ and $\boldsymbol{u}^{(2)}$ are identical.

Proof. According to Definition 1.13, the Hamming distance between $\boldsymbol{u}^{(1)}$ and $\boldsymbol{u}^{(2)}$, denoted by $d_\mathrm{H}\{\boldsymbol{u}^{(1)}, \boldsymbol{u}^{(2)}\}$, is defined as the number of places where they differ [15, p. 76]. If $d_\mathrm{H}\{\boldsymbol{u}^{(1)}, \boldsymbol{u}^{(2)}\}$ is equal to zero, $\boldsymbol{u}^{(1)}$ and $\boldsymbol{u}^{(2)}$ differ in no places, which immediately leads to

$$\boldsymbol{u}^{(1)} = \boldsymbol{u}^{(2)}. \tag{1.58}$$

Now, let $\boldsymbol{u}^{(1)}$ and $\boldsymbol{u}^{(2)}$ be identical. In this case, $\boldsymbol{u}^{(1)}$ and $\boldsymbol{u}^{(2)}$ differ in no places, and thus we find

$$d_\mathrm{H}\left\{\boldsymbol{u}^{(1)}, \boldsymbol{u}^{(2)}\right\} = d_\mathrm{H}\left\{\boldsymbol{u}^{(1)}, \boldsymbol{u}^{(1)}\right\} = d_\mathrm{H}\left\{\boldsymbol{u}^{(2)}, \boldsymbol{u}^{(2)}\right\} = w_\mathrm{H}\{\boldsymbol{0}_k\} = 0, \quad k \in \mathbb{N}^*. \tag{1.59}$$

□

Corollary 1.2 (Hamming Distance Greater Than or Equal to 0). *Let $\boldsymbol{u}^{(1)}$, having the components $u_i^{(1)} \in \mathbb{F}_2$, $i \in \{0, 1 \cdots (k-1)\}$, $k \in \mathbb{N}^*$ and $\boldsymbol{u}^{(2)}$, having the components $u_i^{(2)} \in \mathbb{F}_2$, $i \in \{0, 1 \cdots (k-1)\}$, $k \in \mathbb{N}^*$, be two arbitrary vectors of $\mathbb{F}_2^k$, $k \in \mathbb{N}^*$. Their Hamming distance $d_H\{\cdot, \cdot\}$ has the following property:*

$$d_H\left\{\boldsymbol{u}^{(1)}, \boldsymbol{u}^{(2)}\right\} \geq 0. \tag{1.60}$$

Proof. It follows from (1.53) that the Hamming distance $d_\mathrm{H}\{\boldsymbol{u}^{(1)}, \boldsymbol{u}^{(2)}\}$ is equal to 0 if and only if $\boldsymbol{u}^{(1)}$ is equal to $\boldsymbol{u}^{(2)}$. In this case, (1.51) leads to

$$d_\mathrm{H}\left\{u_i^{(1)}, u_i^{(2)}\right\} = u_i^{(1)} \oplus u_i^{(2)} = 0, \quad i \in \left\{0, 1 \cdots (k-1)\right\}, \quad k \in \mathbb{N}^*, \tag{1.61}$$

and thus

$$d_\mathrm{H}\left\{\boldsymbol{u}^{(1)}, \boldsymbol{u}^{(2)}\right\} = \sum_{i=0}^{k-1} d_\mathrm{H}\left\{u_i^{(1)}, u_i^{(2)}\right\} = \sum_{i=0}^{k-1} 0 = 0, \quad k \in \mathbb{N}^*. \tag{1.62}$$

Now, let us assume that there are $d \in \mathbb{N}^*$ components with the indexes $\{i_1, i_2 \cdots i_d\} \subseteq \{0, 1 \cdots (k-1)\}$, at which $\boldsymbol{u}^{(1)}$ and $\boldsymbol{u}^{(2)}$ differ. In this case,

- $d_\mathrm{H}\{u_i^{(1)}, u_i^{(2)}\}$ is equal to 0 for all indexes $i \notin \{i_1, i_2 \cdots i_d\}$ and
- $d_\mathrm{H}\{u_i^{(1)}, u_i^{(2)}\}$ is equal to 1 for all indexes $i \in \{i_1, i_2 \cdots i_d\}$.

Then we obtain

$$d_\mathrm{H}\left\{\boldsymbol{u}^{(1)}, \boldsymbol{u}^{(2)}\right\} = \sum_{i=0}^{k-1} d_\mathrm{H}\left\{u_i^{(1)}, u_i^{(2)}\right\} = d > 0. \tag{1.63}$$

□

Corollary 1.3 (Hamming Distance is Symmetric). *Let $\boldsymbol{u}^{(1)}$, having the components $u_i^{(1)} \in \mathbb{F}_2$, $i \in \{0, 1 \cdots (k-1)\}$, $k \in \mathbb{N}^*$ and $\boldsymbol{u}^{(2)}$, having the components $u_i^{(2)} \in \mathbb{F}_2$, $i \in \{0, 1 \cdots (k-1)\}$, $k \in \mathbb{N}^*$, be two arbitrary vectors of $\mathbb{F}_2^k$, $k \in \mathbb{N}^*$. Their Hamming distance $d_H\{\boldsymbol{u}^{(1)}, \boldsymbol{u}^{(2)}\}$ has the following property:*

$$d_H\left\{\boldsymbol{u}^{(1)}, \boldsymbol{u}^{(2)}\right\} = d_H\left\{\boldsymbol{u}^{(2)}, \boldsymbol{u}^{(1)}\right\}. \tag{1.64}$$

Proof. With (1.53), we find

$$d_H\{\boldsymbol{u}^{(1)}, \boldsymbol{u}^{(2)}\} = w_H\{\boldsymbol{u}^{(1)} \oplus \boldsymbol{u}^{(2)}\} = w_H\{\boldsymbol{u}^{(2)} \oplus \boldsymbol{u}^{(1)}\} = d_H\{\boldsymbol{u}^{(2)}, \boldsymbol{u}^{(1)}\}. \tag{1.65}$$

□

Theorem 1.4 (Triangle Inequality of the Hamming Distance). *Let $\boldsymbol{u}^{(1)}$, having the components $u_i^{(1)} \in \mathbb{F}_2$, $i \in \{0, 1 \cdots (k-1)\}$, $k \in \mathbb{N}^*$, $\boldsymbol{u}^{(2)}$, having the components $u_i^{(2)} \in \mathbb{F}_2$, $i \in \{0, 1 \cdots (k-1)\}$, $k \in \mathbb{N}^*$ and $\boldsymbol{u}^{(3)}$, having the components $u_i^{(3)} \in \mathbb{F}_2$, $i \in \{0, 1 \cdots (k-1)\}$, $k \in \mathbb{N}^*$, be three arbitrary vectors of $\mathbb{F}_2^k$, $k \in \mathbb{N}^*$. The following relation, termed* triangle inequality *[15, p. 13], [21, p. 3], holds*

$$d_H\{\boldsymbol{u}^{(1)}, \boldsymbol{u}^{(2)}\} + d_H\{\boldsymbol{u}^{(2)}, \boldsymbol{u}^{(3)}\} \geq d_H\{\boldsymbol{u}^{(1)}, \boldsymbol{u}^{(3)}\}. \tag{1.66}$$

Proof. We have

$$0 \leq d_H\{u_i^{(1)}, u_i^{(2)}\} \leq 1, \quad 0 \leq d_H\{u_i^{(2)}, u_i^{(3)}\} \leq 1, \quad 0 \leq d_H\{u_i^{(1)}, u_i^{(3)}\} \leq 1, \\ i \in \{0, 1 \cdots (k-1)\}, \quad k \in \mathbb{N}^*. \tag{1.67}$$

Furthermore, we have

$$d_H\{\boldsymbol{u}^{(1)}, \boldsymbol{u}^{(2)}\} = \sum_{i=0}^{k-1} d_H\{u_i^{(1)}, u_i^{(2)}\}, \tag{1.68}$$

$$d_H\{\boldsymbol{u}^{(2)}, \boldsymbol{u}^{(3)}\} = \sum_{i=0}^{k-1} d_H\{u_i^{(2)}, u_i^{(3)}\}, \tag{1.69}$$

$$d_H\{\boldsymbol{u}^{(1)}, \boldsymbol{u}^{(3)}\} = \sum_{i=0}^{k-1} d_H\{u_i^{(1)}, u_i^{(3)}\}. \tag{1.70}$$

Now, we yield

$$d_H\{\boldsymbol{u}^{(1)}, \boldsymbol{u}^{(2)}\} + d_H\{\boldsymbol{u}^{(2)}, \boldsymbol{u}^{(3)}\} = \sum_{i=0}^{k-1} \left[d_H\{u_i^{(1)}, u_i^{(2)}\} + d_H\{u_i^{(2)}, u_i^{(3)}\} \right], \\ k \in \mathbb{N}^*. \tag{1.71}$$

We have to prove

$$\sum_{i=0}^{k-1} \left[d_H\{u_i^{(1)}, u_i^{(2)}\} + d_H\{u_i^{(2)}, u_i^{(3)}\} \right] \geq \sum_{i=0}^{k-1} d_H\{u_i^{(1)}, u_i^{(3)}\}. \tag{1.72}$$

There are five cases altogether that we have to look at in the proof.

Case 1: $u_i^{(1)} = u_i^{(2)} = u_i^{(3)}$ for all $i \in \{0, 1 \cdots (k-1)\}$

We have

$$d_H\{\boldsymbol{u}^{(1)}, \boldsymbol{u}^{(2)}\} = d_H\{\boldsymbol{u}^{(2)}, \boldsymbol{u}^{(3)}\} = d_H\{\boldsymbol{u}^{(1)}, \boldsymbol{u}^{(3)}\} = 0, \tag{1.73}$$

yielding the true expression

$$\underbrace{d_H\{\boldsymbol{u}^{(1)}, \boldsymbol{u}^{(2)}\}}_{=0} + \underbrace{d_H\{\boldsymbol{u}^{(2)}, \boldsymbol{u}^{(3)}\}}_{=0} = \underbrace{d_H\{\boldsymbol{u}^{(1)}, \boldsymbol{u}^{(3)}\}}_{=0}. \tag{1.74}$$

Case 2: $u_i^{(1)} = u_i^{(2)}$ for all $i \in \{0, 1 \cdots (k-1)\}$ and $u_i^{(1)} = u_i^{(2)} \neq u_i^{(3)}$ for at least one index i

We have

$$d_H\{\boldsymbol{u}^{(1)}, \boldsymbol{u}^{(2)}\} = 0 \tag{1.75}$$

and

$$d_{\mathrm{H}}\left\{\boldsymbol{u}^{(2)}, \boldsymbol{u}^{(3)}\right\} = d_{\mathrm{H}}\left\{\boldsymbol{u}^{(1)}, \boldsymbol{u}^{(3)}\right\} > 0, \tag{1.76}$$

which immediately yields the true expression

$$\underbrace{d_{\mathrm{H}}\left\{\boldsymbol{u}^{(1)}, \boldsymbol{u}^{(2)}\right\}}_{=0} + \underbrace{d_{\mathrm{H}}\left\{\boldsymbol{u}^{(2)}, \boldsymbol{u}^{(3)}\right\}}_{>0} = \underbrace{d_{\mathrm{H}}\left\{\boldsymbol{u}^{(1)}, \boldsymbol{u}^{(3)}\right\}}_{=d_{\mathrm{H}}\{\boldsymbol{u}^{(2)},\boldsymbol{u}^{(3)}\}>0}. \tag{1.77}$$

<u>*Case 3*: $u_i^{(1)} = u_i^{(3)}$ for all $i \in \{0, 1 \cdots (k-1)\}$ and $u_i^{(1)} = u_i^{(3)} \neq u_i^{(2)}$ for at least one index i</u>

We have

$$d_{\mathrm{H}}\left\{\boldsymbol{u}^{(1)}, \boldsymbol{u}^{(3)}\right\} = 0 \tag{1.78}$$

and

$$d_{\mathrm{H}}\left\{\boldsymbol{u}^{(1)}, \boldsymbol{u}^{(2)}\right\} = \underbrace{d_{\mathrm{H}}\left\{\boldsymbol{u}^{(2)}, \boldsymbol{u}^{(3)}\right\}}_{=d_{\mathrm{H}}\{\boldsymbol{u}^{(3)},\boldsymbol{u}^{(2)}\}} > 0, \tag{1.79}$$

which immediately yields the true expression

$$\underbrace{d_{\mathrm{H}}\left\{\boldsymbol{u}^{(1)}, \boldsymbol{u}^{(2)}\right\}}_{>0} + \underbrace{d_{\mathrm{H}}\left\{\boldsymbol{u}^{(2)}, \boldsymbol{u}^{(3)}\right\}}_{=d_{\mathrm{H}}\{\boldsymbol{u}^{(1)},\boldsymbol{u}^{(2)}\}>0} > \underbrace{d_{\mathrm{H}}\left\{\boldsymbol{u}^{(1)}, \boldsymbol{u}^{(3)}\right\}}_{=0}. \tag{1.80}$$

<u>*Case 4*: $u_i^{(2)} = u_i^{(3)}$ for all $i \in \{0, 1 \cdots (k-1)\}$ and $u_i^{(2)} = u_i^{(3)} \neq u_i^{(1)}$ for at least one index i</u>

We have

$$d_{\mathrm{H}}\left\{\boldsymbol{u}^{(2)}, \boldsymbol{u}^{(3)}\right\} = 0 \tag{1.81}$$

and

$$d_{\mathrm{H}}\left\{\boldsymbol{u}^{(1)}, \boldsymbol{u}^{(2)}\right\} = d_{\mathrm{H}}\left\{\boldsymbol{u}^{(1)}, \boldsymbol{u}^{(3)}\right\} > 0, \tag{1.82}$$

which immediately yields the true expression

$$\underbrace{d_{\mathrm{H}}\left\{\boldsymbol{u}^{(1)}, \boldsymbol{u}^{(2)}\right\}}_{>0} + \underbrace{d_{\mathrm{H}}\left\{\boldsymbol{u}^{(2)}, \boldsymbol{u}^{(3)}\right\}}_{=0} = \underbrace{d_{\mathrm{H}}\left\{\boldsymbol{u}^{(1)}, \boldsymbol{u}^{(3)}\right\}}_{=d_{\mathrm{H}}\{\boldsymbol{u}^{(1)},\boldsymbol{u}^{(2)}\}>0}. \tag{1.83}$$

<u>*Case 5*: All three vectors differ</u>

If $\boldsymbol{u}^{(1)}$ and $\boldsymbol{u}^{(2)}$ differ in $u_i^{(1)}$, respectively, $u_i^{(2)}$, then $u_i^{(3)}$ must be identical to either $u_i^{(1)}$ or $u_i^{(2)}$. Then

$$\underbrace{\underbrace{d_{\mathrm{H}}\left\{u_i^{(1)}, u_i^{(2)}\right\}}_{=1} + \underbrace{d_{\mathrm{H}}\left\{u_i^{(2)}, u_i^{(3)}\right\}}_{\geq 0}}_{\geq 1} \geq \underbrace{d_{\mathrm{H}}\left\{u_i^{(1)}, u_i^{(3)}\right\}}_{\geq 0} \geq 0. \tag{1.84}$$

If $\boldsymbol{u}^{(2)}$ and $\boldsymbol{u}^{(3)}$ differ in $u_{\bar{i}}^{(2)}$, respectively, $u_{\bar{i}}^{(3)}$, then $u_{\bar{i}}^{(1)}$ must be identical to either $u_{\bar{i}}^{(2)}$ or $u_{\bar{i}}^{(3)}$. Then

$$\underbrace{\underbrace{d_{\mathrm{H}}\left\{u_{\bar{i}}^{(1)}, u_{\bar{i}}^{(2)}\right\}}_{\geq 0} + \underbrace{d_{\mathrm{H}}\left\{u_{\bar{i}}^{(2)}, u_{\bar{i}}^{(3)}\right\}}_{=1}}_{\geq 1} \geq \underbrace{d_{\mathrm{H}}\left\{u_{\bar{i}}^{(1)}, u_{\bar{i}}^{(3)}\right\}}_{\geq 0} \geq 0. \tag{1.85}$$

Therefore,

$$\underbrace{d_H\{\boldsymbol{u}^{(1)},\boldsymbol{u}^{(2)}\}}_{\geq 1} + \underbrace{d_H\{\boldsymbol{u}^{(2)},\boldsymbol{u}^{(3)}\}}_{\geq 1} > \underbrace{d_H\{\boldsymbol{u}^{(1)},\boldsymbol{u}^{(3)}\}}_{\geq 0}. \tag{1.86}$$

□

Remark 1.7 (On Theorem 1.4). Let $\boldsymbol{u}^{(1)}$, having the components $u_i^{(1)} \in \mathbb{F}_2, i \in \{0, 1 \cdots (k-1)\}, k \in \mathbb{N}^*$, $\boldsymbol{u}^{(2)}$, having the components $u_i^{(2)} \in \mathbb{F}_2, i \in \{0, 1 \cdots (k-1)\}, k \in \mathbb{N}^*$ and $\boldsymbol{u}^{(3)}$, having the components $u_i^{(3)} \in \mathbb{F}_2, i \in \{0, 1 \cdots (k-1)\}, k \in \mathbb{N}^*$ be three arbitrary vectors of $\mathbb{F}_2^k, k \in \mathbb{N}^*$.

Using (1.53), Theorem 1.4 can be written in the following form:

$$w_H\{\boldsymbol{u}^{(1)} \oplus \boldsymbol{u}^{(2)}\} + w_H\{\boldsymbol{u}^{(2)} \oplus \boldsymbol{u}^{(3)}\} \geq w_H\{\boldsymbol{u}^{(1)} \oplus \boldsymbol{u}^{(3)}\}. \tag{1.87}$$

Taking Corollary 1.1, Corollary 1.2, Corollary 1.3 and Theorem 1.4 into account, we find that $d_H\{\cdot,\cdot\}$ of Definition 1.13 fulfills the following three *axioms of the metric space* [21, p. 3]:

- $d_H\{\boldsymbol{u}^{(1)},\boldsymbol{u}^{(2)}\} \geq 0$; $d_H\{\boldsymbol{u}^{(1)},\boldsymbol{u}^{(1)}\} = 0$; $d_H\{\boldsymbol{u}^{(1)},\boldsymbol{u}^{(2)}\} = 0$ yields $\boldsymbol{u}^{(1)} = \boldsymbol{u}^{(2)}$.
- $d_H\{\boldsymbol{u}^{(1)},\boldsymbol{u}^{(2)}\} = d_H\{\boldsymbol{u}^{(2)},\boldsymbol{u}^{(1)}\}$.
- $d_H\{\boldsymbol{u}^{(1)},\boldsymbol{u}^{(2)}\} + d_H\{\boldsymbol{u}^{(2)},\boldsymbol{u}^{(3)}\} \geq d_H\{\boldsymbol{u}^{(1)},\boldsymbol{u}^{(3)}\}$.

Since the Hamming distance $d_H\{\cdot,\cdot\}$ of Definition 1.13 applies to $\mathbb{F}_2^k, k \in \mathbb{N}^*$, the vector space $\mathbb{F}_2^k, k \in \mathbb{N}^*$, is a *metric space* [21, p. 3].

Example 1.3. Let us choose

$$\boldsymbol{u}^{(1)} = (1,0,0,1,0,1,1), \quad \boldsymbol{u}^{(2)} = (0,1,0,0,0,1,1), \quad \boldsymbol{u}^{(3)} = (0,1,1,1,0,1,1). \tag{1.88}$$

We find

$$d_H\{\boldsymbol{u}^{(1)},\boldsymbol{u}^{(2)}\} = 3, \quad d_H\{\boldsymbol{u}^{(2)},\boldsymbol{u}^{(3)}\} = 2, \quad d_H\{\boldsymbol{u}^{(1)},\boldsymbol{u}^{(3)}\} = 3. \tag{1.89}$$

Obviously,

$$d_H\{\boldsymbol{u}^{(1)},\boldsymbol{u}^{(2)}\} + d_H\{\boldsymbol{u}^{(2)},\boldsymbol{u}^{(3)}\} = 5 \geq 3 = d_H\{\boldsymbol{u}^{(1)},\boldsymbol{u}^{(3)}\}. \tag{1.90}$$

Let us look at another example.

Example 1.4. Consider the triangle depicted in Figure 1.4, which has the side lengths

$$a = \overline{BC}, \quad b = \overline{AC}, \quad c = \overline{AB}. \tag{1.91}$$

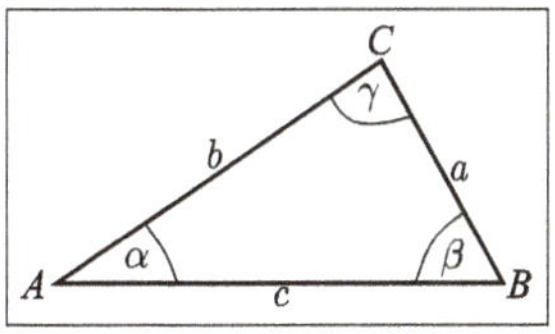

Figure 1.4: Triangle.

From Figure 1.4, we find the *triangle inequalities:*

$$a + b \geq c, \quad a + c \geq b, \quad b + c \geq a. \tag{1.92}$$

Furthermore, in a triangle, we have

$$\alpha + \beta + \gamma = \pi. \tag{1.93}$$

Let

- $\boldsymbol{a}$ be the vector, which originates at B and terminates at C,
- $\boldsymbol{b}$ be the vector, which originates at A and terminates at C and
- $\boldsymbol{c}$ be the vector, which originates at A and terminates at B.

In the two-dimensional Euclidean vector space, let $\|\boldsymbol{a}\|$ and $\|\boldsymbol{b}\|$ be the norms of $\boldsymbol{a}$ and $\boldsymbol{b}$, i. e., the "lengths" of $\boldsymbol{a}$ and $\boldsymbol{b}$. Therefore, $\|\boldsymbol{a}\|$ and $\|\boldsymbol{b}\|$ are the respective distances between $\boldsymbol{a}$ and $\boldsymbol{b}$, and the origin, which is represented by the zero vector $\mathbf{0}$.

Let us denote the angle that $\boldsymbol{a}$ and $\boldsymbol{b}$ enclose by $\measuredangle(\boldsymbol{a}, \boldsymbol{b})$. Clearly, $\measuredangle(\boldsymbol{a}, \boldsymbol{b})$ is lower than or equal to π. In this example, let us define the inner product $\langle \boldsymbol{a} \mid \boldsymbol{b} \rangle$ between $\boldsymbol{a}$ and $\boldsymbol{b}$ as the product of their norms, $\|\boldsymbol{a}\|$ and $\|\boldsymbol{b}\|$ and the cosine of $\measuredangle(\boldsymbol{a}, \boldsymbol{b})$, i. e.,

$$\langle \boldsymbol{a} \mid \boldsymbol{b} \rangle = \|\boldsymbol{a}\| \cdot \|\boldsymbol{b}\| \cdot \cos\left\{\measuredangle(\boldsymbol{a}, \boldsymbol{b})\right\}. \tag{1.94}$$

Clearly,

$$\langle \boldsymbol{a} \mid \boldsymbol{a} \rangle = \|\boldsymbol{a}\|^2, \quad \langle \boldsymbol{b} \mid \boldsymbol{b} \rangle = \|\boldsymbol{b}\|^2, \tag{1.95}$$

and thus

$$\|\boldsymbol{a}\| = \sqrt{\langle \boldsymbol{a} \mid \boldsymbol{a} \rangle}, \quad \|\boldsymbol{b}\| = \sqrt{\langle \boldsymbol{b} \mid \boldsymbol{b} \rangle}, \tag{1.96}$$

respectively. According to Figure 1.4, we yield

$$\boldsymbol{b} - (-\boldsymbol{a}) = \boldsymbol{a} + \boldsymbol{b} = \boldsymbol{c}. \tag{1.97}$$

Using (1.96), the squared distance between $(-\boldsymbol{a})$ and $\boldsymbol{b}$ is given by

$$\begin{aligned} \|\boldsymbol{c}\|^2 = \|\boldsymbol{a} + \boldsymbol{b}\|^2 &= \big\langle (\boldsymbol{a} + \boldsymbol{b}) \mid (\boldsymbol{a} + \boldsymbol{b}) \big\rangle \\ &= \langle \boldsymbol{a} \mid \boldsymbol{a} \rangle + 2\langle \boldsymbol{a} \mid \boldsymbol{b} \rangle + \langle \boldsymbol{b} \mid \boldsymbol{b} \rangle & (1.98) \\ &= \|\boldsymbol{a}\|^2 + \|\boldsymbol{b}\|^2 + 2\|\boldsymbol{a}\|\|\boldsymbol{b}\| \cos\left\{\measuredangle(\boldsymbol{a}, \boldsymbol{b})\right\}. & (1.99) \end{aligned}$$

Since

$$\cos\left\{\measuredangle(\boldsymbol{a}, \boldsymbol{b})\right\} \leq 1, \tag{1.100}$$

we obtain

$$2\|\boldsymbol{a}\|\|\boldsymbol{b}\| \cos\left\{\measuredangle(\boldsymbol{a}, \boldsymbol{b})\right\} \leq 2\|\boldsymbol{a}\|\|\boldsymbol{b}\|, \tag{1.101}$$

leading to the *Cauchy-Schwarz inequality* [11, equation (1.114a), p. 31]

$$\left|\langle \boldsymbol{a} \mid \boldsymbol{b} \rangle\right| \leq \|\boldsymbol{a}\|\|\boldsymbol{b}\|, \tag{1.102}$$

and thus

$$\|\boldsymbol{a}+\boldsymbol{b}\|^2 \le \|\boldsymbol{a}\|^2 + \|\boldsymbol{b}\|^2 + 2\|\boldsymbol{a}\|\|\boldsymbol{b}\|. \tag{1.103}$$

With

$$\|\boldsymbol{a}\|^2 + \|\boldsymbol{b}\|^2 + 2\|\boldsymbol{a}\|\|\boldsymbol{b}\| = \left(\|\boldsymbol{a}\| + \|\boldsymbol{b}\|\right)^2, \tag{1.104}$$

we obtain the *triangle inequality* [11, equation (4.46), p. 266]

$$\|\boldsymbol{c}\| = \|\boldsymbol{a}+\boldsymbol{b}\| \le \|\boldsymbol{a}\| + \|\boldsymbol{b}\|. \tag{1.105}$$

Using Theorem 1.4, we can prove the *reverse triangle inequality*.

Theorem 1.5 (Reverse Triangle Inequality of the Hamming Distance). *Let $\boldsymbol{u}^{(1)}$, having the components $u_i^{(1)} \in \mathbb{F}_2$, $i \in \{0,1\cdots(k-1)\}$, $k \in \mathbb{N}^*$ and $\boldsymbol{u}^{(2)}$, having the components $u_i^{(2)} \in \mathbb{F}_2$, $i \in \{0,1\cdots(k-1)\}$, $k \in \mathbb{N}^*$, be two arbitrary vectors of $\mathbb{F}_2^k$, $k \in \mathbb{N}^*$. The Hamming weight $w_H\{\cdot\}$ has the following property (reverse triangle inequality* [4, equation (15), p. 12]*):*

$$w_H\{\boldsymbol{u}^{(1)}\} - w_H\{\boldsymbol{u}^{(2)}\} \le w_H\{\boldsymbol{u}^{(1)} \oplus \boldsymbol{u}^{(2)}\}, \tag{1.106}$$

$w_H\{\boldsymbol{u}^{(1)} \oplus \boldsymbol{u}^{(2)}\}$ being the Hamming distance $d_H\{\boldsymbol{u}^{(1)}, \boldsymbol{u}^{(2)}\}$.

Proof. The weight of $w_{\mathrm{H}}\{\boldsymbol{u}^{(1)}\}$ of $\boldsymbol{u}^{(1)}$ is given by

$$\begin{aligned} w_{\mathrm{H}}\{\boldsymbol{u}^{(1)}\} &= w_{\mathrm{H}}\Big\{\boldsymbol{u}^{(1)} \oplus \underbrace{\left(\boldsymbol{u}^{(2)} \oplus \boldsymbol{u}^{(2)}\right) \oplus \boldsymbol{0}_k}_{=\boldsymbol{0}_k}\Big\} \\ &= w_{\mathrm{H}}\left\{\left(\boldsymbol{u}^{(1)} \oplus \boldsymbol{u}^{(2)}\right) \oplus \left(\boldsymbol{u}^{(2)} \oplus \boldsymbol{0}_k\right)\right\} \\ &= d_{\mathrm{H}}\left\{\left(\boldsymbol{u}^{(1)} \oplus \boldsymbol{u}^{(2)}\right), \left(\boldsymbol{u}^{(2)} \oplus \boldsymbol{0}_k\right)\right\}. \end{aligned} \tag{1.107}$$

Using Theorem 1.4, we find

$$w_{\mathrm{H}}\{\boldsymbol{u}^{(1)}\} \le d_{\mathrm{H}}\{\boldsymbol{u}^{(1)}, \boldsymbol{u}^{(2)}\} + \underbrace{d_{\mathrm{H}}\{\boldsymbol{u}^{(2)}, \boldsymbol{0}_k\}}_{=w_{\mathrm{H}}\{\boldsymbol{u}^{(2)}\}} = d_{\mathrm{H}}\{\boldsymbol{u}^{(1)}, \boldsymbol{u}^{(2)}\} + w_{\mathrm{H}}\{\boldsymbol{u}^{(2)}\}, \tag{1.108}$$

which yields

$$w_{\mathrm{H}}\{\boldsymbol{u}^{(1)}\} - w_{\mathrm{H}}\{\boldsymbol{u}^{(2)}\} \le d_{\mathrm{H}}\{\boldsymbol{u}^{(1)}, \boldsymbol{u}^{(2)}\} = w_{\mathrm{H}}\{\boldsymbol{u}^{(1)} \oplus \boldsymbol{u}^{(2)}\}. \tag{1.109}$$

If $\boldsymbol{u}^{(1)}$ and $\boldsymbol{u}^{(2)}$ differ, then $w_{\mathrm{H}}\{\boldsymbol{u}^{(1)} \oplus \boldsymbol{u}^{(2)}\}$ is positive, however, $(w_{\mathrm{H}}\{\boldsymbol{u}^{(1)}\} - w_{\mathrm{H}}\{\boldsymbol{u}^{(2)}\})$ will only be positive if $w_{\mathrm{H}}\{\boldsymbol{u}^{(1)}\}$ is greater than $w_{\mathrm{H}}\{\boldsymbol{u}^{(2)}\}$, which means that the equality in (1.106) will not always be reached for distinct $\boldsymbol{u}^{(1)}$ and $\boldsymbol{u}^{(2)}$. If, however, $\boldsymbol{u}^{(1)}$ and $\boldsymbol{u}^{(2)}$ are identical, we always yield the equality in (1.106). □

Letting $\boldsymbol{u}^{(2)} \in \mathbb{F}_2^k$ be the zero vector $\boldsymbol{0}_k \in \mathbb{F}_2^k$, $k \in \mathbb{N}^*$, (1.66) becomes

$$d_{\mathrm{H}}\{\boldsymbol{u}^{(1)}, \boldsymbol{u}^{(3)}\} \le \underbrace{d_{\mathrm{H}}\{\boldsymbol{u}^{(1)}, \boldsymbol{0}_k\}}_{=w_{\mathrm{H}}\{\boldsymbol{u}^{(1)}\}} + \underbrace{d_{\mathrm{H}}\{\boldsymbol{0}_k, \boldsymbol{u}^{(3)}\}}_{=w_{\mathrm{H}}\{\boldsymbol{u}^{(3)}\}}, \quad k \in \mathbb{N}^*. \tag{1.110}$$

Renaming $\boldsymbol{u}^{(3)}$ by $\boldsymbol{u}^{(2)}$ and combining (1.110) with (1.109) yields

$$w_{\mathrm{H}}\{\boldsymbol{u}^{(1)}\} - w_{\mathrm{H}}\{\boldsymbol{u}^{(2)}\} \le \underbrace{d_{\mathrm{H}}\{\boldsymbol{u}^{(1)}, \boldsymbol{u}^{(2)}\}}_{=w_{\mathrm{H}}\{\boldsymbol{u}^{(1)} \oplus \boldsymbol{u}^{(2)}\}} \le w_{\mathrm{H}}\{\boldsymbol{u}^{(1)}\} + w_{\mathrm{H}}\{\boldsymbol{u}^{(2)}\} \tag{1.111}$$

for any $\boldsymbol{u}^{(1)}, \boldsymbol{u}^{(2)} \in \mathbb{F}_2^k$.

In what follows, we will use the notation of the type

$$\bigoplus_{i=0}^{m-1} a_i \odot \boldsymbol{u}^{(i)}, \quad m \in \{2, 3 \cdots 2^k\}, \quad k \in \mathbb{N}^*, \tag{1.112}$$

or, alternatively, $\bigoplus_{i=0}^{m-1} a_i \odot \boldsymbol{u}^{(i)}$, which is the abbreviation of the summation using the additive operation, i. e., $\oplus$, m times, $m \in \{2, 3 \cdots 2^k\}$, $k \in \mathbb{N}^*$.

Let $\boldsymbol{u}^{(0)} \in \mathbb{F}_2^k$, $\boldsymbol{u}^{(1)} \in \mathbb{F}_2^k, \cdots \boldsymbol{u}^{(m-1)} \in \mathbb{F}_2^k$, be $m \in \{2, 3 \cdots 2^k\}$, $k \in \mathbb{N}^*$, vectors of $\mathbb{F}_2^k$, which are not all identical and let $a_0 \in \mathbb{F}_2$, $a_1 \in \mathbb{F}_2, \cdots a_{m-1} \in \mathbb{F}_2$, be arbitrarily chosen elements of $\mathbb{F}_2$. Then taking (1.43) into consideration, leads us to the conclusion that

$$a_0 \odot \boldsymbol{u}^{(0)} \oplus a_1 \odot \boldsymbol{u}^{(1)} \oplus \cdots \oplus a_{m-1} \odot \boldsymbol{u}^{(m-1)} = \bigoplus_{i=0}^{m-1} a_i \odot \boldsymbol{u}^{(i)}, \tag{1.113}$$
$$a_i \in \mathbb{F}_2, \quad \boldsymbol{u}^{(i)} \in \mathbb{F}_2^k, \quad i \in \{0, 1 \cdots m\}, \quad m \in \{2, 3 \cdots 2^k\}, \quad k \in \mathbb{N}^*,$$

is also an element of $\mathbb{F}_2^k$. Equation (1.113) is called a *linear combination of the vectors* $\boldsymbol{u}^{(i)} \in \mathbb{F}_2^k$ [14, p. 411], [22, equation (1.96), p. 70], [21, p. 40]. Equation (1.113) immediately leads us to *linear independence* [11, p. 619], [22, p. 71].

Definition 1.14 (Linearly Independent and Linearly Dependent Vectors of $\mathbb{F}_2^k$). The set $\{\boldsymbol{u}^{(0)}, \boldsymbol{u}^{(1)} \cdots \boldsymbol{u}^{(m-1)}\}$ of vectors, $m \in \{2, 3 \cdots 2^k\}$, $k \in \mathbb{N}^*$, which are taken from $\mathbb{F}_2^k$ and which are mutually distinct, i. e., they pairwise differ from each other, is termed *linearly independent*, if there is no set of elements $\{a_0, a_1 \cdots a_{m-1}\}$, which are taken from $\mathbb{F}_2$ and which are not all equal to 0, such that [11, Section 5.3.7.2, p. 327], [14, p. 411], [22, p. 71], [21, p. 40],

$$\bigoplus_{i=0}^{m-1} a_i \odot \boldsymbol{u}^{(i)} = \boldsymbol{0}_k, \quad a_i \in \mathbb{F}_2, \quad \boldsymbol{u}^{(i)} \in \mathbb{F}_2^k, \quad i \in \{0, 1 \cdots m\}, \quad m \in \left\{2, 3 \cdots 2^k\right\}, \quad k \in \mathbb{N}^*. \tag{1.114}$$

Otherwise, the set $\{\boldsymbol{u}^{(0)}, \boldsymbol{u}^{(1)} \cdots \boldsymbol{u}^{(m-1)}\}$ of vectors, $m \in \{2, 3 \cdots 2^k\}$, $k \in \mathbb{N}^*$, which are taken from $\mathbb{F}_2^k$ and which are mutually distinct, are termed *linearly dependent* [11, Section 5.3.7.2, p. 327], [22, p. 71], [21, p. 40].

An important example of linear dependence is given by *collinear vectors* [11, p. 189], [22, p. 71].

Definition 1.15 (Collinear Vectors of $\mathbb{F}_2^k$). Two vectors $\boldsymbol{u}^{(0)}$ and $\boldsymbol{u}^{(1)}$ of vectors of $\mathbb{F}_2^k$, $k \in \mathbb{N}^*$, are said to be *collinear* if and only if [22, p. 71]

$$\boldsymbol{u}^{(0)} = a \odot \boldsymbol{u}^{(1)}, \quad a \in \mathbb{F}_2, \quad \boldsymbol{u}^{(1)} \in \mathbb{F}_2^k, \tag{1.115}$$

holds.

Remark 1.8 (On Collinear Vectors of $\mathbb{F}_2^k$). Since $a \in \mathbb{F}_2$ used in (1.115) can only assume the values 0 and 1, we can distinguish the following two cases of collinear vectors. First, let a be equal to 0. Then $a \odot \boldsymbol{u}^{(1)}$ is equal to the zero vector $\mathbf{0}_k$, $k \in \mathbb{N}^*$, i. e., $\boldsymbol{u}^{(0)}$ is equal to $\mathbf{0}_k$, $k \in \mathbb{N}^*$. Thus, (1.115) means that every vector $\boldsymbol{u}^{(1)}$ taken from $\mathbb{F}_2^k$, $k \in \mathbb{N}^*$ and the zero vector $\mathbf{0}_k$, $k \in \mathbb{N}^*$ are collinear.

Second, let a be equal to 1. Then $a \odot \boldsymbol{u}^{(1)}$ is equal to $\boldsymbol{u}^{(1)}$, i. e., $\boldsymbol{u}^{(0)}$ must be equal to $\boldsymbol{u}^{(1)}$, which means that $\boldsymbol{u}^{(0)}$ and $\boldsymbol{u}^{(1)}$ are identical; cf. Definition 1.11.

To put it in a nutshell, any vector $\boldsymbol{u} \in \mathbb{F}_2^k$, $k \in \mathbb{N}^*$ is collinear

- to the zero vector $\mathbf{0}_k$, $k \in \mathbb{N}^*$,
- to itself and
- to no other vector of $\mathbb{F}_2^k$, $k \in \mathbb{N}^*$, which is not the zero vector.

Looking at $\boldsymbol{u}$ of (1.23), we find that the components $u_i \in \mathbb{F}_2$, $i \in \{0, 1 \cdots (k-1)\}$, $k \in \mathbb{N}^*$ are the coordinates in the vector space $\mathbb{F}_2^k$, which clearly is k-dimensional. Let us choose the k *unit vectors*

$$\boldsymbol{e}^{(i)} = (e_0^{(i)}, e_1^{(i)} \cdots e_{k-1}^{(i)}) = \Big(0, 0 \cdots \underbrace{1}_{i\text{th position}} \cdots 0\Big) \in \mathbb{F}_2^k, \quad i \in \{0, 1 \cdots (k-1)\}, \quad k \in \mathbb{N}^*, \tag{1.116}$$

we can represent any vector $\boldsymbol{u} \in \mathbb{F}_2^k$, $k \in \mathbb{N}^*$, given by (1.23), in the following form:

$$\boldsymbol{u} = u_0 \odot \boldsymbol{e}^{(0)} \oplus u_1 \odot \boldsymbol{e}^{(1)} \oplus \cdots \oplus u_{k-1} \odot \boldsymbol{e}^{(k-1)} = \bigoplus_{i=0}^{k-1} u_i \odot \boldsymbol{e}^{(i)}, \quad k \in \mathbb{N}^*, \tag{1.117}$$

which is the linear superposition of the k unit vectors $\boldsymbol{e}^{(i)}$, $i \in \{0, 1 \cdots (k-1)\}$, $k \in \mathbb{N}^*$, weighted by the components $u_i \in \mathbb{F}_2$, $i \in \{0, 1 \cdots (k-1)\}$, $k \in \mathbb{N}^*$ of $\boldsymbol{u} \in \mathbb{F}_2^k$, $k \in \mathbb{N}^*$, given by (1.23). Therefore, (1.116) and (1.117) lead us to the conclusion that the k-unit vectors $\boldsymbol{e}^{(i)}$, $i \in \{0, 1 \cdots (k-1)\}$, $k \in \mathbb{N}^*$, form an *algebraic basis* [11, p. 619], [21, p. 40] of $\mathbb{F}_2^k$, which we can use to represent any vector $\boldsymbol{u} \in \mathbb{F}_2^k$, $k \in \mathbb{N}^*$, given by (1.23). An algebraic basis is also called *Hamel basis* [11, p. 619].

Now, it is about time to define what *dimension* means.

Definition 1.16 (Dimension of a Vector Space). The number of elements of an algebraic basis of a vector space $\mathbb{V}$, i. e., the number of the basis vectors is termed the *dimension* $\dim\{\mathbb{V}\}$ *of the vector space* [11, p. 619], [21, p. 41].

Remark 1.9 (On an Algebraic Basis of $\mathbb{F}_2^k$). Since $\mathbb{F}_2^k$, $k \in \mathbb{N}^*$, is k-dimensional, there exist exactly k linearly independent vectors, which span $\mathbb{F}_2^k$, $k \in \mathbb{N}^*$. There cannot be more than k linear independent vectors that span $\mathbb{F}_2^k$, $k \in \mathbb{N}^*$. Also, there cannot be less than k linear independent vectors, which span $\mathbb{F}_2^k$, $k \in \mathbb{N}^*$.

According to (1.116) and (1.117), we can use the set $\{\boldsymbol{e}^{(0)}, \boldsymbol{e}^{(1)} \cdots \boldsymbol{e}^{(k-1)}\}$, $k \in \mathbb{N}^*$, of the k-unit vectors $\boldsymbol{e}^{(i)}$, $i \in \{0, 1 \cdots (k-1)\}$, $k \in \mathbb{N}^*$, to represent any vector $\boldsymbol{u} \in \mathbb{F}_2^k$, $k \in \mathbb{N}^*$, given by (1.23) as a linear combination of these k-unit vectors. The set $\{\boldsymbol{e}^{(0)}, \boldsymbol{e}^{(1)} \cdots \boldsymbol{e}^{(k-1)}\}$, $k \in \mathbb{N}^*$, is also called a *system of vectors* [21, p. 40].

The set of all vectors that can be formed by the linear combination of the elements of the set $\{\boldsymbol{e}^{(0)}, \boldsymbol{e}^{(1)} \cdots \boldsymbol{e}^{(k-1)}\}$, $k \in \mathbb{N}^*$ is sometimes written as $\mathcal{L}(\{\boldsymbol{e}^{(0)}, \boldsymbol{e}^{(1)} \cdots \boldsymbol{e}^{(k-1)}\})$. Clearly, according to (1.116) and (1.117), $\mathcal{L}(\{\boldsymbol{e}^{(0)}, \boldsymbol{e}^{(1)} \cdots \boldsymbol{e}^{(k-1)}\})$ is identical with $\mathbb{F}_2^k$, $k \in \mathbb{N}^*$, i. e.,

$$\mathcal{L}\left(\left\{e^{(0)}, e^{(1)} \cdots e^{(k-1)}\right\}\right) = \mathbb{F}_2^k, \quad k \in \mathbb{N}^*, \tag{1.118}$$

holds. Therefore, the system $\{e^{(0)}, e^{(1)} \cdots e^{(k-1)}\}$, $k \in \mathbb{N}^*$ is an algebraic basis of $\mathbb{F}_2^k$, $k \in \mathbb{N}^*$, [21, p. 40].

Using the $k \times k$ square *identity matrix*,

$$I_k = \begin{pmatrix} e^{(0)} \\ e^{(1)} \\ \vdots \\ e^{(k-1)} \end{pmatrix} = \begin{pmatrix} 1 & 0 & \cdots & 0 \\ 0 & 1 & \cdots & 0 \\ \vdots & \vdots & \ddots & \vdots \\ 0 & 0 & \cdots & 1 \end{pmatrix}, \tag{1.119}$$

(1.117) becomes

$$u = (u_0, u_1 \cdots u_{k-1}) \begin{pmatrix} 1 & 0 & \cdots & 0 \\ 0 & 1 & \cdots & 0 \\ \vdots & \vdots & \ddots & \vdots \\ 0 & 0 & \cdots & 1 \end{pmatrix} = uI_k, \quad k \in \mathbb{N}^*. \tag{1.120}$$

Since the $k \times k$ square identity matrix I_k of (1.119) has k, $k \in \mathbb{N}^*$, rows and k, $k \in \mathbb{N}^*$, columns and, therefore, $k \cdot k$ components, we write

$$I_k \in \mathbb{F}_2^{k \times k} \tag{1.121}$$

to express that I_k, $k \in \mathbb{N}^*$ is a matrix; cf., e. g., [5, p. 487] for a similar notation.

Of course, all that has been found in the case of the vectors $u \in \mathbb{F}_2^k$, $k \in \mathbb{N}^*$ of (1.23) in Section 1.4 also applies to the vectors $v \in \mathbb{F}_2^n$, $n \in \mathbb{N}^*$, introduced in (1.24).

1.5 Linear Channel Codes

According to Section 1.3, a channel code implements the functional mapping of any *message* $u \in \mathbb{F}_2^k$, $k \in \mathbb{N}^*$, of (1.23) onto the corresponding *codeword* $v \in \mathbb{F}_2^n$, $n \in \mathbb{N}^*$ introduced in (1.24). To facilitate proper decoding at the receiver, we will need a *bijective mapping*, i. e., a *one-to-one mapping*, which is "invertible." Hence, a particular message $u \in \mathbb{F}_2^k$, $k \in \mathbb{N}^*$ shall cause a particular, i. e., distinct, codeword $v \in \mathbb{F}_2^n$, $n \in \mathbb{N}^*$, which differs from all other $(2^k - 1)$ possible codewords. We already pointed this fact out at the end of Section 1.3 by stating

> "For a block code to be useful, the 2^k codewords must be distinct [15, p. 66]. Therefore, a one-to-one correspondence between a particular message u and its codeword v is required [15, p. 66]."

Clearly, we need to define what *decoding* means.

Definition 1.17 (Decoding). *Decoding* means converting data by reversing the effect of previous encoding [13, p. 317].

This immediately leads us to the following definition.

Definition 1.18 (Channel Decoding). *Channel decoding* means converting data by reversing the effect of previous encoding by a channel code.

The expression *decoder* has the following meaning.

Definition 1.19 (Decoder). A *decoder* is a device that performs the decoding [13, p. 317].

Then a *channel decoder* is the following.

Definition 1.20 (Channel Decoder). A *channel decoder* is a device that performs the decoding of a channel code.

Considering Section 1.4, in particular, looking at (1.120), we will certainly be led to the question *"what if we replaced the $k \times k$ square identity matrix $\boldsymbol{I}_k$, $k \in \mathbb{N}^*$, in (1.120) by some other nonzero matrix, $\boldsymbol{G}$ say, which maps $\boldsymbol{u} \in \mathbb{F}_2^k$, $k \in \mathbb{N}^*$, onto $\boldsymbol{v} \in \mathbb{F}_2^n$, $n \in \mathbb{N}^*$?"*

Of course, $\boldsymbol{G}$ must be a $k \times n$, $k, n \in \mathbb{N}^*$ matrix [15, p. 67]. Furthermore, the elements of the $k \times n$ matrix $\boldsymbol{G}$ must be taken from $\mathbb{F}_2$. As already mentioned, the $k \times n$ matrix $\boldsymbol{G}$ must not be equal to the $k \times n$ zero matrix $\boldsymbol{0}_{k \times n}$. In this case, (1.120) changes to the following form [15, equation (3.3), p. 67]:

$$\boldsymbol{v} = \boldsymbol{u}\boldsymbol{G}, \quad \boldsymbol{G} \in \mathbb{F}_2^{k \times n} \setminus \{\boldsymbol{0}_{k \times n}\}, \quad \boldsymbol{v} \in \mathbb{F}_2^n, \quad \boldsymbol{u} \in \mathbb{F}_2^k, \quad k, n \in \mathbb{N}^*. \tag{1.122}$$

Defining the k, $k \in \mathbb{N}^*$ row vectors [15, p. 67],

$$\begin{aligned} \boldsymbol{g}^{(i)} &= (g_0^{(i)}, g_1^{(i)} \cdots g_{n-1}^{(i)}), \\ &\boldsymbol{g}^{(i)} \in \mathbb{F}_2^n \setminus \{\boldsymbol{0}_n\} \setminus \{\boldsymbol{0}_n\}, \quad g_j^{(i)} \in \mathbb{F}_2, \\ &i \in \{0, 1 \cdots (k-1)\}, \quad j \in \{0, 1 \cdots (n-1)\}, \quad k, n \in \mathbb{N}^*, \end{aligned} \tag{1.123}$$

the $k \times n$ matrix $\boldsymbol{G}$ takes the form [15, equation (3.2), p. 67]

$$\begin{aligned} \boldsymbol{G} = \begin{pmatrix} \boldsymbol{g}^{(0)} \\ \boldsymbol{g}^{(1)} \\ \vdots \\ \boldsymbol{g}^{(k-1)} \end{pmatrix} &= \begin{pmatrix} g_0^{(0)} & g_1^{(0)} & \cdots & g_{n-1}^{(0)} \\ g_0^{(1)} & g_1^{(1)} & \cdots & g_{n-1}^{(1)} \\ \vdots & \vdots & \ddots & \vdots \\ g_0^{(k-1)} & g_1^{(k-1)} & \cdots & g_{n-1}^{(k-1)} \end{pmatrix}, \\ \boldsymbol{G} \in \mathbb{F}_2^{k \times n} \setminus \{\boldsymbol{0}_{k \times n}\}, \quad &\boldsymbol{g}^{(i)} \in \mathbb{F}_2^n \setminus \{\boldsymbol{0}_n\} \setminus \{\boldsymbol{0}_n\}, \quad g_j^{(i)} \in \mathbb{F}_2, \\ i \in \{0, 1 \cdots (k-1)\}, \quad &j \in \{0, 1 \cdots (n-1)\}, \quad k, n \in \mathbb{N}^*. \end{aligned} \tag{1.124}$$

Using the notation of (1.117) and taking (1.124) into consideration, we may write (1.122) in the following form [15, equations (3.1) and (3.3), p. 67]:

$$\boldsymbol{v} = (u_0, u_1 \cdots u_{k-1}) \begin{pmatrix} \boldsymbol{g}^{(0)} \\ \boldsymbol{g}^{(1)} \\ \vdots \\ \boldsymbol{g}^{(k-1)} \end{pmatrix} = \bigoplus_{i=0}^{k-1} u_i \odot \boldsymbol{g}^{(i)}, \tag{1.125}$$

$$\boldsymbol{v} \in \mathbb{F}_2^n, \quad \boldsymbol{g}^{(i)} \in \mathbb{F}_2^n \setminus \{\boldsymbol{0}_n\}, \quad u_i \in \mathbb{F}_2, \quad i \in \{0, 1 \cdots (k-1)\}, \quad k, n \in \mathbb{N}^*.$$

The codeword $\boldsymbol{v} \in \mathbb{F}_2^n$, $n \in \mathbb{N}^*$ of (1.125) is the linear combination of the k, $k \in \mathbb{N}^*$ row vectors $\boldsymbol{g}^{(i)} \in \mathbb{F}_2^n \setminus \{\boldsymbol{0}_n\}$, $i \in \{0, 1 \cdots (k-1)\}$, $k, n \in \mathbb{N}^*$. To be a useful block code, the k, $k \in \mathbb{N}^*$ row vectors $\boldsymbol{g}^{(i)} \in \mathbb{F}_2^n \setminus \{\boldsymbol{0}_n\}$, $i \in \{0, 1 \cdots (k-1)\}$, $k, n \in \mathbb{N}^*$, must be linearly independent [15, p. 67].

Since the k, $k \in \mathbb{N}^*$, distinct and linearly independent row vectors $\boldsymbol{g}^{(i)} \in \mathbb{F}_2^n \setminus \{\boldsymbol{0}_n\}$, $i \in \{0, 1 \cdots (k-1)\}$, $k, n \in \mathbb{N}^*$ are taken from $\mathbb{F}_2^n$, $n \in \mathbb{N}^*$, they form an algebraic basis of a k-dimensional subspace $\mathbb{V}$ of $\mathbb{F}_2^n$, $n \in \mathbb{N}^*$ [21, p. 40]. Using the notation of (1.118), the k-dimensional subspace $\mathbb{V}$ of $\mathbb{F}_2^n$, $n \in \mathbb{N}^*$, can be written by [21, p. 40]

$$\mathcal{L}(\{\boldsymbol{g}^{(0)}, \boldsymbol{g}^{(1)} \cdots \boldsymbol{g}^{(k-1)}\}) = \mathbb{V}, \quad \dim\{\mathbb{V}\} = k, \tag{1.126}$$

$$\mathbb{V} \subseteq \mathbb{F}_2^n, \quad \boldsymbol{g}^{(i)} \in \mathbb{F}_2^n \setminus \{\boldsymbol{0}_n\}, \quad i \in \{0, 1 \cdots (k-1)\}, \quad k, n \in \mathbb{N}^*$$

$\dim\{\mathbb{V}\}$ denoting the dimension of $\mathbb{V}$; cf. Definition 1.16. Let us therefore call the row vectors $\boldsymbol{g}^{(i)} \in \mathbb{F}_2^n \setminus \{\boldsymbol{0}_n\}$, $i \in \{0, 1 \cdots (k-1)\}$, $k, n \in \mathbb{N}^*$ the *basis vectors*.

Setting out from (1.125), we may define the *inner product*, respectively, the *scalar product* of two arbitrary vectors $\boldsymbol{u}^{(1)} \in \mathbb{F}_2^k$, having the components $u_i^{(1)} \in \mathbb{F}_2$, $i \in \{0, 1 \cdots (k-1)\}$, $k \in \mathbb{N}^*$ and $\boldsymbol{u}^{(2)} \in \mathbb{F}_2^k$ having the components $u_i^{(2)} \in \mathbb{F}_2$, $i \in \{0, 1 \cdots (k-1)\}$, $k \in \mathbb{N}^*$ [23, Definition 1.4, p. 9].

Definition 1.21 (Inner Product/Scalar Product). Let $\boldsymbol{u}^{(1)}$, having the components $u_i^{(1)} \in \mathbb{F}_2$, $i \in \{0, 1 \cdots (k-1)\}$, $k \in \mathbb{N}^*$, $\boldsymbol{u}^{(2)}$ having the components $u_i^{(2)} \in \mathbb{F}_2$, $i \in \{0, 1 \cdots (k-1)\}$, $k \in \mathbb{N}^*$ and $\boldsymbol{u}^{(3)}$, having the components $u_i^{(3)} \in \mathbb{F}_2$, $i \in \{0, 1 \cdots (k-1)\}$, $k \in \mathbb{N}^*$ be three arbitrary vectors of $\mathbb{F}_2^k$, $k \in \mathbb{N}^*$. The *inner product*, respectively, the *scalar product* of $\boldsymbol{u}^{(1)} \in \mathbb{F}_2^k$, $k \in \mathbb{N}^*$ and $\boldsymbol{u}^{(2)} \in \mathbb{F}_2^k$, $k \in \mathbb{N}^*$ is defined by [23, Definition 1.4, p. 9]

$$\langle \boldsymbol{u}^{(1)} \mid \boldsymbol{u}^{(2)} \rangle = \boldsymbol{u}^{(1)} \boldsymbol{u}^{(2)\mathrm{T}} = \bigoplus_{i=0}^{k-1} u_i^{(1)} \odot u_i^{(2)}, \quad \boldsymbol{u}^{(1)}, \boldsymbol{u}^{(2)} \in \mathbb{F}_2^k, \quad k \in \mathbb{N}^*. \tag{1.127}$$

The k-dimensional subspace $\mathbb{V} \subseteq \mathbb{F}_2^n$, $n \in \mathbb{N}^*$ consists of all admissible codewords $\boldsymbol{v}$, which are formed by (1.122), respectively (1.125), is called the *code* $\mathbb{V}$. Clearly, the code $\mathbb{V}$ has the *dimension* k, $k \in \mathbb{N}^*$. Furthermore, the code $\mathbb{V}$ generates codewords, which have n components. Therefore, we say the code $\mathbb{V}$ has *length* n. The code $\mathbb{V}$ is also termed (n, k) *block code* [15, p. 3], respectively, (n, k) *code*.

Since the code $\mathbb{V}$ is the set of all admissible codewords, we can now give an updated definition, accompanying Definition 1.3.

Definition 1.22 (Channel Code—An Update). A *channel code* is the set $\mathbb{V}$ of codewords $\boldsymbol{v}$ [15, p. 3].

We may now state what a *binary code* is.

Definition 1.23 (Binary Code). An (n, k) binary code is the set $\mathbb{V}$ of codewords $\boldsymbol{v}$, all taken from the vector space $\mathbb{F}_2^n$, $n \in \mathbb{N}^*$; see also [7, p. 220]. The components of the codewords $\boldsymbol{v} \in \mathbb{F}_2^n$, $n \in \mathbb{N}^*$, are hence binary, taken from the Galois field $\mathbb{F}_2$; see also [7, p. 220].

Also, a *binary linear code* can be defined.

Definition 1.24 (Binary Linear Code/Binary Linear Block Code). A code, i. e., a block code, $\mathbb{V}$ of dimension k, $k \in \mathbb{N}^*$ and length n, $n \geq k$, $n \in \mathbb{N}^*$ and, therefore, with 2^k codewords is called an (n, k) *binary linear code*, if and only if its n codewords form a k-dimensional subspace of the vector space $\mathbb{F}_2^n$ of all the n-tuples over the finite field $\mathbb{F}_2$; see also [15, Definition 3.1, p. 66].

Owing to their easy mathematical description and, consequently, their efficient implementability, paired with a promising capability to approximate the ideal transmission, linear channel codes have gained a lot of attendance [15, p. 66]. We will therefore restrict ourselves to linear channel codes.

Remark 1.10 (On Binary Linear Codes//Binary Linear Block Codes). Taking (1.122) and (1.125) into consideration, we come to the conclusion that binary linear codes, respectively, binary linear block codes, are based on messages $\boldsymbol{u} \in \mathbb{F}_2^k$, $k \in \mathbb{N}^*$, with components taken from the Galois field $\mathbb{F}_2$. Thus, binary linear codes, respectively, binary linear block codes, have binary inputs and binary outputs.

Example 1.5. Table 1.1 illustrates a (3, 1) binary linear block code, which is the shortest *Hamming code* (cf., e. g., [23, pp. 20–22]), being a repetition code at the same time.

A very popular example is the given in Table 1.2, which represents a (7, 4) binary linear block code with dimension k equal to 4 and length n equal to 7; cf., e. g., [15, Table 3.1, p. 68], [23, pp. 20–22]. This (7, 4) binary linear block code of Table 1.2 is the second shortest *Hamming code* and at the same time the shortest Hamming code [15, p. 100], [23, pp. 20–22], which is not a repetition code.

Table 1.1: Example of a (3, 1) binary linear block code; cf., e. g., [23, pp. 20–22].

message u	codeword v
(0)	(0, 0, 0)
(1)	(1, 1, 1)

Table 1.2: Example of a $(7,4)$ binary linear block code, adapted according to [15, Table 3.1, p. 68] and [23, pp. 20–22].

message u	codeword v
(0, 0, 0, 0)	(0, 0, 0, 0, 0, 0, 0)
(1, 0, 0, 0)	(0, 1, 1, 1, 0, 0, 0)
(0, 1, 0, 0)	(1, 0, 1, 0, 1, 0, 0)
(1, 1, 0, 0)	(1, 1, 0, 1, 1, 0, 0)
(0, 0, 1, 0)	(1, 1, 0, 0, 0, 1, 0)
(1, 0, 1, 0)	(1, 0, 1, 1, 0, 1, 0)
(0, 1, 1, 0)	(0, 1, 1, 0, 1, 1, 0)
(1, 1, 1, 0)	(0, 0, 0, 1, 1, 1, 0)
(0, 0, 0, 1)	(1, 1, 1, 0, 0, 0, 1)
(1, 0, 0, 1)	(1, 0, 0, 1, 0, 0, 1)
(0, 1, 0, 1)	(0, 1, 0, 0, 1, 0, 1)
(1, 1, 0, 1)	(0, 0, 1, 1, 1, 0, 1)
(0, 0, 1, 1)	(0, 0, 1, 0, 0, 1, 1)
(1, 0, 1, 1)	(0, 1, 0, 1, 0, 1, 1)
(0, 1, 1, 1)	(1, 0, 0, 0, 1, 1, 1)
(1, 1, 1, 1)	(1, 1, 1, 1, 1, 1, 1)

Since there are k, $k \in \mathbb{N}^*$, distinct and linearly independent row vectors $\boldsymbol{g}^{(i)} \in \mathbb{F}_2^n \setminus \{\mathbf{0}_n\}$, $i \in \{0, 1 \cdots (k-1)\}$, $k, n \in \mathbb{N}^*$, the "rank" rank$\{\boldsymbol{G}\}$ [11, p. 264] of the $k \times n$ matrix $\boldsymbol{G}$ is equal to k, i. e.,

$$\operatorname{rank}\{\boldsymbol{G}\} = \dim\{\mathbb{V}\} = k, \quad k \in \mathbb{N}^*. \tag{1.128}$$

Clearly, the rows of $\boldsymbol{G}$ of (1.124) generate the (n, k) linear code $\mathbb{V}$ [15, p. 67]. For this reason, the matrix $\boldsymbol{G}$ of (1.124) is called a *generator matrix for* $\mathbb{V}$ [15, p. 67]. Note that any k linearly independent codewords $\boldsymbol{g}^{(i)} \in \mathbb{F}_2^n \setminus \{\mathbf{0}_n\}$, $i \in \{0, 1 \cdots (k-1)\}$, $k, n \in \mathbb{N}^*$, of an (n, k) binary linear block code $\mathbb{V}$ can be used to form a generator matrix for that particular code $\mathbb{V}$ [15, p. 67], [23, p. 22]. This means that the generator matrix $\boldsymbol{G}$ of (1.124) is not unique; see also [15, p. 67], [23, p. 22].

Nevertheless, an (n, k) binary linear block code $\mathbb{V}$ is completely specified by the k, $k \in \mathbb{N}^*$ rows of a generator matrix $\boldsymbol{G}$ of (1.124) [15, p. 67]. Therefore, the encoder has only to store the k, $k \in \mathbb{N}^*$ rows of $\boldsymbol{G}$ given by (1.124) and to form a linear combination of these k, $k \in \mathbb{N}^*$ rows based on the input message $\boldsymbol{u}$ of (1.49) [15, p. 67].

Example 1.6. Let us continue with Example 1.5.

The generator matrix $\boldsymbol{G}$ used to generate the $(3, 1)$ binary linear block code of Table 1.1 is given by

$$\boldsymbol{G} = \begin{pmatrix} 1 & 1 & 1 \end{pmatrix}. \tag{1.129}$$

Obviously, the generator matrix $\boldsymbol{G}$ of (1.129) is composed of the second codeword of Table 1.1, and we find

$$g^{(0)} = (1, 1, 1). \tag{1.130}$$

The generator matrix $\boldsymbol{G}$ used to generate the (7, 4) binary linear block code of Table 1.2 is given by

$$\boldsymbol{G} = \begin{pmatrix} 0 & 1 & 1 & 1 & 0 & 0 & 0 \\ 1 & 0 & 1 & 0 & 1 & 0 & 0 \\ 1 & 1 & 0 & 0 & 0 & 1 & 0 \\ 1 & 1 & 1 & 0 & 0 & 0 & 1 \end{pmatrix}. \tag{1.131}$$

Obviously, the generator matrix $\boldsymbol{G}$ of (1.131) is composed of the second codeword, the third codeword, the fifth codeword and the ninth codeword of Table 1.2. The corresponding basis vectors are

$$g^{(0)} = (0, 1, 1, 1, 0, 0, 0), \tag{1.132}$$

$$g^{(1)} = (1, 0, 1, 0, 1, 0, 0), \tag{1.133}$$

$$g^{(2)} = (1, 1, 0, 0, 0, 1, 0), \tag{1.134}$$

$$g^{(3)} = (1, 1, 1, 0, 0, 0, 1). \tag{1.135}$$

Setting out from (1.122) and (1.125) and bearing in mind that $\boldsymbol{u} \in \mathbb{F}_2^k$, $k \in \mathbb{N}^*$ may be the zero vector $\boldsymbol{0}_k \in \mathbb{F}_2^k$, $k \in \mathbb{N}^*$, the corresponding codeword $\boldsymbol{v} \in \mathbb{F}_2^n$, $n \in \mathbb{N}^*$ is the zero vector $\boldsymbol{0}_n \in \mathbb{F}_2^n$, $n \in \mathbb{N}^*$ [24, p. 129]. In other words, the zero vector $\boldsymbol{0}_n \in \mathbb{F}_2^n$, $n \in \mathbb{N}^*$ is an admissible codeword of an (n, k) linear code [24, p. 129].

The (n, k) binary linear block code $\mathbb{V}$ is the row space of $\boldsymbol{G}$ given by (1.124) [25, p. 53]. Therefore, the (n, k) binary linear block code $\mathbb{V}$ remains unchanged under *elementary row operations* applied to $\boldsymbol{G}$ introduced in (1.124) [25, p. 53]. Such elementary row operations are [5, p. 109]

a) the permutation of rows,
b) the multiplication of a row by a nonzero scalar,
c) the summation of two rows.

The (n, k) binary linear block code $\mathbb{V}$ generated by $\boldsymbol{G}(n, k)$ binary linear block code $\mathbb{V}'$ generated by $\boldsymbol{G}'$, which was formed by elementary row operations upon $\boldsymbol{G}$, are called *identical* [5, pp. 18, 109].

Setting out from an (n, k) binary linear block code $\mathbb{V}$, we can compute the Hamming distance between any two distinct codewords [15, p. 76]. The *minimum distance* $d_{\min}$ of the (n, k) binary linear block code $\mathbb{V}$ is given by the following definition.

Definition 1.25 (Minimum Distance of a Binary Linear Block Code). Let $\boldsymbol{v} \in \mathbb{V}$, and $\boldsymbol{w} \in \mathbb{V}$, be codewords of an (n, k) binary linear block code $\mathbb{V}$. The minimum distance of the code C is given by [15]

$$d_{\min} = \min\{d_{\mathrm{H}}\{\boldsymbol{v}, \boldsymbol{w}\}\}, \quad \boldsymbol{v}, \boldsymbol{w} \in \mathbb{V}, \quad \boldsymbol{v} \neq \boldsymbol{w}. \tag{1.136}$$

Since the (n, k) binary linear block code $\mathbb{V}$ is linear, the sum of two codewords is also a codeword [15, pp. 67, 76]. Thus, the Hamming distance between any two codewords

$\boldsymbol{v} \in \mathbb{V}$ and $\boldsymbol{w} \in \mathbb{V}$ is equal to the Hamming weight of a third codeword $(\boldsymbol{v} \oplus \boldsymbol{w}) \in \mathbb{V}$, yielding

$$d_{\mathrm{min}} = \min\{d_{\mathrm{H}}\{\boldsymbol{v}, \boldsymbol{w}\}\} = \min\{w_{\mathrm{H}}\{\boldsymbol{v} \oplus \boldsymbol{w}\}\} = w_{\mathrm{min}}, \quad \boldsymbol{v}, \boldsymbol{w} \in \mathbb{V}, \quad \boldsymbol{v} \neq \boldsymbol{w}, \tag{1.137}$$

which is equal to the *minimum weight* w_{min} *of all nonzero codewords* $(\boldsymbol{v} \oplus \boldsymbol{w}) \in \mathbb{V}$ [15, p. 77]. Therefore, w_{min} is called the *minimum weight of the* (n, k) *binary linear block code* $\mathbb{V}$ [15, p. 77].

Theorem 1.6 (Minimum Distance of a Binary Linear Block Code). *The minimum distance* d_{min} *of an* (n, k) *binary linear block code* $\mathbb{V}$ *is equal to the minimum weight* w_{min} *of its nonzero codewords and vice versa* [15, *p. 77*].

Proof. The proof is given by (1.137). □

Example 1.7. Let us continue with the Examples 1.5 and 1.6.

Consider Table 1.2. The weights of the nonzero codewords are taken from the set $\{3, 4, 7\}$. The weight 3 occurs seven times, the weight 4 occurs seven times and the weight 7 occurs once. Clearly, the minimum weight w_{min} is equal to 3, which is also equal to the minimum distance d_{min}.

Using the minimum distance d_{min}, often the (n, k) binary linear block code $\mathbb{V}$ is denoted by (n, k, d_{min}) binary linear block code $\mathbb{V}$ [23, Definition 1.9, p. 11], [25, p. 60], [5, Satz 3.4, p. 74], [26, Definition 3.7.2, p. 68].

It is quite clear that the minimum distance d_{min} cannot exceed the length n, $n \in \mathbb{N}^*$, of an (n, k, d_{min}) binary linear block code $\mathbb{V}$. This fact leads us to the following bound:

$$d_{\mathrm{min}} \leq n \quad \Leftrightarrow \quad \frac{d_{\mathrm{min}}}{n} = \delta \leq 1, \quad n \in \mathbb{N}^*. \tag{1.138}$$

Nevertheless, the question remains whether there can be found any relationship between the code rate R and the minimum distance d_{min}. Striving for answers to this question, several bounds, which relate R and d_{min} or δ used in (1.138) were found, some of which we will illustrate in what follows. The first bound that we will consider is termed the *Singleton bound* [23, Satz 6.9, p. 139], [5, Satz 3.7, equation (3.3.1), p. 80].

Theorem 1.7 (Singleton Bound). *For an* (n, k, d_{min}) *binary linear block code* $\mathbb{V}$,

$$d_{\mathrm{min}} \leq n - k + 1, \quad k, n \in \mathbb{N}^*, \tag{1.139}$$

holds [23, *Satz 6.9, p. 139*], [5, *Satz 3.7, equation (3.3.1), p. 80*]. *The expression* $(n - k + 1)$, $k, n \in \mathbb{N}^*$ *in (*1.139*) is termed the* Singleton bound [5, *Satz 3.7, equation (3.3.1), p. 80*]. *In the case of equality in (*1.139*), the* (n, k, d_{min}) *binary linear block code* $\mathbb{V}$ *is termed* maximum distance separable (MDS) code [23, *Satz 6.10, p. 139*], [5, *Satz 3.7, p. 80*].

Proof. All 2^k, $k \in \mathbb{N}^*$ codewords $\boldsymbol{v} \in \mathbb{V} \subseteq \mathbb{F}_2^n$, $n \in \mathbb{N}^*$ differ in at least d_{min} components [5, Satz 3.7, p. 80]. If the first $(d_{\mathrm{min}} - 1)$ components are deleted from all the 2^k, $k \in \mathbb{N}^*$ codewords, then the shortened codewords of the length $(n - d_{\min} + 1)$ are still distinct [5, Satz 3.7, p. 80]. So, there are distinct shortened codewords in the vector space $\mathbb{F}_2^{n-d_{\min}+1}$. This is only possible if and only if $k \le (n - d_{\min} + 1)$ holds [5, Satz 3.7, p. 80]. Hence, (1.139) follows immediately. □

Remark 1.11 (On the Singleton Bound).

$$\frac{d_{\min}}{n} \le 1 - \underbrace{\frac{k}{n}}_{=R} + \frac{1}{n} \quad \Leftrightarrow \quad \frac{d_{\min}}{n} \le R + \frac{1}{n}, \quad k, n \in \mathbb{N}^*, \tag{1.140}$$

follows immediately from (1.139), which becomes the following upper bound on the code rate:

$$R \le 1 - \frac{d_{\min}}{n}, \quad n \gg 1, \quad n \in \mathbb{N}^*. \tag{1.141}$$

Furthermore, the Singleton bound of (1.139) can be expressed as an upper bound on the number of codewords 2^k, $k \in \mathbb{N}^*$ in the following way [27, p. 17f.]:

$$2^k \le 2^{n-d_{\min}+1}, \quad k, n \in \mathbb{N}^*. \tag{1.142}$$

Each codeword of length $n \in \mathbb{N}^*$ is generated by processing k message bits u_i, $i \in \{0, 1 \cdots (k-1)\}$, $k \in \mathbb{N}^*$, and hence contains $(n - k)$, $k, n \in \mathbb{N}^*$ *parity check bits*, i. e., *redundancy bits*, also termed *parity check digits* [15, p. 68], respectively, *redundancy digits* (in German *"Redundanzzeichen"*) [23, p. 8]. The number of possible parity check sequences is thus equal to 2^{n-k}, $k, n \in \mathbb{N}^*$. The Singleton bound (1.139) can be expressed as a lower bound on the number of possible parity check sequences

$$2^{n-k} \ge 2^{d_{\min}-1}, \quad k, n \in \mathbb{N}^*. \tag{1.143}$$

Theorem 1.8 (Plotkin Bound). *For an $(n, k, d_{\min})$ binary linear block code $\mathbb{V}$,*

$$d_{\min} \le \frac{n}{2(1 - 2^{-k})} \approx \frac{n}{2}, \quad k, n \in \mathbb{N}^*, \tag{1.144}$$

holds [14, *equation (3.4-5), p. 109*], [5, *Satz 3.11, equation (3.3.4), p. 82*].

Proof. Each component $v_j \in \mathbb{F}_2$, $j \in \{0, 1 \cdots (n-1)\}$, $n \in \mathbb{N}^*$, of a codeword $\boldsymbol{v} \in \mathbb{V} \subseteq \mathbb{F}_2^n$, $n \in \mathbb{N}^*$, assumes the values 0 and 1 with equal probability 1/2 [5, Satz 3.11, p. 82]. Hence, the average Hamming weight of each component $v_j \in \mathbb{F}_2$, $j \in \{0, 1 \cdots (n-1)\}$, $n \in \mathbb{N}^*$, is equal to 1/2 [5, Satz 3.11, p. 82].

Thus, the average Hamming weight of a codeword $\boldsymbol{v} \in \mathbb{V} \subseteq \mathbb{F}_2^n$, $n \in \mathbb{N}^*$, is equal to $n/2$, $n \in \mathbb{N}^*$ [5, Satz 3.11, p. 82]. If the (all-) zero codeword is omitted, the average Hamming weight of a codeword is increased to

$$\frac{n}{2} \cdot \frac{2^k}{2^k - 1} = \frac{n}{2(1 - 2^{-k})} \approx \frac{n}{2}, \quad k \in \mathbb{N}^*, \tag{1.145}$$

which must be greater than the minimum weight, i. e., the minimum distance $d_{\min}$ [5, Satz 3.11, p. 82]. □

Remark 1.12 (On the Plotkin Bound—Going Once). Setting out from (1.144), we obtain

$$d_{\min} \le \frac{n}{2(1 - 2^{-k})} \quad \Leftrightarrow \quad d_{\min} \le \frac{n}{2} \frac{2^k}{(2^k - 1)} \quad \Leftrightarrow \quad 2^k d_{\min} - d_{\min} \le 2^k \frac{n}{2}, \quad k, n \in \mathbb{N}^*, \tag{1.146}$$

thus yielding

$$2^k d_{\min} - 2^k \frac{n}{2} - d_{\min} \leq 0, \quad \Leftrightarrow \quad 2^k \left(d_{\min} - \frac{n}{2} \right) \leq d_{\min}, \quad k, n \in \mathbb{N}^*. \tag{1.147}$$

Setting out from an $(n, k, d_{\min})$ binary linear block code $\mathbb{V}$, 2^k, $k \in \mathbb{N}^*$ is the *cardinal number* $|\mathbb{V}|$, which denotes the number of distinct elements of the $(n, k, d_{\min})$ binary linear block code $\mathbb{V}$ [11, p. 300], i. e., the number of codewords. Then (1.147) becomes [28, Lemma 3.10, p. 34]

$$|\mathbb{V}| \left(d_{\min} - \frac{n}{2} \right) \leq d_{\min}, \quad k, n \in \mathbb{N}^*, \tag{1.148}$$

which can also be written in the form of an upper bound on the number of codewords [27, p. 16]

$$2^k \leq \frac{d_{\min}}{d_{\min} - n/2}, \quad k, n \in \mathbb{N}^*. \tag{1.149}$$

Theorem 1.9 (Plotkin Bound Revisited). *Setting out from an $(n, k, d_{\min})$ binary linear block code $\mathbb{V}$ with $d_{\min} \leq n/2$, $n \in \mathbb{N}^*$, then the number of codewords 2^k, $k \in \mathbb{N}^*$ is upper bounded by* [28, *Theorem 3.12, p. 35*]

$$2^k \leq d_{\min} \cdot 2^{n - 2d_{\min} + 2}, \quad d_{\min} \leq \frac{n}{2}, \quad k, n \in \mathbb{N}^*. \tag{1.150}$$

Proof. With [28, Theorem 3.12, p. 35]

$$m = n - 2d_{\min} + 1, \quad m, n \in \mathbb{N}^* \tag{1.151}$$

and for each m-tuple $\boldsymbol{x} \in \mathbb{F}_2^m$, let $\mathbb{V}^{(x)}$ be a subset of the $(n, k, d_{\min})$ binary linear block code $\mathbb{V}$, whose first m binary components form the said m-tuple $\boldsymbol{x} \in \mathbb{F}_2^m$. Next, let us delete the aforementioned m binary components from each element of $\mathbb{V}^{(x)} \subset \mathbb{V}$ [28, Theorem 3.12, p. 35], forming the binary linear block code $\tilde{\mathbb{V}}^{(x)}$ with the length $(2d_{\min} - 1)$ and the minimum distance $(d_{\min} + \epsilon)$ for some $\epsilon \geq 0$ [28, Theorem 3.12, p. 35]. Using Theorem 1.8, the left-hand side of (1.148) becomes [28, Theorem 3.12, p. 36]

$$\left| \tilde{\mathbb{V}}^{(x)} \right| \cdot \left([d_{\min} + \epsilon] - \frac{2d_{\min} - 1}{2} \right) = \left| \tilde{\mathbb{V}}^{(x)} \right| \cdot \left(\epsilon + \frac{1}{2} \right) = \frac{1}{2} \cdot \left| \tilde{\mathbb{V}}^{(x)} \right| \cdot (2\epsilon + 1). \tag{1.152}$$

Hence, (1.150) yields [28, Theorem 3.12, p. 36]

$$\frac{1}{2} \cdot \left| \tilde{\mathbb{V}}^{(x)} \right| \cdot (2\epsilon + 1) \leq d_{\min} + \epsilon \quad \Leftrightarrow \quad \left| \tilde{\mathbb{V}}^{(x)} \right| \leq \frac{2(d_{\min} + \epsilon)}{2\epsilon + 1} \leq 2d_{\min}. \tag{1.153}$$

Clearly, [28, Theorem 3.12, p. 36]

$$\begin{aligned} 2^k = |\mathbb{V}| &= \sum_{x \in \mathbb{F}_2^m} \left| \tilde{\mathbb{V}}^{(x)} \right| \leq 2^m \cdot 2d_{\min} = d_{\min} \cdot 2^{m+1} = d_{\min} \cdot 2^{n - 2d_{\min} + 1 + 1} \\ &\leq d_{\min} \cdot 2^{n - 2d_{\min} + 2}, \quad m, n \in \mathbb{N}^*. \end{aligned} \tag{1.154}$$

□

Remark 1.13 (On the Plotkin Bound—Going Twice). The Plotkin bound (1.150) can be expressed as a lower bound on the number of possible parity check sequences

$$2^{n-k} \geq \frac{2^{2(d_{\min} - 1)}}{d_{\min}} = \frac{4^{d_{\min} - 1}}{d_{\min}}, \quad d_{\min} \leq \frac{n}{2}, \quad k, n \in \mathbb{N}^*. \tag{1.155}$$

Theorem 1.10 (Improved Plotkin Bound). *The* improved Plotkin bound *states an upper bound on the code rate*

$$R \leq 1 - \frac{2d_{\min}}{n} = 1 - 2\delta, \quad \frac{d_{\min}}{n} = \delta \leq \frac{1}{2}, \quad d_{\min}, n \in \mathbb{N}^*. \tag{1.156}$$

Proof. Using $|\mathbb{V}|$ equal to 2^k, $k \in \mathbb{N}^*$, (1.154) becomes [28, Theorem 3.14, p. 37]

$$\log_2\{2^k\} \leq \log_2\{d_{\min} \cdot 2^{n-2d_{\min}+2}\} \quad \Leftrightarrow \quad k \leq \log_2\{d_{\min}\} + n - 2d_{\min} + 2, \quad n \in \mathbb{N}^*, \tag{1.157}$$

thus

$$R = \frac{k}{n} \leq 1 - \frac{2d_{\min}}{n} + \underbrace{\frac{\log_2\{d_{\min}\} + 2}{n}}_{\to 0 \text{ for } n \gg 1}, \quad n \in \mathbb{N}^*. \tag{1.158}$$

Therefore, we obtain [28, Theorem 3.14, p. 37]

$$R \leq 1 - \frac{2d_{\min}}{n} = 1 - 2\delta, \quad \frac{d_{\min}}{n} = \delta \leq \frac{1}{2}, \quad d_{\min}, n \in \mathbb{N}^*. \tag{1.159}$$

□

Permutations of the components, i. e., the coordinates of the codeword $\boldsymbol{v}$ of the (n, k) binary linear block code $\mathbb{V}$ correspond to permutations of the columns of its generator matrix $\boldsymbol{G}$ [25, p. 53], leading to a new generator matrix $\boldsymbol{G}''$. The new generator matrix $\boldsymbol{G}''$ generates a (n, k) binary linear block code $\mathbb{V}''$, which has the same minimum distance, the same minimum weight and the same weight distribution [4, p. 229], however, which is usually not identical to the (n, k) binary linear block code $\mathbb{V}$. Since the (n, k) binary linear block code $\mathbb{V}$ and the (n, k) binary linear block code $\mathbb{V}''$ only differ in the order of the components, i. e., the coordinates of the codeword $\boldsymbol{v}$, they are called *equivalent codes* [4, p. 24], [5, pp. 18, 109].

Definition 1.26 (Equivalent Code). Two codes are called equivalent if they differ only in the order of the components, i. e., the coordinates of the codeword $\boldsymbol{v}$, [4, p. 24], [5, pp. 18, 109].

Let us look at an example [4, p. 24].

Example 1.8. The code

$$\mathbb{V} = \{(0,0,0,0), (0,0,1,1), (1,1,0,0), (1,1,1,1)\} \tag{1.160}$$

and the code

$$\mathbb{V}'' = \{(0,0,0,0), (0,1,0,1), (1,0,1,0), (1,1,1,1)\} \tag{1.161}$$

are equivalent, because one needs to permute the components, i. e., the coordinates of the codeword $\boldsymbol{v}$, #2 and #3 of $\mathbb{V}$ to obtain $\mathbb{V}''$ [4, p. 24].

Generally speaking, two (n, k) binary linear block codes $\mathbb{V}$ and $\mathbb{V}''$ are equivalent if and only if their generator matrices can be transformed into each other by elementary row operations and columns permutations [25, p. 53].

Since there are several generator matrices $\boldsymbol{G}$ of (1.124), which generate the identical (n,k) binary linear block code $\mathbb{V}$, let us determine the most favorable version of the generator matrix $\boldsymbol{G}$ by

- choosing those k, $k \in \mathbb{N}^*$, codewords $\boldsymbol{v} \in \mathbb{V}$, which have the *k-unit vectors* $\boldsymbol{e}^{(i)} \in \mathbb{F}_2^k$, $i \in \{0,1\cdots(k-1)\}$, $k \in \mathbb{N}^*$, of (1.116) at their respective ends,
- selecting that particular codeword $\boldsymbol{v} \in \mathbb{V}$, which has $\boldsymbol{e}^{(0)} \in \mathbb{F}_2^k$, $k \in \mathbb{N}^*$ at its end as the first basis vector $\boldsymbol{g}^{(0)}$, then selecting that particular codeword $\boldsymbol{v} \in \mathbb{V}$, which has $\boldsymbol{e}^{(1)} \in \mathbb{F}_2^k$, $k \in \mathbb{N}^*$, at its end as the first basis vector $\boldsymbol{g}^{(1)}$, and so on until selecting that particular codeword $\boldsymbol{v} \in \mathbb{V}$, which has $\boldsymbol{e}^{(k-1)} \in \mathbb{F}_2^k$, $k \in \mathbb{N}^*$ at its end as the first basis vector $\boldsymbol{g}^{(k-1)}$ and, finally,
- setting up the generator matrix $\boldsymbol{G}$ of (1.124).

The so-determined generator matrix $\boldsymbol{G}$ takes the following form [15, equation (3.4), p. 69]

$$\boldsymbol{G} = \begin{pmatrix} p_0^{(0)} & p_1^{(0)} & \cdots & p_{n-k-1}^{(0)} & 1 & 0 & 0 & \cdots & 0 \\ p_0^{(1)} & p_1^{(1)} & \cdots & p_{n-k-1}^{(1)} & 0 & 1 & 0 & \cdots & 0 \\ p_0^{(2)} & p_1^{(2)} & \cdots & p_{n-k-1}^{(2)} & 0 & 0 & 1 & \cdots & 0 \\ \vdots & \vdots & \ddots & \vdots & \vdots & \vdots & \vdots & \ddots & \vdots \\ p_0^{(k-1)} & p_1^{(k-1)} & \cdots & p_{n-k-1}^{(k-1)} & 0 & 0 & 0 & \cdots & 1 \end{pmatrix}, \tag{1.162}$$
$$\boldsymbol{G} \in \mathbb{F}_2^{k\times n} \setminus \{\boldsymbol{0}_{k\times n}\}, \quad p_m^{(i)} \in \mathbb{F}_2,$$
$$i \in \{0,1\cdots(k-1)\}, \quad m \in \{0,1\cdots(n-k-1)\}, \quad k,n \in \mathbb{N}^*,$$

which is composed of a $k \times (n-k)$ *parity matrix* [15, equation (3.4), p. 69]

$$\boldsymbol{P} = \begin{pmatrix} p_0^{(0)} & p_1^{(0)} & \cdots & p_{n-k-1}^{(0)} \\ p_0^{(1)} & p_1^{(1)} & \cdots & p_{n-k-1}^{(1)} \\ p_0^{(2)} & p_1^{(2)} & \cdots & p_{n-k-1}^{(2)} \\ \vdots & \vdots & \ddots & \vdots \\ p_0^{(k-1)} & p_1^{(k-1)} & \cdots & p_{n-k-1}^{(k-1)} \end{pmatrix}, \quad \boldsymbol{P} \in \mathbb{F}_2^{k\times(n-k)} \setminus \{\boldsymbol{0}_{k\times(n-k)}\}, \quad p_m^{(i)} \in \mathbb{F}_2, \tag{1.163}$$
$$i \in \{0,1\cdots(k-1)\}, \quad m \in \{0,1\cdots(n-k-1)\}, \quad k,n \in \mathbb{N}^*,$$

with elements $p_m^{(i)} \in \mathbb{F}_2$, $i \in \{0,1\cdots(k-1)\}$, $m \in \{0,1\cdots(n-k-1)\}$, $k,n \in \mathbb{N}^*$ assuming the values 0 or 1, and the $k \times k$ square *identity matrix* $\boldsymbol{I}_k$, $k \in \mathbb{N}^*$ of (1.119) [15, equation (3.4), p. 69]. With (1.163) and with (1.119), (1.162) becomes [15, p. 69]

$$\boldsymbol{G} = \begin{pmatrix} \boldsymbol{P} & \boldsymbol{I}_k \end{pmatrix}, \quad \boldsymbol{G} \in \mathbb{F}_2^{k\times n} \setminus \{\boldsymbol{0}_{k\times n}\}, \quad \boldsymbol{P} \in \mathbb{F}_2^{k\times(n-k)} \setminus \{\boldsymbol{0}_{k\times(n-k)}\}, \quad \boldsymbol{I}_k \in \mathbb{F}_2^{k\times k}, \tag{1.164}$$
$$k,n \in \mathbb{N}^*.$$

With a message $\boldsymbol{u} \in \mathbb{F}_2^k$, $k \in \mathbb{N}^*$ and with (1.120), the corresponding codeword $\boldsymbol{v} \in \mathbb{V}$ of (1.122) becomes

$$\begin{aligned} \boldsymbol{v} = \boldsymbol{u}\boldsymbol{G} = \boldsymbol{u}\begin{pmatrix} \boldsymbol{P} & \boldsymbol{I}_k \end{pmatrix} = \begin{pmatrix} \boldsymbol{u}\boldsymbol{P} & \boldsymbol{u}\boldsymbol{I}_k \end{pmatrix} = \begin{pmatrix} \boldsymbol{u}\boldsymbol{P} & \boldsymbol{u} \end{pmatrix}, \\ \boldsymbol{G} \in \mathbb{F}_2^{k\times n} \setminus \{\boldsymbol{0}_{k\times n}\}, \quad \boldsymbol{P} \in \mathbb{F}_2^{k\times(n-k)} \setminus \{\boldsymbol{0}_{k\times(n-k)}\}, \quad \boldsymbol{I}_k \in \mathbb{F}_2^{k\times k}, \\ \boldsymbol{v} \in \mathbb{V} \subseteq \mathbb{F}_2^n, \quad \boldsymbol{u} \in \mathbb{F}_2^k, \quad k, n \in \mathbb{N}^*. \end{aligned} \tag{1.165}$$

According to (1.165), the codeword $\boldsymbol{v} \in \mathbb{V}$ contains the message $\boldsymbol{u} \in \mathbb{F}_2^k$, $k \in \mathbb{N}^*$, which is located at the end of $\boldsymbol{v} \in \mathbb{V}$, whereas the first $(n-k)$, $k, n \in \mathbb{N}^*$, components of $\boldsymbol{v} \in \mathbb{V}$ are determined by the product of the message $\boldsymbol{u} \in \mathbb{F}_2^k$, $k \in \mathbb{N}^*$ and the $k \times (n-k)$ parity matrix $\boldsymbol{P}$, $k, n \in \mathbb{N}^*$, [15, equation (3.4), p. 69]. These first $(n-k)$, $k, n \in \mathbb{N}^*$, components of $\boldsymbol{v} \in \mathbb{V}$ are, e. g., termed *parity check digits* [15, p. 68], *redundancy digits* (in German "*Redundanzzeichen*") [23, p. 8] or *redundant bits* [15, p. 3] and the like. Clearly, since $\boldsymbol{v} \in \mathbb{V}$ already contains the message $\boldsymbol{u} \in \mathbb{F}_2^k$, $k \in \mathbb{N}^*$, at its end, the first $(n-k)$, $k, n \in \mathbb{N}^*$, components of $\boldsymbol{v} \in \mathbb{V}$, which are determined by $\boldsymbol{u}\boldsymbol{P}$, are *redundant* and are intended to overcome the equivocation $\mathrm{H}\{X \mid Y\}$; cf. Section 1.2, especially Figure 1.3.

An (n, k) binary linear block code $\mathbb{V}$, which consists only of codewords that have the structure of (1.162), respectively (1.165), containing the message $\boldsymbol{u} \in \mathbb{F}_2^k$, $k \in \mathbb{N}^*$ is termed a *systematic* (n, k) *binary linear block code* $\mathbb{V}$ [15, p. 68]. The resulting *systematic structure* of $\boldsymbol{v} \in \mathbb{V}$ is illustrated in Figure 1.5 [15, Figure 3.1, p. 68].

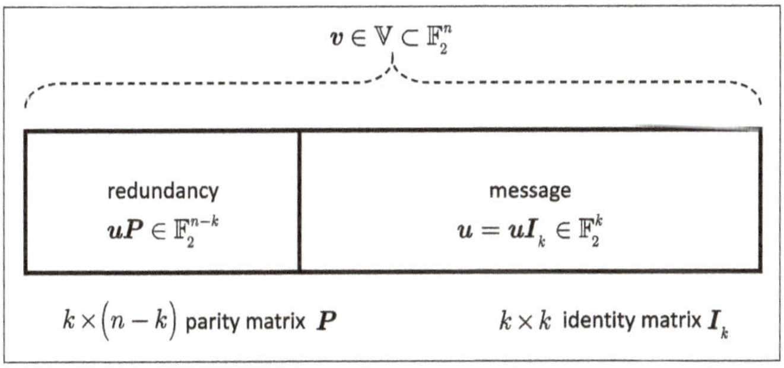

Figure 1.5: Systematic structure of a codeword $\boldsymbol{v} \in \mathbb{V}$ (adapted according to [15, Figure 3.1, p. 68]).

Example 1.9. The two Hamming codes considered in, e. g., the Examples 1.5 and 1.6 are systematic binary linear block codes.

Appropriately applying the aforementioned elementary row operations [5, p. 109] to a predetermined generator matrix $\boldsymbol{G}$ given by (1.124), which has not the form of (1.162), will yield the desired systematic structure, also called the *systematic form*.

We find [15, equation (3.6a) p. 69]

$$v_{n-k+i} = u_i, \quad i \in \{0, 1 \cdots (k-1)\}, \quad k \in \mathbb{N}^*, \tag{1.166}$$

and using (1.125), [15, equation (3.6b) p. 69]

$$v_j = \bigoplus_{i=0}^{k-1} u_i \odot p_j^{(i)}, \quad j \in \{0, 1 \cdots (n-k-1)\}, \quad n \geq k, \quad k, n \in \mathbb{N}^*. \tag{1.167}$$

The $(n-k)$, $k, n \in \mathbb{N}^*$ equations given by (1.167) are called *parity check equations* of the (n, k) binary linear block code $\mathbb{V}$ [15, p. 69].

Every generator matrix $\boldsymbol{G}$ can be converted to a generator matrix in *row echelon form*, also called the *standard form* [25, p. 53]. The row echelon form of a matrix is defined as follows [25, p. 41].

Definition 1.27 (Row-Echelon Form/Standard Form of a Matrix). The row echelon form or standard form of a matrix is defined as follows [25, p. 41]:

a) The leading nonzero term of every nonzero row is one.
b) Every column containing such a leading term has a zero for each of its other entries.
c) The leading term of any row is to the right of the leading term in every higher row.
d) All zero rows are below all nonzero rows.

Let us look at an example.

Example 1.10. The matrix

$$\begin{pmatrix} \mathbf{1} & 2 & \underline{0} & 3 & 4 & \underline{0} \\ \underline{0} & 0 & \mathbf{1} & 5 & 0 & \underline{0} \\ \underline{0} & 0 & \underline{0} & 0 & 0 & \mathbf{1} \\ \underline{0} & 0 & \underline{0} & 0 & 0 & \underline{0} \end{pmatrix} \tag{1.168}$$

is in row echelon form or standard form. The leading nonzero terms are printed in bold face. The zeros in each column containing such a leading term are underlined. Leading terms are in the first, the third and the sixth columns. These three columns do not have any other nonzero elements. The leading term of the first row is in the first column. The leading term of the second row is in the third column and, therefore, to the right of the leading terms of the higher rows, i. e., row 1. The leading term of the third row is in the sixth column and therefore to the right of the leading terms of the higher rows, i. e., rows 1 and 2. The fourth row does not have any leading term. It is a zero row and comes below all nonzero rows.

Since the rows are linearly independent, there is no row which is only composed of zeros [25, p. 53].

Every generator matrix $\boldsymbol{G}$ defined by (1.124), respectively (1.162), can be converted to a generator matrix for an equivalent code with a $k \times k$ identity matrix $\boldsymbol{I}_k$ in the first or in the last k columns by column permutations. We then yield the *systematic form of the generator matrix* [25, p. 53]

$$\boldsymbol{G} = \begin{pmatrix} \boldsymbol{I}_k & \boldsymbol{P} \end{pmatrix}, \quad \boldsymbol{G} \in \mathbb{F}_2^{k\times n} \setminus \{\mathbf{0}_{k\times n}\}, \quad \boldsymbol{P} \in \mathbb{F}_2^{k\times(n-k)} \setminus \{\mathbf{0}_{k\times(n-k)}\}, \quad \boldsymbol{I}_k \in \mathbb{F}_2^{k\times k}, \tag{1.169}$$

$$k, n \in \mathbb{N}^*. \tag{1.170}$$

Remark 1.14 (On the Systematic Form of the Generator Matrix $\boldsymbol{G}$). An encoder of a code is termed a systematic encoder if and only if all the $k \in \mathbb{N}^*$ message bits of a message $\boldsymbol{u}$ are mapped onto $k \leq n$, $k,n \in \mathbb{N}^*$, bits contained in the codeword $\boldsymbol{v}$ without any change [4, Definition, p. 302]. This means that there are distinct, i. e., mutually different coordinates $i_0, i_1 \cdots i_{k-1}, k \in \mathbb{N}^*$, such that [4, Definition, p. 302]

$$c_{i_0} = u_0, \quad c_{i_1} = u_1 \quad \cdots \quad c_{i_{k-1}} = u_{k-1}, \quad k \in \mathbb{N}^*. \tag{1.171}$$

Hence, the message is found unchanged in the codeword [4, Definition, p. 302]. Regrettably, the above very liberal definition (1.171) of a systematic encoder yields a lack of agreement in how the systematic form of a generator matrix should be defined.

Consequently, there is no common understanding about the exact shape of the generator matrix in systematic form. For instance, Lin and Costello use the form of (1.162); cf., e. g., [15, p. 69], whereas for instance Blahut, cf., e. g., [25, p. 53], [29, p. 49], Bossert, cf., e. g., [5, Definition 1.13, p. 14] and Friedrichs, cf., e. g., [5, equation (109), p. 109], seem to prefer (1.169), which is the *row echelon form* or *standard form* (in German *"Gaußsche Normalform"* or *"kanonische Staffelform"* [5, p. 109]); cf., e. g., Definition 1.27 and Example 1.10. MacWilliams and Sloane leave the exact structure completely flexible [4, Definition, p. 302], however, use the row echelon form in some examples; cf., e. g., [4, p. 6f.].

There is another useful matrix associated with every (n, k) binary linear block code $\mathbb{V}$ [15, p. 70]. For any $k \times n$, $k,n \in \mathbb{N}^*$, generator matrix $\boldsymbol{G}$ with k, $k \in \mathbb{N}^*$, nonzero and linearly independent rows $\boldsymbol{g}^{(i)} \in \mathbb{F}_2^n \setminus \{\boldsymbol{0}_n\}$, $i \in \{0,1 \cdots (k-1)\}$, $k \in \mathbb{N}^*$, there exists an $(n-k) \times n$ matrix [4, p. 2], [5, p. 111],

$$\boldsymbol{H} = \begin{pmatrix} \boldsymbol{h}^{(0)} \\ \boldsymbol{h}^{(1)} \\ \vdots \\ \boldsymbol{h}^{(n-k-1)} \end{pmatrix} = \begin{pmatrix} h_0^{(0)} & h_1^{(0)} & \cdots & h_{n-1}^{(0)} \\ h_0^{(1)} & h_1^{(1)} & \cdots & h_{n-1}^{(1)} \\ \vdots & \vdots & \ddots & \vdots \\ h_0^{(n-k-1)} & h_1^{(n-k-1)} & \cdots & h_{n-1}^{(n-k-1)} \end{pmatrix}, \quad \operatorname{rank}\{\boldsymbol{H}\} = n-k, \tag{1.172}$$

$$\begin{aligned} &\boldsymbol{H} \in \mathbb{F}_2^{(n-k)\times n} \setminus \{\boldsymbol{0}_{(n-k)\times n}\}, \quad \boldsymbol{h}^{(m)} \in \mathbb{F}_2^n \setminus \{\boldsymbol{0}_n\}, \quad h_j^{(m)} \in \mathbb{F}_2, \\ &m \in \{0,1 \cdots (n-k-1)\}, \quad j \in \{0,1 \cdots (n-1)\}, \quad k,n \in \mathbb{N}^*, \end{aligned}$$

with the $(n-k)$ linearly independent rows

$$\begin{aligned} \boldsymbol{h}^{(m)} &= (h_0^{(m)}, h_1^{(m)} \cdots h_{n-1}^{(m)}) \in \mathbb{F}_2^n \setminus \{\boldsymbol{0}_n\}, \\ &h_j^{(m)} \in \mathbb{F}_2, \quad m \in \{0,1 \cdots (n-k-1)\}, \quad j \in \{0,1 \cdots (n-1)\}, \quad k,n \in \mathbb{N}^*, \end{aligned} \tag{1.173}$$

such that any of the k, $k \in \mathbb{N}^*$, basis vectors $\boldsymbol{g}^{(i)} \in \mathbb{F}_2^n \setminus \{\boldsymbol{0}_n\}$, $i \in \{0,1 \cdots (k-1)\}$, $k \in \mathbb{N}^*$ (cf. (1.123)) in the row space of $\boldsymbol{G}$ is orthogonal to each one of the $(n-k)$, $k,n \in \mathbb{N}^*$, row vectors $\boldsymbol{h}^{(m)}$, $m \in \{0,1 \cdots (n-k-1)\}$, $k,n \in \mathbb{N}^*$ of $\boldsymbol{H}$ defined by (1.172) and any vector that is orthogonal to the rows of $\boldsymbol{H}$ is in the row space of $\boldsymbol{G}$ [15, p. 70]. This $(n-k) \times n$ matrix $\boldsymbol{H}$ introduced in (1.172) is termed the *parity check matrix* [4, p. 2].

Remark 1.15 (The $k \times (n-k)$ Parity Matrix $\boldsymbol{P}$ of (1.163) and the $(n-k) \times n$ Parity Check Matrix $\boldsymbol{H}$ of (1.172) are Different Matrices). Note that the $k \times (n-k)$ parity matrix $\boldsymbol{P}$ of (1.163) and the $(n-k) \times n$ parity check matrix $\boldsymbol{H}$ of (1.172) are different matrices although their names sound "similar."

Hence, we can describe the (n, k) binary linear block code $\mathbb{V}$ generated by $\boldsymbol{G}$ in an alternative way. Let the $n \times (n-k)$ matrix

$$\boldsymbol{H}^{\mathrm{T}} = \begin{pmatrix} \boldsymbol{h}^{(0)\mathrm{T}} & \boldsymbol{h}^{(1)\mathrm{T}} & \dots & \boldsymbol{h}^{(n-k-1)\mathrm{T}} \end{pmatrix} \tag{1.174}$$

be the transpose of $\boldsymbol{H}$ given by (1.172). The vector $\boldsymbol{v} \in \mathbb{F}_2^n$ is a codeword of the (n, k) binary linear block code $\mathbb{V}$ if and only if [15, p. 70]

$$\boldsymbol{v}\boldsymbol{H}^{\mathrm{T}} = \mathbf{0}_{n-k}, \quad \boldsymbol{v} \in \mathbb{F}_2^n, \quad k, n \in \mathbb{N}^*. \tag{1.175}$$

Consequently, [5, Definition 4.2, p. 111]

$$\boldsymbol{w}\boldsymbol{H}^{\mathrm{T}} \neq \mathbf{0}_{n-k}, \quad \boldsymbol{w} \in \mathbb{F}_2^n \setminus \mathbb{V}, \quad k, n \in \mathbb{N}^*, \tag{1.176}$$

holds for all n-tuples $\boldsymbol{w} \in \mathbb{F}_2^n \setminus \mathbb{V}$, which are not elements of the (n, k) binary linear block code $\mathbb{V}$.

Definition 1.28 (Parity Check Matrix). An $(n-k) \times n$, $k, n \in \mathbb{N}^*$, matrix $\boldsymbol{H}$ is termed *parity check matrix* of the (n, k) binary linear block code $\mathbb{V}$ if the (n, k) binary linear block code $\mathbb{V}$ is the *null space of* $\boldsymbol{H}$ [5, Definition 4.2, p. 111]

$$\mathbb{V} = \left\{ \boldsymbol{v} \in \mathbb{F}_2^n \mid \boldsymbol{v}\boldsymbol{H}^{\mathrm{T}} = \mathbf{0}_{n-k} \right\}, \quad k, n \in \mathbb{N}^*, \tag{1.177}$$

respectively, the *row space of* $\boldsymbol{G}$ [5, Definition 4.2, p. 111]; cf. (1.175).

Theorem 1.11 (Rank of the Parity Check Matrix). *The rank of the $(n-k) \times n$, $k, n \in \mathbb{N}^*$, parity check matrix $\boldsymbol{H}$, rank$\{\boldsymbol{H}\}$ is equal to $(n-k)$ [5, Satz 4.2, p. 111].*

Proof. If rank$\{\boldsymbol{H}\} < (n-k)$, then the $(n-k)$ rows of $(n-k) \times k$, $k, n \in \mathbb{N}^*$, the parity check matrix $\boldsymbol{H}$ given by (1.172), respectively (1.184), are linearly dependent. Applying the elementary row operations to the $(n-k) \times k$, $k, n \in \mathbb{N}^*$, the parity check matrix $\boldsymbol{H}$ yields at least one zero row and the null space of $\boldsymbol{H}$ would have dimension $\geq (k+1)$. Then the null space of $\boldsymbol{H}$ is not the (n, k) binary linear block code $\mathbb{V}$ and the definition of $\boldsymbol{H}$ is not fulfilled. □

The $(n-k)$ rows of $(n-k) \times k$, $k, n \in \mathbb{N}^*$, the parity check matrix $\boldsymbol{H}$ given by (1.172), respectively (1.184), are linearly independent.

Then we find

$$\boldsymbol{u}\boldsymbol{G}\boldsymbol{H}^{\mathrm{T}} = \mathbf{0}_{n-k}, \quad \boldsymbol{u} \in \mathbb{F}_2^k, \quad \boldsymbol{G} \in \mathbb{F}_2^{k \times n} \setminus \{\mathbf{0}_{k \times n}\}, \quad \boldsymbol{H} \in \mathbb{F}_2^{(n-k) \times n} \setminus \{\mathbf{0}_{(n-k) \times n}\}, \quad k, n \in \mathbb{N}^*, \tag{1.178}$$

for all messages $\boldsymbol{u} \in \mathbb{F}_2^k$. (1.178) can only be fulfilled if $\boldsymbol{G}\boldsymbol{H}^{\mathrm{T}}$ is the $k \times (n-k)$ *zero matrix* $\boldsymbol{0}_{k\times(n-k)}$, i. e.,

$$\boldsymbol{G}\boldsymbol{H}^{\mathrm{T}} = \boldsymbol{0}_{k\times(n-k)}, \quad \boldsymbol{G} \in \mathbb{F}_2^{k\times n} \setminus \{\boldsymbol{0}_{k\times n}\}, \quad \boldsymbol{H} \in \mathbb{F}_2^{(n-k)\times n} \setminus \{\boldsymbol{0}_{(n-k)\times n}\}, \quad k,n \in \mathbb{N}^*. \tag{1.179}$$

Theorem 1.12 (Parity Check Matrix Fulfills (1.179)). *The $(n-k)\times n$, $k,n \in \mathbb{N}^*$, matrix $\boldsymbol{H}$ is a parity check matrix for the (n,k) binary linear block code $\mathbb{V}$ if (1.179) is fulfilled* [5, *Satz 4.3 (1), p. 111f.*].

Proof. The proof is given in (1.178). □

Using (1.124) and (1.174), the left-hand side of (1.179) becomes

$$\begin{pmatrix} \boldsymbol{g}^{(0)} \\ \boldsymbol{g}^{(1)} \\ \vdots \\ \boldsymbol{g}^{(k-1)} \end{pmatrix} \begin{pmatrix} \boldsymbol{h}^{(0)\mathrm{T}} & \boldsymbol{h}^{(1)\mathrm{T}} & \cdots & \boldsymbol{h}^{(n-k-1)\mathrm{T}} \end{pmatrix}$$
$$= \begin{pmatrix} \boldsymbol{g}^{(0)}\boldsymbol{h}^{(0)\mathrm{T}} & \boldsymbol{g}^{(0)}\boldsymbol{h}^{(1)\mathrm{T}} & \cdots & \boldsymbol{g}^{(0)}\boldsymbol{h}^{(n-k-1)\mathrm{T}} \\ \boldsymbol{g}^{(1)}\boldsymbol{h}^{(0)\mathrm{T}} & \boldsymbol{g}^{(1)}\boldsymbol{h}^{(1)\mathrm{T}} & \cdots & \boldsymbol{g}^{(1)}\boldsymbol{h}^{(n-k-1)\mathrm{T}} \\ \vdots & \vdots & \ddots & \vdots \\ \boldsymbol{g}^{(k-1)}\boldsymbol{h}^{(0)\mathrm{T}} & \boldsymbol{g}^{(k-1)}\boldsymbol{h}^{(1)\mathrm{T}} & \cdots & \boldsymbol{g}^{(k-1)}\boldsymbol{h}^{(n-k-1)\mathrm{T}} \end{pmatrix}, \tag{1.180}$$

yielding

$$\begin{pmatrix} \boldsymbol{g}^{(0)}\boldsymbol{h}^{(0)\mathrm{T}} & \boldsymbol{g}^{(0)}\boldsymbol{h}^{(1)\mathrm{T}} & \cdots & \boldsymbol{g}^{(0)}\boldsymbol{h}^{(n-k-1)\mathrm{T}} \\ \boldsymbol{g}^{(1)}\boldsymbol{h}^{(0)\mathrm{T}} & \boldsymbol{g}^{(1)}\boldsymbol{h}^{(1)\mathrm{T}} & \cdots & \boldsymbol{g}^{(1)}\boldsymbol{h}^{(n-k-1)\mathrm{T}} \\ \vdots & \vdots & \ddots & \vdots \\ \boldsymbol{g}^{(k-1)}\boldsymbol{h}^{(0)\mathrm{T}} & \boldsymbol{g}^{(k-1)}\boldsymbol{h}^{(1)\mathrm{T}} & \cdots & \boldsymbol{g}^{(k-1)}\boldsymbol{h}^{(n-k-1)\mathrm{T}} \end{pmatrix} = \underbrace{\begin{pmatrix} 0 & 0 & \cdots & 0 \\ 0 & 0 & \cdots & 0 \\ \vdots & \vdots & \ddots & \vdots \\ 0 & 0 & \cdots & 0 \end{pmatrix}}_{=\boldsymbol{0}_{k\times(n-k)}}, \tag{1.181}$$

or, equivalently, using (1.127) of Definition 1.21,

$$\boldsymbol{g}^{(i)}\boldsymbol{h}^{(m)\mathrm{T}} = \langle \boldsymbol{g}^{(i)} \mid \boldsymbol{h}^{(m)} \rangle = [\boldsymbol{G}\boldsymbol{H}^{\mathrm{T}}]_{i,m} = 0, \tag{1.182}$$
$$i \in \{0,1\cdots(k-1)\}, \quad m \in \{0,1\cdots(n-k-1)\}, \quad k,n \in \mathbb{N}^*. \tag{1.183}$$

According to (1.182), the matrix element $[\boldsymbol{G}\boldsymbol{H}^{\mathrm{T}}]_{i,m}$, $i \in \{0,1\cdots(k-1)\}$, $m \in \{0,1\cdots(n-k-1)\}$, $k,n \in \mathbb{N}^*$, in the ith row and the mth column, given by the inner product $\langle \boldsymbol{g}^{(i)} \mid \boldsymbol{h}^{(m)} \rangle$ equal to $\boldsymbol{g}^{(i)}\boldsymbol{h}^{(m)\mathrm{T}}$, $i \in \{0,1\cdots(k-1)\}$, $m \in \{0,1\cdots(n-k-1)\}$, $k,n \in \mathbb{N}^*$, is zero, which means that any of the k, $k \in \mathbb{N}^*$, basis vectors $\boldsymbol{g}^{(i)} \in \mathbb{F}_2^n \setminus\{\boldsymbol{0}_n\}$, $i \in \{0,1\cdots(k-1)\}$, $k \in \mathbb{N}^*$ (cf. (1.123)), in the row space of $\boldsymbol{G}$ is orthogonal to each one of the $(n-k)$, $k,n \in \mathbb{N}^*$, row vectors $\boldsymbol{h}^{(m)}$, $m \in \{0,1\cdots(n-k-1)\}$, $k,n \in \mathbb{N}^*$, of $\boldsymbol{H}$ defined by (1.172) [15, p. 70]. The (n,k) binary linear block code $\mathbb{V}$ is therefore called the *null space of* $\boldsymbol{H}$ [15, p. 70].

If the $k \times n$, $k, n \in \mathbb{N}^*$, generator matrix $\boldsymbol{G}$ of the (n, k) binary linear block code $\mathbb{V}$ is in the systematic form given by (1.162), the $(n-k) \times n$, $k, n \in \mathbb{N}^*$, parity check matrix $\boldsymbol{H}$ may take the following form [15, equation (3.7), p. 70]:

$$\boldsymbol{H} = \begin{pmatrix} \boldsymbol{I}_{n-k} & \boldsymbol{P}^{\mathrm{T}} \end{pmatrix} = \begin{pmatrix} 1 & 0 & 0 & \cdots & 0 & p_0^{(0)} & p_0^{(1)} & \cdots & p_0^{(k-1)} \\ 0 & 1 & 0 & \cdots & 0 & p_1^{(0)} & p_1^{(1)} & \cdots & p_1^{(k-1)} \\ 0 & 0 & 1 & \cdots & 0 & p_2^{(0)} & p_2^{(1)} & \cdots & p_2^{(k-1)} \\ \vdots & \vdots & \vdots & \ddots & \vdots & \vdots & \vdots & \ddots & \vdots \\ 0 & 0 & 0 & \cdots & 1 & p_{n-k-1}^{(0)} & p_{n-k-1}^{(1)} & \cdots & p_{n-k-1}^{(k-1)} \end{pmatrix}, \quad k, n \in \mathbb{N}^*, \tag{1.184}$$

the $(n-k) \times k$, $k, n \in \mathbb{N}^*$, matrix $\boldsymbol{P}^{\mathrm{T}}$ being the transpose of the $k \times (n-k)$, $k, n \in \mathbb{N}^*$, parity matrix $\boldsymbol{P}$. The form of (1.184) of the parity check matrix is sometimes called the *standard form* [4, equation (2), p. 2; equation (39), p. 24]. Therefore, we obtain

$$\boldsymbol{H}^{\mathrm{T}} = \begin{pmatrix} \boldsymbol{I}_{n-k} \\ \boldsymbol{P} \end{pmatrix} = \begin{pmatrix} 1 & 0 & 0 & \cdots & 0 \\ 0 & 1 & 0 & \cdots & 0 \\ 0 & 0 & 1 & \cdots & 0 \\ \vdots & \vdots & \vdots & \ddots & \vdots \\ 0 & 0 & 0 & \cdots & 1 \\ p_0^{(0)} & p_1^{(0)} & p_2^{(0)} & \cdots & p_{n-k-1}^{(0)} \\ p_0^{(1)} & p_1^{(1)} & p_2^{(1)} & \cdots & p_{n-k-1}^{(1)} \\ \vdots & \vdots & \vdots & \ddots & \vdots \\ p_0^{(k-1)} & p_1^{(k-1)} & p_2^{(k-1)} & \cdots & p_{n-k-1}^{(k-1)} \end{pmatrix}, \quad k, n \in \mathbb{N}^*. \tag{1.185}$$

Using (1.162) and (1.184), the left-hand side of (1.179) takes the form

$$\boldsymbol{G}\boldsymbol{H}^{\mathrm{T}} = \begin{pmatrix} \boldsymbol{P} & \boldsymbol{I}_k \end{pmatrix} \begin{pmatrix} \boldsymbol{I}_{n-k} \\ \boldsymbol{P} \end{pmatrix} = \boldsymbol{P}\boldsymbol{I}_{n-k} \oplus \boldsymbol{I}_k\boldsymbol{P} = \boldsymbol{P} \oplus \boldsymbol{P} = \boldsymbol{0}_{k \times (n-k)}, \quad k, n \in \mathbb{N}^*, \tag{1.186}$$

which we can easily verify as follows:

$$\begin{pmatrix} p_0^{(0)} & p_1^{(0)} & \cdots & p_{n-k-1}^{(0)} & 1 & 0 & 0 & \cdots & 0 \\ p_0^{(1)} & p_1^{(1)} & \cdots & p_{n-k-1}^{(1)} & 0 & 1 & 0 & \cdots & 0 \\ p_0^{(2)} & p_1^{(2)} & \cdots & p_{n-k-1}^{(2)} & 0 & 0 & 1 & \cdots & 0 \\ \vdots & \vdots & \ddots & \vdots & \vdots & \vdots & \vdots & \ddots & \vdots \\ p_0^{(k-1)} & p_1^{(k-1)} & \cdots & p_{n-k-1}^{(k-1)} & 0 & 0 & 0 & \cdots & 1 \end{pmatrix}$$

$$
\begin{pmatrix}
1 & 0 & 0 & \cdots & 0 \\
0 & 1 & 0 & \cdots & 0 \\
0 & 0 & 1 & \cdots & 0 \\
\vdots & \vdots & \vdots & \ddots & \vdots \\
0 & 0 & 0 & \cdots & 1 \\
p_0^{(0)} & p_1^{(0)} & p_2^{(0)} & \cdots & p_{n-k-1}^{(0)} \\
p_0^{(1)} & p_1^{(1)} & p_2^{(1)} & \cdots & p_{n-k-1}^{(1)} \\
\vdots & \vdots & \vdots & \ddots & \vdots \\
p_0^{(k-1)} & p_1^{(k-1)} & p_2^{(k-1)} & \cdots & p_{n-k-1}^{(k-1)}
\end{pmatrix}
$$

$$
= \begin{pmatrix}
p_0^{(0)} \oplus p_0^{(0)} & p_1^{(0)} \oplus p_1^{(0)} & \cdots & p_{n-k-1}^{(0)} \oplus p_{n-k-1}^{(0)} \\
p_0^{(1)} \oplus p_0^{(1)} & p_1^{(1)} \oplus p_1^{(1)} & \cdots & p_{n-k-1}^{(1)} \oplus p_{n-k-1}^{(1)} \\
p_0^{(2)} \oplus p_0^{(2)} & p_1^{(2)} \oplus p_1^{(2)} & \cdots & p_{n-k-1}^{(2)} \oplus p_{n-k-1}^{(2)} \\
\vdots & \vdots & \ddots & \vdots \\
p_0^{(k-1)} \oplus p_0^{(k-1)} & p_1^{(k-1)} \oplus p_1^{(k-1)} & \cdots & p_{n-k-1}^{(k-1)} \oplus p_{n-k-1}^{(k-1)}
\end{pmatrix}
$$

$$
= \boldsymbol{P} \oplus \boldsymbol{P} = \boldsymbol{0}_{k\times(n-k)}, \quad k, n \in \mathbb{N}^*. \tag{1.187}
$$

Theorem 1.13 (Parity Check Matrix Fulfills (1.184) for a Generator Matrix in Systematic Form (1.162)). *The $(n-k)\times n$, $k, n \in \mathbb{N}^*$, parity check matrix $\boldsymbol{H}$ of the (n, k) binary linear block code $\mathbb{V}$ takes the form given by (1.184) when the $k \times n$, $k, n \in \mathbb{N}^*$, generator matrix $\boldsymbol{G}$ is in the systematic form of (1.162)* [5, *Satz 4.3 (2), p. 111f.*].

Proof. The proof is given by (1.186). □

Remark 1.16 (On the Alternative Systematic Structure). This result of (1.187), respectively (1.186), does not change in the case of $\boldsymbol{G}$ being equal to $\begin{pmatrix} \boldsymbol{I}_k & \boldsymbol{P} \end{pmatrix}$ and $\boldsymbol{H}$ being equal to $\begin{pmatrix} \boldsymbol{P}^{\mathrm{T}} & \boldsymbol{I}_{n-k} \end{pmatrix}$. Then we have

$$
\boldsymbol{G}\boldsymbol{H}^{\mathrm{T}} = \begin{pmatrix} \boldsymbol{I}_k & \boldsymbol{P} \end{pmatrix} \begin{pmatrix} \boldsymbol{P} \\ \boldsymbol{I}_{n-k} \end{pmatrix} = \boldsymbol{I}_k \boldsymbol{P} \oplus \boldsymbol{P}\boldsymbol{I}_{n-k} = \boldsymbol{P} \oplus \boldsymbol{P} = \boldsymbol{0}_{k\times(n-k)}, \quad k, n \in \mathbb{N}^*. \tag{1.188}
$$

Corollary 1.4 (Minimum Weight of a Code Revisited). *Let $\mathbb{V}$ be a (n, k) binary linear block code with the $(n-k)\times n$, $k, n \in \mathbb{N}^*$, parity check matrix $\boldsymbol{H}$. If no $(d_{\min} - 1)$ or fewer columns of $\boldsymbol{H}$ add to $\boldsymbol{0}_{n-k}$, the code has minimum weight at least $d_{\min}$* [15, *Corollary 3.2.1, p. 78*], [5, *Satz 4.4, p. 113*].

Proof. Let $\boldsymbol{v}$ be a codeword having the Hamming weight $w_{\mathrm{H}}\{\boldsymbol{v}\}$ equal to $d_{\min} \in \mathbb{N}^*$. Then we have

$$
\boldsymbol{0}_{n-k} = \boldsymbol{v}\mathbf{H}^{\mathrm{T}}, \quad k, n \in \mathbb{N}^*, \tag{1.189}
$$

which requires that $d_{\min}$ columns of $\boldsymbol{H}$ add to the zero vector $\boldsymbol{0}_{n-k}$, $k, n \in \mathbb{N}^*$. This means that the minimum number of linearly dependent columns of $\boldsymbol{H}$ is given by $d_{\min}$ [5, Satz 4.4, p. 113].

If only $(d_{\min} - 1)$ or less columns added to the zero vector $\boldsymbol{0}_{n-k}$, $k, n \in \mathbb{N}^*$, i. e., if the minimum number of linearly dependent columns of $\boldsymbol{H}$ was $\leq (d_{\min} - 1)$, there was at least one codeword, which would have the weight $(d_{\min} - 1)$ or less. However, this situation violates the definition of the minimum distance $d_{\min}$; cf. Theorem 1.6. Therefore, the minimum number of linearly dependent columns of $\boldsymbol{H}$, i. e., the minimum number of columns of $\boldsymbol{H}$ that add to the zero vector $\boldsymbol{0}_{n-k}$, $k, n \in \mathbb{N}^*$ is given by $d_{\min}$. □

If $d \geq d_{\mathrm{min}}$ columns of $\boldsymbol{H}$ add to the zero vector $\mathbf{0}_{n-k}$, $k, n \in \mathbb{N}^*$, there exists at least one codeword $\boldsymbol{v}$ with the Hamming weight $w_{\mathrm{H}}\{\boldsymbol{v}\}$ equal to d.

Corollary 1.5 (Minimum Weight of a Code Revisited a Second Time). *Let $\mathbb{V}$ be a (n, k) binary linear block code with the $(n-k) \times n$, $k, n \in \mathbb{N}^*$, parity check matrix $\boldsymbol{H}$. The minimum weight (or the minimum distance) d_{min} of $\mathbb{V}$ is equal to the smallest number of columns of $\boldsymbol{H}$ that are linearly dependent* [15, *Corollary 3.2.2, p. 78*], [5, *Satz 4.4, p. 113*], *and hence sum to $\mathbf{0}_{n-k}$.*

Proof. Let $\boldsymbol{v} \in \mathbb{V}$ be a codeword with the Hamming weight $w_{\mathrm{H}}\{\boldsymbol{v}\}$ equal to $d_{\mathrm{min}} \in \mathbb{N}^*$. Hence, (1.189) holds for $\boldsymbol{v} \in \mathbb{V}$, which requires that d_{min} columns of $\boldsymbol{H}$ add to the zero vector $\mathbf{0}_{n-k}$, $k, n \in \mathbb{N}^*$.

According to Corollary 1.4, the minimum number of columns of $\boldsymbol{H}$, that add to the zero vector $\mathbf{0}_{n-k}$ is d_{min} if there are no fewer arbitrary distinct columns of $\boldsymbol{H}$ that add to the zero vector $\mathbf{0}_{n-k}$, $k, n \in \mathbb{N}^*$.

According to Theorem 1.6, d_{min} is the minimum weight w_{min} of a codeword $\boldsymbol{v}$, which differs from the zero vector $\mathbf{0}_n$, $n \in \mathbb{N}^*$. Therefore, if the smallest number of columns of $\boldsymbol{H}$, which sum to the zero vector $\mathbf{0}_{n-k}$, is equal to d_{min}, which is the minimum weight w_{min} of a codeword $\boldsymbol{v}$, according to Theorem 1.6, d_{min} is the minimum weight (or the minimum distance) of $\mathbb{V}$.

Therefore, the minimum number of linearly dependent columns of the $(n-k) \times n$, $k, n \in \mathbb{N}^*$, parity check matrix $\boldsymbol{H}$ is equal to d_{min}. □

Let us look at an example.

Example 1.11. Consider the $(7, 4, 3)$ binary linear block code $\mathbb{V}$ given in Table 1.2. Its 3×7 parity check matrix $\boldsymbol{H}$ is

$$\boldsymbol{H} = \begin{pmatrix} 1 & 0 & 0 & 0 & 1 & 1 & 1 \\ 0 & 1 & 0 & 1 & 0 & 1 & 1 \\ 0 & 0 & 1 & 1 & 1 & 0 & 1 \end{pmatrix}. \tag{1.190}$$

We see that all columns of $\boldsymbol{H}$ are nonzero and distinct [15, p. 78]. Therefore, no two or fewer columns sum to the zero vector $\mathbf{0}_3$ [15, p. 78]. Thus, the minimum weight of this code is at least 3 [15, p. 78].

However, e. g., the zeroth, the second and the fourth columns sum to $\mathbf{0}_3$ [15, p. 78], i. e.,

$$\begin{pmatrix} 1 \\ 0 \\ 0 \end{pmatrix} \oplus \begin{pmatrix} 0 \\ 0 \\ 1 \end{pmatrix} \oplus \begin{pmatrix} 1 \\ 0 \\ 1 \end{pmatrix} = \begin{pmatrix} 1 \oplus 0 \oplus 1 \\ 0 \oplus 0 \oplus 0 \\ 0 \oplus 1 \oplus 1 \end{pmatrix} = \begin{pmatrix} 0 \\ 0 \\ 0 \end{pmatrix}. \tag{1.191}$$

Thus, according to Corollary 1.4, the minimum weight of the code is 3 [15, p. 78].

From Table 1.2, we see that the minimum weight of the code is indeed 3; cf. Example 1.7. It follows from Corollary 1.5 that the minimum distance is 3.

The further analysis of an (n, k, d_{min}) binary linear block code $\mathbb{V}$ leads us to the following insight.

Theorem 1.14 (At Least $(d_{\mathrm{min}} - 1)$ Linearly Independent Columns of the Parity Check Matrix). *Let $\boldsymbol{H} \in \mathbb{F}_2^{(n-k)\times n}$, $k, n \in \mathbb{N}^*$, be a parity check matrix of an (n, k, d_{min}) binary linear block code $\mathbb{V}$* [5, *Satz 4.4, p. 113*]. *Then the minimum distance d_{min} is the minimum number of linearly dependent columns in $\boldsymbol{H} \in \mathbb{F}_2^{(n-k)\times n}$, $k, n \in \mathbb{N}^*$* [5, *Satz 4.4, p. 113*]. *Therefore, each selection of $(d_{\mathrm{min}} - 1)$ columns of $\boldsymbol{H} \in \mathbb{F}_2^{(n-k)\times n}$, $k, n \in \mathbb{N}^*$ is linearly independent, and there is at least one selection of d_{min} linearly dependent columns* [5, *Satz 4.4, p. 113*].

Proof. Let $\boldsymbol{\eta}^{(j)} \in \mathbb{F}_2^{n-k}, j \in \{0,1\cdots(n-1)\}, k,n \in \mathbb{N}^*$ be the column vectors of the parity check matrix $\boldsymbol{H} \in \mathbb{F}_2^{(n-k)\times n}, k,n \in \mathbb{N}^*$ [5, Satz 4.4, p. 113]

$$\boldsymbol{H} = \begin{pmatrix} \boldsymbol{\eta}^{(0)\mathrm{T}} & \boldsymbol{\eta}^{(1)\mathrm{T}} & \cdots & \boldsymbol{\eta}^{(n-1)\mathrm{T}} \end{pmatrix} \quad \boldsymbol{H} \in \mathbb{F}_2^{(n-k)\times n}, \quad k,n \in \mathbb{N}^*. \tag{1.192}$$

Let us prove the first part of this theorem. Assume that there is a selection of only $(d_{\min} - 1)$ linearly dependent column vectors $\boldsymbol{\eta}^{(r_1)} \in \mathbb{F}_2^{n-k}$, $\boldsymbol{\eta}^{(r_2)} \in \mathbb{F}_2^{n-k}, \cdots \boldsymbol{\eta}^{(r_{[d_{\min}-1]})} \in \mathbb{F}_2^{n-k}$ [5, Satz 4.4, p. 113]. Hence, there is an n-tuple

$$\boldsymbol{y} = (y_0, y_1 \cdots y_{n-1}), \quad \boldsymbol{y} \in \mathbb{F}_2^n, \quad n \in \mathbb{N}^*, \tag{1.193}$$

with the Hamming weight $w_{\mathrm{H}}\{\boldsymbol{y}\} \leq (d_{\min} - 1)$, which complies with the following relation [5, Satz 4.4, p. 113]:

$$\mathbf{0}_{n-k} = \sum_{a=1}^{d_{\min}-1} y_{r_a} \boldsymbol{\eta}^{(r_a)} = \sum_{j=0}^{n-1} y_j \boldsymbol{\eta}^{(j)} = \boldsymbol{y}\boldsymbol{H}^{\mathrm{T}}. \tag{1.194}$$

Hence, $\boldsymbol{y}$ of (1.193) is a codeword, i. e., $\boldsymbol{y} \in \mathbb{V} \subseteq \mathbb{F}_2^n, n \in \mathbb{N}^*$ [5, Satz 4.4, p. 113]. However, this contradicts the above assumption $w_{\mathrm{H}}\{\boldsymbol{y}\} \leq (d_{\min} - 1)$. Therefore, also the assumption that there is a selection of only $(d_{\min} - 1)$ linearly dependent column vectors $\boldsymbol{\eta}^{(r_1)} \in \mathbb{F}_2^{n-k}, \boldsymbol{\eta}^{(r_2)} \in \mathbb{F}_2^{n-k}, \cdots \boldsymbol{\eta}^{(r_{[d_{\min}-1]})} \in \mathbb{F}_2^{n-k}$ is wrong. Hence, the parity check matrix $\boldsymbol{H} \in \mathbb{F}_2^{(n-k)\times n}, k,n \in \mathbb{N}^*$ has at least $(d_{\min}-1)$ linearly independent column vectors.

Now, let us prove the second part of this theorem. Let $\boldsymbol{v} \in \mathbb{V} \subseteq \mathbb{F}_2^n, n \in \mathbb{N}^*$ be a codeword with the Hamming weight $w_{\mathrm{H}}\{\boldsymbol{v}\}$ equal to $d_{\min}$ [5, Satz 4.4, p. 113]. Owing to

$$\mathbf{0}_{n-k} = \boldsymbol{v}\boldsymbol{H}^{\mathrm{T}}, \quad k,n \in \mathbb{N}^*, \tag{1.195}$$

the $d_{\min}$ components of $\boldsymbol{v} \in \mathbb{V} \subseteq \mathbb{F}_2^n, n \in \mathbb{N}^*$, which are equal to 1 determine the selection of $d_{\min}$ linearly dependent column vectors of the parity check matrix $\boldsymbol{H} \in \mathbb{F}_2^{(n-k)\times n}, k,n \in \mathbb{N}^*$. □

The binomial theorem states [11, equations (1.36a), (1.36b) and (1.36c), p. 12]

$$(a+b)^m = \sum_{i=0}^{m} \binom{m}{i} a^{m-i} b^i, \quad a,b \in \mathbb{C}, \quad m \in \mathbb{N}^*. \tag{1.196}$$

Using the Singleton bound (1.139) introduced in Theorem 1.7, we yield

$$d_{\min} - 2 \leq n - k - 1, \quad k,n \in \mathbb{N}^*, \tag{1.197}$$

and letting a and b be equal to 1 and replacing m by $(n-1)$, (1.196) immediately leads us to

$$2^{n-1} = \sum_{i=0}^{n-1} \binom{n-1}{i} = \sum_{i=0}^{d_{\min}-2} \binom{n-1}{i} + \sum_{i=d_{\min}-1}^{n-1} \binom{n-1}{i}, \quad n \in \mathbb{N}^*. \tag{1.198}$$

With

$$2^{n-1} = 2^{n-k} \cdot 2^{k-1}, \quad k,n \in \mathbb{N}^*, \tag{1.199}$$

we obtain

$$2^{n-k} > \frac{1}{2^{k-1}} \sum_{i=0}^{d_{\min}-2} \binom{n-1}{i}, \quad k, n \in \mathbb{N}^*, \tag{1.200}$$

from (1.198). However, the bound of (1.200) can be further improved in the case of an $(n, k, d_{\min})$ binary linear block code $\mathbb{V}$: Setting out from Theorem 1.14, we yield the *Gilbert–Varshamov bound* [5, Satz 3.12, p. 83].

Theorem 1.15 (Gilbert-Varshamov Bound). *If*

$$\sum_{i=0}^{d_{\min}-2} \binom{n-1}{i} < 2^{n-k}, \quad d_{\min}, k, n \in \mathbb{N}^*, \tag{1.201}$$

holds, there exists an $(n, k, d_{\min})$ binary linear block code $\mathbb{V}$ [14, equations (3.4-6) and (3.4-7), p. 109f.], [5, Satz 3.12, p. 83]. Equation (1.201) is termed the Gilbert-Varshamov bound *[14, equations (3.4-6) and (3.4-7), p. 109f.], [5, Satz 3.12, p. 83]. The Gilbert-Varshamov bound is an upper bound on the number of parity check sequences 2^{n-k}, $k, n \in \mathbb{N}^*$.*

Proof. Taking Theorem 1.14 into account, it has to be shown that the parity check matrix $\boldsymbol{H} \in \mathbb{F}_2^{(n-k)\times n}$, $k, n \in \mathbb{N}^*$, with n, $n \in \mathbb{N}^*$, columns of length $(n-k)$, $k, n \in \mathbb{N}^*$ can be constructed in such a way that each choice of $(d_{\min}-1)$ columns is linearly independent [5, Satz 3.12, p. 83].

The first column vector $\boldsymbol{\eta}^{(0)} \in \mathbb{F}_2^{n-k}$, $k, n \in \mathbb{N}^*$, must not be unequal to the zero vector $\mathbf{0}_{n-k}$, $k, n \in \mathbb{N}^*$ [5, Satz 3.12, p. 83]. The second column vector $\boldsymbol{\eta}^{(1)} \in \mathbb{F}_2^{n-k}$, $k, n \in \mathbb{N}^*$, must not be a multiple of the first column [5, Satz 3.12, p. 83]. The third column vector $\boldsymbol{\eta}^{(2)} \in \mathbb{F}_2^{n-k}$, $k, n \in \mathbb{N}^*$, must not be a linear combination of the first two columns, and so on [5, Satz 3.12, p. 83].

Let us assume that $(n-1)$, $n \in \mathbb{N}^*$, column vectors were constructed in the aforementioned way [5, Satz 3.12, p. 83]. Of course, the nth column vector $\boldsymbol{\eta}^{(n-1)} \in \mathbb{F}_2^{n-k}$, $k, n \in \mathbb{N}^*$ must not be a linear combination of $(d_{\min}-2)$ arbitrarily chosen column vectors taken from the previous $(n-1)$, $n \in \mathbb{N}^*$, column vectors [5, Satz 3.12, p. 83].

The number of linear combinations of exactly i column vectors taken from $(n-1)$, $n \in \mathbb{N}^*$ column vectors is equal to

$$\binom{n-1}{i} \tag{1.202}$$

and, therefore, there are

$$\mathfrak{L} = \sum_{i=1}^{d_{\min}-2} \binom{n-1}{i} \tag{1.203}$$

linear combinations of up to $(d_{\min}-2)$ arbitrarily chosen column vectors taken from the first $(n-1)$, $n \in \mathbb{N}^*$, column vectors of the parity check matrix $\boldsymbol{H} \in \mathbb{F}_2^{(n-k)\times n}$, $k, n \in \mathbb{N}^*$ [5, Satz 3.12, p. 83].

Since there are 2^{n-k}, $k, n \in \mathbb{N}^*$ possible realizations of the nth, $n \in \mathbb{N}^*$, column vector, we must make sure that the nth, $n \in \mathbb{N}^*$ column vector is neither equal to the zero vector $\mathbf{0}_{n-k}$, $k, n \in \mathbb{N}^*$, nor to any of the $\mathfrak{L}$ linear combinations; cf. (1.203). Thus,

$$2^{n-k} > 1 + \sum_{i=1}^{d_{\min}-2} \binom{n-1}{i} = \sum_{i=0}^{d_{\min}-2} \binom{n-1}{i}, \quad k, n \in \mathbb{N}^*, \tag{1.204}$$

must hold. □

Remark 1.17 (On the Gilbert-Varshamov Bound). The Gilbert-Varshamov bound of (1.201) can be expressed as an upper bound on the number of codewords

$$2^k < \frac{2^n}{\sum_{i=0}^{d_{\min}-2} \binom{n-1}{i}}, \quad d_{\min}, k, n \in \mathbb{N}^*. \tag{1.205}$$

The 2^{n-k}, $k, n \in \mathbb{N}^*$ linear combinations of the rows of the $(n-k) \times n$, $k, n \in \mathbb{N}^*$, parity check matrix $\boldsymbol{H}$ form an $(n, [n-k])$ binary linear block code $\mathbb{V}^{\perp}$, which is termed the *dual code of* $\mathbb{V}$ [15, p. 70], [4, p. 26], [5, Definition 4.3, p. 115].

Definition 1.29 (Dual Code). Let $\boldsymbol{G}$ be the $k \times n$, $k, n \in \mathbb{N}^*$, generator matrix of the (n, k) binary linear block code $\mathbb{V}$, which has the $(n-k) \times n$, $k, n \in \mathbb{N}^*$ parity check matrix $\boldsymbol{H}$. Its *dual code* $\mathbb{V}^{\perp}$, also called *orthogonal code* $\mathbb{V}^{\perp}$, is the set of 2^{n-k}, $k, n \in \mathbb{N}^*$, vectors $\boldsymbol{x} \in \mathbb{F}_2^n$, $n \in \mathbb{N}^*$, which are orthogonal to all codewords of $\mathbb{V}$ [4, equation (42), p. 26]

$$\mathbb{V}^{\perp} = \left\{\boldsymbol{x} \in \mathbb{F}_2^n \mid \langle \boldsymbol{x} \mid \boldsymbol{v} \rangle = \boldsymbol{x}\boldsymbol{v}^{\mathrm{T}} = \boldsymbol{0}_n \; \forall\, \boldsymbol{v} \in \mathbb{V}\right\}, \quad n \in \mathbb{N}^*. \tag{1.206}$$

$\mathbb{V}^{\perp}$ is an $(n, [n-k])$ binary linear block code [4, p. 26].

$\mathbb{V}^{\perp}$ is the *orthogonal subspace to* $\mathbb{V}$ [4, p. 26].

Therefore, a $(n-k) \times n$, $k, n \in \mathbb{N}^*$, parity check matrix $\boldsymbol{H}$ for a (n, k) binary linear block code $\mathbb{V}$ is a generator matrix for its dual $(n, [n-k])$ binary linear block code $\mathbb{V}^{\perp}$, [15, p. 70], [4, p. 26], [5, p. 115].

Taking (1.206) into consideration, the dual $(n, [n-k])$ binary linear block code $\mathbb{V}^{\perp}$ is the *null space of the* (n, k) *binary linear block code* $\mathbb{V}$ [15, p. 70], also termed the *orthogonal subspace to the* (n, k) *binary linear block code* $\mathbb{V}$ [4, p. 26].

A general discussion of the minimum weight of the dual code goes beyond the scope of this textbook. The reader is therefore referred to the excellent, if not to say perfect, textbook by Florence Jessie *MacWilliams* and Neil James Alexander *Sloane* [4, Chapter 5].

1.6 Syndrome and Error Detection

We will set our from the (n, k) binary linear block code $\mathbb{V}$, having the $k \times n$, $k, n \in \mathbb{N}^*$, generator matrix $\boldsymbol{G}$ and the $(n-k) \times n$, $k, n \in \mathbb{N}^*$, parity check matrix $\boldsymbol{H}$. Let $\boldsymbol{v} \in \mathbb{V}$ be a codeword that was transmitted over a noisy channel. Let the *receive vector* at the input of the decoder be

$$\boldsymbol{r} = (r_0, r_1 \cdots r_{n-1}), \quad \boldsymbol{r} \in \mathbb{F}_2^n, \quad r_j \in \mathbb{F}_2, \quad j \in \{0, 1 \cdots (n-1)\}, \quad n \in \mathbb{N}^*. \tag{1.207}$$

Owing to the channel noise, $\boldsymbol{r}$ of (1.207) may be different from $\boldsymbol{v}$ of (1.24). The vector

$$\begin{aligned} \boldsymbol{e} &= (e_0, e_1 \cdots e_{n-1}) \\ &= \boldsymbol{r} \oplus \boldsymbol{v} \end{aligned} \tag{1.208}$$

$$= (r_0 \oplus v_0\,,\, r_1 \oplus v_1 \cdots r_{n-1} \oplus v_{n-1}),$$
$$\boldsymbol{e}, \boldsymbol{r} \in \mathbb{F}_2^n, \quad \boldsymbol{v} \in \mathbb{V} \subseteq \mathbb{F}_2^n, \quad e_j = \begin{cases} 0 & \text{if } r_j = v_j, \\ 1 & \text{if } r_j \neq v_j, \end{cases}$$
$$e_j, r_j, v_j \in \mathbb{F}_2, \quad j \in \{0, 1 \cdots (n-1)\}, \quad n \in \mathbb{N}^*,$$

is called the *error vector* or *error pattern*. The 1's in $\boldsymbol{e}$ of (1.208) are the *transmission errors* caused by the transmission channel and the noise.

Equations (1.207) and (1.208) yield

$$\boldsymbol{r} = \boldsymbol{v} \oplus \boldsymbol{e}, \quad \boldsymbol{e}, \boldsymbol{r} \in \mathbb{F}_2^n, \quad \boldsymbol{v} \in \mathbb{V} \subseteq \mathbb{F}_2^n, \quad n \in \mathbb{N}^*. \tag{1.209}$$

The decoder will first determine whether $\boldsymbol{r}$ contains transmission errors [15, p. 72]. It may therefore compute [15, equation (3.10), p. 72]

$$\boldsymbol{s} = (s_0, s_1 \cdots s_{n-k-1}) = \boldsymbol{r}\boldsymbol{H}^{\mathrm{T}}, \quad \boldsymbol{s} \in \mathbb{F}_2^{n-k}, \quad \boldsymbol{r} \in \mathbb{F}_2^n, \quad \boldsymbol{H}^{\mathrm{T}} \in \mathbb{F}_2^{n\times(n-k)} \setminus \{\boldsymbol{0}_{n\times(n-k)}\}, \tag{1.210}$$
$$s_m \in \mathbb{F}_2, \quad m \in \{0, 1 \cdots (n-k-1)\}, \quad n \in \mathbb{N}^*,$$

termed the *syndrome of* $\boldsymbol{r}$. If and only if $\boldsymbol{r}$ is a codeword, i. e., $\boldsymbol{e}$ is equal to the zero vector $\boldsymbol{0}_n$, $n \in \mathbb{N}^*$, the syndrome $\boldsymbol{s}$ will be the zero vector $\boldsymbol{0}_{n-k}$, $k, n \in \mathbb{N}^*$. Then the receiver considers $\boldsymbol{r}$ free of transmission errors. If $\boldsymbol{r}$ is not a codeword, we find [15, p. 72]

$$\boldsymbol{s} \neq \boldsymbol{0}_{n-k}, \quad \boldsymbol{s} \in \mathbb{F}_2^{n-k}, \quad k, n \in \mathbb{N}^*. \tag{1.211}$$

If transmission errors were detected, the decoder will either attempt to locate the errors and correct them *(FEC, Forward Error Correction)* or request a retransmission of $\boldsymbol{v}$ *(ARQ, Automatic Repeat Request)* [15, p. 72].

Furthermore, it is possible that the transmission errors in certain error vectors are not detectable, i. e., $\boldsymbol{r}$ contains transmission errors although $\boldsymbol{r}\boldsymbol{H}^{\mathrm{T}}$ is the zero vector $\boldsymbol{0}_{n-k}$, $k, n \in \mathbb{N}^*$, which occurs when the error vector $\boldsymbol{e}$ of (1.208) is a nonzero codeword [15, p. 73]. Such error patterns are termed *undetectable error patterns* [15, p. 73]. Since there are $(2^k - 1)$ nonzero codewords, $\boldsymbol{v} \in \mathbb{V}$ in the (n, k) binary linear block code $\mathbb{V}$, there are $(2^k - 1)$ undetectable error patterns [15, p. 73]. In the occurrence of such an undetectable error pattern, a *decoding error* is unavoidable [15, p. 73].

The syndrome $\boldsymbol{s}$ of the receive vector $\boldsymbol{r}$ (cf. (1.210)) depends only on the error pattern $\boldsymbol{e}$ and not on the transmitted codeword $\boldsymbol{v}$ [15, p. 73]. We yield [15, p. 73]:

$$\boldsymbol{s} = (\boldsymbol{v} \oplus \boldsymbol{e})\,\boldsymbol{H}^{\mathrm{T}} = \underbrace{\boldsymbol{v}\boldsymbol{H}^{\mathrm{T}}}_{=\boldsymbol{0}_{n-k}} \oplus \boldsymbol{e}\boldsymbol{H}^{\mathrm{T}} = \boldsymbol{0}_{n-k} \oplus \boldsymbol{e}\boldsymbol{H}^{\mathrm{T}} = \boldsymbol{e}\boldsymbol{H}^{\mathrm{T}}, \tag{1.212}$$
$$\boldsymbol{s} \in \mathbb{F}_2^{n-k}, \quad \boldsymbol{v}, \boldsymbol{e} \in \mathbb{F}_2^n, \quad \boldsymbol{H}^{\mathrm{T}} \in \mathbb{F}_2^{n\times(n-k)} \setminus \{\boldsymbol{0}_{n\times(n-k)}\}, \quad k, n \in \mathbb{N}^*.$$

1.7 Error-Detecting and Error-Correcting Capabilities of an (n, k) Binary Linear Block Code

1.7.1 Undetectable and Detectable Transmission Errors

When a codeword $\boldsymbol{v} \in \mathbb{V} \subseteq \mathbb{F}_2^n, n \in \mathbb{N}^*$ is transmitted over a transmission channel, an error pattern containing $l, l \in \{0, 1 \cdots n\}, n \in \mathbb{N}^*$, transmission errors will occur in the receive vector $\boldsymbol{r} \in \mathbb{F}_2^n, n \in \mathbb{N}^*$ that differs from the transmitted codeword $\boldsymbol{v} \in \mathbb{V} \subseteq \mathbb{F}_2^n$, $n \in \mathbb{N}^*$, in $l, l \in \{0, 1 \cdots n\}, n \in \mathbb{N}^*$ components. The Hamming distance $d_{\mathrm{H}}\{\boldsymbol{v}, \boldsymbol{r}\}$ between $\boldsymbol{v} \in \mathbb{V} \subseteq \mathbb{F}_2^n, n \in \mathbb{N}^*$ and $\boldsymbol{r} \in \mathbb{F}_2^n, n \in \mathbb{N}^*$ is hence given by [15, p. 78]

$$d_{\mathrm{H}}\{\boldsymbol{v}, \boldsymbol{r}\} = l, \quad l \in \{0, 1 \cdots n\}, \quad n \in \mathbb{N}^*. \tag{1.213}$$

If the minimum distance of the deployed (n, k) binary linear block code $\mathbb{V}$ is $d_{\min}$, any two distinct codewords of the (n, k) binary linear block code $\mathbb{V}$ differ in at least $d_{\min}$ components [15, p. 78]. Hence, the error pattern, which is required to change one admissible codeword into another, must contain $d_{\min}$ transmission errors [15, p. 78]. Therefore, the deployed (n, k) binary linear block code $\mathbb{V}$ can detect all the error patterns of $(d_{\min} - 1)$ or fewer errors [15, p. 78]. However, it cannot detect all the error patterns of $d_{\min}$ transmission errors because there exists at least one pair of codewords, that differ in $d_{\min}$ components, and there is an error pattern of $d_{\min}$ transmission errors that will falsify the one codeword into the other [15, p. 78]. This kind of falsification will also occur for $> d_{\min}$ transmission errors [15, p. 78]. Thus, the error detection capability of the deployed (n, k) binary linear block code $\mathbb{V}$ with the minimum distance $d_{\min}$ is $(d_{\min} - 1)$ [15, p. 78].

Every (n, k) binary linear block code $\mathbb{V}$ is however capable of detecting a large number of error patterns with $d_{\min}$ or more transmission errors [15, p. 78]. It can even detect $(2^n - 2^k)$ error patterns of length n [15, p. 78].

Surprised? Well, let us see.

$\mathbb{F}_2^n, n \in \mathbb{N}^*$, consists of $2^n, n \in \mathbb{N}^*$ vectors, of which $(2^n - 1), n \in \mathbb{N}^*$ are nonzero vectors. This means that there are $(2^n - 1), n \in \mathbb{N}^*$, possible nonzero error patterns $\boldsymbol{e} \in \mathbb{F}_2^n \setminus \{\boldsymbol{0}_n\}, n \in \mathbb{N}^*$. Furthermore, using the notation introduced by the great German mathematician Felix *Hausdorff* in his seminal textbook about *set theory* (in German *Mengenlehre*) [30, pp. 14, 17, 18]

$$\mathbb{F}_2^n = \mathbb{V} + (\mathbb{F}_2^n - \mathbb{V}) = \mathbb{V} \cup (\mathbb{F}_2^n \setminus \mathbb{V}), \quad \mathbb{V}(\mathbb{F}_2^n - \mathbb{V}) = \mathbb{V} \cap (\mathbb{F}_2^n \setminus \mathbb{V}) = \emptyset, \quad n \in \mathbb{N}^*, \tag{1.214}$$

holds. Note that $\boldsymbol{0}_n \notin (\mathbb{F}_2^n - \mathbb{V})$, respectively $\boldsymbol{0}_n \notin (\mathbb{F}_2^n \setminus \mathbb{V}), n \in \mathbb{N}^*$, because $\boldsymbol{0}_n \in \mathbb{V}$, $n \in \mathbb{N}^*$.

Taking (1.214) into consideration, there are $(2^k - 1), k \in \mathbb{N}^*$, error patterns $\boldsymbol{e} \in \mathbb{V} \subseteq \mathbb{F}_2^n$, that are identical to the $(2^k - 1), k \in \mathbb{N}^*$, nonzero codewords $\boldsymbol{v} \in \mathbb{V} \setminus \{\boldsymbol{0}_n\}, n \in \mathbb{N}^*$. These $(2^k - 1), k \in \mathbb{N}^*$, error patterns $\boldsymbol{e} \in \mathbb{V} \subseteq \mathbb{F}_2^n, k \in \mathbb{N}^*$ are *undetectable* because in this case $(\boldsymbol{v} \oplus \boldsymbol{e}) \in \mathbb{V} \subseteq \mathbb{F}_2^n, k \in \mathbb{N}^*$, and thus $\boldsymbol{r} \in \mathbb{V} \subseteq \mathbb{F}_2^n, k \in \mathbb{N}^*$ hold [15, p. 78f.].

However, if an error pattern $\boldsymbol{e}$ is an element of $(\mathbb{F}_2^n - \mathbb{V})$, respectively $(\mathbb{F}_2^n \setminus \mathbb{V})$, it is not identical to a nonzero codeword, and the receive vector $\boldsymbol{r}$ will not be an admissible codeword, i. e., $\boldsymbol{r} \notin \mathbb{V}$, respectively $(\boldsymbol{v} \oplus \boldsymbol{e}) \notin \mathbb{V}$, [15, p. 79]. In this case, a transmission error will be detected [15, p. 79].

The difference of the number $(2^n - 1)$, $n \in \mathbb{N}^*$ of possible nonzero error patterns $\boldsymbol{e} \in \mathbb{F}_2^n \setminus \{\mathbf{0}_n\}$, and the number $(2^k - 1)$, $k \in \mathbb{N}^*$ of undetectable possible nonzero error patterns is the number of *detectable* error patterns [15, p. 79]

$$(2^n - 1) - (2^k - 1) = 2^n - 2^k, \quad k, n \in \mathbb{N}^*. \tag{1.215}$$

Therefore, only an insignificantly small number of error patterns remain undetected for $n \gg k$, $k, n \in \mathbb{N}^*$ [15, p. 79].

1.7.2 Transmission Error Probability Without Decoding

In order to evaluate the *transmission error probability*, we require a model of the transmission channel. To achieve quick insights, this channel model should be easy at hand and simple, however, not unrealistic. A popular choice is the *binary symmetric (noisy) channel (BSC)* model [14, Figure 3.1-1, p. 79f.], [4, Figure 1.1, p. 1].

The discussion of the binary symmetric (noisy) channel (BSC) will benefit from the definition of the *binary entropy function* [5, equation (A.2.3), p. 432], [31, equation (1), p. 64].

Definition 1.30 (Binary Entropy Function). The *binary entropy function* is defined by [5, equation (A.2.3), p. 432]

$$\mathrm{H}_2\{p\} = -p\log_2\{p\} - (1-p)\log_2\{1-p\} = -\log_2\left\{p^p(1-p)^{1-p}\right\}, \quad p \in [0,1]. \tag{1.216}$$

Therefore, we obtain

$$2^{-\mathrm{H}_2\{p\}} = p^p(1-p)^{1-p}, \quad p \in [0,1]. \tag{1.217}$$

Remark 1.18 (On the Binary Entropy Function). The first derivative of the binary entropy function $\mathrm{H}_2\{p\}$ defined by (1.216) with respect to p is given by

$$\frac{\mathrm{dH}_2\{p\}}{\mathrm{d}p} = \log_2\left\{\frac{1-p}{p}\right\}. \tag{1.218}$$

Setting $\mathrm{dH}_2\{p\}/\mathrm{d}p$ to zero, we yield

$$\log_2\left\{\frac{1-p_0}{p_0}\right\} = 0 \quad \Leftrightarrow \quad \frac{1-p_0}{p_0} = 1 \quad \Leftrightarrow \quad p_0 = \frac{1}{2}. \tag{1.219}$$

Furthermore, the second derivative yields

$$\frac{d^2 H_2\{p\}}{dp^2} = -\frac{1}{p(1-p)}, \tag{1.220}$$

which is negative for p_0 equal to 1/2. Hence, the binary entropy function $H_2\{p\}$ defined by (1.216) has the global maximum

$$H_2\{p_0\} = H_2\left\{\frac{1}{2}\right\} = -\frac{1}{2}\log_2\left\{\frac{1}{2}\right\} - \frac{1}{2}\log_2\left\{\frac{1}{2}\right\} = \log_2\{2\} = 1. \tag{1.221}$$

In Figure 1.6, the binary entropy function $H_2\{p\}$ is depicted versus $p \in [0,1]$ [5, equation (A.2.3), Bild A.1, p. 432]. It is illustrated that $H_2\{p\}$ increases monotonically for $p \in [0,1/2]$.

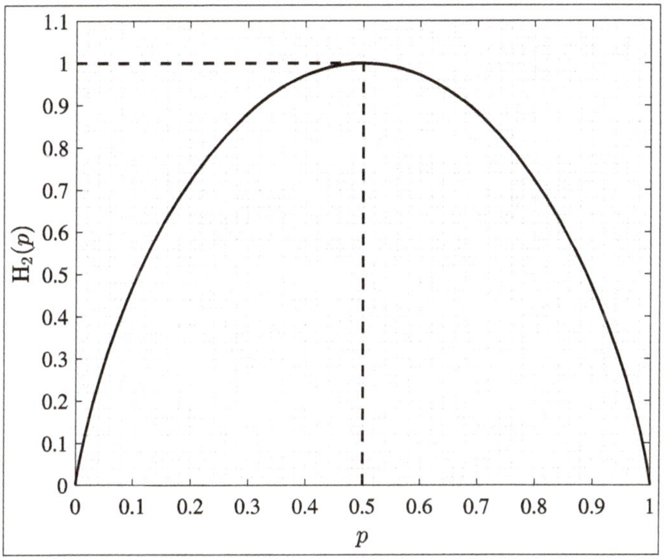

Figure 1.6: Binary entropy function $H_2\{p\}$ versus $p \in [0,1]$; cf. [5, equation (A.2.3), p. 432], [31, equation (1), p. 64].

The inputs of the binary symmetric (noisy) channel (BSC) are the bits $v_j \in \mathbb{F}_2 = \{0,1\}$, $j \in \{0,1\cdots(n-1)\}$, $n \in \mathbb{N}^*$, which are the components of the *transmit codeword* $\mathbf{v} \in \mathbb{V} \subseteq \mathbb{F}_2^n$, $n \in \mathbb{N}^*$ given by (1.24). Each input bit $v_j \in \mathbb{F}_2, j \in \{0,1\cdots(n-1)\}$, $n \in \mathbb{N}^*$, which is transmitted via the channel, is received by the receiver either [14, Figure 3.1-1, p. 79f.], [4, Figure 1.1, p. 1]

- correctly with the probability $(1-p)$, $p \in [0,1]$, or
- erroneously with the probability p, $p \in [0,1]$.

The probability p is termed the *crossover probability* [14, p. 79] or *error probability* [4, Figure 1.1, p. 1]. Consequently, the outputs of the BSC are the bits $r_j \in \mathbb{F}_2, j \in \{0,1\cdots(n-1)\}$,

$n \in \mathbb{N}^*$, which are the components of the *receive vector* $\boldsymbol{r} \in \mathbb{V} \subseteq \mathbb{F}_2^n$, $n \in \mathbb{N}^*$, given by (1.207).

Being taken from $\mathbb{F}_2 = \{0,1\}$, each component $v_j \in \mathbb{F}_2, j \in \{0,1\cdots(n-1)\}$, $n \in \mathbb{N}^*$ of the transmit codeword $\boldsymbol{v} \in \mathbb{V} \subseteq \mathbb{F}_2^n$, $n \in \mathbb{N}^*$, given by (1.24), which is fed into the binary symmetric (noisy) channel (BSC), can assume either the value a_0 equal to 0 or the value a_1 equal to 1. Accordingly, each component $r_j \in \mathbb{F}_2, j \in \{0,1\cdots(n-1)\}$, $n \in \mathbb{N}^*$ of the *receive vector* $\boldsymbol{r} \in \mathbb{V} \subseteq \mathbb{F}_2^n$, $n \in \mathbb{N}^*$, which prevails at the output of the binary symmetric (noisy) channel (BSC), can assume either the value b_0 equal to 0 or the value b_1 equal to 1. The corresponding transitions between the inputs and the outputs of the binary symmetric (noisy) channel (BSC) are illustrated in Figure 1.7; cf., e. g., [14, Figure 3.1-1(a), p. 80].

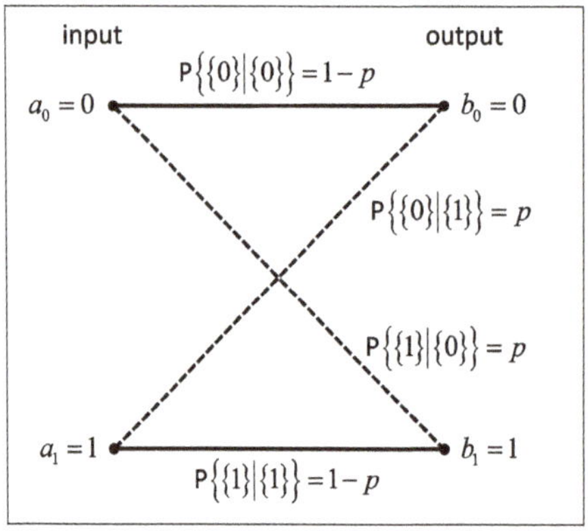

Figure 1.7: Schematic of the binary symmetric (noisy) channel (BSC), adapted according to [14, Figure 3.1-1(a), p. 80]. The events are $\{a_i\}, i \in \{0,1\}$ and $\{b_j\}, j \in \{0,1\}$.

Let the noise vector $\boldsymbol{n} \in \mathbb{V} \subseteq \mathbb{F}_2^n$, $n \in \mathbb{N}^*$, associated with the binary symmetric (noisy) channel (BSC) have the components $n_j \in \mathbb{F}_2, j \in \{0,1\cdots(n-1)\}$, $n \in \mathbb{N}^*$. With the additive operation $\oplus$ in $\mathbb{F}_2 = \{0,1\}$, the transmission via the binary symmetric (noisy) channel (BSC) can be described by

$$r_j = v_j \oplus n_j, \quad n_j, r_j, v_j \in \mathbb{F}_2, \quad j \in \{0,1\cdots(n-1)\}, \quad n \in \mathbb{N}^*. \tag{1.222}$$

Hence, the noise associated with the binary symmetric (noisy) channel (BSC) is additive [31, Figure 1(a), p. 62]. In this regard, the binary symmetric (noisy) channel (BSC) resembles the additive white Gaussian noise (AWGN) channel.

Remark 1.19 (On the Capacity of the Binary Symmetric (Noisy) Channel (BSC)). Let the *a priori probability of the source* be $\mathrm{P}\{\{a_i\}\}$, $i \in \{0,1\}$. Let the *crossover probability* of the binary symmetric (noisy) channel (BSC) be p. Let the *transition probability* of the binary symmetric (noisy) channel (BSC) be $\mathrm{P}\{\{b_j\} \mid \{a_i\}\}$, $i,j \in \{0,1\}$. Let the *a posteriori probability of the sink* be $\mathrm{P}\{\{b_j\}\}$, $j \in \{0,1\}$.

Table 1.3: Events and probabilities in the case of the transmission via the binary symmetric (noisy) channel (BSC).

events		**probabilities**			
$\{a_i\}$	$\{b_j\}$	$\mathrm{P}\{\{a_i\}\}$	$\mathrm{P}\{\{b_j\} \mid \{a_i\}\}$	$\mathrm{P}\{\{b_j\}\{a_i\}\}$	$\mathrm{P}\{\{b_j\}\} = \sum_{i=0}^{1} \mathrm{P}\{\{b_j\}\{a_i\}\}$
$\{0\}$	$\{0\}$	q	$1-p$	$q(1-p)$	$q(1-p)+(1-q)p$
$\{0\}$	$\{1\}$	q	p	qp	$qp+(1-q)(1-p)$
$\{1\}$	$\{0\}$	$1-q$	p	$(1-q)p$	$q(1-p)+(1-q)p$
$\{1\}$	$\{1\}$	$1-q$	$1-p$	$(1-q)(1-p)$	$qp+(1-q)(1-p)$

With the *binary entropy function* $\mathrm{H}_2\{p\}$ defined by (1.216) [5, equation (A.2.3), p. 432] and using the results of Table 1.3, the *entropy of the source* is given by

$$\begin{aligned}
\mathrm{H}\{X\} &= -\sum_{i=0}^{1} \mathrm{P}\big\{\{a_i\}\big\} \log_2\big\{\mathrm{P}\big\{\{a_i\}\big\}\big\} \\
&= -\mathrm{P}\big\{\{a_0\}\big\} \log_2\big\{\mathrm{P}\big\{\{a_0\}\big\}\big\} - \mathrm{P}\big\{\{a_1\}\big\} \log_2\big\{\mathrm{P}\big\{\{a_1\}\big\}\big\} \\
&= -q \log_2\{q\} - (1-q)\log_2\{1-q\} \\
&= \mathrm{H}_2\{q\}.
\end{aligned} \tag{1.223}$$

Furthermore, the *irrelevance* can be determined as

$$\begin{aligned}
\mathrm{H}\{Y \mid X\} &= -\sum_{i=0}^{1}\sum_{j=0}^{1} \mathrm{P}\big\{\{b_j\}\{a_i\}\big\} \log_2\big\{\mathrm{P}\big\{\{b_j\} \mid \{a_i\}\big\}\big\} \\
&= -q(1-p)\log_2\{1-p\} - qp\log_2\{p\} \\
&\quad -(1-q)p\log_2\{p\} - (1-q)(1-p)\log_2\{1-p\} \\
&= \big[(1-p)\log_2\{1-p\}\big] \cdot \underbrace{(-q-1+q)}_{=-1} + \big[p\log_2\{p\}\big] \cdot \underbrace{(-q-1+q)}_{=-1} \\
&= -p\log_2\{p\} - (1-p)\log_2\{1-p\} \\
&= \mathrm{H}_2\{p\}.
\end{aligned} \tag{1.224}$$

As expected, the irrelevance does only depend on p but not on q, and hence not on the source.

With

$$\begin{aligned}
\sum_{j=0}^{1} \mathrm{P}\big\{\{b_j\}\big\} \log_2\big\{\mathrm{P}\big\{\{b_j\}\big\}\big\} &= \mathrm{P}\big\{\{b_0\}\big\} \log_2\big\{\mathrm{P}\big\{\{b_0\}\big\}\big\} + \mathrm{P}\big\{\{b_1\}\big\} \log_2\big\{\mathrm{P}\big\{\{b_1\}\big\}\big\} \\
&= \big[q(1-p)+(1-q)p\big]\log_2\big\{q(1-p)+(1-q)p\big\} \\
&\quad + \big[qp+(1-q)(1-p)\big]\log_2\big\{qp+(1-q)(1-p)\big\} \\
&= -\mathrm{H}_2\big\{qp+(1-q)(1-p)\big\},
\end{aligned} \tag{1.225}$$

the *entropy of the sink* is given by

$$\mathrm{H}\{Y\} = \mathrm{H}_2\big\{qp+(1-q)(1-p)\big\}. \tag{1.226}$$

Furthermore, we yield the *average mutual information* [14, equation (3.1-5), p. 81]

$$I\{X;Y\} = \mathrm{H}\{Y\} - \mathrm{H}\{Y \mid X\}$$

$$= \mathrm{H}_2\{qp + (1-q)(1-p)\} - \underbrace{\{-p\log_2\{p\} - (1-p)\log_2\{1-p\}\}}_{=\mathrm{H}_2\{p\}}$$
$$= \mathrm{H}_2\{qp + (1-q)(1-p)\} - \mathrm{H}_2\{p\}. \tag{1.227}$$

The *equivocation* is hence obtained as

$$\mathrm{H}\{X \mid Y\} = \mathrm{H}\{X\} - I\{X;Y\} = \mathrm{H}_2\{q\} + \mathrm{H}_2\{p\} - \mathrm{H}_2\{qp + (1-q)(1-p)\}. \tag{1.228}$$

$\mathrm{H}_2\{qp + (1-q)(1-p)\}$ in (1.227) achieves its maximum $\mathrm{H}_2\{1/2\}$ equal to 1 for

$$qp + (1-q)(1-p) = \frac{1}{2}. \tag{1.229}$$

The *channel capacity* of the binary symmetric (noisy) channel (BSC) is hence given by [14, p. 82], [31, equation (2), p. 64]

$$C = \max_q I\{X;Y\} = 1 - \mathrm{H}_2\{p\}. \tag{1.230}$$

Figure 1.8 illustrates the channel capacity C of the binary symmetric (noisy) channel (BSC) according to (1.230) versus $p \in [0,1]$; cf. [32, Figure 7.1-4, p. 383]. As expected, C of (1.230) decreases with increasing crossover probability p as long as $p \le 1/2$ holds. The worst case of C equal to 0 is achieved for p equal to 1/2. In this case, we only know that on average half of the input symbols are in error. However, since we do not have any idea about which half is correct, there is no means of error correction, and hence also no means of information transfer. At first sight surprisingly, C of (1.230) increases again with increasing p for $1/2 < p \le 1$. The reason for this effect is that errors imposed on the input symbols become increasingly predictable. So, error correction is feasible again. Just look at the case of p equal to 1 as an example. Then each input is in error with certainty. So, just take the inverse of the received symbols and all is well and the information transfer works superbly. Therefore, C of (1.230) is axis symmetrical with respect the axis $p = 1/2$.

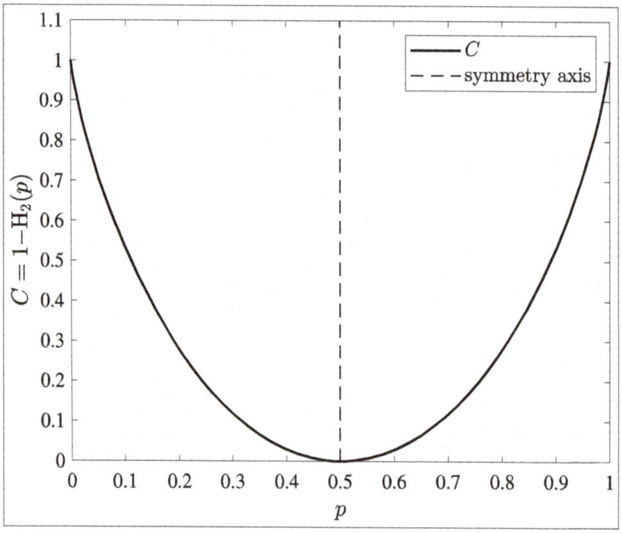

Figure 1.8: Channel capacity C of the binary symmetric (noisy) channel (BSC) according to (1.230) versus $p \in [0,1]$ (adapted according to [32, Figure 7.1-4, p. 383]).

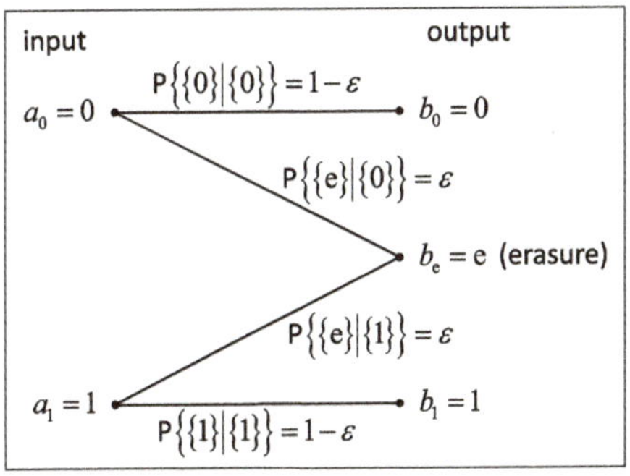

Figure 1.9: Schematic of the binary erasure channel (BEC), adapted according to [33, Figure 2.47, p. 56]. The events are $\{a_i\}, i \in \{0, 1\}$ and $\{b_j\}, j \in \{0, 1, \mathrm{e}\}$.

Remark 1.20 (Binary Erasure Channel (BEC)). Besides the binary symmetric (noisy) channel (BSC), there is a second important channel model, which we will use in the context of polar codes, namely the *binary erasure channel (BEC)*; cf. [31, p. 62]. Let us now analyze it.

Being taken from $\mathbb{F}_2 = \{0, 1\}$, each component $v_j \in \mathbb{F}_2, j \in \{0, 1 \cdots (n-1)\}, n \in \mathbb{N}^*$ of the transmit codeword $\boldsymbol{v} \in \mathbb{V} \subseteq \mathbb{F}_2^n, n \in \mathbb{N}^*$, given by (1.24), which is fed into the binary erasure channel (BEC), can assume either the value a_0 equal to 0 or the value a_1 equal to 1. However, each component $r_j, j \in \{0, 1 \cdots (n-1)\}, n \in \mathbb{N}^*$, of the receive vector $\boldsymbol{r} \in \mathbb{V} \subseteq \mathbb{F}_2^n, n \in \mathbb{N}^*$, which prevails at the output of the binary erasure channel (BEC), can assume either the value b_0 equal to 0 or the value b_1 equal to 1 or it is erased, represented by the value b_{e} equal to e. The noise associated with the binary erasure channel (BEC) is multiplicative [31, Figure 1 (b), p. 62]. In this regard, the binary erasure channel (BEC) resembles a flat Rayleigh fading channel.

The corresponding transitions between the inputs and the outputs of the binary erasure channel (BEC) are illustrated in Figure 1.9; cf., e. g., [33, Figure 2.47, p. 56].

Let us consider the channel capacity of the binary erasure channel (BEC). Let the *a priori probability of the source* be $\mathrm{P}\{\{a_i\}\}, i \in \{0, 1\}$. Let the *erasure probability* of the binary erasure channel (BEC) be ϵ. Let the *transition probability* of the binary erasure channel (BEC) be $\mathrm{P}\{\{b_j\} \mid \{a_i\}\}, i \in \{0, 1\}, j \in \{0, 1, \mathrm{e}\}$. Let the *a posteriori probability of the sink* be $\mathrm{P}\{\{b_j\}\}, j \in \{0, 1\}$.

With the *binary entropy function* $\mathrm{H}_2\{p\}$ defined by (1.216) [5, equation (A.2.3), p. 432] and using the results of Table 1.4, the entropy of the source is given by

$$\begin{aligned}
\mathrm{H}\{X\} &= -\sum_{i=0}^{1} \mathrm{P}\big\{\{a_i\}\big\} \log_2\big\{\mathrm{P}\big\{\{a_i\}\big\}\big\} \\
&= -\mathrm{P}\big\{\{a_0\}\big\} \log_2\big\{\mathrm{P}\big\{\{a_0\}\big\}\big\} - \mathrm{P}\big\{\{a_1\}\big\} \log_2\big\{\mathrm{P}\big\{\{a_1\}\big\}\big\} \\
&= -q \log_2\{q\} - (1-q) \log_2\{1-q\} \\
&= \mathrm{H}_2\{q\}.
\end{aligned} \tag{1.231}$$

Furthermore, the *irrelevance* can be determined as

$$\begin{aligned}
\mathrm{H}\{Y \mid X\} &= -\sum_{i=0}^{1} \sum_{j \in \{0,1,\mathrm{e}\}} \mathrm{P}\big\{\{b_j\}\{a_i\}\big\} \log_2\big\{\mathrm{P}\big\{\{b_j\} \mid \{a_i\}\big\}\big\} \\
&= q\mathrm{H}_2\{\epsilon\} + (1-q)\mathrm{H}_2\{\epsilon\} \\
&= \mathrm{H}_2\{\epsilon\}.
\end{aligned} \tag{1.232}$$

As expected, the irrelevance does only depend on ϵ but not on q, and hence not on the source.

With

$$\begin{aligned}
\sum_{j\in\{0,1,e\}} \mathrm{P}\{\{b_j\}\}\log_2\{\mathrm{P}\{\{b_j\}\}\} &= \mathrm{P}\{\{b_0\}\}\log_2\{\mathrm{P}\{\{b_0\}\}\} \\
&\quad + \mathrm{P}\{\{b_1\}\}\log_2\{\mathrm{P}\{\{b_1\}\}\} \\
&\quad + \mathrm{P}\{\{e\}\}\log_2\{\mathrm{P}\{e\}\} \\
&= q(1-\epsilon)\log_2\{q(1-\epsilon)\} \\
&\quad + \epsilon\log_2\{\epsilon\} \\
&\quad + (1-q)(1-\epsilon)\log_2\{(1-q)(1-\epsilon)\} \\
&= \epsilon\log_2\{\epsilon\} \\
&\quad + q(1-\epsilon)\log_2\{q\} + q(1-\epsilon)\log_2\{1-\epsilon\} \\
&\quad + (1-q)(1-\epsilon)\log_2\{1-q\} + (1-q)(1-\epsilon)\log_2\{1-\epsilon\} \\
&= \underbrace{\epsilon\log_2\{\epsilon\} + (1-\epsilon)\log_2\{1-\epsilon\}}_{=-\mathrm{H}_2\{\epsilon\}}\underbrace{[q+(1-q)]}_{=1} \\
&\quad + (1-\epsilon)\underbrace{[q\log_2\{q\} + (1-q)\log_2\{1-q\}]}_{=-\mathrm{H}_2\{q\}} \\
&= -(1-\epsilon)\mathrm{H}_2\{q\} - \mathrm{H}_2\{\epsilon\},
\end{aligned} \tag{1.233}$$

the *entropy of the sink* yields

$$\mathrm{H}\{Y\} = (1-\epsilon)\mathrm{H}_2\{q\} + \mathrm{H}_2\{\epsilon\}. \tag{1.234}$$

Furthermore, we yield the *average mutual information* [33, equation (2.59), p. 56]

$$\begin{aligned}
I\{X;Y\} &= \mathrm{H}\{Y\} - \mathrm{H}\{Y\,|\,X\} \\
&= (1-\epsilon)\mathrm{H}_2\{q\} + \mathrm{H}_2\{\epsilon\} - \mathrm{H}_2\{\epsilon\} \\
&= (1-\epsilon)\mathrm{H}_2\{q\}.
\end{aligned} \tag{1.235}$$

The *equivocation* is hence obtained as

$$\mathrm{H}\{X\,|\,Y\} = \mathrm{H}\{X\} - I\{X;Y\} = \mathrm{H}_2\{q\} - (1-\epsilon)\mathrm{H}_2\{q\} = \epsilon\mathrm{H}_2\{q\}. \tag{1.236}$$

$\mathrm{H}_2\{q\}$ in (1.235) achieves its maximum $\mathrm{H}_2\{1/2\}$ equal to 1 for

$$q = \frac{1}{2}. \tag{1.237}$$

The *channel capacity* is hence given by [31, equation (2”), p. 65], [33, equation (2.62), p. 56]

$$C = \max_q I\{X;Y\} = 1-\epsilon. \tag{1.238}$$

Figure 1.10 illustrates the channel capacity C of the binary erasure channel (BEC) according to (1.238) versus $\epsilon \in [0,1]$.

Table 1.4: Events and probabilities in the case of the transmission via the binary erasure channel (BEC).

events		probabilities			
$\{a_i\}$	$\{b_j\}$	$\mathrm{P}\{\{a_i\}\}$	$\mathrm{P}\{\{b_j\} \mid \{a_i\}\}$	$\mathrm{P}\{\{b_j\}\{a_i\}\}$	$\mathrm{P}\{\{b_j\}\} = \sum_{i=0}^{1} \mathrm{P}\{\{b_j\}\{a_i\}\}$
{0}	{0}	q	$1-\epsilon$	$q(1-\epsilon)$	$q(1-\epsilon)$
{0}	{e}	q	ϵ	$q\epsilon$	ϵ
{0}	{1}	q	0	0	$(1-q)(1-\epsilon)$
{1}	{0}	$1-q$	0	0	$q(1-\epsilon)$
{1}	{e}	$1-q$	ϵ	$(1-q)\epsilon$	ϵ
{1}	{1}	$1-q$	$1-\epsilon$	$(1-q)(1-\epsilon)$	$(1-q)(1-\epsilon)$

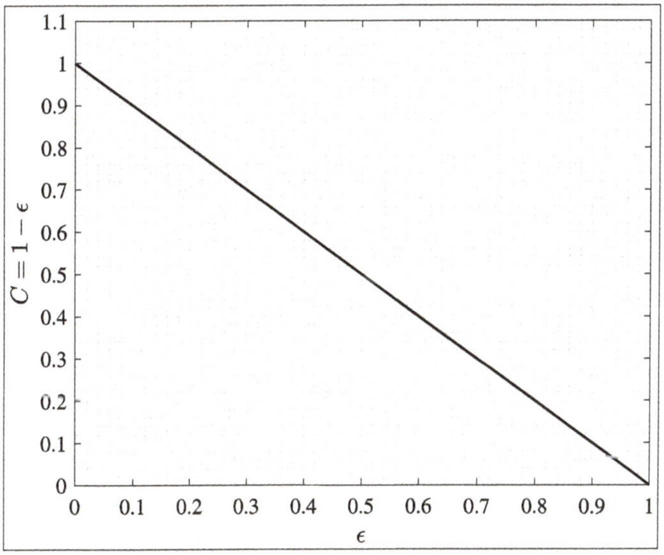

Figure 1.10: Channel capacity C of the binary erasure channel (BEC) according to (1.238) versus $\epsilon \in [0, 1]$.

Remark 1.21 (Bhattacharyya Parameter). The performance evaluation in terms of error probabilities under maximum-likelihood (ML) detection or maximum-likelihood (ML) decoding contains a term, which we call *Bhattacharyya parameter* or *Bhattacharyya bound* (in German “Bhattacharyya-Schranke”) [5, p. 44f.].

Considering binary transmission and causing the channel output b_j, $j \in \{0, 1\}$, in the case of the binary symmetric (noisy) channel (BSC), respectively $j \in \{0, 1, \mathrm{e}\}$, in the case of the binary erasure channel (BEC), the probability of a bit error P_b that the transmitted binary message a_i, $i \in \{0, 1\}$ is received in error is upper bounded by [34, p. 63]

$$P_\mathrm{b} \leq \sum_{j \in \{0,1,\mathrm{e}\}} \sqrt{\mathrm{P}\{\{b_j\} \mid \{a_0\}\} \mathrm{P}\{\{b_j\} \mid \{a_1\}\}}, \tag{1.239}$$

which becomes

$$P_{\mathrm{b}} \leq \int_{-\infty}^{+\infty} \sqrt{\mathrm{p}(b \mid a_0)\mathrm{p}(b \mid a_1)}\, \mathrm{d}b \tag{1.240}$$

in the case of a continuous channel output $b \in \mathbb{R}$ [34, p. 63]. (1.239), respectively (1.240), is called the *Bhattacharyya bound* and

discrete channel output	continuous channel output
$-\ln\left\{\sum_{j\in\{0,1,e\}} \sqrt{\mathrm{P}\{\{b_j\} \mid \{a_0\}\}\mathrm{P}\{\{b_j\} \mid \{a_1\}\}}\right\}$	$-\ln\left\{\int_{-\infty}^{+\infty} \sqrt{\mathrm{p}(b \mid a_0)\mathrm{p}(b \mid a_1)}\, \mathrm{d}b\right\}$

(1.241)

is called the *Bhattacharyya distance* [34, pp. 63, 88].

In the case of a symmetric *discrete memoryless channel (DMC)* such as the binary symmetric (noisy) channel (BSC), the binary erasure channel (BEC) and the binary input additive white Gaussian noise (AWGN) channel, the *Bhattacharyya parameter* is defined by

discrete channel output	continuous channel output
$Z = \sum_{j\in\{0,1,e\}} \sqrt{\mathrm{P}\{\{b_j\} \mid \{a_0\}\}\mathrm{P}\{\{b_j\} \mid \{a_1\}\}}$	$Z = \int_{-\infty}^{+\infty} \sqrt{\mathrm{p}(b \mid a_0)\mathrm{p}(b \mid a_1)}\, \mathrm{d}b$

(1.242)

In (1.242), $\mathrm{P}\{\{b_j\} \mid \{a_0\}\}$ and $\mathrm{P}\{\{b_j\} \mid \{a_1\}\}$ are the transition probabilities, whereas $\mathrm{p}(b \mid a_0)$ and $\mathrm{p}(b \mid a_1)$ are the transition probability densities and the infinitesimal probabilities are given by $(\mathrm{p}(b \mid a_0)\, \mathrm{d}b)$ and $(\mathrm{p}(b \mid a_1)\, \mathrm{d}b)$.

In the case of the binary erasure channel (BEC) having the transition probabilities,

$$\begin{aligned} \mathrm{P}\big\{\{0\} \mid \{0\}\big\} &= 1-\epsilon, \mathrm{P}\big\{\{0\} \mid \{1\}\big\} = 0, \\ \mathrm{P}\big\{\{1\} \mid \{0\}\big\} &= 0, \mathrm{P}\big\{\{1\} \mid \{1\}\big\} = 1-\epsilon, \\ \mathrm{P}\big\{\{e\} \mid \{0\}\big\} &= \epsilon, \mathrm{P}\big\{\{e\} \mid \{1\}\big\} = \epsilon, \end{aligned} \tag{1.243}$$

we yield [33, p. 83]

$$\begin{aligned} Z = {} & \underbrace{\sqrt{\underbrace{\mathrm{P}\big\{\{0\} \mid \{0\}\big\}}_{=1-\epsilon} \underbrace{\mathrm{P}\big\{\{0\} \mid \{1\}\big\}}_{0}}}_{=0} \\ & + \underbrace{\sqrt{\underbrace{\mathrm{P}\big\{\{1\} \mid \{0\}\big\}}_{=0} \underbrace{\mathrm{P}\big\{\{1\} \mid \{1\}\big\}}_{1-\epsilon}}}_{=0} \\ & + \sqrt{\underbrace{\mathrm{P}\big\{\{e\} \mid \{0\}\big\}}_{=\epsilon} \underbrace{\mathrm{P}\big\{\{e\} \mid \{1\}\big\}}_{=\epsilon}} \\ = {} & \epsilon. \end{aligned} \tag{1.244}$$

The Bhattacharyya parameter Z of the binary erasure channel (BEC) is depicted in Figure 1.11 versus the erasure probability ϵ; cf. (1.244).

Strictly speaking, the binary erasure channel (BEC) does not suffer from any bit errors. Rather, it suffers from the erasing of bits. If we considered this a bit error, the "bit error probability" of the binary erasure channel (BEC) was given by

$$P_{\mathrm{b}} = \epsilon \leq Z, \tag{1.245}$$

which is certainly upper bounded by the Bhattacharyya parameter Z of (1.244).

In the case of the binary symmetric (noisy) channel (BSC) having the transition probabilities

$$\begin{aligned} \mathrm{P}\big\{\{0\}\mid\{0\}\big\} &= 1-p, \mathrm{P}\big\{\{0\}\mid\{1\}\big\} = p, \\ \mathrm{P}\big\{\{1\}\mid\{0\}\big\} &= p, \mathrm{P}\big\{\{1\}\mid\{1\}\big\} = 1-p, \end{aligned} \tag{1.246}$$

we find [33, p. 83] (see also [34, p. 88])

$$\begin{aligned} Z &= \sqrt{\underbrace{\mathrm{P}\big\{\{0\}\mid\{0\}\big\}}_{=1-p}\underbrace{\mathrm{P}\big\{\{0\}\mid\{1\}\big\}}_{=p}} + \sqrt{\underbrace{\mathrm{P}\big\{\{1\}\mid\{0\}\big\}}_{=p}\underbrace{\mathrm{P}\big\{\{1\}\mid\{1\}\big\}}_{=1-p}} = 2\sqrt{p(1-p)} \\ &= \sqrt{4p(1-p)}. \end{aligned} \tag{1.247}$$

The Bhattacharyya parameter Z of the binary symmetric (noisy) channel (BSC) is depicted in Figure 1.12 versus the crossover probability p; cf. (1.247).

Keeping in mind the situation depicted in Figure 1.8 and discussed in Remark 1.19, we determine the bit error probability in the case of $p \in [0, 1/2]$ first. Let q be the a priori probability for the occurrence of a_0 at the input of the binary symmetric (noisy) channel (BSC). We obtain

$$P_\mathrm{b} = q\mathrm{P}\big\{\{1\}\mid\{0\}\big\} + (1-q)\mathrm{P}\big\{\{0\}\mid\{1\}\big\} = qp + (1-q)p = p, \quad p \in [0, 1/2]. \tag{1.248}$$

In the case of $p \in [0, 1/2]$, we obtain

$$P_\mathrm{b} = q(1-p) + (1-q)(1-p) = 1-p, \quad p \in \,]1/2, 1].$$

Hence, we also find $P_\mathrm{b} \le Z$ in the case of the binary symmetric (noisy) channel (BSC). Furthermore, we see from Figure 1.8 that the bit error ratio (BER) can never be larger than 1/2, which is true for every binary transmission system.

With the average symbol energy E_b equal to 1, the binary input additive white Gaussian noise (AWGN) channel has the input given by

$$\sqrt{E_\mathrm{b}} \cdot (1-2a_i) = 1 - 2a_i, \quad a_i \in \{0, 1\}, \quad i \in \{0, 1\}. \tag{1.249}$$

With the two-sided noise power spectral density $N_0/2$, i. e., with E_b/N_0 being the signal-to-noise ratio (SNR) at the receiver input, the transition probability densities are given by (cf. [32, equation (5.1-11), p. 235])

$$\mathrm{p}(b \mid a_0) = \frac{1}{\sqrt{\pi N_0}} \exp\left\{-\frac{(b-\sqrt{E_\mathrm{b}})^2}{N_0}\right\}, \quad \mathrm{p}(b \mid a_1) = \frac{1}{\sqrt{\pi N_0}} \exp\left\{-\frac{(b+\sqrt{E_\mathrm{b}})^2}{N_0}\right\}, \tag{1.250}$$

and the Bhattacharyya parameter is given by [11, no. 25, p. 1088]

$$\begin{aligned} Z &= \int_{-\infty}^{+\infty} \sqrt{\mathrm{p}(b \mid a_0)\mathrm{p}(b \mid a_1)}\, \mathrm{d}b \\ &= \int_{-\infty}^{+\infty} \sqrt{\frac{1}{\sqrt{\pi N_0}} \exp\left\{-\frac{(b-\sqrt{E_\mathrm{b}})^2}{N_0}\right\} \frac{1}{\sqrt{\pi N_0}} \exp\left\{-\frac{(b+\sqrt{E_\mathrm{b}})^2}{N_0}\right\}}\, db \\ &= \frac{1}{\sqrt{\pi N_0}} \int_{-\infty}^{+\infty} \sqrt{\exp\left\{-\frac{(b-\sqrt{E_\mathrm{b}})^2}{N_0} - \frac{(b+\sqrt{E_\mathrm{b}})^2}{N_0}\right\}}\, db \end{aligned}$$

$$
\begin{aligned}
&= \frac{1}{\sqrt{\pi N_0}} \int_{-\infty}^{+\infty} \sqrt{\exp\left\{-\frac{b^2 - 2b\sqrt{E_b} + E_b + b^2 + 2b\sqrt{E_b} + E_b}{N_0}\right\}}\, db \\
&= \frac{1}{\sqrt{\pi N_0}} \int_{-\infty}^{+\infty} \exp\left\{-\frac{2b^2 + 2E_b}{2N_0}\right\} db = \exp\left\{-\frac{E_b}{N_0}\right\} \frac{1}{\sqrt{\pi N_0}} \int_{-\infty}^{+\infty} \exp\left\{-\frac{b^2}{N_0}\right\} db \\
&= \exp\left\{-\frac{E_b}{N_0}\right\} \frac{2\sqrt{N_0}}{\sqrt{\pi N_0}} \underbrace{\int_{0}^{+\infty} \exp\{-y^2\}\, dy}_{=\sqrt{\pi}/2} \\
&= \exp\left\{-\frac{E_b}{N_0}\right\}.
\end{aligned}
\tag{1.251}
$$

The Bhattacharyya parameter Z of the binary input additive white Gaussian noise (AWGN) channel is depicted in Figure 1.13. versus the signal-to-noise ratio E_b/N_0; cf. (1.251).

The bit error ratio (BER) in the case of the binary input additive white Gaussian noise (AWGN) channel is given by [32, equation (5.2-4), p. 255], [11, p. 478]

$$
\begin{aligned}
P_b &= \int_{-\infty}^{0} \mathrm{P}\{\{a_0\}\}\mathrm{p}(b \mid a_0)\, db + \int_{0}^{+\infty} \mathrm{P}\{\{a_1\}\}\mathrm{p}(b \mid a_1)\, db \\
&= \frac{q}{\sqrt{\pi N_0}} \int_{-\infty}^{0} \exp\left\{-\frac{(b - \sqrt{E_b})^2}{N_0}\right\} db + \frac{1-q}{\sqrt{\pi N_0}} \int_{0}^{+\infty} \exp\left\{-\frac{(b + \sqrt{E_b})^2}{N_0}\right\} db \\
&= \frac{q\sqrt{N_0}}{\sqrt{\pi N_0}} \int_{-\infty}^{-\sqrt{E_b/N_0}} \exp\{-y^2\}\, dy + \frac{(1-q)\sqrt{N_0}}{\sqrt{\pi N_0}} \underbrace{\int_{\sqrt{E_b/N_0}}^{+\infty} \exp\{-y^2\}\, dy}_{=\sqrt{\pi}\cdot\mathrm{erfc}\{\sqrt{E_b/N_0}\}/2} \\
&= \frac{q}{\sqrt{\pi}} \underbrace{\int_{\sqrt{E_b/N_0}}^{+\infty} \exp\{-y^2\}\, dy}_{=\sqrt{\pi}\cdot\mathrm{erfc}\{\sqrt{E_b/N_0}\}/2} + \frac{(1-q)}{2} \mathrm{erfc}\left\{\sqrt{\frac{E_b}{N_0}}\right\} \\
&= \frac{1}{2}\mathrm{erfc}\left\{\sqrt{\frac{E_b}{N_0}}\right\} \cdot (q + 1 - q) \\
&= \frac{1}{2}\mathrm{erfc}\left\{\sqrt{\frac{E_b}{N_0}}\right\}.
\end{aligned}
\tag{1.252}
$$

Since (cf., e. g., [35, p. 85])

$$
\frac{1}{2}\mathrm{erfc}\left\{\sqrt{\frac{E_b}{N_0}}\right\} \le \exp\left\{-\frac{E_b}{N_0}\right\},
\tag{1.253}
$$

we have $P_b \le Z$ also in the case of the binary input additive white Gaussian noise (AWGN) channel.

The *transmission error probability without decoding*, i. e., the *probability of an undetected transmission error without decoding* can be evaluated by computing the probability of the occurrence of a particular error pattern $\boldsymbol{e} \in \mathbb{F}_2^n$ [4, p. 9] that is not a codeword. In the case of an (n, k) binary linear block code $\mathbb{V}$, the length of the error pattern $\boldsymbol{e} \in \mathbb{F}_2^n$ is equal to n.

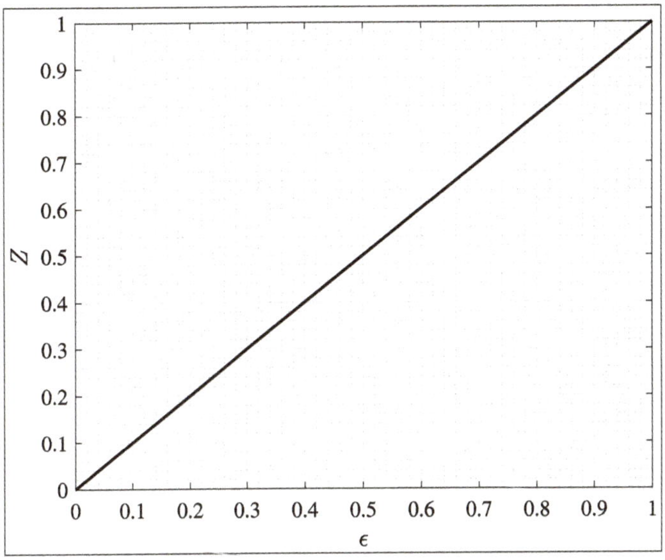

Figure 1.11: Bhattacharyya parameter Z of the binary erasure channel (BEC) versus the erasure probability ε.

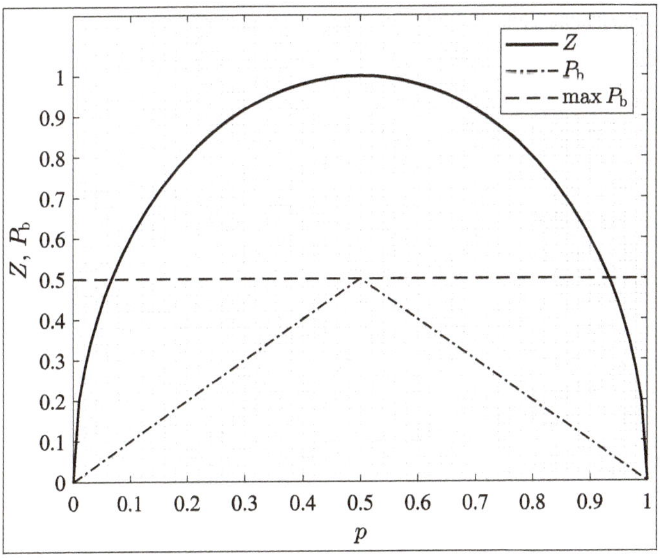

Figure 1.12: Bhattacharyya parameter Z of the binary symmetric (noisy) channel (BSC) versus the crossover probability p.

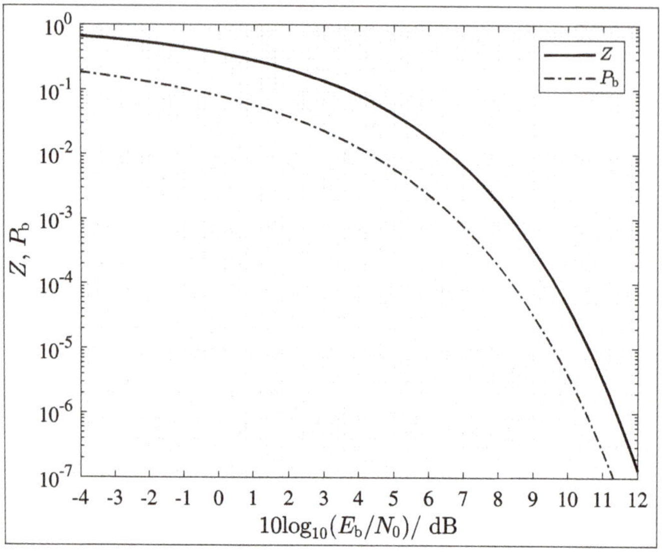

Figure 1.13: Bhattacharyya parameter Z of the binary input additive white Gaussian noise (AWGN) channel versus the signal-to-noise ratio E_b/N_0; cf. (1.251).

First, assume that the error pattern $\boldsymbol{e} \in \mathbb{F}_2^n$ is the zero vector $\mathbf{0}_n$, $n \in \mathbb{N}^*$, having the Hamming weight

$$w_H\{\boldsymbol{e}\} = w_H\{\mathbf{0}_n\} = 0, \quad n \in \mathbb{N}^*. \tag{1.254}$$

The probability $\mathsf{P}\{w_H\{\mathbf{0}_n\} = 0\}$, $n \in \mathbb{N}^*$, that the error pattern $\boldsymbol{e} \in \mathbb{F}_2^n$ is the zero vector $\mathbf{0}_n$, $n \in \mathbb{N}^*$, at the output of the BSC is given by [5, p. 11]

$$\mathsf{P}\{w_H\{\boldsymbol{e}\} = 0\} = (1-p)^n, \quad p \in [0,1], \quad \boldsymbol{e} \in \mathbb{F}_2^n, \quad n \in \mathbb{N}^*, \tag{1.255}$$

which means that none of the components $v_j \in \mathbb{F}_2, j \in \{0, 1 \cdots (n-1)\}, n \in \mathbb{N}^*$ of the transmit codeword $\boldsymbol{v} \in \mathbb{V} \subseteq \mathbb{F}_2^n$, $n \in \mathbb{N}^*$, given by (1.24) is changed during the transmission. Hence, no bit error occurs.

Now, let us assume that a single component $v_j \in \mathbb{F}_2, j \in \{0, 1 \cdots (n-1)\}, n \in \mathbb{N}^*$ of the transmit codeword $\boldsymbol{v} \in \mathbb{V} \subseteq \mathbb{F}_2^n$, $n \in \mathbb{N}^*$, given by (1.24) is changed. Then the error pattern $\boldsymbol{e} \in \mathbb{F}_2^n$ has the weight

$$w_H\{\boldsymbol{e}\} = 1, \quad \boldsymbol{e} \in \mathbb{F}_2^n, n \in \mathbb{N}^*, \tag{1.256}$$

and the probability $\mathsf{P}\{w_H\{\boldsymbol{e}\} = 1\}$ is given by

$$\mathsf{P}\{w_H\{\boldsymbol{e}\} = 1\} = p \cdot (1-p)^{n-1}, \quad p \in [0,1], \quad \boldsymbol{e} \in \mathbb{F}_2^n, \quad n \in \mathbb{N}^*. \tag{1.257}$$

Since there exist n such error patterns with $w_{\mathrm{H}}\{\boldsymbol{e}\}$ equal to 1, the contribution to the transmission error probability without decoding P_{nd}, i. e., the probability of an undetected transmission error without decoding is

$$n \cdot \mathsf{P}\{w_{\mathrm{H}}\{\boldsymbol{e}\} = 1\} = n \cdot p \cdot (1-p)^{n-1}, \quad p \in [0,1], \quad \boldsymbol{e} \in \mathbb{F}_2^n, \quad n \in \mathbb{N}^*. \tag{1.258}$$

Now, let us assume that l, $l \in \{2, 3 \cdots n\}$, $n \in \mathbb{N}^*$, components $v_j \in \mathbb{F}_2$, $j \in \{0, 1 \cdots (n-1)\}$, $n \in \mathbb{N}^*$ of the transmit codeword $\boldsymbol{v} \in \mathbb{V} \subseteq \mathbb{F}_2^n$, $n \in \mathbb{N}^*$ given by (1.24) are changed during the transmission. Then the error pattern $\boldsymbol{e} \in \mathbb{F}_2^n$ has the weight

$$w_{\mathrm{H}}\{\boldsymbol{e}\} = l, \quad l \in \{2, 3 \cdots n\}, \quad \boldsymbol{e} \in \mathbb{F}_2^n, n \in \mathbb{N}^*, \tag{1.259}$$

and the probability $\mathsf{P}\{w_{\mathrm{H}}\{\boldsymbol{e}\} = l\}$ is given by

$$\mathsf{P}\{w_{\mathrm{H}}\{\boldsymbol{e}\} = 1\} = p^l \cdot (1-p)^{n-l}, \qquad p \in [0,1], \quad l \in \{2, 3 \cdots n\}, \quad \boldsymbol{e} \in \mathbb{F}_2^n, \quad n \in \mathbb{N}^*. \tag{1.260}$$

There exist [5, equation (1.3.9), p. 11], [36, p. 391]

$$\binom{n}{l} = \frac{n!}{l!(n-l)!}, \quad l \in \{2, 3 \cdots n\}, \quad n \in \mathbb{N}^*, \tag{1.261}$$

distinct error patterns with $w_{\mathrm{H}}\{\boldsymbol{e}\}$ equal to l, $l \in \{2, 3 \cdots n\}$, $n \in \mathbb{N}^*$ and the contribution to the transmission error probability without decoding P_{nd}, i. e., the probability of an undetected transmission error without decoding is [5, equation (1.3.9), p. 11]

$$\binom{n}{l} \cdot \mathsf{P}\{w_{\mathrm{H}}\{\boldsymbol{e}\} = l\} = \binom{n}{l} \cdot p^l \cdot (1-p)^{n-l}, \tag{1.262}$$

$$p \in [0,1], \quad l \in \{2, 3 \cdots n\}, \quad \boldsymbol{e} \in \mathbb{F}_2^n, \quad n \in \mathbb{N}^*. \tag{1.263}$$

Using the *binomial theorem* [11, equation (1.36c), p. 12],

$$1 = 1^n = ([1-p]+p)^n \equiv \sum_{l=0}^{n} \binom{n}{l} \cdot p^l (1-p)^{n-l}, \quad p \in [0,1], \quad n \in \mathbb{N}^*, \tag{1.264}$$

the transmission error probability without decoding P_{nd}, i. e., the probability of an undetected transmission error without decoding is given by [5, equation (1.3.6), p. 11]

$$\begin{aligned} \mathsf{P}_{\mathrm{nd}} &= \sum_{l=1}^{n} \binom{n}{l} \cdot p^l (1-p)^{n-l} = \underbrace{\sum_{l=0}^{n} \binom{n}{l} \cdot p^l (1-p)^{n-l}}_{=([1-p]+p)^n} - \underbrace{\binom{n}{0}}_{=1} \cdot \underbrace{p^0}_{=1} \cdot (1-p)^n, \\ &= 1 - (1-p)^n, \quad p \in [0,1], \quad n \in \mathbb{N}^*. \end{aligned} \tag{1.265}$$

Remark 1.22 (On the Transmission Error Probability Without Decoding P_{nd}). Using the binomial theorem (cf., e. g., (1.264)), we find [5, p. 11]

$$(1-p)^n = \sum_{l=0}^{n} \binom{n}{l} \cdot 1^{n-l} \cdot (-p)^l$$

$$= \sum_{l=0}^{n} \binom{n}{l} \cdot (-p)^l, \tag{1.266}$$

$$= \binom{n}{0} + \binom{n}{1} \cdot (-p) + \binom{n}{2} \cdot (-p)^2 + \cdots + \binom{n}{n} \cdot (-p)^n$$

$$= 1 - np + \frac{(n-1)n}{2} p^2 + \cdots + (-p)^n, \quad p \in [0,1], \quad n \in \mathbb{N}^*, \tag{1.267}$$

and, therefore, (1.265) becomes [5, equation (1.3.7), p. 11]

$$P_{nd} = np - \frac{(n-1)n}{2} p^2 - \cdots - (-p)^n \approx np, \quad \text{for } np \ll 1, \quad p \in [0,1], \quad n \in \mathbb{N}^*. \tag{1.268}$$

1.7.3 Transmission Error Probability With Decoding

Again, we will consider an (n, k) binary linear block code $\mathbb{V}$ and the transmission via the binary symmetric (noisy) channel (BSC) with crossover probability p, $p \in [0,1]$; cf. Section 1.7.2.

Let $A_l \in \mathbb{N}$, $l \in \{0, 1 \cdots n\}$, $n \in \mathbb{N}^*$ be the number of codewords $\boldsymbol{v} \in \mathbb{V} \subseteq \mathbb{F}_2^n$, $n \in \mathbb{N}^*$ given by (1.24), having the Hamming weight $w_{\mathrm{H}}\{\boldsymbol{v}\}$ equal to l, $l \in \{0, 1 \cdots n\}$, $n \in \mathbb{N}^*$ in the considered (n, k) binary linear block code $\mathbb{V}$. The numbers $A_0, A_1 \cdots A_n$, $n \in \mathbb{N}^*$ are called the *weight distribution of the* (n, k) binary linear block code $\mathbb{V}$ [15].

Example 1.12. Let us continue with the Examples 1.5, 1.6 and 1.7. Considering Table 1.2 and Example 1.7, the weight distribution of the (7, 4, 3) binary linear block code with dimension k equal to 4, length n equal to 7 and minimum distance d_{min} equal to 3, given in Table 1.2 (cf., e. g., [15, Table 3.1, p. 68], [23, pp. 20–22]) is given by [15, p. 79]

$$\begin{aligned} A_0 &= 1, \\ A_1 &= 0, \\ A_2 &= 0, \\ A_3 &= 7, \\ A_4 &= 7, \\ A_5 &= 0, \\ A_6 &= 0, \\ A_7 &= 1. \end{aligned} \tag{1.269}$$

If the (n, k) binary linear block code $\mathbb{V}$ is used only for error detection in the case of the transmission over the aforementioned BSC, the probability that the decoder will fail to detect transmission errors can be computed by exploiting the weight distribution

$A_0, A_1 \cdots A_n$, $n \in \mathbb{N}^*$ of the (n, k) binary linear block code $\mathbb{V}$ [15, p. 79]. Since the (n, k) binary linear block code $\mathbb{V}$ with minimum distance d_{min} will detect all the transmission errors with error patterns $\boldsymbol{e} \in \mathbb{F}_2^n$, $n \in \mathbb{N}^*$, having the Hamming weight $(d_{\text{min}} - 1)$, only such error patterns $\boldsymbol{e} \in \mathbb{F}_2^n$, $n \in \mathbb{N}^*$ with Hamming weights $\geq d_{\text{min}}$ will contribute to the *probability of an undetected error with decoding* P_{d} [15, equation (3.19), p. 79], [5, equation (3.6.1), p. 91],

$$A_l \cdot \mathsf{P}\{w_{\text{H}}\{\boldsymbol{e}\} = l\} = A_l \cdot p^l \cdot (1-p)^{n-l}, \qquad A_l \in \mathbb{N}, \quad p \in [0,1], \quad l \in \{d_{\text{min}}, (d_{\text{min}}+1) \cdots n\}, \quad n \in \mathbb{N}^*, \tag{1.270}$$

therefore, we yield [15, equation (3.19), p. 79], [5, equation (3.6.1), p. 91],

$$\mathsf{P}_{\text{d}} = \sum_{l=d_{\text{min}}}^{n} A_l \cdot p^l \cdot (1-p)^{n-l}, \qquad A_l \in \mathbb{N}, \quad p \in [0,1], \quad l \in \{d_{\text{min}}, (d_{\text{min}}+1) \cdots n\}, \quad n \in \mathbb{N}^*. \tag{1.271}$$

Since in general

$$A_l \leq \binom{n}{l}, \quad A_l \in \mathbb{N}, \quad l \in \{d_{\text{min}}, (d_{\text{min}}+1) \cdots n\}, \quad n \in \mathbb{N}^* \tag{1.272}$$

holds with equality only in rare cases, we yield

$$\mathsf{P}_{\text{d}} \leq \mathsf{P}_{\text{nd}}. \tag{1.273}$$

For $p \ll 1$, which means that $(1-p)^{n-l}$ is approximately equal to 1, (1.271) can be approximated by [5, equation (3.6.3), p. 91]

$$\mathsf{P}_{\approx\text{d}} = A_{d_{\text{min}}} \cdot p^{d_{\text{min}}}, \quad A_{d_{\text{min}}} \in \mathbb{N}. \tag{1.274}$$

Remark 1.23 (On the Probability of an Undetected Error With Decoding P_{d}). Let p be equal to 1/2. Then (1.271) becomes

$$\mathsf{P}_{\text{d}} = \sum_{l=d_{\text{min}}}^{n} A_l \cdot \left(\frac{1}{2}\right)^l \cdot \left(\frac{1}{2}\right)^{n-l} = \sum_{l=d_{\text{min}}}^{n} A_l \cdot \left(\frac{1}{2}\right)^n$$
$$= \frac{1}{2^n} \sum_{l=d_{\text{min}}}^{n} A_l, \quad A_l \in \mathbb{N}, \quad l \in \{d_{\text{min}}, (d_{\text{min}}+1) \cdots n\}, \quad n \in \mathbb{N}^*. \tag{1.275}$$

Since $(\sum_{l=d_{\text{min}}}^{n} A_l)$ is equal to the number $(2^k - 1)$, $k \in \mathbb{N}^*$ of nonzero codewords, (1.275) becomes [23, p. 19]

$$\mathsf{P}_{\text{d}} = \frac{2^k - 1}{2^n} = \frac{1}{2^{n-k}} - \frac{1}{2^n} \leq \frac{1}{2^{n-k}}, \quad k, n \in \mathbb{N}^*. \tag{1.276}$$

Example 1.13. Let us continue with the Examples 1.5, 1.6, 1.7 and 1.12. In the case of the (n,k) binary linear block code $\mathbb{V}$ of Table 1.2, we find

$$\begin{aligned} \mathrm{P_d} &= \sum_{l=d_{\min}}^{n} A_l p^l (1-p)^{n-l} = \sum_{l=3}^{7} A_l p^l (1-p)^{7-l} \\ &= A_3 p^3 (1-p)^{7-3} + A_4 p^4 (1-p)^{7-4} + A_7 p^7 (1-p)^{7-7} \\ &= 7p^3(1-p)^3 + p^7. \end{aligned} \tag{1.277}$$

Furthermore, we find

$$\mathrm{P_{\approx d}} = A_{d_{\min}} \cdot p^{d_{\min}} = 7p^3. \tag{1.278}$$

Figure 1.14 shows $\mathrm{P_{nd}}$ equal to $[1-[1-p]^7]$, $\mathrm{P_d}$ equal to $[7p^3(1-p)^3+p^7]$ and $\mathrm{P_{\approx d}}$ equal to $7p^3$ versus $p \in [0,1]$ for the (n,k) binary linear block code $\mathbb{V}$ of Table 1.2.

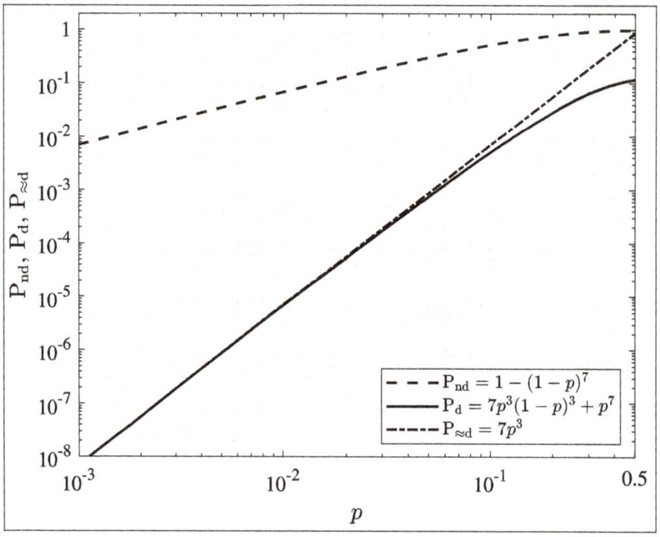

Figure 1.14: $\mathrm{P_{nd}}$ equal to $[1-[1-p]^7]$, $\mathrm{P_d}$ equal to $[7p^3(1-p)^3+p^7]$ and $\mathrm{P_{\approx d}}$ equal to $7p^3$ versus p for the (n,k) binary linear block code $\mathbb{V}$ of Table 1.2.

Still there remains the question about how many errors can an (n,k) binary linear block code $\mathbb{V}$ with the minimum distance $d_{\min}$ correct [15, p. 79]. We can find $t \in \mathbb{N}^*$ such that [15, equation (3.20), p. 79].

$$t \le \frac{d_{\min}-1}{2} \le t + \frac{1}{2}. \tag{1.279}$$

Let $\boldsymbol{v} \in \mathbb{V} \subseteq \mathbb{F}_2^n$, $n \in \mathbb{N}^*$ be the transmitted codeword and $\boldsymbol{r} \in \mathbb{F}_2^n$, $n \in \mathbb{N}^*$ be the receive vector, respectively [15, p. 79]. Furthermore, let $\boldsymbol{w} \in \mathbb{V} \subseteq \mathbb{F}_2^n$, $n \in \mathbb{N}^*$ be any other codeword in the considered (n, k) binary linear block code $\mathbb{V}$. Then the triangle inequality of Theorem 1.4 teaches us [15, equation (3.21), p. 79]

$$d_\mathrm{H}\{\boldsymbol{v}, \boldsymbol{r}\} + d_\mathrm{H}\{\boldsymbol{r}, \boldsymbol{w}\} \geq d_\mathrm{H}\{\boldsymbol{v}, \boldsymbol{w}\}. \tag{1.280}$$

If an error pattern $\boldsymbol{e} \in \mathbb{F}_2^n$, $n \in \mathbb{N}$ representing $t' \in \mathbb{N}^*$ transmission errors occurs, the receive vector $\boldsymbol{r} \in \mathbb{F}_2^n$, $n \in \mathbb{N}^*$ differs from $\boldsymbol{v} \in \mathbb{V} \subseteq \mathbb{F}_2^n$, $n \in \mathbb{N}^*$ in t' components, and we find

$$d_\mathrm{H}\{\boldsymbol{v}, \boldsymbol{r}\} = t', \quad \boldsymbol{v} \in \mathbb{V} \subseteq \mathbb{F}_2^n, \quad \boldsymbol{r} \in \mathbb{F}_2^n, \quad t' \in \mathbb{N}^*, \quad n \in \mathbb{N}^*. \tag{1.281}$$

Since $\boldsymbol{v}, \boldsymbol{w} \in \mathbb{V} \subseteq \mathbb{F}_2^n$, $n \in \mathbb{N}^*$ are codewords using (1.279), we obtain [15, equation (3.22), p. 79]

$$d_\mathrm{H}\{\boldsymbol{v}, \boldsymbol{w}\} \geq d_\mathrm{min} \geq 2t + 1, \quad \boldsymbol{v}, \boldsymbol{w} \in \mathbb{V} \subseteq \mathbb{F}_2^n, \quad \boldsymbol{r} \in \mathbb{F}_2^n, \quad t \in \mathbb{N}^*, \quad n \in \mathbb{N}^*. \tag{1.282}$$

With (1.282), (1.280) becomes [15, p. 80]

$$t' + d_\mathrm{H}\{\boldsymbol{r}, \boldsymbol{w}\} \geq 2t + 1 \quad \Leftrightarrow \quad d_\mathrm{H}\{\boldsymbol{r}, \boldsymbol{w}\} \geq 2t - t' + 1, \qquad t, t' \in \mathbb{N}^*, \quad \boldsymbol{w} \in \mathbb{V} \subseteq \mathbb{F}_2^n, \quad \boldsymbol{r} \in \mathbb{F}_2^n, \quad n \in \mathbb{N}^*. \tag{1.283}$$

In the case of $t' \leq t$, $t, t' \in \mathbb{N}^*$, we yield [15, p. 80]

$$d_\mathrm{H}\{\boldsymbol{r}, \boldsymbol{w}\} > t, \quad t \in \mathbb{N}^*, \quad \boldsymbol{w} \in \mathbb{V} \subseteq \mathbb{F}_2^n, \quad \boldsymbol{r} \in \mathbb{F}_2^n, \quad n \in \mathbb{N}^*. \tag{1.284}$$

This means that if an error pattern $\boldsymbol{e} \in \mathbb{F}_2^n$, $n \in \mathbb{N}$ with $t \in \mathbb{N}^*$ or fewer transmission errors occurs, the receive vector $\boldsymbol{r} \in \mathbb{F}_2^n$, $n \in \mathbb{N}$ is closer to the transmit codeword $\boldsymbol{v} \in \mathbb{V} \subseteq \mathbb{F}_2^n$, $n \in \mathbb{N}^*$ than to any other admissible codeword $\boldsymbol{w} \in \mathbb{V} \subseteq \mathbb{F}_2^n$, $n \in \mathbb{N}^*$ [15, p. 80]. The decoder will then decode $\boldsymbol{v} \in \mathbb{V} \subseteq \mathbb{F}_2^n$, $n \in \mathbb{N}^*$, which means that there are no transmission errors after decoding. Therefore, the considered (n, k) binary linear block code $\mathbb{V}$ can correct t or fewer transmission errors [15, p. 80].

However, the (n, k) binary linear block code $\mathbb{V}$ is not capable of correcting all the error patterns $\boldsymbol{e} \in \mathbb{F}_2^n$, $n \in \mathbb{N}$ with $l > t$, $l, t \in \mathbb{N}^*$ transmission errors because there is at least one case in which an error pattern $\boldsymbol{e} \in \mathbb{F}_2^n$, $n \in \mathbb{N}$ of $l \in \mathbb{N}^*$ transmission errors results in a receive vector $\boldsymbol{r} \in \mathbb{F}_2^n$, $n \in \mathbb{N}$ that is closer to an incorrect codeword than to the transmit codeword $\boldsymbol{v} \in \mathbb{V} \subseteq \mathbb{F}_2^n$, $n \in \mathbb{N}^*$ [15, p. 80]. Let us consider a proof.

Let $\boldsymbol{v}, \boldsymbol{w} \in \mathbb{V} \subseteq \mathbb{F}_2^n$, $n \in \mathbb{N}^*$ be two codewords such that [15, p. 80]

$$d_\mathrm{H}\{\boldsymbol{v}, \boldsymbol{w}\} = d_\mathrm{min}. \tag{1.285}$$

Furthermore, let $\boldsymbol{e}^{(1)} \in \mathbb{F}_2^n$, $n \in \mathbb{N}^*$ and $\boldsymbol{e}^{(2)} \in \mathbb{F}_2^n$, $n \in \mathbb{N}^*$ be two error patterns, which comply with the following [15, p. 80]:

a) $\boldsymbol{e}^{(1)} \oplus \boldsymbol{e}^{(2)} = \boldsymbol{v} \oplus \boldsymbol{w}$, $n \in \mathbb{N}^*$.

b) $\boldsymbol{e}^{(1)} \in \mathbb{F}_2^n, n \in \mathbb{N}^*$ and $\boldsymbol{e}^{(2)} \in \mathbb{F}_2^n, n \in \mathbb{N}^*$ do not have nonzero components in common places.

Thus, we yield [15, equation (3.23), p. 80]

$$w_\mathrm{H}\{\boldsymbol{e}^{(1)}\} + w_\mathrm{H}\{\boldsymbol{e}^{(2)}\} = w_\mathrm{H}\{\boldsymbol{v} \oplus \boldsymbol{w}\} = d_\mathrm{H}\{\boldsymbol{v}, \boldsymbol{w}\} = d_\mathrm{min}, \quad \boldsymbol{e}^{(1)}, \boldsymbol{e}^{(2)} \in \mathbb{F}_2^n, \quad \boldsymbol{v}, \boldsymbol{w} \in \mathbb{V} \subseteq \mathbb{F}_2^n, \quad n \in \mathbb{N}^*. \tag{1.286}$$

Assume that $\boldsymbol{v} \in \mathbb{V} \subseteq \mathbb{F}_2^n, n \in \mathbb{N}^*$ is transmitted and corrupted by the error pattern $\boldsymbol{e}^{(1)} \in \mathbb{F}_2^n, n \in \mathbb{N}^*$ leading to [15, p. 80]

$$\boldsymbol{r} = \boldsymbol{v} \oplus \boldsymbol{e}^{(1)}, \quad \boldsymbol{e}^{(1)}, \boldsymbol{r} \in \mathbb{F}_2^n, \quad \boldsymbol{v} \in \mathbb{V} \subseteq \mathbb{F}_2^n, \quad n \in \mathbb{N}^*, \tag{1.287}$$

and [15, equation (3.24), p. 80]

$$d_\mathrm{H}\{\boldsymbol{v}, \boldsymbol{r}\} = w_\mathrm{H}\{\boldsymbol{v} \oplus \boldsymbol{r}\} = w_\mathrm{H}\{\boldsymbol{e}^{(1)}\}, \quad \boldsymbol{e}^{(1)}, \boldsymbol{r} \in \mathbb{F}_2^n, \quad \boldsymbol{v} \in \mathbb{V} \subseteq \mathbb{F}_2^n, \quad n \in \mathbb{N}^*. \tag{1.288}$$

In this case, we find [15, equation (3.25), p. 80]

$$d_\mathrm{H}\{\boldsymbol{w}, \boldsymbol{r}\} = w_\mathrm{H}\{\boldsymbol{w} \oplus \boldsymbol{r}\} = w_\mathrm{H}\{\boldsymbol{w} \oplus \boldsymbol{v} \oplus \boldsymbol{e}^{(1)}\} = w_\mathrm{H}\{\boldsymbol{e}^{(1)} \oplus \boldsymbol{e}^{(2)} \oplus \boldsymbol{e}^{(1)}\} = w_\mathrm{H}\{\boldsymbol{e}^{(2)}\}, \quad \boldsymbol{e}^{(1)}, \boldsymbol{e}^{(2)}, \boldsymbol{r} \in \mathbb{F}_2^n, \quad \boldsymbol{v}, \boldsymbol{w} \in \mathbb{V} \subseteq \mathbb{F}_2^n, \quad n \in \mathbb{N}^*. \tag{1.289}$$

Let us now assume [15, p. 80]

$$w_\mathrm{H}\{\boldsymbol{e}^{(1)}\} \geq t + 1, \quad \boldsymbol{e}^{(1)} \in \mathbb{F}_2^n, \quad n, t \in \mathbb{N}^*, \tag{1.290}$$

immediately leading to

$$w_\mathrm{H}\{\boldsymbol{e}^{(2)}\} \leq t + 1, \quad \boldsymbol{e}^{(2)} \in \mathbb{F}_2^n, \quad n, t \in \mathbb{N}^*, \tag{1.291}$$

because the maximum of $(t + 1 - w_\mathrm{H}\{\boldsymbol{e}^{(1)}\})$ is zero. Thus, we obtain [15, p. 80]

$$d_\mathrm{H}\{\boldsymbol{v}, \boldsymbol{r}\} \geq d_\mathrm{H}\{\boldsymbol{w}, \boldsymbol{r}\}, \quad \boldsymbol{r} \in \mathbb{F}_2^n, \quad \boldsymbol{v}, \boldsymbol{w} \in \mathbb{V} \subseteq \mathbb{F}_2^n, \quad n \in \mathbb{N}^*. \tag{1.292}$$

Hence, there is an error pattern $\boldsymbol{e}^{(1)} \in \mathbb{F}_2^n, n \in \mathbb{N}^*$ with $l > t$, $l, t \in \mathbb{N}^*$ transmission errors, which results in a receive vector $\boldsymbol{r} \in \mathbb{F}_2^n, n \in \mathbb{N}^*$ that is closer to an incorrect codeword $\boldsymbol{w} \in \mathbb{V} \subseteq \mathbb{F}_2^n, n \in \mathbb{N}^*$ than to the transmit codeword $\boldsymbol{v} \in \mathbb{V} \subseteq \mathbb{F}_2^n, n \in \mathbb{N}^*$ [15, p. 81]. Hence, an incorrect decoding would result [15, p. 81]. Therefore, an (n, k) binary linear block code $\mathbb{V}$ with minimum distance d_min guarantees the correction of all the error patterns consisting of

$$t = \left\lfloor \frac{d_\mathrm{min} - 1}{2} \right\rfloor \tag{1.293}$$

or fewer errors, $\lfloor(d_{\min}-1)/2\rfloor$ denoting the largest integer no greater than $(d_{\min}-1)/2$ [15, p. 81]. The parameter $t \in \mathbb{N}^*$ equal to $\lfloor(d_{\min}-1)/2\rfloor$ according to (1.293) is called the *random-error-correcting capability* of the (n,k) binary linear block code $\mathbb{V}$ [15, p. 81]. The (n,k) binary linear block code $\mathbb{V}$ is thus referred to as a *t-error-correcting code* [15, p. 81].

Definition 1.31 (Gaußklammer / Untere Gaußklammer / Floor Function). Let $x \in \mathbb{R}$. The function [37, Definition 1.12, p. 9]

$$\begin{aligned} f: \quad \mathbb{R} &\mapsto \mathbb{Z} \\ x &\mapsto \lfloor x\rfloor = \max\{k \in \mathbb{Z} \mid k \leq x\} \end{aligned} \tag{1.294}$$

is termed "floor function" by Iverson [38, p. 12] and carries out *"rounding towards minus infinity."*

However, $f:\mathbb{R}\mapsto\mathbb{Z}$ of (1.294) was first introduced by the "princeps mathematicorum," the prince of the mathematicians, Carl Friedrich Gauß in 1808 and is termed *Gaußklammer*, or more modern, *untere Gaußklammer*.

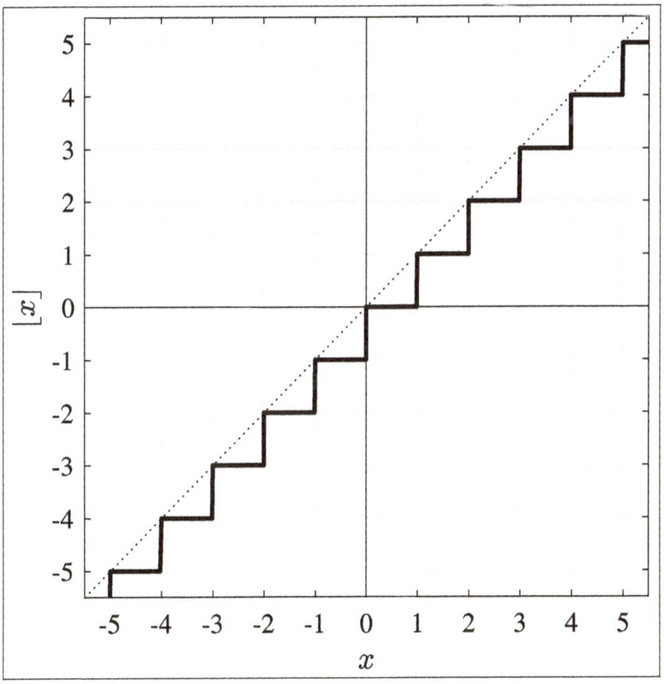

Figure 1.15: Detail of the *(untere) Gaußklammer* (floor function) $f:\mathbb{R}\mapsto\mathbb{Z}$ defined by (1.294) with the mapping $\lfloor x\rfloor$ equal to $\max\{k\in\mathbb{Z}\mid k\leq x\}$.

Figure 1.15 illustrates the *(untere) Gaußklammer (floor function)* $f:\mathbb{R}\mapsto\mathbb{Z}$ defined by (1.294) with the mapping $\lfloor x\rfloor$ equal to $\max\{k\in\mathbb{Z}\mid k\leq x\}$. Obviously, the Gaußklammer (floor function) performs the quantization of the values of any real function such as the *identity function*

$$\begin{aligned} \mathrm{id}: \quad \mathbb{R} &\mapsto \mathbb{R} \\ x &\mapsto \mathrm{id}(x) = x \end{aligned} \tag{1.295}$$

input to $f : \mathbb{R} \mapsto \mathbb{Z}$ defined by (1.294).

Another important way toward quantization is the *obere Gaußklammer (ceiling function)*, which we will illustrate in Definition 1.32.

Definition 1.32 (Obere Gaußklammer / Ceiling Function). Let $x \in \mathbb{R}$. The function [37, Definition 1.12, p. 9]

$$\begin{aligned} c: \quad \mathbb{R} &\mapsto \mathbb{Z} \\ x &\mapsto \lceil x \rceil = \min\{k \in \mathbb{Z} \,|\, k \geq x\} \end{aligned} \tag{1.296}$$

is termed "ceiling function" by Iverson [38, p. 12] as well as *obere Gaußklammer* [37, Definition 1.12, p. 9] and carries out "*rounding towards plus infinity.*"

Figure 1.16 illustrates the *obere Gaußklammer* (ceiling function) $c : \mathbb{R} \mapsto \mathbb{Z}$ defined by (1.296) with the mapping $\lceil x \rceil$ equal to $\min\{k \in \mathbb{Z} \,|\, k \geq x\}$.

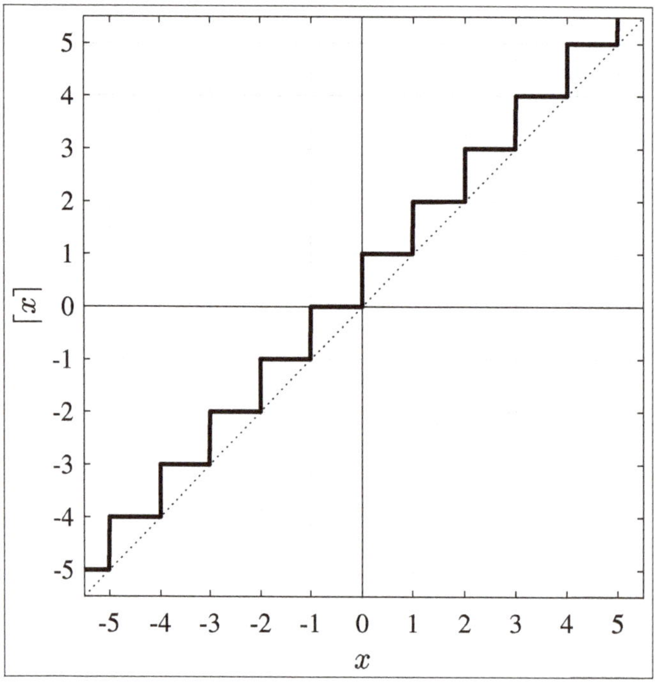

Figure 1.16: Detail of the *obere Gaußklammer* (ceiling function) $c : \mathbb{R} \mapsto \mathbb{Z}$ defined by (1.296) with the mapping $\lceil x \rceil$ equal to $\min\{k \in \mathbb{Z} \,|\, k \geq x\}$.

A third important approach toward quantization is the *rounding function*, which we will illustrate in Definition 1.33.

Definition 1.33 (Rounding Function). Let $x \in \mathbb{R}$. The function [36, p. 5]

$$\begin{aligned} \mathrm{rd}: \quad \mathbb{R} &\mapsto \mathbb{Z} \\ x &\mapsto \mathrm{rd}(x) = \lfloor x + 1/2 \rfloor = \lceil x - 1/2 \rceil \end{aligned} \tag{1.297}$$

carries out the rounding.

Figure 1.17 illustrates the *rounding function* $\mathrm{rd} : \mathbb{R} \mapsto \mathbb{Z}$ defined by (1.297) with the mapping $\mathrm{rd}(x)$ equal to $\lfloor x + 1/2 \rfloor = \lceil x - 1/2 \rceil$.

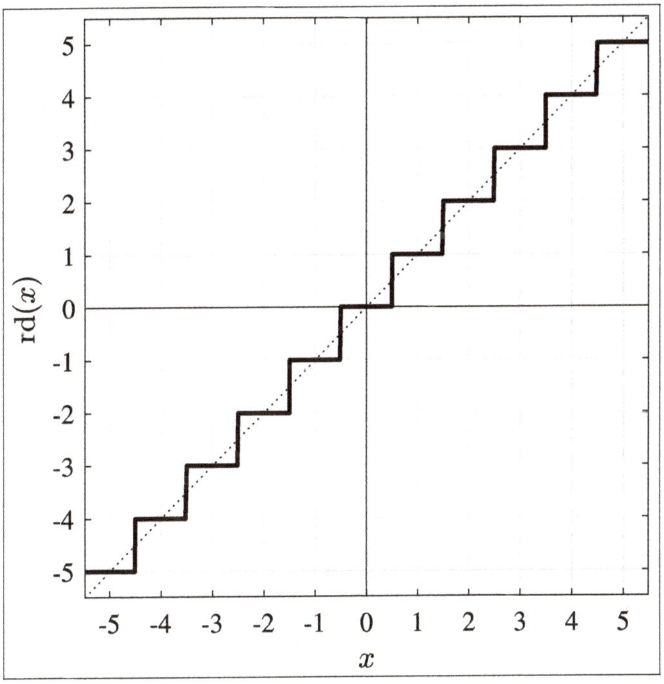

Figure 1.17: Detail of the *rounding function* $\mathrm{rd} : \mathbb{R} \mapsto \mathbb{Z}$ defined by (1.297) with the mapping $\mathrm{rd}(x)$ equal to $\lfloor x + 1/2 \rfloor = \lceil x - 1/2 \rceil$.

An (n, k) binary linear block code $\mathbb{V}$ with random-error-correcting capability t is usually capable of correcting many error patterns of $(t+1)$ or more transmission errors [15, p. 81]. A t-error-correcting (n, k) binary linear block code $\mathbb{V}$ is capable of correcting a total of 2^{n-k} error patterns including those with t or fewer transmission errors [15, p. 81]. If a t-error-correcting (n, k) binary linear block code $\mathbb{V}$ is used strictly for error correction on a BSC with crossover probability p, the error probability that the decoder commits an erroneous decoding of a codeword is upper bounded by [15, equation (3.26), p. 81], [5, equation (3.7.1), p. 93], [23, p. 18],

$$\mathrm{P}(E) \leq \sum_{l=t+1}^{n} \binom{n}{l} p^{l}(1-p)^{n-l}, \tag{1.298}$$

$$p \in [0,1], \quad t = \left\lfloor \frac{(d_{\min}-1)}{2} \right\rfloor, \quad d_{\min}, n \in \mathbb{N}^{*}.$$

Remark 1.24 (On P(E) of (1.298)). Using the *binomial theorem* of (1.264) [11, equation (1.36c), p. 12], we yield

$$1 = \sum_{l=0}^{t} \binom{n}{l} p^{l}(1-p)^{n-l} + \sum_{l=t+1}^{n} \binom{n}{l} p^{l}(1-p)^{n-l}, \tag{1.299}$$

which immediately yields

$$\sum_{l=t+1}^{n} \binom{n}{l} p^{l}(1-p)^{n-l} = 1 - \sum_{l=0}^{t} \binom{n}{l} p^{l}(1-p)^{n-l}. \tag{1.300}$$

Therefore, (1.298) becomes [5, equation (3.7.1), p. 93], [23, p. 18]

$$\mathrm{P}(E) \leq 1 - \sum_{l=0}^{t} \binom{n}{l} p^{l}(1-p)^{n-l}, \quad p \in [0,1], \quad t = \left\lfloor \frac{(d_{\min}-1)}{2} \right\rfloor, \quad d_{\min}, n \in \mathbb{N}^{*}. \tag{1.301}$$

Setting out from the above discussion and using (1.293), we obtain the *sphere packing bound*, which is also termed the *Hamming bound* [5, Satz 3.9, p. 81].

Theorem 1.16 (Hamming Bound / Sphere Packing Bound). *An* $(n, k, d_{\min})$ *binary linear block code* $\mathbb{V}$ *with the random-error-correcting capability t given by (*1.293*), i. e., a t-error-correcting code, must comply with* [5, *Satz 3.9, equations (3.3.2) and (3.3.3), p. 81*]

$$2^{n-k} \geq \sum_{i=0}^{t} \binom{n}{i} = 1 + n + \binom{n}{2} + \cdots + \binom{n}{t}, \quad k, n \in \mathbb{N}^{*}. \tag{1.302}$$

*The inequality (*1.302*) is termed the* sphere packing bound, *respectively, the* Hamming bound [5, *Satz 3.9, p. 81*]. *The Hamming bound is an upper bound on the number of parity check sequences* 2^{n-k}, $k, n \in \mathbb{N}^{*}$.

Proof. Each codeword $\boldsymbol{v} \in \mathbb{V} \subseteq \mathbb{F}_2^n$, $n \in \mathbb{N}^*$, forms the center of a sphere. Let us choose the radii of all 2^k, $k \in \mathbb{N}^*$, maximally large, however, making sure that none of the spheres intersect with any other spheres [23, Satz 1.10, p. 12]. This means that the said spheres are disjoint [5, p. 81]. It is vividly clear that the radius of such a sphere is upper bounded by $d_{\min}/2$ [23, Satz 1.10, p. 12].

We can uniquely assign any receive vector $\boldsymbol{r} \in \mathbb{F}_2^n$, $n \in \mathbb{N}^*$ to a particular codeword $\hat{\boldsymbol{v}} \in \mathbb{V} \subseteq \mathbb{F}_2^n$, $n \in \mathbb{N}^*$, say, as long as the Hamming distance $d_{\mathrm{H}}\{\boldsymbol{r}, \hat{\boldsymbol{v}}\}$ is lower than or the radius of the associated sphere, i. e., as long as the receive vector $\boldsymbol{r} \in \mathbb{F}_2^n$, $n \in \mathbb{N}^*$ is inside the sphere around its center $\hat{\boldsymbol{v}} \in \mathbb{V} \subseteq \mathbb{F}_2^n$, $n \in \mathbb{N}^*$ [23, Satz 1.10, p. 12]. This means that we will decode the transmit codeword correctly. For this reason, the said sphere is termed *decoding sphere* (in German “*Decodierkugel*”) [5, p. 81], *correcting sphere* (in German “*Korrekturkugel*”) [23, Satz 1.10, p. 12] or *Hamming sphere* [25, p. 59f.].

Correct decoding hence requires that $d_H\{\boldsymbol{r}, \hat{\boldsymbol{v}}\}$ is lower than or equal to the random-error-correcting capability t equal to $\lfloor(d_{min} - 1)/2\rfloor$. The number $V_n(t)$ of distinct receive vectors $\boldsymbol{r} \in \mathbb{F}_2^n$, $n \in \mathbb{N}^*$, which fulfill this requirement for a particular decoding sphere is given by [31, equation (6), p. 65]

$$V_n(t) = \sum_{i=0}^{t} \binom{n}{i} \quad t = \left\lfloor \frac{d_{min} - 1}{2} \right\rfloor, \quad n, t \in \mathbb{N}^*. \tag{1.303}$$

To the best knowledge of the author, the denotation $V_n(t)$ (precisely "$V_n(k_1)$") was first introduced by Peter *Elias* in his seminal paper of 1955. A shortened reprint can be found in [31, pp. 61–74].

Since there are 2^k, $k \in \mathbb{N}^*$ such decoding spheres, the total number of receive vectors $\boldsymbol{r} \in \mathbb{F}_2^n$, $n \in \mathbb{N}^*$, which fulfill the requirement of correct decoding is equal to $2^k \cdot V_n(t)$, $k \in \mathbb{N}^*$, and thus

$$2^k \cdot V_n(t) = 2^k \sum_{i=0}^{t} \binom{n}{i} \quad t = \left\lfloor \frac{d_{min} - 1}{2} \right\rfloor, \quad k, n, t \in \mathbb{N}^*. \tag{1.304}$$

Clearly, $2^k \cdot V_n(t)$, $k \in \mathbb{N}^*$ cannot be greater than the total number 2^n, $n \in \mathbb{N}^*$ of receive vectors contained in $\mathbb{F}_2^n$, $n \in \mathbb{N}^*$. Therefore, we obtain

$$2^n \geq 2^k \cdot V_n(t) = 2^k \sum_{i=0}^{t} \binom{n}{i} \quad t = \left\lfloor \frac{d_{min} - 1}{2} \right\rfloor, \quad k, n, t \in \mathbb{N}^*, \tag{1.305}$$

which immediately leads to (1.302). □

Remark 1.25 (On the Hamming Bound). Equation (1.302) can be written in the form of an upper bound on the number of codewords 2^k, $k \in \mathbb{N}^*$, [27, p. 15]

$$2^k \leq \frac{2^n}{\sum_{i=0}^{t} \binom{n}{i}}, \quad k, n \in \mathbb{N}^*. \tag{1.306}$$

Setting out from (1.302), we obtain

$$n - k \geq \log_2 \left\{ \sum_{i=0}^{t} \binom{n}{i} \right\} \quad \Leftrightarrow \quad 1 - \underbrace{\frac{k}{n}}_{=R} \geq \frac{1}{n} \log_2 \left\{ \sum_{i=0}^{t} \binom{n}{i} \right\}, \quad k, n \in \mathbb{N}^*. \tag{1.307}$$

Equation (1.307) means that the code rate R is upper bounded by

$$R \leq 1 - \frac{1}{n} \log_2 \left\{ \sum_{i=0}^{t} \binom{n}{i} \right\}, \quad k, n \in \mathbb{N}^*. \tag{1.308}$$

Remark 1.26 (On the Equivocation). It is about time to revisit what we discussed in Section 1.2, and especially elaborate on the equivocation.

Theorem 1.16 teaches us that $V_n(t)$, $n \in \mathbb{N}^*$ is the number of distinct receive vectors $\boldsymbol{r} \in \mathbb{F}_2^n$, $n \in \mathbb{N}^*$, which are all inside a decoding sphere of diameter t equal to $\lfloor(d_{min}-1)/2\rfloor$, having a permissible codeword $\boldsymbol{v} \in \mathbb{V} \subseteq \mathbb{F}_2^n$, $n \in \mathbb{N}^*$ at its center. In other words, error-free decoding means that all the $V_n(t)$ receive vectors inside the said decoding sphere are mapped onto the one permissible codeword $\boldsymbol{v} \in \mathbb{V} \subseteq \mathbb{F}_2^n$, $n \in \mathbb{N}^*$. Does this ring a bell?

Let us take a different viewpoint. Suppose that the source, i. e., the transmitter, transmits messages of length $n \in \mathbb{N}^*$. Each transmitted message, i. e., each transmit signal has the transmission period T_S.

There are 2^n, $n \in \mathbb{N}^*$ possible transmit signals available for transmission within the transmission period T_S. Let us assume that all these possible transmit signals are equiprobable. Then the source entropy $\mathrm{H}\{X\}$ reaches its maximum, and we yield

$$2^{\{T_S \cdot \mathrm{H}\{X\}\}} \approx 2^n \quad \Leftrightarrow \quad \mathrm{H}\{X\} \approx \frac{n}{T_S}, \quad n \in \mathbb{N}^*, \tag{1.309}$$

"$\approx$" becoming "=" for $T_S \to \infty$.

Now, assume that the perturbations of the transmission channel cause the loss of an average number of $\log_2\{V_n(t)\}$ bits during the transmission. This means that there is an average number $V_n(t)$ of messages, which all "sound the same to the receiver" and are thus mapped onto the same receive sequence. Obviously, the equivocation $\mathrm{H}\{X \mid Y\}$, which is a property of the transmission channel, can be approximated by

$$2^{\{T_S \cdot \mathrm{H}\{X \mid Y\}\}} \approx V_n(t) \quad \Leftrightarrow \quad \mathrm{H}\{X \mid Y\} \approx \frac{\log_2\{V_n(t)\}}{T_S}, \quad n \in \mathbb{N}^*. \tag{1.310}$$

Therefore, only the average number of approximately $T_S \cdot (\mathrm{H}\{X\} - \mathrm{H}\{X \mid Y\})$ bits per receive signal can be decoded error-free.

In order to overcome the ambiguity caused by the perturbations of the transmission channel, only the average number of approximately $T_S \cdot (\mathrm{H}\{X\} - \mathrm{H}\{X \mid Y\})$ information carrying bits per message can be sent, accompanied by the average number of approximately $T_S \cdot \mathrm{H}\{X \mid Y\}$ parity check bits to obtain the average transmission rate $\mathrm{H}\{X\}$, whereas the average information rate is only $(\mathrm{H}\{X\} - \mathrm{H}\{X \mid Y\})$. Since $T_S \cdot \mathrm{H}\{X \mid Y\}$ is not necessarily an integer, the average number of parity check bits, $(n-k)$, should be greater than or equal to $T_S \cdot \mathrm{H}\{X \mid Y\}$, i. e.,

$$n - k \geq T_S \cdot \mathrm{H}\{X \mid Y\} \approx \log_2\left\{V_n(t)\right\}, \quad k, n \in \mathbb{N}^*, \tag{1.311}$$

which is the logarithmic version of the Hamming bound given by (1.302). We immediately yield

$$R \leq 1 - \frac{\log_2\{V_n(t)\}}{n}, \quad k, n \in \mathbb{N}^*. \tag{1.312}$$

Hence, the maximum permissible code rate $R_{\max}$ for a potentially error-free reception is given by $(1 - \log_2\{V_n(t)\}/n)$, $n \in \mathbb{N}^*$ [31, equation (8'), p. 65].

Note that the equivocation $\mathrm{H}\{X \mid Y\}$, being a property of the transmission channel, has been approximately equated with $\log_2\{V_n(t)\}/T_S$, using $V_n(t)$, which is governed by the deployed $(n, k, d_{\min})$ binary linear block code $\mathbb{V}$. This means that in the considered case only such transmission channels with an equivocation $\mathrm{H}\{X \mid Y\}$ not greater than approximately $\log_2\{V_n(t)\}/T_S$ qualify for an error-free transmission, whereas transmission channels with greater equivocation still cause errors, which cannot be corrected.

Finally, it should be mentioned that the model of the transmission channel that must be considered to find the results discussed in Remark 1.26 is the binary symmetric (noisy) channel (BSC); cf., e. g., [31, p. 65].

Furthermore, Theorem 1.16 leads to the following definition.

Definition 1.34 (Perfect Code). A $(n, k, d_{\min})$ binary linear block code $\mathbb{V}$ is said to be a *perfect code* if (1.302) is fulfilled with equality, i. e., if all 2^n, $n \in \mathbb{N}^*$ receive vectors contained in $\mathbb{F}_2^n$, $n \in \mathbb{N}^*$ are inside the $2^k \cdot V_n(t)$, $k \in \mathbb{N}^*$ decoding spheres [14, p. 107], [23, Definition 1.11, p. 12], [5, Satz 3.9, p. 81].

Let us consider the following.

Theorem 1.17 (Gilbert Bound for Binary Linear Codes). *For any $d, n \in \mathbb{N}^*$, which fulfill* [25, *Theorem 12.3.2, p. 387*]

$$2 \le d \le \frac{n}{2}, \tag{1.313}$$

there exists an $(n, k, d_{\min})$ binary linear block code $\mathbb{V}$ with $d_{\min} \ge d$, with the dimension k that satisfies [25, *Theorem 12.3.2, p. 387*]

$$\sum_{i=0}^{d-1} \binom{n}{i} \ge 2^{n-k}, \quad 2 \le d \le \frac{n}{2}, \quad d \le d_{\min}, \quad d, k, n \in \mathbb{N}^*. \tag{1.314}$$

Equation (1.314) is termed the Gilbert bound for binary linear codes [25, *Theorem 12.3.2, p. 387f.*].

Proof. In an $(n, k, d_{\min})$ binary linear block code $\mathbb{V}$, there are 2^k, $k \in \mathbb{N}^*$, codewords, and hence there also exist 2^k, $k \in \mathbb{N}^*$ decoding spheres of radius $(d-1)$, $d \in \mathbb{N}^*$ [25, Theorem 12.3.2, p. 387], each of which contains [25, p. 386]

$$V_n(d-1) = \sum_{i=0}^{d-1} \binom{n}{i}, \quad 2 \le d \le \frac{n}{2}, \quad d \le d_{\min}, \quad d, n \in \mathbb{N}^*, \tag{1.315}$$

points, and hence possible receive vectors.

The number of points, and hence of possible receive vectors, which are contained in all the 2^k, $k \in \mathbb{N}^*$ decoding spheres, is therefore given by [25, Theorem 12.3.2, p. 387]

$$2^k V_n(d-1) = 2^k \sum_{i=0}^{d-1} \binom{n}{i}, \quad 2 \le d \le \frac{n}{2}, \quad d \le d_{\min}, \quad d, k, n \in \mathbb{N}^*, \tag{1.316}$$

which must be lower than the total number of possible receive vectors 2^n, $n \in \mathbb{N}^*$ yielding [25, Theorem 12.3.2, p. 387]

$$2^k \sum_{i=0}^{d-1} \binom{n}{i} < 2^n \quad \Leftrightarrow \quad \sum_{i=0}^{d-1} \binom{n}{i} < 2^{n-k}, \quad 2 \le d \le \frac{n}{2}, \quad d \le d_{\min}, \quad d, k, n \in \mathbb{N}^*, \tag{1.317}$$

so there must be at least one point, $\boldsymbol{w} \in \mathbb{F}_2^n$, $n \in \mathbb{N}^*$, say, that is not in any decoding sphere [25, Theorem 12.3.2, p. 388]. Let $\boldsymbol{v} \in \mathbb{V} \subseteq \mathbb{F}_2^n$, $n \in \mathbb{N}^*$ be a codeword. Then $(\boldsymbol{v} \oplus \boldsymbol{w})$ must be in the same coset as the codeword $\boldsymbol{v} \in \mathbb{V} \subseteq \mathbb{F}_2^n$, $n \in \mathbb{N}^*$ and also cannot be in any decoding sphere because if it were then $\boldsymbol{w} \in \mathbb{F}_2^n$, $n \in \mathbb{N}^*$ had to be in a decoding sphere around another codeword, $\boldsymbol{v}' \in \mathbb{V} \subseteq \mathbb{F}_2^n$, $n \in \mathbb{N}^*$, say [25, Theorem 12.3.2, p. 388]. Then there must be a linear binary block code, which is spanned by $\mathbb{V}$ and $\boldsymbol{w} \in \mathbb{F}_2^n$, $n \in \mathbb{N}^*$, which is larger than $\mathbb{V}$, having the minimum distance of at least d [25, Theorem 12.3.2, p. 388]. Thus, any $(n, k, d_{\min})$ binary linear block code $\mathbb{V}$, which does not comply with (1.314), can be made larger by appending another basis vector [25, Theorem 12.3.2, p. 388]. □

Remark 1.27 (On the Gilbert Bound for Binary Linear Codes). Equation (1.314) can be expressed as a lower bound on the number of codewords [27, p. 19]

$$2^k \ge \frac{2^n}{\sum_{i=0}^{d-1} \binom{n}{i}}, \quad 2 \le d \le \frac{n}{2}, \quad d \le d_{\min}, \quad d, k, n \in \mathbb{N}^*. \tag{1.318}$$

Using the binary entropy function $\mathrm{H}_2\{p\}$ of (1.216), we can consider the following quite useful theorem.

Theorem 1.18 (Bound Using the Binary Entropy Function). *The following bound holds* [5, *Satz A.1, p. 433f.*]

$$\sum_{i=0}^{np} \binom{n}{i} \leq 2^{n\mathrm{H}_2\{p\}}, \quad p \in \left[0, \frac{1}{2}\right], \quad n \in \mathbb{N}^*. \tag{1.319}$$

Proof. Using the binomial theorem given by (1.196) [11, equations (1.36a), (1.36b) and (1.36c), p. 12], we obtain [5, Satz A.1, p. 433f.]

$$1 = (p + 1 - p)^n = \sum_{i=0}^{n} \binom{n}{i} (1-p)^{n-i} p^i \geq (1-p)^n \cdot \sum_{i=0}^{np} \binom{n}{i} \left(\frac{p}{1-p}\right)^i, \tag{1.320}$$

$$p \in \left[0, \frac{1}{2}\right], \quad n \in \mathbb{N}^*.$$

Since $i \leq np$ and [5, Satz A.1, p. 433f.]

$$\frac{p}{1-p} \leq 1 \quad \Rightarrow \quad \left(\frac{p}{1-p}\right)^i \geq \left(\frac{p}{1-p}\right)^{np}, \quad p \in \left[0, \frac{1}{2}\right], \quad n \in \mathbb{N}^*, \tag{1.321}$$

(1.320) becomes

$$1 \geq (1-p)^n \cdot \sum_{i=0}^{np} \binom{n}{i} \left(\frac{p}{1-p}\right)^{np} = \left[p^p (1-p)^{1-p}\right]^n \cdot \sum_{i=0}^{np} \binom{n}{i}, \tag{1.322}$$

and, therefore, we obtain

$$1 \geq 2^{-n\mathrm{H}_2\{p\}} \cdot \sum_{i=0}^{np} \binom{n}{i}, \quad p \in \left[0, \frac{1}{2}\right], \quad n \in \mathbb{N}^*. \tag{1.323}$$

□

Setting out from Theorem 1.18 and (1.18), we obtain

$$\frac{1}{n} \log_2 \left\{ \sum_{i=0}^{np} \binom{n}{i} \right\} \leq \mathrm{H}_2\{p\}, \quad p \in \left[0, \frac{1}{2}\right], \quad n \in \mathbb{N}^*. \tag{1.324}$$

Remark 1.28 (On the Gilbert Bound (1.314)). Using (1.319), the Gilbert bound (1.314) [25, Theorem 12.3.2, p. 387f.] can be given by

$$2^{n\mathrm{H}_2\{\frac{d-1}{n}\}} \geq 2^{n-k}, \quad 2 \leq d \leq \frac{n}{2}, \quad d, k, n \in \mathbb{N}^*, \tag{1.325}$$

which immediately leads to

$$n\mathrm{H}_2\left\{\frac{d-1}{n}\right\} \geq n - k \quad \Leftrightarrow \quad \mathrm{H}_2\left\{\frac{d-1}{n}\right\} \geq 1 - \underbrace{\frac{k}{n}}_{=R}, \quad 2 \leq d \leq \frac{n}{2}, \quad d, k, n \in \mathbb{N}^*, \tag{1.326}$$

thus [25, p. 387]

$$R \geq 1 - \mathrm{H}_2\left\{\frac{d-1}{n}\right\}, \quad 2 \leq d \leq \frac{n}{2}, \quad d, k, n \in \mathbb{N}^*, \tag{1.327}$$

which is also termed the *binary Gilbert bound* [25, p. 386f.].

With $d_{\min} \geq d$ (cf. Theorem 1.17) and, consequently, $d_{\min} \leq n/2$, we can use the following approximation:

$$\mathrm{H}_2\left\{\frac{d-1}{n}\right\} \leq \mathrm{H}_2\left\{\frac{d_{\min}}{n}\right\}, \quad n \in \mathbb{N}^*, \tag{1.328}$$

because $\mathrm{H}_2\{x\}$ is monotonically increasing to $0 \leq x \leq 1/2$. Therefore, (1.327) becomes [5, equation (3.4.5), p. 85]

$$R \geq 1 - \mathrm{H}_2\left\{\frac{d_{\min}}{n}\right\}, \quad n \in \mathbb{N}^*. \tag{1.329}$$

Remark 1.29 (On the Hamming Bound (1.302)). Setting

$$np = t \quad \Leftrightarrow \quad p = \frac{t}{n}, \quad \frac{t}{n} \leq \frac{1}{2}, \quad n, t \in \mathbb{N}^*, \tag{1.330}$$

and using (1.324), we obtain

$$\mathrm{H}_2\left\{\frac{t}{n}\right\} \geq \frac{1}{n} \log_2\left\{\sum_{i=0}^{t} \binom{n}{i}\right\}, \quad \frac{t}{n} \leq \frac{1}{2}, \quad n, t \in \mathbb{N}^*. \tag{1.331}$$

In the case of $n \to \infty$, the bound of (1.331) becomes an equality [5, Satz A.1, equation (A.2.7), p. 433].

Therefore, the logarithmic form (1.307) of the Hamming bound discussed in Theorem 1.16 can be written as

$$1 - \underbrace{\frac{k}{n}}_{=R} \geq \mathrm{H}_2\left\{\frac{t}{n}\right\} \quad \Leftrightarrow \quad 1 - R \geq \mathrm{H}_2\left\{\frac{t}{n}\right\}, \quad \frac{t}{n} \leq \frac{1}{2}, \quad n \gg 1 \quad n, t \in \mathbb{N}^*, \tag{1.332}$$

which immediately leads to

$$R \leq 1 - \mathrm{H}_2\left\{\frac{t}{n}\right\}, \quad \frac{t}{n} \leq \frac{1}{2}, \quad n \gg 1, \quad n, t \in \mathbb{N}^*. \tag{1.333}$$

Since

$$t = \left\lfloor \frac{d_{\min} - 1}{2} \right\rfloor \approx \frac{d_{\min}}{2} \tag{1.334}$$

and, therefore,

$$\mathrm{H}_2\left\{\frac{t}{n}\right\} \approx \mathrm{H}_2\left\{\frac{d_{\min}}{2n}\right\}, \quad \frac{t}{n} \leq \frac{1}{2}, \quad n \gg 1, \quad n, t \in \mathbb{N}^* \tag{1.335}$$

hold, (1.333) becomes [5, equation (3.4.2), p. 84]

$$R \leq 1 - \mathrm{H}_2\left\{\frac{d_{\min}}{2n}\right\}, \quad n \gg 1, \quad d_{\min}, n \in \mathbb{N}^*. \tag{1.336}$$

Table 1.5: Summary of the bounds on the number 2^{n-k}, $k, n \in \mathbb{N}^*$ of parity check sequences.

Singleton bound (1.143)	$2^{n-k} \geq 2^{d_{\min}-1}$
Plotkin bound (1.155)	$2^{n-k} \geq 4^{d_{\min}-1}/d_{\min}$
Improved Plotkin bound (1.159)	$R \leq 1 - \frac{2d_{\min}}{n} = 1 - 2\delta$
Hamming bound (1.302)	$2^{n-k} \geq \sum_{i=0}^{t} \binom{n}{i}, t = \lfloor (d_{\min} - 1)/2 \rfloor$
Gilbert–Varshamov bound (1.201), combined with Gilbert bound for binary linear codes (1.318)	$\sum_{i=0}^{d-1} \binom{n}{i} \geq 2^{n-k} > \sum_{i=0}^{d_{\min}-2} \binom{n-1}{i}$, $2 \leq d \leq \frac{n}{2}, d \leq d_{\min}$

Table 1.5 summarizes

- the Singleton bound (1.143),
- the Plotkin bound (1.155) and the improved Plotkin bound in the form of (1.159),
- the Hamming bound (1.302) and
- the Gilbert–Varshamov bound (1.201), combined with the Gilbert bound for binary linear codes (1.318)

on the number 2^{n-k}, $k, n \in \mathbb{N}^*$ of parity check sequences.

Figure 1.18 illustrates

- the Singleton upper bound in the form of (1.141),
- the improved Plotkin upper bound in the form of (1.159),

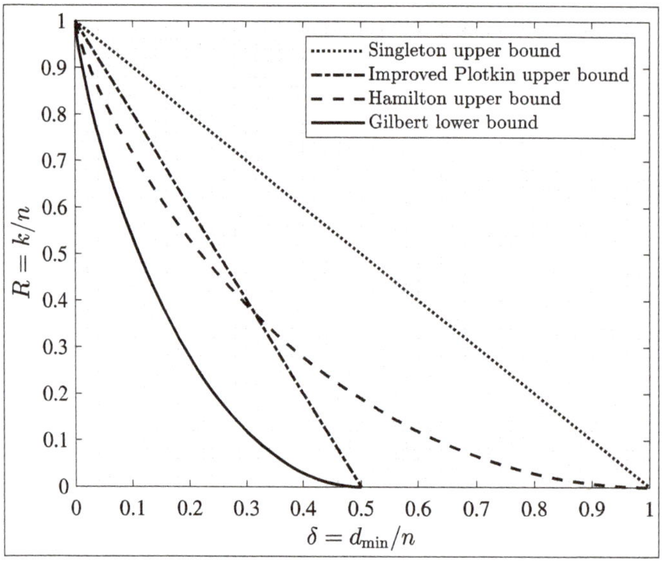

Figure 1.18: Singleton upper bound in the form of (1.141), improved Plotkin upper bound in the form of (1.159), Hamming upper bound in the form of (1.336) and Gilbert lower bound, also termed Gilbert–Varshamov lower bound, in the form of (1.329).

- the Hamming upper bound in the form of (1.336) and
- the Gilbert lower bound, also termed Gilbert–Varshamov lower bound, in the form of (1.329)

versus δ equal to $d_{\min}/n$.

1.8 Syndrome Decoding of an $(n, k, d_{\min})$ Binary Linear Block Code $\mathbb{V}$

Over the past decades, a multitude of decoding schemes for $(n, k, d_{\min})$ binary linear block codes $\mathbb{V}$ have been proposed; cf., e. g., [15, pp. 82–90, 395–447], [14, pp. 96–99], [23, pp. 161–225], [5, pp. 124f., 393–395]. In the case of a general $(n, k, d_{\min})$ binary linear block code $\mathbb{V}$, none of the proposed decoding schemes proves to be efficient [5, p. 124]. Therefore, we will restrict ourselves to a straightforward but simple decoding scheme termed *syndrome decoding*, although syndrome decoding has however also only limited practical relevance, because its realization is still too complex for a powerful $(n, k, d_{\min})$ binary linear block code $\mathbb{V}$ [5, p. 124].

Syndrome decoding sets out from the receive vector $\boldsymbol{r} \in \mathbb{F}_2^n, n \in \mathbb{N}^*$. Therefore, the components $r_j \in \mathbb{F}_2, j \in \{0, 1 \cdots (n-1)\}, n \in \mathbb{N}^*$ of the receive vector $\boldsymbol{r} \in \mathbb{F}_2^n, n \in \mathbb{N}^*$ can only assume the values 0 or 1. This means that the data detector, which carries out the demodulation prior to the channel decoding, restricts its output samples to values taken from $\mathbb{F}_2$ [5, Bild 1.2, p. 8f.]. The said output samples of the data detector are the components $r_j \in \mathbb{F}_2, j \in \{0, 1 \cdots (n-1)\}, n \in \mathbb{N}^*$ of the receive vector $\boldsymbol{r} \in \mathbb{F}_2^n, n \in \mathbb{N}^*$ [5, Bild 1.2, p. 8f.].

Obviously, the components $r_j \in \mathbb{F}_2, j \in \{0, 1 \cdots (n-1)\}, n \in \mathbb{N}^*$ generated by the data detector are taken from the same set $\mathbb{F}_2$ as the components $v_j \in \mathbb{F}_2, j \in \{0, 1 \cdots (n-1)\}, n \in \mathbb{N}^*$ of the transmit codeword $\boldsymbol{v} \in \mathbb{V} \subseteq \mathbb{F}_2^n, n \in \mathbb{N}^*$. Since v_j and $r_j, j \in \{0, 1 \cdots (n-1)\}, n \in \mathbb{N}^*$ are taken from the same set, in our case from $\mathbb{F}_2$, the data detector is said to perform *hard decision detection* [5, pp. 9, 124]. Therefore, syndrome decoding is considered to be a *hard decision decoding* scheme [5, p. 124].

Definition 1.35 (Hard Decision Decoding). A decoding scheme is termed a *hard decision decoding* scheme if and only if the components $v_j \in \mathbb{F}_2, j \in \{0, 1 \cdots (n-1)\}, n \in \mathbb{N}^*$ of the transmit codeword $\boldsymbol{v} \in \mathbb{V} \subseteq \mathbb{F}_2^n$, $n \in \mathbb{N}^*$ and the components $r_j \in \mathbb{F}_2, j \in \{0, 1 \cdots (n-1)\}, n \in \mathbb{N}^*$ of the receive vector $\boldsymbol{r} \in \mathbb{F}_2^n, n \in \mathbb{N}^*$ are taken from the same set, e. g., from $\mathbb{F}_2$ [5, Bild 1.2, p. 8f.].

In the case of hard decision decoding, that particular codeword $\hat{\boldsymbol{v}} \in \mathbb{V} \subseteq \mathbb{F}_2^n, n \in \mathbb{N}^*$ shall be *decoded*, i. e., found and output by the channel decoder, which has the least Hamming distance $d_{\mathrm{H}}\{\hat{\boldsymbol{v}}, \boldsymbol{r}\}$ from the receive vector $\boldsymbol{r} \in \mathbb{F}_2^n, n \in \mathbb{N}^*$, [5, p. 124]. This is *maximum-likelihood decoding (MLD)* [5, Satz 1.2, Satz 1.3, pp. 23, 124].

Theorem 1.19 (Maximum-Likelihood Decoding (MLD)). *Let $\mathbf{v} \in \mathbb{V} \subseteq \mathbb{F}_2^n$, $n \in \mathbb{N}^*$ be the transmit codeword and let $\hat{\mathbf{v}} \in \mathbb{V} \subseteq \mathbb{F}_2^n$, $n \in \mathbb{N}^*$ be the decoded codeword put out by the channel decoder. The error probability $\mathsf{P}\{\hat{\mathbf{v}} \neq \mathbf{v}\}$ after the decoding becomes minimal if decoding is done as follows* [5, *Satz 1.2, p. 23*]. *Setting out from the receive vector $\boldsymbol{r} \in \mathbb{F}_2^n$, $n \in \mathbb{N}^*$ that particular codeword $\hat{\mathbf{v}} \in \mathbb{V} \subseteq \mathbb{F}_2^n$, $n \in \mathbb{N}^*$ is decoded, for which the likelihood function $\mathsf{P}\{\boldsymbol{r} \mid \hat{\mathbf{v}}\}$ assumes its maximum* [5, *Satz 1.2, p. 23*].

Proof. Let $\mathsf{P}\{\mathbf{v}\}$ the a priori probability for the transmission of the transmit codeword $\mathbf{v} \in \mathbb{V} \subseteq \mathbb{F}_2^n, n \in \mathbb{N}^*$. Furthermore, let us assume that this a priori probability $\mathsf{P}\{\mathbf{v}\}$ does not depend on the transmit codeword $\mathbf{v} \in \mathbb{V} \subseteq \mathbb{F}_2^n, n \in \mathbb{N}^*$, i. e.,

$$\mathsf{P}\{\mathbf{v}\} = \frac{1}{2^k} = 2^{-k}, \quad k \in \mathbb{N}^*. \tag{1.337}$$

This means that all admissible codewords of the $(n, k, d_{\min})$ binary linear block code $\mathbb{V}$ have the same probability of occurrence at the transmitter [5, p. 22].

The error probability $\mathsf{P}\{\hat{\mathbf{v}} \neq \mathbf{v}\}$ can be expressed by [5, p. 22]

$$\begin{aligned}
\mathsf{P}\{\hat{\mathbf{v}} \neq \mathbf{v}\} &= \sum_{\mathbf{v}\in\mathbb{V}} \mathsf{P}\big\{\{\hat{\mathbf{v}} \neq \mathbf{v}\}\{\mathbf{v}\}\big\} \\
&= \sum_{\mathbf{v}\in\mathbb{V}} \mathsf{P}\{\hat{\mathbf{v}} \neq \mathbf{v} \mid \mathbf{v}\}\mathsf{P}\{\mathbf{v}\} \\
&= \frac{1}{2^k} \sum_{\mathbf{v}\in\mathbb{V}} \underbrace{\Bigg(\sum_{\substack{\boldsymbol{r}\\ \hat{\mathbf{v}}\neq\mathbf{v}}} \mathsf{P}\{\boldsymbol{r} \mid \mathbf{v}\}\Bigg)}_{=\mathsf{P}\{\hat{\mathbf{v}}\neq\mathbf{v} \mid \mathbf{v}\}} \\
&= \underbrace{\frac{1}{2^k} \sum_{\mathbf{v}\in\mathbb{V},\boldsymbol{r}} \mathsf{P}\{\boldsymbol{r} \mid \mathbf{v}\}}_{=1} - \frac{1}{2^k} \sum_{\substack{\mathbf{v}\in\mathbb{V},\boldsymbol{r}\\ \hat{\mathbf{v}}\equiv\mathbf{v}}} \mathsf{P}\{\boldsymbol{r} \mid \mathbf{v}\} \\
&= 1 - \frac{1}{2^k} \sum_{\boldsymbol{r}} \mathsf{P}\{\boldsymbol{r} \mid \hat{\mathbf{v}}\}, \quad \boldsymbol{r} \in \mathbb{F}_2^n, \quad \mathbf{v}, \hat{\mathbf{v}} \in \mathbb{V} \subseteq \mathbb{F}_2^n, \quad k, n \in \mathbb{N}^*.
\end{aligned} \tag{1.338}$$

Minimizing $\mathsf{P}\{\hat{\mathbf{v}} \neq \mathbf{v}\}$ hence requires the maximization of $\sum_{\boldsymbol{r}} \mathsf{P}\{\boldsymbol{r} \mid \hat{\mathbf{v}}\}$, which is certainly obtained for every sum term $\mathsf{P}\{\boldsymbol{r} \mid \hat{\mathbf{v}}\}$ being maximal for its prevailing receive vector $\boldsymbol{r} \in \mathbb{F}_2^n$, $n \in \mathbb{N}^*$ [5, p. 22]. The maximization of $\mathsf{P}\{\boldsymbol{r} \mid \hat{\mathbf{v}}\}$ is obtained by appropriately choosing $\hat{\mathbf{v}} \in \mathbb{V} \subseteq \mathbb{F}_2^n, n \in \mathbb{N}^*$. Therefore, we obtain

$$\hat{\mathbf{v}} = \arg\max_{\mathbf{v}\in\mathbb{V}}\big\{\mathsf{P}\{\boldsymbol{r} \mid \mathbf{v}\}\big\}, \quad \boldsymbol{r} \in \mathbb{F}_2^n, \quad \mathbf{v}, \hat{\mathbf{v}} \in \mathbb{V} \subseteq \mathbb{F}_2^n, \quad k, n \in \mathbb{N}^*. \tag{1.339}$$

□

At the output of the data detector, the prevailing components $r_j \in \mathbb{F}_2, j \in \{0, 1 \cdots (n-1)\}$, $n \in \mathbb{N}^*$ of the receive vector $\boldsymbol{r} \in \mathbb{F}_2^n$, $n \in \mathbb{N}^*$ can be considered independent [23, p. 173], [5, p. 23]. Therefore, the likelihood function takes the form [5, equation (1.6.7), p. 23]

$$\begin{aligned}
\mathsf{P}\{\boldsymbol{r} \mid \hat{\mathbf{v}}\} &= \prod_{j=0}^{n-1} \mathsf{P}\{r_j \mid \hat{v}_j\}, \\
& \boldsymbol{r} \in \mathbb{F}_2^n, \quad \hat{\mathbf{v}} \in \mathbb{V} \subseteq \mathbb{F}_2^n, \quad r_j, \hat{v}_j \in \mathbb{F}_2, \quad j \in \{0, 1 \cdots (n-1)\}, \quad n \in \mathbb{N}^*.
\end{aligned} \tag{1.340}$$

Assuming that the received vector $\boldsymbol{r} \in \mathbb{F}_2^n$, $n \in \mathbb{N}^*$ and the decoded codeword $\hat{\mathbf{v}} \in \mathbb{V} \subseteq \mathbb{F}_2^n$, $n \in \mathbb{N}^*$ differ in exactly

$$d_{\mathrm{H}}\{\boldsymbol{r}, \hat{\mathbf{v}}\} = t, \quad \boldsymbol{r} \in \mathbb{F}_2^n, \quad \hat{\mathbf{v}} \in \mathbb{V} \subseteq \mathbb{F}_2^n, \quad n, t \in \mathbb{N}^*, \tag{1.341}$$

components and taking the transmission over the binary symmetric channel (BSC) with crossover probability p into account, the likelihood function of (1.340) can be written as [23, p. 173], [5, equation (1.6.7), p. 23]

$$\begin{aligned} \mathsf{P}\{\boldsymbol{r} \mid \hat{\boldsymbol{v}}\} &= (1-p)^{n-t} \cdot p^t \\ &= (1-p)^n \cdot \frac{p^t}{(1-p)^t} \\ &= (1-p)^n \cdot \left(\frac{p}{1-p}\right)^{d_{\mathrm{H}}\{\boldsymbol{r},\hat{\boldsymbol{v}}\}}, \qquad (1.342) \\ \boldsymbol{r} \in \mathbb{F}_2^n,\quad \hat{\boldsymbol{v}} &\in \mathbb{V} \subseteq \mathbb{F}_2^n,\quad p \in [0, \tfrac{1}{2}[,\quad t = d_{\mathrm{H}}\{\boldsymbol{r},\hat{\boldsymbol{v}}\},\quad n, t \in \mathbb{N}^*. \end{aligned}$$

Considering (1.342), $\mathsf{P}\{\boldsymbol{r} \mid \hat{\boldsymbol{v}}\}$ assumes its maximum for a minimum Hamming distance $d_{\mathrm{H}}\{\boldsymbol{r},\hat{\boldsymbol{v}}\}$ between the receive vector $\boldsymbol{r} \in \mathbb{F}_2^n, n \in \mathbb{N}^*$ and the decoded codeword $\hat{\boldsymbol{v}} \in \mathbb{V} \subseteq \mathbb{F}_2^n, n \in \mathbb{N}^*$ [5, Satz 1.3, equation (1.6.8), p. 23].

Theorem 1.20 (Hard Decision Maximum-Likelihood Decoding). *The error probability* $\mathsf{P}\{\hat{\boldsymbol{v}} \neq \boldsymbol{v}\}$ *after the decoding becomes minimal if decoding is done as follows* [5, *Satz 1.3, p. 23*]. *Setting out from the receive vector* $\boldsymbol{r} \in \mathbb{F}_2^n, n \in \mathbb{N}^*$ *choose that particular decoded codeword* $\hat{\boldsymbol{v}} \in \mathbb{V} \subseteq \mathbb{F}_2^n, n \in \mathbb{N}^*$, *for which the Hamming distance* $d_H\{\boldsymbol{r},\hat{\boldsymbol{v}}\}$ *is lower than or equal to* $d_H\{\boldsymbol{r},\boldsymbol{w}\}$ *for any other codeword* $\boldsymbol{w} \in \mathbb{V} \subseteq \mathbb{F}_2^n, n \in \mathbb{N}^*$ [5, *Satz 1.3, equation (1.6.8), p. 23*]

$$d_H\{\boldsymbol{r},\hat{\boldsymbol{v}}\} \le d_H\{\boldsymbol{r},\boldsymbol{w}\}\ \forall\, \boldsymbol{w} \in \mathbb{V} \subseteq \mathbb{F}_2^n,\quad \boldsymbol{r} \in \mathbb{F}_2^n,\quad \hat{\boldsymbol{v}} \in \mathbb{V} \subseteq \mathbb{F}_2^n,\quad n \in \mathbb{N}^*. \qquad (1.343)$$

Proof. The proof is given by (1.340), (1.341) and (1.342). □

Therefore, syndrome decoding is considered to be a *hard decision maximum-likelihood decoding* scheme [5, Satz 4.12, p. 124].

The first step taken by syndrome decoding is the partitioning of the $2^n, n \in \mathbb{N}^*$, possible receive vectors $\boldsymbol{r} \in \mathbb{F}_2^n, n \in \mathbb{N}^*$, into $2^{n-k}, k, n \in \mathbb{N}^*$, disjoint subsets $\mathbb{D}^{(0)}, \mathbb{D}^{(1)} \cdots \mathbb{D}^{(2^{n-k}-1)}, k, n \in \mathbb{N}^*$, each subset $\mathbb{D}^{(m)}, m \in \{0, 1 \cdots (2^{n-k}-1)\}, k, n \in \mathbb{N}^*$, containing $2^k, k \in \mathbb{N}^*$, n-tuples [15, p. 82], [14, p. 103–107], [23, p. 17f.], [5, p. 122f.]. A subset $\mathbb{D}^{(m)}, m \in \{0, 1 \cdots (2^{n-k}-1)\}, k, n \in \mathbb{N}^*$, is termed *coset* (in German "*Nebenklasse*") [5, p. 122f.].

Let us define the $2^{n-k}, k, n \in \mathbb{N}^*$, cosets $\mathbb{D}^{(m)}, m \in \{0, 1 \cdots (2^{n-k}-1)\}, k, n \in \mathbb{N}^*$. Equation (1.212) gives rise to the fact that there are $2^{n-k}, k, n \in \mathbb{N}^*$, different syndromes $\boldsymbol{s}^{(m)} \in \mathbb{F}_2^{n-k}, m \in \{0, 1 \cdots (2^{n-k}-1)\}, k, n \in \mathbb{N}^*$. With all the error patterns $\boldsymbol{e} \in \mathbb{F}_2^n, n \in \mathbb{N}^*$, which lead to the same syndrome $\boldsymbol{s}^{(m)}$ taken from $\mathbb{F}_2^{n-k}, m \in \{0, 1 \cdots (2^{n-k}-1)\}, k, n \in \mathbb{N}^*$, the coset $\mathbb{D}^{(m)}, m \in \{0, 1 \cdots (2^{n-k}-1)\}, k, n \in \mathbb{N}^*$, is given by [5, equation (4.6.3), p. 122]

$$\mathbb{D}^{(m)} = \{\boldsymbol{e} \in \mathbb{F}_2^n \mid \boldsymbol{e}\boldsymbol{H}^{\mathrm{T}} = \boldsymbol{s}^{(m)}\},\quad m \in \{0, 1 \cdots (2^{n-k}-1)\},\quad k, n \in \mathbb{N}^*. \qquad (1.344)$$

Setting

$$\boldsymbol{s}^{(0)} = \boldsymbol{0}_{n-k} \in \mathbb{F}_2^{n-k}, \quad k, n \in \mathbb{N}^*, \tag{1.345}$$

which is the syndrome of every codeword, we yield [5, p. 122]

$$\mathbb{D}^{(0)} = \mathbb{V}, \quad \mathbb{V} \subseteq \mathbb{F}_2^n, \quad n \in \mathbb{N}^*. \tag{1.346}$$

This means that $\mathbb{D}^{(0)}$ of (1.346) consists of all admissible codewords of the $(n, k, d_{\min})$ binary linear block code $\mathbb{V}$ [5, p. 122]. Furthermore, all the other $(2^{n-k} - 1)$, $k, n \in \mathbb{N}^*$, cosets $\mathbb{D}^{(m)}$, $m \in \{1, 2 \cdots (2^{n-k} - 1)\}$, $k, n \in \mathbb{N}^*$, do not contain any admissible codewords [5, p. 122].

Remark 1.30 (Disjoint Cosets). Setting out from (1.344), it will become clear why we claimed that the 2^{n-k}, $k, n \in \mathbb{N}^*$, cosets $\mathbb{D}^{(m)}$ with $m \in \{0, 1 \cdots (2^{n-k} - 1)\}$, $k, n \in \mathbb{N}^*$, are disjoint. The reason for this claim is the fact that a particular error pattern $\boldsymbol{e} \in \mathbb{F}_2^n$, $n \in \mathbb{N}^*$, cannot have more than one syndrome.

Consider two distinct error patterns, $\boldsymbol{e} \in \mathbb{F}_2^n$, $n \in \mathbb{N}^*$ and $\tilde{\boldsymbol{e}} \in \mathbb{F}_2^n$, $n \in \mathbb{N}^*$, say, that are both elements of an arbitrarily chosen coset $\mathbb{D}^{(m)}$, $m \in \{0, 1 \cdots (2^{n-k} - 1)\}$, $k, n \in \mathbb{N}^*$, i. e., [5, p. 122]

$$\boldsymbol{e}, \tilde{\boldsymbol{e}} \in \mathbb{D}^{(m)}, \quad m \in \{0, 1 \cdots (2^{n-k} - 1)\}, \quad k, n \in \mathbb{N}^*. \tag{1.347}$$

Equation (1.344) requires [5, p. 122]

$$\boldsymbol{e}\boldsymbol{H}^{\mathrm{T}} = \tilde{\boldsymbol{e}}\boldsymbol{H}^{\mathrm{T}}, \quad \boldsymbol{e}, \tilde{\boldsymbol{e}} \in \mathbb{D}^{(m)}, \quad m \in \{0, 1 \cdots (2^{n-k} - 1)\}, \quad k, n \in \mathbb{N}^*, \tag{1.348}$$

which yields [5, p. 122]

$$(\boldsymbol{e} - \tilde{\boldsymbol{e}})\boldsymbol{H}^{\mathrm{T}} = \boldsymbol{0}_{n-k}, \quad \boldsymbol{e}, \tilde{\boldsymbol{e}} \in \mathbb{D}^{(m)}, \quad m \in \{0, 1 \cdots (2^{n-k} - 1)\}, \quad k, n \in \mathbb{N}^*. \tag{1.349}$$

Obviously, the difference of two elements $\boldsymbol{e}, \tilde{\boldsymbol{e}} \in \mathbb{D}^{(m)}$, $m \in \{0, 1 \cdots (2^{n-k} - 1)\}$, $k, n \in \mathbb{N}^*$, taken from any $\mathbb{D}^{(m)}$, $m \in \{0, 1 \cdots (2^{n-k} - 1)\}$, $k, n \in \mathbb{N}^*$ is always an admissible codeword [5, p. 122]. Selecting a single but arbitrary error pattern $\boldsymbol{e} \in \mathbb{F}_2^n$, $n \in \mathbb{N}^*$, which leads to the syndrome $\boldsymbol{s}^{(m)}$ taken from $\mathbb{F}_2^{n-k}$, $m \in \{0, 1 \cdots (2^{n-k} - 1)\}$, $k, n \in \mathbb{N}^*$ and using all the 2^k, $k \in \mathbb{N}^*$, codewords $\boldsymbol{v} \in \mathbb{V} \subseteq \mathbb{F}_2^n$, $n \in \mathbb{N}^*$, we can thus rewrite (1.344) in the following way [5, equation (4.6.4), p. 122]

$$\mathbb{D}^{(m)} = \{\boldsymbol{e} \oplus \boldsymbol{v} \,|\, \boldsymbol{v} \in \mathbb{V},\ \boldsymbol{e}\boldsymbol{H}^{\mathrm{T}} = \boldsymbol{s}^{(m)}\}, \quad m \in \{0, 1 \cdots (2^{n-k} - 1)\}, \quad k, n \in \mathbb{N}^*. \tag{1.350}$$

That particular error pattern $\boldsymbol{e}^{(m)}$, $m \in \{0, 1 \cdots (2^{n-k} - 1)\}$, $k, n \in \mathbb{N}^*$, with the minimum Hamming weight of all 2^k, $k \in \mathbb{N}^*$, elements of the coset $\mathbb{D}^{(m)}$, $m \in \{0, 1 \cdots (2^{n-k} - 1)\}$, $k, n \in \mathbb{N}^*$, is called the *coset leader* (in German "*Anführer der Nebenklasse*"). Note that $\boldsymbol{e}^{(m)}$, $m \in \{0, 1 \cdots (2^{n-k} - 1)\}$, of a coset $\mathbb{D}^{(m)}$, $m \in \{0, 1 \cdots (2^{n-k} - 1)\}$, $k, n \in \mathbb{N}^*$, does not have to be unique [5, p. 123]. Using $\boldsymbol{e}^{(m)}$, $m \in \{0, 1 \cdots (2^{n-k} - 1)\}$, (1.350) is given by [5, equation (4.6.7), p. 123]

$$\mathbb{D}^{(m)} = \{\boldsymbol{e}^{(m)} \oplus \boldsymbol{v} \mid \boldsymbol{v} \in \mathbb{V},\ \boldsymbol{e}^{(m)}\boldsymbol{H}^{\mathrm{T}} = \boldsymbol{s}^{(m)}\}, \quad m \in \{0, 1 \cdots (2^{n-k} - 1)\}, \quad k, n \in \mathbb{N}^*. \tag{1.351}$$

Clearly,

$$\boldsymbol{e}^{(0)} = \boldsymbol{0}_n, \quad n \in \mathbb{N}^*, \tag{1.352}$$

holds [5, p. 123]. However, in some cases, there are several coset leader candidates, i. e., error patterns with the same smallest Hamming weight, in a coset [5, p. 123]. In this case, an arbitrary selection of one of these coset leader candidates will do [5, p. 123].

Let us clarify the definitions (1.344), (1.350) and (1.351) in what follows.

Theorem 1.21 (Syndrome of the 2^k Elements of a Coset $\mathbb{D}^{(m)}$). *All the 2^k n-tuples of any coset $\mathbb{D}^{(m)}$, $m \in \{0, 1 \cdots (2^{n-k} - 1)\}$, $k, n \in \mathbb{N}^*$ have the same syndrome* [15, *Theorem 3.6, p. 86*]. *The syndromes for different cosets are different* [15, *Theorem 3.6, p. 86*].

Proof. Using (1.351), each element of the coset $\mathbb{D}^{(m)}$, $m \in \{0, 1 \cdots (2^{n-k} - 1)\}$, $k, n \in \mathbb{N}^*$ has the syndrome

$$\left(\boldsymbol{e}^{(m)} \oplus \boldsymbol{v}\right)\boldsymbol{H}^{\mathrm{T}} = \boldsymbol{e}^{(m)}\boldsymbol{H}^{\mathrm{T}}, \quad \boldsymbol{v} \in \mathbb{V}, \quad m \in \left\{0, 1 \cdots \left(2^{n-k} - 1\right)\right\}, \quad k, n \in \mathbb{N}^*. \tag{1.353}$$

Hence, the syndrome of any element in the coset $\mathbb{D}^{(m)}$, $m \in \{0, 1 \cdots (2^{n-k} - 1)\}$, $k, n \in \mathbb{N}^*$ is equal to the syndrome of the coset leader [15, Theorem 3.6, p. 86]. Therefore, all the vectors of a coset have the same syndrome [15, Theorem 3.6, p. 86].

Now, let $\boldsymbol{e}^{(m)}$ be the coset leader of $\mathbb{D}^{(m)}$ and let $\boldsymbol{e}^{(\mu)}$, $m < \mu$, be the coset leader of $\mathbb{D}^{(\mu)}$, respectively [15, Theorem 3.6, p. 86]. Hence, $\boldsymbol{e}^{(m)}$ and $\boldsymbol{e}^{(\mu)}$ are not in the same coset. Let us assume that the syndromes of these two cosets $\mathbb{D}^{(m)}$ and $\mathbb{D}^{(\mu)}$ are equal [15]. Then

$$\boldsymbol{e}^{(m)}\boldsymbol{H}^{\mathrm{T}} = \boldsymbol{e}^{(\mu)}\boldsymbol{H}^{\mathrm{T}}, \tag{1.354}$$

which yields [15, Theorem 3.6, p. 86]

$$\left(\boldsymbol{e}^{(m)} \oplus \boldsymbol{e}^{(\mu)}\right)\boldsymbol{H}^{\mathrm{T}} = \boldsymbol{0}_{n-k}, \quad k, n \in \mathbb{N}^* \tag{1.355}$$

implying that $(\boldsymbol{e}^{(m)} \oplus \boldsymbol{e}^{(\mu)})$ is a codeword of the $(n, k, d_{\min})$ binary linear block code $\mathbb{V}$ [15, Theorem 3.6, p. 86]. Thus, [15, Theorem 3.6, p. 86]

$$\boldsymbol{e}^{(\mu)} = \boldsymbol{e}^{(m)} \oplus \underbrace{\left(\boldsymbol{e}^{(m)} \oplus \boldsymbol{e}^{(\mu)}\right)}_{\in \mathbb{V}} \tag{1.356}$$

and, therefore, $\boldsymbol{e}^{(m)}$ must be in $\mathbb{D}^{(\mu)}$, which contradicts the construction rule of the cosets [15, Theorem 3.6, p. 86]. □

Using the notation introduced by the great German mathematician Felix *Hausdorff* in his seminal textbook about *set theory* (in German *Mengenlehre*) [30, pp. 14, 17, 18], the family of all the 2^{n-k}, $k, n \in \mathbb{N}^*$, disjoint cosets $\mathbb{D}^{(m)}$, $m \in \{0, 1 \cdots (2^{n-k} - 1)\}$, $k, n \in \mathbb{N}^*$, with

$$\mathbb{D}^{(0)} + \mathbb{D}^{(1)} + \cdots + \mathbb{D}^{(2^{n-k}-1)} = \mathop{\mathfrak{S}}_{m=0}^{2^{n-k}-1} \mathbb{D}^{(m)} = \mathbb{F}_2^n, \quad k, n \in \mathbb{N}^*, \tag{1.357}$$

is called *standard array* (in German "*Nebenklassenzerlegung*") [5, Definition 4.8, p. 123]. Since the 2^{n-k} cosets $\mathbb{D}^{(m)}$, $m \in \{0, 1 \cdots (2^{n-k} - 1)\}$, $k, n \in \mathbb{N}^*$, and their linear superposition results in $\mathbb{F}_2^n$, $n \in \mathbb{N}^*$, according to (1.357), are mutually disjoint they represent a unique *partition*, respectively, a unique *decomposition*, (in German "Zerlegung") of $\mathbb{F}_2^n$, $n \in \mathbb{N}^*$ [5, equation (4.6.6), p. 122].

Let us summarize our findings in the following definitions.

Definition 1.36 (Coset). The disjoint subsets $\mathbb{D}^{(m)}$, $m \in \{0, 1 \cdots (2^{n-k} - 1)\}$, $k, n \in \mathbb{N}^*$, defined by (1.344), (1.350) and (1.351), which fulfill (1.357), are termed *cosets* (in German "*Nebenklassen*") [5, p. 122f.].

Definition 1.37 (Coset Leader). Let $\boldsymbol{e}^{(m)} \in \mathbb{D}^{(m)}$, $m \in \{0, 1 \cdots (2^{n-k} - 1)\}$, $k, n \in \mathbb{N}^*$ be an error pattern with the smallest Hamming weight of all 2^k, $k \in \mathbb{N}^*$, elements of the coset $\mathbb{D}^{(m)}$, $m \in \{0, 1 \cdots (2^{n-k} - 1)\}$, $k, n \in \mathbb{N}^*$. When selected, this particular error pattern is termed the *coset leader* (in German "*Anführer der Nebenklasse*") [5, p. 123]. With this coset leader, the coset $\mathbb{D}^{(m)}$, $m \in \{0, 1 \cdots (2^{n-k} - 1)\}$, $k, n \in \mathbb{N}^*$ is represented by (1.351) [5, equation (4.6.7), p. 123].

Definition 1.38 (Standard Array). The family of all 2^{n-k}, $k, n \in \mathbb{N}^*$, disjoint subsets $\mathbb{D}^{(m)}$, $m \in \{0, 1 \cdots (2^{n-k} - 1)\}$, $k, n \in \mathbb{N}^*$ defined by (1.344), (1.350) and (1.351), which fulfill (1.357), is termed *standard array* (in German "*Nebenklassenzerlegung*") [5, p. 122f.].

The standard array has the following structure (cf., e. g., [15, Figure 3.6, p. 82]):

coset, syndrome	coset leader, column matrix $\boldsymbol{D}^{(0)}$	$\boldsymbol{D}^{(1)}$	$\cdots$	$\boldsymbol{D}^{(l)}$	$\cdots$	$\boldsymbol{D}^{(2^k-1)}$
$\mathbb{D}^{(0)} = \mathbb{V}$, $\boldsymbol{s}^{(0)}$	$\boldsymbol{e}^{(0)} = \boldsymbol{v}^{(0)} = \boldsymbol{0}_n$	$\boldsymbol{v}^{(1)}$	$\cdots$	$\boldsymbol{v}^{(l)}$	$\cdots$	$\boldsymbol{v}^{(2^k-1)}$
$\mathbb{D}^{(1)}$, $\boldsymbol{s}^{(1)}$	$\boldsymbol{e}^{(1)}$	$\boldsymbol{e}^{(1)} \oplus \boldsymbol{v}^{(1)}$	$\cdots$	$\boldsymbol{e}^{(1)} \oplus \boldsymbol{v}^{(l)}$	$\cdots$	$\boldsymbol{e}^{(1)} \oplus \boldsymbol{v}^{(2^k-1)}$
$\vdots$	$\vdots$	$\vdots$	$\vdots$	$\vdots$	$\vdots$	$\vdots$
$\mathbb{D}^{(m)}$, $\boldsymbol{s}^{(m)}$	$\boldsymbol{e}^{(m)}$	$\boldsymbol{e}^{(m)} \oplus \boldsymbol{v}^{(1)}$	$\cdots$	$\boldsymbol{e}^{(m)} \oplus \boldsymbol{v}^{(l)}$	$\cdots$	$\boldsymbol{e}^{(m)} \oplus \boldsymbol{v}^{(2^k-1)}$
$\vdots$	$\vdots$	$\vdots$	$\vdots$	$\vdots$	$\vdots$	$\vdots$
$\mathbb{D}^{(2^{n-k}-1)}$, $\boldsymbol{s}^{(2^{n-k}-1)}$	$\boldsymbol{e}^{(2^{n-k}-1)}$	$\boldsymbol{e}^{(2^{n-k}-1)} \oplus \boldsymbol{v}^{(1)}$	$\cdots$	$\boldsymbol{e}^{(2^{n-k}-1)} \oplus \boldsymbol{v}^{(l)}$	$\cdots$	$\boldsymbol{e}^{(2^{n-k}-1)} \oplus \boldsymbol{v}^{(2^k-1)}$

(1.358)

Theorem 1.22 (Unique Elements in a Standard Array). *No two n-tuples in the same row of a standard array are identical. Every n-tuple appears in one and only one row* [15].

Proof. A particular row, m, $m \in \{0,1\cdots(2^{n-k}-1)\}$, $k,n \in \mathbb{N}^*$, say, forms the coset $\mathbb{D}^{(m)}$, $m \in \{0,1\cdots(2^{n-k}-1)\}$, $k,n \in \mathbb{N}^*$. Let $\boldsymbol{e}^{(m)} \in \mathbb{D}^{(m)}$, $m \in \{0,1\cdots(2^{n-k}-1)\}$, $k,n \in \mathbb{N}^*$, be the coset leader of this coset $\mathbb{D}^{(m)}$, $m \in \{0,1\cdots(2^{n-k}-1)\}$, $k,n \in \mathbb{N}^*$.

Furthermore, let $\boldsymbol{v}^{(l)} \in \mathbb{V} \subseteq \mathbb{F}_2^n$ and $\boldsymbol{v}^{(\lambda)} \in \mathbb{V} \subseteq \mathbb{F}_2^n$, $l,\lambda \in \{0,1\cdots(2^k-1)\}$, $k,n \in \mathbb{N}^*$, be two distinct codewords, i. e.,

$$\boldsymbol{v}^{(l)} \neq \boldsymbol{v}^{(\lambda)} \text{ for } l \neq \lambda, \quad \boldsymbol{v}^{(l)},\boldsymbol{v}^{(\lambda)} \in \mathbb{V} \subseteq \mathbb{F}_2^n, \quad l,\lambda \in \left\{0,1\cdots\left(2^k-1\right)\right\}, \quad k,n \in \mathbb{N}^*. \tag{1.359}$$

Now, let us assume that two arbitrarily chosen n-tuples $(\boldsymbol{e}^{(m)} \oplus \boldsymbol{v}^{(l)}) \in \mathbb{D}^{(m)}$ and $(\boldsymbol{e}^{(m)} \oplus \boldsymbol{v}^{(\lambda)}) \in \mathbb{D}^{(m)}$ say, are identical [15, Theorem 3.3, p. 82f.]

$$\begin{aligned} \boldsymbol{e}^{(m)} \oplus \boldsymbol{v}^{(l)} = \boldsymbol{e}^{(m)} \oplus \boldsymbol{v}^{(\lambda)}, \\ \boldsymbol{e}^{(m)} \in \mathbb{D}^{(m)} \subseteq \mathbb{F}_2^n, \quad m \in \left\{0,1\cdots\left(2^{n-k}-1\right)\right\}, \\ \boldsymbol{v}^{(l)},\boldsymbol{v}^{(\lambda)} \in \mathbb{V} \subseteq \mathbb{F}_2^n, \quad l,\lambda \in \left\{0,1\cdots\left(2^k-1\right)\right\}, \quad k,n \in \mathbb{N}^*, \end{aligned} \tag{1.360}$$

which implies

$$\boldsymbol{v}^{(l)} = \boldsymbol{v}^{(\lambda)}, \quad \boldsymbol{v}^{(l)},\boldsymbol{v}^{(\lambda)} \in \mathbb{V} \subseteq \mathbb{F}_2^n, \quad l,\lambda \in \left\{0,1\cdots\left(2^k-1\right)\right\}, \quad k,n \in \mathbb{N}^*. \tag{1.361}$$

This contradicts the assumption that $\boldsymbol{v}^{(l)} \in \mathbb{V} \subseteq \mathbb{F}_2^n$ and $\boldsymbol{v}^{(\lambda)} \in \mathbb{V} \subseteq \mathbb{F}_2^n$, $l,\lambda \in \{0,1\cdots(2^k-1)\}$, $k,n \in \mathbb{N}^*$ are distinct; cf. (1.359) [15, p. 83].

Now, let us assume that one and the same n-tuple appears at least twice in the coset $\mathbb{D}^{(m)}$, $m \in \{0,1\cdots(2^{n-k}-1)\}$, $k,n \in \mathbb{N}^*$. This, however, contradicts the definition of a coset; cf., e. g., (1.351). □

Furthermore, let us consider the selection of the coset leaders in more detail.

Theorem 1.23 (Selection of the Coset Leaders). *For an $(n,k,d_{\min})$ binary linear block code $\mathbb{V}$, all the n-tuples of weight t equal to $\lfloor (d_{\min}-1)/2 \rfloor$ or less can be used as coset leaders of a standard array of the $(n,k,d_{\min})$ binary linear block code $\mathbb{V}$* [15]. *If all the n-tuples of weight t or less are used as coset leaders, there is at least one n-tuple of weight $(t+1)$ that cannot be used as a coset leader* [15, *Theorem 3.5, p. 85f.*].

Proof. Since the minimum distance of the $(n,k,d_{\min})$ binary linear block code $\mathbb{V}$ is $d_{\min}$, the minimum weight of the $(n,k,d_{\min})$ binary linear block code $\mathbb{V}$ is also $d_{\min}$ [15, Theorem 3.5, p. 85f.].

Let $\boldsymbol{x}$ and $\boldsymbol{y}$ be two n-tuples of weight t or less [15, Theorem 3.5, pp. 85f.], i. e.,

$$w_{\mathrm{H}}\{\boldsymbol{x}\} \leq t \leq \left\lfloor \frac{d_{\min}-1}{2} \right\rfloor, \quad w_{\mathrm{H}}\{\boldsymbol{y}\} \leq t \leq \left\lfloor \frac{d_{\min}-1}{2} \right\rfloor. \tag{1.362}$$

Clearly, the weight $w_{\mathrm{H}}\{\boldsymbol{x}\oplus\boldsymbol{y}\}$ of $(\boldsymbol{x}\oplus\boldsymbol{y})$ is upper bounded by the triangle inequality (cf. Theorem 1.4) [15, Theorem 3.5, p. 85f.]

$$w_{\mathrm{H}}\{\boldsymbol{x}\oplus\boldsymbol{y}\} \leq w_{\mathrm{H}}\{\boldsymbol{x}\} + w_{\mathrm{H}}\{\boldsymbol{y}\} \leq 2\left\lfloor \frac{d_{\min}-1}{2} \right\rfloor \leq d_{\min}-1 < d_{\min}. \tag{1.363}$$

If $\boldsymbol{x}$ and $\boldsymbol{y}$ are in the same coset, $(\boldsymbol{x}\oplus\boldsymbol{y})$ has to be a nonzero codeword of the $(n,k,d_{\min})$ binary linear block code $\mathbb{V}$ [15, Theorem 3.5, p. 85f.], which, however, is impossible because according to (1.363) the Hamming weight $w_{\mathrm{H}}\{\boldsymbol{x}\oplus\boldsymbol{y}\}$ of $(\boldsymbol{x}\oplus\boldsymbol{y})$ is less than the minimum weight $d_{\min}$ [15, Theorem 3.5, p. 85f.].

Therefore, no two n-tuples of weight t or less can be in the same coset of an (n, k, d_{min}) binary linear block code $\mathbb{V}$. Thus, all the n-tuples of weight t or less can be used as coset leaders [15, Theorem 3.5, p. 85f.].

Now, let $\boldsymbol{v}$ be a minimum weight codeword of the (n, k, d_{min}) binary linear block code $\mathbb{V}$, i. e., [15, Theorem 3.5, p. 85f.]

$$w_{\mathrm{H}}\{\boldsymbol{v}\} = d_{\mathrm{min}}. \tag{1.364}$$

Furthermore, let $\boldsymbol{x}$ and $\boldsymbol{y}$ be two n-tuples that satisfy the following conditions [15, Theorem 3.5, p. 85f.]:

a) $\boldsymbol{x} \oplus \boldsymbol{y} = \boldsymbol{v}$;
b) $\boldsymbol{x}$ and $\boldsymbol{y}$ do not have nonzero components in common places.

Obviously, $\boldsymbol{x}$ and $\boldsymbol{y}$ must be in the same coset with

$$w_{\mathrm{H}}\{\boldsymbol{x}\} + w_{\mathrm{H}}\{\boldsymbol{y}\} = w_{\mathrm{H}}\{\boldsymbol{v}\} = d_{\mathrm{min}} \tag{1.365}$$

[15, Theorem 3.5, p. 85f.]. Now, assume that $\boldsymbol{y}$ has $w_{\mathrm{H}}\{\boldsymbol{y}\}$ equal to $(t+1)$ [15, Theorem 3.5, p. 85f.]. Owing to [15, Theorem 3.5, p. 85f.]

$$2t + 1 \leq d_{\mathrm{min}} \leq 2t + 2, \tag{1.366}$$

$w_{\mathrm{H}}\{\boldsymbol{x}\}$ must be either equal to t or $(t+1)$ [15, Theorem 3.5, p. 85f.]. Therefore, if $\boldsymbol{x}$ is chosen to be a coset leader, then $\boldsymbol{y}$ must not be a coset leader [15, Theorem 3.5, p. 85f.]. □

The standard array (cf. (1.358)) is composed of 2^k, $k \in \mathbb{N}^*$ disjoint columns, each column containing 2^{n-k}, $k, n \in \mathbb{N}^*$ n-tuples [15, p. 83]. That particular n-tuple in the topmost row of the standard array (cf. (1.358)) is a codeword of the (n, k, d_{min}) binary linear block code $\mathbb{V}$ [15, p. 83]. Hence, the lth, $l \in \{0, 1 \cdots (2^k - 1)\}$, $k \in \mathbb{N}^*$ column of the standard array (cf. (1.358)) is represented by the column matrix

$$\boldsymbol{D}^{(l)} = \begin{pmatrix} \boldsymbol{v}^{(l)} \\ \boldsymbol{e}^{(1)} \oplus \boldsymbol{v}^{(l)} \\ \boldsymbol{e}^{(2)} \oplus \boldsymbol{v}^{(l)} \\ \vdots \\ \boldsymbol{e}^{(m)} \oplus \boldsymbol{v}^{(l)} \\ \vdots \\ \boldsymbol{e}^{(2^{n-k}-1)} \oplus \boldsymbol{v}^{(l)} \end{pmatrix}, \tag{1.367}$$

$$l \in \{0, 1 \cdots (2^k - 1)\}, \quad m \in \{0, 1 \cdots (2^{n-k} - 1)\}, \quad k, n \in \mathbb{N}^*,$$

the error patterns $\boldsymbol{e}^{(m)}$, $m \in \{0, 1 \cdots (2^{n-k} - 1)\}$, $k, n \in \mathbb{N}^*$ being the coset leaders [15, equation (3.27), p. 83f.].

Let us assume that the transmit codeword is denoted by $\boldsymbol{v}^{(l)}$ $l \in \{0, 1 \cdots (2^k - 1)\}$, $k \in \mathbb{N}^*$, [15, p. 84]. At the output of the noisy transmission channel, we obtain the receive vector $\boldsymbol{r} \in \mathbb{F}_2^n$, $n \in \mathbb{N}^*$, which represents a row of the column matrix $\boldsymbol{D}^{(l)}$, $l \in \{0, 1 \cdots (2^k - 1)\}$, $k \in \mathbb{N}^*$, if the error pattern caused by the noisy transmission channel is a coset leader [15, p. 84]. Then the receive vector $\boldsymbol{r}$ will be decoded correctly.

However, if the error pattern caused by the noisy transmission channel is not a coset leader, the decoding will be erroneous [15, p. 84]. Assume that the noisy transmission channel causes the error pattern $\boldsymbol{e}_x$, which must be in some coset and in some column matrix, however, not in $\boldsymbol{D}^{(l)}$, $l \in \{0, 1 \cdots (2^k - 1)\}$, $k \in \mathbb{N}^*$ [15, p. 84]. Assume that $\boldsymbol{e}_x$ is in the jth coset $\mathbb{D}^{(j)}$, $j \in \{0, 1 \cdots (2^{n-k} - 1)\}$, $k, n \in \mathbb{N}^*$ and in the ith column matrix $\boldsymbol{D}^{(i)}$, $i \in \{0, 1 \cdots (2^k - 1)\}$, $k \in \mathbb{N}^*$, [15, p. 84]. Then we have [15, p. 84]

$$\boldsymbol{e}_x = \boldsymbol{e}^{(j)} \oplus \boldsymbol{v}^{(i)}, \quad i \in \{0, 1 \cdots (2^k - 1)\}, \quad j \in \{0, 1 \cdots (2^{n-k} - 1)\}, \quad k \in \mathbb{N}^*, \tag{1.368}$$

and the receive vector becomes [15, p. 84]

$$\begin{aligned} \boldsymbol{r} = \boldsymbol{v}^{(l)} \oplus \boldsymbol{e}_x &= \boldsymbol{v}^{(l)} \oplus (\boldsymbol{e}^{(j)} \oplus \boldsymbol{v}^{(i)}) = \boldsymbol{e}^{(j)} \oplus (\boldsymbol{v}^{(l)} \oplus \boldsymbol{v}^{(i)}), \\ &i, l \in \{0, 1 \cdots (2^k - 1)\}, \quad j \in \{0, 1 \cdots (2^{n-k} - 1)\}, \quad k \in \mathbb{N}^*, \end{aligned} \tag{1.369}$$

$(\boldsymbol{v}^{(l)} \oplus \boldsymbol{v}^{(i)})$ being a codeword, which is not equal to $\boldsymbol{v}^{(l)}$ if $\boldsymbol{v}^{(i)}$ is nonzero [15, p. 84].

Therefore, the decoding is correct if and only if the error pattern caused by the channel is a coset leader [15, p. 84]. The 2^{n-k}, $k, n \in \mathbb{N}^*$ coset leaders $\boldsymbol{e}^{(m)}$, $m \in \{0, 1 \cdots (2^{n-k} - 1)\}$, $k, n \in \mathbb{N}^*$, including the zero vector $\boldsymbol{0}_n$ are therefore called the *correctable error patterns* [15, p. 84].

Theorem 1.24 (Correction Capability of an $(n, k, d_{\min})$ binary linear block code $\mathbb{V}$). *Every $(n, k, d_{\min})$ binary linear block code $\mathbb{V}$ is capable of correcting 2^{n-k} error patterns* [15, *Theorem 3.4, p. 84*].

Proof. The proof is given in the discussion above. □

Theorem 1.25 (Hard Decision Maximum-Likelihood Decoding Revisited). *Let us set out from an $(n, k, d_{\min})$ binary linear block code $\mathbb{V}$ having the 2^{n-k} cosets $\mathbb{D}^{(m)}$, $m \in \{0, 1 \cdots (2^{n-k} - 1)\}$, $k, n \in \mathbb{N}^*$, with the coset leaders $\boldsymbol{e}^{(m)}$, $m \in \{0, 1 \cdots (2^{n-k} - 1)\}$, $k, n \in \mathbb{N}^*$, each coset leader having the minimum Hamming weight of all the 2^k, $k \in \mathbb{N}^*$ elements in its respective coset. Furthermore, let $\boldsymbol{r} \in \mathbb{F}_2^n$, $n \in \mathbb{N}^*$ be the receive vector.*

If the receive vector $\boldsymbol{r} \in \mathbb{F}_2^n$, $n \in \mathbb{N}^*$ is an element of a particular coset $\mathbb{D}^{(m)}$, $m \in \{0, 1 \cdots (2^{n-k} - 1)\}$, $k, n \in \mathbb{N}^*$, with the coset leader $\boldsymbol{e}^{(m)} \in \mathbb{D}^{(m)} \subseteq \mathbb{F}_2^n$, $m \in \{0, 1 \cdots (2^{n-k} - 1)\}$, $k, n \in \mathbb{N}^*$ the decoded codeword $\hat{\boldsymbol{v}} \in \mathbb{V} \subseteq \mathbb{F}_2^n$, $n \in \mathbb{N}^*$, obtained from hard decision maximum-likelihood decoding, *is given by* [5, *Satz 4.12, p. 124*]

$$\begin{aligned} &\hat{\boldsymbol{v}} = \boldsymbol{r} \oplus \boldsymbol{e}^{(m)}, \\ &\quad \hat{\boldsymbol{v}} \in \mathbb{V} \subseteq \mathbb{F}_2^n, \quad \boldsymbol{r}, \boldsymbol{e}^{(m)} \in \mathbb{D}^{(m)} \subseteq \mathbb{F}_2^n, \quad m \in \left\{0, 1 \cdots \left(2^{n-k} - 1\right)\right\}, \quad n \in \mathbb{N}^*. \end{aligned} \tag{1.370}$$

Proof. Let $\boldsymbol{w} \in \mathbb{V} \subseteq \mathbb{F}_2^n$, $n \in \mathbb{N}^*$ be an arbitrary codeword [5, Satz 4.12, p. 124]. Since $\hat{\boldsymbol{v}} \in \mathbb{V} \subseteq \mathbb{F}_2^n$, $n \in \mathbb{N}^*$ also $(\hat{\boldsymbol{v}} \oplus \boldsymbol{w})$ is a codeword [5, Satz 4.12, p. 124].

Therefore, we have [5, Satz 4.12, p. 124]

$$\boldsymbol{e}^{(m)} \oplus (\hat{\boldsymbol{v}} \oplus \boldsymbol{w}) \in \mathbb{D}^{(m)} \subseteq \mathbb{F}_2^n, \quad m \in \left\{0, 1 \cdots \left(2^{n-k} - 1\right)\right\}, \quad n \in \mathbb{N}^*. \tag{1.371}$$

Since the Hamming weight $w_{\mathrm{H}}\{\boldsymbol{e}^{(m)}\}$, $m \in \{0, 1 \cdots (2^{n-k} - 1)\}$, $k, n \in \mathbb{N}^*$ of the coset leader is minimum, we yield [5, Satz 4.12, p. 124]

$$w_H\{e^{(m)}\} \le w_H\{e^{(m)} \oplus (\hat{v} \oplus w)\}, \quad m \in \{0, 1 \cdots (2^{n-k} - 1)\}, \quad n \in \mathbb{N}^*. \tag{1.372}$$

Since for the coset leader, the following identity holds [5, Satz 4.12, p. 124]:

$$e^{(m)} = r \oplus \hat{v}, \tag{1.373}$$
$$e^{(m)}, r \in \mathbb{D}^{(m)} \subseteq \mathbb{F}_2^n, \quad \hat{v} \in \mathbb{V} \subseteq \mathbb{F}_2^n, \quad m \in \{0, 1 \cdots (2^{n-k} - 1)\}, \quad n \in \mathbb{N}^*,$$

Equation (1.372) becomes [5, Satz 4.12, p. 124]

$$d_H\{r, \hat{v}\} = w_H\{r \oplus \hat{v}\} \le w_H\{r \oplus \hat{v} \oplus (\hat{v} \oplus w)\} = w_H\{r \oplus w\} = d_H\{r, w\}. \tag{1.374}$$

Therefore, hard decision maximum-likelihood decoding is obtained; cf. Theorem 1.20 [5, Satz 4.12, p. 124]. □

The *standard array decoding* can be accomplished as follows [5, p. 124]:
1. Find that particular coset $\mathbb{D}^{(m)}$, $m \in \{0, 1 \cdots (2^{n-k} - 1)\}$, $k, n \in \mathbb{N}^*$, which contains the received vector $r \in \mathbb{F}_2^n$, $n \in \mathbb{N}^*$.
2. Use the coset leader $e^{(m)} \in \mathbb{D}^{(m)} \subseteq \mathbb{F}_2^n$, $m \in \{0, 1 \cdots (2^{n-k} - 1)\}$, $k, n \in \mathbb{N}^*$ to compute (1.370).

The *syndrome decoding* is a computationally simplified version of the standard array decoding, which is achieved as follows [5, p. 124]:
1. Create a table with 2^{n-k}, $k, n \in \mathbb{N}^*$, elements $(s^{(m)}, e^{(m)})$.
2. Compute the syndrome rH^{T} of the receive vector $r \in \mathbb{F}_2^n$, $n \in \mathbb{N}^*$.
3. Find the coset leader $e^{(m)} \in \mathbb{D}^{(m)} \subseteq \mathbb{F}_2^n$, $m \in \{0, 1 \cdots (2^{n-k} - 1)\}$, $k, n \in \mathbb{N}^*$, which corresponds to the syndrome rH^{T} in the aforementioned table.
4. Use this coset leader $e^{(m)} \in \mathbb{D}^{(m)} \subseteq \mathbb{F}_2^n$, $m \in \{0, 1 \cdots (2^{n-k} - 1)\}$, $k, n \in \mathbb{N}^*$ to compute (1.370).

1.9 Some Examples of $(n, k, d_{\min})$ Binary Linear Block Codes

1.9.1 Binary Linear Single-Parity Check Code

A binary linear *single-parity check (SPC) code* is a $(k + 1, k, 2)$ binary linear block code $\mathbb{V}$ with a single parity check digit [15, p. 94], which is capable of the detection of a single transmission error. Since the minimum distance $d_{\min}$ is 2, the random-error-correcting capability t is 0, and error correction is not possible. SPC codes are often used for simple error detection [15, p. 94].

The code rate of the single-parity check (SPC) code is

$$R = \frac{k}{k+1}, \quad k \in \mathbb{N}^*. \tag{1.375}$$

The $k \times (k + 1)$ generator matrix in systematic form is given by [15, equation (3.44), p. 94]

$$\boldsymbol{G} = \underbrace{\begin{pmatrix} 1 & 1 & 0 & 0 & 0 & \cdots & 0 \\ 1 & 0 & 1 & 0 & 0 & \cdots & 0 \\ 1 & 0 & 0 & 1 & 0 & \cdots & 0 \\ 1 & 0 & 0 & 0 & 1 & \cdots & 0 \\ \vdots & \vdots & \vdots & \vdots & \vdots & \ddots & \vdots \\ 1 & 0 & 0 & 0 & 0 & & 1 \end{pmatrix}}_{=\mathbf{1}_k^{\mathrm{T}} \quad = \boldsymbol{I}_k} = \left(\mathbf{1}_k^{\mathrm{T}}\ \boldsymbol{I}_k\right), \quad \mathbf{1}_k = \underbrace{(1, 1 \cdots 1)}_{k \text{ components}}, \quad k \in \mathbb{N}^*. \tag{1.376}$$

With the parity digit [15, equation (3.43), p. 94]

$$p = \bigoplus_{i=0}^{k-1} u_i, \quad p, u_i \in \mathbb{F}_2, \quad i \in \{0, 1 \cdots (k-1)\}, \quad k \in \mathbb{N}^*, \tag{1.377}$$

each codeword is of the form [15, p. 94]

$$\boldsymbol{v} = (p, u_0, u_1, u_2, \cdots, u_{k-1}), \quad \boldsymbol{v} \in \mathbb{V} \subset \mathbb{F}_2^{k+1}, \quad p, u_i \in \mathbb{F}_2, \quad i \in \{0, 1 \cdots (k-1)\}, \quad k \in \mathbb{N}^*. \tag{1.378}$$

Since all the codewords of the SPC code have even Hamming weights, the minimum weight, i. e., the minimum distance, $d_{\min}$ is equal to 2 [15, p. 94]. Since all the codewords have even weights, an SPC code is also called an *even parity check code* [15, p. 94].

The $1 \times (k+1)$ parity check matrix of the SPC code is given by [15, equation (3.45), p. 94]

$$\boldsymbol{H} = \mathbf{1}_{k+1}, \quad k \in \mathbb{N}^*. \tag{1.379}$$

1.9.2 Binary Linear Repetition Code

A binary linear *repetition code* is an $(n, 1, n)$ binary linear block code $\mathbb{V}$ that consists of only two codewords, the (all-) zero codeword [15, p. 94]

$$\mathbf{0}_n = \underbrace{(0, 0 \cdots 0)}_{n \text{ zeros}}, \quad n \in \mathbb{N}^*, \tag{1.380}$$

and the (all-) one codeword [15, p. 94]

$$\mathbf{1}_n = \underbrace{(1, 1 \cdots 1)}_{n \text{ ones}}, \quad n \in \mathbb{N}^*. \tag{1.381}$$

The repetition code simply repeats a single message bit $u_0(n-1)$, $n \in \mathbb{N}^*$, times [15, p. 94]. Its minimum distance $d_{\min}$ is equal to the length n, $n \in \mathbb{N}^*$ of the repetition code. Therefore, the repetition code can detect $(n-1), n \in \mathbb{N}^*$ transmission errors. Its random-error-correcting capability t is $\lfloor (n-1)/2 \rfloor$.

The code rate of the repetition code is

$$R = \frac{1}{n}, \quad n \in \mathbb{N}^*. \tag{1.382}$$

The generator matrix of the repetition code is the $1 \times n$ matrix [15, equation (3.46), p. 94],

$$\boldsymbol{G} = \mathbf{1}_n, \quad n \in \mathbb{N}^*. \tag{1.383}$$

Consequently, the $(n-1) \times n$ parity check matrix is given by

$$\boldsymbol{H} = \Bigg(\underbrace{\begin{matrix} 1 \\ 1 \\ 1 \\ 1 \\ \vdots \\ 1 \end{matrix}}_{=\mathbf{1}_{n-1}^{\mathrm{T}}} \; \underbrace{\begin{matrix} 1 & 0 & 0 & 0 & \cdots & 0 \\ 0 & 1 & 0 & 0 & \cdots & 0 \\ 0 & 0 & 1 & 0 & \cdots & 0 \\ 0 & 0 & 0 & 1 & \cdots & 0 \\ \vdots & \vdots & \vdots & \vdots & \ddots & \vdots \\ 0 & 0 & 0 & 0 & & 1 \end{matrix}}_{=\boldsymbol{I}_{n-1}} \Bigg) = \left(\mathbf{1}_{n-1}^{\mathrm{T}} \; \boldsymbol{I}_{n-1}\right), \quad n \in \mathbb{N}^*. \tag{1.384}$$

We readily see from (1.376) and (1.383) as well as from (1.379) and (1.384) that the repetition code and the single-parity check code are dual codes to each other [15, p. 94].

1.9.3 Binary Maximum Distance Separable (MDS) Codes

Theorem 1.26 (Binary Maximum Distance Separable (MDS) Codes). *The only binary MDS codes (cf. Theorem 1.7) are trivial codes, namely*

- *the* $([k+1], k, 2)$ *binary linear single-parity check (SPC) code,* $k \in \mathbb{N}^*$*; cf. Section 1.9.1,*
- *the* $(n, 1, n)$ *binary linear repetition code,* $n \in \mathbb{N}^*$*, with dimension* k *equal to 1 and length* $n \in \mathbb{N}^*$ *(cf. Sect. 1.9.2), and*
- *the* $(k, k, 1)$ *binary linear code without any redundancy,* $k \in \mathbb{N}^*$*,*

[23, *Satz 6.11, p. 139f.*].

Proof. According to of Theorem 1.7, (1.139) becomes

$$d_{\min} = n - k + 1, \quad \Leftrightarrow \quad n - k = d_{\min} - 1, \quad k, n \in \mathbb{N}^*, \tag{1.385}$$

in the case of an MDS code [23, Satz 6.11, p. 140].

In the case of $d_{\min}$ equal to 1, (1.385) yields the correct mathematical statement

$$k - k = 1 - 1, \quad k \in \mathbb{N}^*, \tag{1.386}$$

only in the case of the $(k, k, 1)$ binary linear code without any redundancy, $k \in \mathbb{N}^*$.

Furthermore, in the case of $d_{\min}$ equal to 2, (1.385) yields the correct mathematical statement

$$n - k = 2 - 1, \quad k, n \in \mathbb{N}^*, \tag{1.387}$$

which can only be fulfilled for n equal to $(k+1)$. This is the case for the $([k+1], k, 2)$ binary linear single-parity check (SPC) code, $k \in \mathbb{N}^*$; cf. Section 1.9.1.

In addition, in the case of $d_{\min}$ equal to n, (1.385) yields the correct mathematical statement

$$n - k = n - 1, \quad k, n \in \mathbb{N}^*, \tag{1.388}$$

only for k equal to 1. This can only be fulfilled in the case of the $(n, 1, n)$ binary linear repetition code, $n \in \mathbb{N}^*$, with dimension k equal to 1 and length $n \in \mathbb{N}^*$; cf. Section 1.9.2.

Hence,

- the $([k+1], k, 2)$ binary linear single-parity check (SPC) code, $k \in \mathbb{N}^*$ (cf. Section 1.9.1),
- the $(n, 1, n)$ binary linear repetition code, $n \in \mathbb{N}^*$, with dimension k equal to 1 and length $n \in \mathbb{N}^*$ (cf. Sect. 1.9.2), and
- the $(k, k, 1)$ binary linear code without any redundancy, $k \in \mathbb{N}^*$,

are MDS codes.

Using (1.385), the Gilbert–Varshamov bound (1.201) of Theorem 1.15 becomes

$$\sum_{i=0}^{d_{\min}-2} \binom{n-1}{i} = \sum_{i=0}^{d_{\min}-2} \binom{d_{\min}+k-2}{i} < 2^{d_{\min}-1} = 2^{n-k}, \quad d_{\min}, k, n \in \mathbb{N}^*. \tag{1.389}$$

In the case of $d_{\min}$ equal to 1, (1.389) becomes

$$\underbrace{\sum_{i=0}^{-1} \binom{k-1}{i}}_{=0} < 2^0 = 1, \quad k \in \mathbb{N}^*, \tag{1.390}$$

which is a correct mathematical statement. Hence, the $(k, k, 1)$ binary linear code without any redundancy, $k \in \mathbb{N}^*$, is not only an MDS code but also meets the Gilbert–Varshamov bound (1.201) of Theorem 1.15.

Now, let us consider the case of $d_{\min}$ equal to 2. Equation (1.389) becomes

$$\underbrace{\sum_{i=0}^{0} \binom{k}{i}}_{=1} < 2^1 = 2, \quad k \in \mathbb{N}^*, \tag{1.391}$$

which is a correct mathematical statement. Hence, the $([k+1], k, 2)$ binary linear single-parity check (SPC) code, $k \in \mathbb{N}^*$ (cf. Sect. 1.9.1) is not only an MDS code but also meets the Gilbert–Varshamov bound (1.201) of Theorem 1.15.

Next, let us consider the case of $d_{\min}$ equal to n, $n \in \mathbb{N}^*$. In this case, (1.389) yields

$$\sum_{i=0}^{n-2} \binom{n-1}{i} < \underbrace{\sum_{i=0}^{n-1} \binom{n-1}{i} = 2^{n-1}}_{\text{binomial theorem [11, equations (1.36a), (1.36b) and (1.36c), p. 12]}}, \quad n \in \mathbb{N}^*, \tag{1.392}$$

which is a correct mathematical statement. Hence, the $(n, 1, n)$ binary linear repetition code, $n \in \mathbb{N}^*$, with dimension k equal to 1 and length $n \in \mathbb{N}^*$ (cf. Section 1.9.2) is not only an MDS code but also meets the Gilbert–Varshamov bound (1.201) of Theorem 1.15.

Now, let us consider the case of $d_{\min} \in \{3, 4 \cdots (n-1)\}$, $n \in \mathbb{N}^*$. We will use the binomial theorem [11, equations (1.36a), (1.36b) and (1.36c), p. 12] in the form

$$2^{d_{\min}-1} = (1+1)^{d_{\min}-1} = \sum_{i=0}^{d_{\min}-1} \binom{d_{\min}-1}{i}, \quad d_{\min} \in \{3, 4 \cdots (n-1)\}, \quad n \in \mathbb{N}^*. \tag{1.393}$$

Since

$$\begin{aligned}\binom{d_{\min}-1}{i} &= \frac{(d_{\min}-1)!}{i!(d_{\min}-1-i)!}\\ &= \frac{(d_{\min}-1)!d_{\min}}{i!(d_{\min}-1-i)!d_{\min}}\\ &\le \frac{d_{\min}!}{i!(d_{\min}-1-i)!(d_{\min}-i)}\\ &\le \frac{d_{\min}!}{i!(d_{\min}-i)!}\\ &\le \frac{d_{\min}!(d_{\min}+1)\cdot\dots\cdot(d_{\min}+k-1)}{i!(d_{\min}-i)!(d_{\min}+1)\cdot\dots\cdot(d_{\min}+k-1)}\\ &\le \frac{(d_{\min}+k-1)!}{i!(d_{\min}-i)!(d_{\min}-i+1)\cdot\dots\cdot(d_{\min}-i+k-1)}\\ &\le \frac{(d_{\min}+k-1)!}{i!(d_{\min}+k-1-i)!}\\ &\le \binom{d_{\min}-1+k}{i},\end{aligned} \tag{1.394}$$

$$i \in \{0,1\cdots(d_{\min}-1)\}, \quad d_{\min} \in \{3,4\cdots(n-1)\}, \quad k,n \in \mathbb{N}^*,$$

with equality only for i equal to 0, we can safely state [23, p. 140]

$$2^{d_{\min}-1} = \sum_{i=0}^{d_{\min}-1}\binom{d_{\min}-1}{i} = \sum_{i=0}^{d_{\min}-2}\binom{d_{\min}-1}{i}+1 < \sum_{i=0}^{d_{\min}-2}\binom{d_{\min}-1+k}{i}, \tag{1.395}$$

$$d_{\min} \in \{3,4\cdots(n-1)\}, \quad k,n \in \mathbb{N}^*.$$

Therefore, (1.389) becomes

$$2^{n-k} < \sum_{i=0}^{d_{\min}-2}\binom{d_{\min}+k-1}{i}, \quad d_{\min} \in \{3,4\cdots(n-1)\}, \quad k,n \in \mathbb{N}^*, \tag{1.396}$$

which requires $(n-k) < (d_{\min}-1)$. Hence, the codes with $d_{\min} \in \{3,4\cdots(n-1)\}$, $n \in \mathbb{N}^*$ cannot be trivial. □

1.9.4 Self-Dual Codes

An (n, k) binary linear block code $\mathbb{V}$ that is equal to its dual $(n, [n-k])$ binary linear block code $\mathbb{V}^\perp$ is called a *self-dual code* [15, p. 94]. Clearly, the dimensions of (n, k) binary linear block code $\mathbb{V}$ and $(n, [n-k])$ binary linear block code $\mathbb{V}^\perp$ must be equal [15, p. 94]

$$k = n - k, \quad \Leftrightarrow \quad k = \frac{n}{2}, \quad k,n \in \mathbb{N}^*, \tag{1.397}$$

which requires the length n to be an even number. The code rate of the self-dual code is [15, p. 94]

$$R = \frac{1}{2}. \tag{1.398}$$

Let [15, p. 95]

$$G = (P\ I_{n/2}), \quad n \in \mathbb{N}^*, \tag{1.399}$$

be an $(n/2) \times n$ generator matrix of a self-dual code $\mathbb{V}$. Then G is also a generator of its dual code $\mathbb{V}^{\perp}$, and thus a parity check matrix of $\mathbb{V}$ [15, p. 94]. Then [15, equation (3.47), p. 94]

$$\mathbf{0}_{n/2\times n/2} = G\, G^{\mathrm{T}}, \quad n \in \mathbb{N}^*, \tag{1.400}$$

immediately leading to

$$\mathbf{0}_{n/2\times n/2} = (P\ I_{n/2}) \begin{pmatrix} P^{\mathrm{T}} \\ I_{n/2} \end{pmatrix} = P\, P^{\mathrm{T}} \oplus I_{n/2}\, I_{n/2} = P\, P^{\mathrm{T}} \oplus I_{n/2}, \quad n \in \mathbb{N}^*, \tag{1.401}$$

and, consequently, [15, equation (3.48), p. 95]

$$P\, P^{\mathrm{T}} = I_{n/2}, \quad n \in \mathbb{N}^*.$$

Thus, if an (n, k) binary linear block code $\mathbb{V}$ with code rate R equal to 1/2 satisfies the condition of (1.400), then it is a self-dual code [15, p. 95].

1.9.5 Hamming Codes

Definition 1.39 (Hamming Code). The parity check matrix $H \in \mathbb{F}_2^{h\times[2^h-1]}$, $h \in \{2, 3 \cdots\}$ of a *Hamming code* has $(2^h - 1)$ column vectors $\eta^{(0)} \in \mathbb{F}_2^h, \eta^{(1)} \in \mathbb{F}_2^h, \cdots \eta^{(2^h-2)} \in \mathbb{F}_2^h$ consisting of h elements taken from $\mathbb{F}_2$, which represent all the vectors of $\mathbb{F}_2^h \setminus \{\mathbf{0}_h\}$ [23, Definition 1.16, p. 20], [4, p. 191].

The parameter h, $h \in \{2, 3 \cdots\}$ is termed the *order of the Hamming code* [5, p. 118].

Any two distinct column vectors of $H \in \mathbb{F}_2^{h\times[2^h-1]}$, $h \in \{2, 3 \cdots\}$ are linearly independent and any three distinct column vectors of $H \in \mathbb{F}_2^{h\times[2^h-1]}$, $h \in \{2, 3 \cdots\}$ are linearly dependent [23, Satz 1.17, p. 20]. Therefore, $d_{\min}$ is equal to 3 [23, Satz 1.18, p. 20]. Hence, with $h \in \{2, 3 \cdots\}$, a Hamming code is a $([2^h - 1], [2^h - h - 1], 3)$ binary linear block code $\mathbb{V}$ [23, p. 20], [5, p. 118].

A Hamming code can thus detect 2 transmission errors, and it can correct 1 transmission error [23, p. 20].

Theorem 1.27 (Hamming Codes are Perfect Codes). *Hamming codes are perfect codes* [23, *Satz 1.18, p. 20*].

Proof. Using the parameters

$$n = 2^h - 1, \quad k = \underbrace{2^h - 1}_{=n} - h \ \Rightarrow n - k = h, \quad d_{\min} = 3 \ \Rightarrow t = \left\lfloor \frac{d_{\min} - 1}{2} \right\rfloor = 1, \tag{1.402}$$

of the $([2^h-1],[2^h-h-1],3)$ binary linear block code $\mathbb{V}$ and setting out from (1.302) of Theorem 1.16, we obtain

$$2^h = 2^{n-k} \geq \sum_{i=0}^{1} \binom{n}{i} = 1 + \binom{2^h-1}{1} = 1 + 2^h - 1 = 2^h \quad k,n \in \mathbb{N}^*. \tag{1.403}$$

Therefore, each Hamming code of the order h, $h \in \{2,3\cdots\}$ fulfills (1.302) with equality and is therefore a perfect code according to Definition 1.34. □

1.9.6 Simplex Codes

Definition 1.40 (Simplex Code). The dual code of a Hamming code, being a $([2^h-1],[2^h-h-1],3)$ binary linear block code $\mathbb{V}$, is a *simplex code*, which is a $([2^h-1],h)$ binary linear block code $\mathbb{V}^{\perp}$ with the parity check matrix $\boldsymbol{H} \in \mathbb{F}_2^{h\times[2^h-1]}$, $h \in \{2,3\cdots\}$, of the Hamming code as its generator matrix [23, p. 110], [4, pp. 30, 221], [5, Definition 4.5, p. 119].

Theorem 1.28 (Minimum Distance $d_{\min}$ of a Simplex Code). *Every nonzero codeword of the simplex code with dimension h, $h \in \{2,3\cdots\}$ has the Hamming weight 2^{h-1}, $h \in \{2,3\cdots\}$, [23, p. 110], [5, Definition 4.5, p. 119], [39, Theorem 3.13, p. 33]. Therefore, the minimum distance $d_{\min}$ is also equal to 2^{h-1}, $h \in \{2,3\cdots\}$ [23, p. 110], [5, Definition 4.5, p. 119], [39, Theorem 3.13, p. 33].*

Proof. Let $\boldsymbol{G} \in \mathbb{F}_2^{h\times[2^h-1]}$, $h \in \{2,3\cdots\}$ be a generator matrix of the h-dimensional, $h \in \{2,3\cdots\}$, simplex code [39, Theorem 3.13, p. 33]

$$\mathbb{V}^{\perp} = \left\{\boldsymbol{uG} \mid \boldsymbol{u} \in \mathbb{F}_2^h\right\}, \quad \boldsymbol{G} \in \mathbb{F}_2^{h\times[2^h-1]}, \quad h \in \{2,3\cdots\}. \tag{1.404}$$

The Hamming weight $w_{\mathrm{H}}\{\boldsymbol{uG}\}$ of each nonzero codeword $\boldsymbol{uG} \in \mathbb{V}^{\perp} \setminus \boldsymbol{0}_{2^h-1}$, $h \in \{2,3\cdots\}$ is given by [39, Theorem 3.13, p. 33]

$$w_{\mathrm{H}}\{\boldsymbol{uG}\} = 2^h - 1 - c, \quad h \in \{2,3\cdots\}, \quad c \in \mathbb{N}^*, \tag{1.405}$$

with $c \in \mathbb{N}^*$ being the number of column vectors $\boldsymbol{\eta} \in \mathbb{F}_2^h$, for which

$$\boldsymbol{u\eta}^{\mathrm{T}} = 0 \tag{1.406}$$

holds [39, Theorem 3.13, p. 33].

The vector space $\mathbb{F}_2^h$, $h \in \{2,3\cdots\}$ contains exactly $(h-1)$, $h \in \{2,3\cdots\}$ vectors, which are orthogonal to $\boldsymbol{u} \in \mathbb{F}_2^h$, $h \in \{2,3\cdots\}$ [39, Theorem 3.13, p. 33]. Since those columns $\boldsymbol{\eta} \in \mathbb{F}_2^h$, $h \in \{2,3\cdots\}$, of $\boldsymbol{G} \in \mathbb{F}_2^{h\times[2^h-1]}$, $h \in \{2,3\cdots\}$, which fulfill (1.406) are linear combinations of the said $(h-1)$, $h \in \{2,3\cdots\}$ vectors orthogonal to $\boldsymbol{u} \in \mathbb{F}_2^h$, $h \in \{2,3\cdots\}$, there are exactly c equal to $(2^{h-1}-1)$, $h \in \{2,3\cdots\}$ such columns [39, Theorem 3.13, p. 33].

Using (1.405), we thus find

$$w_{\mathrm{H}}\{\boldsymbol{uG}\} = 2^h - 1 - \left(2^{h-1} - 1\right) = 2^h - 2^{h-1} = 2^{h-1}(2-1) = 2^{h-1}, \quad h \in \{2,3\cdots\}, \tag{1.407}$$

yielding the minimum distance $d_{\min}$ equal to 2^{h-1}, $h \in \{2,3\cdots\}$ [23, p. 110], [5, Definition 4.5, p. 119], [39, Theorem 3.13, p. 33]. □

Hence, every simplex code is a $([2^h-1],h,2^{h-1})$ binary linear block code $\mathbb{V}^{\perp}$.

The expression “simplex” refers to the fact that the $(2^h - 1)$, $h \in \{2, 3 \cdots\}$, nonzero codewords have equal and constant “lengths” d_{min} equal to 2^{h-1}, $h \in \{2, 3 \cdots\}$, and together with the zero vector $\mathbf{0}_{2^h-1}$, $h \in \{2, 3 \cdots\}$, thus form a $(2^h - 1)$-*simplex* in the 2^{2^h-1}-dimensional vector space of all $(2^h - 1)$-tuples [4, p. 31], [40, p. 149].

Just for the record: The simplex codes meet the Griesmer bound, which is a generalization of the Singleton bound [39, pp. 31–33].

In Figure 1.19, the code rate R versus δ equal to d_{min}/n, $n \in \mathbb{N}^*$ is shown for

- the $([k + 1], k, 2)$ binary linear single-parity check (SPC) code, $k \in \mathbb{N}^*$ (cf. Section 1.9.1),
- the $(n, 1, n)$ binary linear repetition code, $n \in \mathbb{N}^*$, with dimension k equal to 1 and length $n \in \mathbb{N}^*$ (cf. Section 1.9.2) and
- the $(k, k, 1)$ binary linear code without any redundancy, $k \in \mathbb{N}^*$.
- the $([2^h - 1], [2^h - h - 1], 3)$ Hamming code, $h \in \{2, 3 \cdots\}$ and
- the $([2^h - 1], h, 2^{h-1})$ simplex code

considered in Section 1.9.

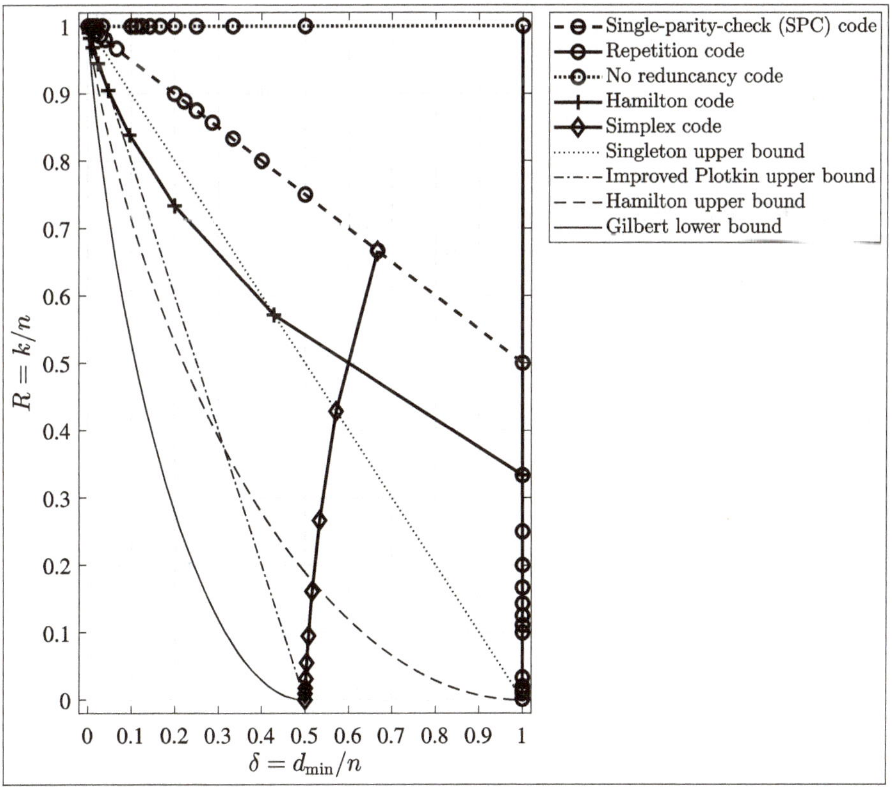

Figure 1.19: Code rate R versus $\delta = d_{min}/n$, $n \in \mathbb{N}^*$ for various binary linear block codes considered in Section 1.9.

1.10 Six Strategies to Construct New Codes from Old

Evolved versions of advanced mobile communications systems must provide backward compatibility to previous and existing systems variants. Meeting this necessity also requires adaptations of such channel coding schemes, which have already been deployed in the predecessors of the evolved versions of advanced mobile communications systems. Therefore, the maxim is the construction of new codes from old. In Section 1.10, six simple strategies to construct new codes from old codes shall be illustrated [4, Chapter 1, Section 9, pp. 27–32].

The first strategy is *adding an overall parity check* [4, Chapter 1, Section 9, p. 27]. Consider an (n, k) binary linear block code $\mathbb{V}$, in which some codewords have odd weight [4, Chapter 1, Section 9, p. 27]. We form a new $([n + 1], k)$ binary linear block code $\mathbb{W}$ by adding

- a 0 at the end of every codeword $\boldsymbol{v} \in \mathbb{W}$ with even weight, and
- a 1 at the end of every codeword $\boldsymbol{v} \in \mathbb{W}$ with odd weight.

[4, Chapter 1, Section 9, p. 27f.] Thus, the new $([n+1], k)$ binary linear block code $\mathbb{W}$ has the property that every codeword has even weight, i. e., it satisfies the new parity check equation

$$v_0 \oplus v_1 \oplus v_2 \oplus \cdots \oplus v_{n-1} \oplus v_n = 0, \quad n \in \mathbb{N}^*, \tag{1.408}$$

which is also called the *"overall" parity check* [4, Chapter 1, Section 9, p. 27]. Since $w_{\mathrm{H}}\{\boldsymbol{v}\}$ is even, the Hamming distance between every pair of codewords is also even [4, Chapter 1, Section 9, p. 27]. If the minimum distance of the (n, k) binary linear block code $\mathbb{V}$ was odd, the minimum distance of the new $([n + 1], k)$ binary linear block code $\mathbb{W}$ is equal to $(d_{\min} + 1)$, which leads to a new $([n + 1], k, [d_{\min} + 1])$ binary linear block code $\mathbb{W}$ [4, Chapter 1, Section 9, p. 27].

Example 1.14. Consider the $(2, 2, 1)$ binary linear block code $\mathbb{V}$ with the 2×2 generator matrix

$$\boldsymbol{G} = \begin{pmatrix} 1 & 0 \\ 0 & 1 \end{pmatrix} = \boldsymbol{I}_k, \tag{1.409}$$

in systematic form. The $(2, 2, 1)$ binary linear block code $\mathbb{V}$ has the following four codewords:

$$\begin{aligned}
\boldsymbol{v}^{(1)} &= \begin{pmatrix} 0 & 0 \end{pmatrix} \begin{pmatrix} 1 & 0 \\ 0 & 1 \end{pmatrix} = \begin{pmatrix} 0 & 0 \end{pmatrix}, \\
\boldsymbol{v}^{(2)} &= \begin{pmatrix} 0 & 1 \end{pmatrix} \begin{pmatrix} 1 & 0 \\ 0 & 1 \end{pmatrix} = \begin{pmatrix} 0 & 1 \end{pmatrix}, \\
\boldsymbol{v}^{(3)} &= \begin{pmatrix} 1 & 0 \end{pmatrix} \begin{pmatrix} 1 & 0 \\ 0 & 1 \end{pmatrix} = \begin{pmatrix} 1 & 0 \end{pmatrix}, \\
\boldsymbol{v}^{(4)} &= \begin{pmatrix} 1 & 1 \end{pmatrix} \begin{pmatrix} 1 & 0 \\ 0 & 1 \end{pmatrix} = \begin{pmatrix} 1 & 1 \end{pmatrix}.
\end{aligned} \tag{1.410}$$

Clearly, d_{min} is equal to 1, which is odd.

Adding an overall parity check yields the (3, 2, 2) binary linear block code $\mathbb{W}$, which is a single-parity check (SPC) code

$$\mathbb{W} = \{(0,0,0), (1,0,1), (1,1,0), (0,1,1)\}, \tag{1.411}$$

which can be generated by using the 2 × 3 generator matrix

$$G^{(\mathbb{W})} = \begin{pmatrix} 1 & 1 & 0 \\ 1 & 0 & 1 \end{pmatrix} = (P\ G) = (P\ I_2), \quad P = 1_2^{\mathrm{T}} = \begin{pmatrix} 1 \\ 1 \end{pmatrix}. \tag{1.412}$$

The new minimum distance is (d_{min} + 1) equal to 2, which is even, just as expected.

Adding check symbols to codewords is generally called *extending a code* [4, Chapter 1, Section 9, p. 27].

Let the (n, k, d_{min}) binary linear block code $\mathbb{V}$ have the parity check matrix $\boldsymbol{H}$. Then the new $([n+1], k, [d_{min}+1])$ binary linear block code $\mathbb{W}$ has the parity check matrix [4, Chapter 1, Section 9, p. 27]

$$H^{(\mathbb{W})} = \begin{pmatrix} 1 & 1 & \cdots & 1 \\ & & & 0 \\ & H & & \vdots \\ & & & 0 \end{pmatrix}. \tag{1.413}$$

Example 1.15. Let us consider the (2, 2, 1) binary linear block code $\mathbb{V}$ of Example 1.14. Its 1 × 3 parity check matrix is given by

$$H^{(\mathbb{W})} = \begin{pmatrix} I_1 & P^{\mathrm{T}} \end{pmatrix} = \begin{pmatrix} 1 & 1 & 1 \end{pmatrix}. \tag{1.414}$$

The dual code spanned by $H^{(\mathbb{W})}$ is the (3, 1, 3) repetition code [15, p. 94], which is also a Hamming code [23, p. 20f.].

The second strategy is *puncturing a code by deleting coordinates*, i. e., components, which is the inverse process to extending a code [4, Chapter 1, Section 9, p. 28f.]. The new punctured code will be denoted by $\mathbb{V}^*$ and is hence a punctured $([n-1], k)$ binary linear block code $\mathbb{V}^*$ [4, Chapter 1, Section 9, p. 29].

Example 1.16. Consider the (3, 2, 2) binary linear block code $\mathbb{V}$, which is a single-parity check (SPC) code having the generator matrix

$$G = \begin{pmatrix} 1 & 1 & 0 \\ 1 & 0 & 1 \end{pmatrix}. \tag{1.415}$$

The (3, 2, 2) binary linear block code $\mathbb{V}$ has the following four codewords:

$$\begin{aligned} \boldsymbol{v}^{(1)} &= \begin{pmatrix} 0 & 0 \end{pmatrix} \begin{pmatrix} 1 & 1 & 0 \\ 1 & 0 & 1 \end{pmatrix} = \begin{pmatrix} 0 & 0 & 0 \end{pmatrix}, \\ \boldsymbol{v}^{(2)} &= \begin{pmatrix} 0 & 1 \end{pmatrix} \begin{pmatrix} 1 & 1 & 0 \\ 1 & 0 & 1 \end{pmatrix} = \begin{pmatrix} 1 & 0 & 1 \end{pmatrix}, \\ \boldsymbol{v}^{(3)} &= \begin{pmatrix} 1 & 0 \end{pmatrix} \begin{pmatrix} 1 & 1 & 0 \\ 1 & 0 & 1 \end{pmatrix} = \begin{pmatrix} 1 & 1 & 0 \end{pmatrix}, \\ \boldsymbol{v}^{(4)} &= \begin{pmatrix} 1 & 1 \end{pmatrix} \begin{pmatrix} 1 & 1 & 0 \\ 1 & 0 & 1 \end{pmatrix} = \begin{pmatrix} 0 & 1 & 1 \end{pmatrix}. \end{aligned} \tag{1.416}$$

Clearly, d_{min} is equal to 2.

Furthermore, we immediately see from (1.416) that the nonzero codewords all have the same "lengths," i. e., the same Hamming weight, d_{min} equal to 2. Therefore, the $(3,2,2)$ binary linear block code $\mathbb{V}$ forms a 3-*simplex*, also called a *tetrahedron*, and is hence a simplex code [4, Figure 1.12, p. 30f.], [40, p. 148f.].

Puncturing this $(3,2,2)$ binary linear block code $\mathbb{V}$ by, e. g., deleting the first coordinate will result in the new $(2,1,1)$ binary linear block code $\mathbb{V}^*$, given by

$$\mathbb{V}^* = \{(0,0),(0,1),(1,0),(1,1)\} \tag{1.417}$$

with the generator matrix

$$\boldsymbol{G}^{(\mathbb{V}^*)} = \begin{pmatrix} 1 & 0 \\ 0 & 1 \end{pmatrix}. \tag{1.418}$$

In general, each time a coordinate is deleted the length n reduces to $(n-1)$ the number of codewords k remains the same and unless we are very lucky the minimum distance d_{min} becomes $(d_{\mathrm{min}}-1)$ [4, Chapter 1, Section 9, p. 29].

The third strategy is *expurgating by throwing away codewords* [4, Chapter 1, Section 9, p. 29]. We will illustrate the most common way in what follows. We will set out from an (n,k,d_{min}) binary linear block code $\mathbb{V}$, containing codewords of both odd and even Hamming weights [4, Chapter 1, Section 9, p. 29]. Half of the codewords have even Hamming weight and the other half have odd Hamming weight [4, Chapter 1, Section 9, p. 29]. We expurgate the (n,k,d_{min}) binary linear block code $\mathbb{V}$ by throwing away those codewords of odd Hamming weight to obtain an $(n,[k-1],d'_{\mathrm{min}})$ binary linear block code $\mathbb{V}'$ [4, Chapter 1, Section 9, p. 29]. Often $d'_{\mathrm{min}} > d_{\mathrm{min}}$, for instance, if d_{min} is odd [4, Chapter 1, Section 9, p. 29].

The fourth strategy is *augmenting by adding new codewords* [4, Chapter 1, Section 9, p. 29]. The most common way to augment a code is by adding the (all-) ones vector $\mathbf{1}_n$, provided it is not already an element of the code [4, Chapter 1, Section 9, p. 29]. This is the same as adding a row of 1's to the generator matrix [4, Chapter 1, Section 9, p. 29]. Let $\boldsymbol{v} \in \mathbb{V} \subseteq \mathbb{F}_2^n$, $n \in \mathbb{N}^*$ be a codeword of the (n,k,d_{min}) binary linear block code $\mathbb{V}$. Assume that the (n,k,d_{min}) binary linear block code $\mathbb{V}$ does not contain the (all-) ones codeword $\mathbf{1}_n$, $n \in \mathbb{N}^*$ [4, Chapter 1, Section 9, p. 29]. Then let all n-tuples of the form $(\mathbf{1}_n \oplus \boldsymbol{v})$, $n \in \mathbb{N}^*$, with $\boldsymbol{v} \in \mathbb{V} \subseteq \mathbb{F}_2^n$, $n \in \mathbb{N}^*$ be the elements of the set $\{\mathbf{1}_n \oplus \mathbb{V}\}$, $n \in \mathbb{N}^*$ [4, Chapter 1, Section 9, p. 29]. The set $\{\mathbf{1}_n \oplus \mathbb{V}\}$, $n \in \mathbb{N}^*$, hence contains all the complements of the codewords $\boldsymbol{v} \in \mathbb{V} \subseteq \mathbb{F}_2^n, n \in \mathbb{N}^*$ [4, Chapter 1, Section 9, p. 29]. Then the augmented version of the $(n, k, d_{\min})$ binary linear block code $\mathbb{V}$ is the augmented $(n, [k+1], d_{\min}^{(\mathbb{W})})$ binary linear block code $\mathbb{W}$ given by [4, Chapter 1, Section 9, p. 29]

$$\mathbb{W} = \mathbb{V} + \{\mathbf{1}_n \oplus \mathbb{V}\} = \mathbb{V} \cup \{\mathbf{1}_n \oplus \mathbb{V}\}, \quad n \in \mathbb{N}^*. \tag{1.419}$$

The minimum distance $d_{\min}^{(\mathbb{W})}$ of the $(n, [k+1], d_{\min}^{(\mathbb{W})})$ binary linear block code $\mathbb{W}$ given by (1.419) is [4, Chapter 1, Section 9, p. 29]

$$d_{\min}^{(\mathbb{W})} = \min\{d_{\min}, (n - d')\}, \quad n \in \mathbb{N}^*, \tag{1.420}$$

d' being the largest weight of any codeword of the $(n, k, d_{\min})$ binary linear block code $\mathbb{V}$ [4, Chapter 1, Section 9, p. 29]. [4, Chapter 1, Section 9, p. 29].

The fifth strategy is *lengthening by adding message symbols* [4, Chapter 1, Section 9, p. 29]. Lengthening a code is readily obtained by [4, Chapter 1, Section 9, p. 29]
- first augmenting it, i. e., by adding the codeword $\mathbf{1}_n$, $n \in \mathbb{N}^*$ and
- then extending it by adding an overall parity check.

This has the effect of adding one more message symbol [4, Chapter 1, Section 9, p. 29].

The sixth and final strategy is *shortening by taking a cross-section* [4, Chapter 1, Section 9, p. 29]. An inverse operation to the lengthening process, illustrated above, is to take the codewords with v_0 equal to 0 and delete the v_0 coordinate, termed *taking a cross-section of the code* [4, Chapter 1, Section 9, p. 29].

1.11 Cyclic $(n, k, d_{\min})$ Binary Linear Block Codes

Cyclic $(n, k, d_{\min})$ binary linear block codes are the most studied of all binary linear block codes, because they are easy to encode and, furthermore, they include the important family of *Bose–Chaudhuri–Hocquenghem (BCH)* codes [4, p. 188]. Furthermore, they are building blocks for many other codes [4, p. 188].

Definition 1.41 (Cyclic $(n, k, d_{\min})$ binary linear block code $\mathbb{V}$). An $(n, k, d_{\min})$ binary linear block code $\mathbb{V}$ is cyclic if it is linear and if any cyclic shift of a codeword is also a codeword [15, Definition 5.1, p. 136], [23, Definition 1.20, p. 23], [5, Definition 5.1, p. 129], [4, p. 188].

According to Definition 1.41, if the n-tuple $\boldsymbol{v} \in \mathbb{V} \subseteq \mathbb{F}_2^n, n \in \mathbb{N}^*$, with

$$\boldsymbol{v} = (v_0, v_1, v_2 \cdots v_{n-2}, v_{n-1}), \quad \boldsymbol{v} \in \mathbb{V} \subseteq \mathbb{F}_2^n, \quad n \in \mathbb{N}^*, \tag{1.421}$$

is a codeword in the cyclic $(n, k, d_{\min})$ binary linear block code $\mathbb{V}$, so is [15, p. 136], [5, p. 129]

$$w = (v_{n-1}, v_0, v_1, v_2 \cdots v_{n-2}), \quad v \in \mathbb{V} \subseteq \mathbb{F}_2^n, \quad v_j \in \mathbb{F}_2, \quad j \in \{0, 1 \cdots (n-1)\}, \quad n \in \mathbb{N}^*. \tag{1.422}$$

Furthermore,

$$\begin{matrix} (v_{n-2}, & v_{n-1}, & v_0 & \cdots & v_{n-4}, & v_{n-3}) & \in \mathbb{V}, \\ (v_{n-3}, & v_{n-2}, & v_{n-1}, & v_0 & \cdots & v_{n-4}) & \in \mathbb{V}, \\ & & & \vdots & & & \\ (v_1 & \cdots & v_{n-3}, & v_{n-2}, & v_{n-1}, & v_0) & \in \mathbb{V}, \end{matrix} \tag{1.423}$$

are codewords in the cyclic $(n, k, d_{\min})$ binary linear block code $\mathbb{V}$ [5, p. 129]. Although (1.421), (1.422) and (1.423) seem to suggest shifting to the right, it is quite obvious that shifting to the left yields the same cyclic $(n, k, d_{\min})$ binary linear block code $\mathbb{V}$ [5, p. 129].

In the case of a cyclic $(n, k, d_{\min})$ binary linear block code $\mathbb{V}$, we may guess that the $k \times n$ generator matrix has the following shape:

$$\begin{aligned} G &= \begin{pmatrix} g_0 & g_1 & \cdots & g_{n-k} & g_{n-k+1} & \cdots & g_{n-1} \\ g_{n-1} & g_0 & \cdots & g_{n-k-1} & g_{n-k} & \cdots & g_{n-2} \\ \vdots & \vdots & \ddots & \vdots & \vdots & \ddots & \vdots \\ g_{n-(k-1)} & g_{n-(k-1)+1} & \cdots & g_0 & g_1 & \cdots & g_{n-k} \end{pmatrix} \\ &= \begin{pmatrix} g_0 & g_1 & \cdots & g_{n-k} & g_{n-k+1} & \cdots & g_{n-1} \\ g_{n-1} & g_0 & \cdots & g_{n-k-1} & g_{n-k} & \cdots & g_{n-2} \\ \vdots & \vdots & \ddots & \vdots & \vdots & \ddots & \vdots \\ g_{n-k+1} & g_{n-k+2} & \cdots & g_0 & g_1 & \cdots & g_{n-k} \end{pmatrix}. \end{aligned} \tag{1.424}$$

Let us now look at two examples.

Example 1.17. The (3, 2, 2) single-parity check (SPC) code

$$\mathbb{V} = \{(0,0,0), (1,1,0), (0,1,1), (1,0,1)\} \tag{1.425}$$

is cyclic [4, p. 189]. Let the message be

$$u = (u_0, u_1), \quad u \in \mathbb{F}_2^2, \quad u_i \in \mathbb{F}_2, \quad i \in \{0, 1\}, \tag{1.426}$$

and the 2×3 generator matrix be

$$G = \begin{pmatrix} 1 & 1 & 0 \\ 0 & 1 & 1 \end{pmatrix}, \quad G \in \mathbb{F}_2^{2\times 3}. \tag{1.427}$$

Then the codeword is given by

$$v = (v_0, v_1, v_2) = uG = (u_0, u_1)\begin{pmatrix} 1 & 1 & 0 \\ 0 & 1 & 1 \end{pmatrix} = (u_0, [u_0 \oplus u_1], u_1), \tag{1.428}$$
$$v \in \mathbb{F}_2^3, \quad u \in \mathbb{F}_2^2, \quad u_i, v_j \in \mathbb{F}_2, \quad i \in \{0,1\}, \quad j \in \{0,1,2\}.$$

We thus obtain the following relationships:

u	v
(0, 0)	(0, 0, 0)
(1, 0)	(1, 1, 0)
(0, 1)	(0, 1, 1)
(1, 1)	(1, 0, 1)

Example 1.18. Let us once more look at the (7, 4, 3) Hamming code

$$\mathbb{V} = \left\{ \begin{array}{llll} (0,0,0,0,0,0,0), & (0,1,1,1,0,0,0), & (0,0,0,1,1,1,0), & (1,0,1,0,1,0,0), \\ (0,1,0,0,1,0,1), & (0,0,1,1,1,0,1), & (1,1,0,1,1,0,0), & (0,1,1,0,1,1,0), \\ (1,1,0,0,0,1,0), & (1,0,1,1,0,1,0), & (1,0,0,1,0,0,1), & (0,0,1,0,0,1,1), \\ (0,1,0,1,0,1,1), & (1,1,1,0,0,0,1), & (1,0,0,0,1,1,1), & (1,1,1,1,1,1,1) \end{array} \right\} \tag{1.429}$$

discussed in, e. g., Examples 1.5, 1.6, 1.7, 1.9 and 1.11. Obviously, this (7, 4, 3) Hamming code is not cyclic, because, for instance, $(0, 1, 1, 1, 0, 0, 0) \in \mathbb{V}$ is a code word, but its cyclic shift $(0, 0, 1, 1, 1, 0, 0) \notin \mathbb{V}$ is not.

The systematic generator matrix G of $\mathbb{V}$ given by (1.429) is presented in (1.131). Setting out from (1.131), however, by moving the fourth column of G of (1.131) in the first position, we find

$$G^{(1)} = \begin{pmatrix} 1 & 0 & 1 & 1 & 0 & 0 & 0 \\ 0 & 1 & 0 & 1 & 1 & 0 & 0 \\ 0 & 1 & 1 & 0 & 0 & 1 & 0 \\ 0 & 1 & 1 & 1 & 0 & 0 & 1 \end{pmatrix}, \quad G^{(1)} \in \mathbb{F}_2^{4\times 7}. \tag{1.430}$$

Now moving the fourth column of $G^{(1)}$ of (1.430) into the second position, we obtain [5, Beispiel 5.1, p. 130]

$$G^{(2)} = \begin{pmatrix} 1 & 1 & 0 & 1 & 0 & 0 & 0 \\ 0 & 1 & 1 & 0 & 1 & 0 & 0 \\ 0 & 0 & 1 & 1 & 0 & 1 & 0 \\ 0 & 1 & 1 & 1 & 0 & 0 & 1 \end{pmatrix} \quad G^{(2)} \in \mathbb{F}_2^{4\times 7}. \tag{1.431}$$

Finally, adding the second row to the fourth row, we yield [5, Beispiel 5.1, p. 130]

$$G^{(3)} = \begin{pmatrix} 1 & 1 & 0 & 1 & 0 & 0 & 0 \\ 0 & 1 & 1 & 0 & 1 & 0 & 0 \\ 0 & 0 & 1 & 1 & 0 & 1 & 0 \\ 0 & 0 & 0 & 1 & 1 & 0 & 1 \end{pmatrix} \quad G^{(3)} \in \mathbb{F}_2^{4\times 7}. \tag{1.432}$$

Table 1.6 illustrates the correspondence of messages u and codewords $v \in \mathbb{V}^{(3)}$ of the (7, 4, 3) Hamming code $\mathbb{V}^{(3)}$ [5, Beispiel 5.1, p. 130]. Clearly, $G^{(3)}$ given by (1.432) is the generator matrix of a cyclic (7, 4, 3) Hamming code $\mathbb{V}^{(3)}$, which is equivalent to the (7, 4, 3) Hamming code generated by G of (1.131) [5, Beispiel 5.1, p. 130].

Table 1.6: Cyclic (7, 4, 3) Hamming code $\in \mathbb{V}^{(3)}$ equivalent to the Hamming code $\mathbb{V}$ generated by $\boldsymbol{G}$ of (1.131), adapted according to [5, Beispiel 5.1, p. 130].

message u	**codeword** $v \in \mathbb{V}^{(3)}$
(0, 0, 0, 0)	(0, 0, 0, 0, 0, 0, 0)
(1, 0, 0, 0)	(1, 1, 0, 1, 0, 0, 0)
(0, 1, 0, 0)	(0, 1, 1, 0, 1, 0, 0)
(0, 0, 1, 0)	(0, 0, 1, 1, 0, 1, 0)
(0, 0, 0, 1)	(0, 0, 0, 1, 1, 0, 1)
(1, 1, 1, 0)	(1, 0, 0, 0, 1, 1, 0)
(0, 1, 1, 1)	(0, 1, 0, 0, 0, 1, 1)
(1, 1, 0, 1)	(1, 0, 1, 0, 0, 0, 1)
(1, 0, 1, 0)	(1, 1, 1, 0, 0, 1, 0)
(0, 1, 0, 1)	(0, 1, 1, 1, 0, 0, 1)
(1, 1, 0, 0)	(1, 0, 1, 1, 1, 0, 0)
(0, 1, 1, 0)	(0, 1, 0, 1, 1, 1, 0)
(0, 0, 1, 1)	(0, 0, 1, 0, 1, 1, 1)
(1, 1, 1, 1)	(1, 0, 0, 1, 0, 1, 1)
(1, 0, 0, 1)	(1, 1, 0, 0, 1, 0, 1)
(1, 0, 1, 1)	(1, 1, 1, 1, 1, 1, 1)

It seems that our first guess of the shape of $\boldsymbol{G}$ of a cyclic $(n, k, d_{\min})$ binary linear block code $\mathbb{V}$ (cf. (1.424)) was right. However, when looking at (1.427) and (1.432), we realize that in both cases the components of the generator matrix $\boldsymbol{G}$ fulfill the conditions (cf., e. g., [15, Theorem 5.2, p. 138f.])

$$g_j \begin{cases} = 1 & \text{for } j \in \{0, (n-k)\}, \\ \in \mathbb{F}_2 & \text{for } j \in \{1, 2 \cdots (n-k-1)\}, \\ = 0 & \text{for } j \in \{(n-k+1), (n-k+2) \cdots (n-1)\}, \end{cases} \qquad k, n \in \mathbb{N}^*. \tag{1.433}$$

Therefore, our second guess of the shape of $\boldsymbol{G}$ of a cyclic $(n, k, d_{\min})$ binary linear block code $\mathbb{V}$ becomes [5, Satz 5.3, p. 135]

$$\boldsymbol{G} = \begin{pmatrix} \overbrace{g_0}^{=1} & g_1 & \cdots & \overbrace{g_{n-k}}^{=1} & 0 & \cdots & 0 \\ 0 & \underbrace{g_0}_{=1} & \cdots & g_{n-k-1} & \underbrace{g_{n-k}}_{=1} & \cdots & 0 \\ \vdots & \vdots & \ddots & \vdots & \vdots & \ddots & \vdots \\ 0 & 0 & \cdots & \underbrace{g_0}_{=1} & g_1 & \cdots & \underbrace{g_{n-k}}_{=1} \end{pmatrix}, \qquad k, n \in \mathbb{N}^*. \tag{1.434}$$

Remark 1.31 (On the Generator Matrix $\boldsymbol{G}$ of (1.434)). In [1, Chapter 4], we introduced the $(N+P-1) \times N$ *time domain channel matrix*

$$\underline{\boldsymbol{A}}^{(t)} = \begin{pmatrix} \underline{h}_0^{(t)} & 0 & \cdots & 0 & 0 \\ \underline{h}_1^{(t)} & \underline{h}_0^{(t)} & \cdots & 0 & 0 \\ \underline{h}_2^{(t)} & \underline{h}_1^{(t)} & \cdots & 0 & 0 \\ \vdots & \vdots & \ddots & \vdots & \vdots \\ \underline{h}_{P-2}^{(t)} & \underline{h}_{P-3}^{(t)} & \cdots & \underline{h}_1^{(t)} & \underline{h}_0^{(t)} \\ \underline{h}_{P-1}^{(t)} & \underline{h}_{P-2}^{(t)} & \cdots & \underline{h}_2^{(t)} & \underline{h}_1^{(t)} \\ 0 & \underline{h}_{P-1}^{(t)} & \cdots & \vdots & \underline{h}_2^{(t)} \\ 0 & 0 & \ddots & \vdots & \vdots \\ \vdots & \vdots & \cdots & \underline{h}_{P-1}^{(t)} & \underline{h}_{P-2}^{(t)} \\ 0 & 0 & \cdots & 0 & \underline{h}_{P-1}^{(t)} \end{pmatrix}, \quad N, P \in \mathbb{N}^*, \tag{1.435}$$

to describe the multipath reception of a transmit signal, which reaches the receiver via a P path mobile radio channel with the complex channel impulse response vector $\underline{\boldsymbol{h}} \in \mathbb{C}^P$ equal to $(\underline{h}_0, \underline{h}_1 \cdots \underline{h}_{P-1})$, $P \in \mathbb{N}^*$. The $(N+P-1) \times N$ time domain channel matrix $\underline{\boldsymbol{A}}^{(t)}$ (1.435) is sometimes termed a *convolution matrix*, because applying $\underline{\boldsymbol{A}}^{(t)}$ to a complex data vector $\underline{\boldsymbol{d}}$ equal to $(\underline{d}_0, \underline{d}_1 \cdots \underline{d}_{N-1})$, $P \in \mathbb{N}^*$, results in the discrete-time convolution of the components of the channel impulse response vector $\underline{\boldsymbol{h}}$ and the complex data symbols contained in the complex data vector $\underline{\boldsymbol{d}}$.

It is remarkable that the transpose $\boldsymbol{G}^{\mathrm{T}}$ of the generator matrix $\boldsymbol{G}$ of (1.434) has the same structure as $\underline{\boldsymbol{A}}^{(t)}$, identifying

$$k \leftrightarrow N, \quad n \leftrightarrow (N+P-1), \quad (n-k) \leftrightarrow (P-1). \tag{1.436}$$

This allows some interesting conclusions:

a) The generator matrix of a cyclic $(n, k, d_{\min})$ binary linear block code $\mathbb{V}$ is a convolution matrix.
b) The encoding of a cyclic $(n, k, d_{\min})$ binary linear block code $\mathbb{V}$ can be regarded as the convolution of the message $\boldsymbol{u}$ and the generator polynomial in the form of the row vector $\boldsymbol{g}^{(0)}$; cf. (1.124).
c) The transmission of a transmit signal over a mobile radio channel can be regarded as the encoding of a cyclic $(n, k, d_{\min})$ binary linear block code $\mathbb{V}$ with the complex data vector $\underline{\boldsymbol{d}}$ corresponding to the message $\boldsymbol{u}$ and the complex channel impulse response vector $\underline{\boldsymbol{h}} \in \mathbb{C}^P$ corresponding to the "generator polynomial in the form of the row vector $\boldsymbol{g}^{(0)}$"; cf. (1.124).

Requiring g_0 and g_{n-k}, $k, n \in \mathbb{N}^*$ to be equal to 1 (cf. (1.433) [15, Theorem 5.2, p. 138f.]) guarantees that no column of (1.434) has the Hamming weight 0, thus avoiding that the binary linear block code had a length lower than n. This is also what we desire in the case of $\boldsymbol{G}$ in systematic form of (1.162). The resulting fact that the maximum of nonzero components per row of the generator matrix $\boldsymbol{G}$ of (1.434) is equal to $(n-k+1)$ is not surprising, bearing in mind that this is also the case for the generator matrix $\boldsymbol{G}$ in systematic form of (1.162).

Remark 1.32 (On Circulants). Now, let us carry out a little thought experiment. In Remark 1.32, let us consider the "regular" calculus defined for the fields $\mathbb{R}$ and $\mathbb{C}$, respectively,

What if we added another $(n-k)$ rows to the generator matrix $\boldsymbol{G}$ of (1.434), each row being identical with the previous one except for a shift to the right by a single location step? Using the n-tuple,

$$\boldsymbol{g} = (g_0, g_1 \cdots g_{n-k}, 0 \cdots 0) = (1, g_1 \cdots 1, 0 \cdots 0), \quad k, n \in \mathbb{N}^*, \tag{1.437}$$

we would yield the $n \times n$ square matrix

$$\begin{aligned}\mathbf{Circ}\,(\boldsymbol{g}) &= \mathbf{Circ}\,(g_0, g_1 \cdots g_{n-k}, 0 \cdots 0)\\ &= \begin{pmatrix} g_0 & g_1 & \cdots & g_{n-k} & 0 & \cdots & 0 \\ 0 & g_0 & \cdots & g_{n-k-1} & g_{n-k} & \cdots & 0 \\ \vdots & \vdots & \ddots & \vdots & \vdots & \ddots & \vdots \\ 0 & 0 & \cdots & g_0 & g_1 & \cdots & g_{n-k} \\ \hdashline g_{n-k} & 0 & \cdots & 0 & g_0 & \cdots & g_{n-k-1} \\ g_{n-k-1} & g_{n-k} & \cdots & 0 & 0 & \cdots & g_{n-k-2} \\ \vdots & \vdots & \ddots & \vdots & \vdots & \ddots & \vdots \\ g_1 & g_2 & \cdots & g_{n-k} & 0 & \cdots & g_0 \end{pmatrix}\\ &= \begin{pmatrix} & & & \boldsymbol{G} & & & \\ \hdashline g_{n-k} & 0 & \cdots & 0 & g_0 & \cdots & g_{n-k-1} \\ g_{n-k-1} & g_{n-k} & \cdots & 0 & 0 & \cdots & g_{n-k-2} \\ \vdots & \vdots & \ddots & \vdots & \vdots & \ddots & \vdots \\ g_1 & g_2 & \cdots & g_{n-k} & 0 & \cdots & g_0 \end{pmatrix}, \quad k, n \in \mathbb{N}^*,\end{aligned} \tag{1.438}$$

which is termed *circulant matrix of order n* or *circulant of order n* [41, p. 66]. Obviously, the generator matrix $\boldsymbol{G}$ of (1.434) is the submatrix of this circulant given by (1.438).

Although we may not use all the nice features of circulants such as being diagonalized by the discrete Fourier transform [41, Theorem 3.2.2, p. 72f.], we may benefit from the knowledge about a further way to represent a circulant [41, pp. 27, 68]. Using the $n \times n$, permutation matrix [41, equations (2.4.14), (2.4.15) and (2.4.16), p. 27]

$$\boldsymbol{\Pi} = \begin{pmatrix} 0 & 1 & 0 & 0 & \cdots & 0 \\ 0 & 0 & 1 & 0 & \cdots & 0 \\ \vdots & \vdots & \vdots & \vdots & \ddots & \vdots \\ 1 & 0 & 0 & 0 & \cdots & 0 \end{pmatrix}, \quad \boldsymbol{\Pi}^k = \underbrace{\boldsymbol{\Pi} \cdot \boldsymbol{\Pi} \cdot \cdots \cdot \boldsymbol{\Pi}}_{k \text{ factors}}, \quad \boldsymbol{\Pi}^n = \boldsymbol{\Pi}^0 = \boldsymbol{I}_n, \quad k, n \in \mathbb{N}^*, \tag{1.439}$$

we find [41, equation (3.1.4), p. 68]

$$\mathrm{Circ}(\boldsymbol{g}) = g_0 \boldsymbol{I}_n + g_1 \boldsymbol{\Pi} + \cdots + g_{n-k} \boldsymbol{\Pi}^{n-k} = \sum_{m=0}^{n-k} g_m \boldsymbol{\Pi}^m, \quad k, n \in \mathbb{N}^*. \tag{1.440}$$

Associating the *polynomial,*

$$\begin{aligned} g(z) &= g_0 \underbrace{z^0}_{=1} + g_1 z^1 + \cdots + g_{n-k} z^{n-k} = \sum_{m=0}^{n-k} g_m z^m \\ &= z^0 + g_1 z^1 + \cdots + z^{n-k}, \quad k, n \in \mathbb{N}^*, \end{aligned} \tag{1.441}$$

with the n-tuple $\boldsymbol{g}$ of (1.437), (1.440) is given by [41, p. 68]

$$\mathrm{Circ}(\boldsymbol{g}) = g(\boldsymbol{\Pi}). \tag{1.442}$$

The leading coefficient g_{n-k} of the polynomial $g(x)$ of (1.441) is equal to 1. Polynomials with the leading coefficient being equal to 1 are termed *monic polynomials* [4, p. 99] or *normalized polynomial* [5, p. 132].

The polynomial $g(x)$ introduced in (1.441) obviously represents the circulant of order n, $n \in \mathbb{N}^*$ given by, e. g., (1.442) and is therefore termed the *representer of the circulant* [41, p. 68].

Motivated by Remark 1.32 and returning to the calculus in $\mathbb{F}_2$, let us define the correspondence of an n-tuple taken from $\mathbb{F}_2^n$, $n \in \mathbb{N}^*$, with a polynomial [5, Definition 5.2, p. 131].

Definition 1.42 (Correspondence of an n-tuple With a Polynomial). The n-tuple, $n \in \mathbb{N}^*$,

$$\boldsymbol{v} = (v_0, v_1 \cdots v_{n-1}), \quad \boldsymbol{v} \in \mathbb{F}_2^n, \quad v_j \in \mathbb{F}_2, \quad j \in \{0, 1 \cdots (n-1)\}, \quad n \in \mathbb{N}^*, \tag{1.443}$$

corresponds with the polynomial of degree $(n-1)$, $n \in \mathbb{N}^*$ in the following way [5, Definition 5.2, p. 131]

$$\boldsymbol{v} \in \mathbb{F}_2^n \quad \leftrightarrow \quad v(x) = v_0 \odot x^0 \oplus v_1 \odot x^1 \oplus \cdots \oplus v_{n-1} \odot x^{n-1} = \bigoplus_{j=0}^{n-1} v_j \odot x^j, \quad n \in \mathbb{N}^*. \tag{1.444}$$

Furthermore, corresponding to the vector space $\mathbb{F}_2^n$, $n \in \mathbb{N}^*$ let $\mathbb{F}_2[x]_{n-1}$, $n \in \mathbb{N}^*$ denote the set of all polynomials of degree $(n-1)$, $n \in \mathbb{N}^*$, or lower with coefficients taken from $\mathbb{F}_2$ [5, Definition 5.2, p. 131].

$\mathbb{F}_2[x]$ denotes the set of all polynomials of arbitrary degree with coefficients taken from $\mathbb{F}_2$ [5, Definition 5.2, p. 131].

Remark 1.33 (On x in (1.444)). The quantity x used in (1.444) must not be mistaken with any input variable of a transmission channel [5, p. 131]. Rather, x is a "wild card." Instead of x, we could have used any other quantity, e. g., z^{-1} or D [5, p. 131].

Remark 1.34 (On Multiplying $v(x)$ With $(x \oplus 1)$). Using

$$\boldsymbol{v} = (1, 1, 1, 1 \cdots 1) \in \mathbb{F}_2^n \tag{1.445}$$

and multiplying it with $(x \oplus 1)$ yields

$$\begin{aligned}(x \oplus 1) \odot v(x) &= (x \oplus 1) \odot \left(1 \oplus x \oplus \cdots \oplus x^{n-1}\right) \\ &= x \oplus x^2 \oplus \cdots \oplus x^{n-1} \oplus x^n \oplus 1 \oplus x \oplus x^2 \oplus \cdots \oplus x^{n-1} \\ &= 1 \oplus \underbrace{(1 \oplus 1)}_{=0} \odot x \oplus \underbrace{(1 \oplus 1)}_{=0} \odot x^2 \oplus \cdots \oplus \underbrace{(1 \oplus 1)}_{=0} \odot x^{n-1} \oplus x^n \\ &= x^n \oplus 1, \quad n \in \mathbb{N}^*. \end{aligned} \tag{1.446}$$

This leads us to the following identity:

$$v(x) = \left(1 \oplus x \oplus \cdots \oplus x^{n-1}\right) = (x \oplus 1)^{-1} \odot \left(x^n \oplus 1\right), \quad n \in \mathbb{N}^*, \tag{1.447}$$

which we will also write in the following form [5, p. 131]:

$$1 \oplus x \oplus \cdots \oplus x^{n-1} = \bigoplus_{j=0}^{n-1} x^j = \frac{x^n \oplus 1}{x \oplus 1} = \left(x^n \oplus 1\right)/(x \oplus 1), \quad n \in \mathbb{N}^*. \tag{1.448}$$

We find the assignments of the row vectors $\boldsymbol{v}$ in $\mathbb{F}_2^n$ and the polynomials $v(x)$ in $\mathbb{F}_2[x]_{n-1}$ given in Table 1.7 [5, p. 131].

Table 1.7: Assignments of the row vectors in $\mathbb{F}_2^n$ and the polynomials in $\mathbb{F}_2[x]_{n-1}$, adapted from [5, p. 131].

$\boldsymbol{v} \in \mathbb{F}_2^n$	$\leftrightarrow$	$v(x) \in \mathbb{F}_2[x]_{n-1}$
$(0, 0, 0, 0 \cdots 0)$	$\leftrightarrow$	0
$(1, 0, 0, 0 \cdots 0)$	$\leftrightarrow$	$x^0 = 1$
$(0, 1, 0, 0 \cdots 0)$	$\leftrightarrow$	x^1
$(0, 0, 1, 0 \cdots 0)$	$\leftrightarrow$	x^2
$(0, 0, 0, 1 \cdots 0)$	$\leftrightarrow$	x^3
$\vdots$	$\vdots$	$\vdots$
$(0, 0, 0, 0 \cdots 1)$	$\leftrightarrow$	x^{n-1}
$\vdots$	$\vdots$	$\vdots$
$(1, 1, 1, 1 \cdots 1)$	$\leftrightarrow$	$\bigoplus_{j=0}^{n-1} x^j = (x^n \oplus 1)/(x \oplus 1)$

Setting out from (1.433) and from Remark 1.32 as well as using Definition 1.42, let us define the *monic generator polynomial over* $\mathbb{F}_2$ as follows [5, Definition 5.3, p. 133], [4, p. 190], [15, Theorems 5.4 and 5.5, p. 139f.], [25, p. 101]:

$$g(x) = \bigoplus_{j=0}^{n-k} g_j \odot x^j, \quad g(x) \in \mathbb{F}_2[x]_{n-k}, \quad g_j \begin{cases} = 1 & \text{for } j \in \{0, (n-k)\}, \\ \in \mathbb{F}_2 & \text{for } j \in \{1, 2 \cdots (n-k-1)\}, \end{cases} \quad k, n \in \mathbb{N}^*, \tag{1.449}$$

which yields

$$g(x) = 1 \oplus g_1 \odot x \oplus g_2 \odot x^2 \oplus \cdots \oplus x^{n-k}, \quad g(x) \in \mathbb{F}_2[x]_{n-k}, \quad g_j \in \mathbb{F}_2 \text{ for } j \in \{1, 2 \cdots (n-k-1)\}, \quad k, n \in \mathbb{N}^*. \tag{1.450}$$

Setting out from a cyclic $(n, k, d_{\min})$ binary linear block code $\mathbb{V}$ and taking Definition 1.42 into account, let the *message polynomial over* $\mathbb{F}_2$ be $u(x) \in \mathbb{F}_2[x]_{k-1}$, $k \in \mathbb{N}^*$, having the degree $(k-1)$, $k \in \mathbb{N}^*$. We yield

$$u(x) = \bigoplus_{i=0}^{k-1} u_i \odot x^i, \quad u(x) \in \mathbb{F}_2[x]_{k-1}, \quad k \in \mathbb{N}^*. \tag{1.451}$$

With the generator polynomial $g(x) \in \mathbb{F}_2[x]_{n-k}$, $k, n \in \mathbb{N}^*$ over $\mathbb{F}_2$ (cf. (1.449)), respectively (1.450), we obtain the *code polynomial over* $\mathbb{F}_2$ [15, p. 139] $v(x) \in \mathbb{F}_2[x]_{n-1}$, $n \in \mathbb{N}^*$ of (1.444), having the degree $(n-1)$, $n \in \mathbb{N}^*$, in the following way:

$$\begin{aligned}
v(x) &= u(x) \odot g(x) \\
&= \left(\bigoplus_{i=0}^{k-1} u_i \odot x^i\right) \odot \left(\bigoplus_{j=0}^{n-k} g_j \odot x^j\right) \\
&= \bigoplus_{i=0}^{k-1}\bigoplus_{j=0}^{n-k} (u_i \odot x^i) \odot (g_j \odot x^j) \\
&= \bigoplus_{i=0}^{k-1}\bigoplus_{j=0}^{n-k} (u_i \odot g_j) \odot x^{i+j} \qquad (1.452)\\
&= (u_0 \odot g_0) \odot x^0 \\
&\quad \oplus (u_0 \odot g_1 \oplus u_1 \odot g_0) \odot x^1 \\
&\quad \oplus (u_0 \odot g_2 \oplus u_1 \odot g_1 \oplus u_2 \odot g_0) \odot x^2 \\
&\quad \oplus (u_0 \odot g_3 \oplus u_1 \odot g_2 \oplus u_2 \odot g_1 \oplus u_3 \odot g_0) \odot x^2 \\
&\quad \oplus \cdots \\
&\quad \oplus (u_{k-1} \odot g_{n-k}) \odot x^{n-1} \\
&= u_0 \odot x^0 \\
&\quad \oplus (u_0 \odot g_1 \oplus u_1) \odot x^1 \\
&\quad \oplus (u_0 \odot g_2 \oplus u_1 \odot g_1 \oplus u_2) \odot x^2 \\
&\quad \oplus (u_0 \odot g_3 \oplus u_1 \odot g_2 \oplus u_2 \odot g_1 \oplus u_3) \odot x^2 \\
&\quad \oplus \cdots \\
&\quad \oplus u_{k-1} \odot x^{n-1} \\
&= \bigoplus_{j=0}^{n-1} \underbrace{\bigoplus_{i=\max\{0,j-(n-k)\}}^{\min\{j,k-1\}} (u_i \odot g_{j-i})}_{=v_j} \odot x^j, \quad n \in \mathbb{N}^*. \qquad (1.453)
\end{aligned}$$

The expression $v_j, j \in \{0, \cdots (n-1)\}$, equal to $\bigoplus_{i=\max\{0,j-(n-k)\}}^{\min\{j,k-1\}} (u_i \odot g_{j-i}), j \in \{0, \cdots (n-1)\}$ is the *discrete convolution* of the message bits u_i, $i \in \{0, \cdots, (k-1)\}$, and the coefficients g_j, $j \in \{0, \cdots (n-k)\}$ of the generator polynomial $g(x)$ [5, Satz 5.3, p. 135]. The results $v_j, j \in \{0, \cdots (n-1)\}$ of $\bigoplus_{i=\max\{0,j-(n-k)\}}^{\min\{j,k-1\}} (u_i \odot g_{j-i}), j \in \{0, \cdots (n-1)\}$ are obtained from the vector-matrix multiplication $\boldsymbol{uG}$ using the generator matrix $\boldsymbol{G}$ of (1.434) [5, Satz 5.3, p. 135].

The above discussion hence leads us to the following first theorem.

Theorem 1.29 (($(n, k, d_{\min})$ Binary Linear Block Code $\mathbb{V}$ and its Monic Generator Polynomial). *The* monic generator polynomial $g(x) \in \mathbb{F}_2[x]_{n-k}$, $k, n \in \mathbb{N}^*$ *over* $\mathbb{F}_2$, *having the degree* $(n-k)$, $k, n \in \mathbb{N}^*$ *generates the* $(n, k, d_{\min})$ *binary linear block code* $\mathbb{V}$ [5, *Definition 5.3, p. 133*]

$$\mathbb{V} = \left\{u(x) \odot g(x) \mid u(x) \in \mathbb{F}_2[x]_{k-1}\right\}, \quad k \in \mathbb{N}^*. \tag{1.454}$$

Proof. Setting out from (1.453), we know that the degree of the code polynomial $v(x) \in \mathbb{F}_2[x]_{n-1}$, $n \in \mathbb{N}^*$ is equal to $(n-1)$, which means that the length of the codeword, and thus the length of the $(n, k, d_{\min})$ binary linear block code $\mathbb{V}$ is equal to n, $n \in \mathbb{N}^*$ [5, Definition 5.3, p. 133].

Of course, considering (1.451), there are 2^k, $k \in \mathbb{N}^*$, message polynomials [5, Definition 5.3, p. 133]. If any two message polynomials $u^{(1)}(x) \in \mathbb{F}_2[x]_{k-1}$ and $u^{(2)}(x) \in \mathbb{F}_2[x]_{k-1}$, $k \in \mathbb{N}^*$ are not identical, we have [5, Definition 5.3, p. 133]

$$\underbrace{u^{(1)}(x) \odot g(x)}_{=v^{(1)} \in \mathbb{F}_2[x]_{n-1}} \neq \underbrace{u^{(2)}(x) \odot g(x)}_{=v^{(2)} \in \mathbb{F}_2[x]_{n-1}}, \tag{1.455}$$
$$u^{(1)}(x) \neq u^{(2)}(x), \quad u^{(1)}(x), u^{(2)}(x) \in \mathbb{F}_2[x]_{k-1}, \quad k, n \in \mathbb{N}^*.$$

Therefore, we yield [5, Definition 5.3, p. 133]

$$|\mathbb{V}| = 2^k, \quad k \in \mathbb{N}^*. \tag{1.456}$$

Hence, the dimension of the $(n, k, d_{\min})$ binary linear block code $\mathbb{V}$ is equal to k, $k \in \mathbb{N}^*$ [5, Definition 5.3, p. 133].

The $(n, k, d_{\min})$ binary linear block code $\mathbb{V}$ is linear because [5, Definition 5.3, p. 133]

$$u^{(1)}(x) \odot g(x) \oplus u^{(2)}(x) \odot g(x) = \left(u^{(1)}(x) \oplus u^{(2)}(x)\right) \odot g(x), \quad k \in \mathbb{N}^*, \tag{1.457}$$

holds. □

The previous few pages already showed that the mathematical treatment of a cyclic $(n, k, d_{\min})$ binary linear block code $\mathbb{V}$ requires some additional knowledge. Therefore, obtaining this additional mathematical know-how seems a necessity before we will be able to continue decently. Therefore, we will next look at a bit of number theory.

1.12 A Primer on Number Theory

1.12.1 Nulls, Units, Irreducible Numbers, Prime Numbers and Composite Numbers

Number theory started as the study of the *positive natural numbers*, which are the elements of the set [42, p. 1]

$$\mathbb{N}^* = \{1, 2, 3 \cdots\}. \tag{1.458}$$

In the recent past, it has been the permanent urging that originated in the set theory and in the computer sciences [19, p. 221], which resulted in the inclusion of the *"null,"* also known as the *"zero,"* in the natural numbers, which have formed the set

$$\mathbb{N} = \{\mathbf{0}, 1, 2, 3 \cdots\}. \tag{1.459}$$

A touching comment on the establishment of $\mathbb{N}$ given by (1.459) can be found in [43, Section 1.1, p. 1]:

> *When God created the numbers, He gave them a straight line as their home. He put a wheel on a point called "zero" and set it in motion. A labeling device left a mark after each full revolution. Since then, the number ray spreads out. Its nonexisting end disappears in the fog of infinity.*
> (translation by the author)

Using this beautiful picture, we term the line segment originating at 0 and terminating at 1 the *unit line segment*, respectively, the *unit distance* [19, p. 220].

Taking the negative variants of the positive natural numbers into account, which arise from the mirroring at zero, we yield the set of integers [42, p. 1]

$$\mathbb{Z} = \{\cdots - 3, -2, -1, 0, 1, 2, 3 \cdots\}, \tag{1.460}$$

which forms the basis of number theory. So, we can rightly claim that *number theory studies integers and integer-valued functions.*

There exist four kinds of such numbers to be studied, namely

- the "*null*," e. g., the number $0 \in \mathbb{N}$,
- the *units*, e. g., the number $1 \in \mathbb{N}$,
- the *prime numbers* [11, p. 330], [44, pp. 55–84], or at least *irreducible numbers* [44, pp. 55, 64–66], and
- the *composite numbers* [11, p. 330], [44, p. 55].

The number 0 is the smallest nonnegative integer. It is even because the division by 2 has no remainder. Also, the number 0 quantifies the amount of null size, of "" (in German "*das Nichts*"), e. g., the cardinality of the empty set $\varnothing$. It is the neutral element with regard to addition. It is *idempotent* for exponents $n \in \mathbb{N}^*$ greater than 0 [11, p. 296], [44, p. 94], i. e.,

$$0^n = 0 \quad \forall n \in \mathbb{N}^*. \tag{1.461}$$

"Wow," might the reader think, another new expression "*Got A Hold On Me*" (Christine McVie, 1984—we miss you and your beautiful music so much). So, enter *idempotence*, and not Sandman this time ("*Enter Sandman,*" Metallica, 1991—can you hear the iconic guitar riff, Lars Ulrich's heavy drumming, James Hetfield's lead vocals and Kirk Hammett's lead guitar playing?).

Definition 1.43 (Idempotence). An element i of a set $\mathbb{A}$, which is equipped with a binary operation $\circ : \mathbb{A} \times \mathbb{A} \mapsto \mathbb{A}$ is termed *idempotent* under $\circ$, if the following relationship holds [4, p. 217]:

$$i \circ i = i, \quad i \in \mathbb{A}. \tag{1.462}$$

The expression "*idempotent*" is synthesized from the Latin words "idem," which means "the same," and "potentia," which means "the power."

In addition, the number 0 belongs to the *five most important constants in analysis*, if you so wish the *five superstars of number theory* namely [45, p. 96]

- the null 0, i. e., the neutral element of addition, also called the additive identity element,
- the unity 1, i. e., the neutral element of multiplication, also termed the multiplicative identity element,
- the *circular number* or *circular constant* (in German "*Kreiszahl*") π, also called *Ludolf's number*,
- *Euler's constant e* and
- the *imaginary unit* j,

which are related with each other by *Euler's identity* [45, p. 96], [46, p. 67]

$$e^{\mathrm{j}\pi} + 1 = 0. \tag{1.463}$$

The discovery of this most remarkable and, at the same time, so simple relation (1.463) has been a greatest act of mathematics carried out by the rightly celebrated Swiss mathematical genius Leonhard *Euler*. According to, e. g., [46, p. 67], Richard *Feynman* called (1.463) "*the most remarkable formula in math.*"

The "*invention of the null*," especially the finding of a symbol for the "null," was a great act by Indian mathematicians in the sixth century [44, pp. 1, 10].

The number 1 is the *unity* and it is therefore not a prime number. It is odd, because the division by 2 yields a remainder. The number 1 is the first, i. e., the smallest, positive natural number. It is idempotent for all exponents $n \in \mathbb{N}^*$ greater than 0, i. e.,

$$1^n = 1 \quad \forall n \in \mathbb{N}^*. \tag{1.464}$$

The number 1 cannot be decomposed into a factorization of prime numbers, i. e., the factorization yields the empty product. According to the above discussion, the number 1 is also one of the five most important constants in analysis.

A *prime number*, also called a *prime*, is a, e. g., natural number greater than 1, which has the following properties [44, pp. 55, 63–66]:

- A prime number can only be divided without remainder by 1 or by itself [44, Satz 4.2, p. 63], i. e., it is irreducible, which is also called *trivial divisor property* or *irreducibility* [44, p. 65].
- If a prime number divides a product, it divides at least one of the factors of the product [44, Satz 4.2, p. 65f.], which is also called the *prime property*.

Prime numbers cannot be formed by multiplying two smaller natural numbers. Euclid showed that there are infinitely many prime numbers [44, pp. 8, 56–69].

Irreducibility does not lead to the prime property in all cases, which we will show in the following example [44, pp. 64–66].

Example 1.19. Let us consider $\mathbb{Z}[\sqrt{-5}]$, which is defined by [44, p. 64]

$$\mathbb{Z}[\sqrt{-5}] := \{a + b\sqrt{-5} \mid a, b \in \mathbb{Z}\}. \tag{1.465}$$

Let us look at the number 6, which is an element of $\mathbb{Z}[\sqrt{-5}]$. In $\mathbb{Z}[\sqrt{-5}]$, the number 6 has two different factorizations, namely

$$6 = 2 \cdot 3, \tag{1.466}$$

and

$$6 = (1 + \sqrt{-5}) \cdot (1 - \sqrt{-5}). \tag{1.467}$$

In $\mathbb{Z}[\sqrt{-5}]$, the elements $2, 3 \in \mathbb{Z}[\sqrt{-5}]$ cannot be further reduced, i. e., they cannot be written as a product of elements of $\mathbb{Z}[\sqrt{-5}]$, which are "smaller in size" than the elements 2 and 3. Also, $(1 + \sqrt{-5})$ and $(1 - \sqrt{-5})$ cannot be further reduced in $\mathbb{Z}[\sqrt{-5}]$ because neither $\pm 2, \pm 3$ nor $\pm\sqrt{-5}$ divide $(1 + \sqrt{-5})$ or $(1 - \sqrt{-5})$. Therefore, in $\mathbb{Z}[\sqrt{-5}]$, the elements $2, 3, (1 + \sqrt{-5})$ and $(1 - \sqrt{-5})$ are irreducible [44, p. 64].

Furthermore, the number 2 divides $(1 + \sqrt{-5}) \cdot (1 - \sqrt{-5})$. However, 2 does not divide $(1 + \sqrt{-5})$ or $(1 - \sqrt{-5})$ [44, p. 64]. In addition, the number 3 divides $(1 + \sqrt{-5}) \cdot (1 - \sqrt{-5})$. However, 3 does not divide $(1 + \sqrt{-5})$ or $(1 - \sqrt{-5})$. Therefore, neither 2 nor 3 are prime numbers although that they are irreducible in $\mathbb{Z}[\sqrt{-5}]$.

In the case of polynomials, the expression "prime" is usually replaced by the term "irreducible" [17, p. 14], even though one or the other polynomial might fulfill both properties.

A natural number greater than 1 that is not a prime number is called a composite number [44, p. 55].

In order to provide a concise algebraic description of cyclic codes, we will rely on the concept of finite fields; cf. Definition 1.8. Finite fields are used in most of the known construction of codes and for decoding [4, pp. 81, 93], [17, p. 1].

1.12.2 Rings, Integral Domains and Euclidean Domains

Our way to the analysis of finite fields leads us directly to the algebraic structures of *rings*, *integral domains* and *Euclidean domains*. In particular, the Euclidean domains are most important for coding theory (cf., e. g., [17, Preface]), especially for the mathematical description of cyclic codes. Therefore, it is advisable to develop a solid understanding of these interesting mathematical structures.

A first step is to consider algebras. A *universal algebra* consists of a set $\mathbb{A}$, also called the support set, and at least one operation defined on $\mathbb{A}$ [11, p. 353].

Definition 1.44 (Algebra). An *algebra* on a set $\mathbb{A}$ is a linear vector space with an associative and distributive multiplication [19, p. 680].

Let us use only binary operations $\circ : \mathbb{A}^2 \mapsto \mathbb{A}$ [11, p. 308], which are unique, and let the set $\mathbb{A}$ be closed under $\circ : \mathbb{A}^2 \mapsto \mathbb{A}$. Using such a distinct binary operation $\circ : \mathbb{A}^2 \mapsto \mathbb{A}$, $(\mathbb{A}, \circ)$ is termed *magma, binar* or *groupoid* [47, p. 147].

Furthermore, $(\mathbb{A}, \circ)$ is termed *semigroup* if $\circ : \mathbb{A}^2 \mapsto \mathbb{A}$ is associative [11, p. 308]

$$(\alpha \circ \beta) \circ \gamma = (\alpha \circ \beta \circ \gamma) = \alpha \circ \beta \circ \gamma, \quad \alpha, \beta, \gamma \in \mathbb{A}. \tag{1.468}$$

A semigroup with an identity element, also called neutral element, $\epsilon \in \mathbb{A}$, which complies with

$$\epsilon \circ \alpha = \alpha \circ \epsilon = \alpha, \quad \alpha, \epsilon \in \mathbb{A}, \tag{1.469}$$

is called a *monoid* [48, p. 6], [49, p. 5].

A monoid that has an inverse element $\alpha^{-1} \in \mathbb{A}$ to every element $\alpha \in \mathbb{A}$ with

$$\alpha^{-1} \circ \alpha = \alpha \circ \alpha^{-1} = \epsilon, \quad \alpha, \alpha^{-1}, \epsilon \in \mathbb{A}, \tag{1.470}$$

is called a *group* [11, p. 309], [48, p. 17].

If the commutative law

$$\alpha \circ \beta = \beta \circ \alpha, \quad \alpha, \epsilon \in \mathbb{A}, \tag{1.471}$$

also holds, a group is termed a *commutative group* or an *Abelian group* [48, p. 18], [49, p. 7] and a monoid is termed a *commutative monoid* [49, p. 5].

Table 1.8 summarizes the above discussion.

Table 1.8: Variants of $(\mathbb{A}, \circ)$.

$(\mathbb{A}, \circ)$	**magma, binar, groupoid**	**semigroup**	**monoid**	**group**	**commutative monoid**	**commutative group, Abelian group**
$\circ$ unique, $\mathbb{A}$ closed under $\circ$	√	√	√	√	√	√
associative law		√	√	√	√	√
identity element			√	√	√	√
inverse element				√		√
commutative law					√	√

So, what is a *ring*?

Definition 1.45 (Ring). A *ring* $(\mathbb{A}, \oplus, \odot)$ [19, p. 700] is a set $\mathbb{A}$ together with two binary operations, $\oplus$, i. e., "add," and $\odot$, i. e., "multiply," denoted by $(\mathbb{A}, \oplus, \odot)$, satisfying the following propositions, or axioms:

a) $(\mathbb{A}, \oplus)$ is an Abelian group.
b) $(\mathbb{A}, \odot)$ is a semigroup.
c) The distributive laws

$$\alpha \odot (\beta \oplus \gamma) = \alpha \odot \beta \,\oplus\, \alpha \odot \gamma \tag{1.472}$$

and

$$(\beta \oplus \gamma) \odot \alpha = \beta \odot \alpha \,\oplus\, \gamma \odot \alpha \tag{1.473}$$

hold for $\alpha, \beta, \gamma \in \mathbb{A}$ [11, p. 323], [19, p. 700].

In the case the *cancellation law* (in German "*nullteilerfrei*") holds also, i. e., if

$$\alpha \odot \beta = \alpha \odot \gamma \quad \Rightarrow \quad \beta = \gamma, \quad \alpha \neq 0 \quad \alpha, \beta, \gamma \in \mathbb{A}, \tag{1.474}$$

Definition 1.45 leads us to the *integral domain* (in German "*Integritätsbereich*") [19, p. 700].

Definition 1.46 (Integral Domain). An *integral domain* $(\mathbb{D}, \oplus, \odot)$ [19, p. 700] is a set $\mathbb{D}$ together with two binary operations, $\oplus$, i. e., "add," and $\odot$, i. e., "multiply," denoted by $(\mathbb{A}, \oplus, \odot)$, satisfying the following propositions, or axioms:

a) $(\mathbb{D}, \oplus)$ is an Abelian group with additive identity element denoted by 0.
b) $(\mathbb{D}, \odot)$ is a commutative monoid (commutative semigroup with multiplicative identity element 1).
c) The distributive laws

$$\alpha \odot (\beta \oplus \gamma) = \alpha \odot \beta \,\oplus\, \alpha \odot \gamma \tag{1.475}$$

and

$$(\beta \oplus \gamma) \odot \alpha = \beta \odot \alpha \,\oplus\, \gamma \odot \alpha \tag{1.476}$$

hold for $\alpha, \beta, \gamma \in \mathbb{D}$.
d) The cancellation law holds for $\alpha, \beta, \gamma \in \mathbb{D}$ [17, p. 3], [19, p. 700].

An integral domain is a commutative ring with cancellation law [19, p. 700].

Example 1.20. Let us look at $(\mathbb{Z}, \oplus, \odot)$. The addition "$\oplus$" is the conventional addition "+" that we all know. The multiplication "$\odot$" is the conventional multiplication "$\cdot$" that we all know.

We first need to check whether $(\mathbb{Z}, \oplus)$ is an Abelian group with additive identity element denoted by 0.

- $(\mathbb{Z}, \oplus)$ is closed, and the addition is unique.
- Identity element is $0 \in \mathbb{Z}$ because

$$\alpha + 0 = \alpha, \quad 0, \alpha \in \mathbb{Z}, \tag{1.477}$$

holds.

- The inverse element is $(-\alpha) \in \mathbb{Z}$ because

$$(-\alpha) + \alpha = 0, \quad \alpha \in \mathbb{Z} \tag{1.478}$$

  holds.
- The associative law

$$\alpha + (\beta + \gamma) = (\alpha + \beta) + \gamma, \quad \alpha, \beta, \gamma \in \mathbb{Z} \tag{1.479}$$

  holds.
- The commutative law

$$\alpha + \beta = \beta + \alpha, \quad \alpha, \beta \in \mathbb{Z} \tag{1.480}$$

  holds.

Next, we need to check that $(\mathbb{Z}, \odot)$ is a commutative monoid, i. e., a commutative semigroup with multiplicative identity element 1.
- $(\mathbb{Z}, \odot)$ is closed, and the multiplication is unique.
- The identity element is 1 because

$$\alpha \cdot 1 = \alpha, \quad 1, \alpha \in \mathbb{Z} \tag{1.481}$$

  holds.
- The associative law

$$\alpha \cdot (\beta \cdot \gamma) = (\alpha \cdot \beta) \cdot \gamma, \quad \alpha, \beta, \gamma \in \mathbb{Z} \tag{1.482}$$

  holds.
- The commutative law

$$\alpha \cdot \beta = \beta \cdot \alpha, \quad \alpha, \beta \in \mathbb{Z} \tag{1.483}$$

  holds.

Furthermore, the distributive laws

$$\alpha \cdot (\beta + \gamma) = \alpha \cdot \beta + \alpha \cdot \gamma, \quad \alpha, \beta, \gamma \in \mathbb{Z}, \tag{1.484}$$

and

$$(\beta + \gamma) \cdot \alpha = \beta \cdot \alpha + \gamma \cdot \alpha, \quad \alpha, \beta, \gamma \in \mathbb{Z}, \tag{1.485}$$

hold.

Finally, we need to check the cancellation law. Since $\alpha \cdot \beta$ and $\alpha \cdot \gamma$, for $\alpha \neq 0$ and $\alpha, \beta, \gamma \in \mathbb{Z}$, are unique,

$$\alpha \cdot \beta = \alpha \cdot \gamma, \quad \alpha, \beta, \gamma \in \mathbb{Z}, \tag{1.486}$$

require

$$\beta = \gamma, \quad \beta, \gamma \in \mathbb{Z}. \tag{1.487}$$

Therefore, $(\mathbb{Z}, \oplus, \odot)$ is an integral domain.

Let us define

$$\mathbb{Z}[\sqrt{-1}] := \{a + b\sqrt{-1} \mid a, b \in \mathbb{Z}\}, \tag{1.488}$$

which is an extension to the integers $\mathbb{Z}$ [44, p. 65]. $\mathbb{Z}[\sqrt{-1}]$ has been studied in detail by the greatest mathematician of all times, the German Carl Friedrich *Gauß*, who has earned himself the honorary title *"princeps mathematicorum,"* the Prince of the Mathematicians [44, p. 65]. To honor Carl Friedrich Gauß, $\mathbb{Z}[\sqrt{-1}]$ is called the *Gaussian integers* (in German *"Gaußsche Zahlen"*) [17, p. 4], [44, p. 65]. Setting out from Example 1.20, we can immediately see that $\mathbb{Z}[\sqrt{-1}]$ is also an integral domain [17, p. 4].

A *Euclidean domain* (in German *"euklidischer Ring"*) $(\mathbb{E}, \oplus, \odot)$ [19, p. 751], [48, pp. 189–194] is an integral domain with the feature of *"size"* among its elements [17, p. 3]. Let us denote the *"size"* of $\alpha \in \mathbb{E}$, $\alpha \neq 0$, by $h\{\alpha\} \in \mathbb{R}_0^+$, i. e., $h\{\alpha\} \geq 0$.

Definition 1.47 (Euclidean Domain). An integral domain with an identity element is called a *Euclidean domain*, if there exists a *Euclidean function* $h : \mathbb{E} \setminus \{0\} \rightarrow \mathbb{N}$, which assigns a nonnegative integer $h\{\alpha\} \in \mathbb{N}$ to each element $\alpha \in \mathbb{E}$, $\alpha \neq 0$, having the following properties [17, p. 4], [19, p. 751]:

- $h\{\alpha \odot \beta\} \geq h\{\alpha\}$, $\quad \forall \alpha, \beta \in \mathbb{E}$, $\quad \alpha \neq 0, \beta \neq 0$.
- For all $\alpha \neq 0$ and $\beta \neq 0$, there exist $q \in \mathbb{E}$ ("quotient") and $r \in \mathbb{E}$ ("remainder") such that

$$\alpha = q \odot \beta \oplus r, \quad \alpha, \beta \in \mathbb{E} \setminus \{0\}, \quad q, r \in \mathbb{E}, \tag{1.489}$$

for which $r \in \mathbb{E}$ is either equal to $0 \in \mathbb{E}$ or $h\{r\} < h\{\beta\}$.

This means that Euclidean domains are *integral domains that have a division algorithm.*

The Euclidean function $h : \mathbb{E} \setminus \{0\} \rightarrow \mathbb{N}$ is also called the *Euclidean norm function* (in German *"euklidische Normfunktion"*) or the *Euclidean absolute value function* (in German *"euklidischer Betrag"*).

In some definitions, one finds $h : \mathbb{E} \rightarrow \mathbb{N}$ instead of $h : \mathbb{E} \setminus \{0\} \rightarrow \mathbb{N}$, setting $h\{0\}$ to 0. Table 1.9 summarizes the above discussion.

Table 1.9: Ring, integral domain and Euclidean domain.

	ring $(\mathbb{A}, \oplus, \odot)$	integral domain $(\mathbb{D}, \oplus, \odot)$	Euclidean domain $(\mathbb{E}, \oplus, \odot)$
$(\mathbb{A}, \oplus) / (\mathbb{D}, \oplus) / (\mathbb{E}, \oplus)$	Abelian group	Abelian group	Abelian group
$(\mathbb{A}, \odot) / (\mathbb{D}, \odot) / (\mathbb{E}, \odot)$	semigroup	commutative monoid	commutative monoid
distributive laws	✓	✓	✓
cancellation law		✓	✓
division algorithm			✓

Before looking at a first example, let us first consider the following theorem.

Theorem 1.30 (Difference of Rational Numbers and Integers). *For every $\lambda \in \mathbb{Q}$, there is a unique $l \in \mathbb{Z}$ with*

$$|\lambda - l| \leq \frac{1}{2}, \quad \lambda \in \mathbb{Q}, \quad l \in \mathbb{Z}. \tag{1.490}$$

Proof. Using the "*Gaußklammer*," we find that

$$\left\lfloor \lambda + \frac{1}{2} \right\rfloor \leq \lambda + \frac{1}{2} < \left\lfloor \lambda + \frac{1}{2} \right\rfloor + 1, \quad \lambda \in \mathbb{Q}, \quad l \in \mathbb{Z}. \tag{1.491}$$

This yields

$$0 \leq \lambda + \frac{1}{2} - \left\lfloor \lambda + \frac{1}{2} \right\rfloor < 1 \quad \Leftrightarrow \quad -\frac{1}{2} \leq \lambda - \left\lfloor \lambda + \frac{1}{2} \right\rfloor < \frac{1}{2}, \quad \lambda \in \mathbb{Q}, \quad l \in \mathbb{Z}. \tag{1.492}$$

Setting

$$l = \left\lfloor \lambda + \frac{1}{2} \right\rfloor, \quad \lambda \in \mathbb{Q}, \quad l \in \mathbb{Z}, \tag{1.493}$$

which is unique, and we yield

$$-\frac{1}{2} \leq \lambda - l < \frac{1}{2} \quad \Rightarrow \quad |\lambda - l| \leq \frac{1}{2} \quad \lambda \in \mathbb{Q}, \quad l \in \mathbb{Z}. \tag{1.494}$$

□

Equation (1.490) tells us that $l \in \mathbb{Z}$ is that particular integer, which is nearest to $\lambda \in \mathbb{Q}$. This is illustrated in Figure 1.20 for λ being taken from the interval $[-3.5, +3.5]$. The selection of $l \in \mathbb{Z}$ being the nearest integer to $\lambda \in \mathbb{Q}$ is made according to (1.490), which results in

$$l = \mathrm{rd}(\lambda), \quad \lambda \in \mathbb{Q}, \quad l \in \mathbb{Z}. \tag{1.495}$$

Figure 1.20 shows a detail of $|\lambda - l| \in \mathbb{Q}$ given by (1.490) with $l \in \mathbb{Z}$ equal to $\mathrm{rd}(\lambda)$, $\lambda \in \mathbb{Q}$, being the nearest integer to λ. In Figure 1.20, the selected $l \in \mathbb{Z}$ are indicated by vertical dotted lines, whereas $|\lambda - l| \in \mathbb{Q}$ is shown as the solid line versus $\lambda \in \mathbb{Q}$. As required, $|\lambda - l| \leq 1/2$ in all cases. The domain in which a particular selection of $l \in \mathbb{Z}$ prevails is shown in the form as a gray shaded rectangle, which changes its shading when $l \in \mathbb{Z}$ is changed.

Example 1.21. Let us look at the integers $(\mathbb{Z}, +, \cdot)$. Let us define

$$h\{a\} = |a| \geq 1, \quad \forall a \in \mathbb{Z}, \quad a \neq 0. \tag{1.496}$$

Using (1.483), we find

$$h\{a \cdot \beta\} = |a\beta| = |a| \cdot \underbrace{|\beta|}_{\geq 1} = h\{a\} \cdot \underbrace{h\{\beta\}}_{\geq 1} \geq h\{a\}, \quad \forall a, \beta \in \mathbb{E}, \quad a \neq 0, \beta \neq 0, \tag{1.497}$$

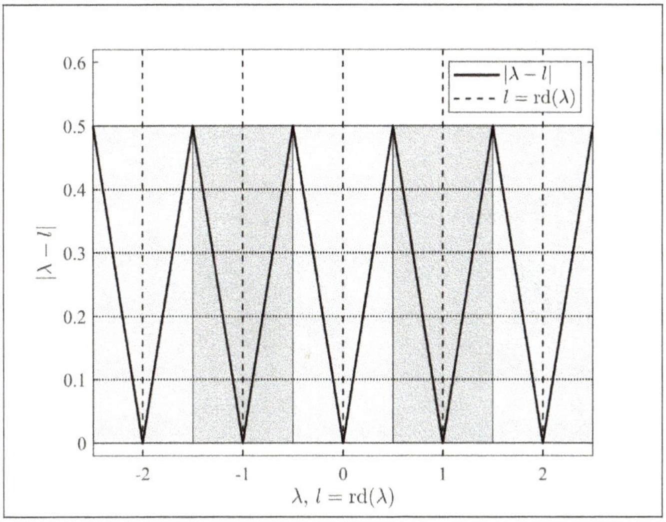

Figure 1.20: Detail of $|\lambda - l| \in \mathbb{Q}$ given by (1.490) with $l \in \mathbb{Z}$ equal to $\mathrm{rd}(\lambda)$, $\lambda \in \mathbb{Q}$, being the nearest integer to λ.

which shows that $h\{\alpha\}$ is multiplicative and that the first property of Definition 1.47 is fulfilled.

Let us now consider the second property of Definition 1.47:

$$\alpha = q\beta + r, \quad h\{r\} < h\{\beta\}, \quad \alpha, \beta \in \mathbb{Z}\backslash\{0\}, \quad q, r \in \mathbb{Z}. \tag{1.498}$$

This means that α and β, being nonzero elements of $\mathbb{Z}$, can be approximated by $q \in \mathbb{Z}$ with $r \in \mathbb{E}$ as small as possible.

Let us assume that β divides α, which is denoted by $\beta|\alpha$, read "β divides α." In this case, $h\{\beta\} \geq 1$. Furthermore, r vanishes.

Let us now consider the case in which r does not vanish. We know that

$$\frac{\alpha}{\beta} = \lambda, \quad \alpha, \beta \in \mathbb{Z}\backslash\{0\}, \quad \lambda \in \mathbb{Q}. \tag{1.499}$$

According to Theorem 1.30, there are unique choices $l \in \mathbb{Z}$ with

$$|\lambda - l| \leq \frac{1}{2}, \quad l \in \mathbb{Z}, \quad \lambda \in \mathbb{Q}, \tag{1.500}$$

i. e., $l \in \mathbb{Z}$ being closest to $\lambda \in \mathbb{Q}$. Then

$$\frac{\alpha}{\beta} = \underbrace{l}_{=q} + \underbrace{\lambda - l}_{=r/\beta} = q + \frac{r}{\beta}, \quad r, \alpha, \beta \in \mathbb{Z}\backslash\{0\}, \quad l \in \mathbb{Z}, \quad \lambda \in \mathbb{Q}. \tag{1.501}$$

With

$$r = (\lambda - l)\beta, \quad r, \beta \in \mathbb{Z}\backslash\{0\}, \quad l \in \mathbb{Z}, \quad \lambda \in \mathbb{Q}, \tag{1.502}$$

we yield

$$h\{r\} = |r| = \left|(\lambda - l)\beta\right| = \underbrace{|\lambda - l|}_{\leq 1/2} \cdot |\beta| \leq \frac{1}{2} \cdot |\beta| < |\beta| = h\{\beta\}, \tag{1.503}$$

$$r, \beta \in \mathbb{Z} \setminus \{0\}, \quad l \in \mathbb{Z}, \quad \lambda \in \mathbb{Q}.$$

Hence, $(\mathbb{Z}, +, \cdot)$ is a Euclidean domain [17, p. 4].

Example 1.22. Let us look at $\mathbb{Z}[\sqrt{-1}]$, the Gaussian integers, which are defined by

$$\mathbb{Z}[\sqrt{-1}] := \{a + b\sqrt{-1} \mid a, b \in \mathbb{Z}\}. \tag{1.504}$$

Let us verify whether $(\mathbb{Z}[\sqrt{-1}], +, \cdot)$ is a Euclidean domain [17, p. 4]. We set

$$\underline{a} = a + b\sqrt{-1} = a + \mathrm{j}b, \quad \underline{a} \in \mathbb{Z}[\sqrt{-1}], \quad a, b \in \mathbb{Z}. \tag{1.505}$$

Let us define

$$h\{\underline{a}\} = |a + \mathrm{j}b|^2 = \underline{a}\,\underline{a}^* = (a + \mathrm{j}b) \cdot (a - \mathrm{j}b) = a^2 + b^2, \quad \underline{a} \in \mathbb{Z}[\sqrt{-1}], \quad a, b \in \mathbb{Z}. \tag{1.506}$$

Clearly, $h\{\underline{a}\} \in \mathbb{N}$, and it can only be zero if $\underline{a}$ was zero, however, for this case, $h\{\underline{a}\}$ is not defined. Hence, $h\{\underline{a}\} \geq 1$ for all $\underline{a} \in \mathbb{Z}[\sqrt{-1}]$.

With

$$\underline{\beta} = c + d\sqrt{-1} = c + \mathrm{j}d, \quad h\{\underline{\beta}\} = c^2 + d^2, \quad \underline{\beta} \in \mathbb{Z}[\sqrt{-1}], \quad c, d \in \mathbb{Z}, \tag{1.507}$$

and with

$$\underline{a} \cdot \underline{\beta} = (a + \mathrm{j}b) \cdot (c + \mathrm{j}d) = ac - bd + \mathrm{j}(ad + bc), \quad \underline{a}, \underline{\beta} \in \mathbb{Z}[\sqrt{-1}], \quad a, b, c, d \in \mathbb{Z}, \tag{1.508}$$

we find

$$\begin{aligned} h\{\underline{a} \cdot \underline{\beta}\} &= (ac - bd)^2 + (ad + bd)^2 \\ &= a^2c^2 - 2abcd + b^2d^2 + a^2d^2 + 2abcd + b^2c^2 \\ &= a^2c^2 + a^2d^2 + b^2c^2 + b^2d^2 \\ &= a^2\left(c^2 + d^2\right) + b^2\left(c^2 + d^2\right) \\ &= \left(a^2 + b^2\right)\left(c^2 + d^2\right) \\ &= h\{\underline{a}\} \cdot h\{\underline{\beta}\}, \quad \underline{a}, \underline{\beta} \in \mathbb{Z}[\sqrt{-1}], \quad a, b, c, d \in \mathbb{Z}, \end{aligned} \tag{1.509}$$

which shows that $h\{\underline{a}\}$ is multiplicative.

We will first have to check the first property given in Definition 1.47, which requires

$$h\{\underline{a} \cdot \underline{\beta}\} \geq h\{\underline{a}\}, \quad \forall\, \underline{a}, \underline{\beta} \in \mathbb{Z}[\sqrt{-1}], \quad \underline{a} \neq 0, \underline{\beta} \neq 0. \tag{1.510}$$

Since for $\underline{\beta} \neq 0$,

$$h\{\underline{\beta}\} \geq 1, \quad \underline{\beta} \in \mathbb{Z}[\sqrt{-1}] \tag{1.511}$$

holds, we yield

$$h\{\underline{a} \cdot \underline{\beta}\} = h\{\underline{a}\} \cdot \underbrace{h\{\underline{\beta}\}}_{\geq 1} \geq h\{\underline{a}\}. \tag{1.512}$$

So, the first property of Definition 1.47 is fulfilled.

Now, will first have to check the second property given in Definition 1.47, according to which for all $\underline{\alpha} \neq 0$ and $\underline{\beta} \neq 0$ there exists $\underline{q}$ ("quotient") and $\underline{r}$ ("remainder") such that

$$\underline{\alpha} = \underline{q}\,\underline{\beta} + \underline{r}, \quad \underline{q}, \underline{r}, \underline{\alpha}, \underline{\beta} \in \mathbb{Z}[\sqrt{-1}], \tag{1.513}$$

with $\underline{r}$ being either equal to 0 or $h\{\underline{r}\} < h\{\underline{\beta}\}$. Let us first assume that $\underline{\beta}|\underline{\alpha}$. In this case, $\underline{r}$ vanishes.

Let us now consider the second part of the second property, i. e., the case in which $\underline{r}$ does not vanish. We know that

$$\frac{\underline{\alpha}}{\underline{\beta}} = \lambda + \mathrm{j}\xi = \frac{\underline{\alpha} \cdot \underline{\beta}^*}{|\underline{\beta}|^2} = \frac{\underline{\alpha} \cdot \underline{\beta}^*}{h\{\underline{\beta}\}}, \quad \underline{\alpha}, \underline{\beta} \in \mathbb{Z}[\sqrt{-1}], \quad \lambda, \xi \in \mathbb{Q}. \tag{1.514}$$

There are unique choices $l, m \in \mathbb{Z}$ with

$$|\lambda - l| \leq \frac{1}{2}, \quad \lambda \in \mathbb{Q}, \quad l \in \mathbb{Z}, \tag{1.515}$$

i. e., $l \in \mathbb{Z}$ being closest to $\lambda \in \mathbb{Q}$, and

$$|\xi - m| \leq \frac{1}{2}, \quad \xi \in \mathbb{Q}, \quad m \in \mathbb{Z}, \tag{1.516}$$

i. e., $m \in \mathbb{Z}$ being closest to $\xi \in \mathbb{Q}$. Then

$$\begin{aligned} \frac{\underline{\alpha}}{\underline{\beta}} &= \lambda + \mathrm{j}\xi \\ &= \lambda + (l - l) + \mathrm{j}\big[\xi + (m - m)\big] \\ &= \underbrace{l + \mathrm{j}m}_{=\underline{q}} + \underbrace{(\lambda - l) + \mathrm{j}(\xi - m)}_{=\underline{r}/\underline{\beta}} \\ &= \underline{q} + \frac{\underline{r}}{\underline{\beta}}, \quad \lambda, \xi \in \mathbb{Q}, \quad l, m \in \mathbb{Z}. \end{aligned} \tag{1.517}$$

We find

$$\begin{aligned} \underline{r} &= (\lambda - l)\underline{\beta} + \mathrm{j}(\xi - m)\underline{\beta} \\ &= (\lambda - l)\big(\mathrm{Re}\{\underline{\beta}\} + \mathrm{j}\,\mathrm{Im}\{\underline{\beta}\}\big) + \mathrm{j}(\xi - m)\big(\mathrm{Re}\{\underline{\beta}\} + \mathrm{j}\,\mathrm{Im}\{\underline{\beta}\}\big) \\ &= \mathrm{Re}\{\underline{\beta}\}(\lambda - l) - \mathrm{Im}\{\underline{\beta}\}(\xi - m) + \mathrm{j}\big\{\mathrm{Re}\{\underline{\beta}\}(\xi - m) + \mathrm{Im}\{\underline{\beta}\}(\lambda - l)\big\} \end{aligned} \tag{1.518}$$

and, therefore,

$$\begin{aligned} h\{\underline{r}\} &= |\underline{r}|^2 \\ &= \big\{\mathrm{Re}\{\underline{\beta}\}(\lambda - l) - \mathrm{Im}\{\underline{\beta}\}(\xi - m)\big\}^2 \\ &\quad + \big\{\mathrm{Re}\{\underline{\beta}\}(\xi - m) + \mathrm{Im}\{\underline{\beta}\}(\lambda - l)\big\}^2, \end{aligned} \tag{1.519}$$

which leads to

$$\begin{aligned} h\{\underline{r}\} &= \left\{\mathrm{Re}\{\underline{\beta}\}\right\}^2(\lambda - l)^2 + \left\{\mathrm{Im}\{\underline{\beta}\}\right\}^2(\xi - m)^2 \\ &\quad - 2\,\mathrm{Re}\{\underline{\beta}\}\,\mathrm{Im}\{\underline{\beta}\}(\lambda - l)(\xi - m) \\ &\quad + \left\{\mathrm{Re}\{\underline{\beta}\}\right\}^2(\xi - m)^2 + \left\{\mathrm{Im}\{\underline{\beta}\}\right\}^2(\lambda - l)^2 \\ &\quad + 2\,\mathrm{Re}\{\underline{\beta}\}\,\mathrm{Im}\{\underline{\beta}\}(\lambda - l)(\xi - m) \\ &= \left(\mathrm{Re}\{\underline{\beta}\}\right)^2\left\{(\lambda - l)^2 + (\xi - m)^2\right\} + \left(\mathrm{Im}\{\underline{\beta}\}\right)^2\left\{(\lambda - l)^2 + (\xi - m)^2\right\} \\ &= \underbrace{\left[\left(\mathrm{Re}\{\underline{\beta}\}\right)^2 + \left(\mathrm{Im}\{\underline{\beta}\}\right)^2\right]}_{=h\{\underline{\beta}\}} \cdot \underbrace{\left[(\lambda - l)^2 + (\xi - m)^2\right]}_{=h\{\underline{r}/\underline{\beta}\}} \\ &= h\{\underline{\beta}\} \cdot h\left\{\frac{\underline{r}}{\underline{\beta}}\right\}. \end{aligned} \tag{1.520}$$

With

$$h\left\{\frac{\underline{r}}{\underline{\beta}}\right\} = (\lambda - l)^2 + (\xi - m)^2 \le \frac{1}{4} + \frac{1}{4} = \frac{1}{2}, \quad \lambda, \xi \in \mathbb{Q}, \quad l, m \in \mathbb{Z}, \tag{1.521}$$

we find

$$h\{\underline{r}\} = h\{\underline{\beta}\} \cdot h\left\{\frac{\underline{r}}{\underline{\beta}}\right\} \le \frac{1}{2} h\{\underline{\beta}\} < h\{\underline{\beta}\}. \tag{1.522}$$

Hence, $(\mathbb{Z}[\sqrt{-1}], +, \cdot)$ is a Euclidean domain [17, p. 4].

Example 1.23. Let us look at $\mathbb{Z}[\sqrt{-p}]$ defined by

$$\mathbb{Z}[\sqrt{-p}] := \left\{a + b\sqrt{-p} \mid a, b \in \mathbb{Z},\ p \in \mathbb{N}^*\right\} \tag{1.523}$$

with $p \in \mathbb{N}^*$ being a positive prime number, i. e., $p \in \{2, 3, 5, 7 \cdots\}$. Let us evaluate whether $(\mathbb{Z}[\sqrt{-p}], +, \cdot)$ is a Euclidean domain [17, p. 4]. We set

$$\underline{\alpha} = a + b\sqrt{-p} = a + \mathrm{j}b\sqrt{p}, \quad \underline{\alpha} \in \mathbb{Z}[\sqrt{-p}], \quad a, b \in \mathbb{Z}, \quad p \in \{2, 3, 5, 7 \cdots\}. \tag{1.524}$$

Let us define

$$h\{\underline{\alpha}\} = |a + \mathrm{j}b\sqrt{p}|^2 = a^2 + pb^2, \quad \underline{\alpha} \in \mathbb{Z}[\sqrt{-p}], \quad a, b \in \mathbb{Z}, \quad p \in \{2, 3, 5, 7 \cdots\}. \tag{1.525}$$

Clearly, $h\{\underline{\alpha}\} \in \mathbb{N}^*$, and it can only be zero if $\underline{\alpha}$ was zero. However, for this case, $h\{\underline{\alpha}\}$ is not defined. Hence, $h\{\underline{\alpha}\} \ge 1$ for all $\underline{\alpha} \in \mathbb{Z}[\sqrt{-p}]$. With

$$\underline{\beta} = c + d\sqrt{-p} = c + \mathrm{j}d\sqrt{p}, \quad \underline{\beta} \in \mathbb{Z}[\sqrt{-p}], \quad c, d \in \mathbb{Z}, \quad p \in \{2, 3, 5, 7 \cdots\}, \tag{1.526}$$

with

$$h\{\underline{\beta}\} = c^2 + pd^2 \quad \underline{\beta} \in \mathbb{Z}[\sqrt{-p}], \quad c, d \in \mathbb{Z}, \quad p \in \{2, 3, 5, 7 \cdots\}, \tag{1.527}$$

with

$$\underline{\alpha} \cdot \underline{\beta} = (a + \mathrm{j}b\sqrt{p}) \cdot (c + \mathrm{j}d\sqrt{p}) = ac - pbd + \mathrm{j}(ad + bc)\sqrt{p}, \tag{1.528}$$

$$\underline{\alpha}, \underline{\beta} \in \mathbb{Z}[\sqrt{-p}], \quad a, b, c, d \in \mathbb{Z}, \quad p \in \{2, 3, 5, 7 \cdots\} \tag{1.529}$$

and with

$$
\begin{aligned}
h\{\underline{\alpha}\} \cdot h\{\underline{\beta}\} &= \left(a^2 + pb^2\right) \cdot \left(c^2 + pd^2\right) \\
&= a^2c^2 + pa^2d^2 + pb^2c^2 + p^2b^2d^2 \\
&= a^2c^2 + p^2b^2d^2 + p\left(a^2d^2 + b^2c^2\right), \qquad (1.530)\\
&\underline{\alpha}, \underline{\beta} \in \mathbb{Z}[\sqrt{-p}], \quad a,b,c,d \in \mathbb{Z}, \quad p \in \{2,3,5,7\cdots\},
\end{aligned}
$$

we find

$$
\begin{aligned}
h\{\underline{\alpha} \cdot \underline{\beta}\} &= (ac - pbd)^2 + p(ad + bc)^2 \\
&= a^2c^2 - 2pabcd + p^2b^2d^2 + pa^2d^2 + 2pabcd + pb^2c^2 \\
&= a^2c^2 + p^2b^2d^2 + p\left(a^2d^2 + b^2c^2\right) \\
&= h\{\underline{\alpha}\} \cdot h\{\underline{\beta}\}, \qquad (1.531)\\
&\underline{\alpha}, \underline{\beta} \in \mathbb{Z}[\sqrt{-p}], \quad a,b,c,d \in \mathbb{Z}, \quad p \in \{2,3,5,7\cdots\},
\end{aligned}
$$

which shows that $h\{\underline{\alpha}\}$ is multiplicative.

We will first have to check the first property given in Definition 1.47: For $\underline{\beta} \neq 0$ and $h\{\underline{\beta}\} \geq 1$, we obtain

$$
h\{\underline{\alpha} \cdot \underline{\beta}\} = h\{\underline{\alpha}\} \cdot \underbrace{h\{\underline{\beta}\}}_{\geq 1} \geq h\{\underline{\alpha}\}, \quad \underline{\alpha}, \underline{\beta} \in \mathbb{Z}[\sqrt{-p}]. \qquad (1.532)
$$

So, the first property of Definition 1.47 is fulfilled.

Now, we will have to check the second property given in Definition 1.47. Let us first assume that $\underline{\beta} | \underline{\alpha}$. In this case, $\underline{r}$ vanishes.

Let us now consider the second part of the second property, i. e., the case in which $\underline{r}$ does not vanish. We know that

$$
\frac{\underline{\alpha}}{\underline{\beta}} = \lambda + \mathrm{j}\xi\sqrt{p}, \quad \lambda, \xi \in \mathbb{Q}, \quad p \in \{2,3,5,7\cdots\}. \qquad (1.533)
$$

There are unique choices $l, m \in \mathbb{Z}$ with

$$
|\lambda - l| \leq \frac{1}{2}, \quad \lambda \in \mathbb{Q}, \quad l \in \mathbb{Z}, \qquad (1.534)
$$

$l \in \mathbb{Z}$ being closest to $\lambda \in \mathbb{Q}$, and

$$
|\xi - m| \leq \frac{1}{2}, \quad \xi \in \mathbb{Q}, m \in \mathbb{Z}, \qquad (1.535)
$$

i. e., $m \in \mathbb{Z}$ being closest to $\xi \in \mathbb{Q}$. Then

$$
\begin{aligned}
\frac{\underline{\alpha}}{\underline{\beta}} &= \lambda + \mathrm{j}\xi\sqrt{p} \\
&= \lambda + l - l + \mathrm{j}(\xi - m + m)\sqrt{p} \\
&= \underbrace{l + \mathrm{j}m\sqrt{p}}_{=\underline{q}} + \underbrace{(\lambda - l) + \mathrm{j}(\xi - m)\sqrt{p}}_{=\underline{r}/\underline{\beta}} \qquad (1.536)\\
&= \underline{q} + \frac{\underline{r}}{\underline{\beta}}, \quad \underline{q}, \underline{r}, \underline{\alpha}, \underline{\beta} \in \mathbb{Z}[\sqrt{-p}], \quad \lambda, \xi \in \mathbb{Q}, \quad l, m \in \mathbb{Z}, \quad p \in \{2,3,5,7\cdots\}.
\end{aligned}
$$

We thus require that the quotient is composed of those integers $l, m \in \mathbb{Z}$, which are nearest to $\lambda, \xi \in \mathbb{Q}$, respectively. We find

$$\frac{\underline{r}}{\underline{\beta}} = (\lambda - l) + \mathrm{j}(\xi - m)\sqrt{p}, \quad \underline{r}, \underline{\beta} \in \mathbb{Z}[\sqrt{-p}], \quad \lambda, \xi \in \mathbb{Q}, \quad l, m \in \mathbb{Z}, \quad p \in \{2, 3, 5, 7 \cdots\} \tag{1.537}$$

and, therefore,

$$h\left\{\frac{\underline{r}}{\underline{\beta}}\right\} = (\lambda - l)^2 + p(\xi - m)^2 \le \frac{1}{4} + \frac{p}{4} = \frac{p+1}{4}, \quad \underline{r}, \underline{\beta} \in \mathbb{Z}[\sqrt{-p}], \tag{1.538}$$
$$\lambda, \xi \in \mathbb{Q}, \quad l, m \in \mathbb{Z}, \quad p \in \{2, 3, 5, 7 \cdots\}.$$

We thus find

$$h\{\underline{r}\} = h\left\{\underline{\beta} \cdot \frac{\underline{r}}{\underline{\beta}}\right\} = h\{\underline{\beta}\} \cdot \frac{p+1}{4}, \quad \underline{r}, \underline{\beta} \in \mathbb{Z}[\sqrt{-p}], \tag{1.539}$$
$$\lambda, \xi \in \mathbb{Q}, \quad l, m \in \mathbb{Z}, \quad p \in \{2, 3, 5, 7 \cdots\},$$

which must be lower than $h\{\underline{\beta}\}$. Therefore, we yield

$$\frac{p+1}{4} < 1 \quad \Leftrightarrow \quad p < 3, \quad p \in \{2, 3, 5, 7 \cdots\}. \tag{1.540}$$

Therefore, $(\mathbb{Z}[\sqrt{-2}], +, \cdot)$ is a Euclidean domain just like the Gaussian integers $(\mathbb{Z}[\sqrt{-1}], \oplus, \odot)$ [17, p. 4].

However, $(\mathbb{Z}[\sqrt{-3}], +, \cdot)$, $(\mathbb{Z}[\sqrt{-5}], +, \cdot)$, $(\mathbb{Z}[\sqrt{-7}], +, \cdot)$, etc. are no Euclidean domains; cf., e. g., [17, p. 5].

Example 1.24. Let us consider the set of monic polynomials $(\mathbb{F}_2[x]_n, \oplus, \odot)$ of maximum degree $n \in \mathbb{N}^*$ over the finite field $\mathbb{F}_2$, i. e., the monic polynomials with coefficients in $\mathbb{F}_2$. $(\mathbb{F}_2[x]_n, \oplus, \odot)$ is an integral domain because $\mathbb{F}_2$ is a field [17, p. 4]. $\mathbb{F}_2[x]_n$ is defined by

$$\mathbb{F}_2[x]_n := \left\{a(x) = x^n \oplus a_{n-1} \odot x^{n-1} \oplus \cdots \oplus a_1 \odot x \oplus a_0 \mid a_i \in \mathbb{F}_2, i \in \left\{0, 1 \cdots (n-1)\right\}, n \in \mathbb{N}\right\} \tag{1.541}$$

with

$$h\left\{a(x)\right\} = \deg\left(a(x)\right) = n \ge 0, \quad a(x) \in \mathbb{F}_2[x]_n, \quad a(x) \ne 0, \quad n \in \mathbb{N}^*, \tag{1.542}$$

i. e., the degree of the polynomial $a(x)$.

Let us evaluate whether $(\mathbb{F}_2[x]_n, \oplus, \odot)$ is a Euclidean domain [17, p. 4]. Therefore, let $m \le n$, $m, n \in \mathbb{N}^*$, and define

$$\beta(x) = x^m \oplus b_{m-1} \odot x^{m-1} \oplus \cdots \oplus b_1 \odot x \oplus b_0, \tag{1.543}$$
$$\beta(x) \in \mathbb{F}_2[x]_n, \quad \beta(x) \ne 0, \quad b_k \in \mathbb{F}_2, \quad k \in \left\{0, 1 \cdots (m-1)\right\}, \quad m, n \in \mathbb{N}^*,$$

with

$$h\left\{\beta(x)\right\} = m \ge 0, \quad \beta(x) \ne 0, \quad n \ge m, \quad m, n \in \mathbb{N}^*. \tag{1.544}$$

We yield

$$\begin{aligned}
\alpha(x) \odot \beta(x) &= \left(\bigoplus_{i=0}^{n} a_i \odot x^i \right) \odot \left(\bigoplus_{k=0}^{m} b_k \odot x^k \right) \\
&= \bigoplus_{i=0}^{n} \bigoplus_{k=0}^{m} a_i \odot b_k \odot x^{i+k} \\
&= a_0 \odot b_0 \\
&\quad \oplus \ (a_1 \odot b_0 \ \oplus \ a_0 \odot b_1) \odot x^1 \\
&\quad \oplus \ (a_2 \odot b_0 \ \oplus \ a_1 \odot b_1 \ \oplus \ a_0 \odot b_2) \odot x^2 \\
&\quad \dots \\
&\quad \oplus \ x^{n+m} \\
&= \bigoplus_{i=0}^{n+m} \ \bigoplus_{k=\max\{0,i-m\}}^{\min\{n,i\}} a_k \odot b_{i-k} \odot x^i, \\
&\quad \alpha(x), \beta(x) \in \mathbb{F}_2[x]_n, \quad \alpha(x) \neq 0, \quad \beta(x) \neq 0, \\
&\quad a_i \in \mathbb{F}_2, \quad i \in \{0, 1 \cdots (n-1)\}, \quad b_k \in \mathbb{F}_2, \quad k \in \{0, 1 \cdots (m-1)\}, \\
&\quad a_n = b_m = 1, \quad n \geq m, \quad m, n \in \mathbb{N}^*.
\end{aligned} \tag{1.545}$$

Furthermore,

$$h\{\alpha(x) \odot \beta(x)\} = h\{\alpha(x)\} + h\{\beta(x)\} = n + m \geq 0, \tag{1.546}$$
$$\alpha(x), \beta(x) \in \mathbb{F}_2[x]_n, \quad \alpha(x) \neq 0, \quad \beta(x) \neq 0,$$
$$a_n = b_m = 1, \quad n \geq m, \quad m, n \in \mathbb{N}^*,$$

which shows that $h\{\alpha(x)\}$ is additive.

In addition, we have

$$h\{\alpha(x) \oplus \beta(x)\} = \begin{cases} \max\{h\{\alpha(x)\}, h\{\beta(x)\}\} & \text{if } h\{\alpha(x)\} \neq h\{\beta(x)\}, \\ h\{\alpha(x)\} - 1 & \text{if } h\{\alpha(x)\} = h\{\beta(x)\}. \end{cases} \tag{1.547}$$

We will first have to check the first property given in Definition 1.47. Equation (1.546) leads to

$$h\{\alpha(x) \odot \beta(x)\} = h\{\alpha(x)\} + h\{\beta(x)\} = n + \underbrace{m}_{\geq 0} \geq n = h\{\alpha(x)\}, \tag{1.548}$$
$$\alpha(x), \beta(x) \in \mathbb{F}_2[x]_n, \quad \alpha(x) \neq 0, \quad \beta(x) \neq 0,$$
$$a_n = b_m = 1, \quad n \geq m, \quad m, n \in \mathbb{N}^*.$$

So, the first property of Definition 1.47 is fulfilled.

Now, let us look at the second property given in Definition 1.47. Let us first assume that $\beta(x)|\alpha(x)$. In this case, the remainder polynomial $r(x)$ vanishes, i. e., $r(x)$ is zero. So, this case fulfills the first part of the second property given in Definition 1.47.

Let us now consider the second part of the second property, i. e., the case in which $r(x)$ does not vanish and which requires $h\{\beta(x)\} \geq 1$. If we admitted $h\{\beta(x)\} > h\{\alpha(x)\}$, we can choose

$$q(x) = 0 \quad \Rightarrow \quad r(x) = \alpha(x), \tag{1.549}$$
$$q(x), r(x), \alpha(x) \in \mathbb{F}_2[x]_n, \quad \alpha(x) \neq 0, \quad a_n = 1, \quad n \in \mathbb{N}^*,$$

clearly, with $h\{\beta(x)\} > h\{r(x)\}$. So, the second property given in Definition 1.47 is fulfilled.

Now, suppose $h\{\beta(x)\} \leq h\{\alpha(x)\}$, yielding

$$\begin{aligned} \alpha(x) = q(x) \odot \beta(x) \oplus r(x) \quad &\Leftrightarrow \quad r(x) = \alpha(x) \oplus q(x) \odot \beta(x), \\ q(x), r(x), \alpha(x), \beta(x) \in \mathbb{F}_2[x]_n, \quad &\alpha(x) \neq 0, \quad \beta(x) \neq 0, \\ a_n = b_m = 1, \quad &m, n \in \mathbb{N}^*. \end{aligned} \tag{1.550}$$

We know that this is true for the trivial cases of first part of the second property given in Definition 1.47. So, we can use these as a Base Case. Let us now move on to the induction step. Suppose that

$$\alpha(x) = q(x) \odot \beta(x) \oplus r(x), \quad h\{r(x)\} < h\{\beta(x)\}, \tag{1.551}$$

hold for

$$h\{\alpha(x)\} = n, \quad n \in \mathbb{N}^*. \tag{1.552}$$

We will now have to consider the case

$$h\{\alpha(x)\} = n + 1, \quad n \in \mathbb{N}^*. \tag{1.553}$$

Let us hence suppose (1.553) and choose

$$\gamma(x) = x^{n+1-m}, \quad m, n \in \mathbb{N}^*, \tag{1.554}$$

which leads to

$$\gamma(x)\beta(x) = x^{n+1-m} \odot \beta(x) = x^{n+1} \oplus \cdots, \quad m, n \in \mathbb{N}^*, \tag{1.555}$$

with

$$h\{\gamma(x) \odot \beta(x)\} = h\{\gamma(x)\} + h\{\beta(x)\} = n + 1 - m + m = n + 1, \quad m, n \in \mathbb{N}^*. \tag{1.556}$$

Set

$$p(x) = \alpha(x) \oplus \gamma(x) \odot \beta(x), \tag{1.557}$$

which has degree at most n because $(x^{n+1} \oplus x^{n+1})$ cancels out

$$h\{p(x)\} = h\{\alpha(x) \oplus \gamma(x) \odot \beta(x)\} \leq n < n + 1, \quad n \in \mathbb{N}^*. \tag{1.558}$$

With the induction hypothesis for degree n, we find

$$p(x) = \tilde{q}(x) \odot \beta(x) \oplus \tilde{r}(x) \tag{1.559}$$

because $h\{p(x)\}$ is equal to n, as required and, therefore, we yield

$$\alpha(x) \oplus \gamma(x) \odot \beta(x) = \tilde{q}(x) \odot \beta(x) \oplus \tilde{r}(x), \quad h\{\tilde{r}(x)\} < h\{\beta(x)\}, \tag{1.560}$$

which immediately becomes

$$\begin{aligned} \alpha(x) &= \tilde{q}(x) \odot \beta(x) \oplus \tilde{r}(x) \oplus \gamma(x) \odot \beta(x) \\ &= \underbrace{\left[\tilde{r}(x) \oplus \gamma(x)\right]}_{=q(x)} \odot \beta(x) \oplus \underbrace{\tilde{r}(x)}_{=r(x)} \\ &= q(x) \odot \beta(x) \oplus r(x), \quad h\{\tilde{r}(x)\} = h\{r(x)\} < h\{\beta(x)\}. \end{aligned} \tag{1.561}$$

Now, we can even consider the uniqueness of the solution. Suppose that there were two distinct solutions using $q(x)$ and $r(x)$ as well as $\tilde{q}(x)$ and $\tilde{r}(x)$, i. e.,

$$a(x) = q(x) \odot \beta(x) \oplus r(x) = \tilde{q}(x) \odot \beta(x) \oplus \tilde{r}(x) \tag{1.562}$$

with

$$h\{r(x)\} < h\{\beta(x)\}, \quad h\{\tilde{r}(x)\} < h\{\beta(x)\} \tag{1.563}$$

and

$$[q(x) \oplus \tilde{q}(x)] \odot \beta(x) = r(x) \oplus \tilde{r}(x). \tag{1.564}$$

Now, we find

$$h\{[q(x) \oplus \tilde{q}(x)] \odot \beta(x)\} = h\{q(x) \oplus \tilde{q}(x)\} + h\{\beta(x)\}, \tag{1.565}$$

which leads to

$$h\{[q(x) \oplus \tilde{q}(x)]\} + h\{\beta(x)\} = \underbrace{h\{r(x) \oplus \tilde{r}(x)\}}_{<h\{\beta(x)\}} < h\{\beta(x)\}, \tag{1.566}$$

and thus

$$h\{[q(x) \oplus \tilde{q}(x)]\} < 0, \tag{1.567}$$

which requires $[q(x) \oplus \tilde{q}(x)]$ to be the null polynomial. Using this result in (1.564) immediately leads to

$$r(x) = \tilde{r}(x). \tag{1.568}$$

Therefore, the second property given in Definition 1.47 is fulfilled. □

Consequently, $(\mathbb{F}_2[x]_n, \oplus, \odot)$ is a Euclidean domain [17, p. 4].

1.12.3 Euclidean Algorithm

Euclidean domains are most important in the mathematical concept of cyclic block codes; cf., e. g., [17, Preface]. Therefore, we illustrated them in Section 1.12.2. We found that a Euclidean domain requires some division algorithm. It does not require much fantasy to guess that such a division algorithm will be most helpful when it comes to the decoding of a cyclic $(n, k, d_{\min})$ binary linear block code $\mathbb{V}$ [15, pp. 155–162], [5, Definition 5.4, pp. 144–161]. Until now, we did not illustrate how such a division algorithm might look like. We will make up for this in Section 1.12.3.

The second property of Definition 1.47 leads us to the search for the *greatest common divisor (gcd)* (in German “größter gemeinsamer Teiler, ggT”) [11, p. 333], [17, p. 4]. If $\alpha|\beta_i$, $\beta_i \in \{\beta_1, \beta_2, \beta_3 \cdots \beta_n\}$, then α is said to be a common divisor of all the β_i, $i \in \{1, 2 \cdots n\}$, $n \in \mathbb{N}^*$ [11, p. 333], [17, p. 4]. If δ is a common divisor of all the β_i, $i \in \{1, 2 \cdots n\}$, $n \in \mathbb{N}^*$, and if every other common divisor of all these β_i, $i \in \{1, 2 \cdots n\}$, $n \in \mathbb{N}^*$ divides δ, then δ is said to be the *greatest common divisor (gcd)* denoted by [11, p. 333], [17, p. 4]

$$\delta = \gcd\{\beta_1, \beta_2, \beta_3 \cdots \beta_n\}, \quad n \in \mathbb{N}^*. \tag{1.569}$$

If the greatest common divisor $\gcd\{\beta_1, \beta_2, \beta_3 \cdots \beta_n\}$ of the set $\{\beta_1, \beta_2, \beta_3 \cdots \beta_n\}$ is equal to 1 or any other unit, the β_i, $i \in \{1, 2 \cdots n\}$, $n \in \mathbb{N}^*$ are termed *coprime, relatively prime* or *mutually prime* (in German "*teilerfremd*") [11, p. 333], [17, p. 14].

Let us look at an example.

Example 1.25. Let us consider $\mathbb{N}$. We are interested in finding the gcd of

$$\alpha = 64 \tag{1.570}$$

and

$$\beta = 48. \tag{1.571}$$

We know that

$$\begin{array}{ccccccc} \alpha & = & 64 & = & 2\cdot2\cdot2\cdot2 & \cdot & 2\cdot2, \\ \beta & = & 48 & = & 2\cdot2\cdot2\cdot2 & \cdot & 3. \end{array} \tag{1.572}$$

Obviously, 64 and 48 have

$$\delta = 2^4 = 16 \tag{1.573}$$

in common, which indeed is their greatest common divisor (gcd).

The "brute force" strategy used in Example 1.25 does however not seem to be very efficient. Therefore, we will have to look for something better if not to say much better.

A first step toward an efficient algorithm for finding the gcd is the following theorem.

Theorem 1.31 (Greatest Common Divisor as a Linear Combination). *If*

$$\mathbb{B} = \{\beta_1, \beta_2, \beta_3 \cdots \beta_n\} \subseteq \mathbb{E}, \quad n \in \mathbb{N}^*, \tag{1.574}$$

is a finite subset of a Euclidean domain $(\mathbb{E}, \oplus, \odot)$, then $\mathbb{E}$ has a greatest common divisor (gcd) δ, which can be expressed as a linear combination of all the β_i, $i \in \{1, 2 \cdots n\}$, $n \in \mathbb{N}^*$, i. e., [17, p. 5]

$$\delta = \bigoplus_{i=1}^{n} \lambda_i \odot \beta_i, \quad \beta_i \in \mathbb{B} \subseteq \mathbb{E}, \quad \lambda_i \in \mathbb{E}, \quad i \in \{1, 2 \cdots n\}, \quad n \in \mathbb{N}^*. \tag{1.575}$$

Proof. Let

$$\mathbb{S} = \left\{ \bigoplus_{i=1}^{n} \mu_i \odot \beta_i \,\middle|\, \mu_i \in \mathbb{E} \right\} \tag{1.576}$$

and let $\delta \in \mathbb{S}$, $\delta \neq 0$, with the smallest possible value of $h\{\delta\}$ [17, p. 5]. As an element of $\mathbb{S}$, $\delta \in \mathbb{S}$ can be given as a linear combination of all the β_i, $i \in \{1, 2 \cdots n\}$, $n \in \mathbb{N}^*$; cf. (1.576) [17, p. 5].

We now claim that $\delta \in \mathbb{S}$ is a gcd of the β_i's [17, p. 5].

We will first show that $\delta | \beta_i$, $i \in \{1, 2 \cdots n\}$, $n \in \mathbb{N}^*$ [17, p. 5]. Since $\delta \neq 0$, according to Definition 1.47, we may write

$$\beta_i = q_i \odot \delta \oplus r_i, \quad \beta_i, \delta \in \mathbb{S}, \quad q_i \in \mathbb{E}, \quad i \in \{1, 2 \cdots n\}, \quad n \in \mathbb{N}^*, \tag{1.577}$$

with either r_i equal to 0 or $h\{r_i\} < h\{\delta\}$ for all $i \in \{1, 2 \cdots n\}, n \in \mathbb{N}^*$ [17, p. 5].

Since $\beta_i, \delta \in \mathbb{S}$ and $q_i \in \mathbb{E}$, r_i is also an element of $\mathbb{S}$, because it is a linear combination of elements of $\mathbb{S}$ [17, p. 5]. Denoting the additive inverse $\oplus(q_i \odot \delta)^{-1}$ of $(q_i \odot \delta)$, $i \in \{1, 2 \cdots n\}, n \in \mathbb{N}^*$, by $\ominus(q_i \odot \delta)$ for the sake of an easier notation, we obtain [17, p. 5]

$$r_i = \beta_i \ominus (q_i \odot \delta), \quad r_i, \beta_i, \delta \in \mathbb{S}, \quad q_i \in \mathbb{E}, \quad i \in \{1, 2 \cdots n\}, \quad n \in \mathbb{N}^*. \tag{1.578}$$

Note that in the case of Euclidean domains formed over $\mathbb{F}_2$, $\ominus$ and $\oplus$ are identical; cf., e. g., [4, equation (14) and below, p. 197]. However, to allow a more general view, we will use $\ominus$ where appropriate.

Since $\delta \in \mathbb{S}$ was chosen to have a smallest possible value of $h\{\delta\}$ among the nonzero elements of $\mathbb{S}$, and since Definition 1.47 requires $h\{r_i\} < h\{\delta\}$, r_i must vanish, i. e.,

$$r_i = 0, \quad i \in \{1, 2 \cdots n\}, \quad n \in \mathbb{N}^* \tag{1.579}$$

and, therefore, (1.577) becomes [17, p. 5]

$$\beta_i = q_i \odot \delta, \quad \beta_i, \delta \in \mathbb{S}, \quad q_i \in \mathbb{E}, \quad i \in \{1, 2 \cdots n\}, \quad n \in \mathbb{N}^*. \tag{1.580}$$

Thus, $\delta \in \mathbb{S}$ is a common divisor of the β_i, $i \in \{1, 2 \cdots n\}, n \in \mathbb{N}^*$ [17, p. 5].

Now, let $\varepsilon \in \mathbb{S}$ be another common divisor of the β_i, $i \in \{1, 2 \cdots n\}, n \in \mathbb{N}^*$, fulfilling [17, p. 5]

$$\beta_i = q_i' \odot \varepsilon, \quad \beta_i, \varepsilon \in \mathbb{S}, \quad q_i' \in \mathbb{E}, \quad i \in \{1, 2 \cdots n\}, \quad n \in \mathbb{N}^*. \tag{1.581}$$

Due to $\delta \in \mathbb{S}$, we can write

$$\begin{aligned} \delta &= \bigoplus_{i=1}^{n} \lambda_i \odot \beta_i \\ &= \bigoplus_{i=1}^{n} \lambda_i \odot q_i' \odot \varepsilon \\ &= \varepsilon \odot \bigoplus_{i=1}^{n} \lambda_i \odot q_i', \\ &\beta_i, \delta, \varepsilon \in \mathbb{S}, \quad q_i', \lambda_i \in \mathbb{E}, \quad i \in \{1, 2 \cdots n\}, \quad n \in \mathbb{N}^*. \end{aligned} \tag{1.582}$$

Equation (1.582) shows that $\delta \in \mathbb{S}$ is a multiple of $\varepsilon \in \mathbb{S}$, with $\varepsilon | \delta$. Therefore, $\delta \in \mathbb{S}$ is the greatest common divisor (gcd) for all the β_i, $i \in \{1, 2 \cdots n\}, n \in \mathbb{N}^*$ just as claimed before [17, p. 5]. □

Theorem 1.31 assures us that greatest common divisors exist [17, p. 5].

In what follows, we will again set out from the Euclidean domain $(\mathbb{E}, \oplus, \odot)$. Furthermore, we will denote the additive inverse $\oplus a^{-1} \in \mathbb{E}$ of $a \in \mathbb{E}$ by $\ominus a \in \mathbb{E}$ for the sake of an easier notation.

Before being able to find greatest common divisors efficiently, the following fact will be helpful.

Theorem 1.32 (Greatest Common Divisor of Linear Functions). *We claim* [17, p. 6]

$$gcd\{s, t\} = gcd\{s, t \ominus r \odot s\} \quad \forall r, s, t \in \mathbb{E}. \tag{1.583}$$

Proof. Let δ be the $\gcd\{s,t\}$. Since $\delta|s$ and $\delta|t$, $\delta|rs$, and thus $\delta|(t \ominus r \odot s)$ [17, p. 6]. So, every common divisor of s and t is also a common divisor of $(t \ominus r \odot s)$ [17, p. 6]. Similarly, a common divisor of s and of $(t \ominus r \odot s)$ must be a common divisor of s and $(t \ominus r \odot s) \oplus r \odot s$, for which, of course,

$$(t \ominus r \odot s) \oplus r \odot s = t \tag{1.584}$$

holds [17, p. 6]. Therefore, a common divisor of $s \in \mathbb{E}$ and of $(t \ominus r \odot s) \in \mathbb{E}$ must be a common divisor of $s \in \mathbb{E}$ and $t \in \mathbb{E}$ [17, p. 6]. □

In what follows, let us use the following notation:

$$t^n = \underbrace{t \odot t \odot t \odot \cdots \odot t}_{n \text{ terms}}, \quad t \in \mathbb{E}, \quad n \in \mathbb{N}^*. \tag{1.585}$$

With Theorem 1.32, we can prove the following.

Theorem 1.33 (Greatest Common Divisor of Polynomials). *Let* $1, t \in \mathbb{E}$ *be elements in any domain* $(\mathbb{E}, \oplus, \odot)$, *in which greatest common divisors exist* [17, p. 6]. *Then, if* $m, n \in \mathbb{N}^*$ *are positive integers, we have* [17, p. 6]

$$gcd\left\{t^n \ominus 1, t^m \ominus 1\right\} = t^{gcd\{n,m\}} \ominus 1, \quad 1, t \in \mathbb{E}, \quad m, n \in \mathbb{N}^*. \tag{1.586}$$

Proof. We will prove (1.586) by induction.

Base Case

Let $m = n = 1$, i. e., $\max\{m, n\} = 1$. Clearly, $\gcd\{1, 1\} = 1$. Then (1.586) takes the form

$$\gcd\{t \ominus 1, t \ominus 1\} = t \ominus 1, \quad 1, t \in \mathbb{E}, \tag{1.587}$$

which is a true claim.

Now, let $m = n$. Clearly, $\gcd\{n, n\} = n$. Then (1.586) takes the form

$$\gcd\left\{t^n \ominus 1, t^n \ominus 1\right\} = t^n \ominus 1, \quad 1, t \in \mathbb{E}, \quad n \in \mathbb{N}^*, \tag{1.588}$$

which is also a true claim.

Induction Step

We suppose that

$$\gcd\left\{t^n \ominus 1, t^m \ominus 1\right\} = t^{\gcd\{n,m\}} \ominus 1, \quad 1, t \in \mathbb{E}, \quad m, n \in \mathbb{N}^*, \tag{1.589}$$

is true [17, p. 6]. We will call this the *induction hypothesis* in what follows.

Without loss of generality, let us assume that $m < n$, $m, n \in \mathbb{N}^*$ [17, p. 6]. Also, we have [17, p. 6]

$$\begin{aligned} t^{n-m} \ominus 1 &= \left(t^n \ominus 1\right) \ominus \left(t^n \ominus 1\right) \oplus \left(t^{n-m} \ominus 1\right) \\ &= \left(t^n \ominus 1\right) \ominus \left\{\left(t^n \ominus 1\right) \ominus \left(t^{n-m} \ominus 1\right)\right\} \\ &= \left(t^n \ominus 1\right) \ominus \left(t^n \ominus t^{n-m}\right) \\ &= \left(t^n \ominus 1\right) \ominus t^{n-m} \odot \left(t^m \ominus 1\right), \quad 1, t \in \mathbb{E}, \quad m, n \in \mathbb{N}^*. \end{aligned} \tag{1.590}$$

Using Theorem 1.32, using (1.589) and using (1.590) as well as setting

$$\text{“}s\text{”} = \left(t^m \ominus 1\right), \quad \text{“}t\text{”} = \left(t^n \ominus 1\right), \quad \text{“}r\text{“} = t^{n-m}, \tag{1.591}$$

we obtain [17, p. 6]

$$\begin{aligned}\gcd\{t^m \ominus 1, t^n \ominus 1\} &= \gcd\{t^m \ominus 1, (t^n \ominus 1) \ominus t^{n-m} \odot (t^m \ominus 1)\}\\ &= \gcd\{t^m \ominus 1, t^{n-m} \ominus 1\}\\ &= t^{\gcd\{m,n-m\}} \ominus 1, \quad 1, t \in \mathbb{E}, \quad m, n \in \mathbb{N}^*.\end{aligned} \tag{1.592}$$

Now, let us consider $\gcd\{m, n-m\}$. Using Theorem 1.32 and setting

$$\text{“}s\text{”} = m, \quad \text{“}t\text{”} = n, \quad \text{“}r\text{”} = 1, \tag{1.593}$$

we yield [17, p. 6]

$$\gcd\{m, n-m\} = \gcd\{m, n\}. \tag{1.594}$$

With (1.594), (1.592) yields [17, p. 6]

$$\gcd\{t^m \ominus 1, t^n \ominus 1\} = t^{\gcd\{m,n\}} \ominus 1, \quad 1, t \in \mathbb{E}, \quad m, n \in \mathbb{N}^*. \tag{1.595}$$

□

Theorem 1.33 leads us to the following obvious conclusion.

Corollary 1.6 (Greatest Common Divisor of Polynomials of Degree q^m). *Let $x \in \mathbb{E}$ be an element in any domain $(\mathbb{E}, \oplus, \odot)$, in which greatest common divisors exist* [17, p. 7]. *Then, if $m, n \in \mathbb{N}$ are integers and $q \in \mathbb{N}^*$ is a positive integer, we claim* [17, p. 7]

$$gcd\{x^{q^m} \ominus x, x^{q^n} \ominus x\} = x^{q^{gcd\{m,n\}}} \ominus x, \quad x \in \mathbb{E}, \quad m, n \in \mathbb{N}, \quad q \in \mathbb{N}^*. \tag{1.596}$$

Proof. Since

$$\gcd\{x^{q^m} \ominus x, x^{q^n} \ominus x\} = x \odot \gcd\{x^{q^m-1} \ominus 1, x^{q^n-1} \ominus 1\},$$
$$x \in \mathbb{E}, \quad m, n \in \mathbb{N}, \quad q \in \mathbb{N}^*, \tag{1.597}$$

we will consider

$$\gcd\{x^{q^m-1} \ominus 1, x^{q^n-1} \ominus 1\}, \quad x \in \mathbb{E}, \quad m, n \in \mathbb{N}, \quad q \in \mathbb{N}^*, \tag{1.598}$$

in what follows. Setting

$$\text{“}n\text{”} = q^m - 1, \quad \text{“}m\text{”} = q^n - 1, \quad m, n \in \mathbb{N}, \quad q \in \mathbb{N}^*, \tag{1.599}$$

and using Theorem 1.33, we find

$$\gcd\{x^{q^m-1} \ominus 1, x^{q^n-1} \ominus 1\} = x^{\gcd\{[q^m-1],[q^n-1]\}} \ominus 1, \quad x \in \mathbb{E}, \quad m, n \in \mathbb{N}, \quad q \in \mathbb{N}^*. \tag{1.600}$$

Using Theorem 1.33 again, we yield

$$\gcd\{q^m - 1, q^n - 1\} = q^{\gcd\{m,n\}} - 1, \quad m, n \in \mathbb{N}, \quad q \in \mathbb{N}^*. \tag{1.601}$$

Hence, we yield

$$\gcd\{x^{q^m-1} \ominus 1, x^{q^n-1} \ominus 1\} = x^{q^{\gcd\{m,n\}}-1} \ominus 1, \quad x \in \mathbb{E}, \quad m, n \in \mathbb{N}, \quad q \in \mathbb{N}^*, \tag{1.602}$$

and thus

$$\gcd\left\{x^{q^m} \ominus x, x^{q^n} \ominus x\right\} = x \odot \gcd\left\{x^{q^m-1} \ominus 1, x^{q^n-1} \ominus 1\right\} = x^{q^{\gcd\{m,n\}}} \ominus x, \tag{1.603}$$
$$x \in \mathbb{E}, \quad m,n \in \mathbb{N}, \quad q \in \mathbb{N}^*.$$

□

Now, we are prepared for a division algorithm, and not just an ordinary one—no, Madam, no Sir, oh no! We take the best that we can get! We take the *Euclidean algorithm* [11, p. 333f.], [17, p. 7f.], [19, pp. 718–721], developed by the one and only *Euclid*. The Euclidean algorithm is the most efficient way to find greatest common divisors [17, p. 5], [19, pp. 718–721].

In what follows, we will consider an extended version of the Euclidean algorithm. The extension of the "conventional" Euclidean algorithm is the computation of a special type of coefficients, called the *Bézout coefficients* s_i and t_i, $i \in \{-1, 0, 1, 2, 3 \cdots (n+1)\}$, $n \in \mathbb{N}^*$. The Bézout coefficients are named after the French mathematician Etienne *Bézout* (1730–1783), whose findings are closely related to the work of the also French mathematician Claude Gaspar Bachet *de Méziriac* (1581–1638) [44, p. 47].

We will set out from a Euclidean domain $(\mathbb{E}, \oplus, \odot)$. Suppose that we are given two elements $\alpha, \beta \in \mathbb{E}$, which shall both be nonzero, i. e., $\alpha \neq 0$ and $\beta \neq 0$. We want to find $\gcd\{\alpha, \beta\} \in \mathbb{E}$ [17, p. 7f.]. For definiteness, let as assume [17, pp. 7f.]

$$h\{\alpha\} \geq h\{\beta\}, \quad \alpha \neq 0, \beta \neq 0, \quad \alpha, \beta \in \mathbb{E}. \tag{1.604}$$

Let us term q_i with $i \in \{-1, 0, 1, 2, 3 \cdots (n+1)\}$, $n \in \mathbb{N}^*$, the "quotient," and let r_i with $i \in \{-1, 0, 1, 2, 3 \cdots (n+1)\}$, $n \in \mathbb{N}^*$, be the "remainder."

Now, it is time to define the *extended Euclidean algorithm.*

Definition 1.48 (Extended Euclidean Algorithm).

Initialization

First, we initialize the "remainders" r_{-1} and r_0 as well as the Bézout coefficients s_{-1}, t_{-1}, s_0 and t_0,

$$\begin{aligned} r_{-1} &= \alpha, \quad s_{-1} = 1, \quad t_{-1} = 0, \\ r_0 &= \beta, \quad s_0 = 0, \quad t_0 = 1. \end{aligned} \tag{1.605}$$

Iteration

For $i \in \{1, 2, 3 \cdots (n+1)\}$, $n \in \mathbb{N}^*$ compute

$$\begin{aligned} r_i &= r_{i-2} \ominus q_i \odot r_{i-1}, \\ s_i &= s_{i-2} \ominus q_i \odot s_{i-1}, \\ t_i &= t_{i-2} \ominus q_i \odot t_{i-1}, \end{aligned} \tag{1.606}$$

until the remainder r_{n+1} equal to 0 is reached, making sure that $h\{r_i\} < h\{r_{i-1}\}$ holds [17, p. 7].

Output

The $\gcd\{\alpha,\beta\} \in \mathbb{E}$ is given by [17, p. 7]

$$\gcd\{\alpha,\beta\} = r_n, \tag{1.607}$$

which can be easily verified by determining [17, pp. 7, 9]

$$s_n \odot \alpha \oplus t_n \odot \beta \stackrel{!}{=} r_n = \gcd\{\alpha,\beta\}. \tag{1.608}$$

Let us prove (1.607).

Proof 1.1 (Proof of (1.607)). We will set out from (1.606), in particular, we will use

$$r_i = r_{i-2} \ominus q_i \odot r_{i-1}. \tag{1.609}$$

Using Theorem 1.32, we yield

$$\gcd\{r_{i-1},r_i\} = \gcd\{r_{i-1},r_{i-2} \ominus q_i \odot r_{i-1}\} = \gcd\{r_{i-1},r_{i-2}\}, \tag{1.610}$$

i. e., [17, p. 8]

$$\gcd\{r_{i-1},r_i\} = \gcd\{r_{i-2},r_{i-1}\}. \tag{1.611}$$

Since

$$\gcd\{r_{-1},r_0\} = \gcd\{\alpha,\beta\}, \quad r_{n+1} = 0, \tag{1.612}$$

hold, using $i \in \{0,1,2,3\cdots(n+1)\}$, $n \in \mathbb{N}^*$, leads to

$$\gcd\{\alpha,\beta\} = \gcd\{r_0,r_1\} = \gcd\{r_1,r_2\} = \cdots = \gcd\{r_n,r_{n+1}\} = \gcd\{r_n,0\} = r_n, \tag{1.613}$$

which proves (1.607). □

Now, let us take a closer look at (1.608). We will prove [17, problem 8d, pp. 9, 11]

$$s_i \odot \alpha \oplus t_i \odot \beta = r_i, \quad i \in \{-1,0,1,2,3\cdots(n+1)\}, \quad n \in \mathbb{N}^*. \tag{1.614}$$

Proof 1.2 (Proof of (1.614)). We will prove (1.614) by induction.

Base Case

For $i = -1$, we have

$$s_{-1} \odot \alpha \oplus t_{-1} \odot \beta = r_{-1}, \tag{1.615}$$

and with (1.605), we obtain

$$1 \odot \alpha \oplus 0 \odot \beta = \alpha, \tag{1.616}$$

which is true [17, p. 9].

For $i = 0$, we have

$$s_0 \odot \alpha \oplus t_0 \odot \beta = r_0 \tag{1.617}$$

and with (1.605)

$$0 \odot \alpha \oplus 1 \odot \beta = \beta, \tag{1.618}$$

which is also true [17, p. 9].

Induction Step

The induction hypothesis is given by [17, p. 9]

$$s_{i-2} \odot \alpha \oplus t_{i-2} \odot \beta = r_{i-2}, \quad s_{i-1} \odot \alpha \oplus t_{i-1} \odot \beta = r_{i-1}. \tag{1.619}$$

Now, we need to prove

$$s_i \odot \alpha \oplus t_i \odot \beta = r_i. \tag{1.620}$$

We will do so by using (1.609) and the above induction hypothesis, which yields

$$\begin{aligned} r_i &= r_{i-2} \ominus q_i \odot r_{i-1} \\ &= s_{i-2} \odot \alpha \oplus t_{i-2} \odot \beta \ominus q_i \odot (s_{i-1} \odot \alpha \oplus t_{i-1} \odot \beta) \\ &= (s_{i-2} \ominus q_i \odot s_{i-1}) \odot \alpha \oplus (t_{i-2} \ominus q_i \odot t_{i-1}) \odot \beta. \end{aligned} \tag{1.621}$$

Since according to (1.606)

$$s_i = s_{i-2} \ominus q_i \odot s_{i-1}, \quad t_i = t_{i-2} \ominus q_i \odot t_{i-1}, \tag{1.622}$$

holds, (1.621) becomes

$$r_i = s_i \odot \alpha \oplus t_i \odot \beta, \tag{1.623}$$

which proves (1.620) [17, p. 9]. □

Note that the Bézout coefficients are not unique [17, p. 11], [44, p. 47], because setting out from (1.608) and denoting the $\gcd\{\alpha, \beta\}$ by δ yield

$$\begin{aligned} \delta &= s_n \odot \alpha \oplus t_n \odot \beta \\ &= s_n \odot \alpha \oplus t_n \odot \beta \oplus \underbrace{k \odot \beta \odot \alpha \ominus k \odot \beta \odot \alpha}_{=0} \\ &= s_n \odot \alpha \oplus k \odot \beta \odot \alpha \oplus t_n \odot \beta \ominus k \odot \alpha \odot \beta \\ &= \underbrace{(s_n \oplus k \odot \beta)}_{=s_{n,k}} \odot \alpha \oplus \underbrace{(t_n \ominus k \odot \alpha)}_{=t_{n,k}} \odot \beta \\ &= s_{n,k} \odot \alpha \oplus t_{n,k} \odot \beta, \quad k \in \mathbb{Z}, \quad n = \text{const}. \end{aligned} \tag{1.624}$$

Indeed, there is an infinite number of Bézout coefficients $s_{n,k}$ and $t_{n,k}$, $k \in \mathbb{Z}$, which fulfill (1.608). Although this might be puzzling, we will not suffer from this specialty when using the extended Euclidean algorithm. The appealing benefit of using the Bézout coefficients is the opportunity to verify the computed greatest common divisor (gcd). And this is why the author prefers the extended Euclidean algorithm to the regular Euclidean algorithm.

Remark 1.35 (About Number Theory and "All That Jazz"). Bored? Don't have any idea of *"What's Going On"* (Marvin Gaye, 1971; what a fantastic soul artist!)? Let me try to get us all back on track, let us become a bit *"Distracted"* (Al Jarreau, 1980; what a genius!). I do hope that we will find some happiness on our own *"Proud Mary"* (Creedence Clearwater Revival, 1969; and a great cover version by Ike and Tina Turner, 1971).

You might recall your earliest years in grammar school while you were struggling your way through mathematical concepts you did not even know were part of number theory because no one had told you. Maybe you had to determine the greatest common divisors of some arbitrarily chosen natural numbers and you kept constantly asking yourself what in this whole wide world was in so desperate need of knowing the respective results of your endeavors. Am I right?

Maybe at some point in time you dared to ask this "silly question about the deeper meaning of it all." Regrettably, the standard answer has been that you had to learn natural numbers because you must know how to count. That sounds just as cool as if someone told you that you can only have dessert if you eat your meat first. Well then, if you were Garfield you might say "one life down, eight to go."

Maybe you were even heroic enough to demand that someone told you an algorithm to find the greatest common divisors. Often the standard answer has sounded like that there was none and, therefore, you should develop "a feeling for it" and have "the guts" to claim what you believe.

Now, let me try to give my answers. Here comes the first one. Determining the greatest common divisors is a great tool for analyzing, and even more important, for decoding cyclic binary linear block codes, which have been widely used in almost uncountable applications. Being a "communications guy" that sounds good to me. Finding greatest common divisors seem to be worthwhile. And the next answer is *"oh yes, there is a great algorithm, the Euclidean algorithm"!*

Feeling a bit better? Well then, let us get *"Back To Life"* (Soul II Soul, 1989), and let us first look at natural numbers, just for the sake of *"Good Times"* (Chic, 1979) we all have had when we were young.

Example 1.26. Let us compute gcd{64, 48} using the extended Euclidean algorithm. The results of the extended Euclidean algorithm are summarized as follows:

iteration number		Bézout coefficients		remainder	quotient	
	i	s_i	t_i	r_i	q_i	
init	−1	*1*	*0*	$\alpha = 64$	–	
	0	*0*	*1*	$\beta = 48$	–	
$n =$	**1**	**1**	**−1**	**16**	**1**	$= \lfloor r_{-1}/r_0 \rfloor$
$n+1 =$	2	−3	4	0	3	$= \lfloor r_0/r_1 \rfloor$

Therefore,

$$\gcd\{64, 48\} = r_1 = 16, \tag{1.625}$$

and, using (1.608), we find

$$1 \cdot 64 + (-1) \cdot 48 = 16. \tag{1.626}$$

Example 1.27. Let us compute $\gcd\{6711, 831\}$.

iteration number		Bézout coefficients		remainder	quotient	
	i	s_i	t_i	r_i	q_i	
init	−1	*1*	*0*	$\alpha = 6711$	-	
	0	*0*	*1*	$\beta = 831$	-	
	1	1	−8	63	8	
	2	−13	105	12	13	
$n =$	3	**66**	**−533**	**3**	5	$= \lfloor r_1/r_2 \rfloor$
$n+1 =$	4	−277	2237	0	4	$= \lfloor r_2/r_3 \rfloor$

Therefore,

$$\gcd\{6711, 831\} = r_3 = 3, \tag{1.627}$$

and, using (1.608), we find

$$66 \cdot 6711 + (-533) \cdot 831 = 3. \tag{1.628}$$

Example 1.28. Let us consider $(\mathbb{F}_2[x]_8, \oplus, \odot)$. Let us compute $\gcd\{x^8, x^6 \oplus x^4 \oplus x^2 \oplus x \oplus 1\}$.

We begin with the following assignments.

iteration number		Bézout coefficients		remainder	quotient
	i	s_i	t_i	r_i	q_i
init	−1	*1*	*0*	$\alpha = x^8$	-
	0	*0*	*1*	$\beta = x^6 \oplus x^4 \oplus x^2 \oplus x \oplus 1$	-

$\underline{i = 1}$

We set out from

$$\begin{aligned} r_1 &= r_{-1} \oplus q_1 \odot r_0 = x^8 \oplus q_1 \odot \left(x^6 \oplus x^4 \oplus x^2 \oplus x \oplus 1\right), \\ s_1 &= s_{-1} \oplus q_1 \cdot s_0 = 1, \\ t_1 &= t_{-1} \oplus q_1 \cdot t_0 = q_1. \end{aligned} \tag{1.629}$$

Since

$$\begin{array}{lllllllllllll}
x^8 & & & & & & & & & & & & \\
x^8 & \oplus & x^6 & \oplus & x^4 & \oplus & x^3 & \oplus & x^2 & & & & \\
\hline
 & & x^6 & \oplus & x^4 & \oplus & x^3 & \oplus & x^2 & & & & \\
 & & x^6 & \oplus & x^4 & & & \oplus & x^2 & \oplus & x & \oplus & 1 \\
\hline
 & & & & & & x^3 & & & \oplus & x & \oplus & 1
\end{array} \quad : x^6 \oplus x^4 \oplus x^2 \oplus x \oplus 1 \;=\; x^2 \oplus 1, \tag{1.630}$$

holds, we choose

$$q_1 = x^2 \oplus 1, \quad r_1 = x^3 \oplus x \oplus 1, \tag{1.631}$$

and yield

$$s_1 = 1, \quad t_1 = x^2 \oplus 1. \tag{1.632}$$

Furthermore, we find

$$h\{r_1\} = 3 < h\{r_0\} = 6 < h\{r_{-1}\} = 8. \tag{1.633}$$

<u>$i = 2$</u>
We have

$$\begin{aligned} r_2 &= r_0 \oplus q_2 \odot r_1 = x^6 \oplus x^4 \oplus x^2 \oplus x \oplus 1 \oplus q_2 \odot \left(x^3 \oplus x \oplus 1\right), \\ s_2 &= s_0 \oplus q_2 \odot s_1 = q_2, \\ t_2 &= t_0 \oplus q_2 \odot t_1 = 1 \oplus q_2 \odot \left(x^2 \oplus 1\right). \end{aligned} \tag{1.634}$$

Since

$$\begin{array}{ccccccccccccl} x^6 & \oplus & x^4 & & & \oplus & x^2 & \oplus & x & \oplus & 1 & : x^3 \oplus x \oplus 1 & = \; x^3 \oplus 1, \\ x^6 & \oplus & x^4 & \oplus & x^3 & & & & & & & & \\ \hline & & & & x^3 & \oplus & x^2 & \oplus & x & \oplus & 1 & & \\ & & & & x^3 & & & \oplus & x & \oplus & 1 & & \\ \cline{5-11} & & & & & & x^2 & & & & & & \end{array} \tag{1.635}$$

we choose

$$q_2 = x^3 \oplus 1, \quad r_2 = x^2, \tag{1.636}$$

and yield

$$\begin{aligned} r_2 &= x^6 \oplus x^4 \oplus x^2 \oplus x \oplus 1 \oplus \left(x^3 \oplus 1\right) \odot \left(x^3 \oplus x \oplus 1\right) = x^2, \\ s_2 &= x^3 \oplus 1, \\ t_2 &= 1 \oplus \left(x^3 \oplus 1\right) \odot \left(x^2 \oplus 1\right) = x^5 \oplus x^3 \oplus x^2, \\ h\{r_2\} &= 2 < h\{r_1\} = 3 < h\{r_0\} = 6 < h\{r_{-1}\} = 8. \end{aligned} \tag{1.637}$$

<u>$i = 3$</u>
We have

$$\begin{aligned} r_3 &= r_1 \oplus q_3 \odot r_2 = \left(x^3 \oplus x \oplus 1\right) \oplus q_3 \odot x^2, \\ s_3 &= s_1 \oplus q_3 \odot s_2 = 1 \oplus q_3 \odot \left(x^3 \oplus 1\right), \\ t_3 &= t_1 \oplus q_3 \odot t_2 = \left(x^2 \oplus 1\right) \oplus q_3 \odot \left(x^5 \oplus x^3 \oplus x^2\right). \end{aligned} \tag{1.638}$$

Since

$$\begin{array}{cccccl} x^3 & \oplus & x & \oplus & 1 & : x^2 \; = \; x, \\ x^3 & & & & & \\ \hline & & x & \oplus & 1 & \end{array} \tag{1.639}$$

we choose

$$q_3 = x, \quad r_3 = x \oplus 1, \tag{1.640}$$

and yield

$$\begin{aligned} r_3 &= (x^3 \oplus x \oplus 1) \oplus x^3 = x \oplus 1, \\ s_3 &= 1 \oplus x \odot (x^3 \oplus 1) = x^4 \oplus x \oplus 1, \\ t_3 &= (x^2 \oplus 1) \oplus x \odot (x^5 \oplus x^3 \oplus x^2) = x^6 \oplus x^4 \oplus x^3 \oplus x^2 \oplus 1, \\ h\{r_3\} &= 1 < h\{r_2\} = 2 < h\{r_1\} = 3 < h\{r_0\} = 6 < h\{r_{-1}\} = 8. \end{aligned} \tag{1.641}$$

$\underline{i = 4}$
We have

$$\begin{aligned} r_4 &= r_2 \oplus q_4 \odot r_3 = x^2 \oplus q_4 \odot (x \oplus 1), \\ s_4 &= s_2 \oplus q_4 \odot s_3 = (x^3 \oplus 1) \oplus q_4 \odot (x^4 \oplus x \oplus 1), \\ t_4 &= t_2 \oplus q_4 \odot t_3 = (x^5 \oplus x^3 \oplus x^2) \oplus q_4 \odot (x^6 \oplus x^4 \oplus x^3 \oplus x^2 \oplus 1), \end{aligned} \tag{1.642}$$

we choose

$$q_4 = x \oplus 1, \quad r_4 = 1, \tag{1.643}$$

and yield

$$\begin{aligned} r_4 &= x^2 \oplus (x \oplus 1) \cdot (x \oplus 1) = 1, \\ s_4 &= x^3 \oplus 1 \oplus (x \oplus 1) \cdot (x^4 \oplus x \oplus 1) = x^5 \oplus x^4 \oplus x^3 \oplus x^2, \\ t_4 &= x^5 \oplus x^3 \oplus x^2 \oplus (x \oplus 1) \odot (x^6 \oplus x^4 \oplus x^3 \oplus x^2 \oplus 1) \\ &= x^7 \oplus x^6 \oplus x^3 \oplus x \oplus 1, \\ h\{r_4\} &= 0 < h\{r_3\} = 1 < h\{r_2\} = 2 < h\{r_1\} = 3 < h\{r_0\} = 6 < h\{r_{-1}\} = 8. \end{aligned} \tag{1.644}$$

$\underline{i = 5}$
We have

$$\begin{aligned} r_5 &= r_3 \oplus q_5 \odot r_4 = x \oplus 1 \oplus q_5 \odot 1, \\ s_5 &= s_3 \oplus q_5 \odot s_4 = x^4 \oplus x \oplus 1 \oplus q_5 \odot (x^5 \oplus x^4 \oplus x^3 \oplus x^2), \\ t_5 &= t_3 \oplus q_5 \odot t_4 = x^6 \oplus x^4 \oplus x^3 \oplus x^2 \oplus 1 \oplus q_5 \odot (x^7 \oplus x^6 \oplus x^3 \oplus x \oplus 1). \end{aligned} \tag{1.645}$$

Since

$$x \oplus 1 : 1 = x \oplus 1, \tag{1.646}$$

we choose

$$q_5 = x \oplus 1, \quad r_5 = 0, \tag{1.647}$$

and yield

$$\begin{aligned} r_5 &= x \oplus 1 \oplus x \oplus 1 = 0, \\ s_5 &= x^4 \oplus x \oplus 1 \oplus (x \oplus 1) \odot (x^5 \oplus x^4 \oplus x^3 \oplus x^2) \\ &= x^6 \oplus x^4 \oplus x^2 \oplus x \oplus 1, \\ t_5 &= x^6 \oplus x^4 \oplus x^3 \oplus x^2 \oplus 1 \oplus (x \oplus 1) \odot (x^7 \oplus x^6 \oplus x^3 \oplus x \oplus 1) \\ &= x^8, \end{aligned} \tag{1.648}$$

Therefore,

$$\gcd\{x^8, x^6 \oplus x^4 \oplus x^2 \oplus x \oplus 1\} = r_4 = 1. \tag{1.649}$$

Furthermore, using (1.608), we find

$$(x^5 \oplus x^4 \oplus x^3 \oplus x^2) \odot x^8 \oplus (x^7 \oplus x^6 \oplus x^3 \oplus x \oplus 1) \odot (x^6 \oplus x^4 \oplus x^2 \oplus x \oplus 1) = 1. \tag{1.650}$$

The following table summarizes the above results.

i	s_i	t_i	r_i	q_i
−1	1	0	$\alpha = x^8$	–
0	0	1	$\beta = x^6 \oplus x^4 \oplus x^2 \oplus x \oplus 1$	–
1	1	$x^2 \oplus 1$	$x^3 \oplus x \oplus 1$	$x^2 \oplus 1$
2	$x^3 \oplus 1$	$x^5 \oplus x^3 \oplus x^2$	x^2	$x^3 \oplus 1$
3	$x^4 \oplus x \oplus 1$	$x^6 \oplus x^4 \oplus x^3 \oplus x^2 \oplus 1$	$x \oplus 1$	x
4	$x^5 \oplus x^4 \oplus x^3 \oplus x^2$	$x^7 \oplus x^6 \oplus x^3 \oplus x \oplus 1$	1	$x \oplus 1$
5	$x^6 \oplus x^4 \oplus x^2 \oplus x \oplus 1$	x^8	0	$x \oplus 1$

Let us consider important properties of the Bézout coefficients.

Theorem 1.34 (Properties of the Bézout Coefficients). *We claim the following (cf.* [17, *problems 8a-c, problem 9, p. 11*]*):*

1:

$$t_i \odot r_{i-1} \ominus t_{i-1} \odot r_i = (\ominus 1)^i \odot \alpha, \quad i \in \{1, 2 \cdots (n+1)\}, \quad n \in \mathbb{N}^*. \tag{1.651}$$

2:

$$s_i r_{i-1} \odot s_{i-1} r_i = (\ominus 1)^{i+1} \odot \beta, \quad i \in \{1, 2 \cdots (n+1)\}, \quad n \in \mathbb{N}^*. \tag{1.652}$$

3:

$$s_i t_{i-1} \ominus s_{i-1} t_i = (\ominus 1)^{i+1}, \quad i \in \{1, 2 \cdots (n+1)\}, \quad n \in \mathbb{N}^*. \tag{1.653}$$

4:

$$\deg(t_i(x)) + \deg(r_{i-1}(x)) = \deg(\alpha(x)), \quad i \in \{1, 2 \cdots (n+1)\}, \quad n \in \mathbb{N}^*. \tag{1.654}$$

5:

$$\deg(s_i(x)) + \deg(r_{i-1}(x)) = \deg(\beta(x)), \quad i \in \{1, 2 \cdots (n+1)\}, \quad n \in \mathbb{N}^*. \tag{1.655}$$

Proof.
Claim 1
We prove (1.651) by induction.

Base Case
We start with $i = 0$ and yield

$$t_0 \odot r_{-1} \ominus t_{-1} \odot r_0 = 1 \odot r_{-1} \ominus 0 \odot r_0 = r_{-1} = \alpha = (\ominus 1)^0 \odot \alpha, \tag{1.656}$$

which is true.

Induction Step

The induction hypothesis is given by

$$t_{i-1} \odot r_{i-2} \ominus t_{i-2} \odot r_{i-1} = (\ominus 1)^{i-1} \odot \alpha. \tag{1.657}$$

We will now evaluate whether

$$t_i \odot r_{i-1} \ominus t_{i-1} \odot r_i = (\ominus 1)^i \odot \alpha \tag{1.658}$$

holds. We know from (1.606)

$$\begin{aligned} r_i &= r_{i-2} \ominus q_i \odot r_{i-1}, \\ s_i &= s_{i-2} \ominus q_i \odot s_{i-1}, \\ t_i &= t_{i-2} \ominus q_i \odot t_{i-1}. \end{aligned} \tag{1.659}$$

We thus yield

$$\begin{aligned} t_i \odot r_{i-1} \ominus t_{i-1} \odot r_i &= (t_{i-2} \ominus q_i \odot t_{i-1}) \odot r_{i-1} \ominus t_{i-1} \odot (r_{i-2} \ominus q_i \odot r_{i-1}) \\ &= t_{i-2} \odot r_{i-1} \ominus q_i \odot t_{i-1} \odot r_{i-1} \ominus t_{i-1} \odot r_{i-2} \oplus t_{i-1} \odot q_i \odot r_{i-1} \\ &= t_{i-2} \odot r_{i-1} \ominus t_{i-1} \odot r_{i-2} \underbrace{\ominus t_{i-1} \odot q_i \odot r_{i-1} \oplus t_{i-1} \odot q_i \odot r_{i-1}}_{=0} \\ &= (\ominus 1) \odot \underbrace{(t_{i-1} \odot r_{i-2} \ominus t_{i-2} \odot r_{i-1})}_{=(\ominus 1)^{i-1} \odot \alpha} \\ &= (\ominus 1) \odot (\ominus 1)^{i-1} \odot \alpha \\ &= (\ominus 1)^i \odot \alpha, \end{aligned} \tag{1.660}$$

which proves (1.651).

Claim 2

We prove (1.652) by induction.

Base Case

We start with $i = 0$ and yield

$$s_0 \odot r_{-1} \ominus s_{-1} \odot r_0 = 0 \odot r_{-1} \ominus 1 \odot r_0 = \ominus 1 \odot \beta = (\ominus 1)\beta \tag{1.661}$$

which is true.

Induction Step

The induction hypothesis is given by

$$s_{i-1} r_{i-2} \ominus s_{i-2} r_{i-1} = (\ominus 1)^i \odot \beta. \tag{1.662}$$

We will now evaluate whether

$$s_i r_{i-1} \ominus s_{i-1} r_i = (\ominus 1)^{i+1} \odot \beta \tag{1.663}$$

holds. We know from (1.606)

$$\begin{aligned} r_i &= r_{i-2} \ominus q_i \odot r_{i-1}, \\ s_i &= s_{i-2} \ominus q_i \odot s_{i-1}, \\ t_i &= t_{i-2} \ominus q_i \odot t_{i-1}. \end{aligned} \tag{1.664}$$

Now, we yield

$$\begin{aligned} s_i r_{i-1} \ominus s_{i-1} r_i &= (s_{i-2} \ominus q_i \odot s_{i-1}) \odot r_{i-1} \ominus s_{i-1} \odot (r_{i-2} \ominus q_i \odot r_{i-1}) \\ &= s_{i-2} \odot r_{i-1} \ominus q_i \odot s_{i-1} \odot r_{i-1} \ominus s_{i-1} \odot r_{i-2} \oplus s_{i-1} \odot q_i \odot r_{i-1} \\ &= s_{i-2} \odot r_{i-1} \ominus s_{i-1} \odot r_{i-2} \underbrace{\ominus q_i \odot s_{i-1} \odot r_{i-1} \oplus q_i \odot s_{i-1} \odot r_{i-1}}_{=0} \\ &= (\ominus 1) \odot \underbrace{(s_{i-1} \odot r_{i-2} - s_{i-2} \odot r_{i-1})}_{=(\ominus 1)^i \odot \beta} \\ &= (\ominus 1)^{i+1} \odot \beta, \end{aligned} \tag{1.665}$$

which proves (1.652).

Claim 3
We prove (1.653) by induction.

Base Case
We start with $i = 0$ and yield

$$s_0 \odot t_{-1} - s_{-1} \odot t_0 = 0 \odot 0 \ominus 1 \odot 1 = \ominus 1 \tag{1.666}$$

which is true.

Induction Step
The induction hypothesis is given by

$$s_{i-1} \odot t_{i-2} \ominus s_{i-2} \odot t_{i-1} = (\ominus 1)^i. \tag{1.667}$$

We will now evaluate whether

$$s_i \odot t_{i-1} \ominus s_{i-1} \odot t_i = (\ominus 1)^{i+1} \tag{1.668}$$

holds. We know from (1.606)

$$\begin{aligned} r_i &= r_{i-2} \ominus q_i \odot r_{i-1}, \\ s_i &= s_{i-2} \ominus q_i \odot s_{i-1}, \\ t_i &= t_{i-2} \ominus q_i \odot t_{i-1}. \end{aligned} \tag{1.669}$$

Now, we yield

$$\begin{aligned} s_i \odot t_{i-1} \ominus s_{i-1} \odot t_i &= (s_{i-2} \ominus q_i \odot s_{i-1}) \odot t_{i-1} \ominus s_{i-1} \odot (t_{i-2} \ominus q_i \odot t_{i-1}) \\ &= s_{i-2} \odot t_{i-1} \ominus q_i \odot s_{i-1} \odot t_{i-1} \ominus s_{i-1} \odot t_{i-2} \oplus s_{i-1} \odot q_i \odot t_{i-1} \\ &= s_{i-2} \odot t_{i-1} \ominus s_{i-1} \odot t_{i-2} \underbrace{\ominus q_i \odot s_{i-1} \odot t_{i-1} \oplus q_i \odot s_{i-1} \odot t_{i-1}}_{=0} \\ &= (\ominus 1) \odot \underbrace{(s_{i-1} \odot t_{i-2} \ominus s_{i-2} \odot t_{i-1})}_{=(\ominus 1)^i} \\ &= (\ominus 1)^{i+1}, \end{aligned} \tag{1.670}$$

which proves (1.653).

Claim 4
We prove (1.654) by induction.

Base Case

We start with $i = 1$. We know that

$$\deg\big(r_{i-1}(x)\big) > \deg\big(r_i(x)\big), \tag{1.671}$$

which is true because $r_i(x)$ is the remainder that we obtain when dividing $r_{i-2}(x)$ by $r_{i-1}(x)$ and we know that

$$t_1(x) = t_{-1}(x) \ominus q_1(x) \odot t_0(x) = \ominus q_1(x). \tag{1.672}$$

Thus, we yield

$$\deg\big(t_1(x)\big) + \deg\big(r_0(x)\big) = \deg\big(t_1(x) \odot r_0(x)\big) = \deg\big(\ominus q_1(x) \odot r_0(x)\big). \tag{1.673}$$

Since the sign does not influence the degree, we have

$$\deg\big(\ominus q_1(x) \odot r_0(x)\big) = \deg\big(q_1(x) \odot r_0(x)\big). \tag{1.674}$$

We now yield

$$\begin{aligned} \deg\big(t_1(x)\big) + \deg\big(r_0(x)\big) &= \deg\big(q_1(x) \odot r_0(x)\big) \\ &= \deg\big(q_1(x) \odot r_0(x) \oplus r_1(x)\big) \\ &= \deg\big(a(x)\big), \end{aligned} \tag{1.675}$$

where we used the facts that

$$\deg\big(r_0(x)\big) > \deg\big(r_1(x)\big), \tag{1.676}$$

which yields

$$\deg\big(q_1(x) \odot r_0(x) \oplus r_1(x)\big) = \deg\big(q_1(x) \odot r_0(x)\big) \tag{1.677}$$

and

$$r_i(x) = r_{i-2}(x) \ominus q_i(x) \odot r_{i-1}(x). \tag{1.678}$$

Induction Step

The induction hypothesis is given by

$$\deg\big(t_{i-1}(x)\big) + \deg\big(r_{i-2}(x)\big) = \deg\big(a(x)\big). \tag{1.679}$$

With

$$t_i(x) \odot r_{i-1}(x) \ominus t_{i-1}(x) \odot r_i(x) = (\ominus 1)^i \odot a(x) \tag{1.680}$$

we find

$$\deg\big(t_i(x) \odot r_{i-1}(x) \ominus t_{i-1}(x) \odot r_i(x)\big) = \deg\big((\ominus 1)^i \odot a(x)\big). \tag{1.681}$$

Since the sign does not influence the degree, we have

$$\deg\big((\ominus 1)^i \odot a(x)\big) = \deg\big(a(x)\big) \tag{1.682}$$

and, therefore, we obtain

$$\deg\big(t_i(x) \odot r_{i-1}(x) \ominus t_{i-1}(x) \odot r_i(x)\big) = \deg\big(a(x)\big). \tag{1.683}$$

Furthermore, we yield

$$\deg\big(\ominus t_{i-1}(x) \odot r_i(x)\big) = \deg\big(t_{i-1}(x) \odot r_i(x)\big) = \deg\big(t_{i-1}(x)\big) + \deg\big(r_i(x)\big). \tag{1.684}$$

With $\deg(r_i(x)) < \deg(r_{i-1}(x)) < \deg(r_{i-2}(x))$, we obtain

$$\deg\big(t_{i-1}(x) \odot r_i(x)\big) = \deg\big(t_{i-1}(x)\big) + \deg\big(r_i(x)\big) < \deg\big(t_{i-1}(x)\big) + \deg\big(r_{i-2}(x)\big). \tag{1.685}$$

Furthermore, we find

$$\deg\big(t_{i-1}(x) \odot r_i(x)\big) < \deg\big(a(x)\big). \tag{1.686}$$

Now, we yield

$$\begin{aligned}\deg\Big(t_i(x) \odot r_{i-1}(x) \ominus \underbrace{t_{i-1}(x) \odot r_i(x)}_{\deg(t_{i-1}(x)\odot r_i(x))<\deg(a(x))}\Big) &= \deg\big(t_i(x) r_{i-1}(x)\big)\\ &= \deg\big(t_i(x)\big) + \deg\big(r_{i-1}(x)\big).\\ &= \deg\big(a(x)\big)\end{aligned} \tag{1.687}$$

and, therefore,

$$\deg\big(t_i(x)\big) + \deg\big(r_{i-1}(x)\big) = \deg\big(a(x)\big), \tag{1.688}$$

which proves the claim.

Claim 5
We prove (1.655) by induction.

Base Case
We start with $i = 1$. We know that

$$s_1(x) = s_{-1}(x) \ominus \underbrace{q_1(x)}_{=1} \odot s_0(x) \tag{1.689}$$

and

$$s_{-1}(x) = 1, \quad s_0(x) = 0; \tag{1.690}$$

cf. (1.605). Therefore, we yield

$$s_1(x) = 1 \tag{1.691}$$

with

$$\deg\big(s_1(x)\big) = \deg(1) = 0. \tag{1.692}$$

Furthermore, we have

$$\underbrace{\deg\big(s_1(x)\big)}_{=0} + \deg\Big(\underbrace{r_0(x)}_{=\beta(x)}\Big) = \deg\big(r_0(x)\big) = \deg\big(\beta(x)\big), \tag{1.693}$$

which is true.

Induction Step

The induction hypothesis is given by

$$\deg\big(s_{i-1}(x)\big) + \deg\big(r_{i-2}(x)\big) = \deg\big(\beta(x)\big). \tag{1.694}$$

Furthermore, we have

$$\deg\big(s_i(x) \odot r_{i-1}(x) \ominus s_{i-1}(x) \odot r_i(x)\big) = \deg\big((\ominus 1)^{i+1} \odot \beta(x)\big) = \deg\big(\beta(x)\big). \tag{1.695}$$

In addition, we yield

$$\deg\big(\ominus s_{i-1}(x) \odot r_i(x)\big) = \deg\big(s_{i-1}(x) \odot r_i(x)\big). \tag{1.696}$$

Also, we know that $\deg(r_i(x)) < \deg(r_{i-1}(x)) < \deg(r_{i-2}(x))$, and hence obtain

$$\deg\big(s_{i-1}(x) \odot r_i(x)\big) = \deg\big(s_{i-1}(x)\big) + \deg\big(r_i(x)\big) < \deg\big(s_{i-1}(x)\big) + \deg\big(r_{i-2}(x)\big). \tag{1.697}$$

Now, we yield

$$\deg\big(s_{i-1}(x) \odot r_i(x)\big) < \deg\big(\beta(x)\big) \tag{1.698}$$

and hence

$$\begin{aligned}\deg\Big(s_i(x) \odot r_{i-1}(x) \ominus \underbrace{s_{i-1}(x) \odot r_i(x)}_{\deg(s_{i-1}(x)\odot r_i(x))<\deg(\beta(x))}\Big) &= \deg\big(s_i(x) \odot r_{i-1}(x)\big)\\ &= \deg\big(s_i(x)\big) + \deg\big(r_{i-1}(x)\big).\\ &= \deg\big(\beta(x)\big).\end{aligned} \tag{1.699}$$

Thus, we find

$$\deg\big(s_i(x)\big) + \deg\big(r_{i-1}(x)\big) = \deg\big(\beta(x)\big), \tag{1.700}$$

which is the claim. □

In 1202, Leonardo *Pisano*, called "*Fibonacci*," employed the recurring series $1, 2, 3, 5, 8, 13 \cdots$ in a problem on the number of offspring of a pair of rabbits [50, p. 393f.]. The resulting most remarkable sequence represents the *golden ratio* (in German "*goldener Schnitt*"), which is omnipresent in nature. For this reason,the Franciscan friar Fra. Luca Bartolomeo de *Pacioli* termed the golden ratio "*Divina proportione*" (in English "*divine proportion*"), in his book on mathematics of 1498. This book was illustrated by the genius Leonardo da *Vinci*, who used to attend Pacioli's lessons.

Owing to their fame, we must have a closer look at the *Fibonacci numbers* [17, p. 11], [18, p. 453]. So, welcome, dear passengers, on our voyage across the wide land of the Euclidean domain $(\mathbb{N}, +, \cdot)$ [17, p. 4].

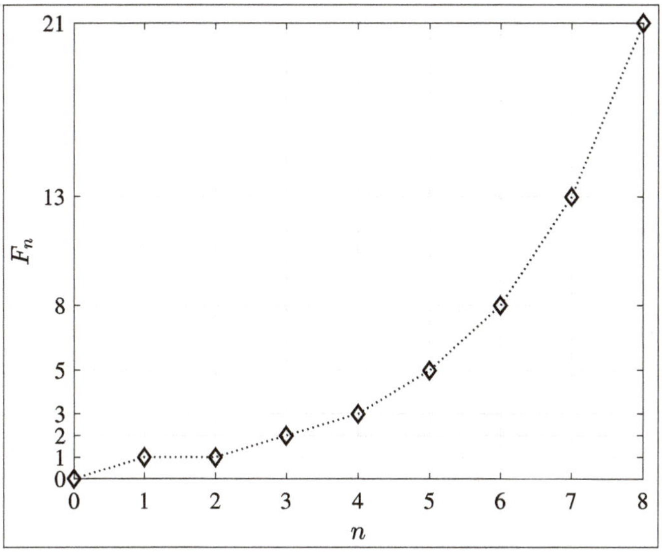

Figure 1.21: First nine Fibonacci numbers 0, 1, 1, 2, 3, 5, 8, 13, 21.

Example 1.29. The *Fibonacci numbers* F_n form a sequence of integers, $\{F_n\}$, whose elements are defined as follows [17, p. 11]:

$$\begin{aligned} F_0 &= 0, \\ F_1 &= 1, \\ F_m &= F_{m-1} + F_{m-2}, \quad m \in \{2, 3, 4 \cdots\}. \end{aligned} \tag{1.701}$$

The corresponding sequence $\{F_n\}$ begins as follows:

$$\{0, 1, 1, 2, 3, 5, 8, 13, 21, 34, 55, 89, 144, 233, 377, 610, 987, 1597, 2584, 4181 \cdots\}. \tag{1.702}$$

The first nine Fibonacci numbers are illustrated in Figure 1.21.

The following has been well-known [17, p. 7]:

$$\gcd\{F_m, F_n\} = F_{\gcd\{m,n\}}, \quad m, n \in \mathbb{N}, \tag{1.703}$$

immediately leading to

$$\gcd\{F_m, F_{m+1}\} = F_{\gcd\{m,m+1\}}, \quad m \in \mathbb{N}. \tag{1.704}$$

Using Theorem 1.32, i. e.,

$$\gcd\{s, t\} = \gcd\{s, t - rs\}, \quad \forall r, s, t \in \mathbb{N}, \tag{1.705}$$

and setting

$$s = m, \quad t = m + 1, \quad r = 1, \quad m \in \mathbb{N}, \tag{1.706}$$

we yield

$$\gcd\{m, m+1\} = \gcd\{m, m+1-1\cdot m\} = \gcd\{m, 1\} = 1, \quad m \in \mathbb{N}. \tag{1.707}$$

Therefore, we yield

$$\gcd\{F_m, F_{m+1}\} = F_{\gcd\{m,m+1\}} = F_{\gcd\{m,1\}} = F_1 = 1, \quad m \in \mathbb{N}. \tag{1.708}$$

We can also evaluate $\gcd\{m, m+1\}$ using the extended Euclidean algorithm, obtaining the following results:

iteration number		Bézout coefficients		remainder	quotient	
	i	s_i	t_i	r_i	q_i	
init	−1	*1*	*0*	$\alpha = m+1$	-	
	0	*0*	*1*	$\beta = m$	-	
$n =$	**1**	1	−1	1	1	$= \lfloor m+1/m \rfloor$
$n+1 =$	2	$-m$	m	0	m	$= \lfloor m/1 \rfloor$

Therefore,

$$\gcd\{m, m+1\} = r_1 = 1, \quad m \in \mathbb{N}. \tag{1.709}$$

Furthermore, using (1.608), we find

$$1 \cdot (m+1) + (-1) \cdot (m) = 1, \quad m \in \mathbb{N}. \tag{1.710}$$

In addition, using Theorem 1.32, i. e.,

$$\gcd\{s, t\} = \gcd\{s, t - rs\}, \quad \forall r, s, t \in \mathbb{N}, \tag{1.711}$$

and setting

$$s = F_m, \quad r = -1, \quad t = F_{m-1}, \quad m \in \mathbb{N}, \tag{1.712}$$

we yield

$$\gcd\{F_m, F_{m-1}\} = \gcd\{F_m, F_m + F_{m-1}\}, \quad m \in \mathbb{N}, \tag{1.713}$$

which becomes

$$\gcd\{F_m, F_{m+1}\} = \gcd\{F_m, F_m + F_{m-1}\} = \gcd\{F_{m-1}, F_m\}, \quad m \in \mathbb{N}. \tag{1.714}$$

Therefore, we obtain

$$\gcd\{F_m, F_{m+1}\} = \gcd\{F_{m-1}, F_m\} = \cdots = \gcd\{F_0, F_1\} = \gcd\{1, 1\} = 1, \quad m \in \mathbb{N}. \tag{1.715}$$

This means that consecutive Fibonacci numbers are relatively prime. Obviously, nature believes that simple symmetries, e. g., doubling or the like, are quite boring. Just like a great sportsman, ,ature wants it a bit more challenging.

Let us revisit Table 1.7. All the 2^n, $n \in \mathbb{N}^*$ polynomials $v(x) \in \mathbb{F}_2[x]_{n-1}$, $n \in \mathbb{N}^*$ given in Table 1.7 have a degree of at most $(n-1)$, $n \in \mathbb{N}^*$. Consequently, dividing these polynomials by $x^n \oplus 1$ gives the quotient 0. The remainders $r(x) \in \mathbb{F}_2[x]_{n-1}$, $n \in \mathbb{N}^*$, of the 2^n, $n \in \mathbb{N}^*$ divisions are the polynomials $v(x) \in \mathbb{F}_2[x]_{n-1}$, $n \in \mathbb{N}^*$ themselves. Therefore, we can write

$$v(x) = \underbrace{a(x)}_{=0} \odot (x^n \oplus 1) \oplus r(x), \quad v(x), r(x) \in \mathbb{F}_2[x]_{n-1}, \quad n \in \mathbb{N}^*. \tag{1.716}$$

We may generalize this cognition by using the second property of a Euclidean domain; cf. Definition 1.47. Since $\mathbb{F}_2[x]_n$, $n \in \mathbb{N}^*$ and $\mathbb{F}_2[x]$ are Euclidean domains, we may write [5, equation (5.1.3), p. 132]

$$a(x) = q(x) \odot g(x) \oplus r(x), \quad h\{r(x)\} < h\{g(x)\}, \quad a(x), g(x), q(x), r(x) \in \mathbb{F}_2[x]. \tag{1.717}$$

Theorem 1.35 (Division Algorithm for Polynomials). *Let $a(x) \in \mathbb{F}_2[x]$ and $g(x) \in \mathbb{F}_2[x] \setminus \{0\}$ be two polynomials of $\mathbb{F}_2[x]$ [29, p. 71]. Then there is a unique pair $q(x), r(x) \in \mathbb{F}_2[x]$ of polynomials, $q(x) \in \mathbb{F}_2[x]$ termed the "quotient polynomial" ("quotient") and $r(x) \in \mathbb{F}_2[x]$ termed the "remainder polynomial" ("remainder"), respectively, the "residue polynomial" ("residue"), such that* [29, p. 72]

$$a(x) = q(x) \odot g(x) \oplus r(x), \quad h\{r(x)\} < h\{g(x)\}, \quad a(x), g(x), q(x), r(x) \in \mathbb{F}_2[x]. \tag{1.718}$$

Proof. Using the usual polynomial division you may have been taught at school, we find the unique quotient $q(x) \in \mathbb{F}_2[x]$ and the unique remainder $r(x) \in \mathbb{F}_2[x]$. In the case $q(x), r(x) \in \mathbb{F}_2[x]$ were not unique, we could write

$$a(x) = q_1(x) \odot g(x) \oplus r_1(x) = q_2(x) \odot g(x) \oplus r_2(x), \tag{1.719}$$

which yields

$$\big(q_1(x) \ominus q_2(x)\big) \odot g(x) = r_2(x) \ominus r_1(x). \tag{1.720}$$

This means that $g(x)$ divides $(r_2(x) \ominus r_1(x))$, i. e., $g(x) | (r_2(x) \ominus r_1(x))$.

Since $h\{r_1(x)\} < h\{g(x)\}$ and $h\{r_2(x)\} < h\{g(x)\}$ If $(r_2(x) \ominus r_1(x))$ is nonzero, $h\{r_2(x) \ominus r_1(x)\} < h\{g(x)\}$ also, which means that $\deg(r_2(x) \ominus r_1(x)) < \deg(g(x))$. Therefore, $g(x) | (r_2(x) \ominus r_1(x))$ if $(r_2(x) \ominus r_1(x))$ is zero.

If $(q_1(x) \ominus q_2(x)) \odot g(x)$ is nonzero, it must have a degree at least $\deg(g(x))$. This contradicts that $\deg(r_2(x) \ominus r_1(x)) < \deg(g(x))$, and hence $(r_2(x) \ominus r_1(x)) = 0$. Therefore, $(q_1(x) \ominus q_2(x))$ must be zero.

Hence, (1.718) holds with unique $q(x), r(x) \in \mathbb{F}_2[x]$. □

Theorem 1.35, also called the *division theorem*, is also discussed in [5, Satz A.4, p. 447].

Setting out from (1.717), it has been common to write [5, p. 132]

$$r(x) = a(x) \bmod g(x), \quad a(x), g(x), r(x) \in \mathbb{F}_2[x], \tag{1.721}$$

read “$r(x)$ *is congruent to* $\alpha(x)$ *modulo* $g(x)$” [18, p. 4], [25, p. 68]. Equation (1.721) is termed a *congruence* [18, p. 4], [25, p. 68], [29, p. 72]. Since $h\{r(x)\} < h\{g(x)\}$, the division by $g(x) \in \mathbb{F}_2[x]$ (see (1.717)) maps all the polynomials $\alpha(x) \in \mathbb{F}_2[x]$ onto the remainders $r(x) \in \mathbb{F}_2[x]$; cf. (1.721). The remainders $r(x)$ form *residue classes* [18, p. 13], sometimes also called *congruence classes* (in German “Restklassen”) [5, p. 132].

A particular remainder $r(x) \in \mathbb{F}_2[x]$ thus represents a residue class, which is often denoted by $R_{g(x)}[\alpha]$ [5, equation (A.6.5), pp. 132, 447], [25, p. 68], [29, p. 72], i. e.,

$$r(x) = \alpha(x) \bmod g(x) = R_{g(x)}[\alpha(x)], \quad \alpha(x), g(x), r(x) \in \mathbb{F}_2[x]. \tag{1.722}$$

Computing mod$g(x)$ means replacing $g(x)$ by 0 [5, p. 132].

Theorem 1.36 (Residue Class Arithmetic). *The following arithmetic rules hold:*

1:

$$R_{g(x)}\left[\alpha(x) \oplus \beta(x)\right] = R_{g(x)}\left[\alpha(x)\right] \oplus R_{g(x)}\left[\beta(x)\right]. \tag{1.723}$$

2:

$$R_{g(x)}\left[\alpha(x) \odot \beta(x)\right] = R_{g(x)}\left[R_{g(x)}\left[\alpha(x)\right] \odot R_{g(x)}\left[\beta(x)\right]\right]. \tag{1.724}$$

3:

$$R_{g(x)}\left[\alpha(x) \odot g(x)\right] = 0. \tag{1.725}$$

4:

$$R_{g(x)}\left[\alpha(x)\right] = R_{g(x)}\left[R_{g(x) \odot h(x)}\left[\alpha(x)\right]\right]. \tag{1.726}$$

5:

$$\deg(\alpha(x)) < \deg(g(x)) \quad \Rightarrow \quad R_{g(x)}\left[\alpha(x)\right] = \alpha(x). \tag{1.727}$$

6:

$$R_{x^n \oplus 1}\left[x^m\right] = x^{m \bmod n} = x^{R_n[m]}. \tag{1.728}$$

Proof.

Claim 1

$$\begin{aligned} R_{g(x)}\left[\alpha(x) \oplus \beta(x)\right] &= \left(\alpha(x) \oplus \beta(x)\right) \bmod g(x) \\ &= \alpha(x) \bmod g(x) \oplus \beta(x) \bmod g(x) \\ &= R_{g(x)}\left[\alpha(x)\right] \oplus R_{g(x)}\left[\beta(x)\right]. \end{aligned} \tag{1.729}$$

Claim 2

$$\begin{aligned} R_{g(x)}\left[\alpha(x) \odot \beta(x)\right] &= \left[\alpha(x) \odot \beta(x)\right] \bmod g(x) \\ &= \Big[\underbrace{\left(q_\alpha(x) \odot g(x) \oplus r_\alpha(x)\right)}_{=\alpha(x)} \odot \underbrace{\left(q_\beta(x) \odot g(x) \oplus r_\beta(x)\right)}_{=\beta(x)}\Big] \bmod g(x) \\ &= \Big[q_\alpha(x) \odot q_\beta(x) \odot \{g(x)\}^2 \oplus q_\alpha(x) \odot r_\beta(x) \odot g(x) \\ &\quad \oplus q_\beta(x) \odot r_\alpha(x) \odot g(x) \oplus r_\alpha(x) \odot r_\beta(x)\Big] \bmod g(x) \\ &= \left[q_\alpha(x) \odot q_\beta(x) \odot \{g(x)\}^2\right] \bmod g(x) \\ &\quad \oplus \left[q_\alpha(x) \odot r_\beta(x) \odot g(x)\right] \bmod g(x) \end{aligned}$$

$$
\begin{aligned}
&\oplus \left[q_\beta(x) \odot r_\alpha(x) \odot g(x)\right] \bmod g(x) \\
&\oplus \left[r_\alpha(x) \odot r_\beta(x)\right] \bmod g(x) \\
&= q_\alpha(x) \odot q_\beta(x) \odot 0 \bmod g(x) \oplus q_\alpha(x) \odot r_\beta(x) \odot 0 \bmod g(x) \\
&\oplus q_\beta(x) \odot r_\alpha(x) \odot 0 \bmod g(x) \oplus \left[r_\alpha(x) \odot r_\beta(x)\right] \bmod g(x) \\
&= \left[r_\alpha(x) \odot r_\beta(x)\right] \bmod g(x) \\
&= R_{g(x)}\left[r_\alpha \odot r_\beta(x)\right] \\
&= \left[\left(\alpha(x) \bmod g(x)\right) \odot \left(\beta(x) \bmod g(x)\right)\right] \bmod g(x) \\
&= R_{g(x)}\left[R_{g(x)}\left[\alpha(x)\right] \odot R_{g(x)}\left[\beta(x)\right]\right]. \qquad (1.730)
\end{aligned}
$$

<u>*Claim 3*</u>

$$
\begin{aligned}
R_{g(x)}\left[\alpha(x) \odot g(x)\right] &= \left[\alpha(x) \odot g(x)\right] \bmod g(x) \\
&= \left[\left(\alpha(x) \bmod g(x)\right) \odot \underbrace{\left(g(x) \bmod g(x)\right)}_{=0}\right] \bmod g(x) \\
&= 0. \qquad (1.731)
\end{aligned}
$$

<u>*Claim 4*</u>

$$
\begin{aligned}
R_{g(x)}\Big[\underbrace{R_{g(x)\odot h(x)}\left[\alpha(x)\right]}_{=r(x)}\Big] &= R_{g(x)}\left[\alpha(x) \ominus q(x) \odot g(x) \odot h(x)\right] \\
&= R_{g(x)}\left[\alpha(x)\right] \ominus \underbrace{R_{g(x)}\left[q(x) \odot g(x) \odot h(x)\right]}_{=0} \\
&= R_{g(x)}\left[\alpha(x)\right]. \qquad (1.732)
\end{aligned}
$$

<u>*Claim 5*</u>
Since

$$
\alpha(x) = \underbrace{q(x)}_{=0} \odot g(x) \oplus r(x) = r(x), \quad \text{for } \deg\left(\alpha(x)\right) < \deg\left(g(x)\right), \qquad (1.733)
$$

we yield

$$
R_{g(x)}\left[\alpha(x)\right] = r(x) = \alpha(x). \qquad (1.734)
$$

<u>*Claim 6*</u>
Computing $\bmod (x^n \ominus 1)$ means replacing x^n by 1 [5, p. 448]. Thus, we obtain

$$
\begin{aligned}
R_{x^n \ominus 1}\left[x^m\right] &= x^m \bmod \left(x^n \ominus 1\right) \\
&= x^{\nu n + m \bmod n} \bmod \left(x^n \ominus 1\right) \\
&= 1^{\nu n} \odot x^{m \bmod n} \\
&= x^{m \bmod n} \qquad (1.735) \\
&= x^{R_n[m]}, \quad m, n, \nu \in \mathbb{N}. \qquad (1.736)
\end{aligned}
$$

□

Having exercised arithmetic (cf. Theorem 1.36), we can easily evaluate the cyclic shifts, which are main sensible processing steps within a cyclic binary linear block code $\mathbb{V}$; cf., e. g., [5, Satz 5.1, p. 132f.].

Theorem 1.37 (Cyclic Shifts of a Codeword). *Let $\mathbf{v} \in \mathbb{V} \subseteq \mathbb{F}_2^n$, $n \in \mathbb{N}^*$, equal to $(v_0, v_1 \cdots v_{n-1})$, $n \in \mathbb{N}^*$, be a codeword, which is associated with the polynomial $v(x) \in \mathbb{F}_2[x]_{n-1}$, $n \in \mathbb{N}^*$, equal to $(v_0 \odot x^0 \oplus v_1 \odot x^1 \odot \cdots \odot v_{n-1} \odot x^{n-1})$, $n \in \mathbb{N}^*$. Then the m-fold cyclic shift yields the codeword $(v_{n-m}, v_{n-m+1} \cdots v_{n-1}, v_0, v_1 \cdots v_{n-m-1})$, $m, n \in \mathbb{N}^*$, which corresponds to the polynomial $R_{x^n \ominus 1}[x^m \odot v(x)]$, $m, n \in \mathbb{N}^*$ [5, Satz 5.1, p. 132f.].*

Proof.

$$\begin{aligned}
R_{x^n\ominus 1}\left[x^m \odot v(x)\right] &= R_{x^n\ominus 1}\left[\left(v_0 \odot x^0 \oplus v_1 \odot x^1 \odot \cdots \odot v_{n-1} \odot x^{n-1}\right) \odot x^m\right] \\
&= \left[v_0 \odot x^m \oplus v_1 \odot x^{m+1} \oplus \cdots \oplus v_{n-m-1} \odot x^{n-1}\right. \\
&\qquad \left. \oplus v_{n-m} \odot x^n \oplus \cdots \odot v_{n-1} \odot x^{m+n-1}\right] \bmod \left(x^n \ominus 1\right) \\
&= v_0 \odot x^m \oplus v_1 \odot x^{m+1} \oplus \cdots \oplus v_{n-m-1} \odot x^{n-1} \\
&\qquad \oplus v_{n-m} \odot x^0 \oplus \cdots \odot v_{n-1} \odot x^{m-1} \\
&= v_{n-m} \odot x^0 \oplus \cdots \odot v_{n-1} \odot x^{m-1} \\
&\qquad \oplus v_0 \odot x^m \oplus v_1 \odot x^{m+1} \oplus \cdots \oplus v_{n-m-1} \odot x^{n-1}.
\end{aligned} \tag{1.737}$$

This polynomial corresponds to the codeword $(v_{n-m}, v_{n-m+1} \cdots v_{n-1}, v_0, v_1 \cdots v_{n-m-1})$, $m, n \in \mathbb{N}^*$. □

1.12.4 Unique Factorization in Euclidean Domains

The primes and irreducible elements as well as the composite elements of a Euclidean domain are related with each other by the *fundamental theorem of arithmetic*, which we will prove in Section 1.12.4. We will start with the formal definition of a *unit*.

Definition 1.49 (Unit). Let $(\mathbb{E}, \oplus, \odot)$ be a Euclidean domain. A unit $u \in \mathbb{E}$ is any divisor of 1, i. e., $u \in \mathbb{E}$ is a unit if and only if there exists $q \in \mathbb{E}$ such that [17, p. 13]

$$u \odot q = 1, \quad u, q \in \mathbb{E}. \tag{1.738}$$

Clearly, a unit has the following property of size:

$$h\{u \odot q\} = h\{1\}, \quad u, q \in \mathbb{E}. \tag{1.739}$$

Example 1.30. Let us consider the integers, i. e., $(\mathbb{Z}, +, \cdot)$, which are a Euclidean domain with [17, p. 4]

$$h\{u\} = |u| \geq 1, \quad u \in \mathbb{Z}\setminus\{0\}, \tag{1.740}$$

which is multiplicative

$$h\{uq\} = h\{u\}h\{q\}, \quad u, q \in \mathbb{Z}\setminus\{0\}. \tag{1.741}$$

We require to find

$$uq = 1, \quad u, q \in \mathbb{Z}\setminus\{0\}. \tag{1.742}$$

Since we require

$$h\{uq\} = |uq| = \underbrace{|u|}_{\geq 1}\,\underbrace{|q|}_{\geq 1} = h\{1\} = 1, \quad u, q \in \mathbb{Z}\setminus\{0\}, \tag{1.743}$$

and

$$h\{u\} = |u| = 1, \quad u, q \in \mathbb{Z}\setminus\{0\}, \tag{1.744}$$

we find

$$u \in \{-1, +1\}. \tag{1.745}$$

Therefore, $u = -1$ and $u = +1$ are the only units in $(\mathbb{Z}, +, \cdot)$ [17, p. 13].

Example 1.31. Let us consider the Euclidean domain of monic polynomials $(\mathbb{F}_2[x]_n, \oplus, \odot)$ over the field $\mathbb{F}_2$, i. e., the polynomials with coefficients in $\mathbb{F}_2$. The units in $(\mathbb{F}_2[x]_n, \oplus, \odot)$ are the scalars, i. e., the polynomials of degree 0 that are units in $\mathbb{F}_2$ [17, p. 13]. In $\mathbb{F}_2$, there is only one element u that fulfills

$$u \odot q = 1, \quad u, q \in \mathbb{F}_2 = \{0, 1\}, \tag{1.746}$$

namely

$$u = 1, \tag{1.747}$$

because

$$u \odot q = \begin{cases} 1 & \text{if } u = 1,\ q = 1, \\ 0 & \text{if } u = 1,\ q = 0, \\ 0 & \text{if } u = 0,\ q = 1, \\ 0 & \text{if } u = 0,\ q = 0, \end{cases} \qquad u, q \in \mathbb{F}_2. \tag{1.748}$$

Therefore, the only unit in $(\mathbb{F}_2[x]_n, \oplus, \odot)$ is

$$u(x) = 1. \tag{1.749}$$

Multiplying any element of $(\mathbb{F}_2[x]_n, \oplus, \odot)$, $p(x)$ say, by $u(x) = 1$ leaves $p(x)$ unchanged, i. e.,

$$p(x) \odot u(x) = p(x). \tag{1.750}$$

Example 1.32. Let us look at the Euclidean domain of the Gaussian integers [44, p. 65]

$$\mathbb{Z}[\sqrt{-1}] := \{a + b\sqrt{-1} \mid a, b \in \mathbb{Z}\}. \tag{1.751}$$

Let $\underline{u}$ be

$$\underline{u} = a + b\sqrt{-1} = a + \mathrm{j}b, \quad a, b \in \mathbb{Z}, \tag{1.752}$$

then

$$h\{\underline{u}\} = a^2 + b^2 \geq 1, \quad \underline{u} \in \mathbb{Z}[\sqrt{-1}]\setminus\{0\}, \tag{1.753}$$

which is multiplicative, i. e.,

$$h\{\underline{u}\underline{q}\} = h\{\underline{u}\}h\{\underline{q}\}, \quad \underline{u}, \underline{q} \in \mathbb{Z}[\sqrt{-1}] \setminus \{0\}. \tag{1.754}$$

We require to find

$$\underline{u}\underline{q} = 1, \quad \underline{u}, \underline{q} \in \mathbb{Z}[\sqrt{-1}] \setminus \{0\}, \tag{1.755}$$

which leads to

$$h\{\underline{u}\underline{q}\} = \underbrace{h\{\underline{u}\}}_{\geq 1} \underbrace{h\{\underline{q}\}}_{\geq 1} = h\{1\} = 1, \quad \underline{u}, \underline{q} \in \mathbb{Z}[\sqrt{-1}] \setminus \{0\}. \tag{1.756}$$

We thus require

$$h\{\underline{u}\} = a^2 + b^2 = 1, \quad \underline{u} \in \mathbb{Z}[\sqrt{-1}] \setminus \{0\}, \tag{1.757}$$

which can be fulfilled if and only if
- $a^2 = 1$, i. e., $a = \pm 1$, and $b^2 = 0$ or
- $a^2 = 0$ and $b^2 = 1$, i. e., $b = \pm 1$.

We hence find four units, namely

$$\underline{u}_1 = 1, \quad \underline{u}_2 = -1, \quad \underline{u}_3 = \mathrm{j}, \quad \underline{u}_4 = -\mathrm{j}. \tag{1.758}$$

Definition 1.50 (Associate). Let $a, b \in \mathbb{E}$ be two elements of the Euclidean domain $(\mathbb{E}, \oplus, \odot)$ and let $u \in \mathbb{E}$ be a unit of $(\mathbb{E}, \oplus, \odot)$. The two elements a, b are called *associates* if [17, p. 13]

$$a = u \odot b, \quad a, b, u \in \mathbb{E}. \tag{1.759}$$

Equation (1.759) is symmetric, which we will prove in what follows. Assume that there is some unit $v \in \mathbb{E}$, which, together with $u \in \mathbb{E}$, fulfills (1.738) of Definition 1.49,

$$v \odot q = 1, \quad u, q \in \mathbb{E}. \tag{1.760}$$

Multiplying both sides of (1.759) by $v \in \mathbb{E}$ leads to

$$v \odot a = \underbrace{v \odot u}_{=1} \odot b = b, \quad a, b, u, v \in \mathbb{E}. \tag{1.761}$$

Also, (1.759) is reflexive because

$$a = u \odot a, \quad a, u \in \mathbb{E}, \tag{1.762}$$

holds for $u = 1$.

Furthermore, (1.759) is transitive because if

$$a = u \odot b, \quad a, b, u \in \mathbb{E}, \tag{1.763}$$

holds for some unit u and if

$$b = v \odot c, \quad b, c, v \in \mathbb{E}, \tag{1.764}$$

is true for some unit v, then

$$a = u \odot v \odot c, \quad a, b, u, v \in \mathbb{E}, \tag{1.765}$$

$u \odot v \in \mathbb{E}$ being a unit.

Since (1.759) is symmetric, reflexive and transitive, it is an *equivalence relation* [17, p. 17].

Example 1.33. Let us consider the integers, i. e., $(\mathbb{Z}, +, \cdot)$, with the unit u taken from $\{-1, +1\}$. For instance, +3 and −3 are associates because [17, p. 13]

$$\underbrace{+3}_{=a} = \underbrace{(-1)}_{=u}\underbrace{(-3)}_{=b}, \quad a, b, u \in \mathbb{Z}. \tag{1.766}$$

Example 1.34. Let us consider $(\mathbb{Z}[\sqrt{-1}], \oplus, \odot)$ with the unit u taken from $\{-1, +1, -\mathrm{j}, +\mathrm{j}\}$. For instance, $(1 + \mathrm{j})$ and $(1 - \mathrm{j})$ are associates because [17, p. 13]

$$\underbrace{1 + \mathrm{j}}_{=a} = \underbrace{\mathrm{j}}_{=u} \odot \underbrace{(1 - \mathrm{j})}_{=b}, \quad a, b, u \in \mathbb{Z}[\sqrt{-1}]. \tag{1.767}$$

Now, let us define the expression *factorization*.

Definition 1.51 (Factorization). Let $b \in \mathbb{E}$ be an element of the Euclidean domain $(\mathbb{E}, \oplus, \odot)$. A *factorization* of $b \in \mathbb{E}$ is an expression of the form [17, p. 13]

$$b = a_1 \odot a_2 \odot a_3 \odot \cdots \odot a_r, \quad a_1, a_2, a_3 \cdots a_r, b \in \mathbb{E}, \quad r \in \mathbb{N}^*. \tag{1.768}$$

Example 1.35. Clearly, (1.738) are a factorization of 1, i. e.,

$$1 = u \odot q, \quad u, q \in \mathbb{E}, \tag{1.769}$$

using the units $u, q \in \mathbb{E}$. Furthermore, multiplying (1.769) by $b \in \mathbb{E}$ yields

$$b = b \odot u \odot q, \quad b, u, q \in \mathbb{E}, \tag{1.770}$$

[17, p. 13], which is trivial.

Generally, we say that the factorization (1.768) is a *trivial factorization* if each of the terms a_i, $i \in \{1, 2 \cdots r\}$, $r \in \mathbb{N}^*$ of (1.768) except for one is a unit and the remaining a_i is an associate of b. This becomes clear from the following consideration. Let us set out from the trivial factorization (1.738)

$$1 = u_1 \cdot q, \quad u_1, q \in \mathbb{E}, \tag{1.771}$$

with the units $u_1 \in \mathbb{E}$ and $q \in \mathbb{E}$. Of course, we can rewrite $q \in \mathbb{E}$ as an associate of a unit $u_2 \in \mathbb{E}$ using the unit $u_3 \in \mathbb{E}$

$$q = u_2 \odot u_3 = 1, \tag{1.772}$$

yielding the trivial factorization

$$1 = u_1 \odot u_2 \odot u_3, \quad u_1, u_2, u_3 \in \mathbb{E}, \tag{1.773}$$

which we can continue to find the trivial factorization

$$1 = u_1 \odot u_2 \odot u_3 \odot \cdots \odot u_r, \quad u_1, u_2, u_3 \cdots u_r \in \mathbb{E}, \quad r \in \mathbb{N}^*. \tag{1.774}$$

Now, multiplying (1.774) by $b \in \mathbb{E}$, we yield the trivial factorization

$$b = b \odot u_1 \odot u_2 \odot u_3 \odot \cdots \odot u_r, \quad b, u_1, u_2, u_3 \cdots u_r \in \mathbb{E}, \quad r \in \mathbb{N}^*. \tag{1.775}$$

Combining an arbitrarily chosen unit in (1.775), u_k say, with $b \in \mathbb{E}$ forms its associate $u_k \odot b$, and we arrive at the trivial factorization

$$b = (u_k \odot b) \odot \underbrace{u_1 \odot u_2 \odot u_3 \odot \cdots \odot u_{k-1} \odot u_{k+1} \odot \cdots \odot u_r}_{r-1 \text{ terms}}, \tag{1.776}$$

$$b, u_1, u_2, u_3 \cdots u_r \in \mathbb{E}, \quad k \in \{1, 2 \cdots r\}, \quad r \in \mathbb{N}^*. \tag{1.777}$$

Comparing (1.777) with (1.768) and identifying

$$a_i = \begin{cases} u_i & \text{for } i \in \{1, 2, 3 \cdots r\} \setminus \{k\}, \\ u_k \odot b & \text{for } i = k, \end{cases}, \quad k \in \{1, 2 \cdots r\}, \quad r \in \mathbb{N}^*, \tag{1.778}$$

shows that *in a trivial factorization exactly one of the factors a_k is an associate of b.*

Let us now look at the elements $p \in \mathbb{E}$ that only have trivial factorizations.

Definition 1.52 (Irreducible Element $p \in \mathbb{E}$). An element $p \in \mathbb{E}$ of a Euclidean domain $(\mathbb{E}, \oplus, \odot)$, which has the property that every possible factorization of $p \in \mathbb{E}$ not being a unit of $(\mathbb{E}, \oplus, \odot)$ is trivial, is called *irreducible* [17, p. 13], respectively, an *irreducible element*.

Definition 1.53 (Prime). If an irreducible element $p \in \mathbb{E}$ of a Euclidean domain $(\mathbb{E}, \oplus, \odot)$ fulfills the prime property [44, Satz 4.2, p. 65f.], it is termed a *prime* [17, p. 13].

Example 1.36. $(\mathbb{Z}, +, \cdot)$ the primes are ±2, ±3, ±5, ±7, ±11, ±13, ±17, ±19, ±23, ±29, ±31, ±37 ⋯, [44, p. 56].

Example 1.37. The primes in $(\mathbb{Z}[\sqrt{-1}], \oplus, \odot)$ are

- the ordinary rational primes congruent to 3 (mod4), e. g.,

$$\begin{aligned}&3, 7, 11, 19, 23, 31, 43, 47, 59, 67, 71, 79, 83, 103,\\&107, 127, 131, 139, 151, 163, 167, 179, 191, 199,\\&211, 223, 227, 239, 251, 263, 271, 283, 307, 311,\\&331, 347, 359, 367, 379, 383, 419, 431, 439, 443,\\&463, 467, 479, 487, 491, 499, 503, 523, 547, 563,\\&571, 587, 599, 607, 619, 631, 643, 647, 659, 683,\\&691, 719, 727, 739, 743, 751, 787, 811, 823, 827,\\&839, 859, 863, 883, 887, 907, 911, 919, 947, 967,\\&971, 983, 991 \cdots,\end{aligned} \tag{1.779}$$

- complex primes of the form $(a + jb)$ with $h\{a + jb\}$ equal to 2, i. e.,

$$(1 + \mathrm{j}), (-1 + \mathrm{j}), (1 - \mathrm{j}) \text{ and } (-1 - \mathrm{j}), \tag{1.780}$$

- complex primes of the form $(a + jb)$ with $h\{a + jb\}$ equal to a rational prime congruent to 1 (mod4), i. e.,

$$\begin{aligned}&(-10 - 9\mathrm{j}), (-10 - 7\mathrm{j}), (-10 - 3\mathrm{j}), (-10 - 1\mathrm{j}), (-10 + 1\mathrm{j}),\\&(-10 + 3\mathrm{j}), (-10 + 7\mathrm{j}), (-10 + 9\mathrm{j}), (-9 - 10\mathrm{j}), (-9 - 4\mathrm{j}),\\&(-9 + 4\mathrm{j}), (-9 + 10\mathrm{j}), (-8 - 7\mathrm{j}), (-8 - 5\mathrm{j}), (-8 - 3\mathrm{j}),\\&(-8 + 3\mathrm{j}), (-8 + 5\mathrm{j}), (-8 + 7\mathrm{j}), (-7 - 10\mathrm{j}), (-7 - 8\mathrm{j}),\\&(-7 - 2\mathrm{j}), (-7 + 2\mathrm{j}), (-7 + 8\mathrm{j}), (-7 + 10\mathrm{j}), (-6 - 5\mathrm{j}),\\&(-6 - 1\mathrm{j}), (-6 + 1\mathrm{j}), (-6 + 5\mathrm{j}), (-5 - 8\mathrm{j}), (-5 - 6\mathrm{j}),\\&(-5 - 4\mathrm{j}), (-5 - 2\mathrm{j}), (-5 + 2\mathrm{j}), (-5 + 4\mathrm{j}), (-5 + 6\mathrm{j}),\\&(-5 + 8\mathrm{j}), (-4 - 9\mathrm{j}), (-4 - 5\mathrm{j}), (-4 - 1\mathrm{j}), (-4 + 1\mathrm{j}),\\&(-4 + 5\mathrm{j}), (-4 + 9\mathrm{j}), (-3 - 10\mathrm{j}), (-3 - 8\mathrm{j}), (-3 - 2\mathrm{j}),\\&(-3 + 2\mathrm{j}), (-3 + 8\mathrm{j}), (-3 + 10\mathrm{j}), (-2 - 7\mathrm{j}), (-2 - 5\mathrm{j}),\\&(-2 - 3\mathrm{j}), (-2 - 1\mathrm{j}), (-2 + 1\mathrm{j}), (-2 + 3\mathrm{j}), (-2 + 5\mathrm{j}),\\&(-2 + 7\mathrm{j}), (-1 - 10\mathrm{j}), (-1 - 6\mathrm{j}), (-1 - 4\mathrm{j}), (-1 - 2\mathrm{j}),\\&(-1 + 2\mathrm{j}), (-1 + 4\mathrm{j}), (-1 + 6\mathrm{j}), (-1 + 10\mathrm{j}), (1 - 10\mathrm{j}),\\&(1 - 6\mathrm{j}), (1 - 4\mathrm{j}), (1 - 2\mathrm{j}), (1 + 2\mathrm{j}), (1 + 4\mathrm{j}), (1 + 6\mathrm{j}),\\&(1 + 10\mathrm{j}), (2 - 7\mathrm{j}), (2 - 5\mathrm{j}), (2 - 3\mathrm{j}), (2 - 1\mathrm{j}), (2 + 1\mathrm{j}),\\&(2 + 3\mathrm{j}), (2 + 5\mathrm{j}), (2 + 7\mathrm{j}), (3 - 10\mathrm{j}), (3 - 8\mathrm{j}), (3 - 2\mathrm{j}),\\&(3 + 2\mathrm{j}), (3 + 8\mathrm{j}), (3 + 10\mathrm{j}), (4 - 9\mathrm{j}), (4 - 5\mathrm{j}), (4 - 1\mathrm{j}),\\&(4 + 1\mathrm{j}), (4 + 5\mathrm{j}), (4 + 9\mathrm{j}), (5 - 8\mathrm{j}), (5 - 6\mathrm{j}), (5 - 4\mathrm{j}),\\&(5 - 2\mathrm{j}), (5 + 2\mathrm{j}), (5 + 4\mathrm{j}), (5 + 6\mathrm{j}), (5 + 8\mathrm{j}), (6 - 5\mathrm{j}),\\&(6 - 1\mathrm{j}), (6 + 1\mathrm{j}), (6 + 5\mathrm{j}), (7 - 10\mathrm{j}), (7 - 8\mathrm{j}), (7 - 2\mathrm{j}),\\&(7 + 2\mathrm{j}), (7 + 8\mathrm{j}), (7 + 10\mathrm{j}), (8 - 7\mathrm{j}), (8 - 5\mathrm{j}), (8 - 3\mathrm{j}),\\&(8 + 3\mathrm{j}), (8 + 5\mathrm{j}), (8 + 7\mathrm{j}), (9 - 10\mathrm{j}), (9 - 4\mathrm{j}), (9 + 4\mathrm{j}),\\&(9 + 10\mathrm{j}), (10 - 9\mathrm{j}), (10 - 7\mathrm{j}), (10 - 3\mathrm{j}), (10 - 1\mathrm{j}), (10 + 1\mathrm{j}),\\&(10 + 3\mathrm{j}), (10 + 7\mathrm{j}), (10 + 9\mathrm{j}) \cdots .\end{aligned} \tag{1.781}$$

Now, what if a divisor $\delta \in \mathbb{E}$ of $b \in \mathbb{E}$ is not an associate of $b \in \mathbb{E}$?

Definition 1.54 (Proper Divisor). Let $b \in \mathbb{E}$ and $\delta \in \mathbb{E}$ be two elements of the Euclidean domain $(\mathbb{E}, \oplus, \odot)$. If $\delta \in \mathbb{E}$ is a divisor of $b \in \mathbb{E}$ and is not an associate of $b \in \mathbb{E}$, then $\delta \in \mathbb{E}$ is called a *proper divisor* [17, p. 14].

In a Euclidean domain $(\mathbb{E}, \oplus, \odot)$, two elements $a \in \mathbb{E}$ and $b \in \mathbb{E}$ are said to be *relatively prime* if their greatest common divisor is 1 or any other unit. This takes us to the next corollary.

Corollary 1.7 (1 as a Linear Combination of Relatively Prime Elements). *In a Euclidean domain $(\mathbb{E}, \oplus, \odot)$, the unit $1 \in \mathbb{E}$ can be expressed as a linear combination of two elements $a \in \mathbb{E}$ and $b \in \mathbb{E}$ that are relatively prime* [17, *p. 14*].

Proof. Theorem 1.31 teaches us that we can express $1 \in \mathbb{E}$ as a linear combination of $a \in \mathbb{E}$ and $b \in \mathbb{E}$, i. e., [17, p. 14]

$$a \odot s \oplus b \odot t = 1 = \gcd\{a, b\}, \quad a, b, s, t \in \mathbb{E}. \tag{1.782}$$

This is also taught by (1.608). □

Corollary 1.8 (Relatively Prime Elements). *Let $p \in \mathbb{E}$ be a prime element in a Euclidean domain $(\mathbb{E}, \oplus, \odot)$. If $p \in \mathbb{E}$ does not divide the element $a \in \mathbb{E}$, then p and a are* relatively prime [17, *p. 14*].

Proof. Let $d \in \mathbb{E}$ be a common divisor of $p \in \mathbb{E}$ and $a \in \mathbb{E}$. Since p is prime, d must either be a unit or an associate of p [17, p. 14]. Since p does not divide a, no associate of p can divide a either [17, p. 14]. Therefore, d must be a unit [17, p. 14]. Thus, the common divisor of p and a is a unit, i. e., $\gcd\{p, a\} = 1$, and hence p and a are relatively prime [17, p. 14]. □

Clearly, if p does not divide a, then there exist $s, t \in \mathbb{E}$ with [17, p. 14]

$$p \odot s \oplus a \odot t = 1 = \gcd\{p, a\}, \quad a, p, s, t \in \mathbb{E}. \tag{1.783}$$

Finally, we can present a well-known fact in the form of a corollary.

Corollary 1.9 (Prime Property in a Euclidean Domain). *Let $a, b, p \in \mathbb{E}$ be three elements in a Euclidean domain $(\mathbb{E}, \oplus, \odot)$. Let p be irreducible. Furthermore, let $p|(a \odot b)$. Then either $p|a$ or $p|b$ or both* [17, *p. 15*].

Proof. If $p|a$, then there is nothing further to prove [17, p. 15].

Otherwise, if p does not divide a, however,

$$p \odot s \oplus a \odot t = 1, \quad a, p, s, t \in \mathbb{E}, \tag{1.784}$$

holds [17, p. 15]. Multiplying by b yields [17, p. 15]

$$p \odot b \odot s \oplus a \odot b \odot t = b, \quad a, b, p, s, t \in \mathbb{E}. \tag{1.785}$$

Since $p|(p \odot b \odot s)$ and $p|(a \odot b)$, which also means that $p|(a \odot b \odot t)$, we have $p|(p \odot b \odot s \oplus a \odot b \odot t)$. Therefore, $p|b$. □

Corollary 1.10 (Size of Proper Divisor). *Let $a \in \mathbb{E}$ and $b \in \mathbb{E}$ be two elements in a Euclidean domain $(\mathbb{E}, \oplus, \odot)$. If $a \in \mathbb{E}$ is a proper divisor of $b \in \mathbb{E}$, then* [17, p. 15]

$$h\{a\} < h\{b\}, \quad a, b \in \mathbb{E}. \tag{1.786}$$

Proof. Suppose

$$b = a \odot c, \quad a, b, c \in \mathbb{E}, \tag{1.787}$$

with $c \in \mathbb{E}$ not a unit [17, p. 15], which means that $a \in \mathbb{E}$ and $b \in \mathbb{E}$ are not associates.

Dividing $a \in \mathbb{E}$ by $b \in \mathbb{E}$ yields [17, p. 15]

$$a = q \odot b \oplus r, \quad h\{r\} < h\{b\}, \quad a, b, q, r \in \mathbb{E}, \tag{1.788}$$

and we yield [17, p. 15]

$$r = a \ominus q \odot b = a \ominus q \odot (a \odot c) = a \odot (1 \ominus q \odot c), \quad a, b, c, q, r \in \mathbb{E}. \tag{1.789}$$

Since $c \in \mathbb{E}$ is not a unit, $(1 \ominus q \odot c)$ does certainly not vanish, and so using the first property of Definition 1.47, we can write [17, p. 15]

$$h\{r\} = h\big\{a \odot (1 \ominus q \odot c)\big\} \geq h\{a\}, \quad a, c, q, r \in \mathbb{E}, \tag{1.790}$$

and obtain

$$h\{a\} \leq h\{r\} < h\{b\}, \quad a, b, r \in \mathbb{E}, \tag{1.791}$$

just as claimed [17, p. 15]. □

Our last two corollaries in this row will come now. Revisiting Corollary 1.9, we find the following.

Corollary 1.11 (Greatest Common Divisor of Two Polynomials). *Let $r_{-1}(x)$ and $r_0(x)$ be polynomials with* [4, *Theorem 14, p. 363*]

$$\deg\big(r_0(x)\big) \leq \deg\big(r_{-1}(x)\big), \tag{1.792}$$

having the greatest common divisor (gcd)

$$gcd\big\{r_{-1}(x), r_0(x)\big\} = h(x). \tag{1.793}$$

Then there exist polynomials $a(x)$ and $b(x)$ such that [4, *Theorem 14, equation (58), p. 363*]

$$a(x) \odot r_{-1}(x) \oplus b(x) \odot r_0(x) = h(x) = gcd\big\{r_{-1}(x), r_0(x)\big\}. \tag{1.794}$$

Proof. We will use the extended Euclidean algorithm (cf. Definition 1.48 and Theorem 1.35) to prove our claim.

Let the polynomials $a_i(x)$ and $b_i(x)$, $i \in \{-1, 0, 1, 2 \cdots\}$, be defined as follows [4, Theorem 14, equation (59), p. 363]:

$$a_{-1}(x) = 0, \quad b_{-1}(x) = 1, \tag{1.795}$$

$$a_0(x) = 1, \quad b_0(x) = 0, \tag{1.796}$$

and [4, Theorem 14, equation (60), p. 363]

$$a_i(x) = q_i(x) \odot a_{i-1}(x) \oplus a_{i-2}(x), \quad i \in \{-1, 0, 1, 2 \cdots\}, \tag{1.797}$$

$$b_i(x) = q_i(x) \odot b_{i-1}(x) \oplus b_{i-2}(x), \quad i \in \{-1, 0, 1, 2 \cdots\}, \tag{1.798}$$

i. e., [4, Theorem 14, equation (61), p. 364]

$$\begin{aligned} \begin{pmatrix} a_i(x) & a_{i-1}(x) \\ b_i(x) & b_{i-1}(x) \end{pmatrix} &= \begin{pmatrix} a_{i-1}(x) & a_{i-2}(x) \\ b_{i-1}(x) & b_{i-2}(x) \end{pmatrix} \odot \begin{pmatrix} q_i(x) & 1 \\ 1 & 0 \end{pmatrix} \\ &= \underbrace{\begin{pmatrix} 1 & 0 \\ 0 & 1 \end{pmatrix}}_{=I_2} \odot \begin{pmatrix} q_1(x) & 1 \\ 1 & 0 \end{pmatrix} \odot \begin{pmatrix} q_2(x) & 1 \\ 1 & 0 \end{pmatrix} \odot \cdots \odot \begin{pmatrix} q_i(x) & 1 \\ 1 & 0 \end{pmatrix}, \end{aligned} \tag{1.799}$$

having the determinant [4, Theorem 14, p. 364]

$$\det \begin{pmatrix} a_i(x) & a_{i-1}(x) \\ b_i(x) & b_{i-1}(x) \end{pmatrix} = (\ominus 1)^i = (\oplus 1)^i = 1, \quad i \in \mathbb{N}^*, \tag{1.800}$$

which is a consequence of the multiplication theorem

$$\det(\boldsymbol{AB}) = \det(\boldsymbol{A}) \cdot \det(\boldsymbol{B}), \tag{1.801}$$

(cf. e. g. [1] for a proof) also exploiting that $\ominus$ and $\oplus$ are identical for the arithmetic over $\mathbb{F}_2$.

Furthermore, we have [4, Theorem 14, equation (62), p. 364]

$$\begin{pmatrix} r_{-1}(x) \\ r_0(x) \end{pmatrix} = \begin{pmatrix} a_i(x) & a_{i-1}(x) \\ b_i(x) & b_{i-1}(x) \end{pmatrix} \odot \begin{pmatrix} r_{i-1}(x) \\ r_i(x) \end{pmatrix}, \tag{1.802}$$

leading to [4, Theorem 14, equation (63), p. 364]

$$\begin{aligned} \begin{pmatrix} r_{i-1}(x) \\ r_i(x) \end{pmatrix} &= (\ominus 1)^i \odot \begin{pmatrix} b_{i-1}(x) & \ominus a_{i-1}(x) \\ \ominus b_i(x) & a_i(x) \end{pmatrix} \odot \begin{pmatrix} r_{-1}(x) \\ r_0(x) \end{pmatrix} \\ &= \begin{pmatrix} b_{i-1}(x) & a_{i-1}(x) \\ b_i(x) & a_i(x) \end{pmatrix} \odot \begin{pmatrix} r_{-1}(x) \\ r_0(x) \end{pmatrix}. \end{aligned} \tag{1.803}$$

In particular, using (1.612) and (1.613), we find [4, Theorem 14, equation (64), p. 364]

$$\begin{aligned} r_i(x) &= b_i(x) \odot r_{-1}(x) \oplus a_i(x) \odot r_0(x) = a_i(x) \odot r_0(x) \oplus b_i(x) \odot r_{-1}(x) \\ &= \gcd\{r_{-1}(x), r_0(x)\}. \end{aligned} \tag{1.804}$$

□

Corollary 1.11 immediately leads to the following.

Corollary 1.12 (Greatest Common Divisor of Two Relatively Prime Polynomials). *Let $r_{-1}(x)$ and $r_0(x)$ be two relatively prime polynomials with* [4, *Theorem 14, p. 363*]

$$\deg(r_0(x)) \leq \deg(r_{-1}(x)), \tag{1.805}$$

having the greatest common divisor (gcd)

$$gcd\{r_{-1}(x), r_0(x)\} = 1. \tag{1.806}$$

Then there exist polynomials $a(x)$ and $b(x)$ such that [4, *Theorem 14, equation (58), p. 363*]

$$a(x) \odot r_{-1}(x) \oplus b(x) \odot r_0(x) = 1. \tag{1.807}$$

Proof. The proof is given in Corollary 1.11. □

Finally, we can give the *fundamental theorem of arithmetic*, which is also known as the *unique factorization theorem*, respectively, the *unique-prime-factorization theorem.*

Theorem 1.38 (Fundamental Theorem of Arithmetic). *Let $b \in \mathbb{E}$ be an element in a Euclidean domain $(\mathbb{E}, \oplus, \odot)$. Then:*

1: *$b \in \mathbb{E}$ can be written as a product of primes $p_1 \in \mathbb{E}, p_2 \in \mathbb{E}, p_3 \in \mathbb{E} \cdots p_r \in \mathbb{E}, r \in \mathbb{N}^*$, i. e.,*

$$b = p_1 \odot p_2 \odot p_3 \odot \cdots \odot p_r, \quad p_i \in \mathbb{E}, \quad i \in \{1, 2 \cdots r\}, \quad r \in \mathbb{N}^*, \tag{1.808}$$

2: *and if $b \in \mathbb{E}$ is written in another way as a product of primes $q_1 \in \mathbb{E}, q_2 \in \mathbb{E}, q_3 \in \mathbb{E} \cdots q_s \in \mathbb{E}, s \in \mathbb{N}^*$, then*

$$r = s, \quad r, s \in \mathbb{N}^*, \tag{1.809}$$

and, after a suitable renumbering, the p_i and the q_i, $i \in \{1, 2 \cdots r\}, r \in \mathbb{N}^*$ are associates [17, *p. 15*].

Proof.

Claim 1

If $b \in \mathbb{E}$ is a prime, the expression $b = b$ satisfies us [17, p. 15].

Otherwise, $b \in \mathbb{E}$ has a nontrivial factorization

$$b = a \odot c, \quad a, b, c \in \mathbb{E}, \tag{1.810}$$

$a \in \mathbb{E}$ and $c \in \mathbb{E}$ being proper divisors of $b \in \mathbb{E}$ [17, p. 15].

Owing to Corollary 1.10, we have

$$h\{a\} < h\{b\}, \quad h\{c\} < h\{b\}, \quad a, b, c \in \mathbb{E}, \tag{1.811}$$

and so, by induction, $a \in \mathbb{E}$ as well as $c \in \mathbb{E}$ can be expressed as a product of primes, i. e.,

$$\begin{aligned} a = p_1 \odot p_2 \odot \cdots \odot p_j, \quad c = p_{j+1} \odot p_{j+2} \odot \cdots \odot p_r, \\ p_i \in \mathbb{E}, \quad i \in \{1, 2 \cdots r\}, \quad j \in \{1, 2 \cdots (r-1)\}, \quad r \in \mathbb{N}^*, \quad a, c \in \mathbb{E}, \end{aligned} \tag{1.812}$$

leading to

$$\begin{aligned} b = a \odot c = p_1 \odot p_2 \odot \cdots \odot p_j \odot p_{j+1} \odot p_{j+2} \odot \cdots \odot p_r, \\ a, b, c, p_i \in \mathbb{E}, \quad i \in \{1, 2 \cdots r\}, \quad r \in \mathbb{N}^*, \end{aligned} \tag{1.813}$$

just as asserted [17, p. 16].

Claim 2

We know that, for instance, $p_1 | (q_1 \odot q_2 \odot q_3 \odot \cdots \odot q_s)$. Hence, p_1 must divide at least one of the $q_i \in \mathbb{E}$, $i \in \{1, 2 \cdots s\}, s \in \mathbb{N}^*$ [17, p. 16]. Let us carry out the renumbering such that $p_1 | q_1$ [17, p. 16].

Since, however, $p_1 \in \mathbb{E}$ and $q_1 \in \mathbb{E}$ are primes and p_1 is not a unit, p_1 and q_1 must be associates, i. e.,

$$q_1 = u_1 \cdot p_1, \quad p_1, q_1, u_1 \in \mathbb{E}, \tag{1.814}$$

for some unit $u_1 \in \mathbb{E}$ [17, p. 16]. In general, let us define

$$q_i = u_i \odot p_i, p_i, q_i, u_i \in \mathbb{E}, \quad i \in \left\{1, 2 \cdots \min\{r, s\}\right\} \quad r, s \in \mathbb{N}^*, \tag{1.815}$$

for some units u_i, $i \in \{1, 2 \cdots \min\{r, s\}\}$, $r, s \in \mathbb{N}^*$ [17, p. 16]. Therefore, we arrive at

$$p_1 \odot p_2 \odot p_3 \odot \cdots \odot p_r = \underbrace{q_1}_{=u_1 \odot p_1} \odot q_2 \odot q_3 \odot \cdots \odot q_s = (u_1 \odot p_1) \odot q_2 \odot q_3 \odot \cdots \odot q_s, \quad r, s \in \mathbb{N}^*, \tag{1.816}$$

and, after dividing both sides by p_1, we obtain

$$\begin{aligned} p_2 \odot p_3 \odot \cdots \odot p_r &= (u_1 \odot q_2) \odot q_3 \odot \cdots \odot q_s \\ &= u_1 \odot \underbrace{q_2}_{=u_2 \odot p_2} \odot q_3 \odot \cdots \odot q_s \\ &= u_1 \odot u_2 \odot p_2 \odot q_3 \odot \cdots \odot q_s. \quad r, s \in \mathbb{N}^*. \end{aligned} \tag{1.817}$$

Dividing by p_2 yields

$$p_3 \odot \cdots \odot p_r = u_1 \odot u_2 \odot q_3 \odot \cdots \odot q_s, \quad r, s \in \mathbb{N}^*. \tag{1.818}$$

In what follows, we will use the notation of the following type:

$$\bigodot_{i=1}^{r} p_i, \quad r \in \mathbb{N}^*, \tag{1.819}$$

or, alternatively, $\bigodot_{i=1}^{r} p_i$, which is the abbreviation of the multiplication using the multiplicative operation, i. e., $\odot$, r times, $r \in \mathbb{N}^*$.

Let us now assume that $r < s$. Furthermore, let us replace $r < s$ terms on the right side and divide both sides by $\bigodot_{i=3}^{r} p_i$. We thereby obtain

$$1 = \left(\bigodot_{i=3}^{r} u_i\right) \odot q_{r+1} \odot \cdots \odot q_s, \quad r, s \in \mathbb{N}^*, \tag{1.820}$$

requiring that $q_{r+1}, q_{r+2} \cdots q_s$, $r, s \in \mathbb{N}^*$ are units, which contradicts the prerequisite. Therefore, the case $r < s$ is impossible.

What about $r > s$? Now, after replacing $s < r$ terms on the right side and dividing both sides by $\bigodot_{i=3}^{s} p_i$, we obtain

$$p_{s+1} \odot \cdots \odot p_r = \bigodot_{i=3}^{s} u_i, \quad r, s \in \mathbb{N}^*, \tag{1.821}$$

requiring that $p_{s+1}, p_{s+2} \cdots p_r$ are units, which contradicts the prerequisite. Therefore, $r > s$ is also impossible.

The only admissible case is obtained for

$$r = s, \quad r, s \in \mathbb{N}^*, \tag{1.822}$$

in which we replace $r = s$ terms on the right side and divide both sides by $\bigodot_{i=3}^{r} p_i$, which leads to the true expression

$$1 = \bigodot_{i=3}^{r} u_i, \quad r \in \mathbb{N}^*. \tag{1.823}$$

This means that expressing $b \in \mathbb{E}$ in two different forms

$$b = p_1 \odot p_2 \odot p_3 \cdots\cdots p_r = q_1 \odot q_2 \odot q_3 \cdots q_s, \quad r, s \in \mathbb{N}^*, \tag{1.824}$$

with $p_i \in \mathbb{E}$ and $q_i \in \mathbb{E}$, $i \in \{1, 2 \cdots r\}$, $r \in \mathbb{N}^*$, being primes, is only possible when after appropriate renumbering, the $p_i \in \mathbb{E}$ and the $q_i \in \mathbb{E}$, $i \in \{1, 2 \cdots r\}$, $r \in \mathbb{N}^*$ are pairwise associates. □

Example 1.38. Let us look at $(\mathbb{Z}[\sqrt{-1}], +, \cdot)$ [17, p. 4]. Let us factor $8 \in \mathbb{Z}[\sqrt{-1}]$. We find

$$8 = (1 + \mathrm{j})^5 \cdot (-1 + \mathrm{j}). \tag{1.825}$$

Let us factor $(8 + 4\mathrm{j}) \in \mathbb{Z}[\sqrt{-1}]$. We find

$$8 + 4\mathrm{j} = -(1 + \mathrm{j})^4 \cdot (2 + \mathrm{j}). \tag{1.826}$$

Let us factor $10 \in \mathbb{Z}[\sqrt{-1}]$. We find

$$10 = -(1 + \mathrm{j})^2 \cdot (2 + \mathrm{j}) \cdot (1 + 2\mathrm{j}). \tag{1.827}$$

Let us factor $12 \in \mathbb{Z}[\sqrt{-1}]$. We find

$$12 = -(1 + \mathrm{j})^4 \cdot 3. \tag{1.828}$$

Let us factor $(45 + 3\mathrm{j}) \in \mathbb{Z}[\sqrt{-1}]$. We find

$$45 + 3\mathrm{j} = (1 + \mathrm{j}) \cdot 3 \cdot (8 - 7\mathrm{j}). \tag{1.829}$$

Let us factor $60 \in \mathbb{Z}[\sqrt{-1}]$. We find

$$60 = -(1 + \mathrm{j})^3 \cdot (1 - \mathrm{j}) \cdot 3 \cdot (2 + \mathrm{j}) \cdot (1 + 2\mathrm{j}). \tag{1.830}$$

Example 1.39. Let us consider the Euclidean domain of monic polynomials $(\mathbb{F}_2[x]_n, \oplus, \odot)$ over the field $\mathbb{F}_2$, i. e., the polynomials with coefficients in $\mathbb{F}_2$.

We already know that the unit is

$$u(x) = 1. \tag{1.831}$$

The polynomial

$$i_1(x) = x, \quad i_1(x) \in \mathbb{F}_2[x]_n, \tag{1.832}$$

is irreducible because it can only be divided by $u(x)$ or by $i_1(x)$.

Furthermore, the polynomial

$$i_2(x) = x \oplus 1, \quad i_2(x) \in \mathbb{F}_2[x]_n, \tag{1.833}$$

is irreducible because it can only be divided by $u(x)$ or by $i_2(x)$.

The polynomial

$$p_1(x) = x^2, \quad p_1(x) \in \mathbb{F}_2[x]_n, \tag{1.834}$$

is not irreducible because

$$p_1(x) = x^2 = x \odot x = \left(i_1(x)\right)^2. \tag{1.835}$$

The polynomial

$$p_2(x) = x^2 \oplus 1, \quad p_2(x) \in \mathbb{F}_2[x]_n, \tag{1.836}$$

is not irreducible. We find

$$\begin{array}{ccccccc} x^2 & \oplus & & & 1 & : x \oplus 1 & = & x \oplus 1 \\ x^2 & \oplus & x & & & & & \\ \hline & & x & \oplus & 1 & & & \\ & & x & \oplus & 1 & & & \\ \cline{3-5} & & & & 0 & & & \end{array} \tag{1.837}$$

and, therefore, we have

$$p_2(x) = x^2 \oplus 1 = \left(i_2(x)\right)^2 = (x \oplus 1)^2. \tag{1.838}$$

The polynomial

$$i_3(x) = x^2 \oplus x \oplus 1, \quad i_3(x) \in \mathbb{F}_2[x]_n, \tag{1.839}$$

is irreducible because neither $i_1(x)$ nor $i_2(x)$ divide it.

The polynomial

$$p_3(x) = x^2 \oplus x, \quad p_3(x) \in \mathbb{F}_2[x]_n, \tag{1.840}$$

is not irreducible, because we find

$$p_3(x) = x^2 \oplus x = x \cdot (x \oplus 1) = i_1(x) \odot i_2(x). \tag{1.841}$$

The polynomial

$$i_4(x) = x^3 \oplus x \oplus 1, \quad i_4(x) \in \mathbb{F}_2[x]_n, \tag{1.842}$$

is irreducible because neither $i_1(x)$ nor $i_2(x)$ nor $i_3(x)$ divide it.

The polynomial

$$i_5(x) = x^3 \oplus x^2 \oplus 1, \quad i_5(x) \in \mathbb{F}_2[x]_n, \tag{1.843}$$

is irreducible because neither $i_1(x)$ nor $i_2(x)$ nor $i_3(x)$ nor $i_4(x)$ divide it.

The polynomials

$$\begin{aligned} i_6(x) &= x^4 \oplus x \oplus 1, \\ i_7(x) &= x^4 \oplus x^3 \oplus x^2 \oplus x \oplus 1, \\ i_8(x) &= x^4 \oplus x^3 \oplus 1, \quad i_6(x), i_7(x), i_8(x) \in \mathbb{F}_2[x]_n, \end{aligned} \tag{1.844}$$

are also irreducible.

1.12.5 Construction of the Finite Field $\mathbb{F}_{2^n}$

Let us consider the vector space $\mathbb{F}_2^n$, $n \in \mathbb{N}^*$. We already know that the n-tuples, i. e., the row vectors with n components taken from $\mathbb{F}_2$, can easily be added by vector addition. Taking the said components from $\mathbb{F}_2$, vector subtraction is the same as vector addition [4, p. 82]. For instance, we yield [4, p. 82]

$$(a_0, a_1, a_2, a_3 \cdots a_{n-1}) \oplus (a_0, a_1, a_2, a_3 \cdots a_{n-1}) = \mathbf{0}_n, \quad n \in \mathbb{N}^*. \tag{1.845}$$

What about multiplication? This does not seem to be illustratively clear by just taking vectors of the above form into account. However, we already learned that we can represent a vector by a polynomial. So, let us choose an abstract "placeholder," i. e., some field element $\alpha \in \mathbb{F}_{2^n}$ say, which we can use to associate any permissible n-tuple, $n \in \mathbb{N}^*$, i. e., row vectors, with a polynomial in α [4, p. 82], for instance, like this:

n-tuple $\in \mathbb{F}_2^n$	polynomial $\in \mathbb{F}_2[x]_{n-1}$ with field element $\alpha \in \mathbb{F}_{2^n}$
$(0, 0, 0, 0 \cdots 0, 0)$	0
$(1, 0, 0, 0 \cdots 0, 0)$	1
$(0, 1, 0, 0 \cdots 0, 0)$	α
$(1, 1, 0, 0 \cdots 0, 0)$	$1 \oplus \alpha$
$(0, 0, 1, 0 \cdots 0, 0)$	α^2
$(1, 0, 1, 0 \cdots 0, 0)$	$1 \oplus \alpha^2$
$(0, 1, 1, 0 \cdots 0, 0)$	$\alpha^2 \oplus \alpha^3$
$\vdots$	$\vdots$
$(1, 1, 1, 1 \cdots 1, 1)$	$\bigoplus_{i=0}^{n-1} \alpha^i$

The multiplication of the n-tuples of $\mathbb{F}_2^n$ corresponds to the multiplication of the above polynomials, for instance, like this [4, p. 82]:

$$(1, 1, 1, 1 \cdots 1, 1) \odot (1, 1, 1, 1 \cdots 1, 1) \leftrightarrow \bigoplus_{i=0}^{n-1} \alpha^i \odot \bigoplus_{j=0}^{n-1} \alpha^j = \bigoplus_{i=0}^{n-1} \bigoplus_{j=0}^{n-1} \alpha^{i+j} \tag{1.846}$$

with degree $2(n-1)$, which is greater than $(n-1)$ for $n > 1$. However, the reader might think "wait a minute, this result does not correspond to any of the permissible n-tuples"! We obviously must find a way to overcome this issue [4, p. 82]. To do this, we agree that α will have to satisfy a certain fixed equation of degree n, $n \in \mathbb{N}^*$. The required equation will have the following form:

$$\underbrace{1 \oplus [\text{polynomial in } \alpha \text{ without the constant 1 and with degree at most } (n-1)] \oplus \alpha^n}_{\text{polynomial } \pi(\alpha) \text{ of degree } n} = 0,$$
$$n \in \mathbb{N}^*, \tag{1.847}$$

yielding

$$\pi(\alpha) = 0, \quad n \in \mathbb{N}^*. \tag{1.848}$$

We see from (1.848) that α is obviously a *zero* of the polynomial $\pi(x)$ [25, p. 78]. The field element α is also called a *root* of the equation (1.848) [25, p. 78].

Equation (1.848) immediately leads to

$$\alpha^n = 1 \oplus [\text{polynomial in } \alpha \text{ without the constant 1 and with degree at most } (n-1)], \quad n \in \mathbb{N}^*.$$

Example 1.40. Let us consider n equal to 4. A suitable version of (1.848) is [4, p. 82]

$$\pi(\alpha) = 1 \oplus \alpha \oplus \alpha^4 = 0, \quad \Leftrightarrow \quad \alpha^4 = 1 \oplus \alpha. \tag{1.849}$$

Then we find

$$\begin{aligned}
\alpha^{-\infty} \quad & = 0, \\
\alpha^0 \quad & = 1, \\
\alpha^1 \quad & = \alpha, \\
\alpha^2 \quad & = \alpha^2, \\
\alpha^3 \quad & = \alpha^3, \\
\alpha^4 \quad & = 1 \oplus \alpha, \\
\alpha^5 \quad & = \alpha \oplus \alpha^2, \\
\alpha^6 \quad & = \alpha^2 \oplus \alpha^3, \\
\alpha^7 = \alpha^3 \oplus \alpha^4 \quad & = 1 \oplus \alpha \oplus \alpha^3, \\
\alpha^8 = \alpha^4 \oplus \alpha^5 \quad & = 1 \oplus \alpha^2, \\
\alpha^9 = \alpha^5 \oplus \alpha^6 \quad & = \alpha \oplus \alpha^3, \\
\alpha^{10} = \alpha^6 \oplus \alpha^7 \quad & = 1 \oplus \alpha \oplus \alpha^2, \\
\alpha^{11} = \alpha^7 \oplus \alpha^8 \quad & = \alpha \oplus \alpha^2 \oplus \alpha^3, \\
\alpha^{12} = \alpha^8 \oplus \alpha^9 \quad & = 1 \oplus \alpha \oplus \alpha^2 \oplus \alpha^3, \\
\alpha^{13} = \alpha^9 \oplus \alpha^{10} \quad & = 1 \oplus \alpha^2 \oplus \alpha^3, \\
\alpha^{14} = \alpha^{10} \oplus \alpha^{11} \quad & = 1 \oplus \alpha^3, \\
\alpha^{15} = \alpha^{11} \oplus \alpha^{12} \quad & = 1, \\
\alpha^{16} = \alpha^{12} \oplus \alpha^{13} \quad & = \alpha.
\end{aligned} \tag{1.850}$$

Thus, we yield the following table [4, p. 85].

4-tuple $\in \mathbb{F}_2^4$	**polynomial** $\in \mathbb{F}_2[x]_3$ **with field element** $\alpha \in \mathbb{F}_2^4$	**power of** $\alpha \in \mathbb{F}_2^4$	**logarithm**
$(0,0,0,0)$	0	$\alpha^{-\infty}$	$-\infty$
$(1,0,0,0)$	1	α^0	0
$(0,1,0,0)$	α	α^1	1
$(0,0,1,0)$	α^2	α^2	2
$(0,0,0,1)$	α^3	α^3	3
$(1,1,0,0)$	$1 \oplus \alpha$	α^4	4
$(0,1,1,0)$	$\alpha \oplus \alpha^2$	α^5	5
$(0,0,1,1)$	$\alpha^2 \oplus \alpha^3$	α^6	6
$(1,1,0,1)$	$1 \oplus \alpha \oplus \alpha^3$	α^7	7
$(1,0,1,0)$	$1 \oplus \alpha^2$	α^8	8
$(0,1,0,1)$	$\alpha \oplus \alpha^3$	α^9	9
$(1,1,1,0)$	$1 \oplus \alpha \oplus \alpha^2$	α^{10}	10
$(0,1,1,1)$	$\alpha \oplus \alpha^2 \oplus \alpha^3$	α^{11}	11
$(1,1,1,1)$	$1 \oplus \alpha \oplus \alpha^2 \oplus \alpha^3$	α^{12}	12
$(1,0,1,1)$	$1 \oplus \alpha^2 \oplus \alpha^3$	α^{13}	13
$(1,0,0,1)$	$1 \oplus \alpha^3$	α^{14}	14

The above table represents the finite field $\mathbb{F}_{2^4}$ equal to $\mathbb{F}_{16}$, which is generated by $1 \oplus \alpha \oplus \alpha^4 = 0$ [4, p. 85]. $\mathbb{F}_{16}$ is called the *Galois field of order 16* [4, p. 84]. The finite field $\mathbb{F}_{16}$ consists of all polynomials in α with binary coefficients and degree at most 3, with calculations performed modulo the polynomial $\pi(\alpha)$ equal to $(1 \oplus \alpha \oplus \alpha^4)$ [4, p. 93].

According to the above table, the field elements can be written in several different ways [4, p. 84]. We note that the nonzero elements of the field form a *cyclic group of order 15 with generator* α, where

$$\alpha^{15} = 1. \tag{1.851}$$

In Example 1.21, we will elaborate that the chosen polynomial $\pi(x)$, which obviously has the generator α as a zero, is an *irreducible polynomial* [4, p. 84]. Moreover, we will define what we mean by irreducible polynomial in Definition 1.55.

The element α or any other generator of this cyclic group is called a *primitive element* of $\mathbb{F}_{2^4}$ [4, p. 84]. For example α and α^2 are primitive but α^5 is not [4, p. 84]. A polynomial having a primitive element as a zero is called a *primitive polynomial* [4, p. 84].

Using

$$\alpha^{15} \oplus 1 = 0, \tag{1.852}$$

we yield

$$\underbrace{\alpha^{14}}_{=1\oplus\alpha^3} \odot \alpha \oplus 1 = 0, \tag{1.853}$$

i. e.,

$$\alpha \oplus \alpha^4 \oplus 1 = 0 \quad \Leftrightarrow \quad 1 \oplus \alpha \oplus \alpha^4 = 0. \tag{1.854}$$

All finite fields can be obtained in the way illustrated in this Example 1.40 [4, Chapter 4].

Not all irreducible polynomials are primitive [4, p. 84]. For instance, $(1 \oplus x \oplus x^2 \oplus x^3 \oplus x^4)$ is irreducible and could hence by used to generate the finite field $\mathbb{F}_{16}$, but it is not a primitive polynomial [4, p. 84].

Using the above table, we may multiply (1, 1, 0, 1) by (1, 0, 0, 1), yielding [4, p. 82]

$$\begin{aligned}\left(1 \oplus a \oplus a^3\right) \odot \left(1 \oplus a^3\right) &= 1 \oplus a \oplus a^4 \oplus a^6 \\ &= \underbrace{1 \oplus a \oplus 1 \oplus a}_{=0} \oplus (1 \oplus a) \odot a^2, \\ &= a^2 \oplus a^3.\end{aligned} \tag{1.855}$$

We now obtain [4, p. 82]

$$a^2 \oplus a^3 \leftrightarrow (0, 0, 1, 1). \tag{1.856}$$

The computation according to (1.855) is equivalent to dividing $(1 \oplus a \oplus a^4 \oplus a^6)$ by $(1 \oplus a \oplus a^4)$ and keeping the remainder [4, p. 82].

Another way of describing this process is that we reduce a product of polynomials modulo $\pi(a)$ [4, p. 83],

$$1 \oplus a \oplus a^4 \oplus a^6 = \underbrace{\left[1 \oplus a^2\right]}_{\text{“quotient”}} \odot \pi(a) \oplus \underbrace{a^2 \oplus a^3}_{\text{“remainder”}} = \left(a^2 \oplus a^3\right) \bmod \pi(a). \tag{1.857}$$

Similarly, we find [4, p. 83]

$$a^4 = (1 \oplus a) \bmod \pi(a). \tag{1.858}$$

Since we require $\mathbb{F}_{2^n}$, $n \in \mathbb{N}^*$ to be a finite field, the multiplication must have an inverse. This will only be possible if the polynomial $\pi(x) \in \mathbb{F}_2[x]_{n-1}$, $n \in \mathbb{N}^*$ is irreducible over $\mathbb{F}_2$.

Definition 1.55 (Irreducible Polynomial). A polynomial is irreducible over a field, if it is not the product of two polynomials of lower degree in the field [4, p. 83].

An irreducible polynomial is like a prime number in that it has no nontrivial factors [4, p. 83]. Similar to Theorem 1.38, apart from a constant factor, any polynomial can be written uniquely as the product of irreducible polynomials [4, p. 83].

Theorem 1.39 (Irreducible Polynomial and Inverse). *If $\pi(x)$ is irreducible, then every nonzero polynomial $B(a)$ has a unique inverse $B(a)^{-1}$ such that* [4, p. 83]

$$B(a) \odot B(a)^{-1} \equiv 1 \bmod \pi(a). \tag{1.859}$$

Proof. Look at all the products $A(a) \odot B(a)$, in which $A(a)$ runs through all the polynomials [4, p. 83], e. g.,

$$1, a, a \oplus 1, a^2 \cdots \left(a^3 \oplus a^2 \oplus a \oplus 1\right), \tag{1.860}$$

e. g., in the case of $\mathbb{F}_{16}$ of degree lower than or equal to 3.

The said products $A(a) \odot B(a)$ must all be distinct mod $\pi(a)$. For instance, if

$$A_1(a) \odot B(a) = A_2(a) \odot B(a) \bmod \pi(a), \tag{1.861}$$

we yield

$$A_1(\alpha) \odot B(\alpha) \oplus A_2(\alpha) \odot B(\alpha) = 0 \bmod \pi(\alpha), \tag{1.862}$$

i. e.,

$$\left[A_1(\alpha) \oplus A_2(\alpha)\right] \odot B(\alpha) = 0 \bmod \pi(\alpha), \tag{1.863}$$

thus $\pi(\alpha)|[A_1(\alpha) \oplus A_2(\alpha)] \odot B(\alpha)$. Since $\pi(\alpha)$ is irreducible either $\pi(\alpha)|(A_1(\alpha) \oplus A_2(\alpha))$ or $\pi(\alpha)|B(\alpha)$ [4, p. 83]. Since the degrees of $A_1(\alpha)$, $A_2(\alpha)$ and $B(\alpha)$ are less than the degree of $\pi(\alpha)$, this situation can only occur if [4, p. 83]

$$A_1(\alpha) = A_2(\alpha). \tag{1.864}$$

Therefore, all the products $A(\alpha) \odot B(\alpha)$ are distinct. These products must also be equal to (1.860) in some order [4, p. 83]. In particular, for just one given $A(\alpha)$, $A(\alpha) \odot B(\alpha)$ is equal to 1. Then $A(\alpha)$ is equal to $B(\alpha)^{-1}$ [4, p. 83]. □

Example 1.41. Let us revisit Example 1.40. We shall see that the polynomial $(1 \oplus x \oplus x^4)$ is irreducible over $\mathbb{F}_2$ [4, p. 83f.]. The polynomial $(1 \oplus x \oplus x^4)$ has degree 4 and so, if not irreducible, contains a factor of degree 1 or 2. In the case of the degree being equal to 1, there exist only two polynomials, namely [4, p. 84]
- x and
- $(1 \oplus x)$.

In the case of the degree being equal to 2, we find exactly four polynomials, namely [4, p. 84]
- x^2,
- $(1 \oplus x^2)$,
- $(x \oplus x^2)$ and
- $(1 \oplus x \oplus x^2)$.

Let us consider whether x divides $\pi(x)$:

$$\begin{array}{cccccccc} x^4 & \oplus & x & \oplus & 1 & : & x & = x^3 \oplus 1 \\ x^4 & & & & & & & \\ \hline & & x & \oplus & 1 & & & \\ & & x & & & & & \\ \hline & & & & 1 & & & \end{array} \tag{1.865}$$

So, we have a nonzero remainder, 1, and, therefore, $x \nmid \pi(x)$, read "x does not divide $\pi(x)$."

Let us consider whether $\pi(x)$ can be divided by $(1 \oplus x)$:

$$\begin{array}{cccccc} x^4 & \oplus & x & \oplus & 1 & : x \oplus 1 \; = \; x^3 \oplus x^2 \oplus x \\ x^4 & \oplus & x^3 & & & \\ \hline x^3 & \oplus & x & \oplus & 1 & \\ x^3 & \oplus & x^2 & & & \\ \hline x^2 & \oplus & x & \oplus & 1 & \\ x^2 & \oplus & x & & & \\ \hline & & & & 1 & \end{array} \tag{1.866}$$

Again, we have a nonzero remainder, 1, and, therefore, $(1 \oplus x) \nmid \pi(x)$.

Let us consider whether $\pi(x)$ can be divided by x^2:

$$\begin{array}{ccccccc} x^4 & \oplus & x & \oplus & 1 & : & x^2 = x^2 \\ x^4 & & & & & & \\ \hline & & x & \oplus & 1 & & \end{array} \tag{1.867}$$

So, we have a nonzero remainder, $(1 \oplus x)$, and, therefore, $x^2 \nmid \pi(x)$.

Let us consider whether $\pi(x)$ can be divided by $(1 \oplus x^2)$:

$$\begin{array}{ccccccc} x^4 & \oplus & x & \oplus & 1 & : & x^2 \oplus 1 = x^2 \oplus 1 \\ x^4 & \oplus & x^2 & & & & \\ \hline x^2 & \oplus & x & \oplus & 1 & & \\ x^2 & & & \oplus & 1 & & \\ \hline & & x & & & & \end{array} \tag{1.868}$$

So, we have a nonzero remainder, x, and, therefore, $(1 \oplus x^2) \nmid \pi(x)$.

Let us consider whether $\pi(x)$ can be divided by $(x \oplus x^2)$:

$$\begin{array}{ccccccc} x^4 & \oplus & x & \oplus & 1 & : & x^2 \oplus x = x^2 \oplus x \\ x^4 & \oplus & x^3 & & & & \\ \hline x^3 & \oplus & x & \oplus & 1 & & \\ x^3 & \oplus & x & & & & \\ \hline & & & & 1 & & \end{array} \tag{1.869}$$

So, we have a nonzero remainder, 1, and, therefore, $(x \oplus x^2) \nmid \pi(x)$.

Let us finally consider whether $\pi(x)$ can be divided by $(1 \oplus x \oplus x^2)$:

$$\begin{array}{ccccccc} x^4 & \oplus & x & \oplus & 1 & : & x^2 \oplus x \oplus 1 = x^2 \oplus x \\ x^4 & \oplus & x^3 & \oplus & x^2 & & \\ \hline x^3 \oplus x^2 & \oplus & x & \oplus & 1 & & \\ x^3 \oplus x^2 & \oplus & x & & & & \\ \hline & & & & 1 & & \end{array} \tag{1.870}$$

So, we have a nonzero remainder, 1, and, therefore, $(1 \oplus x \oplus x^2) \nmid \pi(x)$.

Clearly, $\pi(x)$ is irreducible. Therefore, using $\pi(x)$ equal to $(1 \oplus x \oplus x^4)$ allows to establish $\mathbb{F}_{16}$.

Example 1.42. Let us consider $\mathbb{F}_{2^3}$ equal to $\mathbb{F}_8$. Tables 1.10 and 1.11 contain two versions of $\mathbb{F}_8$, one being obtained using the irreducible polynomial [4, p. 93]

$$\pi_1(x) = 1 \oplus x \oplus x^3, \tag{1.871}$$

the other being obtained using the irreducible polynomial [4, p. 93]

$$\pi_2(x) = 1 \oplus x^2 \oplus x^3. \tag{1.872}$$

Example 1.43. For later reference, we will give $\mathbb{F}_{2^2}$ equal to $\mathbb{F}_4$ generated by the irreducible polynomial

$$\pi(x) = 1 \oplus x \oplus x^2, \tag{1.873}$$

Table 1.10: One version of $\mathbb{F}_{2^3}$ equal to $\mathbb{F}_8$, adapted according to [4, p. 101].

$\pi_1(x) = 1 \oplus x \oplus x^3$ $(\alpha^7 = 1)$			
as a 3-tuple	**as a polynomial**	**as a power of α**	**logarithm**
(0, 0, 0)	0	0	$-\infty$
(1, 0, 0)	1	α^0	0
(0, 1, 0)	α	α^1	1
(0, 0, 1)	α^2	α^2	2
(1, 1, 0)	$1 \oplus \alpha$	α^3	3
(0, 1, 1)	$\alpha \oplus \alpha^2$	α^4	4
(1, 1, 1)	$1 \oplus \alpha \oplus \alpha^2$	α^5	5
(1, 0, 1)	$1 \oplus \alpha^2$	α^6	6

Table 1.11: Another version of $\mathbb{F}_{2^3}$ equal to $\mathbb{F}_8$, adapted according to [4, p. 101].

$\pi_2(x) = 1 \oplus x^2 \oplus x^3$ $(\alpha^7 = 1)$			
as a 3-tuple	**as a polynomial**	**as a power of α**	**logarithm**
(0, 0, 0)	0	0	$-\infty$
(1, 0, 0)	1	α^0	0
(0, 1, 0)	α	α^1	1
(0, 0, 1)	α^2	α^2	2
(1, 0, 1)	$1 \oplus \alpha^2$	α^3	3
(1, 1, 1)	$1 \oplus \alpha \oplus \alpha^2$	α^4	4
(1, 1, 0)	$1 \oplus \alpha$	α^5	5
(0, 1, 1)	$\alpha \oplus \alpha^2$	α^6	6

and $\mathbb{F}_2$ generated by the irreducible polynomial

$$\pi(x) = 1 \oplus x; \qquad (1.874)$$

cf., Table 1.12 and Table 1.13. Both $\mathbb{F}_2$ and $\mathbb{F}_{2^2}$ are unique.

Table 1.12: $\mathbb{F}_{2^2}$ equal to $\mathbb{F}_4$.

$\pi(x) = 1 \oplus x \oplus x^2$ $(\alpha^3 = 1)$			
as a 2-tuple	**as a polynomial**	**as a power of α**	**logarithm**
(0, 0)	0	0	$-\infty$
(1, 0)	1	α^0	0
(0, 1)	α	α^1	1
(1, 1)	$1 \oplus \alpha$	α^2	2

Table 1.13: $\mathbb{F}_2$.

$\pi(x) = 1 \oplus x \; (\alpha^1 = 1)$			
as a 1-tuple	as a polynomial	as a power of α	logarithm
0	0	0	$-\infty$
1	1	α^0	0

1.12.6 Minimal Polynomials, Conjugates and Cyclotomic Cosets

To further evaluate cyclic codes, we will introduce the concept of *minimal polynomials* [4, p. 99], [5, p. 178]. We will start with the cyclic nature of the considered finite fields and move on to *Fermat's theorem.*

Let us choose a maximal set $\mathbb{F}$ of m, $m \in \mathbb{N}^*$, field elements $\beta_0 = 1, \beta_1, \beta_2 \cdots \beta_{m-1}$, $m \in \mathbb{N}^*$, which are linearly independent over $\mathbb{F}_2$ [4, p. 96]. These m elements form a set of basis elements of $\mathbb{F}$. Then $\mathbb{F}$ contains only the elements [4, p. 96]

$$a_0 \odot \beta_0 \oplus a_1 \odot \beta_1 \oplus \cdots \oplus a_{m-1} \odot \beta_{m-1}, \qquad (1.875)$$
$$\beta_i \in \mathbb{F}, \quad a_i \in \mathbb{F}_2, \quad i \in \{0, 1 \cdots (m-1)\}, \quad m \in \mathbb{N}^*.$$

According to (1.875), each element of $\mathbb{F}$ is formed by a linear combination of the m basis elements $\beta_0 = 1, \beta_1, \beta_2 \cdots \beta_{m-1}$, $m \in \mathbb{N}^*$, which are weighted by unique binary weighting factors $a_i \in \mathbb{F}_2$, $i \in \{0, 1 \cdots (m-1)\}$, $m \in \mathbb{N}^*$ [4, p. 96].

Thus, $\mathbb{F}$ is a vector space of dimension m over $\mathbb{F}_2$ and contains 2^m elements, $m \in \mathbb{N}^*$ [4, p. 96]. Hence, $\mathbb{F}$ has the order 2^m [4, p. 96].

Let $\mathbb{F}^*$ stand for the set of $(2^m - 1)$, $m \in \mathbb{N}^*$, nonzero elements of $\mathbb{F}$ [4, p. 96].

Theorem 1.40 ($\mathbb{F}^*$ as a Cyclic Multiplicative Group). $\mathbb{F}^*$ *is a finite* cyclic multiplicative group *of order* r *equal to* $(2^m - 1)$ [4, *p. 96*]. *A finite multiplicative group is cyclic if it consists of the elements* $1, \alpha, \alpha^2, \alpha^3, \cdots, \alpha^{r-1}$ *with* $\alpha^r = 1$ [4, *p. 96*]. *Then* α *is called a generator of the group* [4, *p. 96*].

Proof. It follows from the definition of a finite field [4, p. 96] that $\mathbb{F}^*$ is a multiplicative group. Let $\alpha \in \mathbb{F}^*$ [4, p. 96]. Since $\mathbb{F}^*$ has size $(2^m - 1)$, $m \in \mathbb{N}^*$, the expression α^i, $i \in \{0, 1 \cdots (2^m - 2)\}$ has at most $(2^m - 1)$ distinct values [4, p. 96]. Therefore, there are integers r and i, with $1 \le r \le (2^m - 1)$ such that [4, p. 96]

$$\alpha^{r+i} = \alpha^i \quad \Leftrightarrow \quad \alpha^r = 1, \qquad (1.876)$$
$$i \in \{0, 1 \cdots (2^m - 2)\}, \quad r \in \{1, 2 \cdots (2^m - 1)\}, \quad m \in \mathbb{N}^*.$$

The smallest r, which fulfills (1.876), is called the *order of* $\alpha \in \mathbb{F}^*$ [4, p. 96].

Now choose $\alpha \in \mathbb{F}^*$ so that the order r of $\alpha \in \mathbb{F}^*$ is as large as possible [4, p. 96]. We shall show that the order l of any element $\beta \in \mathbb{F}^*$ divides r [4, p. 96].

For any prime p, we yield the orders r and l as follows [4, p. 96]:

$$r = p^a r', \quad l = p^b l', \quad r \nmid r', \quad r \nmid l'. \qquad (1.877)$$

Using (1.876), we obtain [4, p. 96]

$$\left(\alpha^{p^a}\right)^{r'+i} = \alpha^{p^a(r'+i)} = \alpha^{p^a r' + p^a i} = \alpha^{r+p^a i} = \alpha^{p^a i} = \left(\alpha^{p^a}\right)^i, \qquad (1.878)$$
$$i \in \{0, 1 \cdots (2^m - 2)\}, \quad m \in \mathbb{N}^*.$$

Hence, α^{p^a} has the order r' [4, p. 96].

Using (1.876), we furthermore find [4, p. 96]

$$\beta^{l+i} = \beta^{\pi^b l' + i} = \left(\beta^{l'}\right)^{p^b + \frac{i}{l'}} = \left(\beta^{l'}\right)^{\frac{i}{l'}}, \qquad (1.879)$$
$$i \in \{0, 1 \cdots (2^m - 2)\}, \quad m \in \mathbb{N}^*.$$

Hence, $\beta^{l'}$ has the order p^b [4, p. 96].

Therefore, $\alpha^{p^a} \odot \beta^{l'}$ has the order $r' \cdot p^b$ [4, p. 96]. Hence, $b \le a$ or else r would not be maximal [4, p. 96]. Thus, every prime power that is a divisor of l is also a divisor of r, and so l divides r [4, p. 96]. Therefore, every $\beta \in \mathbb{F}^*$ satisfies the equation $x^r \oplus 1 = 0$, which means that $x^r \oplus 1$ is divisible by $\bigodot_{\beta \in F^*} (x \oplus \beta)$ [4, p. 96]. Since there are $(2^m - 1)$ elements in $\mathbb{F}^*$, we find $r \ge (2^m - 1)$ [4, p. 96]. But the prerequisite requires $r \le (2^m - 1)$. Both conditions can only be fulfilled for [4, p. 96]

$$r = 2^m - 1, \quad m \in \mathbb{N}^*. \qquad (1.880)$$

Thus,

$$\bigodot_{\beta \in F^*} (x \oplus \beta) = x^{2^m - 1} \oplus 1, \quad m \in \mathbb{N}^*, \qquad (1.881)$$

and the nonzero elements of $\mathbb{F}$ form the cyclic group $\alpha, \alpha^2, \alpha^3 \cdots \alpha^{2^m-2}, \alpha^{2^m-1} = 1$. □

Theorem 1.40 is quite interesting because it teaches us that in $\mathbb{F}_{2^m}$, $m \in \mathbb{N}^*$, having the generator $\alpha \in \mathbb{F}_{2^m}$, $m \in \mathbb{N}^*$, we have

$$\alpha^{2^m - 1} = 1, \quad m \in \mathbb{N}^*; \qquad (1.882)$$

(cf. (1.876)) and, therefore, the *inverse* $\alpha^{-1} \in \mathbb{F}_{2^m}$, $m \in \mathbb{N}^*$ *of the generator* $\alpha \in \mathbb{F}_{2^m}$, $m \in \mathbb{N}^*$ is given by

$$\alpha^{-1} = \alpha^{2^m - 2}, \quad m \in \mathbb{N}^*. \qquad (1.883)$$

Corollary 1.13 (Fermat's Theorem). *Every element* β *of a field* $\mathbb{F}$ *of order* 2^m *satisfies the identity*

$$\beta^{2^m} = \beta, \qquad (1.884)$$

or, equivalently, is a root of the equation [4, p. 96]

$$x^{2^m} = x. \qquad (1.885)$$

Thus,

$$x^{2^m} \oplus x = \bigodot_{\beta \in \mathbb{F}} (x \oplus \beta), \tag{1.886}$$

follows directly from Theorem 1.40 [4, *p. 96*].

If $\mathbb{F}$ is a field of order 2^m, a primitive element α of $\mathbb{F}$ has the order $(2^m - 1)$ [4, p. 97]. Then it follows that any nonzero element of $\mathbb{F}$ is a power of α [4, p. 97].

Theorem 1.41 (Primitive Element in a Field). *Any finite field* $\mathbb{F}$ *contains a primitive element* [4, *p. 97*].

Proof. Take α to be a generator of the cyclic group $\mathbb{F}^*$ [4, p. 97]. □

Let $\mathbb{F}$ be an arbitrary finite field of order 2^m, $m \in \mathbb{N}$ [4, p. 95]. $\mathbb{F}$ contains the unit element 1, and since $\mathbb{F}$ is finite, the elements 1, $(1 \oplus 1)$, $(1 \oplus 1 \oplus 1) \cdots$ cannot all be distinct [4, p. 95]. Indeed the smallest number p such that $\underbrace{1 \oplus 1 \oplus \cdots \oplus 1}_{p \text{ times}} = 0$ is 2 in our case [4, p. 95]. This p equal to 2 is a prime, and it is called the *characteristic* of the field [4, p. 95].

Theorem 1.42 (Powers of Elements in a Field). *In any finite field* $\mathbb{F}$ *of characteristic* 2, *we have* [4, *p. 97*]

$$(x \oplus y)^2 = x^2 \oplus y^2. \tag{1.887}$$

Proof. We have

$$(x \oplus y)^2 = \underbrace{x \odot x}_{x^2} \oplus \underbrace{x \odot y \oplus x \odot y}_{=0} \oplus \underbrace{y \odot y}_{y^2} = x^2 \oplus y^2. \tag{1.888}$$

□

Let us come back to the polynomial $(x^{2^m} \oplus x)$, for which we have

$$\beta^{2^m} \oplus \beta = 0, \quad m \in \mathbb{N}^*. \tag{1.889}$$

Definition 1.56 (Minimal Polynomial). The minimal polynomial of β over $\mathbb{F}_2$ is the lowest degree monic polynomial $M(x)$ with coefficients from $\mathbb{F}_2$ such that [4, p. 99]

$$M(\beta) = 0. \tag{1.890}$$

Let us consider important properties of minimal polynomials.

Theorem 1.43 (Property M1: Minimal Polynomials are Irreducible). *$M(x)$ is irreducible* [4, *p. 99*].

Proof. Assume $M(x)$ was not irreducible. Then we can write

$$M(x) = M_1(x) \odot M_2(x) \tag{1.891}$$

with degrees of $M_1(x)$ and $M_2(x)$ both greater than 0 [4, p. 99]. Thus, we have

$$M(\beta) = M_1(\beta) \odot M_2(\beta) = 0, \tag{1.892}$$

which means that at least either $M_1(\beta)$ must be equal to 0 or $M_2(\beta)$ must be equal to 0. However, this contradicts the fact that $M(x)$ has the lowest degree with $M(\beta)$ being equal to 0 [4, p. 99]. □

Theorem 1.44 (Property M2: $M(x)$ divides $f(x)$). *If $f(x)$ is any polynomial with coefficients in $\mathbb{F}_2$ such that $f(\beta)$ is equal to 0, then $M(x)$ divides $f(x)$* [4, *p. 99*].

Proof. By dividing $M(x)$ into $f(x)$, we find

$$f(x) = M(x) \odot a(x) \oplus r(x), \tag{1.893}$$

where the degree of the remainder $r(x)$ is less than the degree of $M(x)$ [4, p. 100]. Let $x = \beta$, yielding

$$0 = 0 \oplus r(\beta), \tag{1.894}$$

and so $r(x)$ is a polynomial with lower degree than $M(x)$ having β as a root [4, p. 100]. This is a contradiction unless $r(x)$ is equal to 0 [4, p. 100]. Then $f(x)$ is divisible by $M(x)$ [4, p. 100]. □

With Fermat's theorem (cf. Corollary 1.13 and Theorem 1.44 as a proof), we find the following important theorem.

Theorem 1.45 (Property M3: $M(x)$ divides $x^{2^m} \oplus x$). *$x^{2^m} \oplus x$ is divisible by $M(x)$* [4, *p. 100*].

Proof. As already stated, the proof is Theorem 1.44. □

Theorem 1.46 (Property M4: Degree of $M(x)$). $\deg M(x) \leq m$ [4, *p. 100*].

Proof. $\mathbb{F}_{2^m}$ is a vector space of dimension m over $\mathbb{F}_2$ [4, p. 100]. Therefore, any $(m+1)$ elements, such as $1, \beta, \beta^2, \beta^3 \cdots \beta^m$ are linearly dependent, i. e., there exist coefficients $a_i \in \mathbb{F}_2$ not all zero such that [4, p. 100]

$$\bigoplus_{i=0}^{m} a_i \odot \beta^i = 0. \tag{1.895}$$

Thus,

$$\bigoplus_{i=0}^{m} a_i \odot x^i \tag{1.896}$$

is a polynomial of degree $\leq m$ having β as a root [4, p. 100]. Therefore, $\deg M(x) \leq m$ [4, p. 100]. □

Theorem 1.47 (Property M5: Primitive Polynomial). *The minimal polynomial of a primitive element of $\mathbb{F}_{2^m}$ has degree m* [4, *p. 100*]. *Such a polynomial is called a* primitive polynomial [4, *p. 100*].

Proof. Let α be a primitive element of $\mathbb{F}_{2^m}$ with the minimal polynomial $M(x)$ of degree d [4, p. 100]. Then we may use $M(x)$ to generate a finite field $\mathbb{F}$ of order 2^m [4, p. 100].

But $\mathbb{F}$ contains α and hence all elements of $\mathbb{F}_{2^m}$, so the degree of $M(x)$ as the generator of $\mathbb{E}$ must fulfill the requirement $d \geq m$ [4, p. 100]. However, since Theorem 1.46 requires $\deg M(x) \leq m$, the degree of the primitive polynomial is exactly m [4, p. 100]. □

Example 1.44. Let $m = 1$. We have [4, p. 107]

$$a^1 = 1, \quad \Leftrightarrow \quad 1 = a^{-1}, \tag{1.897}$$

in $\mathbb{F}_2$. Furthermore, in $\mathbb{F}_2$ defined by $a \oplus 1 = 0$, the minimal polynomials have coefficients equal to 0 or 1. We thus yield the following:

Element of $\mathbb{F}_2$	Minimal Polynomial		
0			x
$1 = a^{-1}$	$M^{(0)}(x)$	$=$	$x \oplus 1$

Both polynomials, x and $M^{(0)}(x)$, both having degree 1, are irreducible [4, p. 107].

Let $m = 2$. We have

$$a^3 = 1, \quad \Leftrightarrow \quad a^2 = a^{-1}, \tag{1.898}$$

in $\mathbb{F}_{2^2}$ equal to $\mathbb{F}_4$. Furthermore, in $\mathbb{F}_{2^2}$ equal to $\mathbb{F}_4$, defined by $a^2 \oplus a \oplus 1 = 0$, the minimal polynomials have coefficients equal to 0 or 1. We thus yield the following [4, p. 108]:

Element of $\mathbb{F}_{2^2}$	Minimal Polynomial		
0			x
1	$M^{(0)}(x)$	$=$	$x \oplus 1$
$a, a^2 = a^{-1}$	$M^{(1)}(x)$	$=$	$x^2 \oplus x \oplus 1$

There is only one irreducible polynomial of degree 2, namely $M^{(1)}(x)$ equal to $x^2 \oplus x \oplus 1$. Since $M^{(1)}(x)$ equal to $x^2 \oplus x \oplus 1$ is the polynomial of the primitive element a, it is the primitive polynomial of $\mathbb{F}_{2^2}$ equal to $\mathbb{F}_4$, defined by $a^2 \oplus a \oplus 1 = 0$.

Let $m = 3$. We have

$$a^7 = 1, \quad \Leftrightarrow \quad a^6 = a^{-1}, \tag{1.899}$$

in $\mathbb{F}_{2^3}$ equal to $\mathbb{F}_8$. Furthermore, in $\mathbb{F}_{2^3}$ equal to $\mathbb{F}_8$, defined by $a^3 \oplus a \oplus 1 = 0$, the minimal polynomials have coefficients equal to 0 or 1. We thus yield the following [4, p. 108]:

Element of $\mathbb{F}_{2^3}$	Minimal Polynomial		
0			x
1	$M^{(0)}(x)$	$=$	$x \oplus 1$
a, a^2, a^4	$M^{(1)}(x)$	$=$	$x^3 \oplus x \oplus 1$
$a^3, a^6 = a^{-1}, a^5$	$M^{(3)}(x)$	$=$	$x^3 \oplus x^2 \oplus 1$

There are two irreducible polynomial of degree 3, namely $M^{(1)}(x)$ equal to $x^3 \oplus x \oplus 1$ and $M^{(3)}(x)$ equal to $x^3 \oplus x^2 \oplus 1$. Since $M^{(1)}(x)$ equal to $x^3 \oplus x \oplus 1$ is the polynomial of the primitive element a, it is the primitive polynomial of $\mathbb{F}_{2^3}$ equal to $\mathbb{F}_8$, defined by $a^3 \oplus a \oplus 1 = 0$.

Let $m = 4$. According to Example 1.40, we have

$$\alpha^{15} = 1, \quad \Leftrightarrow \quad \alpha^{14} = \alpha^{-1}, \tag{1.900}$$

in $\mathbb{F}_{2^4}$ equal to $\mathbb{F}_{16}$. Furthermore, in $\mathbb{F}_{2^4}$ equal to $\mathbb{F}_{16}$, defined by $\alpha^4 \oplus \alpha \oplus 1 = 0$, the minimal polynomials have coefficients equal to 0 or 1. We thus yield the following [4, pp. 99, 109]:

Element of $\mathbb{F}_{2^4}$	Minimal Polynomial		
0			x
1	$M^{(0)}(x)$	=	$x \oplus 1$
$\alpha, \alpha^2, \alpha^4, \alpha^8$	$M^{(1)}(x)$	=	$x^4 \oplus x \oplus 1$
$\alpha^3, \alpha^6, \alpha^{12}, \alpha^9$	$M^{(3)}(x)$	=	$x^4 \oplus x^3 \oplus x^2 \oplus x \oplus 1$
α^5, α^{10}	$M^{(5)}(x)$	=	$x^2 \oplus x \oplus 1$
$\alpha^7, \alpha^{14} = \alpha^{-1}, \alpha^{13}, \alpha^{11}$	$M^{(7)}(x)$	=	$x^4 \oplus x^3 \oplus 1$

There are three irreducible polynomial of degree 4, namely $M^{(1)}(x)$ equal to $x^4 \oplus x \oplus 1$, $M^{(3)}(x)$ equal to $x^4 \oplus x^3 \oplus x^2 \oplus x \oplus 1$ and $M^{(7)}(x)$ equal to $x^4 \oplus x^3 \oplus 1$. Since $M^{(1)}(x)$ equal to $x^4 \oplus x \oplus 1$ is the polynomial of the primitive element α, it is the primitive polynomial of $\mathbb{F}_{2^4}$ equal to $\mathbb{F}_{16}$, defined by $\alpha^4 \oplus \alpha \oplus 1 = 0$.

Let $m = 5$. We have

$$\alpha^{31} = 1, \quad \Leftrightarrow \quad \alpha^{30} = \alpha^{-1}, \tag{1.901}$$

in $\mathbb{F}_{2^5}$ equal to $\mathbb{F}_{32}$. Also, in $\mathbb{F}_{2^5}$ equal to $\mathbb{F}_{32}$, defined by $\alpha^5 \oplus \alpha^2 \oplus 1 = 0$, the minimal polynomials have coefficients equal to 0 or 1. We thus yield the following [4, p. 109]:

Element of $\mathbb{F}_{2^5}$	Minimal Polynomial		
0			x
1	$M^{(0)}(x)$	=	$x \oplus 1$
$\alpha, \alpha^2, \alpha^4, \alpha^8, \alpha^{16}$	$M^{(1)}(x)$	=	$x^5 \oplus x^2 \oplus 1$
$\alpha^3, \alpha^6, \alpha^{12}, \alpha^{24}, \alpha^{17}$	$M^{(3)}(x)$	=	$x^5 \oplus x^4 \oplus x^3 \oplus x^2 \oplus 1$
$\alpha^5, \alpha^{10}, \alpha^{20}, \alpha^9, \alpha^{18}$	$M^{(5)}(x)$	=	$x^5 \oplus x^4 \oplus x^2 \oplus x \oplus 1$
$\alpha^7, \alpha^{14}, \alpha^{28}, \alpha^{25}, \alpha^{19}$	$M^{(7)}(x)$	=	$x^5 \oplus x^3 \oplus x^2 \oplus x \oplus 1$
$\alpha^{11}, \alpha^{22}, \alpha^{13}, \alpha^{26}, \alpha^{21}$	$M^{(11)}(x)$	=	$x^5 \oplus x^4 \oplus x^3 \oplus x \oplus 1$
$\alpha^{15}, \alpha^{30} = \alpha^{-1}, \alpha^{29}, \alpha^{27}, \alpha^{23}$	$M^{(15)}(x)$	=	$x^5 \oplus x^3 \oplus 1$

There are six irreducible polynomial of degree 5, namely $M^{(1)}(x)$ equal to $x^5 \oplus x^2 \oplus 1$, $M^{(3)}(x)$ equal to $x^5 \oplus x^4 \oplus x^3 \oplus x^2 \oplus 1$, $M^{(5)}(x)$ equal to $x^5 \oplus x^4 \oplus x^2 \oplus x \oplus 1$, $M^{(7)}(x)$ equal to $x^5 \oplus x^3 \oplus x^2 \oplus x \oplus 1$, $M^{(11)}(x)$ equal to $x^5 \oplus x^4 \oplus x^3 \oplus x \oplus 1$ and $M^{(15)}(x)$ equal to $x^5 \oplus x^3 \oplus 1$. Since $M^{(1)}(x)$ equal to $x^5 \oplus x^2 \oplus 1$ is the polynomial of the primitive element α, it is the primitive polynomial of $\mathbb{F}_{2^5}$ equal to $\mathbb{F}_{32}$, defined by $\alpha^5 \oplus \alpha^2 \oplus 1 = 0$.

Finally, in $\mathbb{F}_{2^5}$ equal to $\mathbb{F}_{32}$, defined by $\alpha^5 \oplus \alpha^3 \oplus 1 = 0$, the minimal polynomials have coefficients equal to 0 or 1. We thus yield the following [4, p. 109]:

Element of $\mathbb{F}_{2^5}$	Minimal Polynomial		
0			x
1	$M^{(0)}(x)$	=	$x \oplus 1$
$\alpha, \alpha^2, \alpha^4, \alpha^8, \alpha^{16}$	$M^{(1)}(x)$	=	$x^5 \oplus x^3 \oplus 1$
$\alpha^3, \alpha^6, \alpha^{12}, \alpha^{24}, \alpha^{17}$	$M^{(3)}(x)$	=	$x^5 \oplus x^3 \oplus x^2 \oplus x \oplus 1$
$\alpha^5, \alpha^{10}, \alpha^{20}, \alpha^9, \alpha^{18}$	$M^{(5)}(x)$	=	$x^5 \oplus x^4 \oplus x^3 \oplus x \oplus 1$
$\alpha^7, \alpha^{14}, \alpha^{28}, \alpha^{25}, \alpha^{19}$	$M^{(7)}(x)$	=	$x^5 \oplus x^4 \oplus x^3 \oplus x^2 \oplus 1$
$\alpha^{11}, \alpha^{22}, \alpha^{13}, \alpha^{26}, \alpha^{21}$	$M^{(11)}(x)$	=	$x^5 \oplus x^4 \oplus x^2 \oplus x \oplus 1$
$\alpha^{15}, \alpha^{30} = \alpha^{-1}, \alpha^{29}, \alpha^{27}, \alpha^{23}$	$M^{(15)}(x)$	=	$x^5 \oplus x^2 \oplus 1$

Since $M^{(1)}(x)$ equal to $x^5 \oplus x^3 \oplus 1$ is the polynomial of the primitive element α, it is the primitive polynomial of $\mathbb{F}_{2^5}$ equal to $\mathbb{F}_{32}$, defined by $\alpha^5 \oplus \alpha^3 \oplus 1 = 0$.

Remark 1.36 (On Minimal Polynomials). When using an irreducible polynomial $\pi(x)$ to construct the finite field $\mathbb{F}_{2^m}, m \in \mathbb{N}^*$ and $\alpha \in \mathbb{F}_{2^m}$ being a root of $\pi(x)$, then $\pi(x)$ is the minimal polynomial of α [4, p. 100].

Two fields $\mathbb{F}$ and $\mathbb{G}$ are said to be *isomorphic* if there is a one-to-one mapping from $\mathbb{F}$ onto $\mathbb{G}$, which preserves addition and multiplication [4, p. 101].

Theorem 1.48 (Isomorphism of Finite Fields). *All finite fields of order $2^m, m \in \mathbb{N}^*$ are isomorphic* [4, p. 101].

Proof. Let $\mathbb{F}$ and $\mathbb{G}$ be fields of order 2^m, and let α be a primitive element of $\mathbb{F}$ with minimal polynomial $M(x)$ [4, p. 101]. According to Theorem 1.45, $M(x)$ divides $x^{2^m} \oplus x$ [4, p. 101].

Using Fermat's theorem (cf. Corollary 1.13), there is an element of $\mathbb{G}$, β say, which has minimal polynomial $M(x)$ [4, p. 101]. Now, $\mathbb{F}$ can be considered to consist of all polynomials in α of degree $\leq (m-1)$, i. e., $\mathbb{F}$ consists of polynomials modulo $M(x)$ [4, p. 101].

Furthermore, $\mathbb{G}$ consists of all polynomials in β of degree $\leq (m-1)$ [4, p. 101].

Therefore, the mapping $\alpha \leftrightarrow \beta$ is an isomorphism $\mathbb{F} \leftrightarrow \mathbb{G}$ [4, p. 101]. □

Example 1.45. The two versions of $\mathbb{F}_{2^3}$ equal to $\mathbb{F}_8$ given in Table 1.10 and Table 1.11 are obviously isomorphic [4, p. 101].

There are two further properties of minimal polynomials, which will be important in analyzing and designing cyclic codes.

Theorem 1.49 (Property M6: Elements with Identical Minimal Polynomials). *In $\mathbb{F}_{2^m}$, β and β^2 have the same minimal polynomial* [4, p. 103].

Proof. [Proof by Example] Let us consider $\mathbb{F}_{16}$. Assume that β has the minimal polynomial $(x^4 \oplus x \oplus 1)$, i. e., [4, p. 103]

$$\beta^4 \oplus \beta \oplus 1 = 0. \tag{1.902}$$

Now, let us look at β^2, which must clearly fulfill [4, p. 103]

$$\left(\beta^2\right)^4 \oplus \beta^2 \oplus 1 = 0. \tag{1.903}$$

With

$$(x \oplus y)^2 = x^2 \oplus y^2 \tag{1.904}$$

(cf. Theorem 1.42), and with

$$(\beta \oplus 1)^2 = (\beta \oplus 1) \odot (\beta \oplus 1) = \beta^2 \oplus 1, \tag{1.905}$$

we have [4, p. 103]

$$(\beta^2)^4 \oplus \beta^2 \oplus 1 = (\beta^4)^2 \oplus (\beta \oplus 1)^2 = (\beta^4 \oplus \beta \oplus 1)^2 = 0. \tag{1.906}$$

So, by Theorem 1.44, the minimal polynomial of β^2 divides $(x^4 \oplus x \oplus 1)$ [4, p. 104].

But $(\beta^2)^8 = \beta$, so we can use the same argument to show that the minimal polynomial of β divides that of β^2 [4, p. 104]. Therefore, they are equal [4, p. 104]. □

Elements of the field with the same minimal polynomial are called *conjugates* [4, p. 104].

Example 1.46. Let us look at what happens in $\mathbb{F}_{16}$ [4, p. 104]. Setting out from Theorem 1.49, the following elements all have the same minimal polynomial [4, p. 104]:

$$a,\ a^2,\ (a^2)^2 = a^4,\ (a^4)^2 = a^8.\ ((a^8)^2 = a^{16} = a). \tag{1.907}$$

Likewise,

$$a^3,\ (a^3)^2 = a^6,\ (a^6)^2 = a^{12},\ (a^{12})^2 = a^{24} = a^9,\ ((a^9)^2 = a^{18} = a^3) \tag{1.908}$$

all have the same minimal polynomial [4, p. 104]. Also,

$$a^5,\ (a^5)^2 = a^{10},\ ((a^{10})^2 = a^{20} = a^5) \tag{1.909}$$

all have the same minimal polynomial. Finally,

$$\begin{aligned} a^7,\ (a^7)^2 = a^{14},\ (a^{14})^2 = a^{28} = a^{13},\ (a^{13})^2 = a^{26} = a^{11}, \\ ((a^{11})^2 = a^{22} = a^7) \end{aligned} \tag{1.910}$$

all have the same minimal polynomial.

We see that the powers of a fall into disjoint sets, which we call *cyclotomic cosets*.

All a^j where j runs through a cyclotomic coset have the same minimal polynomial [4, p. 104].

Definition 1.57 (Cyclotomic Coset). The operation of multiplying by the characteristic 2 divides the integers $\mathrm{mod}(2^m - 1)$ into sets called the cyclotomic cosets [4, p. 104].

The cyclotomic coset containing s consists of

$$\{s, (2 \cdot s), (2^2 \cdot s), (2^3 \cdot s), (2^4 \cdot s), (2^5 \cdot s), (2^6 \cdot s) \cdots (2^{m_s - 1} \cdot s)\}, \tag{1.911}$$

$m_s \in \mathbb{N}^*$ being the smallest positive natural number such that [4, p. 104]

$$2^{m_s} \cdot s = s \bmod (2^m - 1) = R_{2^m-1}[s], \quad m_s, m \in \mathbb{N}^*. \tag{1.912}$$

Note that the cyclotomic cosets are disjoint [4, p. 104].

Remark 1.37 (On Cyclotomic Cosets). Let us shed some light onto the construction of the cyclotomic cosets. We know that multiplying by 2 divides the nonnegative integers $\bmod(2^m - 1)$, $m \in \mathbb{N}$ into *cyclotomic cosets* [4, Definition, p. 104]. With $s \in \mathbb{N}$ being the *coset representative* [4, p. 104], each cyclotomic coset is given by $\{s, 2s, 2^2s, 2^3s \cdots 2^{m_s-1}s\}$, $m_s \in \mathbb{N}$, with $2^{m_s}s$ being equal to $s \bmod (2^m-1)$, i. e., $R_{2^m-1}[s]$, $m \in \mathbb{N}$.

Clearly, s equal to 0 leads to the cyclotomic coset $\mathbb{C}_0$ equal to $\{0\}$. Furthermore, the discussion above implies that the coset representatives $s > 0$ must be odd integers.

An illustrative way of constructing the said cyclotomic cosets for $s > 0$ is by setting out from the binary representation of the integers > 0. Since $(2^m - 1) \bmod (2^m - 1)$ is zero, all integers are mapped into the set $\{0, 1 \cdots (2^m - 2)\}$, $m \in \mathbb{N}$.

The construction rule illustrated above shows that the possible coset representatives $s > 0$ fulfill

$$0 \le s \le 2^{m-1} - 1, \quad s \text{ odd}, \quad m \in \{2, 3, 4 \cdots\}. \tag{1.913}$$

Then we may write

$$\begin{aligned} s &= 2^{m-2} \cdot s_{m-2} + 2^{m-3} \cdot s_{m-3} + \cdots + 2^3 \cdot s_3 + 2^2 \cdot s_2 + 2^1 \cdot s_1 + 2^0 \cdot \underbrace{1}_{=s_0,\ s \text{ is odd}}, \\ &= 2^{m-1} \cdot s_{m-1} + 2^{m-2} \cdot s_{m-2} + \cdots + 2^3 \cdot s_3 + 2^2 \cdot s_2 + 2^1 \cdot s_1 + 1, \\ & s_j \in \mathbb{F}_2, \quad j \in \{0, 1 \cdots (m-1)\}, \quad m \in \{2, 3, 4 \cdots\}. \end{aligned} \tag{1.914}$$

Therefore, s can be represented by the following binary vector:

	2^{m-1}	2^{m-2}	$\cdots$	2^3	2^2	2^1	$2^0 = 1$	
(	0,	s_{m-2}	$\cdots$	s_3,	s_2,	s_1,	1	)

Let us begin with s equal to 1, which can only occur for $m \ge 2$, and construct $\mathbb{C}_1$. We yield the following table:

construction rule		$\mathbf{2^{m-1}}$	$\mathbf{2^{m-2}}$	$\cdots$	$\mathbf{2^3}$	$\mathbf{2^2}$	$\mathbf{2^1}$	$\mathbf{2^0 = 1}$		**coset element**
s	(	0,	0	$\cdots$	0,	0,	0,	1	)	1
$2s$	(	0,	0	$\cdots$	0,	0,	1,	0	)	2
$4s$	(	0,	0	$\cdots$	0,	1,	0,	0	)	4
$8s$	(	0,	0	$\cdots$	1,	0,	0,	0	)	8
$\vdots$	$\vdots$	$\vdots$	$\vdots$	$\vdots$	$\vdots$	$\vdots$	$\vdots$	$\vdots$	$\vdots$	$\vdots$
$2^{m-2}s$	(	0,	1	$\cdots$	0,	0,	0,	0	)	2^{m-2}
$2^{m-1}s$	(	1,	0	$\cdots$	0,	0,	0,	0	)	2^{m-1}
$2^m s \bmod (2^m - 1)$	(	0,	0	$\cdots$	0,	0,	0,	1	)	1

Therefore, we find

$$\mathbb{C}_1 = \{1, 2, 4 \cdots 2^{m-1}\}, \quad w_H\{s\} = 1, \quad m_s = m, \quad m, m_s \in \mathbb{N}. \tag{1.915}$$

Next, let us look at coset representatives with Hamming weight $w_H\{s\}$ equal to 2. In general, these coset representatives have the form

$$s = 2^j + 1, \quad j \in \{1, 2 \cdots (m-1)\}, \quad m \in \mathbb{N}, \tag{1.916}$$

i. e., $s \in \{3, 5, 9, 17, 33, 65 \cdots\}$.

For example, in the case of s equal to 3, which can only occur for $m \geq 3$, we find the following table:

construction rule		2^{m-1}	2^{m-2}	···	2^3	2^2	2^1	$2^0 = 1$		**coset element**
s	(	0,	0	···	0,	0,	1,	1	)	3
$2s$	(	0,	0	···	0,	1,	1,	0	)	6
$4s$	(	0,	0	···	1,	1,	0,	0	)	12
⋮	⋮	⋮	⋮	⋮	⋮	⋮	⋮	⋮	⋮	⋮
$2^{m-2}s$	(	1,	1	···	0,	0,	0,	0	)	$3 \cdot 2^{m-2}$
$2^{m-1}s$	(	1,	0	···	0,	0,	0,	1	)	$2^{m-1} + 1$
$2^m s \bmod (2^m - 1)$	(	0,	0	···	0,	0,	1,	1	)	3

Therefore, we yield

$$\mathbb{C}_3 = \{3, 6 \cdots 3 \cdot 2^{m-2}, (2^{m-1} + 1)\}, \quad w_H\{s\} = 2, \quad m_s = m, \quad m, m_s \in \mathbb{N}. \tag{1.917}$$

In the case of s equal to 5, which can only occur for $m \geq 4$, we obtain the following table:

construction rule		2^{m-1}	2^{m-2}	···	2^3	2^2	2^1	$2^0 = 1$		**coset element**
s	(	0,	0	···	0,	1,	0,	1	)	5
$2s$	(	0,	0	···	1,	0,	1,	0	)	10
⋮	⋮	⋮	⋮	⋮	⋮	⋮	⋮	⋮	⋮	⋮
$2^{m-3}s$	(	1,	0	···	0,	0,	0,	0	)	$5 \cdot 2^{m-3}$
$2^{m-2}s$	(	0,	1	···	0,	0,	0,	1	)	$2^{m-2} + 1$
$2^{m-1}s$	(	1,	0	···	0,	0,	1,	0	)	$2^{m-1} + 2$
$2^m s \bmod (2^m - 1)$	(	0,	0	···	0,	1,	0,	1	)	5

Therefore, we find

$$\mathbb{C}_5 = \{5, 10 \cdots 5 \cdot 2^{m-3}, (2^{m-2} + 1), (2^{m-1} + 2)\}, \tag{1.918}$$
$$w_H\{s\} = 2, \quad m_s = m, \quad m, m_s \in \mathbb{N}.$$

In the case of s equal to 7, which can only occur for $m \geq 4$, we yield the following table:

construction rule		2^{m-1}	2^{m-2}	$\cdots$	2^3	2^2	2^1	$2^0 = 1$		coset element
s	(	0,	0	$\cdots$	0,	1,	1,	1	)	7
$2s$	(	0,	0	$\cdots$	1,	1,	1,	0	)	14
$\vdots$	$\vdots$	$\vdots$	$\vdots$	$\vdots$	$\vdots$	$\vdots$	$\vdots$	$\vdots$	$\vdots$	$\vdots$
$2^{m-3}s$	(	1,	1	$\cdots$	0,	0,	0,	0	)	$7 \cdot 2^{m-3}$
$2^{m-2}s$	(	1,	1	$\cdots$	0,	0,	0,	1	)	$3 \cdot 2^{m-2} + 1$
$2^{m-1}s$	(	1,	0	$\cdots$	0,	0,	1,	1	)	$2^{m-1} + 3$
$2^m s \bmod (2^m - 1)$	(	0,	0	$\cdots$	0,	1,	1,	1	)	7

Therefore, we obtain

$$\mathbb{C}_7 = \left\{7, 14 \cdots 7 \cdot 2^{m-3}, \left(3 \cdot 2^{m-2} + 1\right), \left(2^{m-1} + 3\right)\right\}, \qquad (1.919)$$
$$w_{\mathrm{H}}\{s\} = 3, \quad m_s = m, \quad m, m_s \in \mathbb{N}.$$

In the case of s equal to 9, which can only occur for $m \geq 6$, because $9 \in \mathbb{C}_5$ for m equal to 5, we find the following table:

construction rule		2^{m-1}	2^{m-2}	$\cdots$	2^3	2^2	2^1	$2^0 = 1$		coset element
s	(	0,	0	$\cdots$	1,	0,	0,	1	)	9
$2s$	(	0,	0	$\cdots$	0,	0,	1,	0	)	18
$\vdots$	$\vdots$	$\vdots$	$\vdots$	$\vdots$	$\vdots$	$\vdots$	$\vdots$	$\vdots$	$\vdots$	$\vdots$
$2^{m-4}s$	(	1,	0	$\cdots$	0,	0,	0,	0	)	$9 \cdot 2^{m-4}$
$2^{m-3}s$	(	0,	0	$\cdots$	0,	0,	0,	1	)	$2^{m-3} + 1$
$2^{m-2}s$	(	0,	1	$\cdots$	0,	0,	1,	0	)	$2^{m-2} + 2$
$2^{m-1}s$	(	1,	0	$\cdots$	0,	1,	0,	0	)	$2^{m-1} + 4$
$2^m s \bmod (2^m - 1)$	(	0,	0	$\cdots$	1,	0,	0,	1	)	9

Therefore, we yield

$$\mathbb{C}_9 = \left\{9, 18 \cdots 9 \cdot 2^{m-4}, \left(2^{m-3} + 1\right), \left(2^{m-2} + 2\right), \left(2^{m-1} + 4\right)\right\}, \qquad (1.920)$$
$$w_{\mathrm{H}}\{s\} = 2, \quad m_s = m, \quad m, m_s \in \mathbb{N}.$$

In the case of s equal to 11, which can only occur for $m \geq 5$, we obtain the following table:

construction rule		2^{m-1}	2^{m-2}	$\cdots$	2^3	2^2	2^1	$2^0 = 1$		coset element
s	(	0,	0	$\cdots$	1,	0,	1,	1	)	11
$2s$	(	0,	0	$\cdots$	0,	1,	1,	0	)	22
$\vdots$	$\vdots$	$\vdots$	$\vdots$	$\vdots$	$\vdots$	$\vdots$	$\vdots$	$\vdots$	$\vdots$	$\vdots$
$2^{m-4}s$	(	1,	0	$\cdots$	0,	0,	0,	0	)	$11 \cdot 2^{m-4}$
$2^{m-3}s$	(	0,	1	$\cdots$	0,	0,	0,	1	)	$3 \cdot 2^{m-3} + 1$
$2^{m-2}s$	(	1,	1	$\cdots$	0,	0,	1,	0	)	$3 \cdot 2^{m-2} + 2$
$2^{m-1}s$	(	1,	0	$\cdots$	0,	1,	0,	1	)	$2^{m-1} + 5$
$2^m s \bmod (2^m - 1)$	(	0,	0	$\cdots$	1,	0,	1,	1	)	11

Therefore, we find

$$\mathbb{C}_{11} = \{11, 22 \cdots 11 \cdot 2^{m-4}, (3 \cdot 2^{m-3} + 1), (3 \cdot 2^{m-2} + 2), (2^{m-1} + 5)\}, \qquad (1.921)$$
$$w_H\{s\} = 3, \quad m_s = m, \quad m, m_s \in \mathbb{N}.$$

In the case of s equal to 13, which can only occur for $m \geq 6$, because $13 \in \mathbb{C}_{11}$ for m equal to 5, we yield the following table:

construction rule		2^{m-1}	2^{m-2}	...	2^3	2^2	2^1	$2^0 = 1$		coset element
s	(	0,	0	...	1,	1,	0,	1	)	13
$2s$	(	0,	0	...	1,	0,	1,	0	)	26
⋮	⋮	⋮	⋮	⋮	⋮	⋮	⋮	⋮	⋮	⋮
$2^{m-4}s$	(	1,	1	...	0,	0,	0,	0	)	$13 \cdot 2^{m-4}$
$2^{m-3}s$	(	1,	0	...	0,	0,	0,	1	)	$5 \cdot 2^{m-3} + 1$
$2^{m-2}s$	(	0,	1	...	0,	0,	1,	1	)	$2^{m-2} + 3$
$2^{m-1}s$	(	1,	0	...	0,	1,	1,	0	)	$2^{m-1} + 6$
$2^m s \bmod (2^m - 1)$	(	0,	0	...	1,	1,	0,	1	)	13

Therefore, we obtain

$$\mathbb{C}_{13} = \{11, 22 \cdots 13 \cdot 2^{m-4}, (5 \cdot 2^{m-3} + 1), (2^{m-2} + 3), (2^{m-1} + 6)\}, \qquad (1.922)$$
$$w_H\{s\} = 3, \quad m_s = m, \quad m, m_s \in \mathbb{N}.$$

In the case of s equal to 15, which can only occur for $m \geq 5$, we yield the following table:

construction rule		2^{m-1}	2^{m-2}	...	2^3	2^2	2^1	$2^0 = 1$		coset element
s	(	0,	0	...	1,	1,	1,	1	)	15
$2s$	(	0,	0	...	1,	1,	1,	0	)	30
⋮	⋮	⋮	⋮	⋮	⋮	⋮	⋮	⋮	⋮	⋮
$2^{m-4}s$	(	1,	1	...	0,	0,	0,	0	)	$15 \cdot 2^{m-4}$
$2^{m-3}s$	(	1,	1	...	0,	0,	0,	1	)	$7 \cdot 2^{m-3} + 1$
$2^{m-2}s$	(	1,	1	...	0,	0,	1,	1	)	$3 \cdot 2^{m-2} + 3$
$2^{m-1}s$	(	1,	0	...	0,	1,	1,	1	)	$2^{m-1} + 7$
$2^m s \bmod (2^m - 1)$	(	0,	0	...	1,	1,	1,	1	)	15

Therefore, we find

$$\mathbb{C}_{15} = \{15, 30 \cdots 15 \cdot 2^{m-4}, (7 \cdot 2^{m-3} + 1), (3 \cdot 2^{m-2} + 3), (2^{m-1} + 7)\}, \qquad (1.923)$$
$$w_H\{s\} = 4, \quad m_s = m, \quad m, m_s \in \mathbb{N}.$$

Example 1.47. Let us consider $m = 1$. We have to compute $\mathrm{mod}(2^1 - 1) = \mathrm{mod}(1)$ in this case. The elements for $s > 0$ are contained in already existing cyclotomic cosets. Therefore, the cyclotomic coset for $m = 1$ is

$$\mathbb{C}_0 = \{0\}. \tag{1.924}$$

Let us consider $m = 2$. We have to compute $\mathrm{mod}(2^2 - 1) = \mathrm{mod}(3)$ in this case. The elements for $s > 1$ are contained in already existing cyclotomic cosets. Therefore, the cyclotomic cosets for $m = 2$ are

$$\mathbb{C}_0 = \{0\}, \quad \mathbb{C}_1 = \{1, 2\}. \tag{1.925}$$

Let us consider $m = 3$. We have to compute $\mathrm{mod}(2^3 - 1) = \mathrm{mod}(7)$ in this case. The elements for $s = 2$ are contained in the cyclotomic coset C_1. The elements for $s > 3$ are contained in already existing cyclotomic cosets. Therefore, the cyclotomic cosets for $m = 3$ are [4, p. 105]

$$\mathbb{C}_0 = \{0\}, \quad \mathbb{C}_1 = \{1, 2, 4\}, \quad \mathbb{C}_3 = \{3, 6, 5\}. \tag{1.926}$$

Our notation is that if s is the smallest number in the coset, the coset is denoted by $\mathbb{C}_s$ [4, p. 104]. The subscripts s are called the *coset representatives* $\mathrm{mod}(2^m - 1)$ [4, p. 104].

Example 1.48. Now, let us consider $m = 4$. We have to compute $\mathrm{mod}(2^4 - 1) = \mathrm{mod}(15)$ in this case. The elements for $s = 2$ and $s = 4$ are contained in the cyclotomic $\mathbb{C}_1$. The elements for $s = 6$ are contained in the cyclotomic coset $\mathbb{C}_3$. The elements for $s > 7$ are contained in already existing cyclotomic cosets. Therefore, the cyclotomic cosets for $m = 4$ are [4, p. 104]

$$\begin{aligned} \mathbb{C}_0 &= \{0\}, \\ \mathbb{C}_1 &= \{1, 2, 4, 8\}, \\ \mathbb{C}_3 &= \{3, 6, 12, 9\}, \\ \mathbb{C}_5 &= \{5, 10\}, \\ \mathbb{C}_7 &= \{7, 14, 13, 11\}. \end{aligned} \tag{1.927}$$

Now, let us consider $m = 5$. We have to compute $\mathrm{mod}(2^5 - 1) = \mathrm{mod}(31)$ in this case. The elements for $s = 2$, $s = 4$ and $s = 8$ are contained in the cyclotomic coset $\mathbb{C}_1$. The elements for $s = 6$ and $s = 12$ are contained in the cyclotomic coset $\mathbb{C}_3$. The elements for $s = 9$ and $s = 10$ are contained in the cyclotomic coset $\mathbb{C}_5$. The elements for $s = 14$ are contained in the cyclotomic coset $\mathbb{C}_7$. The elements for $s = 13$ are contained in the cyclotomic coset $\mathbb{C}_{11}$. The elements for $s > 15$ are contained in already existing cyclotomic cosets. Therefore, the cyclotomic cosets for $m = 5$ are [4, p. 105]

$$\begin{aligned} \mathbb{C}_0 &= \{0\}, \\ \mathbb{C}_1 &= \{1, 2, 4, 8, 16\}, \\ \mathbb{C}_3 &= \{3, 6, 12, 24, 17\}, \\ \mathbb{C}_5 &= \{5, 10, 20, 9, 18\}, \\ \mathbb{C}_7 &= \{7, 14, 28, 25, 19\}, \\ \mathbb{C}_{11} &= \{11, 22, 13, 26, 21\}, \\ \mathbb{C}_{15} &= \{15, 30, 29, 27, 23\}. \end{aligned} \tag{1.928}$$

Now, we can define the minimal polynomial of a coset.

Definition 1.58 (Minimal Polynomial of a Cyclotomic Coset). Let $M^{(i)}(x)$ be the minimal polynomial of $\alpha^i \in \mathbb{F}_{2^m}$ [4, p. 104]. Setting out from Theorem 1.49, we find [4, p. 104]

$$M^{(2i)}(x) = M^{(i)}(x). \tag{1.929}$$

From the preceding discussion, it follows that if i is in the cyclotomic coset $\mathbb{C}_s$, then in $\mathbb{F}_{2^m}$ we yield [4, p. 109]

$$\bigodot_{j \in \mathbb{C}_s} (x \oplus \alpha^j) \text{ divides } M^{(i)}(x). \tag{1.930}$$

This yields the seventh property of minimal polynomials.

Theorem 1.50 (Property M7: Minimal Polynomial as a Product of $(x \oplus \alpha^j)$). *If i is in the cyclotomic coset $\mathbb{C}_s$, then [4, p. 105]*

$$M^{(i)}(x) = \bigodot_{j \in \mathbb{C}_s} (x \oplus \alpha^j). \tag{1.931}$$

Corollary 1.14 (Property M8: Product of Nontrivial Minimal Polynomials). *Setting out from Fermat's theorem (cf. Corollary 1.13), we find*

$$x^{2^m} \oplus x = x \odot \bigodot_s M^{(s)}(x), \tag{1.932}$$

and thus

$$x^{2^m-1} \oplus 1 = \bigodot_s M^{(s)}(x), \tag{1.933}$$

where s runs through the coset representatives $\mathrm{mod}(2^m - 1)$ *[4, p. 105].*

Remark 1.38 (On the Factorization of $(x^{2^m-1} \oplus 1)$ over $\mathbb{F}_{2^m}$). Setting out from (1.931) and (1.933), we find [5, Satz 6.3, equation (6.2.13), p. 174], [4, equation (13), p. 197]

$$x^n \oplus 1 = x^{2^m-1} \oplus 1 = \bigodot_s \bigodot_{j \in \mathbb{C}_s} (x \oplus \alpha^j) = \bigodot_{j=0}^{2^m-2} (x \oplus \alpha^j) = \bigodot_{j=0}^{n-1} (x \oplus \alpha^j), \tag{1.934}$$
$$n = 2^m - 1, \quad m \in \mathbb{N}^*,$$

α being a primitive element of the finite field $\mathbb{F}_{2^m}$, $m \in \mathbb{N}^*$; cf. also Theorem 1.40.

Equation (1.934) is the *factorization of $(x^n \oplus 1)$ equal to $(x^{2^m-1} \oplus 1)$ over $\mathbb{F}_{2^m}$ equal to $\mathbb{F}_{n+1}$* [4, p. 196]. Therefore, $\mathbb{F}_{2^m}$ equal to $\mathbb{F}_{n+1}$ is called the *splitting field of $(x^n \oplus 1)$ equal to $(x^{2^m-1} \oplus 1)$* (in German *"Kreisteilungskörper, Zerfällungskörper"*) [5, p. 174], [4, p. 196].

The terms $\alpha^j, j \in \{0, 1 \cdots (n-1)\} = \{0, 1 \cdots (2^m - 2)\}$ are the zeros of $(x^n \oplus 1)$ equal to $(x^{2^m-1} \oplus 1)$ and are called the *primitive nth roots of unity* [4, p. 196f.]. These primitive nth roots of unity form a cyclic subgroup of $\mathbb{F}^*_{2^m}$ [4, p. 196f.].

Equation (1.934) teaches us that $(x^n \oplus 1)$ equal to $(x^{2^m-1} \oplus 1)$ has exactly n equal to $(2^m - 1)$, $m \in \mathbb{N}^*$, distinct zeros [4, p. 196].

Before coming to a formal proof of (1.934), let us consider the factorization of a primitive polynomial over $\mathbb{F}_{2^m}$, $m \in \mathbb{N}^*$.

Theorem 1.51 (Factorization of a Primitive Polynomial $p(x) \in \mathbb{F}_2[x]_m$). *Let $p(x) \in \mathbb{F}_2[x]_m$, $m \in \mathbb{N}^*$, be a primitive polynomial with $\deg(p(x))$ equal to $m \in \mathbb{N}^*$, let $\alpha \in \mathbb{F}_{2^m}$, $m \in \mathbb{N}^*$, be a primitive element and let n be equal to $(2^m - 1)$, $m \in \mathbb{N}^*$. The primitive polynomial $p(x) \in \mathbb{F}_2[x]_m$, $m \in \mathbb{N}^*$ is irreducible over $\mathbb{F}_2$ [5, Satz 6.3, p. 174]. Furthermore, the primitive polynomial $p(x)$ has m distinct zeros α^{2^i}, $i \in \{0, 1 \cdots (m-1)\}$, $m \in \mathbb{N}^*$, i. e.,*

$$p(x) = \bigodot_{i=0}^{m-1} \left(x \oplus \alpha^{2^i}\right), \quad m \in \mathbb{N}^*, \tag{1.935}$$

hence $p(x)$ can be split into linear factors $(x \oplus \alpha^{2^i})$, $i \in \{0, 1 \cdots (m-1)\}$, $m \in \mathbb{N}^*$, over $\mathbb{F}_{2^m}$ [5, Satz 6.3, p. 174].

Proof. Using Theorem 1.42, we find [5, Satz 6.2, p. 173f.]

$$\left[p(x)\right]^{2^m} = p\left(x^{2^m}\right), \quad m \in \mathbb{N}^*, \tag{1.936}$$

and hence [5, p. 175]

$$p\left(\alpha^{2^i}\right) = \left[p(\alpha)\right]^{2^i} = 0^{2^i} = 0, \quad i \in \left\{0, 1 \cdots (m-1)\right\}, \quad m \in \mathbb{N}^*, \tag{1.937}$$

because $p(\alpha)$ is equal to 0. Therefore, the m values α^{2^i}, $i \in \{0, 1 \cdots (m-1)\}$, $m \in \mathbb{N}^*$ are the zeros of $p(x) \in \mathbb{F}_2[x]_m$.

Using $i \in \{0, 1 \cdots (m-1)\}$, $m \in \mathbb{N}^*$, we yield

$$0 \leq 2^i \leq 2^{m-1} < 2^m - 1, \quad i \in \left\{0, 1 \cdots (m-1)\right\}, \quad m \in \mathbb{N}^*, \tag{1.938}$$

which clearly demonstrates that all the m values α^{2^i}, $i \in \{0, 1 \cdots (m-1)\}$, $m \in \mathbb{N}^*$ are distinct [5, p. 175]. □

Now, let us consider proving (1.934).

Theorem 1.52 (Factorization of $(x^{2^m-1} \oplus 1)$). *Let n be equal to $(2^m - 1)$, $m \in \mathbb{N}^*$ and let $\alpha \in \mathbb{F}_{2^m}$, $m \in \mathbb{N}^*$ be a primitive element.*

Then the polynomial $(x^{2^m-1} \oplus 1)$ has n be equal to $(2^m - 1)$, $m \in \mathbb{N}^*$, zeros α^{2^i}, $i \in \{0, 1 \cdots (2^m - 2)\}$, $m \in \mathbb{N}^*$, which form $\mathbb{F}_{2^m} \backslash \{0\}$ [5, Satz 6.3, p. 174]. This means that the $m \in \mathbb{N}^*$, zeros α^{2^i}, $i \in \{0, 1 \cdots (2^m - 2)\}$, $m \in \mathbb{N}^*$ are the nonzero elements of $\mathbb{F}_{2^m}$, $m \in \mathbb{N}^*$. $\mathbb{F}_{2^m} \backslash \{0\}$, $m \in \mathbb{N}^*$ can also be denoted by $\mathbb{F}_{2^m}^*$ [4, p. 96].

Proof. Rewriting the argument of the proof of Theorem 1.51 proves the claim made. □

Finally, let us revisit the *primitive polynomials*. We already know that

- a polynomial having a primitive element as a zero is called a primitive polynomial [4, p. 84] and,
- according to Theorem 1.47, the minimal polynomial of a primitive element of $\mathbb{F}_{2^m}$ has degree m and is a primitive polynomial [4, p. 100].

Theorem 1.53 (Primitive Polynomial Revisited). *Let n be equal to* $(2^m - 1)$, $m \in \mathbb{N}^*$ *and let* $p(x) \in \mathbb{F}_2[x]_m$, $m \in \mathbb{N}^*$ *be an irreducible polynomial with* $\deg(p(x))$ *equal to* $m \in \mathbb{N}^*$. *Then* [5, *Satz 6.4, p. 175f.*]

$$p(x) \text{ is a primitive polynomial} \quad \Leftrightarrow \quad 2^m - 1 = \min\{l \in \mathbb{N}^* \mid p(x) \mid (x^l \oplus 1)\}. \tag{1.939}$$

Proof. First, suppose that $p(x)$ is a primitive polynomial. Combining Theorem 1.51 and Theorem 1.52, we know that $p(x) \mid (x^n \oplus 1)$, $n = (2^m - 1)$, $m \in \mathbb{N}^*$ [5, Satz 6.4, p. 175f.]. If $p(x)$ also divided $(x^l \oplus 1)$, $l < n$, $n = (2^m - 1)$, $m \in \mathbb{N}^*$, we would have

$$q(x) \odot p(x) = x^l \oplus 1, \quad l < n, \quad n = (2^m - 1), \quad m \in \mathbb{N}^*, \tag{1.940}$$

for some polynomial $q(x) \in \mathbb{F}_2[x]_m$ [5, Satz 6.4, p. 175f.]. Using the primitive element $\alpha \in \mathbb{F}_{2^m}$, we have $p(\alpha)$ equal to 0, and hence [5, Satz 6.4, p. 175f.]

$$\alpha^l = 1, \quad \alpha \in \mathbb{F}_{2^m}, \quad l < n, \quad n = (2^m - 1), \quad m \in \mathbb{N}^*, \tag{1.941}$$

which contradicts the supposition that $p(x)$ is a primitive polynomial. Hence, $p(x)$ does not divide $(x^l \oplus 1)$, $l < n$, $n = (2^m - 1)$, $m \in \mathbb{N}^*$ [5, Satz 6.4, p. 175f.].

Now, suppose that $p(x)$ is a not a primitive polynomial [5, Satz 6.4, p. 175f.]. Then there is an element $\gamma \in \mathbb{F}_{2^m}$ with $p(\gamma)$ equal to 0 and [5, Satz 6.4, p. 175f.]

$$\gamma^l = 1, \quad \gamma \in \mathbb{F}_{2^m}, \quad l < n, \quad n = (2^m - 1), \quad m \in \mathbb{N}^*. \tag{1.942}$$

According to the division theorem, Theorem 1.35, there are polynomials $q(x)$ and $r(x)$ with [5, Satz 6.4, p. 175f.]

$$x^l \oplus 1 = q(x) \odot p(x) \oplus r(x) \quad \deg(r(x)) < m, \quad l < n, \quad n = (2^m - 1), \quad m \in \mathbb{N}^*, \tag{1.943}$$

which leads to the requirement that $r(x)$ must have $m \in \mathbb{N}^*$ distinct zeros, which can only be fulfilled for $r(x) = 0$ [5, Satz 6.4, p. 175f.]. This, however, contradicts the supposition that $p(x)$ does not divide $(x^l \oplus 1)$ [5, Satz 6.4, p. 175f.]. Hence, $p(x)$ must be a primitive polynomial [5, Satz 6.4, p. 175f.]. □

The main properties of minimal polynomials discussed in this Section 1.12.6 are summarized in Table 1.14 [4, p. 99–105].

Table 1.14: Properties of minimal polynomials $M(x)$ and $M^{(i)}(x)$, $i \in \mathbb{C}_s$; for a discussion (cf., e. g., [4, p. 99–105]).

no.	property
M1	$M(x)$ is irreducible.
M2	$M(x)$ with $M(\beta) = 0$ divides any polynomial $f(x)$ with $f(\beta) = 0$
M3	$M(x)$ divides $x^{2^m} \oplus x$
M4	$\deg(M(x)) \leq m$ in $\mathbb{F}_{2^m}$
M5	$\deg(M(x)) = m$ of a minimal polynomial $M(x)$ of a primitive element α in $\mathbb{F}_{2^m}$
M6	$\beta \in \mathbb{F}_{2^m}$ and $\beta^2 \in \mathbb{F}_{2^m}$ have the same minimal polynomial $M(x)$
M7	$M^{(i)}(x) = \bigodot_{j \in \mathbb{C}_s} (x \oplus \alpha^j)$, $i \in \mathbb{C}_s$
M8	$x^{2^m - 1} \oplus 1 = \bigodot_s M^{(s)}(x)$, where s runs through the coset representatives $\bmod (2^m - 1)$

Section 1.12 provided quite a bit of interesting information, especially about polynomials. Plunging into the polynomial sea and swimming to the deeper areas seems quite safe, now. So, why not try?

Alright, we already know that $\mathbb{F}_2[x]_{n-1}, n \in \mathbb{N}^*$ denotes the set of all polynomials of degree $(n-1), n \in \mathbb{N}^*$ or less [5, Definition 5.2, p. 131]; cf. Definition 1.42.

Furthermore, Example 1.24 teaches us that $(\mathbb{F}_2[x]_{n-1}, \oplus, \odot)$ is a Euclidean domain and, therefore, also a *ring*, or more precisely a *commutative ring with identity* [4, p. 189]; cf. Definition 1.46 combined with Definition 1.47. Therefore, we may write [4, p. 189], [25, Definition 4.4.1, p. 79]

$$\mathbb{F}_2[x]_{n-1} = \frac{\mathbb{F}_2[x]}{x^n \oplus 1} = \frac{\mathbb{F}_2[x]}{(x^n \oplus 1)} = \frac{\mathbb{F}_2[x]}{\langle x^n \oplus 1\rangle}, \quad n \in \mathbb{N}^*. \tag{1.944}$$

Letting $p(x) \in \mathbb{F}_2[x]$ be an arbitrary polynomial in $\mathbb{F}_2[x]$, we find

$$R_{x^n \oplus 1}[p(x)] \in \mathbb{F}_2[x]_{n-1}, \quad n \in \mathbb{N}^*. \tag{1.945}$$

Therefore, $\mathbb{F}_2[x]_{n-1}, n \in \mathbb{N}^*$, consists of all the residue classes $R_{x^n \oplus 1}[p(x)]$ with $p(x) \in \mathbb{F}_2[x]$ [4, p. 189].

Definition 1.59 (Ring of Polynomials $\mathbb{F}_2[x]_{n-1}$ Equal to $\mathbb{F}_2[x]/(x^n \oplus 1)$ Equal to $\mathbb{F}_2[x]/\langle x^n \oplus 1\rangle$). The ring of polynomials $p(x)$ modulo $(x^n \oplus 1), n \in \mathbb{N}^*$ denoted by $\mathbb{F}_2[x]_{n-1}$ equal to $\mathbb{F}_2[x]/(x^n \oplus 1)$ equal to $\mathbb{F}_2[x]/\langle x^n \oplus 1\rangle, n \in \mathbb{N}^*$ (cf. (1.944)) is the set of polynomials with a degree smaller than $n \in \mathbb{N}^*$, together with polynomial addition and polynomial multiplication modulo $(x^n \oplus 1), n \in \mathbb{N}^*$.

We know that multiplying any polynomial, $v(x) \in \mathbb{F}_2[x]$ say, by x in $\mathbb{F}_2[x]_{n-1}, n \in \mathbb{N}^*$ is a cyclic shift (cf. Theorem 1.37), because x^n is equal to 1 in $\mathbb{F}_2[x]_{n-1}, n \in \mathbb{N}^*$ [4, p. 189]. This knowledge takes us directly to the next two definitions [5, Definition A.5, p. 442], [4, p. 189].

Definition 1.60 (Ideal). An ideal $\mathcal{I}$ of $\mathbb{F}_2[x]_{n-1}, n \in \mathbb{N}^*$ is a subset, respectively, a linear subspace of $\mathbb{F}_2[x]_{n-1}, n \in \mathbb{N}^*$, such that [5, Definition A.5, p. 442], [4, p. 189].

(i) If $v(x) \in \mathcal{I}$, then so is $r(x) \odot v(x)$ for all $r(x) \in \mathbb{F}_2[x]_{n-1}, n \in \mathbb{N}^*$ [4, p. 189].

Clearly, we can replace (i) by

(ii) if $v(x) \in \mathcal{I}$, then so is $x \odot v(x)$ [4, p. 189].

In Definition 1.60, we required by item (i) that all multiples of an arbitrary polynomial $v(x) \in \mathcal{I}$ are also elements of the ideal $\mathcal{I} \subseteq \mathbb{F}_2[x]_{n-1}, n \in \mathbb{N}^*$.

Now, let us consider a fixed polynomial, $g(x) \in \mathcal{I}$ say. This polynomial $g(x) \in \mathcal{I}$ generates the ideal $\mathcal{I}$.

Definition 1.61 (Principal Ideal). An ideal $\mathcal{I}$, which is generated by a fixed polynomial $g(x)$, denoted by [5, Definition A.5, p. 442]

$$\mathcal{I} = \langle g(x) \rangle = \{g(x) \odot r(x) \mid r(x) \in \mathbb{F}_2[x]_{n-1}\}, \quad g(x) \text{ fixed}, \quad n \in \mathbb{N}^*, \tag{1.946}$$

is termed *principal ideal* [4, p. 190] (in German *"Hauptideal"* [5, Definition A.5, p. 442]). The polynomial $g(x)$ is called a *generator polynomial of the principal ideal* [4, p. 190].

In fact, every ideal $\mathcal{I} \subseteq \mathbb{F}_2[x]_{n-1}$, $n \in \mathbb{N}^*$, in $\mathbb{F}_2[x]_{n-1}$, $n \in \mathbb{N}^*$ is a principal ideal [4, p. 190].

1.12.7 Finding Irreducible Polynomials

In order to generate a cyclic (n, k, d_{min}) binary linear block code $\mathbb{V}$, we require irreducible polynomials. Finding such irreducible polynomials is hence the focus of Section 1.12.7.

Theorem 1.54 (Product of Irreducible Monic Polynomials). *For any field $\mathbb{F}_{2^m}$ [4, p. 107],*

$$x^{2^m} \oplus x = \text{product of all monic polynomials, irreducible over } \mathbb{F}_{2^m}\text{, whose degree divides } m. \tag{1.947}$$

Proof. The first part of the proof sets out from the irreducible polynomial $\pi(x)$ over $\mathbb{F}_2$ [4, Theorem 10, p. 107]. Let $\deg(\pi(x))$ be $d \in \mathbb{N}^*$, and let d be a divisor of m, i. e., $d|m$ [4, Theorem 10, p. 107]. The trivial case is $\pi(x)$ equal to x [4, Theorem 10, p. 107]. So, suppose $\pi(x) \neq x$ [4, Theorem 10, p. 107]. Using $\pi(x) \neq x$ to construct a finite field $\mathbb{F}_{2^m}$, then $\pi(x)$ is the minimal polynomial of one of the field elements [4, Theorem 10, p. 107]. Therefore, $\pi(x)$ divides $(x^{2^d-1} \oplus 1)$, i. e., $\pi(x) \big| (x^{2^d-1} \oplus 1)$ (cf. Theorem 1.45) and, consequently, $\pi(x) \big| (x^{2^d} \oplus x)$ [4, Theorem 10, p. 107].

Now, let us look at the converse part of the proof and assume the polynomial $\pi(x)$ is irreducible and $\deg(\pi(x))$ is equal to $d \in \mathbb{N}^*$ [4, Theorem 10, p. 107]. Furthermore, suppose $\pi(x) \big| (x^{2^d} \oplus x)$ [4, Theorem 10, p. 107]. We need to show $d|m$ [4, Theorem 10, p. 107]. Again, let us consider the nontrivial case $\pi(x) \neq x$ with $\pi(x) \big| (x^{2^d-1} \oplus 1)$; cf. Theorem 1.45 [4, Theorem 10, p. 107]. Let us use $\pi(x) \neq x$ to construct a finite field $\mathbb{F}_{2^m}$ of order 2^d, again [4, Theorem 10, p. 107]. Let us denote a zero of $\pi(x)$ by $\alpha \in \mathbb{F}_{2^m}$ [4, Theorem 10, p. 107]. Furthermore, let $\beta \in \mathbb{F}_{2^m}$ be a primitive element of $\mathbb{F}_{2^m}$ [4, Theorem 10, p. 107]. Then we can write [4, Theorem 10, equation (9), p. 107]

$$\beta = c_0 \oplus c_1 \odot \alpha \oplus \cdots \oplus c_{d-1} \odot \alpha^{d-1}, \quad c_i \in \mathbb{F}_2, \quad i \in \{0, 1 \cdots (d-1)\}, \quad d \in \mathbb{N}^*. \tag{1.948}$$

Since $\pi(\alpha)$ is equal to 0 and

$$\alpha^{2^m} = \alpha, \tag{1.949}$$

Theorem 1.42 and (1.948) lead to [4, Theorem 10, p. 107]

$$\beta^{2^m} = \beta \quad \Leftrightarrow \quad \beta^{2^m-1} = 1. \tag{1.950}$$

The order $(2^d - 1)$ must hence divide $(2^m - 1)$. Therefore, $d|m$. □

Example 1.49. First, consider $m = 1$, i. e., $\mathbb{F}_2$ [4, p. 107]. Using (1.932), we find the equality [4, p. 107]

$$x^2 \oplus x = x \odot M^{(0)}(x) = x \odot (x \oplus 1). \tag{1.951}$$

$M^{(0)}(x)$ can also be found using (1.924) and (1.931),

$$M^{(0)}(x) = \bigodot_{j \in \mathbb{C}_0} \left(x \oplus \alpha^j\right) = \bigodot_{j \in \{0\}} \left(x \oplus \alpha^j\right) = x \oplus 1, \quad \mathbb{C}_0 = \{0\}. \tag{1.952}$$

There are two irreducible polynomials of degree 1, namely x and $(x \oplus 1)$ [4, p. 107]. The minimal polynomials of 0 and 1 in $\mathbb{F}_2$ are respectively x and $(x \oplus 1)$ [4, p. 107]. Hence,

Element	Minimal Polynomial
0	x
1	$M^{(0)}(x) = x \oplus 1$

Example 1.50. Now, let us look at $m = 2$, i. e. $\mathbb{F}_{2^2}$ equal to $\mathbb{F}_4$ [4, p. 107]. Using (1.932), we find the equality

$$\begin{aligned} x^4 \oplus x &= x \odot \bigodot_{s=0}^{1} M^{(s)}(x) \\ &= x \odot M^{(0)}(x) \odot M^{(1)}(x) \\ &= x \odot (x \oplus 1) \odot M^{(1)}(x). \end{aligned} \tag{1.953}$$

In (1.953), we used

$$M^{(0)}(x) = x \oplus 1 \tag{1.954}$$

because we know from (1.925) and (1.931) that

$$M^{(0)}(x) = \bigodot_{j \in \mathbb{C}_0} \left(x \oplus \alpha^j\right) = x \oplus 1, \quad \mathbb{C}_0 = \{0\}. \tag{1.955}$$

$M^{(1)}(x)$ can also be found using (1.925) and (1.931),

$$\begin{aligned} M^{(1)}(x) &= \bigodot_{j \in \mathbb{C}_1} \left(x \oplus \alpha^j\right) \\ &= \bigodot_{j \in \{1,2\}} \left(x \oplus \alpha^j\right) \\ &= (x \oplus \alpha) \odot \left(x \oplus \alpha^2\right) \\ &= x^2 \oplus x \odot \left(\alpha^2 \oplus \alpha\right) \oplus \alpha^3. \end{aligned} \tag{1.956}$$

Since we have

$$\alpha^2 \oplus \alpha = 1, \quad \alpha^3 = 1, \tag{1.957}$$

in $\mathbb{F}_4$, (1.956) becomes

$$M^{(1)}(x) = x^2 \oplus x \oplus 1. \tag{1.958}$$

Another approach to find $M^{(1)}(x)$ is as follows. We first divide $(x^4 \oplus x)$ by x, yielding

$$x^3 \oplus 1 = \underbrace{(x \oplus 1)}_{=M^{(0)}(x)} \odot M^{(1)}(x). \tag{1.959}$$

Now, let us divide $(x^3 \oplus 1)$ by $(x \oplus 1)$,

$$\begin{array}{lllll} x^3 & & & \oplus & 1 \quad : \quad x \oplus 1 \quad = x^2 \oplus x \oplus 1 \\ x^3 & \oplus & x^2 & & \\ \hline & & x^2 & \oplus & 1 \\ & & x^2 & \oplus & x \\ \hline & & & x \oplus & 1 \\ & & & x \oplus & 1 \\ \hline & & & x \oplus & 0 \end{array} \tag{1.960}$$

immediately leading to

$$M^{(1)}(x) = x^2 \oplus x \oplus 1. \tag{1.961}$$

There is one irreducible polynomial of degree 2, namely $x^2 \oplus x \oplus 1$ [4, p. 108]. The minimal polynomials of $\mathbb{F}_4$ are [4, p. 108],

Elements of $\mathbb{F}_4$	**Minimal Polynomial**
0	x
1	$M^{(0)}(x) = x \oplus 1$
a, a^2	$M^{(1)}(x) = M^{(2)}(x) = x^2 \oplus x \oplus 1$

Example 1.51. Let us now consider $m = 3$, i. e. $\mathbb{F}_{2^3}$ equal to $\mathbb{F}_8$, defined by $a^3 \oplus a \oplus 1 = 0$ [4, p. 108]. Using (1.926), we yield

$$M^{(0)}(x) = \bigodot_{j \in \mathbb{C}_0} \left(x \oplus a^j\right) = \bigodot_{j \in \{0\}} \left(x \oplus a^j\right) = x \oplus 1. \tag{1.962}$$

With (1.932), we find the equality

$$x^7 \oplus 1 = \underbrace{(x \oplus 1)}_{=M^{(0)}(x)} \odot M^{(1)}(x) \odot M^{(3)}(x). \tag{1.963}$$

Setting out from (1.448), dividing $(x^7 \oplus 1)$ by $(x \oplus 1)$ yields

$$\left(x^7 \oplus 1\right) = (x \oplus 1) \odot \left(x^6 \oplus x^5 \oplus x^4 \oplus x^3 \oplus x^2 \oplus x \oplus 1\right), \tag{1.964}$$

hence, we obtain

$$M^{(1)}(x) \odot M^{(3)}(x) = x^6 \oplus x^5 \oplus x^4 \oplus x^3 \oplus x^2 \oplus x \oplus 1. \tag{1.965}$$

We can compute $M^{(1)}(x)$ by setting out from (1.926) and (1.931),

$$\begin{aligned}
M^{(1)}(x) &= \bigodot_{j\in\{1,2,4\}} \left(x \oplus \alpha^j\right) \\
&= (x \oplus \alpha) \odot \left(x \oplus \alpha^2\right) \odot \left(x \oplus \alpha^4\right) \\
&= \left(x^2 \oplus \alpha \odot x \oplus \alpha^2 \odot x \oplus \alpha^3\right) \odot \left(x \oplus \alpha^4\right) \\
&= x^3 \oplus \alpha \odot x^2 \oplus \alpha^2 \odot x^2 \oplus \alpha^3 \odot x \oplus \alpha^4 \odot x^2 \oplus \alpha^5 \odot x \oplus \alpha^6 \odot x \oplus \alpha^7 \\
&= x^3 \oplus \alpha \odot x^2 \oplus \alpha^2 \odot x^2 \oplus \alpha^4 \odot x^2 \oplus \alpha^3 \odot x \oplus \alpha^5 \odot x \oplus \alpha^6 \odot x \oplus \alpha^7 \\
&= x^3 \oplus x^2 \odot \left(\alpha \oplus \alpha^2 \oplus \alpha^4\right) \oplus x \odot \left(\alpha^3 \oplus \alpha^5 \oplus \alpha^6\right) \oplus \alpha^7 .
\end{aligned} \tag{1.966}$$

Since we have

$$\alpha^3 = 1 \oplus \alpha, \quad \alpha^7 = 1, \tag{1.967}$$

in $\mathbb{F}_8$, we yield

$$\begin{aligned}
M^{(1)}(x) &= x^3 \oplus x^2 \odot \alpha \odot \underbrace{\left(1 \oplus \alpha \oplus \alpha^3\right)}_{=0} \oplus x \odot \left(\alpha^3 \oplus \alpha^5 \oplus \alpha^6\right) \oplus 1 \\
&= x^3 \oplus x \odot \alpha^3 \odot \left(1 \oplus \alpha^2 \oplus \underbrace{\alpha^3}_{=1\oplus\alpha}\right) \oplus 1 \\
&= x^3 \oplus x \odot \alpha^4 \odot \underbrace{(1 \oplus \alpha)}_{=\alpha^3} \oplus 1 \\
&= x^3 \oplus x \odot \underbrace{\alpha^7}_{=1} \oplus 1,
\end{aligned} \tag{1.968}$$

and hence

$$M^{(1)}(x) = x^3 \oplus x \oplus 1. \tag{1.969}$$

Using (1.926) and (1.931), we find

$$\begin{aligned}
M^{(3)}(x) &= \bigodot_{j\in\{3,6,5\}} \left(x \oplus \alpha^j\right) \\
&= \left(x \oplus \alpha^3\right) \odot \left(x \oplus \alpha^6\right) \odot \left(x \oplus \alpha^5\right) \\
&= \left[x^2 \oplus x \odot \left(\alpha^6 \oplus \alpha^3\right) \oplus \alpha^9\right] \odot \left(x \oplus \alpha^5\right) \\
&= x^3 \oplus x^2 \odot \left(\alpha^6 \oplus \alpha^3\right) \oplus x \odot \alpha^9 \oplus x^2 \odot \alpha^5 \oplus x \odot \alpha^5 \odot \left(\alpha^6 \oplus \alpha^3\right) \oplus \underbrace{\alpha^{14}}_{=1} \\
&= x^3 \oplus x^2 \odot \left(\alpha^6 \oplus \alpha^3 \oplus \alpha^5\right) \oplus x \odot \alpha^5 \odot \left(\alpha^6 \oplus \alpha^3 \oplus \alpha^4\right) \oplus 1 \\
&= x^3 \oplus x^2 \odot \alpha^3 \odot \left(1 \oplus \alpha^2 \oplus \underbrace{\alpha^3}_{=1\oplus\alpha}\right) \oplus x \odot \alpha^8 \odot \underbrace{\left(\alpha^3 \oplus 1 \oplus \alpha\right)}_{=0} \oplus 1 \\
&= x^3 \oplus x^2 \odot \alpha^4 \odot \underbrace{(\alpha \oplus 1)}_{=\alpha^3} \oplus 1 \\
&= x^3 \oplus x^2 \odot \underbrace{\alpha^7}_{=1} \oplus 1,
\end{aligned} \tag{1.970}$$

and hence

$$M^{(3)}(x) = x^3 \oplus x^2 \oplus 1. \tag{1.971}$$

So, we find the following table [4, p. 108].

Elements	Minimal Polynomial
0	x
1	$M^{(0)}(x) = x \oplus 1$
a, a^2, a^4	$M^{(1)}(x) = x^3 \oplus x \oplus 1$
a^3, a^6, a^5	$M^{(3)}(x) = x^3 \oplus x^2 \oplus 1$

The minimal polynomials $M^{(1)}(x)$ equal to $(x^3 \oplus x \oplus 1)$ and $M^{(3)}(x)$ equal to $(x^3 \oplus x^2 \oplus 1)$ are called *reciprocal polynomials* because [4, p. 108]

$$M^{(1)}(x) = x^3 \odot M^{(3)}\left(x^{-1}\right). \tag{1.972}$$

In general, the reciprocal polynomial of the polynomial $f(x)$ is [4, p. 108]

$$x^{\deg f(x)} f\left(x^{-1}\right). \tag{1.973}$$

The zeros of the reciprocal polynomial are the reciprocals of zeros of the original polynomial [4, p. 108]. The reciprocal of an irreducible polynomial is also irreducible [4, p. 108]. So, if a has minimal polynomial $M^{(1)}(x)$, we know immediately that a^{-1} has minimal polynomial $M^{(-1)}(x)$, which is the reciprocal polynomial of $M^{(1)}(x)$ [4, p. 108].

Now, we are prepared to extend Section 1.11.

1.13 Cyclic (n, k, d_{min}) Binary Linear Block Codes Revisited

1.13.1 Ideals and Generator Polynomials

Now, let us "*Get Back*" (The Beatles, 1969) to cyclic (n, k, d_{min}) binary linear block codes and continue with Section 1.11. Using our newly acquired knowledge of number theory (cf. Section 1.12), we will first revisit generator polynomials.

Theorem 1.55 (Generator Polynomial of a Cyclic (n, k, d_{min}) Binary Linear Block Code $\mathbb{V}$ Divides $(x^n \oplus 1)$). *Let $g(x) \in \mathbb{F}_2[x]_{n-k}$, $k, n \in \mathbb{N}^*$, over $\mathbb{F}_2$, having the degree $(n-k)$, $k, n \in \mathbb{N}^*$, be a generator polynomial of an (n, k, d_{min}) binary linear block code $\mathbb{V}$ [5, Satz 5.2, p. 134]. The (n, k, d_{min}) binary linear block code $\mathbb{V}$ is cyclic if and only if $g(x) \in \mathbb{F}_2[x]_{n-k}$, $k, n \in \mathbb{N}^*$, divides $(x^n \oplus 1)$ [5, Satz 5.2, p. 134].*

Proof. According to Theorem 1.35, there are polynomials $q(x)$ and $r(x)$ that fulfill

$$x^n \oplus 1 = q(x) \odot g(x) \oplus r(x), \quad \deg\left(r(x)\right) < \deg\left(g(x)\right) = n - k, \quad k, n \in \mathbb{N}^*. \tag{1.974}$$

Therefore, we yield [5, Satz 5.2, p. 134]

$$r(x) = R_{x^n \oplus 1}\left[r(x)\right] = R_{x^n \oplus 1}\left[x^n \oplus 1 \oplus q(x) \odot g(x)\right] = R_{x^n \oplus 1}\left[q(x) \odot g(x)\right]. \tag{1.975}$$

Since $q(x) \odot g(x) \bmod (x^n \oplus 1)$ is the linear superposition of cyclic shifts of $g(x)$ with $q(x)$ governing these shifts (cf. Theorem 1.37), $r(x)$ equal to $R_{x^n \oplus 1}[q(x) \odot g(x)]$ is also a codeword, i. e., $R_{x^n \oplus 1}[q(x) \odot g(x)] \in \mathbb{V}$, however, with $\deg(r(x)) < (n-k)$, $k, n \in \mathbb{N}^*$. This is only possible for $r(x) = 0$, which means that $g(x) | (x^n \oplus 1)$ [5, Satz 5.2, p. 134].

Using Theorem 1.37, the cyclic shift $R_{x^n \oplus 1}[x^m \odot v(x)] \in \mathbb{V}$ of the codeword

$$v(x) = u(x) \odot g(x) \tag{1.976}$$

is also a codeword, which requires $g(x)|(x^n \oplus 1)$ [5, Satz 5.2, p. 134].

The monic generator polynomial $g(x) \in \mathbb{F}_2[x]_{n-k}$, $k, n \in \mathbb{N}^*$, over $\mathbb{F}_2$, having the degree $(n-k)$, $k, n \in \mathbb{N}^*$, is unique for the following reason. Suppose that there was a second generator, $g(x) \odot y(x)$, then $\deg(y(x))$ is equal to 0 because $\deg(g(x) \odot y(x))$ must be equal to $\deg(g(x))$ [5, Satz 5.2, p. 134]. Hence, $y(x)$ is a constant, namely 1. Hence, $g(x) \odot y(x)$ is equal to $g(x)$. □

Corollary 1.15 (Nontrivial Minimal Polynomials as Generator Polynomials). *Combining Corollary* 1.14 *and Theorem* 1.55 *yield*

$$n = 2^m - 1, \quad m \in \mathbb{N}^* \tag{1.977}$$

(cf., e. g., [4, *p. 192*]*), immediately leading to* [4, *equation (14), p. 197*]

$$x^n \oplus 1 = \bigodot_s M^{(s)}(x) \quad n \in \mathbb{N}^*, \tag{1.978}$$

where s runs through the coset representatives mod *n*, $n \in \mathbb{N}^*$.

This teaches us that the irreducible nontrivial minimal polynomials $M^{(s)}(x)$*, where s runs through the coset representatives* mod *n*, $n \in \mathbb{N}^*$*, as well as the product of arbitrarily chosen irreducible nontrivial and distinct minimal polynomials* $M^{(s)}(x)$*, each one associated with the corresponding coset representative of the cyclotomic coset* $\mathbb{C}_s$*, can be used as generator polynomials of cyclic* $(n, k, d_{\min})$ *binary linear block codes; cf., e. g.,* [4, *Theorem 2, p. 192; Theorem 3, p. 193; equation (15), p. 199; p. 205*].

The latter insight leads to the Bose-Chaudhuri-Hocquenghem (BCH) codes, which form a large family of powerful cyclic codes [15, *pp. 194f.*]. *BCH codes are a generalization of the Hamming codes* [15, *p. 194*]. *Binary BCH codes, which we will solely consider in this textbook, were discovered in 1959 by Hocquenghem and in 1960 by Bose and Chaudhuri* [15, *p. 194*].

Corollary 1.16 (Factors of $(x^n \oplus 1)$ Used to Form Generator Polynomials). *Referring to Remark* 1.38*, and letting* $\mathbb{K}$ *be any union of cyclotomic cosets* $\mathbb{C}_s$*, then each product of* $(x \oplus \alpha^j)$ *with* $j \in \mathbb{K}$ *is can be used as a generator polynomial* [4, *equation (15), p. 199*]

$$g(x) = \bigodot_{j \in \mathbb{K}} \left(x \oplus \alpha^j\right). \tag{1.979}$$

Every cyclic code has a generator polynomial $g(x) \in \mathcal{I} \subseteq \mathbb{F}_2[x]_{n-1}$, $n \in \mathbb{N}^*$ [4, p. 190].

We can define a cyclic code in terms of principal ideals.

Definition 1.62 (Cyclic Code). A cyclic code of length n is a principal ideal $\mathcal{I} \subseteq \mathbb{F}_2[x]_{n-1}$, $n \in \mathbb{N}^*$ [4, p. 189].

The cyclic $(n, k, d_{\min})$ binary linear block code $\mathbb{V}$ is obviously given by [4, Theorem 1, p. 190]

$$\mathcal{I} = \langle g(x) \rangle. \tag{1.980}$$

According to Corollary 1.16 and Definition 1.62, a cyclic $(n, k, d_{\min})$ binary linear block code $\mathbb{V}$ is defined in terms of the zeros of any polynomial $v(x) \in \mathcal{I}$ [4, p. 199].

Wow, looking at number theory was definitely worthwhile.

Often, the length n of a cyclic code is an odd natural number, e. g., n is equal to $(2^m - 1)$, $m \in \mathbb{N}^*$; cf., e. g., our discussions about factorization above.

Example 1.52. The cyclic (3, 2, 3) binary linear block code

$$\mathbb{V} = \{(0,0,0), (1,1,0), (0,1,1), (1,0,1)\} \tag{1.981}$$

is represented by the principal ideal [4, p. 190]

$$\mathcal{I} = \Big\{ \underbrace{0}_{=v^{(0)}(x)}, \underbrace{1 \oplus x}_{=v^{(1)}(x)}, \underbrace{x \oplus x^2}_{=v^{(2)}(x)}, \underbrace{1 \oplus x^2}_{=v^{(3)}(x)} \Big\} \subset \mathbb{F}_2[x]_3. \tag{1.982}$$

Since $\mathbb{F}_2[x]_3$ contains eight elements, the principal ideal $\mathcal{I}$ is its subset.

Proof. $\mathbb{V}$, and, therefore, $\mathcal{I}$, is closed under addition and, therefore, linear [4, p. 190].

$$\begin{aligned}
(0,0,0) \oplus (0,0,0) &= (0,0,0) \in \mathbb{V}, \\
(0,0,0) \oplus (1,1,0) &= (1,1,0) \in \mathbb{V}, \\
(0,0,0) \oplus (0,1,1) &= (0,1,1) \in \mathbb{V}, \\
(0,0,0) \oplus (1,0,1) &= (1,0,1) \in \mathbb{V}, \\
(1,1,0) \oplus (1,1,0) &= (0,0,0) \in \mathbb{V}, \\
(1,1,0) \oplus (0,1,1) &= (1,0,1) \in \mathbb{V}, \\
(1,1,0) \oplus (1,0,1) &= (0,1,1) \in \mathbb{V}, \\
(0,1,1) \oplus (0,1,1) &= (0,0,0) \in \mathbb{V}, \\
(0,1,1) \oplus (1,0,1) &= (1,1,0) \in \mathbb{V}, \\
(1,0,1) \oplus (1,0,1) &= (0,0,0) \in \mathbb{V}.
\end{aligned} \tag{1.983}$$

Since

$$x^3 = 1, \tag{1.984}$$

any multiple $x^i \odot v^{(k)}(x)$, $i \in \mathbb{Z}$, $k \in \{0,1,2,3\}$ is again in $\mathcal{I}$ [4, p. 190]

$$\begin{aligned}
&x^i \odot v^{(0)}(x) = v^{(0)}(x) = 0, \\
&x^{3i+1} \odot v^{(1)}(x) = x^{3i+0} \odot v^{(2)}(x) = x^{3i-1} \odot v^{(3)}(x) = v^{(2)}(x), \\
&x^{3i+2} \odot v^{(1)}(x) = x^{3i+1} \odot v^{(2)}(x) = x^{3i+0} \odot v^{(3)}(x) = v^{(3)}(x), \\
&x^{3i+3} \odot v^{(1)}(x) = x^{3i+2} \odot v^{(2)}(x) = x^{3i+1} \odot v^{(3)}(x) = v^{(1)}(x), \\
&i \in \mathbb{Z}.
\end{aligned} \tag{1.985}$$

□

The next theorem proves this and other basic properties of cyclic codes.

Theorem 1.56 (Properties of Cyclic $(n, k, d_{\min})$ Binary Linear Block Codes). *Let $\mathcal{I}$ be a nonzero ideal in $\mathbb{F}_2[x]_{n-1}$, $n \in \mathbb{N}^*$, i. e., a cyclic code of length n* [4, p. 190].

1: *There is a unique monic polynomial $g(x)$ of minimal degree in $\mathcal{I}$* [4, p. 190].

2: *$\mathcal{I} = \langle g(x) \rangle$, i. e., $g(x)$ is a generator polynomial of $\mathcal{I}$* [4, p. 190].

3: *$g(x)$ is a factor of $(x^n \oplus 1)$* [4, p. 190].

4: *Any $v(x) \in \mathcal{I}$ can be written uniquely as $v(x) = u(x) \odot g(x)$ in $\mathbb{F}_2[x]$, where the message $u(x) \in \mathbb{F}_2[x]$ has degree $< (n-r) = k$, r equal to $(n-k)$ being the degree of $g(x)$* [4, p. 190]. *The dimension of $\mathcal{I}$ is $(n-r) = k$. Thus, the message $u(x) \in \mathbb{F}_2[x]$ becomes the codeword $v(x) = u(x) \odot g(x)$* [4, p. 190].

5: *If $g(x)$ is equal to $(g_0 \oplus g_1 \odot x \oplus g_2 \odot x^2 \oplus \cdots \oplus g_r \odot x^r\}$, then $\mathcal{I}$ is generated by the rows of the generator matrix* [4, pp. 190f.]

$$G = \begin{pmatrix} g_0 & g_1 & \cdots & \cdots & \cdots & \cdots & g_{n-k-1} & g_{n-k} & 0 & 0 & \cdots & 0 \\ 0 & g_0 & g_1 & \cdots & \cdots & \cdots & \cdots & g_{n-k-1} & g_{n-k} & 0 & \cdots & 0 \\ \vdots & \vdots & \ddots & \ddots & \ddots & \ddots & \ddots & \ddots & \ddots & \ddots & \ddots & \vdots \\ \vdots & \vdots & \ddots & \ddots & \ddots & \ddots & \ddots & \ddots & \ddots & \ddots & \ddots & \vdots \\ 0 & 0 & \cdots & 0 & g_0 & g_1 & \cdots & \cdots & \cdots & \cdots & g_{n-k-1} & g_{n-k} \end{pmatrix}, \tag{1.986}$$

i. e., [4, *equation (1), p. 191*]

$$G = \begin{pmatrix} g(x) \\ x \odot g(x) \\ \vdots \\ x^{k-1} \odot g(x) \end{pmatrix}, \quad k \in \mathbb{N}^*. \tag{1.987}$$

Proof.

Claim 1

Suppose $f(x), g(x) \in \mathcal{I}$ are monic and have the minimal degree r equal to $(n-k)$, $k, n \in \mathbb{N}^*$ [4, p. 191]. Since both are monic, $(f(x) \oplus g(x)) \in \mathcal{I}$ has lower degree, which is a contradiction unless $f(x)$ and $g(x)$ are identical [4, p. 191].

Claim 2

Let us assume that $v(x) \in \mathcal{I}$ [4, p. 191]. Then the division Theorem 1.35 suggests that [4, p. 191]

$$v(x) = q(x) \odot g(x) \oplus r(x), \quad \deg\big(r(x)\big) < r = n-k, \quad k, n \in \mathbb{N}^*; \tag{1.988}$$

hence,

$$r(x) = v(x) \oplus q(x) \odot g(x) \in \mathcal{I}, \tag{1.989}$$

because the code is linear [4, p. 191]. Therefore, $r(x)$ must vanish [4, p. 191]. Thus, $v(x) \in \langle g(x) \rangle$ [4, p. 191].

Claim 3

The above Claim 2 and the division Theorem 1.35 suggest that [4, p. 191]

$$x^n \oplus 1 = q(x) \odot g(x) \oplus r(x), \quad \deg\big(r(x)\big) < r = n-k, \quad k, n \in \mathbb{N}^*. \tag{1.990}$$

In $\mathbb{F}_2[x]_{n-1}$, $n \in \mathbb{N}^*$, this says that

$$r(x) = \ominus\, q(x) \odot g(x) = q(x) \odot g(x) \in \mathcal{I}, \tag{1.991}$$

which is a contradiction, unless $r(x)$ vanishes [4, p. 191].

Claim 4, Claim 5

According to the above Claim 2, any $v(x) \in \mathcal{I}$, having $\deg(v(x)) < n$, is equal to $u(x) \odot g(x) \in \mathbb{F}_2[x]_{n-1}$, $n \in \mathbb{N}^*$ [4, p. 191]. In $\mathbb{F}_2[x]$, using $r(x)$ equal to 0 and introducing $e(x)$, we thus yield [4, p. 191]

$$\begin{aligned} v(x) &= w(x) \odot g(x) \oplus e(x) \odot \underbrace{\left(x^n \oplus 1\right)}_{=q(x)\odot g(x)} \in \mathbb{F}_2[x] \\ &= w(x) \odot g(x) \oplus e(x) \odot q(x) \odot g(x) \in \mathbb{F}_2[x] \\ &= \underbrace{\left[w(x) \oplus e(x) \odot q(x)\right]}_{=u(x)} \odot g(x) \in \mathbb{F}_2[x] \\ &= u(x) \odot g(x) \in \mathbb{F}_2[x], \end{aligned} \tag{1.992}$$

where obviously [4, p. 191]

$$\deg\big(u(x)\big) \leq n - r - 1 = k - 1, \quad k, n \in \mathbb{N}^*. \tag{1.993}$$

Thus, the cyclic $(n, k, d_{\min})$ binary linear block code $\mathbb{V}$ consists of multiples of $g(x)$ by polynomials of degree $\leq (n - r - 1)$ equal to $(k - 1)$, $k, n \in \mathbb{N}^*$, evaluated in $\mathbb{F}_2[x]$, but not in $\mathbb{F}_2[x]_{n-1}$, $n \in \mathbb{N}^*$ [4, p. 191].

There are $(n-r)$ equal to k, $k, n \in \mathbb{N}^*$ linearly independent multiples of $g(x)$, namely $g(x)$, $(x \odot g(x))$, $(x^2 \odot g(x)) \cdots (x^{n-r-1} \odot g(x)) = (x^{k-1} \odot g(x))$ [4, p. 191]. The corresponding vectors are the rows of the generator matrix $\boldsymbol{G}$. Thus, the code has dimension $(n - r)$ equal to k, $k, n \in \mathbb{N}^*$, [4, p. 191]. □

Up to now, we have taken the generator polynomial of a code $g(x)$ to be the lowest degree monic polynomial in the code [4, p. 199]. But other generators are possible [4, p. 199].

Theorem 1.57 (Other Generator Polynomials of a Cyclic Code). *If $p(x) \in \mathbb{F}_2[x]_{n-1}$, $n \in \mathbb{N}^*$ does not introduce any new zeros, i. e., if $p(\alpha^i) \neq 0$ for all $i \notin \mathbb{K}$, $\mathbb{K}$ being the union of the cyclotomic cosets $\mathbb{C}_s$, used to determine $g(x)$ (cf. Corollary 1.16), then $g(x)$ and $p(x) \odot g(x)$ generate the same code. For instance, $[g(x)]^2$ generates the same code as $g(x)$ [4, p. 199].*

Proof. Clearly, $\langle g(x)\rangle \supseteq \langle p(x) \odot g(x)\rangle$ [4, p. 199]. Since $g(x)$ divides $(x^n \oplus 1)$, $n \in \mathbb{N}^*$, there is a polynomial $h(x)$, which we term the *check polynomial*, with

$$g(x) \odot h(x) = x^n \oplus 1 \quad \text{in } \mathbb{F}_2[x], \quad n \in \mathbb{N}^*, \tag{1.994}$$

hence complying with

$$g(x) \odot h(x) = 0 \quad \text{in } \mathbb{F}_2[x]_{n-1}, \quad n \in \mathbb{N}^*. \tag{1.995}$$

The check polynomial $h(x)$ and $p(x)$ must be relatively prime if $g(x)$ and $p(x) \odot g(x)$ generate the same code [4, p. 199]. In this case, there are polynomials $a(x)$, $b(x)$ such that [4, p. 199]

$$1 = a(x) \odot p(x) \oplus b(x) \odot h(x) \quad \text{in } \mathbb{F}_2[x], \tag{1.996}$$

which is equivalent to [4, p. 199]

$$g(x) = a(x) \odot p(x) \odot g(x) \oplus b(x) \underbrace{h(x) \odot g(x)}_{=0 \text{ in } \mathbb{F}_2[x]_{n-1}} \quad \text{in } \mathbb{F}_2[x]. \tag{1.997}$$

This is obviously identical to [4, p. 199]

$$g(x) = a(x) \odot p(x) \odot g(x) \quad \text{in } \mathbb{F}_2[x]_{n-1}, \tag{1.998}$$

meaning that [4, p. 199]

$$\langle g(x) \rangle \subseteq \langle p(x) \odot g(x) \rangle. \tag{1.999}$$

Since $\langle g(x) \rangle \supseteq \langle p(x) \odot g(x) \rangle$ also holds, we obtain

$$\langle g(x) \rangle = \langle p(x) \odot g(x) \rangle. \tag{1.1000}$$

□

1.13.2 Check Polynomial

Let the cyclic (n, k, d_{min}) binary linear block code $\mathbb{V}$ have the generator polynomial $g(x)$ [4, p. 194]. It follows from Theorem 1.56 that this generator polynomial $g(x)$ divides $(x^n \oplus 1)$ [4, p. 194]. Therefore, with some nonzero polynomial $h(x)$, which we call the *check polynomial* as we already know from Theorem 1.57, we obtain [4, p. 194]

$$g(x) \odot h(x) = x^n \oplus 1 \quad \text{in } \mathbb{F}_2[x], \quad \deg(g(x) \odot h(x)) = n, \quad n \in \mathbb{N}^*, \tag{1.1001}$$

i. e., [4, p. 194]

$$g(x) \odot h(x) = 0 \quad \text{in } \mathbb{F}_2[x]_{n-1}, \quad n \in \mathbb{N}^*. \tag{1.1002}$$

Since $\deg(g(x) \odot h(x))$ is equal to $n \in \mathbb{N}^*$ in $\mathbb{F}_2[x]$ and $\deg(g(x))$ is equal to $(n - k)$, $k, n \in \mathbb{N}^*$, we yield [4, p. 195]

$$\deg(h(x)) = \deg(g(x) \odot h(x)) - \deg(g(x)) = n - (n - k) = k, \quad k, n \in \mathbb{N}^*, \tag{1.1003}$$

which is the dimension of the cyclic (n, k, d_{min}) binary linear block code $\mathbb{V}$.

Then we find [4, p. 194]

$$h(x) = \frac{x^n \oplus 1}{g(x)} = \bigoplus_{m=0}^{k} h_m \odot x^m \quad \text{in } \mathbb{F}_2[x], \quad h_k \neq 0, \quad k, n \in \mathbb{N}^*. \tag{1.1004}$$

Remark 1.39 (Why Call $h(x)$ the Check Polynomial?). The reason for the name "check polynomial" is the following [4, p. 194]. If the polynomial

$$v(x) = u(x) \odot g(x) = \bigoplus_{j=0}^{n-1} v_j \odot x^j, \quad n \in \mathbb{N}^* \tag{1.1005}$$

is a code polynomial of $\mathcal{I}$, i. e., represents a codeword of the cyclic (n, k, d_{min}) binary linear block code $\mathbb{V}$, then [4, p. 194]

$$v(x) \odot h(x) = \underbrace{u(x) \odot g(x)}_{=v(x)} \odot h(x)$$

$$
\begin{aligned}
&= \bigoplus_{j=0}^{n-1} \bigoplus_{m=0}^{k} v_j \odot h_m \odot x^{j+m} \\
&= \bigoplus_{j=0}^{n-1} \bigoplus_{m=j}^{k+j} v_j \odot h_{m-j} \odot x^{m} \\
&= 0 \quad \text{in } \mathbb{F}_2[x]_{n-1}, \quad n \in \mathbb{N}^*. \qquad (1.1006)
\end{aligned}
$$

Of course, the subscripts must be taken mod n in (1.1006) [4, p. 194]. In (1.1006), the coefficient of x^m is given by [4, p. 194]

$$
\bigoplus_{j=0}^{n-1} v_j \odot h_{m-j} = 0, \quad m \in \{0, 1 \cdots (n-1)\}, \quad n \in \mathbb{N}^*, \tag{1.1007}
$$

where the subscripts must be taken mod n. (1.1007) says that $v(x) \in \mathcal{I}$ must satisfy the $(n-k)$, $k, n \in \mathbb{N}^*$, parity check equations [4, equation (11), p. 195]

$$
\begin{array}{ccccccccccc}
v_{n-k-1} \odot h_k & \oplus & v_{n-k} \odot h_{k-1} & \oplus & v_{n-k+1} \odot h_{k-2} & \oplus & \cdots & \oplus & v_{n-1} \odot h_0 & = & 0, \\
v_{n-k-2} \odot h_k & \oplus & v_{n-k-1} \odot h_{k-1} & \oplus & v_{n-k} \odot h_{k-2} & \oplus & \cdots & \oplus & v_{n-2} \odot h_0 & = & 0, \\
v_{n-k-3} \odot h_k & \oplus & v_{n-k-2} \odot h_{k-1} & \oplus & v_{n-k-1} \odot h_{k-2} & \oplus & \cdots & \oplus & v_{n-3} \odot h_0 & = & 0, \\
\vdots & & \vdots & & \vdots & & & & \vdots & & \vdots \\
v_0 \odot h_k & \oplus & v_1 \odot h_{k-1} & \oplus & v_2 \odot h_{k-2} & \oplus & \cdots & \oplus & v_k \odot h_0 & = & 0,
\end{array} \tag{1.1008}
$$

with $h_k \neq 0$ [4, p. 195], i. e., [4, equation (11), p. 195]

$$
\begin{array}{ccccccccccc}
v_{n-k-1} & \oplus & v_{n-k} \odot h_{k-1} & \oplus & v_{n-k+1} \odot h_{k-2} & \oplus & \cdots & \oplus & v_{n-1} \odot h_0 & = & 0. \\
v_{n-k-2} & \oplus & v_{n-k-1} \odot h_{k-1} & \oplus & v_{n-k} \odot h_{k-2} & \oplus & \cdots & \oplus & v_{n-2} \odot h_0 & = & 0, \\
v_{n-k-3} & \oplus & v_{n-k-2} \odot h_{k-1} & \oplus & v_{n-k-1} \odot h_{k-2} & \oplus & \cdots & \oplus & v_{n-3} \odot h_0 & = & 0, \\
\vdots & & \vdots & & \vdots & & & & \vdots & & \vdots \\
v_0 & \oplus & v_1 \odot h_{k-1} & \oplus & v_2 \odot h_{k-2} & \oplus & \cdots & \oplus & v_k \odot h_0 & = & 0.
\end{array} \tag{1.1009}
$$

Hence, $v(x) \in \mathcal{I}$ satisfies the *linear recurrence* [4, equation (12), p. 195]

$$
\begin{aligned}
&v_j \oplus v_{j+1} \odot h_{k-1} \oplus v_{j+2} \odot h_{k-2} \oplus \cdots \oplus v_{j+k} \odot h_0 = 0, \qquad (1.1010) \\
&j \in \{0, 1 \cdots (n-k-1)\}, \quad n \in \mathbb{N}^*.
\end{aligned}
$$

If $v_{n-k}, v_{n-k+1} \cdots v_{n-2}, v_{n-1}$, $k, n \in \mathbb{N}^*$ are taken as the $k \in \mathbb{N}^*$ message symbols, (1.1007) and (1.1010) successively define the $(n-k)$, $k, n \in \mathbb{N}^*$ parity check symbols $v_0, v_1, \cdots v_{n-k-2}, v_{n-k-1}$ [4, p. 195].

Let the *parity check matrix* be [4, equation (9), p. 195]

$$
\boldsymbol{H} = \begin{pmatrix}
0 & \cdots & 0 & h_k & \cdots & h_2 & h_1 & h_0 \\
0 & \cdots & h_k & \cdots & h_2 & h_1 & h_0 & 0 \\
\vdots & \iddots & \vdots & \iddots & \iddots & \iddots & \iddots & \vdots \\
h_k & \cdots & h_2 & h_1 & h_0 & 0 & \cdots & 0
\end{pmatrix}. \tag{1.1011}
$$

Then (1.1007) lets us know that if $\boldsymbol{v}$ is a codeword, i. e., $\boldsymbol{v} \in \mathbb{V}$, and hence $v(x) \in \mathcal{I}$ is a code polynomial, and we obtain [4, p. 195]

$$\boldsymbol{v}\boldsymbol{H}^{\mathrm{T}} = \boldsymbol{0}_{n-k}, \quad k, n \in \mathbb{N}^*. \tag{1.1012}$$

Now, let $\mathbb{V}$ be a cyclic code with generator polynomial $g(x)$ and let the check polynomial $h(x)$ be $(x^n \oplus 1)/g(x)$ [4, p. 196]. Then (1.1011) proves the following theorem.

Theorem 1.58 (Dual Code of a Cyclic Code). *The dual code $\mathbb{V}^{\perp}$ of a cyclic $(n, k, d_{\min})$ binary linear block code $\mathbb{V}$ is cyclic and has the generator polynomial* [4, *p. 196*],

$$g^{\perp}(x) = x^{\deg(h(x))} \odot h\left(x^{-1}\right). \tag{1.1013}$$

Proof. Equation (1.1011) is the proof. □

According to Theorem 1.58, the code with generator polynomial $h(x)$ equal to $(x^n \oplus 1)/g(x)$ is equivalent to $\mathbb{V}^{\perp}$ [4, p. 196]. In fact, it consists of the codewords of $\mathbb{V}^{\perp}$ written backwards [4, p. 196].

1.13.3 Inversion Formula

Let us now broaden our view on finite fields. First, let us introduce the following theorem that we already know for infinite sets, e. g., the set of complex numbers $\mathbb{C}$.

Theorem 1.59 (Sum of Zeros in $\mathbb{F}_{2^m}$). *Let $\xi \in \mathbb{F}_{2^m}$, $m \in \mathbb{N}^*$, be any zero of $(x^n \oplus 1)$, i. e., $(\xi^n \oplus 1) = 0$* [4, *p. 200*]. *Let n be equal to $(2^m - 1)$, $m \in \mathbb{N}^*$, i. e., n is an odd natural number. Then we have* [4, *p. 200*]

$$\bigoplus_{i=0}^{n-1} \xi^i = \begin{cases} 0 & \text{if } \xi \neq 1, \\ n \bmod 2 & \text{if } \xi = 1, \end{cases} = \begin{cases} 0 & \text{if } \xi \neq 1, \\ 1 & \text{if } \xi = 1. \end{cases} \tag{1.1014}$$

Proof. If $\xi = 1$, the sum $(\bigoplus_{i=0}^{n-1} 1)$ is equal to $n \bmod 2$ [4, p. 200].

Now, assume that $\xi \neq 1$. Then, with [11, equation (1.54c), p. 19], we obtain [4, p. 200]

$$\bigoplus_{i=0}^{n-1} \xi^i = \frac{1}{\xi \oplus 1} \cdot \underbrace{\left(\xi^n \oplus 1\right)}_{=0} = \frac{0}{\xi \oplus 1} = 0. \tag{1.1015}$$

□

Theorem 1.60 (Inversion Formula). *The codeword $\boldsymbol{v}$ equal to $(v_0, v_1 \cdots v_{n-1})$, $n \in \mathbb{N}^*$ may be recovered from*

$$v(x) = v_0 \oplus v_1 \odot x \oplus \cdots \oplus v_{n-1} \odot x^{n-1}, \quad n \in \mathbb{N}^*, \tag{1.1016}$$

by [4, *p. 200*]

$$v_i = \bigoplus_{j=0}^{n-1} v\left(\alpha^j\right) \odot \alpha^{-i \cdot j}, \quad i \in \{0, 1 \cdots (n-1)\}, \quad n \in \mathbb{N}^*, \quad n \text{ odd}. \tag{1.1017}$$

Proof. With

$$v\left(\alpha^j\right) = v_0 \oplus v_1 \odot \alpha^j \oplus \cdots \oplus v_{n-1} \odot \alpha^{j(n-1)} = \bigoplus_{k=0}^{n-1} v_k \odot \alpha^{j \cdot k}, \tag{1.1018}$$

we have [4, p. 200]

$$\bigoplus_{j=0}^{n-1} v(\alpha^j) \odot \alpha^{-i\cdot j} = \bigoplus_{j=0}^{n-1}\bigoplus_{k=0}^{n-1} v_k \alpha^{-i\cdot j} \odot \alpha^{j\cdot k} = \bigoplus_{k=0}^{n-1} v_k \bigoplus_{j=0}^{n-1} \alpha^{-j\cdot i+j\cdot k} = \bigoplus_{k=0}^{n-1} v_k \bigoplus_{j=0}^{n-1} \alpha^{j\cdot(k-i)}. \tag{1.1019}$$

With the Kronecker symbol,

$$\delta_{ik} = \begin{cases} 1 & \text{if } i = k, \\ 0 & \text{if } i \neq k, \end{cases} \tag{1.1020}$$

and using Theorem 1.59, we find

$$\bigoplus_{j=0}^{n-1} \alpha^{j\cdot(k-i)} = \begin{cases} 0 & \text{if } k \neq i \\ 1 & \text{if } k = i \end{cases} = \delta_{ik}. \tag{1.1021}$$

Therefore, we obtain [4, p. 200]

$$\bigoplus_{j=0}^{n-1} v(\alpha^j) \odot \alpha^{-i\cdot j} = \bigoplus_{k=0}^{n-1} v_k \delta_{ik} = v_i, \quad i \in \{0, 1 \cdots (n-1)\}, \quad n \in \mathbb{N}^*, \quad n \text{ odd}. \tag{1.1022}$$

□

Remark 1.40 (On (1.1017)). Letting $\alpha^j, j \in \{0, 1 \cdots (n-1)\}, n \in \mathbb{N}^*$, be the nth roots of unity, which take the form [11, eq. (19.222b), p. 956]

$$\omega_n^j = \left[\exp\left\{\mathrm{j}\frac{2\pi}{n}\right\}\right]^j, \quad \mathrm{j} = \sqrt{-1}, \quad j \in \{0, 1 \cdots (n-1)\}, \quad n \in \mathbb{N}^*, \tag{1.1023}$$

in the set of complex numbers $\mathbb{C}$, it is quite remarkable to see that $\bigoplus_{j=0}^{n-1} v(\alpha^j) \odot \alpha^{-i\cdot j}$ of (1.1017) looks like the *discrete Fourier transform (DFT)* [25, Definition 6.1.1, p. 132].

1.13.4 Idempotents

A most remarkable polynomial associated with a cyclic $(n, k, d_{\min})$ binary linear block code $\mathbb{V}$ is its *idempotent* $E(x)$ [4, p. 216].

Definition 1.63 (Idempotent Polynomial). A polynomial $E(x) \in \mathbb{F}_2[x]_{n-1}$ is an *idempotent*, respectively an *idempotent polynomial*, if [4, p. 217]

$$E(x) = [E(x)]^2 = E(x^2). \tag{1.1024}$$

Example 1.53. Let us look at $\mathbb{F}_2[x]_7$. In $\mathbb{F}_2[x]_7$, we have

$$x^7 = 1. \tag{1.1025}$$

$E(x)$ equal to $(x \oplus x^2 \oplus x^4)$ is an idempotent in $\mathbb{F}_2[x]_7$, because we have

$$\begin{aligned}
[E(x)]^2 &= (x \oplus x^2 \oplus x^4) \odot (x \oplus x^2 \oplus x^4) \\
&= x^2 \oplus x^3 \oplus x^5 \oplus x^3 \oplus x^4 \oplus x^6 \oplus x^5 \oplus x^6 \oplus x^8 \\
&= x^2 \oplus \underbrace{x^3 \oplus x^3}_{=0} \oplus x^4 \oplus \underbrace{x^5 \oplus x^5}_{=0} \oplus \underbrace{x^6 \oplus x^6}_{=0} \oplus x^8 \\
&= x^2 \oplus x^4 \oplus \underbrace{x^8}_{=x} \\
&= E(x^2) \\
&= x \oplus x^2 \oplus x^4 \\
&= E(x).
\end{aligned} \tag{1.1026}$$

Furthermore, $E(x)$ equal to 1 is an idempotent in $\mathbb{F}_2[x]_7$,

$$[E(x)]^2 = 1^2 = 1 = E(x^2) = E(x). \tag{1.1027}$$

Also, $E(x)$ equal to $(x^3 \oplus x^6 \oplus x^5)$ is an idempotent in $\mathbb{F}_2[x]_7$, because we yield

$$\begin{aligned}
[E(x)]^2 &= (x^3 \oplus x^6 \oplus x^5) \odot (x^3 \oplus x^6 \oplus x^5) \\
&= x^6 \oplus x^9 \oplus x^8 \oplus x^9 \oplus x^{12} \oplus x^{11} \oplus x^8 \oplus x^{11} \oplus x^{10} \\
&= x^6 \oplus x^{12} \oplus x^{10} \oplus \underbrace{x^9 \oplus x^9}_{=0} \oplus \underbrace{x^8 \oplus x^8}_{=0} \oplus \underbrace{x^{11} \oplus x^{11}}_{=0} \\
&= x^6 \oplus x^{12} \oplus x^{10} \\
&= E(x^2) \\
&= x^6 \oplus \underbrace{x^7}_{=1} x^5 \oplus \underbrace{x^7}_{=1} x^3 \\
&= x^3 \oplus x^6 \oplus x^5 \\
&= E(x).
\end{aligned} \tag{1.1028}$$

In general,

$$E(x) = \bigoplus_{k=0}^{n-1} \varepsilon_k \odot x^k, \quad \varepsilon_k \in \mathbb{F}_2, \quad k \in \{0, 1 \cdots (n-1)\}, \quad n \in \mathbb{N}^*, \tag{1.1029}$$

is an idempotent if and only if ε_k is equal to ε_{2k} with subscripts mod n [4, p. 217].

If $E(x)$ is an idempotent, so is $(1 \oplus E(x))$ [4, p. 217].

Example 1.54. Let us look at $\mathbb{F}_2[x]_7$ again. We already know that

$$x^7 = 1. \tag{1.1030}$$

$E(x)$ equal to $(1 \oplus x \oplus x^2 \oplus x^4)$ is an idempotent in $\mathbb{F}_2[x]_7$, because we yield

$$
\begin{aligned}
\left[E(x)\right]^2 &= \left(1 \oplus x \oplus x^2 \oplus x^4\right) \odot \left(1 \oplus x \oplus x^2 \oplus x^4\right) \\
&= 1 \oplus x \oplus x^2 \oplus x^4 \oplus x \oplus x^2 \oplus x^3 \oplus x^5 \oplus x^2 \oplus x^3 \oplus x^4 \oplus x^6 \oplus x^4 \oplus x^5 \oplus x^6 \oplus x^8 \\
&= 1 \oplus \underbrace{x \oplus x}_{=0} \oplus \underbrace{x^2 \oplus x^2}_{=0} \oplus x^2 \oplus \underbrace{x^3 \oplus x^3}_{=0} \oplus \underbrace{x^4 \oplus x^4}_{=0} \oplus x^4 \oplus \underbrace{x^5 \oplus x^5}_{=0} \oplus \underbrace{x^6 \oplus x^6}_{=0} \oplus x^8 \\
&= 1 \oplus x^2 \oplus x^4 \oplus x^8 \\
&= E\left(x^2\right) \\
&= 1 \oplus x^2 \oplus x^4 \oplus \underbrace{x^8}_{=x} \\
&= 1 \oplus x \oplus x^2 \oplus x^4 \\
&= E(x).
\end{aligned}
\tag{1.1031}
$$

Furthermore, $E(x)$ equal to 0 is an idempotent in $\mathbb{F}_2[x]_7$, because we yield

$$
\left[E(x)\right]^2 = 0^2 = 0 = E\left(x^2\right) = E(x). \tag{1.1032}
$$

Also, $E(x)$ equal to $(1 \oplus x^3 \oplus x^6 \oplus x^5)$ is an idempotent in $\mathbb{F}_2[x]_7$, because we yield

$$
\begin{aligned}
\left[E(x)\right]^2 &= \left(1 \oplus x^3 \oplus x^6 \oplus x^5\right)\left(1 \oplus x^3 \oplus x^6 \oplus x^5\right) \\
&= 1 \oplus x^3 \oplus x^6 \oplus x^5 \oplus x^3 \oplus x^6 \oplus x^9 \oplus x^8 \oplus x^6 \oplus x^9 \oplus x^{12} \oplus x^{11} \oplus x^5 \oplus x^8 \oplus x^{11} \oplus x^{10} \\
&= 1 \oplus \underbrace{x^3 \oplus x^3}_{==0} \oplus \underbrace{x^5 \oplus x^5}_{=0} \oplus \underbrace{x^6 \oplus x^6}_{=0} \oplus x^6 \oplus \underbrace{x^8 \oplus x^8}_{=0} \oplus \underbrace{x^9 \oplus x^9}_{=0} \oplus \underbrace{11}_{=0} \oplus x^{11} \oplus x^{12} \oplus x^{10} \\
&= 1 \oplus x^6 \oplus x^{12} \oplus x^{10} \\
&= E\left(x^2\right) \\
&= 1 \oplus x^6 \oplus \underbrace{x^7}_{=1} x^5 \oplus \underbrace{x^7}_{=1} x^3 \\
&= 1 \oplus x^3 \oplus x^6 \oplus x^5 \\
&= E(x).
\end{aligned}
\tag{1.1033}
$$

Let us look at a cyclic $(n, k, d_{\min})$ binary linear block code $\mathbb{V}$ again, and prove the following theorem.

Theorem 1.61 (A Cyclic Code has a Unique Idempotent).

1 *A cyclic $(n, k, d_{\min})$ binary linear block code $\mathcal{I}$ equal to $\langle g(x) \rangle$ contains a unique idempotent $E(x)$ such that* [4, *Theorem 1, p. 217*]

$$
\mathcal{I} = \left\langle E(x) \right\rangle. \tag{1.1034}
$$

$E(x)$ is equal to $(p(x) \odot g(x))$ for some polynomial $p(x)$, complying with [4, *Theorem 1, p. 217*]

$$
E\left(\alpha^i\right) = 0 \text{ if and only if } g\left(\alpha^i\right) = 0. \tag{1.1035}
$$

2 *Furthermore, $v(x) \in \mathcal{I}$ if and only if $v(x) \odot E(x)$ is equal to $v(x)$* [4, *Theorem 1, p. 217*].

Proof.
Claim 1
Let

$$x^n \oplus 1 = g(x) \odot h(x), \tag{1.1036}$$

where $g(x)$ and $h(x)$ are relatively prime [4, Theorem 1, p. 217]. According to Corollary 1.12, there exist polynomials $p(x)$ and $q(x)$ such that [4, Theorem 1, equation (1), p. 217]

$$p(x) \odot g(x) \oplus q(x) \odot h(x) = 1 \quad \text{in } \mathbb{F}_2[x]. \tag{1.1037}$$

Set $E(x)$ equal to $p(x) \odot g(x)$ [4, Theorem 1, p. 217]. Then we obtain [4, Theorem 1, p. 217]

$$p(x) \odot g(x) \odot \underbrace{\left[p(x) \odot g(x) \oplus q(x) \odot h(x)\right]}_{=1} = p(x) \odot g(x), \tag{1.1038}$$

which is equal to

$$\begin{aligned} &\underbrace{p(x) \odot g(x)}_{=E(x)} \odot \underbrace{p(x) \odot g(x)}_{=E(x)} \oplus p(x) \odot \underbrace{g(x) \odot h(x)}_{=0 \text{ in } \mathbb{F}_2[x]_{n-1}} \odot q(x) \\ &\quad = \left[E(x)\right]^2 = E(x) \text{ in } \mathbb{F}_2[x]_{n-1}, \quad n \in \mathbb{N}^*. \end{aligned} \tag{1.1039}$$

Thus, $E(x)$ is an idempotent [4, Theorem 1, p. 217].

Since $g(x)$ and $h(x)$ are relatively prime, an nth root of unity is a zero of either $g(x)$ or $h(x)$, but not of both [4, Theorem 1, p. 217]. Therefore, an nth root of unity, which is a zero of $p(x)$, must also be a zero of $g(x)$ [4, p. 217]. Since $p(x)$ does not introduce any new zeros (cf. Theorem 1.57), $E(x)$ and $g(x)$ generate the same code [4, Theorem 1, p. 217].

Claim 2
If $v(x)$ is equal to $(v(x) \odot E(x))$, then clearly $v(x) \in \mathcal{I}$ [4, Theorem 1, p. 217]. Conversely, if $v(x) \in \mathcal{I}$, then [4, Theorem 1, p. 217f.]

$$v(x) = b(x) \odot E(x), \tag{1.1040}$$

and

$$v(x) \odot E(x) = b(x) \odot E(x) \odot E(x) = b(x) \odot \left[E(x)\right]^2 = b(x) \odot E(x) = v(x). \tag{1.1041}$$

Uniqueness
Finally, let us show that $E(x)$ is the unique idempotent, which generates $\mathcal{I}$. Assume there was another idempotent, $F(x)$ say, which generates $\mathcal{I}$ [4, Theorem 1, p. 218]. Then, according to Claim 2, we have [4, Theorem 1, p. 218]

$$F(x) \odot E(x) = F(x), \tag{1.1042}$$

and at the same time [4, Theorem 1, p. 218]

$$F(x) \odot E(x) = E(x), \tag{1.1043}$$

which yields [4, Theorem 1, p. 218]

$$F(x) = E(x). \tag{1.1044}$$

□

Owing to (1.1014) and Theorem 1.59, we have [4, Lemma 2, p. 218]

$$E(\alpha^i) = \bigoplus_{k=0}^{n-1} \varepsilon_k \alpha^{i \cdot k} = 0 \text{ or } 1, \quad j \in \{0, 1 \cdots (n-1)\}, \quad n \in \mathbb{N}^*. \tag{1.1045}$$

Example 1.55. We will set out from the three idempotents of $\mathbb{F}_2[x]_7$,

$$\begin{aligned} E_1(x) &= x \oplus x^2 \oplus x^4, \\ E_2(x) &= 1, \\ E_3(x) &= x^3 \oplus x^6 \oplus x^5 \end{aligned} \tag{1.1046}$$

that we already considered in Example 1.53. Clearly, we have

$$\begin{aligned} E_1(1) &= 1 \oplus 1 \oplus 1 = 1, \\ E_2(1) &= 1, \\ E_3(1) &= 1 \oplus 1 \oplus 1 = 1. \end{aligned} \tag{1.1047}$$

Let us now look at x equal to $\alpha \in \mathbb{F}_8$. Since in $\mathbb{F}_8$, we have, e. g.,

$$\alpha^3 = 1 \oplus \alpha, \tag{1.1048}$$

we find

$$\begin{aligned} E_1(\alpha) &= \alpha \oplus \alpha^2 \oplus \alpha^4 = \alpha \odot \left(1 \oplus \alpha \oplus \alpha^3\right) = \alpha \odot (\underbrace{1 \oplus \alpha \oplus 1 \oplus \alpha}_{=0}) = 0 \\ E_2(\alpha) &= 1, \\ E_3(\alpha) &= \alpha^3 \oplus \alpha^6 \oplus \alpha^5 = \alpha^3(1 \oplus \alpha^3 \oplus \alpha^2) \\ &= (1 \oplus \alpha) \odot \left(\underbrace{1 \oplus 1}_{=0} \oplus \alpha \oplus \alpha^2\right) \\ &= \alpha \oplus \underbrace{\alpha^2 \oplus \alpha^2}_{=0} \oplus \alpha^3 = \underbrace{\alpha \oplus \alpha}_{=0} \oplus 1 = 1. \end{aligned} \tag{1.1049}$$

Owing to Theorem 1.60, we have [4, Lemma 2, p. 218]

$$\varepsilon_i = \bigoplus_{j=0}^{n-1} E(\alpha^j) \odot \alpha^{-i \cdot j} = \bigoplus_{s} \bigoplus_{j \in \mathbb{C}_s} \alpha^{-i \cdot j}, \quad i \in \{0, 1 \cdots (n-1)\}, \quad n \in \mathbb{N}^*, \tag{1.1050}$$

as coefficients of $E(x)$. Hence, we obtain [4, Lemma 4, p. 219]

$$E(x) = \bigoplus_{k=0}^{n-1} \varepsilon_k \odot x^k = \bigoplus_{s} \bigoplus_{j \in \mathbb{C}_s} x^j, \quad n \in \mathbb{N}^*. \tag{1.1051}$$

Let the polynomial $a(x) \in \mathbb{F}_2[x]_{n-1}$, $n \in \mathbb{N}^*$ be [4, p. 218]

$$a(x) = a_0 \oplus a_1 \odot x \oplus a_2 \odot x^2 \oplus \cdots \oplus a_{n-1} \odot x^{n-1} = \bigoplus_{j=0}^{n-1} a_j \odot x^j, \quad n \in \mathbb{N}^*. \tag{1.1052}$$

Note that x^n is replaced by 1 in $\mathbb{F}_2[x]_{n-1}, n \in \mathbb{N}^*$. We now define $a^*(x) \in \mathbb{F}_2[x]_{n-1}, n \in \mathbb{N}^*$, as follows:

$$a^*(x) = a_0 \oplus a_1 \odot x^{n-1} \oplus a_2 \odot x^{n-2} \oplus \cdots \oplus a_{n-1} \odot x = \bigoplus_{j=0}^{n-1} a_j \odot x^{n-j}, \quad n \in \mathbb{N}^*. \tag{1.1053}$$

Lemma 1.1 ($E^*(x)$ is an Idempotent). *If $E(x)$ is an idempotent, then so is $E^*(x)$ [4, Lemma 4, p. 219].*

Proof. Using (1.1051), i. e.,

$$E(x) = \bigoplus_s \bigoplus_{j \in \mathbb{C}_s} x^j, \quad n \in \mathbb{N}^*, \tag{1.1054}$$

we yield

$$E^*(x) = \bigoplus_s \bigoplus_{j \in \mathbb{C}_s} x^{-j} = \bigoplus_s \bigoplus_{j \in \mathbb{C}_{-s}} x^j, \quad n \in \mathbb{N}^*. \tag{1.1055}$$

□

Theorem 1.62 ($(1 \oplus E(x))^*$ is the Idempotent of the Dual Code). *If the cyclic $(n, k, d_{\min})$ binary linear block code $\mathbb{V}$ has the idempotent $E(x)$, then its dual code $\mathbb{V}^\perp$ has the idempotent $(1 \oplus E(x))^*$ [4, Theorem 5, p. 219].*

Proof. Let $\alpha_1, \alpha_2 \cdots \alpha_n$, $n \in \mathbb{N}^*$ be the nth roots of unity [4, Theorem 5, p. 219]. Now, assume that $\alpha_1, \alpha_2 \cdots \alpha_t$, $1 \le t < n, n \in \mathbb{N}^*$ are the zeros of $\mathcal{I}$, and hence of the cyclic $(n, k, d_{\min})$ binary linear block code $\mathbb{V}$, i. e., [4, Theorem 5, p. 219]

$$E(\alpha_i) = 0, \quad i \in \{1, 2 \cdots t\}, \quad 1 \le t < n, \quad n \in \mathbb{N}^*, \tag{1.1056}$$

and [4, Theorem 5, p. 219]

$$E(\alpha_i) = 1, \quad i \in \{(t+1), (t+2) \cdots n\}, \quad 1 \le t < n, \quad n \in \mathbb{N}^*. \tag{1.1057}$$

Clearly, $(1 \oplus E(x))$ has the zeros $\alpha_{t+1}, \alpha_{t+2} \cdots \alpha_n$, $1 \le t < n, n \in \mathbb{N}^*$, and $(1 \oplus E(x))^*$ has the zeros $\alpha_{t+1}^{-1}, \alpha_{t+2}^{-1} \cdots \alpha_n^{-1}$, $1 \le t < n, n \in \mathbb{N}^*$ [4, Theorem 5, p. 219]. Taking Theorem 1.58 into account, these are the zeros of the dual code $\mathbb{V}^\perp$ [4, Theorem 5, p. 219]. □

1.13.5 Minimal Ideals and Primitive Idempotents

A *minimal ideal* is an ideal, which does not contain any smaller nonzero ideal [4, p. 219]. The corresponding cyclic $(n, k, d_{\min})$ binary linear block code $\mathbb{V}$ is called a minimal or irreducible code, and the idempotent of the minimal ideal is called a *primitive idempotent* [4, p. 219]. The nonzeros of a minimal ideal must be $\{\alpha^i \mid i \in \mathbb{C}_s\}$ for some cyclotomic coset $\mathbb{C}_s$ [4, p. 219]. We denote this minimal ideal by $\mathcal{M}_s$, and the corresponding primitive idempotent by $\theta_s(x)$ [4, p. 219],

$$\mathcal{M}_s = \langle \theta_s(x) \rangle. \tag{1.1058}$$

Thus, [4, p. 219]

$$\theta_s(\alpha^j) = \begin{cases} 1 & \text{if } j \in \mathbb{C}_s, \\ 0 & \text{otherwise.} \end{cases} \quad (1.1059)$$

In particular, $\theta_0(x)$ has the single nonzero x equal to 1 and is given by [4, p. 219]

$$\theta_0(x) = \frac{x^n \oplus 1}{x \oplus 1} = \bigoplus_{i=0}^{n-1} x^i, \quad n \in \mathbb{N}^*. \quad (1.1060)$$

Theorem 1.63 (Construction Rule for Primitive Idempotents). *The primitive idempotent $\theta_s(x)$ associated with the cyclotomic coset $\mathbb{C}_s$ is given by* [4, *Theorem 6, p. 220*]

$$\theta_s(x) = \bigoplus_{i=0}^{n-1} \varepsilon_i \odot x^i, \quad (1.1061)$$

where [4, *Theorem 6, p. 220*]

$$\varepsilon_i = \bigoplus_{j \in \mathbb{C}_s} \alpha^{-i \cdot j}, \quad i \in \{0, 1 \cdots (n-1)\}, \quad n \in \mathbb{N}^*. \quad (1.1062)$$

Proof. It follows from Theorem 1.60 that [4, p. 220]

$$\varepsilon_i = \bigoplus_{j=0}^{n-1} \theta_s(\alpha^j) \odot \alpha^{-i \cdot j} = \bigoplus_{j \in \mathbb{C}_s} \alpha^{-i \cdot j}, \quad i \in \{0, 1 \cdots (n-1)\}, \quad n \in \mathbb{N}^*. \quad (1.1063)$$

□

Combining (1.1061) with (1.1062), we obviously obtain [4, p. 221]

$$\theta_s(x) = \bigoplus_{i=0}^{n-1} \underbrace{\left(\bigoplus_{j \in \mathbb{C}_s} \alpha^{-i \cdot j} \right)}_{= \varepsilon_i} \odot x^i, \quad n \in \mathbb{N}^*. \quad (1.1064)$$

Example 1.56. According to Example 1.47, there is one cyclotomic coset for $m = 1$, i. e., for $\mathbb{F}_2$, namely

$$\mathbb{C}_0 = \{0\}. \quad (1.1065)$$

Furthermore, we have n equal to $(2^m - 1)$, which is 1. There is only one primitive idempotent, $\theta_0(x)$, which is given by

$$\theta_0(x) = \bigoplus_{i=0}^{0} x^i = 1 \quad (1.1066)$$

according to (1.1060). Note that

$$\bigoplus_s \theta_s(x) = \theta_0(x) = 1. \quad (1.1067)$$

Looking at (1.1059), we find that $\theta_0(\alpha^0)$ is equal to 1.

In the case of m equal to 2, n is equal to $(2^m - 1)$, which is 3. According to Example 1.47, there are two cyclotomic cosets for $m = 2$, i. e., for $\mathbb{F}_{2^2}$ equal to $\mathbb{F}_4$, namely

$$\mathbb{C}_0 = \{0\}, \quad \mathbb{C}_1 = \{1, 2\}. \tag{1.1068}$$

There are two primitive idempotents, namely

$$\theta_0(x) = \bigoplus_{i=0}^{2} x^i = 1 \oplus x \oplus x^2 \tag{1.1069}$$

according to (1.1060) and $\theta_1(x)$, which we will come to in a moment. According to Table 1.12, we have

$$\begin{aligned} a^3 &= 1, \\ a^2 &= a^{-1}, \\ a^1 &= a^{-2}, \end{aligned} \tag{1.1070}$$

and

$$a^2 = 1 \oplus a. \tag{1.1071}$$

Clearly,

$$\theta_0\left(a^0\right) = \bigoplus_{i=0}^{2} \left(a^0\right)^i = \bigoplus_{i=0}^{2} 1 = 1 \tag{1.1072}$$

because a^0 equal to 1 belongs to the cyclotomic coset $\mathbb{C}_0$.

Furthermore, we obtain

$$\theta_0(a) = 1 \oplus a \oplus a^2 = 1 \oplus a \oplus 1 \oplus a = \underbrace{1 \oplus 1}_{=0} \oplus \underbrace{a \oplus a}_{=0} = 0 \tag{1.1073}$$

and

$$\begin{aligned} \theta_0\left(a^2\right) &= 1 \oplus a^2 \oplus a^4 = 1 \oplus 1 \oplus a \oplus (1 \oplus a)^2 = 1 \oplus (1 \oplus a) \odot (\underbrace{1 \oplus 1}_{=0} \oplus a) \\ &= 1 \oplus a \oplus a^2 = 1 \oplus a \oplus 1 \oplus a = \underbrace{1 \oplus 1}_{=0} \oplus \underbrace{a \oplus a}_{=0} \\ &= 0 \end{aligned} \tag{1.1074}$$

because a and a^2 belong to the cyclotomic coset $\mathbb{C}_1$ but not to the cyclotomic coset $\mathbb{C}_0$.

Also,

$$\theta_0\left(x^2\right) = 1 \oplus x^2 \oplus x^4 = 1 \oplus x^2 \oplus \underbrace{x^3}_{=1} x = 1 \oplus x \oplus x^2 = \theta_0(x) \tag{1.1075}$$

and

$$\begin{aligned} \left[\theta_0(x)\right]^2 &= \left[1 \oplus x \oplus x^2\right]^2 \\ &= \left(1 \oplus x \oplus x^2\right) \odot \left(1 \oplus x \oplus x^2\right) \\ &= 1 \oplus x \oplus x^2 \oplus x \oplus x^2 \oplus x^3 \oplus x^2 \oplus x^3 \oplus x^4 \\ &= 1 \oplus \underbrace{x \oplus x}_{=0} \oplus \underbrace{x^2 \oplus x^2}_{=0} \oplus x^2 \oplus \underbrace{x^3 \oplus x^3}_{=0} \oplus x^4 \\ &= 1 \oplus x^2 \oplus x^4 \\ &= 1 \oplus x^2 \oplus \underbrace{x^3}_{=1} \odot x \\ &= 1 \oplus x \oplus x^2 \\ &= \theta_0(x). \end{aligned} \tag{1.1076}$$

Using (1.1062), we find

$$\varepsilon_0 = \bigoplus_{j \in \mathbb{C}_1} \alpha^0 = \bigoplus_{j \in \{1,2\}} \alpha^0 = 1 \oplus 1 = 0. \tag{1.1077}$$

Furthermore, we find

$$\varepsilon_1 = \bigoplus_{j \in \mathbb{C}_1} \alpha^{-j} = \bigoplus_{j \in \{1,2\}} \alpha^{-j} = \alpha^{-1} \oplus \alpha^{-2} = \alpha^2 \oplus \alpha = 1 \oplus \underbrace{\alpha \oplus \alpha}_{=0} = 1. \tag{1.1078}$$

Since

$$\varepsilon_2 = \varepsilon_1, \tag{1.1079}$$

we have

$$\varepsilon_2 = 1. \tag{1.1080}$$

According to (1.1061), we find

$$\theta_1(x) = \bigoplus_{i=0}^{2} \varepsilon_i x^i = \varepsilon_0 x^0 \oplus \varepsilon_1 x^1 \oplus \varepsilon_2 x^2 = x \oplus x^2. \tag{1.1081}$$

Clearly,

$$\theta_1(\alpha) = \alpha \oplus \alpha^2 = 1 \tag{1.1082}$$

and

$$\begin{aligned} \theta_1\left(\alpha^2\right) &= \alpha^2 \oplus \alpha^4 = 1 \oplus \alpha \oplus (1 \oplus \alpha)^2 = (1 \oplus \alpha) \odot (\underbrace{1 \oplus 1}_{=0} \oplus \alpha) \\ &= \alpha \oplus \alpha^2 = 1, \end{aligned} \tag{1.1083}$$

because α and α^2 belong to the cyclotomic coset $\mathbb{C}_1$. In addition, we find

$$\theta_1\left(\alpha^0\right) = 1 \oplus 1 = 0, \tag{1.1084}$$

because α^0 belongs to the cyclotomic coset $\mathbb{C}_0$ but not to the cyclotomic coset $\mathbb{C}_1$. Also, we yield

$$\begin{aligned} \left(\theta_1(x)\right)^2 &= \left(x \oplus x^2\right) \odot \left(x \oplus x^2\right) \\ &= x^2 \oplus \underbrace{x^3 \oplus x^3}_{=0} \oplus x^4 \\ &= x^2 \oplus x^4 \\ &= \theta_1\left(x^2\right) \\ &= x^2 \oplus \underbrace{x^3}_{=1} \odot x \\ &= x \oplus x^2 \\ &= \theta_1(x). \end{aligned} \tag{1.1085}$$

Note that

$$\bigoplus_{s} \theta_s(x) = \theta_0(x) \oplus \theta_1(x) = \underbrace{1 \oplus x \oplus x^2}_{=\theta_0(x)} \oplus \underbrace{x \oplus x^2}_{=\theta_1(x)} = 1 \oplus \underbrace{x \oplus x}_{=0} \oplus \underbrace{x^2 \oplus x^2}_{=0} = 1. \tag{1.1086}$$

In the case of m equal to 3, n is equal to $(2^m - 1)$, which is 7. According to Example 1.47, there are three cyclotomic cosets for $m = 2$, i. e., for $\mathbb{F}_{2^3}$ equal to $\mathbb{F}_8$, namely

$$\mathbb{C}_0 = \{0\}, \quad \mathbb{C}_1 = \{1, 2, 4\}, \quad \mathbb{C}_3 = \{3, 6, 5\}. \tag{1.1087}$$

Consequently, there are three primitive idempotents, namely

$$\theta_0(x) = \bigoplus_{i=0}^{6} x^i = 1 \oplus x \oplus x^2 \oplus x^3 \oplus x^4 \oplus x^5 \oplus x^6 \tag{1.1088}$$

according to (1.1060) and $\theta_1(x)$ and θ_3, which we will come to in a moment.

According to Table 1.10, we have

$$\begin{aligned} a^7 &= 1 \\ a^6 &= a^{-1} \\ a^5 &= a^{-2} \\ a^4 &= a^{-3} \\ a^3 &= a^{-4} \\ a^1 &= a^{-5} \\ a^1 &= a^{-6} \\ a^0 &= a^{-7} \end{aligned} \tag{1.1089}$$

and

$$a^3 = 1 \oplus a. \tag{1.1090}$$

Let us now consider $\theta_1(x)$. Using (1.1062), we find [4, p. 220]

$$\varepsilon_i = \bigoplus_{j \in \mathbb{C}_1} a^{-ij} = \bigoplus_{j \in \{1,2,4\}} a^{-ij} = a^{-i} \oplus a^{-2i} \oplus a^{-4i}, \quad i \in \{1, 2 \cdots 6\}, \tag{1.1091}$$

which yields

$$\begin{aligned} \varepsilon_0 &= 1 \oplus 1 \oplus 1 = 1, \\ \varepsilon_1 &= a^{-1} \oplus a^{-2} \oplus a^{-4} = a^6 \oplus a^5 \oplus a^3 = 1 \oplus a^2 \oplus 1 \oplus a \oplus a^2 \oplus 1 \oplus a = 1, \\ \varepsilon_2 &= \varepsilon_1 = 1, \\ \varepsilon_3 &= a^{-3} \oplus a^{-6} \oplus a^{-12} = a^{-12}\left(a^9 \oplus a^6 \oplus 1\right) = a^2\left(a^2 \oplus 1 \oplus a^2 \oplus 1\right) = 0, \\ \varepsilon_4 &= \varepsilon_2 = 1, \\ \varepsilon_5 &= a^{-5} \oplus a^{-10} \oplus a^{-20} = a^{-20}\left(a^{15} \oplus a^{10} \oplus 1\right) = a^{-20}(a \oplus 1 \oplus a \oplus 1) = 0, \\ \varepsilon_6 &= \varepsilon_3 = 0. \end{aligned} \tag{1.1092}$$

Hence, we have [4, p. 220]

$$\theta_1(x) = 1 \oplus x \oplus x^2 \oplus x^4. \tag{1.1093}$$

Let us now consider $\theta_3(x)$. Using (1.1062), we find [4, p. 220]

$$\varepsilon_i = \bigoplus_{j\in\mathbb{C}_3} \alpha^{-i\cdot j} = \bigoplus_{j\in\{3,6,5\}} \alpha^{-i\cdot j} = \alpha^{-3i} \oplus \alpha^{-6i} \oplus \alpha^{-5i}, \quad i \in \{1, 2 \cdots 6\}. \tag{1.1094}$$

This yields

$$\begin{aligned}
\varepsilon_0 &= 1 \oplus 1 \oplus 1 = 1, \\
\varepsilon_1 &= \alpha^{-3} \oplus \alpha^{-6} \oplus \alpha^{-5} = \alpha \oplus \alpha^2 \oplus \alpha \oplus \alpha^2 = 0, \\
\varepsilon_2 &= \varepsilon_1 = 0, \\
\varepsilon_3 &= \alpha^{-9} \oplus \alpha^{-15} \oplus \alpha^{-18} = \alpha^5 \oplus \alpha^6 \oplus \alpha^3 = 1 \oplus \alpha^2 \oplus \alpha \oplus 1 \oplus \alpha^2 \oplus 1 \oplus \alpha = 1, \\
\varepsilon_4 &= \varepsilon_2 = 0, \\
\varepsilon_5 &= \alpha^{-15} \oplus \alpha^{-25} \oplus \alpha^{-30} = \alpha^5 \alpha^2 = 1, \\
\varepsilon_6 &= \varepsilon_3 = 1.
\end{aligned} \tag{1.1095}$$

Hence, we have [4, p. 220]

$$\theta_3(x) = 1 \oplus x^3 \oplus x^5 \oplus x^6. \tag{1.1096}$$

Now, when using Table 1.11, we have

$$\alpha^3 = 1 \oplus \alpha^2, \tag{1.1097}$$

which does not change $\theta_0(x)$, however, $\theta_1(x)$ and $\theta_3(x)$ are interchanged, [4, p. 220]

Note that

$$\begin{aligned}
\bigoplus_s \theta_s(x) &= \theta_0(x) \oplus \theta_1(x) \oplus \theta_3(x) \\
&= \underbrace{1 \oplus x \oplus x^2 \oplus x^3 \oplus x^4 \oplus x^5 \oplus x^6}_{=\theta_0(x)} \\
&\oplus \underbrace{1 \oplus x \oplus x^2 \oplus x^4}_{=\theta_1(x)} \\
&\oplus \underbrace{1 \oplus x^3 \oplus x^5 \oplus x^6}_{=\theta_3(x)} \\
&= 1 \oplus \underbrace{1 \oplus 1}_{=0} \oplus \underbrace{x \oplus x}_{=0} \oplus \underbrace{x^2 \oplus x^2}_{=0} \oplus \underbrace{x^3 \oplus x^3}_{=0} \\
&\oplus \underbrace{x^4 \oplus x^4}_{=0} \oplus \underbrace{x^5 \oplus x^5}_{=0} \oplus \underbrace{x^6 \oplus x^6}_{=0} \\
&= 1.
\end{aligned} \tag{1.1098}$$

Thus, using different polynomials to define the field has the effect of relabeling the primitive idempotents [4, p. 220].

Theorem 1.64 (Properties of Primitive Idempotents). *The primitive idempotents have the following properties [4, p. 220]:*

1

$$\bigoplus_s \theta_s(x) = 1. \tag{1.1099}$$

2

$$\theta_i(x) \odot \theta_j(x) = 0, \quad j \neq i. \tag{1.1100}$$

3 *$\mathbb{F}_2[x]_{n-1}$, $n \in \mathbb{N}^*$ is the direct sum of the minimal ideals generated by all the primitive idempotents $\theta_s(x)$. Thus, any polynomial $a(x) \in \mathbb{F}_2[x]_{n-1}$, $n \in \mathbb{N}^*$ can be uniquely written in the form*

$$a(x) = \bigoplus_s a_s(x), \tag{1.1101}$$

where $a_s(x)$ is in the ideal generated by $\theta_s(x)$.

4 *If $E(x)$ is an idempotent, then for some values $a_s \in \mathbb{F}_2$, $E(x)$ can be written as*

$$E(x) = \bigoplus_s a_s \odot \theta_s(x). \tag{1.1102}$$

Conversely, any such expression is an idempotent.

Proof.

Claim 1

Using (1.1064), we find [4, p. 221]

$$\begin{aligned}
\bigoplus_s \theta_s(x) &= \bigoplus_s \bigoplus_{i=0}^{n-1} \bigoplus_{j \in \mathbb{C}_s} \alpha^{-i \cdot j} \odot x^i \\
&= \bigoplus_{i=0}^{n-1} x^i \odot \underbrace{\bigoplus_s \bigoplus_{j \in \mathbb{C}_s} \alpha^{-i \cdot j}}_{= \bigoplus_{j=0}^{n-1} \alpha^{-i \cdot j}} \\
&= \bigoplus_{i=0}^{n-1} x^i \odot \bigoplus_{j=0}^{n-1} \alpha^{-i \cdot j} \\
&= x^0 \odot \bigoplus_{j=0}^{n-1} \left[\alpha^{-0}\right]^j \oplus \bigoplus_{i=1}^{n-1} x^i \odot \bigoplus_{j=0}^{n-1} \alpha^{-i \cdot j} \\
&= \underbrace{\bigoplus_{j=0}^{n-1} 1^j}_{=1 \text{ according to Theorem 1.59}} \oplus \bigoplus_{i=1}^{n-1} x^i \odot \underbrace{\bigoplus_{j=0}^{n-1} \alpha^{-i \cdot j}}_{=0 \text{ according to Theorem 1.59}} \\
&= 1.
\end{aligned} \tag{1.1103}$$

Claim 2

Let $\mathcal{M}_i$ be the minimal ideal with the primitive idempotent $\theta_i(x)$ and let $\mathcal{M}_j$ be the minimal ideal with primitive idempotent $\theta_j(x)$ [4, p. 221]. Let $i \neq j$ [4, p. 221]. Then $\mathcal{M}_i\mathcal{M}_j$ equal to $\mathcal{M}_i \cap \mathcal{M}_j$ is a proper subideal of $\mathcal{M}_i$. Hence, $\mathcal{M}_i\mathcal{M}_j$ is the empty set $\emptyset$ for $i \neq j$ [4, p. 221]. Therefore, $\theta_i(x) \odot \theta_j(x) \in \mathcal{M}_i\mathcal{M}_j$, and thus $\theta_i(x) \odot \theta_j(x)$ must be equal to 0 [4, p. 221].

Claim 3

Using (1.1099) leads us to

$$a(x) = a(x) \odot 1 = a(x) \odot \bigoplus_s \theta_s(x) = \bigoplus_s a_s(x), \tag{1.1104}$$

where $a_s(x)$ is in the ideal generated by $\theta_s(x)$ [4, p. 220].

Claim 4

The nonzeros of $E(x)$ are a union of sets of the nonzeros of minimal idempotents [4, p. 220]. The result follows from (1.1059) and the fact that $E(\alpha^j)$ is either 1 or 0 [4, p. 220]. □

Every idempotent $E(x)$ is hence the sum of primitive idempotents $\theta_s(x)$, and every polynomial $a(x)$ in $\mathbb{F}_2[x]_{n-1}$ can be written uniquely as a sum of polynomials $a_s(x)$ from minimal ideals [4, p. 220f.].

Using the definition (1.1053), the polynomial $\theta_s^*(x)$ given by

$$\theta_s^*(x) = \theta_{s'}(x), \quad s' \in \mathbb{C}_{-s}, \tag{1.1105}$$

s' being unique and the smallest possible integer, is also a primitive idempotent with nonzeros $\{\alpha^i \mid i \in \mathbb{C}_{-s}\}$ [4, p. 221].

Example 1.57. We know from Example 1.56 that for $m = 1$, there is only one cyclotomic coset, $\mathbb{C}_0$. Therefore, $s' \in \mathbb{C}_0$ is equal to 0, and we find

$$\theta_0^*(x) = \theta_0(x) = 1. \tag{1.1106}$$

Example 1.58. In the case of $m = 2$, there are two cyclotomic cosets, namely

$$\mathbb{C}_0 = \{0\}, \quad \mathbb{C}_1 = \{1, 2\}; \tag{1.1107}$$

cf. Example 1.56. In the case of $s = 0$, $s' \in \mathbb{C}_0$ is equal to 0 as well, and we yield

$$\theta_0^*(x) = \theta_0(x) = 1 \oplus x \oplus x^2; \tag{1.1108}$$

cf. Example 1.56. In the case of $s = 1$, $s' \in \mathbb{C}_{-1}$. Since

$$(-1) \bmod \left(2^2 - 1\right) = (-1 + 3) \bmod \left(2^2 - 1\right) = 2 \in \mathbb{C}_1, \tag{1.1109}$$

we readily see that $\mathbb{C}_{-1}$ is equal to $\mathbb{C}_1$ and s' is equal to min(1, 2), which is 1. Therefore, we find

$$\theta_1^*(x) = \theta_1(x) = x \oplus x^2; \tag{1.1110}$$

cf. Example 1.56.

Example 1.59. In the case of $m = 3$, there are three cyclotomic cosets, namely

$$\mathbb{C}_0 = \{0\}, \quad \mathbb{C}_1 = \{1, 2, 4\}, \quad \mathbb{C}_3 = \{3, 6, 5\}; \tag{1.1111}$$

cf. Example 1.56. In the case of $s = 0$, $s' \in \mathbb{C}_0$ is equal to 0 as well, and we find

$$\theta_0^*(x) = \theta_0(x) = 1 \oplus x \oplus x^2 \oplus x^3 \oplus x^4 \oplus x^5 \oplus x^6; \tag{1.1112}$$

cf. Example 1.56. For $s = 1$, $s' \in \mathbb{C}_{-1}$. Since

$$(-1) \bmod \left(2^3 - 1\right) = (-1 + 7) \bmod \left(2^2 - 1\right) = 6 \in \mathbb{C}_3, \tag{1.1113}$$

we readily see that $\mathbb{C}_{-1}$ is equal to $\mathbb{C}_3$ and s' is equal to min(3, 6, 5), which is 3. Therefore, we find

$$\theta_1^*(x) = \theta_3(x) = 1 \oplus x^3 \oplus x^5 \oplus x^6; \tag{1.1114}$$

cf. Example 1.56. In the case of $s = 3$, $s' \in \mathbb{C}_{-3}$. Since

$$(-3) \bmod \left(2^3-1\right) = (-3+7) \bmod \left(2^2-1\right) = 4 \in \mathbb{C}_1, \tag{1.1115}$$

we readily see that $\mathbb{C}_{-3}$ is equal to $\mathbb{C}_1$ and s' is equal to min(1, 2, 4), which is 1. Therefore, we find

$$\theta_1^*(x) = \theta_3(x) = 1 \oplus x \oplus x^2 \oplus x^4; \tag{1.1116}$$

cf. Example 1.56.

Example 1.60. In the case of $m = 4$, there are five cyclotomic cosets, namely [4, p. 104]

$$\begin{aligned} \mathbb{C}_0 &= \{0\}, \\ \mathbb{C}_1 &= \{1,2,4,8\}, \\ \mathbb{C}_3 &= \{3,6,12,9\}, \\ \mathbb{C}_5 &= \{5,10\}, \\ \mathbb{C}_7 &= \{7,14,13,11\}; \end{aligned} \tag{1.1117}$$

cf. Example 1.48. We will consider $\mathbb{F}_{2^4}$ equal to $\mathbb{F}_{16}$ defined by $\alpha^4 \oplus \alpha \oplus 1 = 0$, illustrated in Example 1.44. In the case of $s = 0$, $s' \in \mathbb{C}_0$ is equal to 0 as well, and we find [4, p. 221]

$$\theta_0^*(x) = \theta_0(x) = \bigoplus_{j=0}^{14} x^j = 1 \oplus x \oplus x^2 \oplus \cdots \oplus x^{14}. \tag{1.1118}$$

For $s = 1$, $s' \in \mathbb{C}_{-1}$. Since

$$(-1) \bmod \left(2^4-1\right) = (-1+15) \bmod 15 = 14 \in \mathbb{C}_7, \tag{1.1119}$$

we readily see that $\mathbb{C}_{-1}$ is equal to $\mathbb{C}_7$ and s' is equal to min(7, 14, 13, 11), which is 7. Therefore, we find [4, p. 221]

$$\theta_1^*(x) = \theta_7(x) = x^3 \oplus x^6 \oplus x^7 \oplus x^9 \oplus x^{11} \oplus x^{12} \oplus x^{13} \oplus x^{14}. \tag{1.1120}$$

In the case of $s = 3$, $s' \in \mathbb{C}_{-3}$. Since

$$(-3) \bmod \left(2^4-1\right) = (-3+15) \bmod (15) = 12 \in \mathbb{C}_3, \tag{1.1121}$$

we readily see that $\mathbb{C}_{-3}$ is equal to $\mathbb{C}_3$ and s' is equal to min(3, 6, 12, 9), which is 3. Therefore, we find [4, p. 221]

$$\begin{aligned} \theta_3^*(x) &= \theta_3(x) \\ &= x \oplus x^2 \oplus x^3 \oplus x^4 \oplus x^6 \oplus x^7 \oplus x^8 \oplus x^9 \oplus x^{11} \oplus x^{12} \oplus x^{13} \oplus x^{14}. \end{aligned} \tag{1.1122}$$

In the case of $s = 5$, $s' \in \mathbb{C}_{-5}$. Since

$$(-5) \bmod \left(2^4-1\right) = (-5+15) \bmod (15) = 10 \in \mathbb{C}_5, \tag{1.1123}$$

we readily see that $\mathbb{C}_{-5}$ is equal to $\mathbb{C}_5$ and s' is equal to min(5, 10), which is 5. Therefore, we find [4, p. 221]

$$\theta_5^*(x) = \theta_5(x) = x \oplus x^2 \oplus x^4 \oplus x^5 \oplus x^7 \oplus x^8 \oplus x^{10} \oplus x^{11} \oplus x^{13} \oplus x^{14}. \tag{1.1124}$$

In the case of $s = 7$, $s' \in \mathbb{C}_{-7}$. Since

$$(-7) \bmod \left(2^4 - 1\right) = (-7 + 15) \bmod (15) = 8 \in \mathbb{C}_1, \tag{1.1125}$$

we readily see that $\mathbb{C}_{-7}$ is equal to $\mathbb{C}_1$ and s' is equal to min(1, 2, 4, 8), which is 1. Therefore, we find [4, p. 221]

$$\theta_7^*(x) = \theta_1(x) = x \oplus x^2 \oplus x^3 \oplus x^4 \oplus x^6 \oplus x^8 \oplus x^9 \oplus x^{12}. \tag{1.1126}$$

Example 1.61. In the case of $m = 5$, there are five cyclotomic cosets, namely [4, p. 104]

$$\begin{aligned}
\mathbb{C}_0 &= \{0\},\\
\mathbb{C}_1 &= \{1, 2, 4, 8, 16\},\\
\mathbb{C}_3 &= \{3, 6, 12, 24, 17\},\\
\mathbb{C}_5 &= \{5, 10, 20, 9, 18\},\\
\mathbb{C}_7 &= \{7, 14, 28, 25, 19\},\\
\mathbb{C}_{11} &= \{11, 22, 13, 26, 21\},\\
\mathbb{C}_{15} &= \{15, 30, 29, 27, 23\};
\end{aligned} \tag{1.1127}$$

cf. Example 1.48. We will consider $\mathbb{F}_{2^5}$ equal to $\mathbb{F}_{32}$ defined by $\alpha^5 \oplus \alpha^3 \oplus 1 = 0$, illustrated in Example 1.44. In the case of $s = 0$, $s' \in \mathbb{C}_0$ is equal to 0 as well, and we find [4, p. 222]

$$\theta_0^*(x) = \theta_0(x) = \bigoplus_{j=0}^{30} x^j = 1 \oplus x \oplus x^2 \oplus \cdots \oplus x^{30}. \tag{1.1128}$$

For $s = 1$, $s' \in \mathbb{C}_{-1}$. Since

$$(-1) \bmod \left(2^5 - 1\right) = (-1 + 31) \bmod 31 = 30 \in \mathbb{C}_{15}, \tag{1.1129}$$

we readily see that $\mathbb{C}_{-1}$ is equal to $\mathbb{C}_{15}$ and s' is equal to min(15, 30, 29, 27, 23), which is 15. Therefore, we find [4, p. 222]

$$\begin{aligned}
\theta_1^*(x) &= \theta_{15}(x)\\
&= 1 \oplus x^3 \oplus x^5 \oplus x^6 \oplus x^9 \oplus x^{10} \oplus x^{11} \oplus x^{12} \oplus x^{13}\\
&\quad \oplus x^{17} \oplus x^{18} \oplus x^{20} \oplus x^{21} \oplus x^{22} \oplus x^{24} \oplus x^{26}.
\end{aligned} \tag{1.1130}$$

In the case of $s = 3$, $s' \in \mathbb{C}_{-3}$. Since

$$(-3) \bmod \left(2^5 - 1\right) = (-3 + 31) \bmod 31 = 28 \in \mathbb{C}_7, \tag{1.1131}$$

we readily see that $\mathbb{C}_{-3}$ is equal to $\mathbb{C}_7$ and s' is equal to min(7, 14, 28, 25, 19), which is 7. Therefore, we find [4, p. 222]

$$\begin{aligned}
\theta_3^*(x) &= \theta_7(x)\\
&= 1 \oplus x \oplus x^2 \oplus x^3 \oplus x^4 \oplus x^6 \oplus x^7 \oplus x^8 \oplus x^{12}\\
&\quad \oplus x^{14} \oplus x^{16} \oplus x^{17} \oplus x^{19} \oplus x^{24} \oplus x^{25} \oplus x^{28}.
\end{aligned} \tag{1.1132}$$

In the case of $s = 5$, $s' \in \mathbb{C}_{-5}$. Since

$$(-5) \bmod \left(2^5 - 1\right) = (-5 + 31) \bmod 31 = 26 \in \mathbb{C}_{11}, \tag{1.1133}$$

we readily see that $\mathbb{C}_{-5}$ is equal to $\mathbb{C}_5$ and s' is equal to min(11, 22, 13, 26, 21), which is 11. Therefore, we find [4, p. 222]

$$\begin{aligned}\theta_5^*(x) &= \theta_{11}(x)\\ &= 1 \oplus x \oplus x^2 \oplus x^4 \oplus x^8 \oplus x^{11} \oplus x^{13} \oplus x^{15} \oplus x^{16}\\ &\quad \oplus x^{21} \oplus x^{22} \oplus x^{23} \oplus x^{26} \oplus x^{27} \oplus x^{29} \oplus x^{30}.\end{aligned} \tag{1.1134}$$

In the case of $s = 7$, $s' \in \mathbb{C}_{-7}$. Since

$$(-7) \bmod \left(2^5 - 1\right) = (-7 + 31) \bmod 31 = 24 \in \mathbb{C}_3, \tag{1.1135}$$

we readily see that $\mathbb{C}_{-7}$ is equal to $\mathbb{C}_3$ and s' is equal to min(3, 6, 12, 24, 17), which is 3. Therefore, we find [4, p. 222]

$$\begin{aligned}\theta_7^*(x) &= \theta_3(x)\\ &= 1 \oplus x^3 \oplus x^6 \oplus x^7 \oplus x^{12} \oplus x^{14} \oplus x^{15} \oplus x^{17} \oplus x^{19}\\ &\quad \oplus x^{23} \oplus x^{24} \oplus x^{25} \oplus x^{27} \oplus x^{28} \oplus x^{29} \oplus x^{30}.\end{aligned} \tag{1.1136}$$

In the case of $s = 11$, $s' \in \mathbb{C}_{-11}$. Since

$$(-11) \bmod \left(2^5 - 1\right) = (-11 + 31) \bmod 31 = 20 \in \mathbb{C}_5, \tag{1.1137}$$

we readily see that $\mathbb{C}_{-11}$ is equal to $\mathbb{C}_5$ and s' is equal to min(5, 10, 20, 9, 18), which is 5. Therefore, we find [4, p. 222]

$$\begin{aligned}\theta_{11}^*(x) &= \theta_5(x)\\ &= 1 \oplus x \oplus x^2 \oplus x^3 \oplus x^5 \oplus x^8 \oplus x^9 \oplus x^{10} \oplus x^{15}\\ &\quad \oplus x^{16} \oplus x^{18} \oplus x^{20} \oplus x^{23} \oplus x^{27} \oplus x^{29} \oplus x^{30}.\end{aligned} \tag{1.1138}$$

In the case of $s = 15$, $s' \in \mathbb{C}_{-15}$. Since

$$(-15) \bmod \left(2^5 - 1\right) = (-15 + 31) \bmod 31 = 16 \in \mathbb{C}_1, \tag{1.1139}$$

we readily see that $\mathbb{C}_{-11}$ is equal to $\mathbb{C}_5$ and s' is equal to min(1, 2, 4, 8, 16), which is 1. Therefore, we find [4, p. 222]

$$\begin{aligned}\theta_{15}^*(x) &= \theta_1(x)\\ &= 1 \oplus x^5 \oplus x^7 \oplus x^9 \oplus x^{10} \oplus x^{11} \oplus x^{13} \oplus x^{14} \oplus x^{18} \oplus x^{19}\\ &\quad \oplus x^{20} \oplus x^{21} \oplus x^{22} \oplus x^{25} \oplus x^{26} \oplus x^{28}.\end{aligned} \tag{1.1140}$$

1.14 Some Examples of Cyclic $(n, k, d_{\min})$ Binary Linear Block Codes

1.14.1 Introductory Remark

In Section 1.14, three types of cyclic $(n, k, d_{\min})$ binary linear block codes, namely

- cyclic Hamming codes,
- cyclic simplex codes and

– double-error-correcting Bose–Chaudhuri–Hocquenghem (BCH) codes

will be briefly illustrated. You, dear reader, might ask yourself, why this selection is so sparse. The reason is quite simple. Up to date, only such classical $(n, k, d_{\min})$ binary linear block codes with code lengths up to n equal to 256 bits and with dimensions k lower or equal to 256 bits have been identified [27, Table 1.1, p. 21], [51]. Such classical $(n, k, d_{\min})$ binary linear block codes are much too short for data transmission in advanced mobile communications systems and are hence considered for the error protection of control information, solely. Therefore, the author believes that only a few examples suffice.

1.14.2 Cyclic Hamming Codes

Let us revisit the Hamming codes; cf. Section 1.9.5.

Example 1.62. The (7, 4, 3) Hamming code is a binary linear block code $\mathbb{V}$ with the parity check matrix [4, p. 24]

$$\boldsymbol{H} = \begin{pmatrix} 0 & 0 & 0 & 1 & 1 & 1 & 1 \\ 0 & 1 & 1 & 0 & 0 & 1 & 1 \\ 1 & 0 & 1 & 0 & 1 & 0 & 1 \end{pmatrix}. \tag{1.1141}$$

We have taken the columns in the natural order of increasing binary numbers, the first column representing $(0 \cdot 2^2 + 0 \cdot 2^1 + 1 \cdot 2^0) = 1$, the second column representing $(0 \cdot 2^2 + 1 \cdot 2^1 + 0 \cdot 2^0) = 2$, the third column representing $(0 \cdot 2^2 + 1 \cdot 2^1 + 1 \cdot 2^0) = 3$, the fourth column representing $(1 \cdot 2^2 + 0 \cdot 2^1 + 0 \cdot 2^0) = 4$, the fifth column representing $(1 \cdot 2^2 + 0 \cdot 2^1 + 1 \cdot 2^0) = 5$, the sixth column representing $(1 \cdot 2^2 + 1 \cdot 2^1 + 0 \cdot 2^0) = 6$ and the seventh column representing $(1 \cdot 2^2 + 1 \cdot 2^1 + 1 \cdot 2^0) = 7$.

To obtain the parity check matrix in the standard form [4, equation (2), p. 2; equation (39), p. 24] of (1.184), we take the columns in a different order, namely (4, 2, 1, 3, 5, 6, 7), yielding [4, p. 24]

$$\boldsymbol{H}' = \begin{pmatrix} 1 & 0 & 0 & 0 & 1 & 1 & 1 \\ 0 & 1 & 0 & 1 & 0 & 1 & 1 \\ 0 & 0 & 1 & 1 & 1 & 0 & 1 \end{pmatrix}. \tag{1.1142}$$

$\boldsymbol{H}'$ represents an equivalent (7, 4, 3) Hamming code $\mathbb{V}'$.

A further equivalent (7, 4, 3) Hamming code $\mathbb{V}''$ has the parity check matrix [4, p. 25]

$$\boldsymbol{H}'' = \begin{pmatrix} 1 & 1 & 1 & 0 & 1 & 0 & 0 \\ 0 & 1 & 1 & 1 & 0 & 1 & 0 \\ 0 & 0 & 1 & 1 & 1 & 0 & 1 \end{pmatrix}, \tag{1.1143}$$

which is obviously a right-circulant matrix, the first row being

$$\begin{pmatrix} 1 & 1 & 1 & 0 & 1 & 0 & 0 \end{pmatrix}. \tag{1.1144}$$

Setting out from Definition 1.39, we find that the parity check matrix of a binary Hamming code of length n equal to $(2^h - 1)$, $h \in \{2, 3, 4 \cdots\}$ has as columns all $(2^h - 1)$ distinct nonzero h-tuples [4, p. 191].

Now, if α is a primitive element of $\mathbb{F}_{2^h}$, $h \in \{2,3,4\cdots\}$, then $1, \alpha, \alpha^2, \alpha^3 \cdots \alpha^{2^h-2}$, $h \in \{2,3,4\cdots\}$ are distinct and can be represented by distinct nonzero binary h-tuples [4, p. 191].

So, the cyclic ($n = 2^h - 1, k = 2^h - h - 1, d_{\min} = 3$) Hamming code has a parity check matrix, which can be taken to be

$$\boldsymbol{H} = (1, \alpha, \alpha^2, \alpha^3 \cdots \alpha^{2^h-2}), \tag{1.1145}$$

where each entry is to be replaced by the corresponding column vector of h 0's and 1's [4, p. 191f.].

Example 1.63. Let us consider h equal to 3 and $\mathbb{F}_{2^3}$ in the form of Table 1.10 with the primitive polynomial

$$\pi_1(x) = 1 \oplus x \oplus x^3, \tag{1.1146}$$

which is the minimal polynomial $M^{(1)}(x)$.

With

$$2^h - 2 = 2^3 - 2 = 6 \tag{1.1147}$$

and using (1.1145), we find

$$\boldsymbol{H}_{M^{(1)}(x)} = (1, \alpha, \alpha^2, \alpha^3, \cdots, \alpha^6) = \begin{pmatrix} 0 & 0 & 1 & 0 & 1 & 1 & 1 \\ 0 & 1 & 0 & 1 & 1 & 1 & 0 \\ 1 & 0 & 0 & 1 & 0 & 1 & 1 \end{pmatrix}, \tag{1.1148}$$

where

$$1 \oplus \alpha \oplus \alpha^3 = 0. \tag{1.1149}$$

A codeword $\boldsymbol{v}$ fulfills

$$\boldsymbol{v}\boldsymbol{H}^{\mathrm{T}}_{M^{(1)}(x)} = \boldsymbol{0}_3, \tag{1.1150}$$

i. e.,

$$(v_0, v_1, v_2, v_3, v_4, v_5, v_6, v_7) \begin{pmatrix} 0 & 0 & 1 \\ 0 & 1 & 0 \\ 1 & 0 & 0 \\ 0 & 1 & 1 \\ 1 & 1 & 0 \\ 1 & 1 & 1 \\ 1 & 0 & 1 \end{pmatrix} = (0, 0, 0), \tag{1.1151}$$

which yields the parity check equations

$$\begin{aligned} v_2 \oplus v_4 \oplus v_5 \oplus v_6 &= 0, \\ v_1 \oplus v_3 \oplus v_4 \oplus v_5 &= 0, \\ v_0 \oplus v_3 \oplus v_5 \oplus v_6 &= 0. \end{aligned} \tag{1.1152}$$

Let the codeword be $\boldsymbol{v}$ equal to $(v_0 v_1 \cdots v_{n-1})$, $n \in \mathbb{N}^*$, and its corresponding polynomial shall be $v(x)$ equal to $v_0 \oplus v_1 \odot x \oplus \cdots \oplus v_{n-1} \odot x^{n-1}$, $n \in \mathbb{N}^*$. In general, using $\boldsymbol{H}$ of (1.1145), the vector $\boldsymbol{v}$ equal to $(v_0, v_1 \cdots v_{n-1})$, $n \in \mathbb{N}^*$ is a codeword if and only if

$$\boldsymbol{v}\boldsymbol{H}^{\mathrm{T}} = \boldsymbol{0}_h, \quad h \in \{2, 3, 4 \cdots\}, \tag{1.1153}$$

which yields

$$\boldsymbol{v} \begin{pmatrix} 1 \\ \alpha \\ \alpha^2 \\ \alpha^3 \\ \vdots \\ \alpha^{2^h-2} \end{pmatrix} = \boldsymbol{0}_h, \quad h \in \{2, 3, 4 \cdots\}, \tag{1.1154}$$

or, equivalently

$$\bigoplus_{i=0}^{n-1} v_i \odot \alpha^i = 0 \quad \Leftrightarrow \quad v(\alpha) = 0, \tag{1.1155}$$

which means if and only if the minimal polynomial $M^{(1)}(x)$ divides $v(x)$ [4, p. 192]. Clearly, this proves the following theorem.

Theorem 1.65 (Generator Polynomial of a Cyclic Hamming Code). *The cyclic $(n = 2^h - 1, k = 2^h - h - 1, d_{min} = 3)$ Hamming code has the generator polynomial $g(x)$ equal to the minimal polynomial $M^{(1)}(x)$ [4, p. 192].*

Example 1.64. It shall be noted that an equivalent cyclic (7, 4, 3) Hamming code can also be formed using the minimal polynomial

$$M^{(3)}(x) = 1 \oplus x^2 \oplus x^3. \tag{1.1156}$$

We find

$$\boldsymbol{H}_{M^{(3)}(x)} = \begin{pmatrix} 0 & 0 & 1 & 1 & 1 & 0 & 1 \\ 0 & 1 & 0 & 0 & 1 & 1 & 1 \\ 1 & 0 & 0 & 1 & 1 & 1 & 0 \end{pmatrix}, \tag{1.1157}$$

which can be obtained by permuting the columns of $\boldsymbol{H}_{M^{(1)}(x)}$ in the following way (1, 2, 3, 7, 6, 4, 5).

1.14.3 Cyclic Simplex Codes

If n is equal to $(2^h - 1)$, $h \in \{2, 3, 4 \cdots\}$, the primitive idempotent $\theta_1(x)$ generates the cyclic $(2^h - 1, h, 2^{h-1})$ simplex code [4, p. 221]. The nonzero codewords are the $(2^h - 1)$ cyclic shifts of $\theta_1(x)$ [4, p. 221].

If n is equal to $(2^h - 1)$ and $(2^h - 1)$ is prime to s, α^s is also a primitive element, and so $\theta_s(x)$ generates a code equivalent to the above mentioned cyclic simplex code [4, p. 221].

We will see later that cyclic simplex codes can be generated by using m-sequences [4, p. 89].

1.14.4 Double-error-correcting Bose–Chaudhuri–Hocquenghem (BCH) Codes

In Section 1.14.4, we will illustrate the double-error-correcting Bose–Chaudhuri–Hocquenghem (BCH) code $\mathbb{V}$ of length n equal to $(2^m - 1)$, $m \in \{3, 4, 5 \cdots\}$ can be defined to have the parity check matrix [4, equation (7), p. 192f.]

$$\boldsymbol{H} = \begin{pmatrix} 1 & \alpha & \alpha^2 & \cdots & \alpha^{2^m-2} \\ 1 & \alpha^3 & \alpha^{3\cdot 2} & \cdots & \alpha^{3\cdot(2^m-2)} \end{pmatrix} = \begin{pmatrix} 1 & \alpha & \alpha^2 & \cdots & \alpha^{2^m-2} \\ 1 & \alpha^3 & \alpha^6 & \cdots & \alpha^{3\cdot(2^m-2)} \end{pmatrix}, \quad m \in \{3, 4, 5 \cdots\}, \tag{1.1158}$$

where again each entry is to be replaced by the corresponding binary m-tuple [4, pp. 88, 192f.].

Clearly, [4, p. 193]

$$\boldsymbol{v}\boldsymbol{H}^{\mathrm{T}} = \boldsymbol{0}_{(n-k)} \tag{1.1159}$$

is required for $\boldsymbol{v}$ to be an admissible codeword. This requirement yields two parity check equations, which have to be fulfilled simultaneously [4, p. 193]

$$\bigoplus_{i=0}^{n-1} v_i \alpha^i = 0 \quad \text{and} \quad \bigoplus_{i=0}^{n-1} v_i \alpha^{3i} = 0, \tag{1.1160}$$

which leads to [4, p. 193]

$$v(\alpha) = 0 \quad \text{and} \quad v(\alpha^3) = 0. \tag{1.1161}$$

This means that the minimal polynomial $M^{(1)}(x)$ of α must divide $v(x)$ and at the same time the minimal polynomial $M^{(3)}(x)$ of α^3 must divide $v(x)$, i. e., [4, p. 193]

$$M^{(1)}(x) \big| \, v(x) \quad \text{and} \quad M^{(3)}(x) \big| \, v(x). \tag{1.1162}$$

The two conditions of (1.1162) must be fulfilled simultaneously and, therefore, the generator polynomial $g(x)$ must be [4, p. 193]

$$g(x) = \operatorname{lcm}\{M^{(1)}(x), M^{(2)}(x)\}, \tag{1.1163}$$

$\mathrm{lcm}\{M^{(1)}(x), M^{(2)}(x)\}$ denoting the *least common multiple (LCM)* (in German *"kleinstes gemeinsames Vielfaches (kgV)"* [11, p. 334]) of the minimal polynomials $M^{(1)}(x)$ and $M^{(3)}(x)$ [15, p. 195], [4, p. 193].

As we already know, the minimal polynomial $M^{(1)}(x)$ of α is irreducible (cf. the property M1 of Table 1.14), and corresponds to the cyclotomic coset $\mathbb{C}_1$; cf. the property M7 of Table 1.14. Furthermore, the minimal polynomial $M^{(3)}(x)$ of α^3 is also irreducible; cf. the property M1 of Table 1.14, and corresponds to the cyclotomic coset $\mathbb{C}_3$; cf. the property M7 of Table 1.14. Since both $M^{(1)}(x)$ and $M^{(3)}(x)$ are irreducible and obviously distinct, we have $\mathrm{lcm}\{M^{(1)}(x), M^{(2)}(x)\}$ is equal to $M^{(1)}(x) \odot M^{(2)}(x)$, and thus [4, Theorem 3, p. 193]

$$g(x) = M^{(1)}(x) \odot M^{(2)}(x). \tag{1.1164}$$

Theorem 1.66 (Generator Polynomial of Double-Error-Correcting BCH Codes). *The double-error-correcting BCH code has the parameters* [4, *Theorem 3, p. 193*]

$$\left(n = 2^m - 1, \quad k = 2^m - 1 - 2m, \quad d_{\min} \geq 5\right), \quad m \in \{3, 4, 5 \cdots\}, \tag{1.1165}$$

and is a cyclic code with generator polynomial [4, *Theorem 3, p. 193*]

$$g(x) = M^{(1)}(x) \odot M^{(3)}(x). \tag{1.1166}$$

Proof. The proof is given above. □

1.15 Permutation Matrices

It is beneficial for the determination of generator matrices of *Reed–Muller (RM) codes* (cf. Sections 1.17 and especially 1.17.4 for the analysis of convolutional turbo codes; cf., e. g., Chapter 2, and of *polar codes*; cf. Chapter 4 to review some aspects of matrix calculus). First, we will consider *permutation matrices*, which form an important component when establishing the generator matrices of polar codes. A comprehensive treatment of permutation matrices can, e. g., be found in [41, Section 2.4, pp. 24–31].

A *permutation matrix* $\boldsymbol{\Sigma}$ allows us to permute the rows of the generator matrix $\boldsymbol{G}$ of a polar code. With the $N \times N$ identity matrix $\boldsymbol{I}_N$, an $N \times N$ permutation matrix $\boldsymbol{\Sigma}_N$ has the following property:

$$\boldsymbol{\Sigma}_N^{\mathrm{T}} \boldsymbol{\Sigma}_N = \boldsymbol{\Sigma}_N \boldsymbol{\Sigma}_N^{\mathrm{T}} = \boldsymbol{I}_N, \quad N \in \mathbb{N}^*. \tag{1.1167}$$

$\boldsymbol{\Sigma}_N$ is termed a *permutation matrix of order N* [41, p. 25]. In general, permutation matrices are orthogonal, i. e., unitary in the case of complex calculus [41, equations (2.4.10)–(2.4.12), p. 26], [52, p. 54f.]. Hence, permutation matrices also preserves the distances between pairs of vectors.

Permutations, especially random permutations, have been important in coding theory; cf., e. g., Chapter 2 on convolutional turbo codes. An important random cyclic permutation scheme was introduced by Sandra *Sattolo* [53]. Taking the random part away, this scheme leads us to the "mod(p) perfect shuffle" $\boldsymbol{\Sigma}_{p,q}$ with $p \cdot q$ being equal to N, $N, p, q \in \mathbb{N}^*$. The "mod(p) perfect shuffle" is defined by where it maps the unit row vectors

$$\boldsymbol{e}_k = \Big(\underbrace{0}_{\text{0th position}} \cdots 0 \underbrace{1}_{k\text{th position}} 0 \cdots \underbrace{0}_{(N-1)\text{th position}} \Big), \quad k \in \{0, 1 \cdots (N-1)\}, \quad N \in \mathbb{N}^*. \tag{1.1168}$$

We have

$$\boldsymbol{e}^{\mathrm{T}}_{\tilde{k}(k)} = \boldsymbol{\Sigma}_{p,q} \boldsymbol{e}^{\mathrm{T}}_k, \quad \tilde{k}(k) = \left\lfloor \frac{k}{p} \right\rfloor + (k \bmod p) \cdot q, \quad p \cdot q = N, \tag{1.1169}$$
$$k \in \{0, 1 \cdots (N-1)\}, \quad N, p, q \in \mathbb{N}^*.$$

Let us restrict ourselves to even values of N, which means that $N/2 \in \mathbb{N}^*$. Choosing p equal to 2 and q equal to $N/2$, we find

$$\tilde{k}(k) = \left\lfloor \frac{k}{2} \right\rfloor + (k \bmod 2) \cdot \frac{N}{2}, \quad k \in \{0, 1 \cdots (N-1)\}, \quad N \in \mathbb{N}^*. \tag{1.1170}$$

Hence, we obtain

$$\begin{aligned}
\tilde{k}(0) &= 0 + [0 \bmod 2] \cdot \frac{N}{2} = 0 + 0 \cdot \frac{N}{2} = 0, \\
\tilde{k}(1) &= 0 + [1 \bmod 2] \cdot \frac{N}{2} = 0 + 1 \cdot \frac{N}{2} = \frac{N}{2}, \\
\tilde{k}(2) &= 1 + [2 \bmod 2] \cdot \frac{N}{2} = 1 + 0 \cdot \frac{N}{2} = 1, \\
\tilde{k}(3) &= 1 + [3 \bmod 2] \cdot \frac{N}{2} = 1 + 1 \cdot \frac{N}{2} = \frac{N}{2} + 1, \\
\tilde{k}(4) &= 2 + [4 \bmod 2] \cdot \frac{N}{2} = 2 + 0 \cdot \frac{N}{2} = 2, \\
\tilde{k}(5) &= 2 + [5 \bmod 2] \cdot \frac{N}{2} = 2 + 1 \cdot \frac{N}{2} = \frac{N}{2} + 2, \\
&\vdots \\
\tilde{k}(N-2) &= [\frac{N-2}{2} \rfloor + [(N-2) \bmod 2] \cdot \frac{N}{2} = \frac{N}{2} - 1, \\
\tilde{k}(N-1) &= \left\lfloor \frac{N-1}{2} \right\rfloor + [(N-1) \bmod 2] \cdot \frac{N}{2} = N - 1,
\end{aligned} \tag{1.1171}$$

and thus

$$(v_0, v_2, v_4 \cdots v_{N-2}, v_1, v_3, v_5 \cdots v_{N-1})^{\mathrm{T}} = \boldsymbol{\Sigma}_{2,\frac{N}{2}} (v_0, v_1 \cdots v_{N-1})^{\mathrm{T}}. \tag{1.1172}$$

This means that

$$\boldsymbol{\Sigma}_{2,\frac{N}{2}} = \begin{pmatrix} 1 & 0 & 0 & 0 & 0 & \cdots & 0 & 0 & 0 \\ 0 & 0 & 1 & 0 & 0 & \cdots & 0 & 0 & 0 \\ 0 & 0 & 0 & 0 & 1 & \cdots & 0 & 0 & 0 \\ \vdots & \vdots & \vdots & \vdots & \vdots & \cdots & \vdots & \vdots & \vdots \\ 0 & 0 & 0 & 0 & 0 & \cdots & 0 & 1 & 0 \\ 0 & 1 & 0 & 0 & 0 & \cdots & 0 & 0 & 0 \\ 0 & 0 & 0 & 1 & 0 & \cdots & 0 & 0 & 0 \\ \vdots & \vdots & \vdots & \vdots & \vdots & \cdots & \vdots & \vdots & \vdots \\ 0 & 0 & 0 & 0 & 0 & \cdots & 1 & 0 & 0 \\ 0 & 0 & 0 & 0 & 0 & \cdots & 0 & 0 & 1 \end{pmatrix} \begin{matrix} \leftarrow \text{row} & 0 \\ \leftarrow \text{row} & 1 \\ \leftarrow \text{row} & 2 \\ & \\ \leftarrow \text{row} & N/2-1 \\ \leftarrow \text{row} & N/2 \\ \leftarrow \text{row} & N/2+1 \\ & \\ \leftarrow \text{row} & N-2 \\ \leftarrow \text{row} & N-1 \end{matrix} \tag{1.1173}$$

and

$$\boldsymbol{\Sigma}^{\mathrm{T}}_{2,\frac{N}{2}} = \begin{pmatrix} 1 & 0 & 0 & \cdots & 0 & 0 & 0 & \cdots & 0 & 0 \\ 0 & 0 & 0 & \cdots & 0 & 1 & 0 & \cdots & 0 & 0 \\ 0 & 1 & 0 & \cdots & 0 & 0 & 0 & \cdots & 0 & 0 \\ 0 & 0 & 0 & \cdots & 0 & 0 & 1 & \cdots & 0 & 0 \\ 0 & 0 & 1 & \cdots & 0 & 0 & 0 & \cdots & 0 & 0 \\ \vdots & \vdots & \vdots & \vdots & \vdots & \vdots & \vdots & \vdots & \vdots & \vdots \\ 0 & 0 & 0 & \cdots & 0 & 0 & 0 & \cdots & 1 & 0 \\ 0 & 0 & 0 & \cdots & 1 & 0 & 0 & \cdots & 0 & 0 \\ 0 & 0 & 0 & \cdots & 0 & 0 & 0 & \cdots & 0 & 1 \end{pmatrix}. \tag{1.1174}$$

Clearly,

$$\boldsymbol{\Sigma}_{2,\frac{N}{2}} \boldsymbol{\Sigma}^{\mathrm{T}}_{2,\frac{N}{2}} = \boldsymbol{\Sigma}^{\mathrm{T}}_{2,\frac{N}{2}} \boldsymbol{\Sigma}_{2,\frac{N}{2}} = \boldsymbol{I}_N. \tag{1.1175}$$

Obviously, all the $N^2, N \in \mathbb{N}^*$ components, i. e., the elements of a permutation matrix are taken from $\mathbb{F}_2$. Furthermore, the Hamming weight of each row of a permutation matrix, also termed the *row weight* [15, p. 857], respectively, *row Hamming weight*, is equal to 1 by construction using the unit row vectors $\boldsymbol{e}_k$, $k \in \{0, 1 \cdots (N-1)\}$, $N \in \mathbb{N}^*$ of (1.1168). Since each one of the unit row vectors $\boldsymbol{e}_k$, $k \in \{0, 1 \cdots (N-1)\}$, $N \in \mathbb{N}^*$ of (1.1168) is used only once in a permutation matrix, also the Hamming weight of each column of a permutation matrix, also termed the *column weight* [15, p. 857], respectively, *column Hamming weight*, is equal to 1.

Example 1.65. Let us consider N equal to 2. We have

$$\boldsymbol{\Sigma}_{2,1} = \boldsymbol{\Sigma}_{2,1}^{\mathrm{T}} = \begin{pmatrix} 1 & 0 \\ 0 & 1 \end{pmatrix} = \boldsymbol{I}_2. \tag{1.1176}$$

Now, let us consider N equal to 4. We have

$$\boldsymbol{\Sigma}_{2,2} = \boldsymbol{\Sigma}_{2,2}^{\mathrm{T}} = \begin{pmatrix} 1 & 0 & 0 & 0 \\ 0 & 0 & 1 & 0 \\ 0 & 1 & 0 & 0 \\ 0 & 0 & 0 & 1 \end{pmatrix}. \tag{1.1177}$$

Now, let us consider N equal to 8. We have

$$\boldsymbol{\Sigma}_{2,4} = \begin{pmatrix} 1 & 0 & 0 & 0 & 0 & 0 & 0 & 0 \\ 0 & 0 & 1 & 0 & 0 & 0 & 0 & 0 \\ 0 & 0 & 0 & 0 & 1 & 0 & 0 & 0 \\ 0 & 0 & 0 & 0 & 0 & 0 & 1 & 0 \\ 0 & 1 & 0 & 0 & 0 & 0 & 0 & 0 \\ 0 & 0 & 0 & 1 & 0 & 0 & 0 & 0 \\ 0 & 0 & 0 & 0 & 0 & 1 & 0 & 0 \\ 0 & 0 & 0 & 0 & 0 & 0 & 0 & 1 \end{pmatrix} \tag{1.1178}$$

and

$$\boldsymbol{\Sigma}_{2,4}^{\mathrm{T}} = \begin{pmatrix} 1 & 0 & 0 & 0 & 0 & 0 & 0 & 0 \\ 0 & 0 & 0 & 0 & 1 & 0 & 0 & 0 \\ 0 & 1 & 0 & 0 & 0 & 0 & 0 & 0 \\ 0 & 0 & 0 & 0 & 0 & 1 & 0 & 0 \\ 0 & 0 & 1 & 0 & 0 & 0 & 0 & 0 \\ 0 & 0 & 0 & 0 & 0 & 0 & 1 & 0 \\ 0 & 0 & 0 & 1 & 0 & 0 & 0 & 0 \\ 0 & 0 & 0 & 0 & 0 & 0 & 0 & 1 \end{pmatrix}. \tag{1.1179}$$

In the case of polar codes, we will only need the matrices $\boldsymbol{\Sigma}_{2,N/2}$ and $\boldsymbol{\Sigma}_{2,N/2}^{\mathrm{T}}$. Let us simplify the notation in the following way:

$$\boldsymbol{\Pi}_N = \boldsymbol{\Sigma}_{2,\frac{N}{2}}^{\mathrm{T}}, \quad \boldsymbol{\Pi}_N^{\mathrm{T}} = \boldsymbol{\Sigma}_{2,\frac{N}{2}}. \tag{1.1180}$$

Using (1.1180), (1.1175) becomes

$$\boldsymbol{\Pi}_N^{\mathrm{T}}\boldsymbol{\Pi}_N = \boldsymbol{\Pi}_N\boldsymbol{\Pi}_N^{\mathrm{T}} = \boldsymbol{I}_N. \tag{1.1181}$$

1.16 Kronecker Product

Next, let us look at the *Kronecker product* [15, p. 114], [4, p. 421], [54, equation (1.21), p. 13], [52, p. 107], [55, equation (2.1), p. 21]. The Kronecker product is named after the German mathematician Leopold *Kronecker*. The Kronecker product always operates on matrices [52, p. 107], which are the representations of tensors of order 2 in vector spaces in their given bases. These vector spaces are of arbitrary dimension, which may be infinite, keyword "Hilbert spaces" [52, Definition 4.16, p. 289]. Such tensors represent linear operators acting on the said vector spaces [52, Definition 4.16, p. 289]. In this regard, the tensor product is an extension of the Kronecker product [52, p. viii].

Since

- a scalar number can be regarded as a special matrix with a single row and a single column, also considered as a tensor of order 0,
- a row vector can be regarded as a special matrix with a single row and N columns, $N \in \mathbb{N}^*$, also considered as a tensor of order 1, and
- a column vector can be regarded as a special matrix with N rows, $N \in \mathbb{N}^*$, and a single column, as well considered as a tensor of order 1,

the Kronecker product can be applied to scalars and vectors. In the case of a scalar, i. e., a tensor of order 0, the Kronecker product is identical with the conventional multiplication.

In Section 1.16, we will consider matrices, which have components, i. e., elements taken from the set $\mathbb{R}$ of real numbers or from the set $\mathbb{C}$ of complex numbers. Since $\mathbb{F}_2 \subset \mathbb{R} \subset \mathbb{C}$ and since the multiplicative operation "$\odot$" in $\mathbb{F}_2$ has the same effect as the conventional multiplication of $\mathbb{R}$ and $\mathbb{C}$, we renounce on using the denotation "$\odot$" in what follows.

With the $m \times m$ matrix $\boldsymbol{A}$, having the components a_{ij} $i, j \in \{0, 1 \cdots (m-1)\}$, $m \in \mathbb{N}^*$, and with the $n \times n$ matrix $\boldsymbol{B}$, having the components b_{kl} $k, l \in \{0, 1 \cdots (n-1)\}$, $n \in \mathbb{N}^*$ the Kronecker product of $\boldsymbol{A}$ and $\boldsymbol{B}$ is the $mn \times mn$ matrix $\boldsymbol{A} \otimes \boldsymbol{B}$ given by [15, p. 114], [4, p. 421], [54, equation (1.21), p. 13], [52, p. 107], [55, equation (2.1), p. 21]

$$\boldsymbol{A} \otimes \boldsymbol{B} = \begin{pmatrix} a_{00}\boldsymbol{B} & a_{01}\boldsymbol{B} & \cdots & a_{0(m-1)}\boldsymbol{B} \\ a_{10}\boldsymbol{B} & a_{11}\boldsymbol{B} & \cdots & a_{1(m-1)}\boldsymbol{B} \\ \vdots & \vdots & \ddots & \vdots \\ a_{(m-1)0}\boldsymbol{B} & a_{(m-1)1}\boldsymbol{B} & \cdots & a_{(m-1)(m-1)}\boldsymbol{B} \end{pmatrix}. \tag{1.1182}$$

Note that for a_{ij} equal to 1, we have $a_{ij}\boldsymbol{B}$ equal to $\boldsymbol{B}$ and for a_{ij} equal to 0 we have $a_{ij}\boldsymbol{B}$ equal to the zero matrix $\boldsymbol{0}_{n \times n}$.

Kronecker powers are defined by [52, pp. 116f.]

$$\boldsymbol{A}^{\otimes N} = \boldsymbol{A} \otimes \boldsymbol{A}^{\otimes(N-1)} = \boldsymbol{A}^{\otimes(N-1)} \otimes \boldsymbol{A}, \quad N \in \mathbb{N}^*, \tag{1.1183}$$

i. e.,

$$\begin{aligned} \boldsymbol{A}^{\otimes 2} &= \boldsymbol{A} \otimes \boldsymbol{A}, \\ \boldsymbol{A}^{\otimes 3} &= \boldsymbol{A} \otimes \boldsymbol{A}^{\otimes 2} = \boldsymbol{A} \otimes \boldsymbol{A} \otimes \boldsymbol{A} = \boldsymbol{A}^{\otimes 2} \otimes \boldsymbol{A}, \\ \boldsymbol{A}^{\otimes 4} &= \boldsymbol{A} \otimes \boldsymbol{A}^{\otimes 3} = \boldsymbol{A} \otimes \boldsymbol{A} \otimes \boldsymbol{A}^{\otimes 2} = \boldsymbol{A} \otimes \boldsymbol{A} \otimes \boldsymbol{A} \otimes \boldsymbol{A} = \boldsymbol{A}^{\otimes 3} \otimes \boldsymbol{A}, \end{aligned} \tag{1.1184}$$

and so on.

Now, let

$$A = \begin{pmatrix} a_{00} & a_{01} \\ a_{10} & a_{11} \end{pmatrix}, \quad C = \begin{pmatrix} c_{00} & c_{01} \\ c_{10} & c_{11} \end{pmatrix} \tag{1.1185}$$

be 2×2 matrices and $\boldsymbol{B}$ and $\boldsymbol{D}$ be $N/2 \times N/2$ matrices, respectively. Then we find [52, p. 119]

$$(\boldsymbol{A} \otimes \boldsymbol{B})(\boldsymbol{C} \otimes \boldsymbol{D}) = \begin{pmatrix} a_{00}\boldsymbol{B} & a_{01}\boldsymbol{B} \\ a_{10}\boldsymbol{B} & a_{11}\boldsymbol{B} \end{pmatrix} \begin{pmatrix} c_{00}\boldsymbol{D} & c_{01}\boldsymbol{D} \\ c_{10}\boldsymbol{D} & c_{11}\boldsymbol{D} \end{pmatrix} = \underbrace{\begin{pmatrix} \sum_{j=0}^{1} a_{0j}c_{j0} & \sum_{j=0}^{1} a_{0j}c_{j1} \\ \sum_{j=0}^{1} a_{1j}c_{j0} & \sum_{j=0}^{1} a_{1j}c_{j1} \end{pmatrix}}_{=\boldsymbol{AC}} \otimes \boldsymbol{BD}$$

$$= \boldsymbol{AC} \otimes \boldsymbol{BD}. \tag{1.1186}$$

With the 2×2 matrix $\boldsymbol{A}$ and the $N/2 \times N/2$ matrix $\boldsymbol{B}$, we have

$$\boldsymbol{B} \otimes \boldsymbol{A} = \boldsymbol{\Pi}_N(\boldsymbol{A} \otimes \boldsymbol{B})\boldsymbol{\Pi}_N^{\mathrm{T}}. \tag{1.1187}$$

Let us look deeper.

Theorem 1.67 (Noncommutative Property of the Kronecker Product). *Let $\boldsymbol{A}$ be an $m \times n$ matrix and $\boldsymbol{B}$ be a $p \times q$ matrix. Furthermore, let $\boldsymbol{\Pi}$ be an $nq \times nq$ permutation matrix with the inverse $\boldsymbol{\Pi}^T$ and let $\boldsymbol{\Phi}$ be an $mp \times mp$ permutation matrix. Then we yield*

$$\boldsymbol{B} \otimes \boldsymbol{A} = \boldsymbol{\Phi}(\boldsymbol{A} \otimes \boldsymbol{B})\boldsymbol{\Pi}^T. \tag{1.1188}$$

Proof. Let

$$\boldsymbol{x} = (x_0, x_1 \cdots x_{n-1})^{\mathrm{T}} \tag{1.1189}$$

be a column vector with n components $x_0, x_1 \cdots x_{n-1}$ and

$$\boldsymbol{r} = (r_0, r_1 \cdots r_{m-1})^{\mathrm{T}} = \boldsymbol{Ar}, \tag{1.1190}$$

be a column vector with m components $r_0, r_1 \cdots r_{m-1}$.

Furthermore, let

$$\boldsymbol{y} = (y_0, y_1 \cdots y_{q-1})^{\mathrm{T}} \tag{1.1191}$$

be a column vector with q components $y_0, y_1 \cdots y_{q-1}$ and

$$\boldsymbol{s} = (s_0, s_1 \cdots s_{p-1})^{\mathrm{T}} = \boldsymbol{By}, \tag{1.1192}$$

be a column vector with p components $s_0, s_1 \cdots s_{p-1}$.

Apart from the order of the elements, the Kronecker products

$$\boldsymbol{x}\otimes\boldsymbol{y}=\begin{pmatrix}x_0\\x_1\\\vdots\\x_{n-1}\end{pmatrix}\otimes\boldsymbol{y}=\begin{pmatrix}x_0\boldsymbol{y}\\x_1\boldsymbol{y}\\\vdots\\x_{n-1}\boldsymbol{y}\end{pmatrix}=\begin{pmatrix}x_0y_0\\x_0y_1\\\vdots\\x_0y_{q-1}\\x_1y_0\\x_1y_1\\\vdots\\x_1y_{q-1}\\\vdots\\x_{n-1}y_0\\x_{n-1}y_1\\\vdots\\x_{n-1}y_{q-1}\end{pmatrix}\tag{1.1193}$$

and

$$\boldsymbol{y}\otimes\boldsymbol{x}=\begin{pmatrix}y_0\\y_1\\\vdots\\y_{q-1}\end{pmatrix}\otimes\boldsymbol{x}=\begin{pmatrix}y_0\boldsymbol{x}\\y_1\boldsymbol{x}\\\vdots\\y_{q-1}\boldsymbol{x}\end{pmatrix}=\begin{pmatrix}y_0x_0\\y_0x_1\\\vdots\\y_0x_{n-1}\\y_1x_0\\y_1x_1\\\vdots\\y_1x_{n-1}\\\vdots\\y_{q-1}x_0\\y_{q-1}x_1\\\vdots\\y_1q-1x_{n-1}\end{pmatrix}\tag{1.1194}$$

contain the same elements.

Therefore, there exists an invertible $nq\times nq$ permutation matrix $\boldsymbol{\Pi}$, which transforms $\boldsymbol{x}\otimes\boldsymbol{y}$ into $\boldsymbol{y}\otimes\boldsymbol{x}$, i. e., we yield

$$\boldsymbol{y}\otimes\boldsymbol{x}=\boldsymbol{\Pi}(\boldsymbol{x}\otimes\boldsymbol{y}).\tag{1.1195}$$

Apart from the order of the elements, the Kronecker products

$$\boldsymbol{r} \otimes \boldsymbol{s} = \begin{pmatrix} r_0 \\ r_1 \\ \vdots \\ r_{m-1} \end{pmatrix} \otimes \boldsymbol{s} = \begin{pmatrix} r_0 \boldsymbol{s} \\ r_1 \boldsymbol{s} \\ \vdots \\ r_{m-1} \boldsymbol{s} \end{pmatrix} = \begin{pmatrix} r_0 s_0 \\ r_0 s_1 \\ \vdots \\ r_0 s_{p-1} \\ r_1 s_0 \\ r_1 s_1 \\ \vdots \\ r_1 s_{p-1} \\ \vdots \\ r_{m-1} s_0 \\ r_{m-1} s_1 \\ \vdots \\ r_{m-1} s_{p-1} \end{pmatrix} \tag{1.1196}$$

and

$$\boldsymbol{s} \otimes \boldsymbol{r} = \begin{pmatrix} s_0 \\ s_1 \\ \vdots \\ s_{p-1} \end{pmatrix} \otimes \boldsymbol{r} = \begin{pmatrix} s_0 \boldsymbol{r} \\ s_1 \boldsymbol{r} \\ \vdots \\ s_{p-1} \boldsymbol{r} \end{pmatrix} = \begin{pmatrix} s_0 r_0 \\ s_0 r_1 \\ \vdots \\ s_0 r_{m-1} \\ s_1 r_0 \\ s_1 r_1 \\ \vdots \\ s_1 r_{m-1} \\ \vdots \\ s_{p-1} r_0 \\ s_{p-1} r_1 \\ \vdots \\ s_{p-1} r_{m-1} \end{pmatrix} \tag{1.1197}$$

contain the same elements.

Therefore, there exists an invertible $mp \times mp$ permutation matrix $\boldsymbol{\Phi}$, which transforms $\boldsymbol{r} \otimes \boldsymbol{s}$ into $\boldsymbol{s} \otimes \boldsymbol{r}$, i. e., we have

$$\boldsymbol{s} \otimes \boldsymbol{r} = \boldsymbol{\Phi}(\boldsymbol{r} \otimes \boldsymbol{s}). \tag{1.1198}$$

Setting out from (1.1186) and using (1.1190) as well as (1.1192), we find

$$(\boldsymbol{A} \otimes \boldsymbol{B})(\boldsymbol{x} \otimes \boldsymbol{y}) = \boldsymbol{A}\boldsymbol{x} \otimes \boldsymbol{B}\boldsymbol{y} = \boldsymbol{r} \otimes \boldsymbol{s} \tag{1.1199}$$

and

$$(\boldsymbol{B} \otimes \boldsymbol{A})(\boldsymbol{y} \otimes \boldsymbol{x}) = \boldsymbol{B}\boldsymbol{y} \otimes \boldsymbol{A}\boldsymbol{x} = \boldsymbol{s} \otimes \boldsymbol{r}. \tag{1.1200}$$

With (1.1198), (1.1199) becomes

$$\boldsymbol{\Phi}(\boldsymbol{A} \otimes \boldsymbol{B})(\boldsymbol{x} \otimes \boldsymbol{y}) = \boldsymbol{s} \otimes \boldsymbol{r} \tag{1.1201}$$

and with (1.1195), (1.1200) becomes

$$(\boldsymbol{B} \otimes \boldsymbol{A})\boldsymbol{\Pi}(\boldsymbol{x} \otimes \boldsymbol{y}) = \boldsymbol{s} \otimes \boldsymbol{r}. \tag{1.1202}$$

Since (1.1201) and (1.1202) are equal, we find

$$\boldsymbol{\Phi}(\boldsymbol{A} \otimes \boldsymbol{B})(\boldsymbol{x} \otimes \boldsymbol{y}) = (\boldsymbol{B} \otimes \boldsymbol{A})\boldsymbol{\Pi}(\boldsymbol{x} \otimes \boldsymbol{y}) \tag{1.1203}$$

and, therefore,

$$\boldsymbol{\Phi}(\boldsymbol{A} \otimes \boldsymbol{B}) = (\boldsymbol{B} \otimes \boldsymbol{A})\boldsymbol{\Pi}. \tag{1.1204}$$

Since $\boldsymbol{\Pi}$ is invertible, we find

$$\boldsymbol{B} \otimes \boldsymbol{A} = \boldsymbol{\Phi}(\boldsymbol{A} \otimes \boldsymbol{B})\boldsymbol{\Pi}^{\mathrm{T}}. \tag{1.1205}$$

□

Now, let us consider the case of (1.1187).

Theorem 1.68 (Noncommutative Property of the Kronecker Product Revisited). *Let $\boldsymbol{A}$ be a 2×2 matrix and $\boldsymbol{B}$ be an $N/2 \times N/2$ matrix. Furthermore, let $\boldsymbol{\Pi}_N$ be the $N \times N$ permutation matrix given by* (7.132). *Then we have*

$$\boldsymbol{B} \otimes \boldsymbol{A} = \boldsymbol{\Pi}_N(\boldsymbol{A} \otimes \boldsymbol{B})\boldsymbol{\Pi}_N^T. \tag{1.1206}$$

Proof. Let

$$\boldsymbol{x} = (x_0, x_1)^{\mathrm{T}} \tag{1.1207}$$

with 2 components x_0, x_1 and

$$\boldsymbol{r} = (r_0, r_1)^{\mathrm{T}} = \boldsymbol{A}\boldsymbol{x} \tag{1.1208}$$

with 2 components r_0, r_1. Furthermore, let

$$\boldsymbol{y} = (y_0, y_1 \cdots y_{\frac{N}{2}-1})^{\mathrm{T}} \tag{1.1209}$$

with $N/2$ components $y_0, y_1 \cdots y_{\frac{N}{2}-1}$ and

$$\boldsymbol{s} = (s_0, s_1 \cdots s_{\frac{N}{2}-1})^{\mathrm{T}} = \boldsymbol{B}\boldsymbol{y}, \tag{1.1210}$$

with $N/2$ components $s_0, s_1 \cdots s_{\frac{N}{2}-1}$.

Clearly,

$$\boldsymbol{x} \otimes \boldsymbol{y} = \begin{pmatrix} x_0 \\ x_1 \end{pmatrix} \otimes \boldsymbol{y} = \begin{pmatrix} x_0\boldsymbol{y} \\ x_1\boldsymbol{y} \end{pmatrix} = \begin{pmatrix} x_0y_0 \\ x_0y_1 \\ \vdots \\ x_0y_{\frac{N}{2}-1} \\ x_1y_0 \\ x_1y_1 \\ \vdots \\ x_1y_{\frac{N}{2}-1} \end{pmatrix} \tag{1.1211}$$

and

$$\boldsymbol{y} \otimes \boldsymbol{x} = \begin{pmatrix} y_0 \\ y_1 \\ \vdots \\ y_{\frac{N}{2}-1} \end{pmatrix} \otimes \boldsymbol{x} = \begin{pmatrix} y_0\boldsymbol{x} \\ y_1\boldsymbol{x} \\ \vdots \\ y_{\frac{N}{2}-1}\boldsymbol{x} \end{pmatrix} = \begin{pmatrix} y_0x_0 \\ y_0x_1 \\ y_1x_0 \\ y_1x_1 \\ \vdots \\ y_{\frac{N}{2}-1}x_0 \\ y_{\frac{N}{2}.1}x_1 \end{pmatrix} \tag{1.1212}$$

contain the same elements.

Choosing the permutation matrix $\boldsymbol{\Pi}_N$ appropriately, we find

$$\boldsymbol{y} \otimes \boldsymbol{x} = \boldsymbol{\Pi}_N(\boldsymbol{x} \otimes \boldsymbol{y}). \tag{1.1213}$$

Furthermore,

$$\boldsymbol{r} \otimes \boldsymbol{s} = \begin{pmatrix} r_0 \\ r_1 \end{pmatrix} \otimes \boldsymbol{s} = \begin{pmatrix} r_0\boldsymbol{s} \\ r_1\boldsymbol{s} \end{pmatrix} = \begin{pmatrix} r_0s_0 \\ r_0s_1 \\ \vdots \\ r_0s_{\frac{N}{2}-1} \\ r_1s_0 \\ r_1s_1 \\ \vdots \\ r_1s_{\frac{N}{2}-1} \end{pmatrix} \tag{1.1214}$$

and

$$\boldsymbol{s} \otimes \boldsymbol{r} = \begin{pmatrix} s_0 \\ s_1 \\ \vdots \\ s_{\frac{N}{2}-1} \end{pmatrix} \otimes \boldsymbol{r} = \begin{pmatrix} s_0\boldsymbol{r} \\ s_1\boldsymbol{r} \\ \vdots \\ s_{\frac{N}{2}-1}\boldsymbol{r} \end{pmatrix} = \begin{pmatrix} s_0r_0 \\ s_0r_1 \\ s_1r_0 \\ s_1r_1 \\ \vdots \\ s_{\frac{N}{2}-1}r_0 \\ s_{\frac{N}{2}-1}r_1 \end{pmatrix} \tag{1.1215}$$

contain the same elements. Clearly, choosing $\boldsymbol{\Phi}$ equal to $\boldsymbol{\Pi}_N$, we find

$$\boldsymbol{s} \otimes \boldsymbol{r} = \boldsymbol{\Pi}_N(\boldsymbol{r} \otimes \boldsymbol{s}). \tag{1.1216}$$

Equation (1.1203) becomes

$$\boldsymbol{\Pi}_N(\boldsymbol{A} \otimes \boldsymbol{B})(\boldsymbol{x} \otimes \boldsymbol{y}) = (\boldsymbol{B} \otimes \boldsymbol{A})\boldsymbol{\Pi}_N(\boldsymbol{x} \otimes \boldsymbol{y}) \tag{1.1217}$$

and, therefore,

$$\boldsymbol{\Pi}_N(\boldsymbol{A} \otimes \boldsymbol{B}) = (\boldsymbol{B} \otimes \boldsymbol{A})\boldsymbol{\Pi}_N, \tag{1.1218}$$

which leads to

$$\boldsymbol{B} \otimes \boldsymbol{A} = \boldsymbol{\Pi}_N(\boldsymbol{A} \otimes \boldsymbol{B})\boldsymbol{\Pi}_N^{\mathrm{T}}. \tag{1.1219}$$

□

Now, let us choose special matrices $\boldsymbol{A}$ and $\boldsymbol{B}$. First, we will replace the 2 matrix $\boldsymbol{A}$ by $\boldsymbol{I}_2$. Next, let us denote the $N/2 \times N/2$ matrix $\boldsymbol{B}$ by $\boldsymbol{B}_{N/2}$. Then (1.1187) becomes

$$\boldsymbol{B}_{N/2} \otimes \boldsymbol{I}_2 = \boldsymbol{\Pi}_N(\boldsymbol{I}_2 \otimes \boldsymbol{B}_{N/2})\boldsymbol{\Pi}_N^{\mathrm{T}}. \tag{1.1220}$$

Using (1.1181), multiplying (1.1220) by $\boldsymbol{\Pi}_N^{\mathrm{T}}$ from the left yields

$$\boldsymbol{\Pi}_N^{\mathrm{T}}(\boldsymbol{B}_{N/2} \otimes \boldsymbol{I}_2) = \underbrace{\boldsymbol{\Pi}_N^{\mathrm{T}}\boldsymbol{\Pi}_N}_{=\boldsymbol{I}_N}(\boldsymbol{I}_2 \otimes \boldsymbol{B}_{N/2})\boldsymbol{\Pi}_N^{\mathrm{T}} = (\boldsymbol{I}_2 \otimes \boldsymbol{B}_{N/2})\boldsymbol{\Pi}_N^{\mathrm{T}}, \tag{1.1221}$$

and multiplying (1.1220) by $\boldsymbol{\Pi}_N$ from the right leads to

$$(\boldsymbol{B}_{N/2} \otimes \boldsymbol{I}_2)\boldsymbol{\Pi}_N = \boldsymbol{\Pi}_N(\boldsymbol{I}_2 \otimes \boldsymbol{B}_{N/2})\underbrace{\boldsymbol{\Pi}_N^{\mathrm{T}}\boldsymbol{\Pi}_N}_{=\boldsymbol{I}_N} = \boldsymbol{\Pi}_N(\boldsymbol{I}_2 \otimes \boldsymbol{B}_{N/2}). \tag{1.1222}$$

Since $(\boldsymbol{B}_{N/2} \otimes \boldsymbol{I}_2)\boldsymbol{\Pi}_N$ and $\boldsymbol{\Pi}_N(\boldsymbol{I}_2 \otimes \boldsymbol{B}_{N/2})$ are $N \times N$ matrices, using

$$\boldsymbol{B}_N \overset{\text{def}}{=} \boldsymbol{\Pi}_N(\boldsymbol{I}_2 \otimes \boldsymbol{B}_{N/2}), \quad \boldsymbol{B}_1 = 1, \quad N \in \{2, 4, 8, 16 \cdots 2^{\mu} \cdots\}, \quad \mu \in \mathbb{N}^*, \tag{1.1223}$$

yields the recursive definition of $\boldsymbol{B}_N$.

Example 1.66. Let us consider N equal to 2. Then we have

$$\boldsymbol{B}_2 = \boldsymbol{\Pi}_2(\boldsymbol{I}_2 \otimes \boldsymbol{B}_1) = \boldsymbol{\Sigma}_{2,1}^{\mathrm{T}}\boldsymbol{I}_2 = \begin{pmatrix} 1 & 0 \\ 0 & 1 \end{pmatrix}\boldsymbol{I}_2 = \boldsymbol{I}_2 = \boldsymbol{B}_2^{\mathrm{T}}. \tag{1.1224}$$

Let us consider N equal to 4. Then we have

$$\boldsymbol{B}_4 = \boldsymbol{\Pi}_4(\boldsymbol{I}_2 \otimes \boldsymbol{B}_2) = \begin{pmatrix} 1 & 0 & 0 & 0 \\ 0 & 0 & 1 & 0 \\ 0 & 1 & 0 & 0 \\ 0 & 0 & 0 & 1 \end{pmatrix}\boldsymbol{I}_4 = \begin{pmatrix} 1 & 0 & 0 & 0 \\ 0 & 0 & 1 & 0 \\ 0 & 1 & 0 & 0 \\ 0 & 0 & 0 & 1 \end{pmatrix} = \boldsymbol{\Pi}_4 = \boldsymbol{B}_4^{\mathrm{T}}. \tag{1.1225}$$

Let us consider N equal to 8. Then we have

$$\boldsymbol{B}_8 = \boldsymbol{\Pi}_8(\boldsymbol{I}_2 \otimes \boldsymbol{B}_4) = \begin{pmatrix} 1 & 0 & 0 & 0 & 0 & 0 & 0 & 0 \\ 0 & 0 & 0 & 0 & 1 & 0 & 0 & 0 \\ 0 & 1 & 0 & 0 & 0 & 0 & 0 & 0 \\ 0 & 0 & 0 & 0 & 0 & 1 & 0 & 0 \\ 0 & 0 & 1 & 0 & 0 & 0 & 0 & 0 \\ 0 & 0 & 0 & 0 & 0 & 0 & 1 & 0 \\ 0 & 0 & 0 & 1 & 0 & 0 & 0 & 0 \\ 0 & 0 & 0 & 0 & 0 & 0 & 0 & 1 \end{pmatrix} \begin{pmatrix} 1 & 0 & 0 & 0 & 0 & 0 & 0 & 0 \\ 0 & 0 & 1 & 0 & 0 & 0 & 0 & 0 \\ 0 & 1 & 0 & 0 & 0 & 0 & 0 & 0 \\ 0 & 0 & 0 & 1 & 0 & 0 & 0 & 0 \\ 0 & 0 & 0 & 0 & 1 & 0 & 0 & 0 \\ 0 & 0 & 0 & 0 & 0 & 0 & 1 & 0 \\ 0 & 0 & 0 & 0 & 0 & 1 & 0 & 0 \\ 0 & 0 & 0 & 0 & 0 & 0 & 0 & 1 \end{pmatrix}$$

$$= \begin{pmatrix} 1 & 0 & 0 & 0 & 0 & 0 & 0 & 0 \\ 0 & 0 & 0 & 0 & 1 & 0 & 0 & 0 \\ 0 & 0 & 1 & 0 & 0 & 0 & 0 & 0 \\ 0 & 0 & 0 & 0 & 0 & 0 & 1 & 0 \\ 0 & 1 & 0 & 0 & 0 & 0 & 0 & 0 \\ 0 & 0 & 0 & 0 & 0 & 1 & 0 & 0 \\ 0 & 0 & 0 & 1 & 0 & 0 & 0 & 0 \\ 0 & 0 & 0 & 0 & 0 & 0 & 0 & 1 \end{pmatrix} = \boldsymbol{B}_8^{\mathrm{T}}, \tag{1.1226}$$

and so on.

Obviously, $\boldsymbol{B}_N$ is symmetric, i. e.,

$$\boldsymbol{B}_N = \boldsymbol{B}_N^{\mathrm{T}}. \tag{1.1227}$$

Proof 1.3 (Proof of (1.1227)). Equation (1.1227) can be shown by induction.

Base Case

The cases N equal to 2, 4 and 8 is already shown above, i. e., $\boldsymbol{B}_2 = \boldsymbol{B}_2^{\mathrm{T}}$.

Induction Step

For the induction step, we set out from

$$\boldsymbol{B}_{N/2} = \boldsymbol{B}_{N/2}^{\mathrm{T}} \tag{1.1228}$$

and

$$\begin{aligned} \boldsymbol{\Pi}_N &= \begin{pmatrix} [\boldsymbol{\Pi}_N]_1 & [\boldsymbol{\Pi}_N]_2 \\ [\boldsymbol{\Pi}_N]_3 & [\boldsymbol{\Pi}_N]_4 \end{pmatrix} = \begin{pmatrix} [\boldsymbol{\Pi}_N]_1^{\mathrm{T}} & [\boldsymbol{\Pi}_N]_3^{\mathrm{T}} \\ [\boldsymbol{\Pi}_N]_2^{\mathrm{T}} & [\boldsymbol{\Pi}_N]_4^{\mathrm{T}} \end{pmatrix}, \\ \boldsymbol{\Pi}_N^{\mathrm{T}} &= \begin{pmatrix} [\boldsymbol{\Pi}_N]_1 & [\boldsymbol{\Pi}_N]_2 \\ [\boldsymbol{\Pi}_N]_3 & [\boldsymbol{\Pi}_N]_4 \end{pmatrix}^{\mathrm{T}} = \begin{pmatrix} [\boldsymbol{\Pi}_N]_1^{\mathrm{T}} & [\boldsymbol{\Pi}_N]_3^{\mathrm{T}} \\ [\boldsymbol{\Pi}_N]_2^{\mathrm{T}} & [\boldsymbol{\Pi}_N]_4^{\mathrm{T}} \end{pmatrix} \end{aligned} \tag{1.1229}$$

with the submatrices $[\boldsymbol{\Pi}_N]_1$, $[\boldsymbol{\Pi}_N]_2$, $[\boldsymbol{\Pi}_N]_3$ and $[\boldsymbol{\Pi}_N]_4$. Then we find

$$\begin{aligned} \boldsymbol{B}_N^{\mathrm{T}} &= \left[\boldsymbol{\Pi}_N(\boldsymbol{I}_2 \otimes \boldsymbol{B}_{N/2})\right]^{\mathrm{T}} = (\boldsymbol{I}_2 \otimes \boldsymbol{B}_{N/2})^{\mathrm{T}}\boldsymbol{\Pi}_N^{\mathrm{T}} = \left(\boldsymbol{I}_2 \otimes \boldsymbol{B}_{N/2}^{\mathrm{T}}\right)\boldsymbol{\Pi}_N^{\mathrm{T}} \\ &= (\boldsymbol{I}_2 \otimes \boldsymbol{B}_{N/2})\boldsymbol{\Pi}_N^{\mathrm{T}} \\ &= \begin{pmatrix} \boldsymbol{B}_{N/2} & \boldsymbol{0}_{N/2\times N/2} \\ \boldsymbol{0}_{N/2\times N/2} & \boldsymbol{B}_{N/2} \end{pmatrix}\boldsymbol{\Pi}_N^{\mathrm{T}} \end{aligned}$$

$$= \begin{pmatrix} \boldsymbol{B}_{N/2} & \boldsymbol{0}_{N/2\times N/2} \\ \boldsymbol{0}_{N/2,N/2} & \boldsymbol{B}_{N/2} \end{pmatrix} \begin{pmatrix} [\boldsymbol{\Pi}_N]_1^{\mathrm{T}} & [\boldsymbol{\Pi}_N]_3^{\mathrm{T}} \\ [\boldsymbol{\Pi}_N]_2^{\mathrm{T}} & [\boldsymbol{\Pi}_N]_4^{\mathrm{T}} \end{pmatrix}$$

$$= \begin{pmatrix} \boldsymbol{B}_{N/2}[\boldsymbol{\Pi}_N]_1^{\mathrm{T}} & \boldsymbol{B}_{N/2}[\boldsymbol{\Pi}_N]_3^{\mathrm{T}} \\ \boldsymbol{B}_{N/2}[\boldsymbol{\Pi}_N]_2^{\mathrm{T}} & \boldsymbol{B}_{N/2}[\boldsymbol{\Pi}_N]_4^{\mathrm{T}} \end{pmatrix} \tag{1.1230}$$

$$= \begin{pmatrix} ([\boldsymbol{\Pi}_N]_1\boldsymbol{B}_{N/2})^{\mathrm{T}} & ([\boldsymbol{\Pi}_N]_3\boldsymbol{B}_{N/2})^{\mathrm{T}} \\ ([\boldsymbol{\Pi}_N]_2\boldsymbol{B}_{N/2})^{\mathrm{T}} & ([\boldsymbol{\Pi}_N]_4\boldsymbol{B}_{N/2})^{\mathrm{T}} \end{pmatrix}$$

$$= \begin{pmatrix} ([\boldsymbol{\Pi}_N]_1\boldsymbol{B}_{N/2}) & ([\boldsymbol{\Pi}_N]_2\boldsymbol{B}_{N/2}) \\ ([\boldsymbol{\Pi}_N]_3\boldsymbol{B}_{N/2}) & ([\boldsymbol{\Pi}_N]_4\boldsymbol{B}_{N/2}) \end{pmatrix}$$

$$= \begin{pmatrix} [\boldsymbol{\Pi}_N]_1 & [\boldsymbol{\Pi}_N]_2 \\ [\boldsymbol{\Pi}_N]_3 & [\boldsymbol{\Pi}_N]_4 \end{pmatrix} \begin{pmatrix} \boldsymbol{B}_{N/2} & \boldsymbol{0}_{N/2\times N/2} \\ \boldsymbol{0}_{N/2\times N/2} & \boldsymbol{B}_{N/2} \end{pmatrix}$$

hence yielding

$$\boldsymbol{B}_N^{\mathrm{T}} = \boldsymbol{\Pi}_N \begin{pmatrix} \boldsymbol{B}_{N/2} & \boldsymbol{0}_{N/2,N/2} \\ \boldsymbol{0}_{N/2\times N/2} & \boldsymbol{B}_{N/2} \end{pmatrix} = \boldsymbol{\Pi}_N(\boldsymbol{I}_2 \otimes \boldsymbol{B}_{N/2}) = \boldsymbol{B}_N. \tag{1.1231}$$

□

Summarizing, we have

$$\boldsymbol{B}_N = \boldsymbol{B}_N^{\mathrm{T}} = (\boldsymbol{I}_2 \otimes \boldsymbol{B}_{N/2})\boldsymbol{\Pi}_N^{\mathrm{T}} = \boldsymbol{\Pi}_N(\boldsymbol{I}_2 \otimes \boldsymbol{B}_{N/2}). \tag{1.1232}$$

Now, we are prepared for the Reed–Muller (RM) codes and the polar codes.

1.17 Reed–Muller (RM) Codes

1.17.1 Definition of Reed–Muller (RM) Codes

Reed–Muller (RM) codes [15, pp. 99, 105–119], [5, pp. 399–401] [23, pp. 107–109, 115–119], [25, pp. 166–169, 418–426], [39, pp. 122–126] are $(n, k, d_{\min})$ binary linear block codes and belong to the important linear block codes, which have been deployed in advanced mobile communications systems since the establishment of *Universal Mobile Telecommunications System (UMTS)* [56, Section 4.3.1, pp. 27–29], [57, Table 8, Section 4.3.3, p. 55f.]. To the best knowledge of the author, the most detailed discussion and analysis of Reed–Muller (RM) codes can be found in [4, Chapters 13–15, pp. 370–479].

Reed–Muller (RM) were discovered by David E. *Muller* in 1954 [58], [15, p. 99] and are multiple error-correcting codes [15, p. 99]. The first decoding algorithm for these codes was devised by Irving S. *Reed*, also in 1954 [59], [15, p. 99].

Reed–Muller (RM) codes, being simple in construction and rich in structural properties [15, p. 99], are one of the oldest and best understood families of codes [4, p. 370] and can be decoded using either hard-decision or soft-decision algorithms [15, p. 99].

Reed–Muller (RM) codes can be defined very simply in terms of *Boolean functions* [4, p. 370]. We are going to define codes of length n equal to 2^m, $m, n \in \mathbb{N}^*$, and to do so we shall need m, $m \in \mathbb{N}^*$, variables $x_1, x_2 \cdots x_m$, $m \in \mathbb{N}^*$, which take on the values 0 or 1, i. e., $x_1, x_2 \cdots x_m \in \mathbb{F}_2$ [4, p. 370].

Alternatively, let

$$\boldsymbol{x} = (x_1, x_2 \cdots x_m), \quad \boldsymbol{x} \in \mathbb{F}_2^m, \quad m \in \mathbb{N}^*, \tag{1.1233}$$

be the set of all binary m-tuples [4, p. 370]. Any function

$$f(\boldsymbol{x}) = f(x_1, x_2 \cdots x_m) \in \mathbb{F}_2, \quad m \in \mathbb{N}^*, \tag{1.1234}$$

which takes on the values 0 and 1, is called a *Boolean function* [4, p. 371]. Such a Boolean function can be specified by a truth table, which gives the value of $f(\boldsymbol{x})$ at all of its 2^m, $m \in \mathbb{N}^*$ arguments [4, p. 371].

Example 1.67. For m equal to 3, one such Boolean function is specified by the following truth table [4, p. 371]:

$\sum_{j=1}^{m} 2^{j-1} x_j$	0	1	2	3	4	5	6	7
x_3	0	0	0	0	1	1	1	1
x_2	0	0	1	1	0	0	1	1
x_1	0	1	0	1	0	1	0	1
$f(\boldsymbol{x}) = f(x_1, x_2, x_3)$	0	0	0	1	1	0	0	0

(1.1235)

Note that $x_j^2, j \in \{1, 2, 3\}$ is equal to $x_j, j \in \{1, 2, 3\}$, [4, p. 371].

The columns of the truth table are have the natural ordering, increasing from 0 to 7 Example 1.67 [4, p. 371].

The last row of the above table gives the values taken by $f(x_1, x_2, x_3)$ [4, p. 371]. All these values form the binary vector

$$\boldsymbol{f} = (0, 0, 0, 1, 1, 0, 0, 0) \in \mathbb{F}_2^8 \tag{1.1236}$$

of length n equal to 2^m, $m \in \mathbb{N}^*$, in our case 8 [4, p. 371].

A code will consist of all binary vectors $\boldsymbol{f} \in \mathbb{F}_2^m$, $m \in \mathbb{N}^*$, where the function $f(\boldsymbol{x})$ of (1.1234) belongs to a certain class [4, p. 371].

The last row of the truth table can be filled in arbitrarily. There are 2^{2^m} Boolean functions of m variables [4, p. 371]. A Boolean function can be found immediately from its truth table [4, p. 371].

Definition 1.64 (Logical Operations Applied to Boolean Functions). Using the additive operation $\oplus$ and the multiplicative operation $\odot$ of the finite field $\mathbb{F}_2$, the following logical operations are applied to Boolean Functions [4, p. 371]:

$$\begin{aligned} f(\boldsymbol{x}) \text{ EXCLUSIVE OR } g(\boldsymbol{x}) &= f(\boldsymbol{x}) \oplus g(\boldsymbol{x}), \\ f(\boldsymbol{x}) \text{ AND } g(\boldsymbol{x}) &= f(\boldsymbol{x}) \odot g(\boldsymbol{x}), \\ f(\boldsymbol{x}) \vee g(\boldsymbol{x}) &= f(\boldsymbol{x}) \vee g(\boldsymbol{x}) && = f(\boldsymbol{x}) \oplus g(\boldsymbol{x}) \oplus f(\boldsymbol{x}) \odot g(\boldsymbol{x}), \\ \text{NOT} f(\boldsymbol{x}) &= \bar{f}(\boldsymbol{x}) && = 1 \oplus f(\boldsymbol{x}). \end{aligned} \tag{1.1237}$$

Example 1.68. In the preceding Example 1.67, the following Boolean function was used:

$$f(x_1, x_2, x_3) = x_1 \odot x_2 \odot \overline{x_3} \vee \overline{x_1} \odot \overline{x_2} \odot x_3, \tag{1.1238}$$

called the *disjunctive normal form* of f [4, p. 371].

According to (1.1237), we find [4, p. 371]

$$\begin{aligned} x_1 \odot x_2 \odot \overline{x_3} \vee \overline{x_1} \odot \overline{x_2} \odot x_3 &= x_1 \odot x_2 \odot \overline{x_3} \oplus \overline{x_1} \odot \overline{x_2} \odot x_3 \\ &\oplus x_1 \odot x_2 \odot \overline{x_3} \odot \overline{x_1} \odot \overline{x_2} \odot x_3 \\ &= x_1 \odot x_2 \odot \overline{x_3} \oplus \overline{x_1} \odot \overline{x_2} \odot x_3 \\ &\oplus \underbrace{x_1 \odot \overline{x_1}}_{=0} \odot \underbrace{x_2 \odot \overline{x_2}}_{=0} \odot \underbrace{\overline{x_3} \odot x_3}_{=0} \\ &= x_1 \odot x_2 \odot \underbrace{(1 \oplus x_3)}_{=\overline{x_3}} \oplus \underbrace{(1 \oplus x_1)}_{=\overline{x_1}} \odot \underbrace{(1 \oplus x_2)}_{=\overline{x_2}} \odot x_3 \\ &= x_1 \odot x_2 \oplus x_1 \odot x_2 \odot x_3 \\ &\oplus (1 \oplus x_1) \odot (1 \oplus x_2) \odot x_3 \\ &= x_1 \odot x_2 \oplus x_1 \odot x_2 \odot x_3 \\ &\oplus (1 \oplus x_2) \odot x_3 \oplus x_1 (1 \oplus x_2) \odot x_3 \\ &= x_1 \odot x_2 \oplus x_1 \odot x_2 \odot x_3 \oplus x_3 \\ &\oplus x_2 \odot x_3 \oplus x_1 \odot x_3 \oplus x_1 \odot x_2 \odot x_3 \\ &= x_1 \odot x_2 \oplus x_3 \oplus x_2 \odot x_3 \oplus x_1 \odot x_3 \\ &\oplus \underbrace{x_1 \odot x_2 \odot x_3 \oplus x_1 \odot x_2 \odot x_3}_{=0} \\ &= x_1 \odot x_2 \oplus x_1 \odot x_3 \oplus x_2 \odot x_3 \oplus x_3. \end{aligned}$$

Any Boolean function can obviously be expressed as a sum of the elementary functions

$$\begin{matrix} 1, \\ x_1, x_2 \cdots x_m, \\ x_1 \odot x_2, x_1 \odot x_3 \cdots x_{m-1} \odot x_m, \\ x_1 \odot x_2 \odot x_3, x_1 \odot x_2 \odot x_4 \cdots x_{m-2} \odot x_{m-1} \odot x_m, \\ \cdots \\ x_1 \odot x_2 \odot \cdots \odot x_m, \end{matrix} \quad m \in \mathbb{N}^*, \tag{1.1239}$$

with coefficients 0 or 1 [4, p. 371]. Using the binomial theorem [11, equation (1.36c), p. 12]. There are

$$\begin{aligned} &\binom{m}{0} + \binom{m}{1} + \binom{m}{2} + \cdots + \binom{m}{m} \\ &= \sum_{i=0}^{m} \binom{m}{i} = \sum_{i=0}^{m} \binom{m}{i} 1^{m-i} 1^i = 2^m, \quad m \in \mathbb{N}^*, \end{aligned} \tag{1.1240}$$

such functions (1.1239) [4, p. 371]. Using (1.1233), any Boolean function is hence represented by

$$\begin{aligned} f(\mathbf{x}) = a_0 \\ &\oplus a_1 \odot x_1 \oplus a_2 \odot x_2 \oplus \cdots \oplus a_m \odot x_m \\ &\oplus a_{m+1} \odot x_1 \odot x_2 \oplus a_{m+2} \odot x_1 \odot x_3 \oplus \cdots \oplus a_{(m+1)m/2} \odot x_{m-1} \odot x_m \\ &\oplus a_{(m+1)m/2+1} \odot x_1 \odot x_2 \odot x_3 \\ &\oplus a_{(m+1)m/2+2} \odot x_1 \odot x_2 \odot x_4 \oplus \cdots \\ &\oplus a_{(m^2+5)m/6} \odot x_{m-2} \odot x_{m-1} \odot x_m \\ &\oplus \cdots \\ &\oplus a_{2^m-1} \odot x_1 \odot x_2 \odot \cdots \odot x_m, \quad m \in \mathbb{N}^*. \end{aligned} \tag{1.1241}$$

Since there are 2^{2^m} Boolean functions of m variables altogether, all these sums must be distinct [4, p. 371]. In other words, the 2^m vectors corresponding to the functions (1.1239) are linearly independent [4, p. 371].

If $f(x_1, x_2 \cdots x_m)$ is a Boolean function, then

$$f(x_1, x_2 \cdots x_m) = x_m \odot f(x_1, x_2 \cdots x_{m-1}, 1) \vee \overline{x_m} \odot f(x_1, x_2 \cdots x_{m-1}, 0), \tag{1.1242}$$

because if x_m is equal to 1, $\overline{x_m} f(x_1, x_2 \cdots x_{m-1}, 0)$ is equal to 0 and

$$\begin{aligned} 1 \odot f(x_1, x_2 \cdots x_{m-1}, 1) = a_0 \\ &\oplus a_1 \odot x_1 \oplus a_2 \odot x_2 \oplus \cdots \oplus a_m \odot 1 \\ &\oplus a_{m+1} \odot x_1 \odot x_2 \\ &\oplus a_{m+2} \odot x_1 \odot x_3 \oplus \cdots \\ &\oplus a_{(m+1)m/2} \odot x_{m-1} \odot 1 \\ &\oplus a_{(m+1)m/2+1} \odot x_1 \odot x_2 \odot x_3 \\ &\oplus a_{(m+1)m/2+2} \odot x_1 \odot x_2 \odot x_4 \oplus \cdots \\ &\oplus a_{(m^2+5)m/6} \odot x_{m-2} \odot x_{m-1} \odot 1 \\ &\oplus \cdots \\ &\oplus a_{2^m-1} \odot x_1 \odot x_2 \odot \cdots \odot 1, \quad m \in \mathbb{N}^*, \end{aligned} \tag{1.1243}$$

which yields $f(x_1, x_2 \cdots x_m)$, and if x_m is equal to 0, i. e., $\overline{x_m}$ is equal to 1, $x_m f(x_1, x_2 \cdots x_{m-1}, 1)$ is equal to 0 and

$$\begin{aligned} 1 \odot f(x_1, x_2 \cdots x_{m-1}, 0) = a_0 \\ &\oplus a_1 \odot x_1 \oplus a_2 \odot x_2 \oplus \cdots \oplus a_m \odot 0 \\ &\oplus a_{m+1} \odot x_1 \odot x_2 \end{aligned}$$

$$\begin{aligned}
&\oplus\ a_{m+2} \odot x_1 \odot x_3\ \oplus \cdots \\
&\oplus\ a_{(m+1)m/2} \odot x_{m-1} \odot 0 \\
&\oplus\ a_{(m+1)m/2+1} \odot x_1 \odot x_2 \odot x_3 \\
&\oplus\ a_{(m+1)m/2+2} \odot x_1 \odot x_2 \odot x_4\ \oplus \cdots \\
&\oplus\ a_{(m^2+5)m/6} \odot x_{m-2} \odot x_{m-1} \odot 0 \\
&\oplus \cdots \\
&\oplus\ a_{2^m-1} \odot x_1 \odot x_2 \odot \cdots \odot 0, \quad m \in \mathbb{N}^*, \qquad (1.1244)
\end{aligned}$$

which yields $f(x_1, x_2 \cdots x_m)$ [4, p. 372].

Also, if $f(x_1, x_2 \cdots x_m)$ is a Boolean function, then

$$f(x_1, x_2 \cdots x_m) = x_m \odot g(x_1, x_2 \cdots x_{m-1})\ \oplus\ h(x_1, x_2 \cdots x_{m-1}) \qquad (1.1245)$$

if $g(x_1, x_2 \cdots x_{m-1})$ and $h(x_1, x_2 \cdots x_{m-1})$ are Boolean functions, each using 2^{m-1} elementary functions [4, p. 372]

$$\begin{array}{c}
1, \\
x_1,\ x_2\ \cdots\ x_{m-1}, \\
x_1 \odot x_2,\ x_1 \odot x_3\ \cdots\ x_{m-2} \odot x_{m-1}, \\
x_1 \odot x_2 \odot x_3,\ x_1 \odot x_2 \odot x_4\ \cdots\ x_{m-3} \odot x_{m-2} \odot x_{m-1}, \\
\cdots \\
x_1 \odot x_2 \odot\ \cdots\ \odot x_{m-1},
\end{array} \quad m \in \mathbb{N}^*. \qquad (1.1246)$$

In addition, we have the disjunctive normal form [4, p. 372]

$$f(x_1, x_2 \cdots x_m) = \bigoplus_{i_1=0}^{1} \cdots \bigoplus_{i_m=0}^{1} f(i_1, i_2 \cdots i_m) \odot y_1^{(i_1)} \odot y_2^{(i_2)} \odot \cdots \odot y_m^{(i_m)}, \qquad (1.1247)$$

$$y_j^{(1)} = x_j, \quad y_j^{(0)} = \overline{x_j} = 1 \oplus x_j, \quad j \in \{1, 2 \cdots m\}, \quad m \in \mathbb{N}^*.$$

Theorem 1.69 (Expansion of a Boolean Function). *Any Boolean function $f(x_1, x_2 \cdots x_m) \in \mathbb{F}_2$ can be expanded in powers of $x_j \in \mathbb{F}_2, j \in \{1, 2 \cdots m\}, m \in \mathbb{N}^*$, as* [4, p. 372]

$$f(\boldsymbol{x}) = f(x_1, x_2 \cdots x_m) = \bigoplus_{\boldsymbol{a} \in \mathbb{F}_2^m} g(\boldsymbol{a}) \odot x_1^{a_1} \odot x_2^{a_2} \cdots \odot x_m^{a_m}, \quad \boldsymbol{a}, \boldsymbol{x} \in \mathbb{F}_2^m. \qquad (1.1248)$$

With

$$\boldsymbol{b} = (b_1, b_2 \cdots b_m) \in \mathbb{F}_2^m, \qquad (1.1249)$$

and with $\boldsymbol{b} \lhd \boldsymbol{a}$ meaning that the 1's in $\boldsymbol{b}$ are a subset of the 1's in $\boldsymbol{a}$, the coefficients in (1.1248) are given by [4, p. 372]

$$g(\boldsymbol{a}) = \bigoplus_{\boldsymbol{b} \lhd \boldsymbol{a}} f(b_1, b_2 \cdots b_m). \qquad (1.1250)$$

Proof. For m equal to 1, (1.1241) yields

$$f(x_1) = a_0 \oplus a_1 x_1, \tag{1.1251}$$

and (1.1248) yields

$$f(x_1) = \bigoplus_{\mathbf{a}\in\mathbb{F}_2} g(a_1)\cdot x_1^{a_1} = g(0)\cdot 1 \oplus g(1)\cdot x_1, \tag{1.1252}$$

and the disjunctive normal form for $f(x_1)$ is given by

$$\begin{aligned} f(x_1) &= \bigoplus_{i_1=0}^{1} f(i_1)\cdot y_1^{(i_1)} \\ &= f(0)\cdot y_1^{(0)} \oplus f(1)\cdot y_1^{(1)} \\ &= f(0)\cdot(1 \oplus x_1) \oplus f(1)\cdot x_1 \\ &= f(0)\cdot 1 \oplus \big(f(0) \oplus f(1)\big)\cdot x_1. \end{aligned} \tag{1.1253}$$

Choosing

$$g(0) = a_0 = f(0), \quad g(1) = a_1 = f(0) \oplus f(1) \tag{1.1254}$$

proves (1.1250) and (1.1248) [4, p. 372].

Similarly, for m equal to 2, (1.1241) yields

$$f(x_1, x_2) = a_0 \oplus a_1 x_1 \oplus a_2 x_2 \oplus a_3 x_1 x_2, \tag{1.1255}$$

and (1.1248) yields

$$\begin{aligned} f(x_1, x_2) &= \bigoplus_{\mathbf{a}\in\mathbb{F}_2^2} g(a_1, a_2) \odot x_1^{a_1} \odot x_2^{a_2} \\ &= g(0,0)\odot 1 \oplus g(1,0)\odot x_1 \oplus g(0,1)\odot x_2 \oplus g(1,1)\odot x_1 \odot x_2, \end{aligned} \tag{1.1256}$$

and the disjunctive normal form for $f(x_1, x_2)$ is given by

$$\begin{aligned} f(x_1, x_2) &= \bigoplus_{i_1=0}^{1}\bigoplus_{i_2=0}^{1} f(i_1, i_2) \odot y_1^{(i_1)} \odot y_2^{(i_2)} \\ &= \bigoplus_{i_2=0}^{1} f(0, i_2) \odot y_1^{(0)} \odot y_2^{(i_2)} \odot \bigoplus_{i_2=0}^{1} f(1, i_2) \cdot y_1^{(1)} \odot y_2^{(i_2)} \\ &= f(0,0)\odot y_1^{(0)} \odot y_2^{(0)} \oplus f(0,1)\odot y_1^{(0)} \odot y_2^{(1)} \\ &\quad \oplus f(1,0)\odot y_1^{(1)} \odot y_2^{(0)} \oplus f(1,1)\odot y_1^{(1)} \odot y_2^{(1)} \\ &= f(0,0)\odot(1 \oplus x_1)\odot(1 \oplus x_2) \oplus f(0,1)\odot(1 \oplus x_1)\odot x_2 \\ &\quad \oplus f(1,0)\odot x_1 \odot (1 \oplus x_2) \oplus f(1,1)\odot x_1 \odot x_2 \\ &= f(0,0)\odot[1 \oplus x_1 \oplus x_2 \oplus x_1 \odot x_2] \oplus f(0,1)\odot[x_2 \oplus x_1 \odot x_2] \\ &\quad \oplus f(1,0)\odot[x_1 \oplus x_1 \odot x_2] \oplus f(1,1)\odot x_1 \odot x_2 \\ &= f(0,0)\odot 1 \\ &\quad \oplus \big[f(0,0) \oplus f(1,0)\big]\odot x_1 \end{aligned} \tag{1.1257}$$

$$\oplus \left[f(0,0) \oplus f(0,1)\right] \odot x_2$$
$$\oplus \left[f(0,0) \oplus f(0,1) \oplus f(1,0) \oplus f(1,1)\right] \odot x_1 \odot x_2.$$

Choosing

$$\begin{aligned} g(0,0) &= a_0 = f(0,0), \\ g(1,0) &= a_1 = f(0,0) \oplus f(1,0), \\ g(0,1) &= a_2 = f(0,0) \oplus f(0,1), \\ g(1,1) &= a_3 = f(0,0) \oplus f(0,1) \oplus f(1,0) \oplus f(1,1), \end{aligned} \tag{1.1258}$$

proves (1.1250) and (1.1248) [4, p. 372].

We can easily continue this way. □

For any integers m and r with

$$0 \le r \le m, \quad r, m \in \mathbb{N}, \tag{1.1259}$$

there exists a binary rth-order Reed–Muller (RM) code, denoted by [15, p. 105]

$$\mathcal{RM}(r,m).$$

Let 2 equal to $(x_1, x_2 \cdots x_m)$ be a vector which ranges over $\mathbb{F}_2^m$ and f the vector of length 2^m obtained from a Boolean function $f(x_1, x_2 \cdots x_m)$ as discussed above [4, p. 372].

Definition 1.65 (Reed–Muller (RM) Code). The rth-order Reed–Muller (RM) code $\mathcal{RM}(r,m)$ of length n equal to 2^m for $0 \le r \le m, r, m \in \mathbb{N}$ is the set of all binary vectors $\boldsymbol{f}$, where $f(x_1, x_2 \cdots x_m)$, $m \in \mathbb{N}$ is a Boolean function, which is a polynomial of degree at most r [4, p. 373].

Example 1.69. Let m be equal to 3. Then let us define the Boolean function

$$f(x_1, x_2, x_3) = a_0 \oplus a_1 \odot x_1 \oplus a_2 \odot x_2 \oplus a_3 \odot x_3. \tag{1.1260}$$

The degree of this function is 1. Therefore, according to Definition 1.65, we shall be able to form the first-order $\mathcal{RM}(1,3)$ code using (1.1260). Clearly, we find the following truth table:

$\sum_{j=1}^{m} 2^{j-1} x_j$	0	1	2	3
x_3	0	0	0	0
x_2	0	0	1	1
x_1	0	1	0	1
$f(x_1,x_2,x_3)$	a_0	$a_0 \oplus a_1$	$a_0 \oplus a_2$	$a_0 \oplus a_1 \oplus a_2$
$\sum_{j=1}^{m} 2^{j-1} x_j$	4	5	6	7
x_3	1	1	1	1
x_2	0	0	1	1
x_1	0	1	0	1
$f(x_1,x_2,x_3)$	$a_0 \oplus a_3$	$a_0 \oplus a_1 \oplus a_3$	$a_0 \oplus a_2 \oplus a_3$	$a_0 \oplus a_1 \oplus a_2 \oplus a_3$

(1.1261)

We will find 2^{m+1}, $m \in \mathbb{N}$, i. e., 16, binary vectors $\boldsymbol{f}^{(j)}, j \in \{1, 2 \cdots 2^{m+1}\}$, $m \in \mathbb{N}$.

Let $a_0 = a_1 = a_2 = a_3 = 0$. Then the first binary vector is given by

$$\boldsymbol{f}^{(1)} = \begin{pmatrix} 0 & 0 & 0 & 0 & 0 & 0 & 0 & 0 \end{pmatrix} = \mathbf{0}_8. \tag{1.1262}$$

Now, let $a_3 = 1$ and $a_0 = a_1 = a_2 = 0$. Then the second binary vector is given by

$$\boldsymbol{f}^{(2)} = \begin{pmatrix} 0 & 0 & 0 & 0 & 1 & 1 & 1 & 1 \end{pmatrix}. \tag{1.1263}$$

Now, let $a_2 = 1$ and $a_0 = a_1 = a_3 = 0$. Then the third binary vector is given by

$$\boldsymbol{f}^{(3)} = \begin{pmatrix} 0 & 0 & 1 & 1 & 0 & 0 & 1 & 1 \end{pmatrix}. \tag{1.1264}$$

Now, let $a_1 = 1$ and $a_0 = a_2 = a_3 = 0$. Then the fourth binary vector is given by

$$\boldsymbol{f}^{(4)} = \begin{pmatrix} 0 & 1 & 0 & 1 & 0 & 1 & 0 & 1 \end{pmatrix}. \tag{1.1265}$$

Now, let $a_2 = a_3 = 1$ and $a_0 = a_1 = 0$. Then the fifth binary vector is given by

$$\boldsymbol{f}^{(5)} = \begin{pmatrix} 0 & 0 & 1 & 1 & 1 & 1 & 0 & 0 \end{pmatrix}. \tag{1.1266}$$

Now, let $a_1 = a_3 = 1$ and $a_0 = a_2 = 0$. Then the sixth binary vector is given by

$$\boldsymbol{f}^{(6)} = \begin{pmatrix} 0 & 1 & 0 & 1 & 1 & 0 & 1 & 0 \end{pmatrix}. \tag{1.1267}$$

Now, let $a_1 = a_2 = 1$ and $a_0 = a_3 = 0$. Then the seventh binary vector is given by

$$\boldsymbol{f}^{(7)} = \begin{pmatrix} 0 & 1 & 1 & 0 & 0 & 1 & 1 & 0 \end{pmatrix}. \tag{1.1268}$$

Now, let $a_1 = a_2 = a_3 = 1$ and $a_0 = 0$. Then the eighth binary vector is given by

$$\boldsymbol{f}^{(8)} = \begin{pmatrix} 0 & 1 & 1 & 0 & 1 & 0 & 0 & 1 \end{pmatrix}. \tag{1.1269}$$

Now, let $a_0 = 1$ and $a_1 = a_2 = a_3 = 0$. Then the ninth binary vector is given by

$$\boldsymbol{f}^{(9)} = \begin{pmatrix} 1 & 1 & 1 & 1 & 1 & 1 & 1 & 1 \end{pmatrix} = \mathbf{1}_8. \tag{1.1270}$$

Now, let $a_0 = a_3 = 1$ and $a_1 = a_2 = 0$. Then the tenth binary vector is given by

$$\boldsymbol{f}^{(10)} = \begin{pmatrix} 1 & 1 & 1 & 1 & 0 & 0 & 0 & 0 \end{pmatrix}. \tag{1.1271}$$

Now, let $a_0 = a_2 = 1$ and $a_1 = a_3 = 0$. Then the eleventh binary vector is given by

$$\boldsymbol{f}^{(11)} = \begin{pmatrix} 1 & 1 & 0 & 0 & 1 & 1 & 0 & 0 \end{pmatrix}. \tag{1.1272}$$

Now, let $a_0 = a_1 = 1$ and $a_2 = a_3 = 0$. Then the twelfth binary vector is given by

$$\boldsymbol{f}^{(12)} = \begin{pmatrix} 1 & 0 & 1 & 0 & 1 & 0 & 1 & 0 \end{pmatrix}. \tag{1.1273}$$

Now, let $a_0 = a_2 = a_3 = 1$ and $a_1 = 0$. Then the thirteenth binary vector is given by

$$\boldsymbol{f}^{(13)} = \begin{pmatrix} 1 & 1 & 0 & 0 & 0 & 0 & 1 & 1 \end{pmatrix}. \tag{1.1274}$$

Now, let $a_0 = a_1 = a_3 = 1$ and $a_2 = 0$. Then the fourteenth binary vector is given by

$$f^{(14)} = \begin{pmatrix} 1 & 0 & 1 & 0 & 0 & 1 & 0 & 1 \end{pmatrix}. \tag{1.1275}$$

Now, let $a_0 = a_1 = a_2 = 1$ and $a_3 = 0$. Then the fifteenth binary vector is given by

$$f^{(15)} = \begin{pmatrix} 1 & 0 & 0 & 1 & 1 & 0 & 0 & 1 \end{pmatrix}. \tag{1.1276}$$

Now, let $a_0 = a_1 = a_2 = a_3 = 1$. Then the sixteenth binary vector is given by

$$f^{(16)} = \begin{pmatrix} 1 & 0 & 0 & 1 & 0 & 1 & 1 & 0 \end{pmatrix}. \tag{1.1277}$$

Therefore, the first-order $\mathcal{RM}(1,3)$ code is given by

$$\mathbb{V} = \left\{ \begin{matrix} \begin{pmatrix} 0 & 0 & 0 & 0 & 0 & 0 & 0 & 0 \end{pmatrix}, \begin{pmatrix} 0 & 0 & 0 & 0 & 1 & 1 & 1 & 1 \end{pmatrix}, \\ \begin{pmatrix} 0 & 0 & 1 & 1 & 0 & 0 & 1 & 1 \end{pmatrix}, \begin{pmatrix} 0 & 1 & 0 & 1 & 0 & 1 & 0 & 1 \end{pmatrix}, \\ \begin{pmatrix} 0 & 0 & 1 & 1 & 1 & 1 & 0 & 0 \end{pmatrix}, \begin{pmatrix} 0 & 1 & 0 & 1 & 1 & 0 & 1 & 0 \end{pmatrix}, \\ \begin{pmatrix} 0 & 1 & 1 & 0 & 0 & 1 & 1 & 0 \end{pmatrix}, \begin{pmatrix} 0 & 1 & 1 & 0 & 1 & 0 & 0 & 1 \end{pmatrix}, \\ \begin{pmatrix} 1 & 1 & 1 & 1 & 1 & 1 & 1 & 1 \end{pmatrix}, \begin{pmatrix} 1 & 1 & 1 & 1 & 0 & 0 & 0 & 0 \end{pmatrix}, \\ \begin{pmatrix} 1 & 1 & 0 & 0 & 1 & 1 & 0 & 0 \end{pmatrix}, \begin{pmatrix} 1 & 0 & 1 & 0 & 1 & 0 & 1 & 0 \end{pmatrix}, \\ \begin{pmatrix} 1 & 1 & 0 & 0 & 0 & 0 & 1 & 1 \end{pmatrix}, \begin{pmatrix} 1 & 0 & 1 & 0 & 0 & 1 & 0 & 1 \end{pmatrix}, \\ \begin{pmatrix} 1 & 0 & 0 & 1 & 1 & 0 & 0 & 1 \end{pmatrix}, \begin{pmatrix} 1 & 0 & 0 & 1 & 0 & 1 & 1 & 0 \end{pmatrix} \end{matrix} \right\} \tag{1.1278}$$

Let us define the m be equal to three basis vectors

$$\begin{aligned} v^{(3)} &= \begin{pmatrix} 0 & 0 & 0 & 0 & 1 & 1 & 1 & 1 \end{pmatrix}, \\ v^{(2)} &= \begin{pmatrix} 0 & 0 & 1 & 1 & 0 & 0 & 1 & 1 \end{pmatrix}, \\ v^{(1)} &= \begin{pmatrix} 0 & 1 & 0 & 1 & 0 & 1 & 0 & 1 \end{pmatrix}. \end{aligned} \tag{1.1279}$$

Then, together with

$$v^{(0)} = \mathbf{1}_8 = \begin{pmatrix} 1 & 1 & 1 & 1 & 1 & 1 & 1 & 1 \end{pmatrix}, \tag{1.1280}$$

the $(m+1) \times 2^m$ equal to 4×8 generator matrix of the first-order $\mathcal{RM}(1,3)$ code can be given in the following form:

$$G_{\mathcal{RM}(1,3)} = \begin{pmatrix} v^{(0)} \\ v^{(1)} \\ v^{(2)} \\ v^{(3)} \end{pmatrix} = \begin{pmatrix} 1 & 1 & 1 & 1 & 1 & 1 & 1 & 1 \\ 0 & 1 & 0 & 1 & 0 & 1 & 0 & 1 \\ 0 & 0 & 1 & 1 & 0 & 0 & 1 & 1 \\ 0 & 0 & 0 & 0 & 1 & 1 & 1 & 1 \end{pmatrix}, \tag{1.1281}$$

and we can map the 16 codewords of the first-order $\mathcal{RM}(1,3)$ code to the sixteen functions $f^{(1)}$ through $f^{(16)}$ in the following way [4, p. 373]:

$$
\begin{aligned}
\mathbf{0}_8 &= \begin{pmatrix} 0 & 0 & 0 & 0 & 0 & 0 & 0 & 0 \end{pmatrix} = \boldsymbol{f}^{(1)}, \\
\boldsymbol{v}^{(3)} &= \begin{pmatrix} 0 & 0 & 0 & 0 & 1 & 1 & 1 & 1 \end{pmatrix} = \boldsymbol{f}^{(2)}, \\
\boldsymbol{v}^{(2)} &= \begin{pmatrix} 0 & 0 & 1 & 1 & 0 & 0 & 1 & 1 \end{pmatrix} = \boldsymbol{f}^{(3)}, \\
\boldsymbol{v}^{(1)} &= \begin{pmatrix} 0 & 1 & 0 & 1 & 0 & 1 & 0 & 1 \end{pmatrix} = \boldsymbol{f}^{(4)}, \\
\boldsymbol{v}^{(2)} \oplus \boldsymbol{v}^{(3)} &= \begin{pmatrix} 0 & 0 & 1 & 1 & 1 & 1 & 0 & 0 \end{pmatrix} = \boldsymbol{f}^{(5)}, \\
\boldsymbol{v}^{(1)} \oplus \boldsymbol{v}^{(3)} &= \begin{pmatrix} 0 & 1 & 0 & 1 & 1 & 0 & 1 & 0 \end{pmatrix} = \boldsymbol{f}^{(6)}, \\
\boldsymbol{v}^{(1)} \oplus \boldsymbol{v}^{(2)} &= \begin{pmatrix} 0 & 1 & 1 & 0 & 0 & 1 & 1 & 0 \end{pmatrix} = \boldsymbol{f}^{(7)}, \\
\boldsymbol{v}^{(1)} \oplus \boldsymbol{v}^{(2)} \oplus \boldsymbol{v}^{(3)} &= \begin{pmatrix} 0 & 1 & 1 & 0 & 1 & 0 & 0 & 1 \end{pmatrix} = \boldsymbol{f}^{(8)}, \\
\mathbf{1}_8 &= \begin{pmatrix} 1 & 1 & 1 & 1 & 1 & 1 & 1 & 1 \end{pmatrix} = \boldsymbol{f}^{(9)}, \\
\mathbf{1}_8 \oplus \boldsymbol{v}^{(3)} &= \begin{pmatrix} 1 & 1 & 1 & 1 & 0 & 0 & 0 & 0 \end{pmatrix} = \boldsymbol{f}^{(10)}, \\
\mathbf{1}_8 \oplus \boldsymbol{v}^{(2)} &= \begin{pmatrix} 1 & 1 & 0 & 0 & 1 & 1 & 0 & 0 \end{pmatrix} = \boldsymbol{f}^{(11)}, \\
\mathbf{1}_8 \oplus \boldsymbol{v}^{(1)} &= \begin{pmatrix} 1 & 0 & 1 & 0 & 1 & 0 & 1 & 0 \end{pmatrix} = \boldsymbol{f}^{(12)}, \\
\mathbf{1}_8 \oplus \boldsymbol{v}^{(2)} \oplus \boldsymbol{v}^{(3)} &= \begin{pmatrix} 1 & 1 & 0 & 0 & 0 & 0 & 1 & 1 \end{pmatrix} = \boldsymbol{f}^{(13)}, \\
\mathbf{1}_8 \oplus \boldsymbol{v}^{(1)} \oplus \boldsymbol{v}^{(3)} &= \begin{pmatrix} 1 & 0 & 1 & 0 & 0 & 1 & 0 & 1 \end{pmatrix} = \boldsymbol{f}^{(14)}, \\
\mathbf{1}_8 \oplus \boldsymbol{v}^{(1)} \oplus \boldsymbol{v}^{(2)} &= \begin{pmatrix} 1 & 0 & 0 & 1 & 1 & 0 & 0 & 1 \end{pmatrix} = \boldsymbol{f}^{(15)}, \\
\mathbf{1}_8 \oplus \boldsymbol{v}^{(1)} \oplus \boldsymbol{v}^{(2)} \oplus \boldsymbol{v}^{(3)} &= \begin{pmatrix} 1 & 0 & 0 & 1 & 0 & 1 & 1 & 0 \end{pmatrix} = \boldsymbol{f}^{(16)}.
\end{aligned}
\tag{1.1282}
$$

It is easy to see that the minimum distance $d_{\min}$ of the first-order $\mathcal{RM}(1,3)$ code is equal to 4, i. e., 2^{m-1}, $m \in \mathbb{N}$.

Therefore, the first-order Reed–Muller (RM) code $\mathcal{RM}(1,3)$ is a $(2^m, [m+1], 2^{m-1})$ binary linear block code $\mathbb{V}$ with m equal to 3, i. e., it is a $(8,4,4)$ binary linear block code $\mathbb{V}$.

1.17.2 Properties of Reed–Muller (RM) Codes

The rth-order Reed–Muller code $\mathcal{RM}(r,m)$ has the following parameters [15, p. 105], [4, p. 373ff.]:

(i) The *length* n of the rth-order Reed–Muller (RM) code $\mathcal{RM}(r,m)$ is given by Definition 1.65 [15, p. 105], [4, p. 373]

$$n = 2^m, \quad m, n \in \mathbb{N}. \tag{1.1283}$$

(ii) Taking (1.1240) into account, the *dimension* $k(r,m)$ of the rth-order Reed–Muller (RM) code $\mathcal{RM}(r,m)$ is given by [15, p. 105], [4, pp. 371, 373]

$$k(r,m) = \binom{m}{0} + \binom{m}{1} + \binom{m}{2} + \cdots + \binom{m}{r} = \sum_{i=0}^{r} \binom{m}{i}, \tag{1.1284}$$

$$0 \le r \le m, \quad m, r \in \mathbb{N}.$$

(iii) The *minimum distance* $d_{\min}$ of the rth-order Reed–Muller (RM) code $\mathcal{RM}(r,m)$ is given by [15, p. 105], [4, Theorem 3, p. 375]

$$d_{\min} = 2^{m-r}, \quad 0 \le r \le m, \quad m, r \in \mathbb{N}. \tag{1.1285}$$

The rth-order Reed–Muller (RM) code $\mathcal{RM}(r,m)$ is hence a

$$\left(2^m, \left[\sum_{i=0}^{r} \binom{m}{i}\right], 2^{m-r}\right) \text{ binary linear block code } \mathbb{V}$$

[15, p. 105], [4, p. 376].

(iv) Clearly, with the dimension $k(r,m)$, $0 \le r \le m$, $m, r \in \mathbb{N}$, of the rth-order Reed–Muller (RM) code $\mathcal{RM}(r,m)$, given by (1.1284), the *code rate* R of the rth-order Reed–Muller (RM) code $\mathcal{RM}(r,m)$ is given by

$$R = \frac{k(r,m)}{n} = \frac{\sum_{i=0}^{r} \binom{m}{i}}{2^m}, \quad 0 \le r \le m, \quad r, m \in \mathbb{N}, \tag{1.1286}$$

and we find

$$\underbrace{\frac{1}{2^m}}_{R \text{ of the } \mathcal{RM}(0,m) \text{ code}} \le \underbrace{\frac{1}{2^m} + \frac{m}{2^m}}_{R \text{ of the } \mathcal{RM}(1,m) \text{ code}} \le \underbrace{\frac{1}{2^m} + \frac{m}{2^m} + \frac{(m-1)m}{2^{m+1}}}_{R \text{ of the } \mathcal{RM}(2,m) \text{ code}} \le \underbrace{1}_{R \text{ of the } \mathcal{RM}(m,m) \text{ code}},$$
$$m \in \mathbb{N}. \tag{1.1287}$$

Example 1.70. Let $m = 5$ and $r = 1$. Then

$$n = 2^5 = 32, \quad k(1,5) = 1 + 5 = 6, \quad d_{\min} = 2^{5-1} = 16. \tag{1.1288}$$

Hence, the first-order Reed–Muller (RM) code $\mathcal{RM}(1,5)$ is a $(32, 6, 16)$ binary linear block code $\mathbb{V}$.

The code rate of the first-order Reed–Muller (RM) code $\mathcal{RM}(1,5)$ is remarkably low, given by only

$$R = \frac{\sum_{i=0}^{1} \binom{5}{i}}{2^5} = \frac{3}{16} = 0.1875 < \frac{1}{5}. \tag{1.1289}$$

Example 1.71. Let $m = 5$ and $r = 2$. Then we yield

$$n = 2^5 = 32, \quad k(2,5) = 1 + 5 + 10 = 16, \quad d_{\min} = 2^{5-2} = 8. \tag{1.1290}$$

Hence, the second-order Reed–Muller (RM) code $\mathcal{RM}(2,5)$ is a $(32, 16, 8)$ binary linear block code $\mathbb{V}$.

The code rate of the second-order Reed–Muller (RM) code $\mathcal{RM}(2,5)$ is much more acceptable, given by

$$R = \frac{\sum_{i=0}^{2} \binom{5}{i}}{2^5} = \frac{1}{2}. \tag{1.1291}$$

For $1 \leq i \leq m$, $i, m \in \mathbb{N}$, let $\boldsymbol{v}^{(i)}$ be a 2^m-tuple over $\mathbb{F}_2$ of the following form [15, equation (4.4), p. 105]:

$$\boldsymbol{v}^{(i)} = (\underbrace{0\cdots0}_{2^{i-1}}\underbrace{1\cdots1}_{2^{i-1}}\underbrace{0\cdots0}_{2^{i-1}}\cdots\underbrace{1\cdots1}_{2^{i-1}}), \quad 1 \leq i \leq m, \quad i, m \in \mathbb{N}. \tag{1.1292}$$

Obviously, $\boldsymbol{v}^{(i)}$, $i \in \mathbb{N}$ consists of $2^{m-(i-1)}$, $i, m \in \mathbb{N}$ alternating “all zeros tuples” and “all ones tuples,” each of length 2^{i-1}, $i \in \mathbb{N}$ [15, equation (4.4), p. 105]. The Hamming weight of $\boldsymbol{v}^{(i)}$ is given by

$$w_{\mathrm{H}}\{\boldsymbol{v}^{(i)}\} = 2^{m-(i-1)-1} \cdot 2^{i-1} = 2^{m-1}, \quad 1 \leq i \leq m, \quad i, m \in \mathbb{N}. \tag{1.1293}$$

Example 1.72. Let $m = 4$ [15, p. 105]. Then each $\boldsymbol{v}^{(i)}$, $1 \leq i \leq 4$, $i \in \mathbb{N}$ has 2^4 equal to 16 components, i. e., coordinates. Hence, $\boldsymbol{v}^{(i)}$ is a 16-tuple [15, p. 105]. Since $m = 4$, there are m equal to 4 such 16-tuples $\boldsymbol{v}^{(i)}$, $1 \leq i \leq 4$, $i \in \mathbb{N}$.

The first 16-tuple $\boldsymbol{v}^{(1)}$ has a total of $2^{4-(i-1)}$ equal to 2^4, i. e., 16, alternating “all zeros tuples” and “all ones tuples” of length 2^{i-1} equal to 2^0, i. e., 1:

$$\boldsymbol{v}^{(1)} = (0, 1, 0, 1, 0, 1, 0, 1, 0, 1, 0, 1, 0, 1, 0, 1). \tag{1.1294}$$

As expected, we have

$$w_{\mathrm{H}}\{\boldsymbol{v}^{(1)}\} = 2^{4-1} = 8. \tag{1.1295}$$

The second 16-tuple $\boldsymbol{v}^{(2)}$ has a total of $2^{4-(i-1)}$ equal to 2^3, i. e., 8, alternating “all zeros tuples” and “all ones tuples” of length 2^{i-1} equal to 2^1, i. e., 2:

$$\boldsymbol{v}^{(2)} = (0, 0, 1, 1, 0, 0, 1, 1, 0, 0, 1, 1, 0, 0, 1, 1). \tag{1.1296}$$

As expected, we have

$$w_{\mathrm{H}}\{\boldsymbol{v}^{(2)}\} = 2^{4-1} = 8. \tag{1.1297}$$

The third 16-tuple $\boldsymbol{v}^{(3)}$ has a total of $2^{4-(i-1)}$ equal to 2^2, i. e., 4, alternating “all zeros tuples” and “all ones tuples” of length 2^{i-1} equal to 2^2, i. e., 4:

$$\boldsymbol{v}^{(3)} = (0, 0, 0, 0, 1, 1, 1, 1, 0, 0, 0, 0, 1, 1, 1, 1). \tag{1.1298}$$

As expected, we have

$$w_{\mathrm{H}}\{\boldsymbol{v}^{(3)}\} = 2^{4-1} = 8. \tag{1.1299}$$

The last, i. e., the fourth, 16-tuple $\boldsymbol{v}^{(4)}$ has a total of $2^{4-(i-1)}$ equal to 2^1, i. e., 2, alternating “all zeros tuples” and “all ones tuples” of length 2^{i-1} equal to 2^3, i. e., 8:

$$\boldsymbol{v}^{(4)} = (0, 0, 0, 0, 0, 0, 0, 0, 1, 1, 1, 1, 1, 1, 1, 1). \tag{1.1300}$$

As expected, we have

$$w_{\mathrm{H}}\{\boldsymbol{v}^{(4)}\} = 2^{4-1} = 8. \tag{1.1301}$$

Let

$$a = (a_0, a_1 \cdots a_{2^m-1}) \in \mathbb{F}_2^{2^m}, \quad m \in \mathbb{N}, \tag{1.1302}$$

and

$$b = (b_0, b_1 \cdots b_{2^m-1}) \in \mathbb{F}_2^{2^m}, \quad m \in \mathbb{N}, \tag{1.1303}$$

two binary 2^m-tuples [15, p. 105]. We define the *logic (Boolean) product* of $\boldsymbol{a} \in \mathbb{F}_2^{2^m}$ and $\boldsymbol{b} \in \mathbb{F}_2^{2^m}$ by [15, p. 105]

$$\boldsymbol{a} \boxdot \boldsymbol{b} = (a_0 \odot b_0, a_1 \odot b_1 \cdots a_{2^m-1} \odot b_{2^m-1}) \in \mathbb{F}_2^{2^m}, \quad n \in \mathbb{N}^*. \tag{1.1304}$$

Furthermore, let [15, p. 105]

$$\boldsymbol{v}^{(0)} = \mathbf{1}_{2^m}, \quad n \in \mathbb{N}^* \tag{1.1305}$$

be the all ones vector of length 2^m and Hamming weight 2^m. The product vector

$$\boldsymbol{v}^{(i_1)} \boxdot \boldsymbol{v}^{(i_2)} \boxdot \cdots \boxdot \boldsymbol{v}^{(i_p)}, \quad 1 \le i_1 \le i_2 \le \cdots \le i_p \le m, \quad i_1, i_2 \cdots i_p, m \in \mathbb{N} \tag{1.1306}$$

is said to have degree p [15, p. 105f.]. Since the Hamming weights of $\boldsymbol{v}^{(1)}, \boldsymbol{v}^{(2)}, \boldsymbol{v}^{(3)} \cdots \boldsymbol{v}^{(m)}$ are all equal to 2^{m-1}, the Hamming weight of the product vector $\boldsymbol{v}^{(i_1)} \boxdot \boldsymbol{v}^{(i_2)} \boxdot \cdots \boxdot \boldsymbol{v}^{(i_p)}$, $1 \le i_1 \le i_2 \le \cdots \le i_p \le m$ is also even and a power of 2 [15, p. 106]

$$w_{\mathrm{H}}\{\boldsymbol{v}^{(i_1)} \boxdot \boldsymbol{v}^{(i_2)} \boxdot \cdots \boxdot \boldsymbol{v}^{(i_p)}\} = 2^{m-p}, \quad 1 \le i_1 \le i_2 \le \cdots \le i_p \le m, \quad i_1, i_2 \cdots i_p, m \in \mathbb{N}. \tag{1.1307}$$

Example 1.73. Let $m = 4$ [15, p. 106]. We have

$$\boldsymbol{v}^{(0)} = (1,1,1,1,1,1,1,1,1,1,1,1,1,1,1,1), \quad w_{\mathrm{H}}\{\boldsymbol{v}^{(0)}\} = 16. \tag{1.1308}$$

In Example 1.72, we derived the four 16-tuples

$$\begin{aligned} \boldsymbol{v}^{(1)} &= (0,1,0,1,0,1,0,1,0,1,0,1,0,1,0,1), \\ \boldsymbol{v}^{(2)} &= (0,0,1,1,0,0,1,1,0,0,1,1,0,0,1,1), \\ \boldsymbol{v}^{(3)} &= (0,0,0,0,1,1,1,1,0,0,0,0,1,1,1,1), \\ \boldsymbol{v}^{(4)} &= (0,0,0,0,0,0,0,0,1,1,1,1,1,1,1,1), \end{aligned} \tag{1.1309}$$

with

$$w_{\mathrm{H}}\{\boldsymbol{v}^{(1)}\} = w_{\mathrm{H}}\{\boldsymbol{v}^{(2)}\} = w_{\mathrm{H}}\{\boldsymbol{v}^{(3)}\} = w_{\mathrm{H}}\{\boldsymbol{v}^{(4)}\} = 2^{4-1} = 8. \tag{1.1310}$$

We find

$$\begin{aligned}
\boldsymbol{v}^{(1)} \boxdot \boldsymbol{v}^{(2)} &= (0,1,0,1,0,1,0,1,0,1,0,1,0,1,0,1) \boxdot (0,0,1,1,0,0,1,1,0,0,1,1,0,0,1,1)\\
&= (0,0,0,1,0,0,0,1,0,0,0,1,0,0,0,1),\\
\boldsymbol{v}^{(1)} \boxdot \boldsymbol{v}^{(3)} &= (0,1,0,1,0,1,0,1,0,1,0,1,0,1,0,1) \boxdot (0,0,0,0,1,1,1,1,0,0,0,0,1,1,1,1)\\
&= (0,0,0,0,0,1,0,1,0,0,0,0,0,1,0,1),\\
\boldsymbol{v}^{(1)} \boxdot \boldsymbol{v}^{(4)} &= (0,1,0,1,0,1,0,1,0,1,0,1,0,1,0,1) \boxdot (0,0,0,0,0,0,0,0,1,1,1,1,1,1,1,1)\\
&= (0,0,0,0,0,0,0,0,0,1,0,1,0,1,0,1),\\
\boldsymbol{v}^{(2)} \boxdot \boldsymbol{v}^{(3)} &= (0,0,1,1,0,0,1,1,0,0,1,1,0,0,1,1) \boxdot (0,0,0,0,1,1,1,1,0,0,0,0,1,1,1,1)\\
&= (0,0,0,0,0,0,1,1,0,0,0,0,0,0,1,1),\\
\boldsymbol{v}^{(2)} \boxdot \boldsymbol{v}^{(4)} &= (0,0,1,1,0,0,1,1,0,0,1,1,0,0,1,1) \boxdot (0,0,0,0,0,0,0,0,1,1,1,1,1,1,1,1)\\
&= (0,0,0,0,0,0,0,0,0,0,1,1,0,0,1,1),\\
\boldsymbol{v}^{(3)} \boxdot \boldsymbol{v}^{(4)} &= (0,0,0,0,1,1,1,1,0,0,0,0,1,1,1,1) \boxdot (0,0,0,0,0,0,0,0,1,1,1,1,1,1,1,1)\\
&= (0,0,0,0,0,0,0,0,0,0,0,0,1,1,1,1).
\end{aligned} \tag{1.1311}$$

As expected, we have

$$\begin{aligned}
w_{\mathrm{H}}\{\boldsymbol{v}^{(1)} \boxdot \boldsymbol{v}^{(2)}\} &= w_{\mathrm{H}}\{\boldsymbol{v}^{(1)} \boxdot \boldsymbol{v}^{(3)}\}\\
&= w_{\mathrm{H}}\{\boldsymbol{v}^{(1)} \boxdot \boldsymbol{v}^{(4)}\}\\
&= w_{\mathrm{H}}\{\boldsymbol{v}^{(2)} \boxdot \boldsymbol{v}^{(3)}\}\\
&= w_{\mathrm{H}}\{\boldsymbol{v}^{(2)} \boxdot \boldsymbol{v}^{(4)}\}\\
&= w_{\mathrm{H}}\{\boldsymbol{v}^{(3)} \boxdot \boldsymbol{v}^{(4)}\}\\
&= 2^{4-2} = 4.
\end{aligned} \tag{1.1312}$$

Furthermore, we find

$$\begin{aligned}
\boldsymbol{v}^{(1)} \boxdot \boldsymbol{v}^{(2)} \boxdot \boldsymbol{v}^{(3)} &= (0,0,0,1,0,0,0,1,0,0,0,1,0,0,0,1)\\
&\boxdot (0,0,0,0,1,1,1,1,0,0,0,0,1,1,1,1)\\
&= (0,1,0,1,0,1,0,1,0,1,0,1,0,1,0,1)\\
&\boxdot (0,0,0,0,0,0,1,1,0,0,0,0,0,0,1,1)\\
&= (0,0,0,0,0,0,0,1,0,0,0,0,0,0,0,1),\\
\boldsymbol{v}^{(1)} \boxdot \boldsymbol{v}^{(2)} \boxdot \boldsymbol{v}^{(4)} &= (0,0,0,1,0,0,0,1,0,0,0,1,0,0,0,1)\\
&\boxdot (0,0,0,0,0,0,0,0,1,1,1,1,1,1,1,1)\\
&= (0,1,0,1,0,1,0,1,0,1,0,1,0,1,0,1)\\
&\boxdot (0,0,0,0,0,0,0,0,0,0,1,1,0,0,1,1)\\
&= (0,0,0,0,0,0,0,0,0,0,0,1,0,0,0,1),\\
\boldsymbol{v}^{(1)} \boxdot \boldsymbol{v}^{(3)} \boxdot \boldsymbol{v}^{(4)} &= (0,0,0,0,0,1,0,1,0,0,0,0,0,1,0,1)\\
&\boxdot (0,0,0,0,0,0,0,0,1,1,1,1,1,1,1,1)\\
&= (0,1,0,1,0,1,0,1,0,1,0,1,0,1,0,1)\\
&\boxdot (0,0,0,0,0,0,0,0,0,0,0,0,1,1,1,1)\\
&= (0,0,0,0,0,0,0,0,0,0,0,0,0,1,0,1),
\end{aligned} \tag{1.1313}$$

$$\begin{aligned}\boldsymbol{v}^{(2)} \boxdot \boldsymbol{v}^{(3)} \boxdot \boldsymbol{v}^{(4)} &= (0,0,0,0,0,0,1,1,0,0,0,0,0,0,1,1)\\ &\boxdot (0,0,0,0,0,0,0,0,1,1,1,1,1,1,1,1)\\ &= (0,0,1,1,0,0,1,1,0,0,1,1,0,0,1,1)\\ &\boxdot (0,0,0,0,0,0,0,0,0,0,0,0,1,1,1,1)\\ &= (0,0,0,0,0,0,0,0,0,0,0,0,0,0,1,1).\end{aligned}$$

As expected, we have

$$\begin{aligned}w_{\mathrm{H}}\left\{\boldsymbol{v}^{(1)} \boxdot \boldsymbol{v}^{(2)} \boxdot \boldsymbol{v}^{(3)}\right\} &= w_{\mathrm{H}}\left\{\boldsymbol{v}^{(1)} \boxdot \boldsymbol{v}^{(2)} \boxdot \boldsymbol{v}^{(4)}\right\}\\ &= w_{\mathrm{H}}\left\{\boldsymbol{v}^{(1)} \boxdot \boldsymbol{v}^{(3)} \boxdot \boldsymbol{v}^{(4)}\right\}\\ &= w_{\mathrm{H}}\left\{\boldsymbol{v}^{(2)} \boxdot \boldsymbol{v}^{(3)} \boxdot \boldsymbol{v}^{(4)}\right\}\\ &= 2^{4-3} = 2.\end{aligned} \tag{1.1314}$$

Finally, we find

$$\begin{aligned}\boldsymbol{v}^{(1)} \boxdot \boldsymbol{v}^{(2)} \boxdot \boldsymbol{v}^{(3)} \boxdot \boldsymbol{v}^{(4)} &= (0,0,0,0,0,0,0,1,0,0,0,0,0,0,0,1)\\ &\boxdot (0,0,0,0,0,0,0,0,1,1,1,1,1,1,1,1)\\ &= (0,0,0,0,0,0,0,0,0,0,0,0,0,0,0,1).\end{aligned} \tag{1.1315}$$

As expected, we have

$$w_{\mathrm{H}}\left\{\boldsymbol{v}^{(1)} \boxdot \boldsymbol{v}^{(2)} \boxdot \boldsymbol{v}^{(3)} \boxdot \boldsymbol{v}^{(4)}\right\} = 2^{4-4} = 1. \tag{1.1316}$$

The rth-order Reed–Muller (RM) code $\mathcal{RM}(r,m)$ of length 2^m is generated, i. e., *spanned* by the following set of $k(r,m)$, $0 \le r \le m$, $m, r \in \mathbb{N}$ mutually independent binary vectors, taken from $\mathbb{F}_2^{2^m}$ [15, equation (4.5), p. 106]:

$$\left\{\begin{matrix} \boldsymbol{v}^{(0)}, \\ \boldsymbol{v}^{(1)}, \boldsymbol{v}^{(2)} \dots \boldsymbol{v}^{(m)}, \\ \boldsymbol{v}^{(1)} \boxdot \boldsymbol{v}^{(2)}, \boldsymbol{v}^{(1)} \boxdot \boldsymbol{v}^{(3)} \dots \boldsymbol{v}^{(m-1)} \boxdot \boldsymbol{v}^{(m)}, \\ \boldsymbol{v}^{(1)} \boxdot \boldsymbol{v}^{(2)} \boxdot \boldsymbol{v}^{(3)}, \boldsymbol{v}^{(1)} \boxdot \boldsymbol{v}^{(3)} \boxdot \boldsymbol{v}^{(4)} \dots \boldsymbol{v}^{(m-2)} \boxdot \boldsymbol{v}^{(m-1)} \boxdot \boldsymbol{v}^{(m)}, \\ \vdots \\ \text{up to products of degree } r \end{matrix}\right\}, \quad 0 \le r \le m, \quad m, r \in \mathbb{N}. \tag{1.1317}$$

The generator matrix $\boldsymbol{G}_{\mathcal{RM}(r,m)}$ is established by using these $k(r,m)$, $0 \le r \le m$, $m, r \in \mathbb{N}$ mutually independent binary vectors of (1.1317), arranging these $k(r,m)$, $0 \le r \le m$, $m, r \in \mathbb{N}$ mutually independent binary vectors of (1.1317) as the rows of $\boldsymbol{G}_{\mathcal{RM}(r,m)}$, yielding [15, equation (4.5), p. 106]:

$$\boldsymbol{G}_{\mathcal{RM}(r,m)} = \begin{pmatrix} \boldsymbol{v}^{(0)} \\ \boldsymbol{v}^{(1)} \\ \boldsymbol{v}^{(2)} \\ \vdots \\ \boldsymbol{v}^{(m)} \\ \boldsymbol{v}^{(1)} \boxdot \boldsymbol{v}^{(2)} \\ \boldsymbol{v}^{(1)} \boxdot \boldsymbol{v}^{(3)} \\ \vdots \\ \boldsymbol{v}^{(m-1)} \boxdot \boldsymbol{v}^{(m)} \\ \boldsymbol{v}^{(1)} \boxdot \boldsymbol{v}^{(2)} \boxdot \boldsymbol{v}^{(3)} \\ \boldsymbol{v}^{(1)} \boxdot \boldsymbol{v}^{(3)} \boxdot \boldsymbol{v}^{(4)} \\ \vdots \\ \boldsymbol{v}^{(m-2)} \boxdot \boldsymbol{v}^{(m-1)} \boxdot \boldsymbol{v}^{(m)} \\ \vdots \\ \text{up to products of degree } r \end{pmatrix} = \begin{pmatrix} \boldsymbol{1}_{2^m} \\ \boldsymbol{v}^{(1)} \\ \boldsymbol{v}^{(2)} \\ \vdots \\ \boldsymbol{v}^{(m)} \\ \boldsymbol{v}^{(1)} \boxdot \boldsymbol{v}^{(2)} \\ \boldsymbol{v}^{(1)} \boxdot \boldsymbol{v}^{(3)} \\ \vdots \\ \boldsymbol{v}^{(m-1)} \boxdot \boldsymbol{v}^{(m)} \\ \boldsymbol{v}^{(1)} \boxdot \boldsymbol{v}^{(2)} \boxdot \boldsymbol{v}^{(3)} \\ \boldsymbol{v}^{(1)} \boxdot \boldsymbol{v}^{(3)} \boxdot \boldsymbol{v}^{(4)} \\ \vdots \\ \boldsymbol{v}^{(m-2)} \boxdot \boldsymbol{v}^{(m-1)} \boxdot \boldsymbol{v}^{(m)} \\ \vdots \\ \text{up to products of degree } r \end{pmatrix}, \tag{1.1318}$$

$$0 \le r \le m, \quad m, r \in \mathbb{N}.$$

Therefore, the rth-order Reed–Muller (RM) code $\mathcal{RM}(r,m)$ has dimension $k(r,m)$ [15, p. 106].

For $0 \le p \le r$, $p, r \in \mathbb{N}$, there are exactly

$$\binom{m}{p} = \frac{m!}{p!(m-p)!}, \quad 0 \le p \le r \le m, \quad m, p, r \in \mathbb{N}, \tag{1.1319}$$

rows of Hamming weight 2^{m-p}, $m, p \in \mathbb{N}$ in $\boldsymbol{G}_{\mathcal{RM}(r,m)}$ [15, p. 106].

Let us look at r being equal to 0, first. Then p can only be zero. In the case of the zeroth-order Reed–Muller (RM) code $\mathcal{RM}(0,m)$, we have

$$\binom{m}{0} = \frac{m!}{m!} = 1, \quad m \in \mathbb{N}, \tag{1.1320}$$

row of Hamming weight 2^m, $m \in \mathbb{N}$ in $\boldsymbol{G}_{\mathcal{RM}(0,m)}$, $m \in \mathbb{N}$. There are no further rows in $\boldsymbol{G}_{\mathcal{RM}(0,m)}$, $m \in \mathbb{N}$. Therefore, the zeroth-order Reed–Muller (RM) code $\mathcal{RM}(0,m)$ is the repetition code of length 2^m, generated by the 1×2^m generator matrix

$$\boldsymbol{G}_{\mathcal{RM}(0,m)} = \left(\boldsymbol{v}^{(0)}\right) = (\boldsymbol{1}_{2^m}). \tag{1.1321}$$

Let us now look at r being equal to 1. In this case, p can assume the values 0 and 1. Hence, the generator matrix $\boldsymbol{G}_{\mathcal{RM}(1,m)}$ of the first-order Reed–Muller (RM) code $\mathcal{RM}(1,m)$ has

$$\binom{m}{0} = \frac{m!}{m!} = 1, \quad m \in \mathbb{N}, \tag{1.1322}$$

row with Hamming weight 2^m, $m \in \mathbb{N}$, i. e., the all ones vector $\boldsymbol{v}^{(0)}$ equal to $\mathbf{1}_{2^m}$, and

$$\binom{m}{1} = \frac{m!}{(m-1)!} = m, \quad m \in \mathbb{N}, \tag{1.1323}$$

rows with Hamming weight 2^{m-1}, $m \in \mathbb{N}$, which are the binary vectors $\boldsymbol{v}^{(i)}$, $1 \leq i \leq m$, $i, m \in \mathbb{N}$. There are no further rows in $\boldsymbol{G}_{\mathcal{RM}(1,m)}$. Therefore, the first-order Reed–Muller (RM) code $\mathcal{RM}(1,m)$ is generated by the $(m+1) \times 2^m$, $m \in \mathbb{N}$, generator matrix

$$\boldsymbol{G}_{\mathcal{RM}(1,m)} = \begin{pmatrix} \boldsymbol{v}^{(0)} \\ \boldsymbol{v}^{(1)} \\ \boldsymbol{v}^{(2)} \\ \vdots \\ \boldsymbol{v}^{(m-1)} \\ \boldsymbol{v}^{(m)} \end{pmatrix} = \begin{pmatrix} \mathbf{1}_{2^m} \\ \boldsymbol{v}^{(1)} \\ \boldsymbol{v}^{(2)} \\ \vdots \\ \boldsymbol{v}^{(m-1)} \\ \boldsymbol{v}^{(m)} \end{pmatrix}, \quad m \in \mathbb{N}. \tag{1.1324}$$

Summarizing all the above discussions in the view of the first-order Reed–Muller (RM) code $\mathcal{RM}(1,m)$, we find the following definition [23, Definition 5.1, p. 107f.].

Definition 1.66 (First-order Reed–Muller (RM) Code). The generator matrix $\boldsymbol{G}_{\mathcal{RM}(1,m)}$ of the first-order Reed–Muller (RM) code $\mathcal{RM}(1,m)$ consists of [23, Definition 5.1, p. 107f.]
- a row being the zero vector $\mathbf{0}_{2^m}$, $m \in \mathbb{N}^*$, and
- a submatrix, which has 2^m, $m \in \mathbb{N}^*$ columns, each being a distinct m-tuple, $m \in \mathbb{N}^*$, taken $\mathbb{F}_2^m$, $m \in \mathbb{N}^*$.

The first-order Reed–Muller (RM) code $\mathcal{RM}(1,m)$ is a $(2^m, m+1, 2^{m-1})$ binary linear block code [23, Definition 5.1, p. 107f.].

Example 1.74. The first-order Reed–Muller (RM) code $\mathcal{RM}(1,4)$ is a $(16,5,8)$ binary linear block code with the 5×16 generator matrix

$$\boldsymbol{G}_{\mathcal{RM}(1,4)} = \begin{pmatrix} \boldsymbol{v}^{(0)} \\ \boldsymbol{v}^{(4)} \\ \boldsymbol{v}^{(3)} \\ \boldsymbol{v}^{(2)} \\ \boldsymbol{v}^{(1)} \end{pmatrix} = \begin{pmatrix} 1&1&1&1&1&1&1&1&1&1&1&1&1&1&1&1 \\ 0&0&0&0&0&0&0&0&1&1&1&1&1&1&1&1 \\ 0&0&0&0&1&1&1&1&0&0&0&0&1&1&1&1 \\ 0&0&1&1&0&0&1&1&0&0&1&1&0&0&1&1 \\ 0&1&0&1&0&1&0&1&0&1&0&1&0&1&0&1 \end{pmatrix}, \tag{1.1325}$$

when reordering the rows; cf. also Example 1.75.

Setting out from (1.1325), we find the $(m+1) \times 2^m$, i. e., 5×16, matrix

$$\boldsymbol{W}_{(m+1)\times 2^m} = \mathbf{1}_{(m+1)\times 2^m} - 2 \cdot \boldsymbol{G}_{\mathcal{RM}(1,m)} \tag{1.1326}$$

$$= \boldsymbol{W}_{5\times16}$$
$$= \boldsymbol{1}_{5\times16} - 2\cdot \boldsymbol{G}_{\mathcal{RM}(1,4)}$$
$$= \begin{pmatrix} -1 & -1 & -1 & -1 & -1 & -1 & -1 & -1 & -1 & -1 & -1 & -1 & -1 & -1 & -1 & -1 \\ +1 & +1 & +1 & +1 & +1 & +1 & +1 & +1 & -1 & -1 & -1 & -1 & -1 & -1 & -1 & -1 \\ +1 & +1 & +1 & +1 & -1 & -1 & -1 & -1 & +1 & +1 & +1 & +1 & -1 & -1 & -1 & -1 \\ +1 & +1 & -1 & -1 & +1 & +1 & -1 & -1 & +1 & +1 & -1 & -1 & +1 & +1 & -1 & -1 \\ +1 & -1 & +1 & -1 & +1 & -1 & +1 & -1 & +1 & -1 & +1 & -1 & +1 & -1 & +1 & -1 \end{pmatrix}. \tag{1.1327}$$

Let $[\boldsymbol{W}_{5\times16}]_{ij}$, $i \in \{0,1,2,3,4,5\}$, $j \in \{0,1\cdots15\}$ denote the element of $\boldsymbol{W}_{5\times16}$ in row i and column j, then we immediately find

$$\begin{aligned} \frac{1}{2^m}\sum_{j=0}^{2^m-1}[\boldsymbol{W}_{(m+1)\times2^m}]_{ij}\cdot[\boldsymbol{W}_{(m+1)\times2^m}]_{kj} &= \frac{1}{16}\sum_{j=0}^{15}[\boldsymbol{W}_{5\times16}]_{ij}\cdot[\boldsymbol{W}_{5\times16}]_{kj} \\ &= \begin{cases} 1 & \text{if } i = k, \\ 0 & \text{if } i \neq k, \end{cases} \\ &= \delta_{ik}, \quad i,k \in \{0,1,2,3,4,5\}, \end{aligned} \tag{1.1328}$$

i. e., the row vectors of $\boldsymbol{W}_{5\times16}$ are mutually orthogonal.

To further illustrate the structure of the $([m+1]\times 2^m)$ generator matrix $\boldsymbol{G}_{\mathcal{RM}(1,m)}$, $m \in \mathbb{N}^*$, respectively, the $([m+1]\times 2^m)$ matrix $\boldsymbol{W}_{(m+1)\times2^m}$, $m \in \mathbb{N}^*$ let us define the *Rademacher sine function* [54, equation (3.1), p. 46], which are named after the German mathematician Hans Adolph *Rademacher* [60].

Definition 1.67 (Rademacher Functions). Using the Gaußklammer $\lfloor x \rfloor$, the *Rademacher sine function* sir : $\mathbb{R} \to \{-1,+1\}$ has the functional mapping [54, equation (3.1), p. 46]

$$\operatorname{sir}\{x\} = (-1)^{\lfloor 2\cdot x\rfloor}. \tag{1.1329}$$

Using the Gaußklammer $\lfloor x \rfloor$, the *Rademacher cosine function* cor : $\mathbb{R} \to \{-1,+1\}$ has the functional mapping [54, equation (3.1), p. 46]

$$\operatorname{cor}\{x\} = (-1)^{\lfloor 2\cdot x+\frac{1}{2}\rfloor} = \operatorname{sir}\left\{x+\frac{1}{4}\right\}. \tag{1.1330}$$

Using the Rademacher sine function, we immediately see that
- the first row of the $([m+1]\times 2^m)$ matrix $\boldsymbol{W}_{(m+1)\times2^m}$, $m \in \mathbb{N}^*$ is given by $-\operatorname{sir}\{j/2^{m+1}\}$ equal to $\operatorname{cor}\{j/2^{m+1}+1/4\}$ and equal to $(-1)^{\lfloor j/2^m\rfloor}$, i. e., $+1$, $j \in \{0,1\cdots(2^m-1)\}$, $m \in \mathbb{N}^*$,
- the second row of the $([m+1]\times 2^m)$ matrix $\boldsymbol{W}_{(m+1)\times2^m}$, $m \in \mathbb{N}^*$ is given by $\operatorname{sir}\{j/2^m\}$ equal to $(-1)^{\lfloor j/2^{m-1}\rfloor}$, $j \in \{0,1\cdots(2^m-1)\}$, $m \in \mathbb{N}^*$,
- the third row of the $([m+1]\times 2^m)$ matrix $\boldsymbol{W}_{(m+1)\times2^m}$, $m \in \mathbb{N}^*$ is given by $\operatorname{sir}\{j/2^{m-1}\}$ equal to $(-1)^{\lfloor j/2^{m-2}\rfloor}$, $j \in \{0,1\cdots(2^m-1)\}$, $m \in \mathbb{N}^*$,
- the fourth row of the $([m+1]\times 2^m)$ matrix $\boldsymbol{W}_{(m+1)\times2^m}$, $m \in \mathbb{N}^*$ is given by $\operatorname{sir}\{j/2^{m-2}\}$ equal to $(-1)^{\lfloor j/2^{m-3}\rfloor}$, $j \in \{0,1\cdots(2^m-1)\}$, $m \in \mathbb{N}^*$,

- the fifth row of the $([m+1] \times 2^m)$ matrix $\boldsymbol{W}_{(m+1)\times 2^m}$, $m \in \mathbb{N}^*$ is given by $\mathrm{sir}\{j/2^{m-3}\}$ equal to $(-1)^{\lfloor j/2^{m-4} \rfloor}, j \in \{0, 1 \cdots (2^m - 1)\}, m \in \mathbb{N}^*$,
- ...
- the mth row of the $([m+1] \times 2^m)$ matrix $\boldsymbol{W}_{(m+1)\times 2^m}$, $m \in \mathbb{N}^*$ is given by $\mathrm{sir}\{j/2^2\}$ equal to $(-1)^{\lfloor j/2 \rfloor}, j \in \{0, 1 \cdots (2^m - 1)\}, m \in \mathbb{N}^*$ and
- the $(m+1)$st and last row of the $([m+1] \times 2^m)$ matrix $\boldsymbol{W}_{(m+1)\times 2^m}$, $m \in \mathbb{N}^*$ is given by $\mathrm{sir}\{j/2\}$ equal to $(-1)^{\lfloor j/2^0 \rfloor}$, i. e., $(-1)^{\lfloor j \rfloor}, j \in \{0, 1 \cdots (2^m - 1)\}, m \in \mathbb{N}^*$.

Let us look at r being equal to 2, now. In this case, p can assume the values 0, 1 and 2. Hence, the generator matrix $\boldsymbol{G}_{\mathcal{RM}(2,m)}$ of the second-order Reed–Muller (RM) code $\mathcal{RM}(2, m)$ has

$$\binom{m}{0} = \frac{m!}{m!} = 1, \quad m \in \mathbb{N}, \tag{1.1331}$$

row with Hamming weight 2^m, $m \in \mathbb{N}$, i. e., the all ones vector $\boldsymbol{v}^{(0)}$ equal to $\boldsymbol{1}_{2^m}$,

$$\binom{m}{1} = \frac{m!}{(m-1)!} = m, \quad m \in \mathbb{N}, \tag{1.1332}$$

rows with Hamming weight 2^{m-1}, $m \in \mathbb{N}$, which are the binary vectors $\boldsymbol{v}^{(i)}$, $1 \le i \le m$, $i, m \in \mathbb{N}$ and

$$\binom{m}{2} = \frac{m!}{2(m-2)!} = \frac{(m-1)m}{2}, \quad m \in \mathbb{N}, \tag{1.1333}$$

rows with Hamming weight 2^{m-2}, $m \in \mathbb{N}$, which are the binary vectors formed by the logic (Boolean) products

$$\boldsymbol{v}^{(i_1)} \boxdot \boldsymbol{v}^{(i_2)}, \quad 1 \le i_1 \le i_2 \le m, \quad i_1, i_2, m \in \mathbb{N} \tag{1.1334}$$

of degree 2. There are no further rows in $\boldsymbol{G}_{\mathcal{RM}(2,m)}$. Therefore, the second-order Reed–Muller (RM) code $\mathcal{RM}(2, m)$ is generated by the $[(m^2 + m + 2)/2] \times 2^m$ generator matrix

$$\boldsymbol{G}_{\mathcal{RM}(2,m)} = \begin{pmatrix} \boldsymbol{v}^{(0)} \\ \boldsymbol{v}^{(1)} \\ \boldsymbol{v}^{(2)} \\ \vdots \\ \boldsymbol{v}^{(m-1)} \\ \boldsymbol{v}^{(m)} \\ \boldsymbol{v}^{(1)} \boxdot \boldsymbol{v}^{(2)} \\ \boldsymbol{v}^{(1)} \boxdot \boldsymbol{v}^{(3)} \\ \boldsymbol{v}^{(1)} \boxdot \boldsymbol{v}^{(4)} \\ \vdots \\ \boldsymbol{v}^{(m-2)} \boxdot \boldsymbol{v}^{(m)} \\ \boldsymbol{v}^{(m-1)} \boxdot \boldsymbol{v}^{(m)} \end{pmatrix} = \begin{pmatrix} \mathbf{1}_{2^m} \\ \boldsymbol{v}^{(1)} \\ \boldsymbol{v}^{(2)} \\ \vdots \\ \boldsymbol{v}^{(m-1)} \\ \boldsymbol{v}^{(m)} \\ \boldsymbol{v}^{(1)} \boxdot \boldsymbol{v}^{(2)} \\ \boldsymbol{v}^{(1)} \boxdot \boldsymbol{v}^{(3)} \\ \boldsymbol{v}^{(1)} \boxdot \boldsymbol{v}^{(4)} \\ \vdots \\ \boldsymbol{v}^{(m-2)} \boxdot \boldsymbol{v}^{(m)} \\ \boldsymbol{v}^{(m-1)} \boxdot \boldsymbol{v}^{(m)} \end{pmatrix}, \quad m \in \mathbb{N}. \tag{1.1335}$$

It follows that the $(m-1)$-order Reed–Muller (RM) code $\mathcal{RM}([m-1], m)$ has the minimum distance

$$d_{\min} = 2. \tag{1.1336}$$

This $(m-1)$-order Reed–Muller (RM) code $\mathcal{RM}([m-1], m)$ is a single-parity check code [15, p. 106].

Example 1.75. Let us consider $m = 4$. The zeroth-order Reed–Muller (RM) code $\mathcal{RM}(0,4)$ is generated by the 1×16 generator matrix

$$\boldsymbol{G}_{\mathcal{RM}(0,4)} = (1111111111111111). \tag{1.1337}$$

Its minimum distance $d_{\min}$ is 16.

The first-order Reed–Muller (RM) code $\mathcal{RM}(1,4)$ is generated by the 5×16 generator matrix

$$\boldsymbol{G}_{\mathcal{RM}(1,4)} = \begin{pmatrix} \boldsymbol{v}^{(0)} \\ \boldsymbol{v}^{(1)} \\ \boldsymbol{v}^{(2)} \\ \boldsymbol{v}^{(3)} \\ \boldsymbol{v}^{(4)} \end{pmatrix} = \begin{pmatrix} 1&1&1&1&1&1&1&1&1&1&1&1&1&1&1&1 \\ 0&1&0&1&0&1&0&1&0&1&0&1&0&1&0&1 \\ 0&0&1&1&0&0&1&1&0&0&1&1&0&0&1&1 \\ 0&0&0&0&1&1&1&1&0&0&0&0&1&1&1&1 \\ 0&0&0&0&0&0&0&0&1&1&1&1&1&1&1&1 \end{pmatrix}. \tag{1.1338}$$

Its minimum distance is 8.

Of course, the rows of the 5×16 generator matrix can be reordered without changing the code, e. g.,

$$\boldsymbol{G}_{\mathcal{RM}(1,4)} = \begin{pmatrix} \boldsymbol{v}^{(0)} \\ \boldsymbol{v}^{(4)} \\ \boldsymbol{v}^{(3)} \\ \boldsymbol{v}^{(2)} \\ \boldsymbol{v}^{(1)} \end{pmatrix} = \begin{pmatrix} 1&1&1&1&1&1&1&1&1&1&1&1&1&1&1&1 \\ 0&0&0&0&0&0&0&0&1&1&1&1&1&1&1&1 \\ 0&0&0&0&1&1&1&1&0&0&0&0&1&1&1&1 \\ 0&0&1&1&0&0&1&1&0&0&1&1&0&0&1&1 \\ 0&1&0&1&0&1&0&1&0&1&0&1&0&1&0&1 \end{pmatrix}. \tag{1.1339}$$

In the case of the second-order Reed–Muller (RM) code $\mathcal{RM}(2,4)$, there are

$$\begin{pmatrix} 4 \\ 2 \end{pmatrix} = 6 \tag{1.1340}$$

rows with Hamming weight 2^{4-2} equal to 4, which are the binary vectors formed by the logic (Boolean) products

$$\begin{aligned}
\boldsymbol{v}^{(1)} \boxdot \boldsymbol{v}^{(2)} &= (0,0,0,1,0,0,0,1,0,0,0,1,0,0,0,1), \\
\boldsymbol{v}^{(1)} \boxdot \boldsymbol{v}^{(3)} &= (0,0,0,0,0,1,0,1,0,0,0,0,0,1,0,1), \\
\boldsymbol{v}^{(1)} \boxdot \boldsymbol{v}^{(4)} &= (0,0,0,0,0,0,0,0,0,1,0,1,0,1,0,1), \\
\boldsymbol{v}^{(2)} \boxdot \boldsymbol{v}^{(3)} &= (0,0,0,0,0,0,1,1,0,0,0,0,0,0,1,1), \\
\boldsymbol{v}^{(2)} \boxdot \boldsymbol{v}^{(4)} &= (0,0,0,0,0,0,0,0,0,0,1,1,0,0,1,1), \\
\boldsymbol{v}^{(3)} \boxdot \boldsymbol{v}^{(4)} &= (0,0,0,0,0,0,0,0,0,0,0,0,1,1,1,1).
\end{aligned} \tag{1.1341}$$

The corresponding 11×16 generator matrix is given by

$$\boldsymbol{G}_{\mathcal{RM}(2,4)} = \begin{pmatrix} \boldsymbol{v}^{(0)} \\ \boldsymbol{v}^{(1)} \\ \boldsymbol{v}^{(2)} \\ \boldsymbol{v}^{(3)} \\ \boldsymbol{v}^{(4)} \\ \boldsymbol{v}^{(1)} \boxdot \boldsymbol{v}^{(2)} \\ \boldsymbol{v}^{(1)} \boxdot \boldsymbol{v}^{(3)} \\ \boldsymbol{v}^{(1)} \boxdot \boldsymbol{v}^{(4)} \\ \boldsymbol{v}^{(2)} \boxdot \boldsymbol{v}^{(3)} \\ \boldsymbol{v}^{(2)} \boxdot \boldsymbol{v}^{(4)} \\ \boldsymbol{v}^{(3)} \boxdot \boldsymbol{v}^{(4)} \end{pmatrix} = \begin{pmatrix}
1&1&1&1&1&1&1&1&1&1&1&1&1&1&1&1 \\
0&1&0&1&0&1&0&1&0&1&0&1&0&1&0&1 \\
0&0&1&1&0&0&1&1&0&0&1&1&0&0&1&1 \\
0&0&0&0&1&1&1&1&0&0&0&0&1&1&1&1 \\
0&0&0&0&0&0&0&0&1&1&1&1&1&1&1&1 \\
0&0&0&1&0&0&0&1&0&0&0&1&0&0&0&1 \\
0&0&0&0&0&1&0&1&0&0&0&0&0&1&0&1 \\
0&0&0&0&0&0&0&0&0&1&0&1&0&1&0&1 \\
0&0&0&0&0&0&1&1&0&0&0&0&0&0&1&1 \\
0&0&0&0&0&0&0&0&0&0&1&1&0&0&1&1 \\
0&0&0&0&0&0&0&0&0&0&0&0&1&1&1&1
\end{pmatrix}. \tag{1.1342}$$

Its minimum distance is 4. Also in the case of the second-order Reed–Muller (RM) code $\mathcal{RM}(2,4)$, a reordering of the rows will not change the code:

$$\boldsymbol{G}_{\mathcal{RM}(2,4)} = \begin{pmatrix} \boldsymbol{v}^{(0)} \\ \boldsymbol{v}^{(4)} \\ \boldsymbol{v}^{(3)} \\ \boldsymbol{v}^{(2)} \\ \boldsymbol{v}^{(1)} \\ \boldsymbol{v}^{(3)} \boxdot \boldsymbol{v}^{(4)} \\ \boldsymbol{v}^{(2)} \boxdot \boldsymbol{v}^{(4)} \\ \boldsymbol{v}^{(1)} \boxdot \boldsymbol{v}^{(4)} \\ \boldsymbol{v}^{(2)} \boxdot \boldsymbol{v}^{(3)} \\ \boldsymbol{v}^{(1)} \boxdot \boldsymbol{v}^{(3)} \\ \boldsymbol{v}^{(1)} \boxdot \boldsymbol{v}^{(2)} \end{pmatrix} = \begin{pmatrix}
1&1&1&1&1&1&1&1&1&1&1&1&1&1&1&1 \\
0&0&0&0&0&0&0&0&1&1&1&1&1&1&1&1 \\
0&0&0&0&1&1&1&1&0&0&0&0&1&1&1&1 \\
0&0&1&1&0&0&1&1&0&0&1&1&0&0&1&1 \\
0&1&0&1&0&1&0&1&0&1&0&1&0&1&0&1 \\
0&0&0&0&0&0&0&0&0&0&0&0&1&1&1&1 \\
0&0&0&0&0&0&0&0&0&0&1&1&0&0&1&1 \\
0&0&0&0&0&0&0&0&0&1&0&1&0&1&0&1 \\
0&0&0&0&0&0&1&1&0&0&0&0&0&0&1&1 \\
0&0&0&0&0&1&0&1&0&0&0&0&0&1&0&1 \\
0&0&0&1&0&0&0&1&0&0&0&1&0&0&0&1
\end{pmatrix}. \tag{1.1343}$$

Since all the vectors in $\boldsymbol{G}_{\mathcal{RM}(r,m)}$ are of even Hamming weight, all the codewords in the rth-order Reed–Muller (RM) code $\mathcal{RM}(r,m)$ have even Hamming weight [15, p. 106].

We readily see from the code construction that the $[r-1]$th-order Reed–Muller (RM) code $\mathcal{RM}([r-1], m)$ is a proper subcode of the rth-order Reed–Muller (RM) code $\mathcal{RM}(r, m)$ [15, p. 106]. Hence, we have the following inclusion chain [15, equation (4.6), p. 106]:

$$\mathcal{RM}(0, m) \subset \mathcal{RM}(1, m) \subset \mathcal{RM}(2, m) \subset \cdots \subset \mathcal{RM}(r, m) \subset \cdots \subset \mathcal{RM}(m, m). \quad (1.1344)$$

Example 1.76. It follows from Example 1.75 that

$$\mathcal{RM}(0, 4) \subset \mathcal{RM}(1, 4) \subset \mathcal{RM}(2, 4). \quad (1.1345)$$

1.17.3 Punctured Reed–Muller (RM) Codes

Puncturing means deleting at least one coordinate, i. e., one component of the codeword [4, p. 28f.]. Hence, puncturing means discarding a column of the generator matrix. Of course, puncturing can be applied to an rth-order Reed–Muller (RM) code $\mathcal{RM}(r, m)$, $m, r \in \mathbb{N}$.

Definition 1.68 (Punctured Reed–Muller (RM) Code). For $0 \leq r \leq (m-1)$, the rth-order punctured Reed–Muller (RM) code $\mathcal{RM}(r, m)^*$ is obtained by puncturing, i. e., deleting that particular component, i. e., coordinate, given by the value of the Boolean function $f(x_1, x_2 \cdots x_m)$, at

$$x_1 = x_2 = x_3 = \cdots = x_m = 0, \quad m \in \mathbb{N}, \quad (1.1346)$$

from all the codewords $\boldsymbol{f} \in \mathbb{F}_2^{2^m}$, $m \in \mathbb{N}$, of the rth-order Reed–Muller (RM) code $\mathcal{RM}(r, m)$ [4, p. 377].

The rth-order punctured Reed–Muller (RM) code $\mathcal{RM}(r, m)^*$ has the length n equal to $[2^m - 1]$, $m \in \mathbb{N}$ and the unchanged dimension $k(r, m)$, $0 \leq r \leq m$, $m, r \in \mathbb{N}$; cf. (1.1284) [4, p. 377].

Due to the puncturing, the minimum distance d_{min} of the rth-order punctured Reed–Muller (RM) code $\mathcal{RM}(r, m)^*$ is reduced by 1, i. e., it is $[2^{m-r} - 1]$, $0 \leq r \leq m$, $m, r \in \mathbb{N}$.

Hence, the rth-order punctured Reed–Muller (RM) code $\mathcal{RM}(r, m)^*$ is a [4, p. 377]

$$\left([2^m - 1], \left[\sum_{i=0}^{r} \binom{m}{i}\right], [2^{m-r} - 1]\right) \text{ binary linear block code } \mathbb{V}.$$

For instance, the first-order punctured Reed–Muller (RM) code $\mathcal{RM}(1, m)^*$ has
- the length n equal to $[2^m - 1]$,
- the dimension $k(r, m)$ equal to $(m + 1)$ and
- the minimum distance d_{min} equal to $(2^{m-1} - 1)$,

which makes it a $(2^m - 1, m + 1, 2^{m-1} - 1)$ binary linear block code $\mathbb{V}$ [4, Figure 1.13, p. 31].

The coordinate, i. e., the component, mentioned in Definition 1.68 corresponds to that particular column of the generator matrix $\boldsymbol{G}_{\mathcal{RM}(r,m)}$, which has zeros in all rows except the first one. Often, this is the first column of $G_{\mathcal{RM}(r,m)}$. This column corresponds to the overall parity check. Hence, the puncturing deletes that particular coordinate of the rth-order Reed–Muller (RM) code $\mathcal{RM}(r,m)$, which implements the overall parity check [4, Figure 1.13, p. 31].

Example 1.77. We will set out from Example 1.69. Immediately, we see that the coordinate to be punctured is the first coordinate. Therefore, the first-order punctured Reed–Muller (RM) code $\mathcal{RM}(1,3)^*$ is given by

$$\mathbb{V} = \left\{ \begin{matrix} (0,0,0,0,0,0,0), & (0,0,0,1,1,1,1), & (0,1,1,0,1,1,1), & (1,0,1,0,1,0,1), \\ (0,1,1,1,1,0,0), & (1,0,1,1,0,1,0), & (1,1,0,0,1,1,0), & (1,1,0,1,0,0,1), \\ (1,1,1,1,1,1,1), & (1,1,1,0,0,0,0), & (1,0,0,1,1,0,0), & (0,1,0,1,0,1,0), \\ (1,0,0,0,0,1,1), & (0,1,0,0,1,0,1), & (0,0,1,1,0,0,1), & (0,0,1,0,1,1,0) \end{matrix} \right\}. \tag{1.1347}$$

Let us define the m be equal to three basis vectors:

$$\begin{aligned} v^{(3)*} &= (0,0,0,1,1,1,1), \\ v^{(2)*} &= (0,1,1,0,0,1,1), \\ v^{(1)*} &= (1,0,1,0,1,0,1), \end{aligned} \tag{1.1348}$$

then, together with

$$v^{(0)*} = \mathbf{1}_7 = (1,1,1,1,1,1,1), \tag{1.1349}$$

the $(m+1) \times [2^m - 1]$ equal to 4×7 generator matrix of the first-order punctured Reed–Muller (RM) code $\mathcal{RM}(1,3)^*$ can be given by

$$G_{\mathcal{RM}(1,3)^*} = \begin{pmatrix} v^{(0)*} \\ v^{(1)*} \\ v^{(2)*} \\ v^{(3)*} \end{pmatrix} = \begin{pmatrix} 1 & 1 & 1 & 1 & 1 & 1 & 1 \\ 1 & 0 & 1 & 0 & 1 & 0 & 1 \\ 0 & 1 & 1 & 0 & 0 & 1 & 1 \\ 0 & 0 & 0 & 1 & 1 & 1 & 1 \end{pmatrix}. \tag{1.1350}$$

It is easy to see that the minimum distance $d_{\min}$ of the first-order punctured Reed–Muller (RM) code $\mathcal{RM}(1,3)^*$ is equal to 3, i. e., $(2^{m-1} - 1)$ for $m = 3$. Therefore, the first-order punctured Reed–Muller (RM) code $\mathcal{RM}(1,3)^*$ is a $(7,4,3)$ binary linear block code $\mathbb{V}$.

Let us establish an equivalent version of the first-order punctured Reed–Muller (RM) code $\mathcal{RM}(1,3)^*$ by permuting columns of $G_{\mathcal{RM}(1,3)^*}$. First, we permute columns 2 and 3 to yield the generator matrix

$$\begin{pmatrix} 1 & 1 & 1 & 1 & 1 & 1 & 1 \\ 1 & 1 & 0 & 0 & 1 & 0 & 1 \\ 0 & 1 & 1 & 0 & 0 & 1 & 1 \\ 0 & 0 & 0 & 1 & 1 & 1 & 1 \end{pmatrix}. \tag{1.1351}$$

Now, let us continue with permuting the columns 3 and 7 to yield

$$\begin{pmatrix} 1 & 1 & 1 & 1 & 1 & 1 & 1 \\ 1 & 1 & 1 & 0 & 1 & 0 & 0 \\ 0 & 1 & 1 & 0 & 0 & 1 & 1 \\ 0 & 0 & 1 & 1 & 1 & 1 & 0 \end{pmatrix}. \tag{1.1352}$$

In this next step, let us permute the columns 4 and 6 and yield

$$\begin{pmatrix} 1 & 1 & 1 & 1 & 1 & 1 & 1 \\ 1 & 1 & 1 & 0 & 1 & 0 & 0 \\ 0 & 1 & 1 & 1 & 0 & 0 & 1 \\ 0 & 0 & 1 & 1 & 1 & 1 & 0 \end{pmatrix}. \tag{1.1353}$$

Finally, we permute the columns 6 and 7 and yield the equivalent version of the first-order punctured Reed–Muller (RM) code $\mathcal{RM}(1,3)^*$ with the generator matrix

$$\boldsymbol{G}'_{\mathcal{RM}(1,3)^*} = \begin{pmatrix} 1 & 1 & 1 & 1 & 1 & 1 & 1 \\ 1 & 1 & 1 & 0 & 1 & 0 & 0 \\ 0 & 1 & 1 & 1 & 0 & 1 & 0 \\ 0 & 0 & 1 & 1 & 1 & 0 & 1 \end{pmatrix}. \tag{1.1354}$$

A close look shows that the third and the fourth rows are cyclic shifts of the second row. Also, we see that the $\mathcal{RM}(1,3)$ code is the augmented dual of the cyclic binary (7, 4, 3) Hamming code. Therefore, we may say that the first-order punctured Reed–Muller (RM) code $\mathcal{RM}(1,3)^*$ is a cyclic code.

The second, the third and the fourth row of $\boldsymbol{G}'_{\mathcal{RM}(1,3)^*}$ of (1.1354) are linearly independent. Note that in the case of m equal to 3, there are only three linearly independent shifts. A further shift of the fourth row results in the vector

$$\begin{pmatrix} 1 & 0 & 0 & 1 & 1 & 1 & 0 \end{pmatrix}, \tag{1.1355}$$

which is the sum of the second and the third rows of $\boldsymbol{G}'_{\mathcal{RM}(1,3)^*}$ of (1.1354). Another shift yields

$$\begin{pmatrix} 0 & 1 & 0 & 0 & 1 & 1 & 1 \end{pmatrix}, \tag{1.1356}$$

which is the sum of the third and the fourth rows of $\boldsymbol{G}'_{\mathcal{RM}(1,3)^*}$ of (1.1354). Another shift yields

$$\begin{pmatrix} 1 & 0 & 1 & 0 & 0 & 1 & 1 \end{pmatrix}, \tag{1.1357}$$

which is the sum of the second, the third and the fourth rows of $\boldsymbol{G}'_{\mathcal{RM}(1,3)^*}$ of (1.1354). In addition, another shift yields

$$\begin{pmatrix} 1 & 1 & 0 & 1 & 0 & 0 & 1 \end{pmatrix}, \tag{1.1358}$$

which is the sum of the second and the fourth rows of $\boldsymbol{G}'_{\mathcal{RM}(1,3)^*}$ of (1.1354). The next shift will result in the second row of $\boldsymbol{G}'_{\mathcal{RM}(1,3)^*}$ of (1.1354).

Example 1.78. Let us continue Example 1.77. The sequence

$$\begin{pmatrix} 1 & 1 & 1 & 0 & 1 & 0 & 0 \end{pmatrix}, \tag{1.1359}$$

which forms the second row of $\boldsymbol{G}'_{\mathcal{RM}(1,3)^*}$ of (1.1354) can be represented by the polynomial

$$\theta_1(x) = 1 \oplus x \oplus x^2 \oplus x^4, \tag{1.1360}$$

which is the primitive idempotent derived in Example 1.56. The primitive idempotent $\theta_1(x)$ can be generated by a *linear feedback shift register (LFSR)* with the feedback polynomial

$$M^{(1)}(x) = 1 \oplus x \oplus x^3 \tag{1.1361}$$

when the initial state is chosen to be

$$\begin{pmatrix} 1 & 1 & 1 \end{pmatrix}. \tag{1.1362}$$

A linear feedback shift register (LFSR), which is governed by a polynomial over $\mathbb{F}_2$ with degree $m \in \mathbb{N}^*$, can assume 2^m different states, each state being enumerated by an integer ranging from 0 to $(2^m - 1)$ and being characterized by a distinct binary vector of length $m \in \mathbb{N}^*$, which is the binary representation of the said integer.

Since initializing the linear feedback shift register (LFSR) in the state 0, which is characterized by the binary vector

$$\Big(\underbrace{0, 0 \cdots 0}_{m \text{ components}} \Big), \tag{1.1363}$$

will produce an all zero output sequence without any changes of the state of the linear feedback shift register (LFSR), we must avoid such an initialization in the state 0 [61, p. 43]. Rather, the linear feedback shift register (LFSR) must be initialized in a nonzero state. The number of nonzero states that the linear feedback shift register (LFSR) can exhibit is $(2^m - 1)$, $m \in \mathbb{N}^*$ [61, p. 43].

Initializing the linear feedback shift register (LFSR) in one of these nonzero states will yield a periodic output sequence [61, p. 43], composed of 0's and 1's. Therefore, the shape of the output sequence repeats after a given number of clock cycles. The period of the said output sequence is limited to a maximum of $(2^m - 1)$, $m \in \mathbb{N}^*$, clock cycles, i. e., it is given by the number of nonzero states that the linear feedback shift register (LFSR) can assume [61, p. 43]. Whether or not this maximum period of $(2^m - 1)$, $m \in \mathbb{N}^*$ is obtained depends on the choice of the aforementioned polynomial over $\mathbb{F}_2$ with degree $m \in \mathbb{N}^*$ [61, p. 43].

The feedback polynomial $M^{(1)}(x)$ used in Figure 1.22 is an irreducible polynomial, which is taken from the factorization

$$x^7 \oplus 1 = (1 \oplus x) \odot \left(1 \oplus x \oplus x^3\right) \odot \left(1 \oplus x^2 \oplus x^3\right) \tag{1.1364}$$

of $(x^7 \oplus 1)$, setting

$$\begin{aligned} M^{(0)}(x) &= 1 \oplus x, \\ M^{(1)}(x) &= 1 \oplus x \oplus x^3, \\ M^{(3)}(x) &= 1 \oplus x^2 \oplus x^3 \end{aligned} \tag{1.1365}$$

and using $M^{(1)}(x)$. We can easily show that both sides of (1.1364) are equal:

$$\begin{aligned} & x^7 \oplus 1 = (1 \oplus x) \odot \left(1 \oplus x \oplus x^3\right) \odot \left(1 \oplus x^2 \oplus x^3\right) \\ & = (1 \oplus x) \odot \left(1 \oplus x \oplus x^3 \oplus x^2 \oplus x^3 \oplus x^5 \oplus x^3 \oplus x^4 \oplus x^6\right) \\ & = (1 \oplus x) \odot \Big(1 \oplus x \oplus x^2 \oplus \underbrace{x^3 \oplus x^3 \oplus x^3}_{=x^3} \oplus x^4 \oplus x^5 \oplus x^6\Big) \\ & = (1 \oplus x) \odot \left(1 \oplus x \oplus x^2 \oplus x^3 \oplus x^4 \oplus x^5 \oplus x^6\right) \\ & = 1 \oplus x \oplus x^2 \oplus x^3 \oplus x^4 \oplus x^5 \oplus x^6 \oplus x \oplus x^2 \oplus x^3 \oplus x^4 \oplus x^5 \oplus x^6 \oplus x^7 \\ & = 1 \oplus \underbrace{x \oplus x}_{=0} \oplus \underbrace{x^2 \oplus x^2}_{=0} \oplus \underbrace{x^3 \oplus x^3}_{=0} \oplus \underbrace{x^4 \oplus x^4}_{=0} \oplus \underbrace{x^5 \oplus x^5}_{=0} \oplus \underbrace{x^6 \oplus x^6}_{=0} \oplus x^7 \\ & = x^7 \oplus 1. \end{aligned} \tag{1.1366}$$

Furthermore, the polynomial $M^{(1)}(x)$ equal to $x^3 \oplus x \oplus 1$ is the primitive polynomial [61, Tabelle 3.2, p. 44] $\pi_1(x)$ of $\mathbb{F}_8$ according to Table 1.10. Using primitive polynomials to setup the feedback of a linear feedback shift register (LFSR) always yields output sequences with maximum periods [61, p. 43]. Such output sequences are hence called *maximum-length sequences*, respectively *m-sequences* [61, p. 43]. Such m-sequences resemble random sequences of 0's and 1's and are therefore also termed *pseudo-noise (PN) sequences* [4, Definition, p. 408]. Hence, the output sequence, having a period with polynomial $\theta_1(x) = x^4 \oplus x^2 \oplus x \oplus 1$ [61, Tabelle 3.3, p. 46], which is a primitive idempotent, is an m-sequence.

To put it in a nutshell, a primitive idempotent is the period of an m-sequence generated by a linear feedback shift register (LFSR), using a primitive polynomial as its feedback polynomial.

Setting out from $M^{(3)}(x)$ and establishing a modified version of the LFSR shown in Figure 1.22 (cf. Figure 1.23), using the initial state

$$\begin{pmatrix} 0 & 0 & 1 \end{pmatrix}, \tag{1.1367}$$

we can generate a further sequence, which we term $\theta_3(x)$ and which can also serve as a second row of an equivalent version $G''_{\mathcal{RM}(1,3)^*}$ of $G'_{\mathcal{RM}(1,3)^*}$ given by (1.1354). The corresponding generator matrix is given by

$$G''_{\mathcal{RM}(1,3)^*} = \begin{pmatrix} 1 & 1 & 1 & 1 & 1 & 1 & 1 \\ 1 & 0 & 0 & 1 & 0 & 1 & 1 \\ 1 & 1 & 0 & 0 & 1 & 0 & 1 \\ 1 & 1 & 1 & 0 & 0 & 1 & 0 \end{pmatrix}. \tag{1.1368}$$

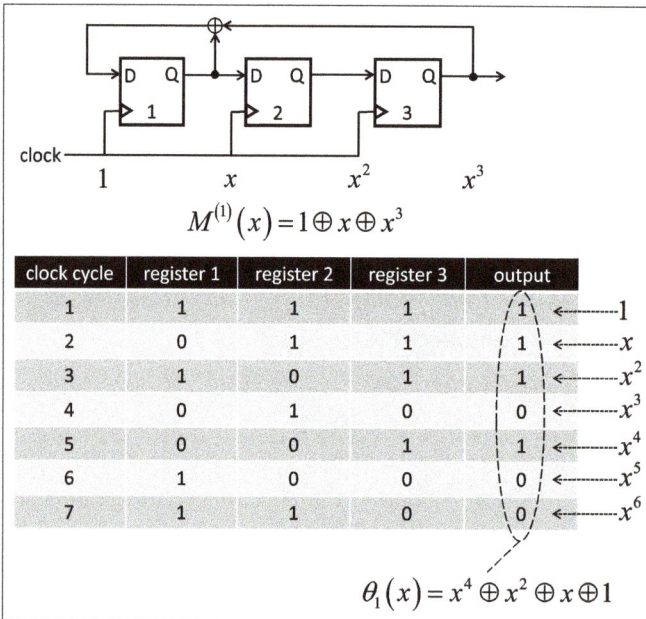

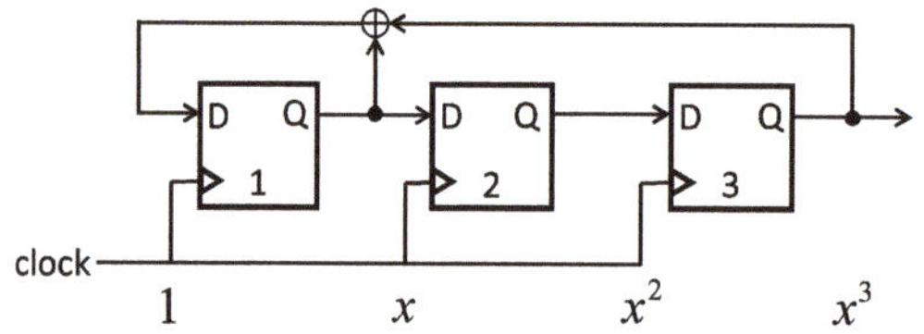

$$M^{(1)}(x) = 1 \oplus x \oplus x^3$$

clock cycle	register 1	register 2	register 3	output	
1	1	1	1	1	$\leftarrow 1$
2	0	1	1	1	$\leftarrow x$
3	1	0	1	1	$\leftarrow x^2$
4	0	1	0	0	$\leftarrow x^3$
5	0	0	1	1	$\leftarrow x^4$
6	1	0	0	0	$\leftarrow x^5$
7	1	1	0	0	$\leftarrow x^6$

$$\theta_1(x) = x^4 \oplus x^2 \oplus x \oplus 1$$

Figure 1.22: *Linear feedback shift register (LFSR)* with polynomial $M^{(1)}(x) = x^3 \oplus x \oplus 1$, which is irreducible in $\mathbb{F}_2$ [61, p. 42] and the primitive polynomial [61, Tabelle 3.2, p. 44] $\pi_1(x)$ of $\mathbb{F}_8$ according to Table 1.10, generating the output sequence with the primitive idempotent $\theta_1(x) = x^4 \oplus x^2 \oplus x \oplus 1$ [61, Tabelle 3.3, p. 46] being a period of this output sequence.

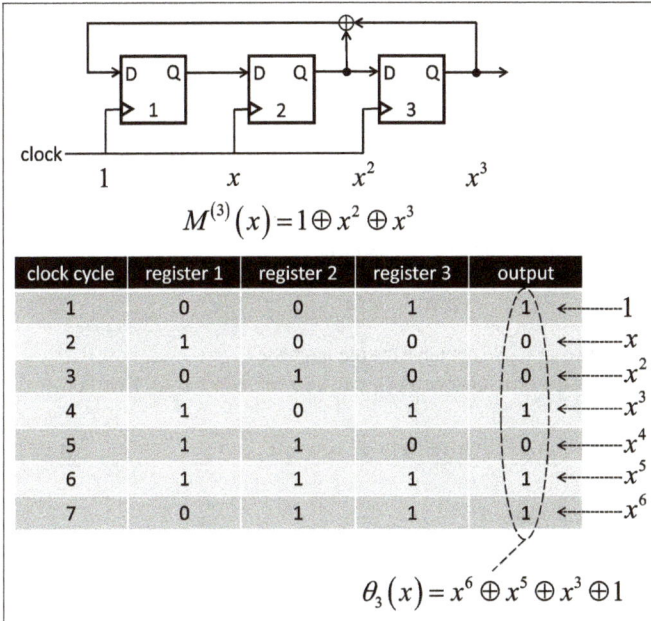

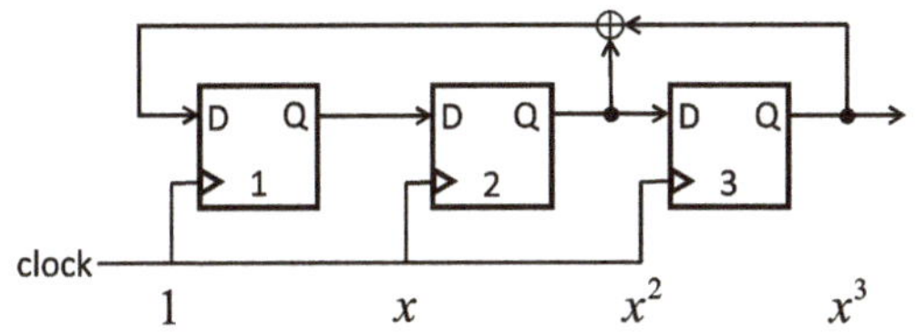

$$M^{(3)}(x) = 1 \oplus x^2 \oplus x^3$$

clock cycle	register 1	register 2	register 3	output	
1	0	0	1	1	$\leftarrow 1$
2	1	0	0	0	$\leftarrow x$
3	0	1	0	0	$\leftarrow x^2$
4	1	0	1	1	$\leftarrow x^3$
5	1	1	0	0	$\leftarrow x^4$
6	1	1	1	1	$\leftarrow x^5$
7	0	1	1	1	$\leftarrow x^6$

$$\theta_3(x) = x^6 \oplus x^5 \oplus x^3 \oplus 1$$

Figure 1.23: *Linear feedback shift register (LFSR)* with polynomial $M^{(3)}(x) = x^3 \oplus x^2 \oplus 1$, which is irreducible in $\mathbb{F}_2$ [61, p. 43] and the primitive polynomial $\pi_2(x)$ of $\mathbb{F}_8$ according to Table 1.11, generating the output sequence with the primitive idempotent $\theta_3(x) = x^6 \oplus x^5 \oplus x^3 \oplus 1$ being a period of this output sequence.

The rth-order punctured Reed–Muller (RM) code $\mathcal{RM}(r, m)^*$ is a cyclic code, which has as zeros α^s for all coset representatives s satisfying

$$1 \leq \mathcal{H}\{s\} \leq m - r - 1 \text{ and } 1 \leq s \leq 2^{m-1} - 1, \tag{1.1369}$$

where $\mathcal{H}\{s\}$ is the number of 1's in the binary expansion of $s \geq 0$ [4, p. 383]. The generator polynomial $g(x)$ and the check polynomial $h(x)$ of the rth-order punctured Reed–Muller (RM) code $\mathcal{RM}(r, m)^*$ are given by

$$g(x) = \bigodot_{\substack{1 \leq \mathcal{H}\{s\} \leq m-r-1 \\ 1 \leq s \leq 2^{m-1}-1}} M^{(s)}(x), \tag{1.1370}$$

$$h(x) = (x \oplus 1) \odot \bigodot_{\substack{m-r \leq \mathcal{H}\{s\} \leq m-1 \\ 1 \leq s \leq 2^{m-1}-1}} M^{(s)}(x), \tag{1.1371}$$

for $0 \leq r \leq m - 1$, $m, r \in \mathbb{N}$, where s runs through the representatives of the cyclotomic cosets and $M^{(s)}(x)$ is the minimal polynomial of α^s [4, p. 383]. Do not forget that an empty product is equal to 1 [4, p. 383].

Example 1.79. Consider the first-order punctured Reed–Muller (RM) code $\mathcal{RM}(1,3)^*$. Using (1.1369), $r = 1$ and $m = 3$, we have

$$\mathcal{H}\{s\} = 1 \text{ and } 1 \le s \le 3, \tag{1.1372}$$

thus yielding

$$s = 1, \quad s = 2, \quad s = 4, \tag{1.1373}$$

i. e., the elements of the cyclotomic coset $\mathbb{C}_1$, fulfill the requirement and

$$s = 3, \quad s = 5, \quad s = 6, \tag{1.1374}$$

i. e., the elements of the cyclotomic coset $\mathbb{C}_3$ do not fulfill the requirement.

Since only the representative of the cyclotomic coset $\mathbb{C}_1$ is to be taken, which is $s = 1$, we find

$$g(x) = \bigodot_{s=1}^{1} M^{(s)}(x) = M^{(1)}(x) = x^3 \oplus x \oplus 1. \tag{1.1375}$$

For the check polynomial, we must consider

$$m - r \le \mathcal{H}\{s\} \le m - 1, \tag{1.1376}$$

which yields

$$2 \le \mathcal{H}\{s\} \le 2, \tag{1.1377}$$

thus leading to

$$s = 3, \quad s = 5, \quad s = 6, \tag{1.1378}$$

i. e., the elements of the cyclotomic coset $\mathbb{C}_3$ fulfill this requirement. Therefore,

$$h(x) = (x \oplus 1) \odot \bigodot_{s=3}^{3} M^{(s)}(x) = (x \oplus 1) \odot M^{(3)}(x), \tag{1.1379}$$

with

$$M^{(3)}(x) = x^3 \oplus x^2 \oplus 1, \tag{1.1380}$$

leading to

$$\begin{aligned} h(x) &= (x \oplus 1) \odot \left(x^3 \oplus x^2 \oplus 1\right) = x^4 \oplus x^3 \oplus x \oplus x^3 \oplus x^2 \oplus 1 \\ &= x^4 \oplus \underbrace{x^3 \oplus x^3}_{=0} \oplus x^2 \oplus x \oplus 1 \\ &= x^4 \oplus x^2 \oplus x \oplus 1. \end{aligned} \tag{1.1381}$$

Now, let us compute $g(x) \odot h(x)$ in $\mathbb{F}_2[x]_7$, which yields

$$\begin{aligned} g(x) \odot h(x) &= \left(x^3 \oplus x \oplus 1\right) \odot \left(x^4 \oplus x^2 \oplus x \oplus 1\right) \\ &= \underbrace{x^7}_{=1} \oplus x^5 \oplus x^4 \oplus x^3 \oplus x^5 \oplus x^3 \oplus x^2 \oplus x \oplus x^4 \oplus x^2 \oplus x \oplus 1 \end{aligned} \tag{1.1382}$$

$$= \underbrace{x^5 \oplus x^5}_{=0} \oplus \underbrace{x^4 \oplus x^4}_{=0} \oplus \underbrace{x^3 \oplus x^3}_{=0} \oplus \underbrace{x^2 \oplus x^2}_{=0} \oplus \underbrace{x \oplus x}_{=0}$$
$$= 0,$$

just as expected.

Example 1.80. Let us consider the first-order punctured Reed–Muller (RM) code $\mathcal{RM}(1,4)^*$. We have

$$1 \leq \mathcal{H}\{s\} \leq m - r - 1 \text{ and } 1 \leq s \leq 2^{m-1} - 1, \tag{1.1383}$$

which yields

$$1 \leq \mathcal{H}\{s\} \leq 2 \text{ and } 1 \leq s \leq 7, \tag{1.1384}$$

thus

$$s = 1, \quad s = 2, \quad s = 4, \quad s = 8, \tag{1.1385}$$

i. e., the elements of the cyclotomic coset $\mathbb{C}_1$,

$$s = 3, \quad s = 6, \quad s = 12, \quad s = 9, \tag{1.1386}$$

i. e., the elements of the cyclotomic coset $\mathbb{C}_3$, and

$$s = 5, \quad s = 10, \tag{1.1387}$$

i. e., the elements of the cyclotomic coset $\mathbb{C}_5$ fulfill the requirement, but

$$s = 7, \quad s = 14, \quad s = 13, \quad s = 11, \tag{1.1388}$$

i. e., the elements of the cyclotomic coset $\mathbb{C}_7$ do not fulfill the requirement. Since only the coset representatives of the cyclotomic cosets $\mathbb{C}_1$, $\mathbb{C}_3$ and $\mathbb{C}_5$ are to be taken, we find

$$\begin{aligned} g(x) &= \bigodot_{s \in \{1,3,5\}} M^{(s)}(x) = M^{(1)}(x) \odot M^{(3)}(x) \odot M^{(5)}(x) \\ &= \left(x^4 \oplus x \oplus 1\right) \odot \left(x^4 \oplus x^3 \oplus x^2 \oplus x \oplus 1\right) \odot \left(x^2 \oplus x \oplus 1\right) \\ &= \left(x^8 \oplus x^7 \oplus x^6 \oplus x^5 \oplus x^4 \oplus x^5 \oplus x^4 \oplus x^3 \oplus x^2 \oplus x \oplus x^4 \oplus x^3 \oplus x^2 \oplus x \oplus 1\right) \\ &\quad \odot \left(x^2 \oplus x \oplus 1\right) \\ &= \left(x^8 \oplus x^7 \oplus x^6 \oplus x^4 \oplus 1\right) \odot \left(x^2 \oplus x \oplus 1\right) \\ &= x^{10} \oplus x^9 \oplus x^8 \oplus x^9 \oplus x^8 \oplus x^7 \oplus x^8 \oplus x^7 \oplus x^6 \oplus x^6 \oplus x^5 \oplus x^4 \oplus x^2 \oplus x \oplus 1 \\ &= x^{10} \oplus x^8 \oplus x^5 \oplus x^4 \oplus x^2 \oplus x \oplus 1. \end{aligned} \tag{1.1389}$$

For the check polynomial, we must consider

$$m - r \leq \mathcal{H}\{s\} \leq m - 1, \tag{1.1390}$$

which yields

$$3 \leq \mathcal{H}\{s\} \leq 3, \tag{1.1391}$$

thus

$$s = 7, \quad s = 14, \quad s = 13, \quad s = 11, \tag{1.1392}$$

i. e., the elements of the cyclotomic coset $\mathbb{C}_7$ fulfill this requirement. Therefore, we yield

$$h(x) = (x \oplus 1) \odot \bigodot_{s=7} M^{(s)}(x) = (x \oplus 1) \odot M^{(7)}(x) \tag{1.1393}$$

with

$$M^{(7)}(x) = x^4 \oplus x^3 \oplus 1, \tag{1.1394}$$

i. e.,

$$\begin{aligned} h(x) &= (x \oplus 1) \odot \left(x^4 \oplus x^3 \oplus 1\right) = x^5 \oplus x^4 \oplus x \oplus x^4 \oplus x^3 \oplus 1 \\ &= x^5 \oplus x^3 \oplus x \oplus 1. \end{aligned} \tag{1.1395}$$

Now, let us compute $g(x) \odot h(x)$ in $\mathbb{F}_2[x]_{15}$, which yields

$$\begin{aligned} g(x) \odot h(x) &= \left(x^{10} \oplus x^8 \oplus x^5 \oplus x^4 \oplus x^2 \oplus x \oplus 1\right) \\ &\quad \odot \left(x^5 \oplus x^3 \oplus x \oplus 1\right) \\ &= \underbrace{x^{15}}_{=1} \oplus x^{13} \oplus x^{11} \oplus x^{10} \oplus x^{13} \oplus x^{11} \oplus x^9 \oplus x^8 \\ &\quad \oplus x^{10} \oplus x^8 \oplus x^6 \oplus x^5 \oplus x^9 \oplus x^7 \oplus x^5 \oplus x^4 \\ &\quad \oplus x^7 \oplus x^5 \oplus x^3 \oplus x^2 \oplus x^6 \oplus x^4 \oplus x^2 \oplus x \\ &\quad \oplus x^5 \oplus x^3 \oplus x \oplus 1 \\ &= 0, \end{aligned} \tag{1.1396}$$

just as expected.

Using the coset representative s (cf. Remark 1.37), the *idempotent* of the rth-order punctured Reed–Muller (RM) code $\mathcal{RM}(r, m)^*$ is given by [4, p. 383]

$$E_{\mathcal{RM}(r,m)^*}(x) = \theta_0(x) \oplus \bigoplus_{\substack{1 \le \mathcal{H}\{s\} \le r \\ 1 \le s \le 2^{m-1}-1}} \theta_s(x), \tag{1.1397}$$

or, equivalently, replacing α by α^{-1}, [4, p. 383]

$$E_{\mathcal{RM}(r,m)^*}(x) = \theta_0(x) \oplus \bigoplus_{\substack{m-r \le \mathcal{H}\{s\} \le m-1 \\ 1 \le s \le 2^{m-1}-1}} \theta_s(x), \tag{1.1398}$$

where s runs through the representatives of the cyclotomic cosets [4, p. 383].

For instance, consider the first-order punctured Reed–Muller (RM) code $\mathcal{RM}(1, m)^*$. In this case, we have to fulfill

$$1 \le \mathcal{H}\{s\} \le 1 \text{ and } 1 \le s \le 2^{m-1} - 1, \tag{1.1399}$$

which can only be fulfilled by the elements of the cyclotomic cosets $\mathbb{C}_1$. Therefore, the idempotent of the first-order punctured Reed–Muller (RM) code $\mathcal{RM}(1,m)^*$ is given by

$$E_{\mathcal{RM}(1,m)^*}(x) = \theta_0(x) \oplus \theta_1(x). \tag{1.1400}$$

Now, consider the second-order punctured Reed–Muller (RM) code $\mathcal{RM}(2,m)^*$. In this case, we have the requirements

$$1 \le \mathcal{H}\{s\} \le 2 \text{ and } 1 \le s \le 2^{m-1} - 1, \tag{1.1401}$$

which can only be fulfilled by the elements of the cyclotomic cosets
- $\mathbb{C}_1$ and
- $\mathbb{C}_{2^i+1}$, $i \in \{1, 2 \cdots (m-1)\}$.

Therefore, the idempotent of the second-order punctured Reed–Muller (RM) code $\mathcal{RM}(2,m)^*$ is given by

$$E_{\mathcal{RM}(2,m)^*}(x) = \theta_0(x) \oplus \bigoplus_{\substack{\mathcal{H}\{s\}=1\\ 1\le s\le 2^{m-1}-1}} \theta_s(x) \oplus \bigoplus_{\substack{\mathcal{H}\{s\}=2\\ 3\le s\le 2^{m-1}-1}} \theta_s(x)$$

$$= \theta_0(x) \oplus \theta_1(x) \bigoplus_{\substack{\mathcal{H}\{s\}=2\\ 3\le s\le 2^{m-1}-1}} \theta_s(x), \quad m \in \mathbb{N}. \tag{1.1402}$$

The idempotents of the rth-order punctured Reed–Muller (RM) codes $\mathcal{RM}(r,m)^*$ can also be represented using the primitive idempotents $\theta_s^*(x)$ [4, p. 383f.]. Then we find the idempotent of the first-order punctured Reed–Muller (RM) code $\mathcal{RM}(1,m)^*$ to be [4, p. 384]

$$\theta_0(x) \oplus \theta_1^*(x), \tag{1.1403}$$

and the idempotent of the second-order punctured Reed–Muller (RM) code $\mathcal{RM}(2,m)^*$ may be taken to be [4, p. 384]

$$\theta_0(x) \oplus \theta_1^*(x) \oplus \bigoplus_{i=\lfloor (m+1)/2 \rfloor}^{m-1} \theta_{1+2^i}^*(x), \quad m \in \mathbb{N}. \tag{1.1404}$$

Example 1.81. Consider the punctured Reed–Muller (RM) code $\mathcal{RM}(1,3)^*$. We have

$$1 \le \mathcal{H}\{s\} \le r \text{ and } 1 \le s \le 3, \tag{1.1405}$$

as a requirement, which becomes

$$1 \le \mathcal{H}\{s\} \le 1 \text{ and } 1 \le s \le 3, \tag{1.1406}$$

thus

$$s = 1, \quad s = 2, \quad s = 4, \tag{1.1407}$$

i. e., the elements of the cyclotomic coset $\mathbb{C}_1$, fulfill the requirement and

$$s = 3, \quad s = 5, \quad s = 6, \tag{1.1408}$$

i. e., the elements of the cyclotomic coset $\mathbb{C}_3$ do not fulfill the requirement.

Since only the coset representative of the cyclotomic coset $\mathbb{C}_1$ is to be taken, which is $s = 1$, we find

$$\begin{aligned} E_{\mathcal{RM}(1,3)^*}(x) &= \theta_0(x) \bigoplus_{s\in\{1\}} \theta_s(x) = \theta_0(x) \oplus \theta_1(x) \\ &= \underbrace{\left(1 \oplus x \oplus x^2 \oplus x^3 \oplus x^4 \oplus x^5 \oplus x^6\right)}_{=\theta_0(x)} \oplus \underbrace{\left(1 \oplus x \oplus x^2 \oplus x^4\right)}_{=\theta_1(x)} \\ &= x^3 \oplus x^5 \oplus x^6. \end{aligned} \tag{1.1409}$$

Note that

$$E_{\mathrm{RM}(1,3)^*}(x) = 1 \oplus \theta_3(x). \tag{1.1410}$$

The general codewords in the zeroth-order punctured Reed–Muller (RM) code $\mathcal{RM}(0, m)^*$ are of the form

$$a_0 \odot \theta_0(x), \quad a_0 \in \mathbb{F}_2, \tag{1.1411}$$

i. e., they are either $\theta_0(x)$ or $0 \odot \theta_0(x)$ [4, p. 384]. Hence, the general codewords in the zeroth-order Reed–Muller (RM) code $\mathcal{RM}(0, m)$ are of the form [4, p. 384]

$$|\, a_0 \,|\, a_0 \odot \theta_0(x) \,|, \quad a_0 \in \mathbb{F}_2, \tag{1.1412}$$

where the first part of the codeword, $a_0 \in \mathbb{F}_2$, corresponds to the *overall parity check* and the second part, $a_0 \odot \theta_0(x)$ is in the cyclic zeroth-order punctured Reed–Muller (RM) code $\mathcal{RM}(0, m)^*$ [4, p. 384]. Clearly, both the cyclic zeroth-order punctured Reed–Muller (RM) code $\mathcal{RM}(0, m)^*$ and the zeroth-order Reed–Muller (RM) code $\mathcal{RM}(0, m)$ are repetition codes.

The general codewords in the first-order punctured Reed–Muller (RM) code $\mathcal{RM}(1, m)^*$ are of the form

$$\begin{gathered} a_0 \odot \theta_0(x) \oplus a_1 \odot x^{i_1} \odot \theta_1^*(x), \\ a_0, a_1 \in \mathbb{F}_2, \quad i_1 \in \{0, 1, 2 \cdots (2^m - 2)\}, \quad m \in \mathbb{N}, \end{gathered} \tag{1.1413}$$

where $(x^{i_1} \odot \theta_1^*(x))$ are the $(2^m - 1)$, $m \in \mathbb{N}$, cyclic shifts of the idempotent $\theta_1^*(x)$ [4, p. 384].

Therefore, the general codewords in the first-order Reed–Muller (RM) code $\mathcal{RM}(1, m)$ are of the form [4, p. 384]

$$|\, a_0 \,|\, a_0 \odot \theta_0(x) \oplus a_1 \odot x^{i_1} \odot \theta_1^*(x) \,|, \tag{1.1414}$$

$$a_0, a_1 \in \mathbb{F}_2, \quad i_1 \in \{0, 1, 2 \cdots (2^m - 2)\}, \quad m \in \mathbb{N}.$$

Correspondingly, the general codewords in the second-order punctured Reed–Muller (RM) code $\mathcal{RM}(2, m)^*$ are of the form [4, p. 384]

$$a_0 \odot \theta_0(x) \oplus a_1 \odot x^{i_1} \odot \theta_1^*(x) \bigoplus_{i=\lfloor (m+1)/2 \rfloor}^{m-1} a_i \odot x^{s_i} \odot \theta_{1+2^i}^*(x) \tag{1.1415}$$

$$a_0, a_1, a_i \in \mathbb{F}_2, \quad i \in \{\lfloor (m+1)/2 \rfloor \cdots (m-1)\}, \quad i_1 \in \{0, 1, 2 \cdots (2^m - 2)\}, \quad m \in \mathbb{N}.$$

Therefore, the general codewords in the second-order Reed–Muller (RM) code $\mathcal{RM}(2, m)$ are of the form [4, p. 384]

$$| \, a_0 \, | \, a_0 \odot \theta_0(x) \oplus a_1 \odot x^{i_1} \odot \theta_1^*(x) \bigoplus_{i=\lfloor (m+1)/2 \rfloor}^{m-1} a_i \odot x^{s_i} \odot \theta_{1+2^i}^*(x) \, |, \tag{1.1416}$$

$$a_0, a_1, a_i \in \mathbb{F}_2, \quad i \in \{\lfloor (m+1)/2 \rfloor \cdots (m-1)\}, \quad i_1, s_i \in \{0, 1, 2 \cdots (2^m - 2)\}, \quad m \in \mathbb{N}.$$

Example 1.82. Let us look at the zeroth-order Reed–Muller (RM) code $\mathcal{RM}(0, 5)$ the first-order Reed–Muller (RM) code $\mathcal{RM}(1, 5)$, and the second-order Reed–Muller (RM) code $\mathcal{RM}(2, 5)$.

According to Example 1.48, the cyclotomic cosets of $\mathbb{F}_{2^5}$ equal to $\mathbb{F}_{32}$ are

$$\begin{aligned}
\mathbb{C}_0 &= \{0\},\\
\mathbb{C}_1 &= \{1, 2, 4, 8, 16\},\\
\mathbb{C}_3 &= \{3, 6, 12, 24, 17\},\\
\mathbb{C}_5 &= \{5, 10, 20, 9, 18\},\\
\mathbb{C}_7 &= \{7, 14, 28, 25, 19\},\\
\mathbb{C}_{11} &= \{11, 22, 13, 26, 21\},\\
\mathbb{C}_{15} &= \{15, 30, 29, 27, 23\}.
\end{aligned} \tag{1.1417}$$

In particular, looking at the cyclotomic cosets, we yield [4, p. 222]

$$\begin{aligned}
\mathbb{C}_0 : \quad & \theta_0^*(x) = \theta_0(x) = \bigoplus_{j=0}^{30} x^j,\\
\mathbb{C}_{15} : \quad & \theta_1^*(x) = \theta_{15}(x)\\
& = 1 \oplus x^3 \oplus x^5 \oplus x^6 \oplus x^9 \oplus x^{10} \oplus x^{11}\\
& \oplus x^{12} \oplus x^{13} \oplus x^{17} \oplus x^{18} \oplus x^{20}\\
& \oplus x^{21} \oplus x^{22} \oplus x^{24} \oplus x^{26},\\
\mathbb{C}_7 : \quad & \theta_3^*(x) = \theta_7(x) = \theta_{17}^*(x) \quad \left(-17 \bmod 31 = 14 \in \mathbb{C}_7 \mapsto s' = 7\right)\\
& = 1 \oplus x \oplus x^2 \oplus x^3 \oplus x^4 \oplus x^6 \oplus x^7\\
& \oplus x^8 \oplus x^{12} \oplus x^{14} \oplus x^{16} \oplus x^{17}\\
& \oplus x^{19} \oplus x^{24} \oplus x^{25} \oplus x^{28},
\end{aligned}$$

$$\begin{aligned}
\mathbb{C}_{11}:\quad \theta_5^*(x) &= \theta_{11}(x) = \theta_9^*(x) \quad \left(-9 \bmod 31 = 22 \in \mathbb{C}_{11} \mapsto s' = 11\right)\\
&= 1 \oplus x \oplus x^2 \oplus x^4 \oplus x^8 \oplus x^{11} \oplus x^{13}\\
&\oplus x^{15} \oplus x^{16} \oplus x^{21} \oplus x^{22} \oplus x^{23}\\
&\oplus x^{26} \oplus x^{27} \oplus x^{29} \oplus x^{30}, \qquad (1.1418)\\
\mathbb{C}_3:\quad \theta_7^*(x) &= \theta_3(x)\\
&= 1 \oplus x^3 \oplus x^6 \oplus x^7 \oplus x^{12} \oplus x^{14}\\
&\oplus x^{15} \oplus x^{17} \oplus x^{19} \oplus x^{23} \oplus x^{24}\\
&\oplus x^{25} \oplus x^{27} \oplus x^{28} \oplus x^{29} \oplus x^{30},\\
\mathbb{C}_5:\quad \theta_{11}^*(x) &= \theta_5(x)\\
&= 1 \oplus x \oplus x^2 \oplus x^3 \oplus x^5 \oplus x^8\\
&\oplus x^9 \oplus x^{10} \oplus x^{15} \oplus x^{16} \oplus x^{18}\\
&\oplus x^{20} \oplus x^{23} \oplus x^{27} \oplus x^{29} \oplus x^{30},\\
\mathbb{C}_1:\quad \theta_{15}^*(x) &= \theta_1(x)\\
&= 1 \oplus x^5 \oplus x^7 \oplus x^9 \oplus x^{10} \oplus x^{11}\\
&\oplus x^{13} \oplus x^{14} \oplus x^{18} \oplus x^{19} \oplus x^{20}\\
&\oplus x^{21} \oplus x^{22} \oplus x^{25} \oplus x^{26} \oplus x^{28}.
\end{aligned}$$

The idempotent of the second-order punctured Reed–Muller (RM) code $\mathcal{RM}(2,5)^*$ is given by [4, p. 384]

$$\begin{aligned}
E_{\mathcal{RM}(2,5)^*}(x) &= \theta_0(x) \oplus \theta_1^*(x) \oplus \bigoplus_{i=\lfloor(5+1)/2\rfloor}^{5-1} \theta_{1+2^i}^*(x)\\
&= \theta_0(x) \oplus \theta_1^*(x) \oplus \bigoplus_{i=3}^{4} \theta_{1+2^i}^*(x)\\
&= \theta_0(x) \oplus \theta_1^*(x) \oplus \theta_{1+2^3}^*(x) \oplus \theta_{1+2^4}^*(x)\\
&= \theta_0(x) \oplus \theta_1^*(x) \oplus \underbrace{\theta_9^*(x)}_{=\theta_5^*(x)} \oplus \underbrace{\theta_{17}^*(x)}_{=\theta_3^*(x)}\\
&= \theta_0(x) \oplus \theta_1^*(x) \oplus \theta_3^*(x) \oplus \theta_5^*(x). \qquad (1.1419)
\end{aligned}$$

Therefore, we find

$$\begin{aligned}
\mathcal{RM}(0,5) &: \left|\, a_0 \mid a_0 \odot \theta_0(x) \,\right|,\\
\mathcal{RM}(1,5) &: \left|\, a_0 \mid a_0 \odot \theta_0(x) \oplus a_1 \odot x^{i_1}\theta_1^*(x) \,\right|,\\
\mathcal{RM}(2,5) &: \left|\, a_0 \mid a_0 \odot \theta_0(x) \oplus a_1 \odot x^{i_1} \odot \theta_1^*(x) \right. \qquad (1.1420)\\
&\quad \left. \oplus\, a_2 \odot x^{s_1} \odot \theta_5^*(x) \oplus a_3 \odot x^{s_2} \odot \theta_3^*(x) \,\right|,\\
&\quad a_0, a_1, a_2, a_3 \in \mathbb{F}_2, \quad i_1, s_1, s_2 \in \{0, 1 \cdots 30\}.
\end{aligned}$$

We now obtain

$$\boldsymbol{G}_{\mathcal{RM}(0,5)} = \begin{pmatrix} 1 & \theta_0(x) \end{pmatrix} = \begin{pmatrix} 1 & \theta_0^*(x) \end{pmatrix}, \tag{1.1421}$$

$$\boldsymbol{G}_{\mathcal{RM}(1,5)} = \begin{pmatrix} 1 & \theta_0(x) \\ 0 & \theta_1^*(x) \\ 0 & x^1 \odot \theta_1^*(x) \\ 0 & x^2 \odot \theta_1^*(x) \\ 0 & x^3 \odot \theta_1^*(x) \\ 0 & x^4 \odot \theta_1^*(x) \end{pmatrix} = \begin{pmatrix} 1 & \theta_0^*(x) \\ 0 & \theta_1^*(x) \\ 0 & x^1 \odot \theta_1^*(x) \\ 0 & x^2 \odot \theta_1^*(x) \\ 0 & x^3 \odot \theta_1^*(x) \\ 0 & x^4 \odot \theta_1^*(x) \end{pmatrix}, \tag{1.1422}$$

$$\boldsymbol{G}_{\mathcal{RM}(2,5)} = \begin{pmatrix} 1 & \theta_0(x) \\ 0 & \theta_1^*(x) \\ 0 & x^1 \odot \theta_1^*(x) \\ 0 & x^2 \odot \theta_1^*(x) \\ 0 & x^3 \odot \theta_1^*(x) \\ 0 & x^4 \odot \theta_1^*(x) \\ 0 & \theta_5^*(x) \\ 0 & x^1 \odot \theta_5^*(x) \\ 0 & x^2 \odot \theta_5^*(x) \\ 0 & x^3 \odot \theta_5^*(x) \\ 0 & x^4 \odot \theta_5^*(x) \\ 0 & \theta_3^*(x) \\ 0 & x^1 \odot \theta_3^*(x) \\ 0 & x^2 \odot \theta_3^*(x) \\ 0 & x^3 \odot \theta_3^*(x) \\ 0 & x^4 \odot \theta_3^*(x) \end{pmatrix} = \begin{pmatrix} 1 & \theta_0^*(x) \\ 0 & \theta_1^*(x) \\ 0 & x^1 \odot \theta_1^*(x) \\ 0 & x^2 \odot \theta_1^*(x) \\ 0 & x^3 \odot \theta_1^*(x) \\ 0 & x^4 \odot \theta_1^*(x) \\ 0 & \theta_5^*(x) \\ 0 & x^1 \odot \theta_5^*(x) \\ 0 & x^2 \odot \theta_5^*(x) \\ 0 & x^3 \odot \theta_5^*(x) \\ 0 & x^4 \odot \theta_5^*(x) \\ 0 & \theta_3^*(x) \\ 0 & x^1 \odot \theta_3^*(x) \\ 0 & x^2 \odot \theta_3^*(x) \\ 0 & x^3 \odot \theta_3^*(x) \\ 0 & x^4 \odot \theta_3^*(x) \end{pmatrix}. \tag{1.1423}$$

Note that the generated codes will not change when rows are permuted.

Furthermore, note that the above given primitive idempotents [4, p. 222] can also be represented as the binary vectors

$$\begin{aligned}
\boldsymbol{\theta}_0^* &= \boldsymbol{\theta}_0 & &= (1\,1\,1\,1\,1\,1\,1\,1\,1\,1\,1\,1\,1\,1\,1\,1\,1\,1\,1\,1\,1\,1\,1\,1\,1\,1\,1\,1\,1\,1\,1), \\
\boldsymbol{\theta}_1^* &= \boldsymbol{\theta}_{15} & &= (1\,0\,0\,1\,0\,1\,1\,0\,0\,1\,1\,1\,1\,1\,0\,0\,0\,1\,1\,0\,1\,1\,1\,0\,1\,0\,1\,0\,0\,0\,0), \\
\boldsymbol{\theta}_3^* &= \boldsymbol{\theta}_7 &= \boldsymbol{\theta}_{17}^* &= (1\,1\,1\,1\,1\,0\,1\,1\,1\,0\,0\,0\,1\,0\,1\,0\,1\,1\,0\,1\,0\,0\,0\,0\,1\,1\,0\,0\,1\,0\,0), \\
\boldsymbol{\theta}_5^* &= \boldsymbol{\theta}_{11} &= \boldsymbol{\theta}_9^* &= (1\,1\,1\,0\,1\,0\,0\,0\,1\,0\,0\,1\,0\,1\,0\,1\,1\,0\,0\,0\,0\,1\,1\,1\,0\,0\,1\,1\,0\,1\,1), \\
\boldsymbol{\theta}_7^* &= \boldsymbol{\theta}_3 & &= (1\,0\,0\,1\,0\,0\,1\,1\,0\,0\,0\,0\,1\,0\,1\,1\,0\,1\,0\,1\,0\,0\,0\,1\,1\,1\,0\,1\,1\,1\,1), \\
\boldsymbol{\theta}_{11}^* &= \boldsymbol{\theta}_5 & &= (1\,1\,1\,0\,1\,1\,0\,0\,1\,1\,1\,0\,0\,0\,0\,1\,1\,0\,1\,0\,1\,0\,0\,1\,0\,0\,0\,1\,0\,1\,1), \\
\boldsymbol{\theta}_{15}^* &= \boldsymbol{\theta}_1 & &= (1\,0\,0\,0\,0\,1\,0\,1\,0\,1\,1\,1\,0\,1\,1\,0\,0\,0\,1\,1\,1\,1\,1\,0\,0\,1\,1\,0\,1\,0\,0).
\end{aligned} \tag{1.1424}$$

These vectors can be directly used when establishing the generator matrices $\boldsymbol{G}_{\mathcal{RM}(0,5)}$, $\boldsymbol{G}_{\mathcal{RM}(1,5)}$ and $\boldsymbol{G}_{\mathcal{RM}(2,5)}$.

1.17.4 Constructing Reed–Muller (RM) Codes Using the Kronecker Product

Using the 2×2 matrix over $\mathbb{F}_2$ [15, equation (4.14), p. 114]

$$\boldsymbol{H}_2 = \begin{pmatrix} 1 & 1 \\ 0 & 1 \end{pmatrix}, \tag{1.1425}$$

the two-fold Kronecker product of $\boldsymbol{H}_2$ is given by the 4×4 matrix [15, equation (4.15), p. 114]

$$\boldsymbol{H}_4 = \boldsymbol{H}_2^{\otimes 2} = \boldsymbol{H}_2 \otimes \boldsymbol{H}_2 = \begin{pmatrix} \boldsymbol{H}_2 & \boldsymbol{H}_2 \\ \boldsymbol{0}_{2\times 2} & \boldsymbol{H}_2 \end{pmatrix} = \begin{pmatrix} 1 & 1 & 1 & 1 \\ 0 & 1 & 0 & 1 \\ 0 & 0 & 1 & 1 \\ 0 & 0 & 0 & 1 \end{pmatrix}. \tag{1.1426}$$

The three-fold Kronecker product of $\boldsymbol{H}_2$ is defined as the 8×8 matrix [15, equation (4.16), p. 115]

$$\begin{aligned} \boldsymbol{H}_8 &= \boldsymbol{H}_2^{\otimes 3} \\ &= \boldsymbol{H}_2 \otimes \boldsymbol{H}_2 \otimes \boldsymbol{H}_2 \\ &= \boldsymbol{H}_2 \otimes \boldsymbol{H}_4 \\ &= \begin{pmatrix} \boldsymbol{H}_4 & \boldsymbol{H}_4 \\ \boldsymbol{0}_{4\times 4} & \boldsymbol{H}_4 \end{pmatrix} \\ &= \begin{pmatrix} 1 & 1 & 1 & 1 & 1 & 1 & 1 & 1 \\ 0 & 1 & 0 & 1 & 0 & 1 & 0 & 1 \\ 0 & 0 & 1 & 1 & 0 & 0 & 1 & 1 \\ 0 & 0 & 0 & 1 & 0 & 0 & 0 & 1 \\ 0 & 0 & 0 & 0 & 1 & 1 & 1 & 1 \\ 0 & 0 & 0 & 0 & 0 & 1 & 0 & 1 \\ 0 & 0 & 0 & 0 & 0 & 0 & 1 & 1 \\ 0 & 0 & 0 & 0 & 0 & 0 & 0 & 1 \end{pmatrix}. \end{aligned} \tag{1.1427}$$

Furthermore, the four-fold Kronecker product of $\boldsymbol{H}_2$ is obtained as the 16×16 matrix [15, p. 115]

$$\begin{aligned} \boldsymbol{H}_{16} = \boldsymbol{H}_2^{\otimes 4} &= \begin{pmatrix} \boldsymbol{H}_8 & \boldsymbol{H}_8 \\ \boldsymbol{0}_{8\times 8} & \boldsymbol{H}_8 \end{pmatrix} \\ &= \begin{pmatrix} 1 & 1 & 1 & 1 & 1 & 1 & 1 & 1 & 1 & 1 & 1 & 1 & 1 & 1 & 1 & 1 \\ 0 & 1 & 0 & 1 & 0 & 1 & 0 & 1 & 0 & 1 & 0 & 1 & 0 & 1 & 0 & 1 \\ 0 & 0 & 1 & 1 & 0 & 0 & 1 & 1 & 0 & 0 & 1 & 1 & 0 & 0 & 1 & 1 \\ 0 & 0 & 0 & 1 & 0 & 0 & 0 & 1 & 0 & 0 & 0 & 1 & 0 & 0 & 0 & 1 \\ 0 & 0 & 0 & 0 & 1 & 1 & 1 & 1 & 0 & 0 & 0 & 0 & 1 & 1 & 1 & 1 \\ 0 & 0 & 0 & 0 & 0 & 1 & 0 & 1 & 0 & 0 & 0 & 0 & 0 & 1 & 0 & 1 \\ 0 & 0 & 0 & 0 & 0 & 0 & 1 & 1 & 0 & 0 & 0 & 0 & 0 & 0 & 1 & 1 \\ 0 & 0 & 0 & 0 & 0 & 0 & 0 & 1 & 0 & 0 & 0 & 0 & 0 & 0 & 0 & 1 \\ 0 & 0 & 0 & 0 & 0 & 0 & 0 & 0 & 1 & 1 & 1 & 1 & 1 & 1 & 1 & 1 \\ 0 & 0 & 0 & 0 & 0 & 0 & 0 & 0 & 0 & 1 & 0 & 1 & 0 & 1 & 0 & 1 \\ 0 & 0 & 0 & 0 & 0 & 0 & 0 & 0 & 0 & 0 & 1 & 1 & 0 & 0 & 1 & 1 \\ 0 & 0 & 0 & 0 & 0 & 0 & 0 & 0 & 0 & 0 & 0 & 1 & 0 & 0 & 0 & 1 \\ 0 & 0 & 0 & 0 & 0 & 0 & 0 & 0 & 0 & 0 & 0 & 0 & 1 & 1 & 1 & 1 \\ 0 & 0 & 0 & 0 & 0 & 0 & 0 & 0 & 0 & 0 & 0 & 0 & 0 & 1 & 0 & 1 \\ 0 & 0 & 0 & 0 & 0 & 0 & 0 & 0 & 0 & 0 & 0 & 0 & 0 & 0 & 1 & 1 \\ 0 & 0 & 0 & 0 & 0 & 0 & 0 & 0 & 0 & 0 & 0 & 0 & 0 & 0 & 0 & 1 \end{pmatrix}. \end{aligned} \tag{1.1428}$$

Similarly, we can define the m-fold Kronecker product of $\boldsymbol{H}_2$ as the $2^m \times 2^m$ matrix [15, p. 115], [4, p. 422]

$$\boldsymbol{H}_{2^m} = \boldsymbol{H}_2^{\otimes m} = \boldsymbol{H}_2 \otimes \boldsymbol{H}_{2^{m-1}} = \underbrace{\boldsymbol{H}_2 \otimes \boldsymbol{H}_2 \otimes \cdots \otimes \boldsymbol{H}_2}_{m \text{ factors}} \tag{1.1429}$$

$$= \begin{pmatrix} \boldsymbol{H}_{2^{m-1}} & \boldsymbol{H}_{2^{m-1}} \\ \boldsymbol{0}_{2^{m-1} \times 2^{m-1}} & \boldsymbol{H}_{2^{m-1}} \end{pmatrix}, \quad m \in \mathbb{N}^*. \tag{1.1430}$$

The matrices $\boldsymbol{H}_{2^m}$, $m \in \mathbb{N}^*$, are closely related with the *Sylvester-type Hadamard matrices* [23, p. 116], [4, Figure 2.3, p. 44].

The rows of $\boldsymbol{H}_{2^m}$ have Hamming weights [15, p. 115]

$$2^0, 2^1, 2^2 \cdots 2^{m-2}, 2^{m-1}, 2^m, \quad m \in \mathbb{N}^*, \tag{1.1431}$$

and the number of rows with Hamming weight [15, p. 115] 2^{m-l}, $0 \le l \le m$, $l, m \in \mathbb{N}^*$ is equal to

$$\binom{m}{l}, \quad 0 \le l \le m, \quad l, m \in \mathbb{N}^*. \tag{1.1432}$$

The generator matrix $\boldsymbol{G}_{\mathcal{RM}(r,m)}$ of the rth-order Reed–Muller (RM) code $\mathcal{RM}(r,m)$ consists of those row vectors of $\boldsymbol{H}_{2^m}$ defined by (1.1429) with Hamming weights equal to or greater than 2^{m-r} [15, p. 115]. These row vectors are the same row vectors used in (1.1318) except they are different permutations [15, p. 115], and hence yield an equivalent code.

Example 1.83. The generator matrix $\boldsymbol{G}_{\mathcal{RM}(2,4)}$ consists of those row vectors of $\boldsymbol{H}_{16}$ given by (1.1428) with Hamming weights 4, 8 and 16 [15, p. 115]

$$\boldsymbol{G}_{\mathcal{RM}(2,4)} = \begin{pmatrix}
1&1&1&1&1&1&1&1&1&1&1&1&1&1&1&1\\
0&1&0&1&0&1&0&1&0&1&0&1&0&1&0&1\\
0&0&1&1&0&0&1&1&0&0&1&1&0&0&1&1\\
0&0&0&1&0&0&0&1&0&0&0&1&0&0&0&1\\
0&0&0&0&1&1&1&1&0&0&0&0&1&1&1&1\\
0&0&0&0&0&1&0&1&0&0&0&0&0&1&0&1\\
0&0&0&0&0&0&1&1&0&0&0&0&0&0&1&1\\
0&0&0&0&0&0&0&0&1&1&1&1&1&1&1&1\\
0&0&0&0&0&0&0&0&0&1&0&1&0&1&0&1\\
0&0&0&0&0&0&0&0&0&0&1&1&0&0&1&1\\
0&0&0&0&0&0&0&0&0&0&0&0&1&1&1&1
\end{pmatrix}. \tag{1.1433}$$

Example 1.84. The generator matrix $\boldsymbol{G}_{\mathcal{RM}(1,4)}$ consists of those row vectors of $\boldsymbol{H}_{16}$ given by (1.1428) with Hamming weights 8 and 16 [15, p. 115]

$$\boldsymbol{G}_{\mathcal{RM}(1,4)} = \begin{pmatrix} 1 & 1 & 1 & 1 & 1 & 1 & 1 & 1 & 1 & 1 & 1 & 1 & 1 & 1 & 1 & 1 \\ 0 & 1 & 0 & 1 & 0 & 1 & 0 & 1 & 0 & 1 & 0 & 1 & 0 & 1 & 0 & 1 \\ 0 & 0 & 1 & 1 & 0 & 0 & 1 & 1 & 0 & 0 & 1 & 1 & 0 & 0 & 1 & 1 \\ 0 & 0 & 0 & 0 & 1 & 1 & 1 & 1 & 0 & 0 & 0 & 0 & 1 & 1 & 1 & 1 \\ 0 & 0 & 0 & 0 & 0 & 0 & 0 & 0 & 1 & 1 & 1 & 1 & 1 & 1 & 1 & 1 \end{pmatrix}. \tag{1.1434}$$

1.17.5 Subcodes of Reed–Muller (RM) Codes

Let us begin with an example.

Example 1.85. In the case of m equal to 5, there are exactly six irreducible polynomials, i. e., minimal polynomials of degree 5 over $\mathbb{F}_2$; cf. Example 1.44 and [4, Figure 7.1, p. 198]. These are factors of $x^{31} \oplus 1$ [4, p. 109]. Let us take $\mathbb{F}_{2^5}$ equal to $\mathbb{F}_{32}$ defined by $\alpha^5 \oplus \alpha^3 \oplus 1 = 0$, illustrated in Example 1.44 into account. Giving the factor of $x^{31} \oplus 1$ in octal with the lowest degree on left (cf. [4, Figure 7.1, p. 198]), and the associated cyclotomic coset in brackets, we find the six minimal polynomials

$$M^{(0)}(x) = x \oplus 1 \qquad (6_8, \text{associated with } \mathbb{C}_0), \tag{1.1435}$$

$$M^{(1)}(x) = x^5 \oplus x^3 \oplus 1 \qquad (45_8, \text{associated with } \mathbb{C}_1), \tag{1.1436}$$

$$M^{(3)}(x) = x^5 \oplus x^3 \oplus x^2 \oplus x \oplus 1 \qquad (75_8, \text{associated with } \mathbb{C}_3), \tag{1.1437}$$

$$M^{(5)}(x) = x^5 \oplus x^4 \oplus x^3 \oplus x \oplus 1 \qquad (67_8, \text{associated with } \mathbb{C}_5), \tag{1.1438}$$

$$M^{(7)}(x) = x^5 \oplus x^4 \oplus x^3 \oplus x^2 \oplus 1 \qquad (57_8, \text{associated with } \mathbb{C}_7), \tag{1.1439}$$

$$M^{(11)}(x) = x^5 \oplus x^4 \oplus x^2 \oplus x \oplus 1 \qquad (73_8, \text{associated with } \mathbb{C}_{11}), \tag{1.1440}$$

$$M^{(15)}(x) = x^5 \oplus x^2 \oplus 1 \qquad (51_8, \text{associated with } \mathbb{C}_{15}). \tag{1.1441}$$

The primitive idempotents [4, p. 222]

$$\begin{aligned} \theta_0^*(x) &= \theta_0(x) = \bigoplus_{j=0}^{30} x^j = 1 \oplus x \oplus x^2 \oplus \cdots \oplus x^{30}, \\ \theta_1^*(x) &= \theta_{15}(x) \\ &= 1 \oplus x^3 \oplus x^5 \oplus x^6 \oplus x^9 \oplus x^{10} \oplus x^{11} \oplus x^{12} \oplus x^{13} \\ &\quad \oplus x^{17} \oplus x^{18} \oplus x^{20} \oplus x^{21} \oplus x^{22} \oplus x^{24} \oplus x^{26}, \\ \theta_3^*(x) &= \theta_7(x) \\ &= 1 \oplus x \oplus x^2 \oplus x^3 \oplus x^4 \oplus x^6 \oplus x^7 \oplus x^8 \oplus x^{12} \\ &\quad \oplus x^{14} \oplus x^{16} \oplus x^{17} \oplus x^{19} \oplus x^{24} \oplus x^{25} \oplus x^{28}, \\ \theta_5^*(x) &= \theta_{11}(x) \\ &= 1 \oplus x \oplus x^2 \oplus x^4 \oplus x^8 \oplus x^{11} \oplus x^{13} \oplus x^{15} \oplus x^{16} \\ &\quad \oplus x^{21} \oplus x^{22} \oplus x^{23} \oplus x^{26} \oplus x^{27} \oplus x^{29} \oplus x^{30}, \\ \theta_7^*(x) &= \theta_3(x) \\ &= 1 \oplus x^3 \oplus x^6 \oplus x^7 \oplus x^{12} \oplus x^{14} \oplus x^{15} \oplus x^{17} \oplus x^{19} \\ &\quad \oplus x^{23} \oplus x^{24} \oplus x^{25} \oplus x^{27} \oplus x^{28} \oplus x^{29} \oplus x^{30}, \end{aligned} \tag{1.1442}$$

$$\begin{aligned}\theta_{11}^*(x) &= \theta_5(x)\\ &= 1 \oplus x \oplus x^2 \oplus x^3 \oplus x^5 \oplus x^8 \oplus x^9 \oplus x^{10} \oplus x^{15}\\ &\quad \oplus x^{16} \oplus x^{18} \oplus x^{20} \oplus x^{23} \oplus x^{27} \oplus x^{29} \oplus x^{30},\\ \theta_{15}^*(x) &= \theta_1(x)\\ &= 1 \oplus x^5 \oplus x^7 \oplus x^9 \oplus x^{10} \oplus x^{11} \oplus x^{13} \oplus x^{14} \oplus x^{18} \oplus x^{19}\\ &\quad \oplus x^{20} \oplus x^{21} \oplus x^{22} \oplus x^{25} \oplus x^{26} \oplus x^{28}.\end{aligned}$$

We make the following observations:
- Using the minimal polynomial $M^{(15)}(x)$ of (1.1441) as the feedback polynomial of a linear feedback shift register (LFSR), the primitive idempotent θ_1^* will be generated as one period of an m-sequence.
- Using the minimal polynomial $M^{(7)}(x)$ of (1.1439) as the feedback polynomial of a linear feedback shift register (LFSR), the primitive idempotent θ_3^* will be generated as one period of an m-sequence.
- Using the minimal polynomial $M^{(11)}(x)$ of (1.1440) as the feedback polynomial of a linear feedback shift register (LFSR), the primitive idempotent θ_5^* will be generated as one period of an m-sequence.
- Using the minimal polynomial $M^{(3)}(x)$ of (1.1437) as the feedback polynomial of a linear feedback shift register (LFSR), the primitive idempotent θ_7^* will be generated as one period of an m-sequence.
- Using the minimal polynomial $M^{(5)}(x)$ of (1.1438) as the feedback polynomial of a linear feedback shift register (LFSR), the primitive idempotent θ_{11}^* will be generated as one period of an m-sequence.
- Using the minimal polynomial $M^{(1)}(x)$ of (1.1436) as the feedback polynomial of a linear feedback shift register (LFSR), the primitive idempotent θ_{15}^* will be generated as one period of an m-sequence.

Following Example 1.82, the second-order punctured Reed–Muller (RM) code $\mathcal{RM}(2,5)^*$ has the idempotent [4, p. 384]

$$\theta_0(x) \oplus \theta_1^*(x) \oplus \theta_5^*(x) \oplus \theta_3^*(x). \tag{1.1443}$$

Setting out from (1.1421), (1.1422) and (1.1423) of Example 1.82 as well as from (1.1443), we can see that
- the first-order Reed–Muller (RM) code $\mathcal{RM}(1,5)$ is based on a single m-sequence, represented by its respective period, which is the primitive idempotent $\theta_1^*(x)$, and
- the second-order Reed–Muller (RM) code $\mathcal{RM}(2,5)$ is based on three distinct m-sequences, represented by their respective periods, which are the primitive idempotents $\theta_1^*(x)$, $\theta_5^*(x)$ and $\theta_3^*(x)$.

In the case of the first-order Reed–Muller (RM) code $\mathcal{RM}(1,5)$ any m-sequence can be selected, provided that we can accept equivalent versions of the code. This means that the first-order Reed–Muller (RM) code $\mathcal{RM}(1,5)$ could, e. g., be based on $\theta_5^*(x)$ or $\theta_3^*(x)$ instead of $\theta_1^*(x)$.

Since the first-order Reed–Muller (RM) code $\mathcal{RM}(1,5)$, being a [4, Figure 1.13, p. 31]

$$(2^m, m+1, 2^{m-1}) = (32, 6, 16) \tag{1.1444}$$

binary linear block code, relies on an m-sequence, the first-order punctured Reed–Muller (RM) code $\mathcal{RM}(1,5)^*$, being a [4, Figure 1.13, p. 31]

$$(2^m - 1, m+1, 2^{m-1} - 1) = (31, 6, 15) \tag{1.1445}$$

binary linear block code, contains the cyclic shifted versions of this m-sequence. Expurgating, i. e., dropping the first row from the generator matrix $\boldsymbol{G}_{\mathcal{RM}(1,5)^*}$ of the first-order punctured Reed–Muller (RM) code $\mathcal{RM}(1,5)^*$, yields the *simplex code*, which is a [4, Figure 1.13, p. 31]

$$(2^m - 1, m, 2^{m-1}) = (31, 5, 16) \tag{1.1446}$$

binary linear block code. Hence, the simplex code contains only cyclic shifted versions of an m-sequence [4, p. 89].

Let's continue with the Example 1.85.

Example 1.86. We can readily see from Example 1.85, that

- the zeroth-order Reed–Muller (RM) code $\mathcal{RM}(0,5)$, using the primitive idempotent $\theta_0(x)$, is a subcode of the first-order Reed–Muller (RM) code $\mathcal{RM}(1,5)$, based on the primitive idempotent $\theta_0(x)$ as well as on any one further primitive idempotent, taken from the set

$$\{\theta_1^*(x), \theta_3^*(x), \theta_5^*(x)\}, \tag{1.1447}$$

- which itself is a subcode of the second-order Reed–Muller (RM) code $\mathcal{RM}(2,5)$, relying on the primitive idempotents $\theta_0(x)$, $\theta_1^*(x)$, $\theta_3^*(x)$ and $\theta_5^*(x)$.

We may now come to the conclusion, that we could construct subcodes, which lie in between the first-order Reed–Muller (RM) code $\mathcal{RM}(1,5)$ and the second-order Reed–Muller (RM) code $\mathcal{RM}(2,5)$ by, e. g., discarding from $\boldsymbol{G}_{\mathcal{RM}(2,5)}$ given by (1.1423) either

- the rows containing the primitive idempotent $\theta_1^*(x)$, leaving those rows containing the primitive idempotents $\theta_0(x)$, $\theta_5^*(x)$ and $\theta_3^*(x)$,
- the rows containing the primitive idempotent $\theta_5^*(x)$, leaving those rows containing the primitive idempotents $\theta_0(x)$, $\theta_1^*(x)$ and $\theta_3^*(x)$, or
- the rows containing the primitive idempotent $\theta_3^*(x)$, leaving those rows containing the primitive idempotents $\theta_0(x)$, $\theta_1^*(x)$ and $\theta_5^*(x)$.

Two distinct m-sequences in a code definitely "*Ring My Bell*" (Anita Ward, 1979). Is this yours also? An important lesson taught by Robert *Gold* is that there is

> *"a class of pairs of maximal linear sequences (remark by the author: m-sequences), for which we can determine precisely the cross correlation function, having three values"*

[62, p. 154]; see also [63, p. 619]. Such a pair of two selected m-sequences is also called a *preferred pair* [61, p. 96], [64, p. 406]. Combining, i. e., adding, the two m-sequences of a preferred pair yields a *Gold code* [64, pp. 404-407], also called *Gold sequence* [61, pp. 95f.]. For instance, [64, Figure 8-19, p. 407] shows a generator of a Gold code that is based on the sum of the primitive idempotent $\theta_{15}^*(x)$; cf. the linear feedback shift register (LFSR) depicted in the upper part of [64, Figure 8-19, p. 407], and the primitive idempotent $\theta_7^*(x)$; cf. the linear feedback shift register (LFSR) depicted in the lower part of [64, Figure 8-19, p. 407].

Let $\theta_i^*(x)$ be a period of the first m-sequence of the preferred pair, having the vector representation

$$\begin{aligned}\boldsymbol{\theta}_i^* &= ([\boldsymbol{\theta}_i^*]_0, [\boldsymbol{\theta}_i^*]_1 \cdots [\boldsymbol{\theta}_i^*]_{2^m-2}), \quad i \in \{1, 2 \cdots (2^{m-1} - 1)\}, \\ &= ([\boldsymbol{\theta}_i^*]_0, [\boldsymbol{\theta}_i^*]_1 \cdots [\boldsymbol{\theta}_i^*]_{30}), \quad i \in \{1, 2 \cdots 15\},\end{aligned} \tag{1.1448}$$

and let $\theta_j^*(x)$, $i \neq j$, be a period of the second m-sequence of the preferred pair, having the vector representation

$$\begin{aligned}\boldsymbol{\theta}_j^* &= ([\boldsymbol{\theta}_j^*]_0, [\boldsymbol{\theta}_j^*]_1 \cdots [\boldsymbol{\theta}_j^*]_{2^m-2}), \quad i \neq j, \quad i,j \in \{1, 2 \cdots (2^{m-1} - 1)\}, \\ &= ([\boldsymbol{\theta}_j^*]_0, [\boldsymbol{\theta}_j^*]_1 \cdots [\boldsymbol{\theta}_j^*]_{30}), \quad i \neq j, \quad i,j \in \{1, 2 \cdots 15\}.\end{aligned} \tag{1.1449}$$

Then, in our considered case of m equal to 5, the *periodic cross correlation function* is given by [61, equation (2.49), p. 24]

$$\phi_{ij}(\tau) = \sum_{n=0}^{2^m-2} (1 - 2 \cdot [\boldsymbol{\theta}_i^*]_n) \cdot (1 - 2 \cdot [\boldsymbol{\theta}_j^*]_{[(n+\tau) \bmod 31]}), \tag{1.1450}$$

$$i \neq j, \quad i,j \in \{1, 2 \cdots (2^{m-1} - 1)\}, \quad \tau \in \{0, 1 \cdots (2^m - 2)\},$$

$$= \sum_{n=0}^{30} (1 - 2 \cdot [\boldsymbol{\theta}_i^*]_n) \cdot (1 - 2 \cdot [\boldsymbol{\theta}_j^*]_{[(n+\tau) \bmod 31]}), \tag{1.1451}$$

$$i \neq j, \quad i,j \in \{1, 2 \cdots 15\}, \quad \tau \in \{0, 1 \cdots 30\},$$

$[\boldsymbol{\theta}_i^*]_n, n \in \{0, 1 \cdots (2^m-2)\}, i \in \{1, 2 \cdots (2^{m-1}-1)\}$, being the nth component of the primitive idempotent in vector form, $\boldsymbol{\theta}_i^*, i \in \{1, 2 \cdots (2^{m-1}-1)\}$ and $(1-2\cdot[\boldsymbol{\theta}_i^*]_n), n \in \{0, 1 \cdots (2^m-2)\}$, $i \in \{1, 2 \cdots (2^{m-1} - 1)\}$, representing the bipolar version of the nth component of the primitive idempotent in vector form, $[\boldsymbol{\theta}_i^*]_n$, $n \in \{0, 1 \cdots (2^m - 2)\}, i \in \{1, 2 \cdots (2^{m-1} - 1)\}$.

The "*three values*" mentioned above are [62, p. 155]

$$\phi_{ij}(\tau) = \begin{cases} (-1) & \text{if } (1 - 2 \cdot [\boldsymbol{\theta}_i^*]_\tau) = +1, \\ (-[2^{(m+1)/2} + 1]) \text{ or } (+[2^{(m+1)/2} - 1]) & \text{if } (1 - 2 \cdot [\boldsymbol{\theta}_i^*]_\tau) = -1, \end{cases} \tag{1.1452}$$

$$\tau \in \{0, 1 \cdots (2^m - 2)\},$$

$$= \begin{cases} (-1) & \text{if } (1 - 2 \cdot [\boldsymbol{\theta}_i^*]_\tau) = +1, \\ (-9) \text{ or } (+7) & \text{if } (1 - 2 \cdot [\boldsymbol{\theta}_i^*]_\tau) = -1, \end{cases} \tag{1.1453}$$

$$\tau \in \{0, 1 \cdots 30\}.$$

Let us continue with Example 1.86.

Example 1.87. Let us now consider the following pairs of m-sequences:
- $\theta_3^*(x)$ and $\theta_5^*(x)$,
- $\theta_1^*(x)$ and $\theta_3^*(x)$ and
- $\theta_1^*(x)$ and $\theta_5^*(x)$

and evaluate whether these pairs are preferred pairs. Evaluating the results shown in Figure 1.24, Figure 1.25 and Figure 1.26, we see that all three pairs of m-sequences are preferred pairs. Hence, the second-order punctured Reed–Muller (RM) code $\mathcal{RM}(2,5)^*$ contains cyclic shifted versions of m-sequences and of Gold codes. Therefore, also the second-order Reed–Muller (RM) code $\mathcal{RM}(2,5)$ relies on both m-sequences and Gold codes.

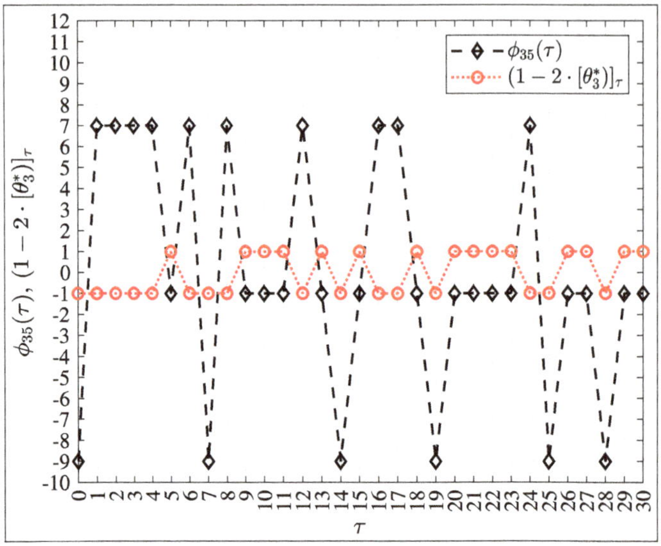

Figure 1.24: Periodic cross correlation function $\phi_{35}(m)$.

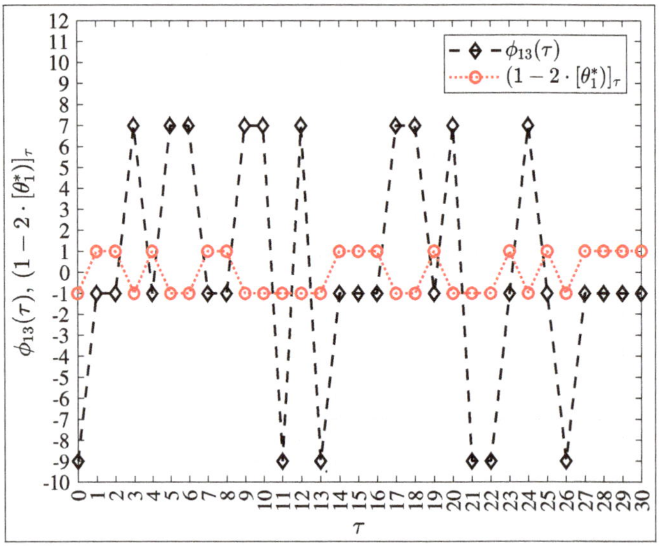

Figure 1.25: Periodic cross-correlation function $\phi_{13}(m)$.

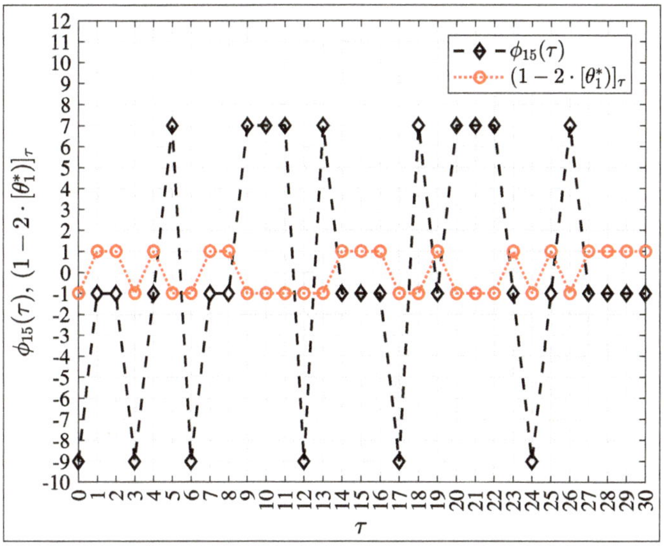

Figure 1.26: Periodic cross-correlation function $\phi_{15}(m)$.

In a nutshell, we may say, that there are subcodes of the second-order Reed–Muller (RM) code $\mathcal{RM}(2,m)$ and, consequently, of the second-order punctured Reed–Muller (RM) code $\mathcal{RM}(2,m)^*$ [4, Chapter 15, pp. 499–455], especially [4, Corollary 17, p. 455]. We will use Corollary 17 of [4, Corollary 17, p. 455] in what follows.

Theorem 1.70 (Subcodes of the Second-Order Reed–Muller (RM) Codes). *Let $m = 2t+1$ be odd, and let d be any number in the range $1 \le d \le t$ [4, Corollary 17, p. 455]. Then there exist two*

$$\left[2^m, m(t-d+2)+1, 2^{m-1}-2^{m-d-1}\right] \tag{1.1454}$$

subcodes of the second-order Reed–Muller (RM) code $\mathcal{RM}(2,m)$, which are obtained by extending the cyclic subcodes of the second-order punctured Reed–Muller (RM) code $\mathcal{RM}(2,m)^*$. The said second-order punctured Reed–Muller (RM) code $\mathcal{RM}(2,m)^*$ has the idempotents [4, Corollary 17, p. 455]

$$\theta_0(x) \oplus \theta_1^*(x) \bigoplus_{i=d}^{t} \theta_{1+2^i}^*(x) \tag{1.1455}$$

and [4, Corollary 17, p. 455]

$$\theta_0(x) \oplus \theta_1^*(x) \bigoplus_{i=1}^{t-d+1} \theta_{1+2^i}^*(x). \tag{1.1456}$$

These subcodes have the minimum distances $d_{\min}$ 2^{m-1} and $(2^{m-1} \pm 2^{m-h-1})$ for all h in the range $d \le h \le t$ [4, Corollary 17, p. 455].

Example 1.88. Let us again consider the second-order Reed–Muller (RM) code $\mathcal{RM}(2,5)$, i. e., $m = 5$. In this case, using $m = 2t + 1$, we have

$$t = \frac{m-1}{2} = \frac{5-1}{2} = 2. \tag{1.1457}$$

With $d = 1$, there are two $(32,16,8)$ binary linear block codes, which are subcodes of the second-order Reed–Muller (RM) code $\mathcal{RM}(2,5)$. However, since the second-order Reed–Muller (RM) code $\mathcal{RM}(2,5)$ already is a $(32,16,8)$ binary linear block code, it does not make any sense to further pursue this path, trying to define any subcodes in between $\mathcal{RM}(1,5)$ and $\mathcal{RM}(2,5)$.

Choosing $d = 2$, there are two $(32,11,12)$ binary linear block codes, which are subcodes $\mathbb{V}_1$ and $\mathbb{V}_2$ of the second-order Reed–Muller (RM) code $\mathcal{RM}(2,5)$ and which lie in between $\mathcal{RM}(1,5)$ and $\mathcal{RM}(2,5)$, i. e.,

$$\mathcal{RM}(1,5) \subset \mathbb{V}_1 \subset \mathcal{RM}(2,5), \tag{1.1458}$$

$$\mathcal{RM}(1,5) \subset \mathbb{V}_2 \subset \mathcal{RM}(2,5). \tag{1.1459}$$

We will follow this path further. The idempotent of $\mathbb{V}_1$ is given by

$$E^{(1)}(x) = \theta_0(x) \oplus \theta_1^*(x) \oplus \theta_5^*(x). \tag{1.1460}$$

The idempotent of $\mathbb{V}_2$ is given by

$$E^{(2)}(x) = \theta_0(x) \oplus \theta_1^*(x) \oplus \theta_3^*(x). \tag{1.1461}$$

With the primitive idempotents $\theta_0(x), \theta_1^*(x), \theta_3^*(x)$ and $\theta_5^*(x)$ [4, p. 222] (cf. (1.1442)), the generator matrices of $\mathbb{V}_1$ and $\mathbb{V}_2$ can be formed as

$$\begin{aligned}
\boldsymbol{G}_{\mathbb{V}_1} &= \begin{pmatrix} 1 & \theta_0(x) \\ 0 & \theta_3^*(x) \\ 0 & x^1 \odot \theta_3^*(x) \\ 0 & x^2 \odot \theta_3^*(x) \\ 0 & x^3 \odot \theta_3^*(x) \\ 0 & x^4 \odot \theta_3^*(x) \\ 0 & \theta_1^*(x) \\ 0 & x^1 \odot \theta_1^*(x) \\ 0 & x^2 \odot \theta_1^*(x) \\ 0 & x^3 \odot \theta_1^*(x) \\ 0 & x^4 \odot \theta_1^*(x) \end{pmatrix}, \\
\boldsymbol{G}_{\mathbb{V}_2} &= \begin{pmatrix} 1 & \theta_0(x) \\ 0 & \theta_5^*(x) \\ 0 & x^1 \odot \theta_5^*(x) \\ 0 & x^2 \odot \theta_5^*(x) \\ 0 & x^3 \odot \theta_5^*(x) \\ 0 & x^4 \odot \theta_5^*(x) \\ 0 & \theta_1^*(x) \\ 0 & x^1 \odot \theta_1^*(x) \\ 0 & x^2 \odot \theta_1^*(x) \\ 0 & x^3 \odot \theta_1^*(x) \\ 0 & x^4 \odot \theta_1^*(x) \end{pmatrix}.
\end{aligned} \tag{1.1462}$$

When appropriately reordering the columns of $G_{\mathbb{V}_2}$ in the following way:

$$\begin{aligned}&(5,23,4,26,22,3,10,25,28,30,21,13,2,9,19,\\&6,24,27,11,29,31,14,20,7,12,32,15,8,16,17,18),\end{aligned} \tag{1.1463}$$

i. e., column 5 of the original $G_{\mathbb{V}_2}$ now becomes column 1, column 23 of the original $G_{\mathbb{V}_2}$ now becomes column 2 and so on. We obtain an equivalent code. This code has the well-known form of the $\mathcal{RM}(1,m)$ generator matrix in the first six rows. Finally, expurgating by deleting the last row of the new generator matrix, we yield the $(32,10,12)$ code used in UMTS for the protection of the *transport format combination indicator (TFCI)* [65, Table 8, Section 4.3.3, p. 55f.]. A punctured version of an equivalent code was already mentioned in [4, p. 451].

The existence of such subcodes of the second-order Reed–Muller (RM) code $\mathcal{RM}(2,m)$ was already mentioned in [62, 2nd footnote, right column, p. 154].

1.18 Convolutional Codes

According to Section 1.14.1, only such classical (n,k,d_{min}) binary linear block codes with code lengths up to n equal to 256 bits and with dimensions k lower or equal to 256 bits have been identified until now [27, Table 1.1, p. 21], [51]. Such code lengths and dimensions have been much too small for many communications applications and, in particular, for advanced mobile communications.

Moreover, the mobile radio channel poses considerable obstacles on the transmission, yielding a large equivocation $\mathrm{H}\{X \mid Y\}$, and hence often requiring code rates of 1/2 or even less. This requires limiting the length k of the transmitted messages to less than 130 bits. This is out of the question for modern communications systems.

We may conclude that codes, which at the same time, facilitate large k and combat the equivocation $\mathrm{H}\{X \mid Y\}$ at the same time have been necessary from the beginning of modern information theory. Today, they are more than urgently needed.

A first promising solution have been *binary convolutional codes*, originally termed *sliding parity-check codes*. Binary convolutional codes were first introduced by Peter *Elias* in 1955 as an alternative to block codes, [15, p. 453]. In his seminal paper, Elias wrote analogously [31, pp. 70–73]:

> "There are *(author's note: the number of codewords)* 2^k, $k \in \mathbb{N}^*$ entries *(author's note: the codewords)* in a codebook and n binary digits in each entry *(author's note: each codeword contains n bits)*. This codebook must be stored at transmitter and receiver, which is impractical for values of *(author's note: the length of the code)* n and *(author's note: the code rate)* R large enough to be useful in greatly reducing the ambiguous detection probability.
> (· · ·)
> It is, of course, possible to devise coding schemes that are systematic (· · ·). However, the only scheme of this kind that has been shown to transmit information at a positive rate does not attain channel capacity, nor does its error probability decrease as rapidly as it should.

(···)
There is an alternative procedure. This is to find a small subset of codes that have a simple encoding and decoding procedure and are typical of the set of all possible codes, in the sense that they have the same average probability of error.
(···)
It is possible to construct a random parity-check code requiring only $(n-1)$ random binary selections by modifying the coefficient matrix (···).
(···)
If in a code of this type, which may be called a *sliding parity-check code*, the check digits are interspersed among the information digits, then the block length n can be made indefinitely large."

This leads us to the following knowledge.

a) Looking at Elias's words, we may guess that the *sliding* mentioned by Elias may be realized by a convolution.
b) Cyclic (n, k, d_{min}) binary linear block codes are governed by the convolution, as well; cf., e. g., Remark 1.31.
c) Convolutional codes can be built on cyclic (n, k, d_{min}) binary linear block codes.

But how can we overcome the length limitation mentioned above? Let us be pragmatic and try the following *brute force* attempt:

a) Let us forget that the number $(n-k+1)$, $k, n \in \mathbb{N}^*$, of possibly nonzero components $g_0, g_1 \cdots g_{n-k}$ of the nonzero row vector $\boldsymbol{g}$ defined in (1.437) governs the code rate R, because as k grows, n will grow accordingly whereas $(k-n)$ will remain constant and finite, thus R will approach 1 for $k, n \to \infty$.
b) Instead, let us replace $(n-k)$ used in $\boldsymbol{g}$ of (1.437) by the finite quantity $(\nu-1) \in \mathbb{N}^*$. Then let us define a nonzero *first generator vector of the convolutional code*

$$\boldsymbol{g}^{(0)} = (g_0^{(0)}, g_1^{(0)} \cdots g_{\nu-1}^{(0)}), \quad g_0^{(0)} = g_{\nu-1}^{(0)} = 1, \; g_m^{(0)} \in \mathbb{F}_2, \; m \in \{1, 2 \cdots (\nu-2)\}, \; \nu \in \mathbb{N}^*, \tag{1.1464}$$

and let us term the finite quantity $\nu \in \mathbb{N}^*$ the *constraint length of the convolutional code* (in German "Einflusslänge" [5, p. 246]); cf., e. g., [34, p. 229], [26, p. 159].

Remark 1.41 (On the Constraint Length). Unlike with block codes, with binary convolutional codes, large minimum distances and low error probabilities are achieved not by increasing k and n, $k, n \in \mathbb{N}^*$, but by increasing the constraint length $\nu \in \mathbb{N}^*$ of the convolutional code [15, p. 453]. It is however remarkable that small $\nu \in \mathbb{N}^*$, typically lower than 18 suffice for persuasive error performance; cf., e. g., [5, Tabelle 8.4, p. 257], [23, Tabelle 8.10, p. 319; Tabelle 8.11, p. 320; Tabelle 8.14 and Tabelle 8.15, p. 321], [32, pp. 492–496], [14, Table 4.2-1, p. 215].

Regrettably, the constraint length of a convolutional code is a quantity that does not have a unique and commonly accepted definition.

Lin and Costello [15, p. 459], Blahut [25, p. 271f.] and Bossert [23, p. 232] call the greatest index $(\nu-1) \in \mathbb{N}^*$, of the components $g_m \in \mathbb{F}_2$, $m \in \{0, 1 \cdots (\nu-1)\}$, $\nu \in \mathbb{N}^*$ of the first generator vector $\boldsymbol{g}^{(0)}$ of the convolutional code; cf. (1.1464), the constraint length of the convolutional code.

However, Friedrichs [5, p. 246], Proakis [32, p. 471f.], Viterbi and Omura [34, p. 229], Anderson and Mohan [14, p. 201], Glover and Grant [66, p. 352] and Biglieri [26, p. 159] define that the constraint length of the convolutional code shall be the length $\nu \in \mathbb{N}^*$ of the first generator vector $\boldsymbol{g}^{(0)}$ of the convolutional code; cf. (1.1464).

c) Let us then establish a first part of the convolutional encoding by using a nonzero *first generator matrix of the convolutional code*

$$\boldsymbol{G}^{(0)} = \begin{pmatrix} g_0^{(0)} & g_1^{(0)} & \cdots & g_{\nu-1}^{(0)} & 0 & \cdots & 0 & \cdots \\ 0 & g_0^{(0)} & \cdots & g_{\nu-2}^{(0)} & g_{\nu-1}^{(0)} & \cdots & 0 & \cdots \\ \vdots & \vdots & \ddots & \vdots & \vdots & \ddots & \vdots & \ddots \\ 0 & 0 & \cdots & g_0^{(0)} & g_1^{(0)} & \cdots & g_{\nu-1}^{(0)} & \cdots \\ \vdots & \vdots & \ddots & \vdots & \ddots & \ddots & \vdots & \ddots \end{pmatrix}, \quad k, \nu \in \mathbb{N}^*, \tag{1.1465}$$

which can become infinitely large.

Remark 1.42 (On Convolutional Codes as Block Codes). In the case of a message $\boldsymbol{u}$ and the generator vector $\boldsymbol{g}^{(0)}$ that are finite in length, we may easily obtain a finite length codeword $\boldsymbol{v}$ also when using a convolutional code. In this case, $\boldsymbol{G}^{(0)}$ of (1.1465) will be a $k \times (k + \nu - 1)$, $k, \nu \in \mathbb{N}^*$ matrix.

d) Let us control the code rate R, and hence the maximum equivocation $\max \mathrm{H}\{X \mid Y\}$ that we can combat by using more than just a single generator vector of the convolutional code, and thus more than just a single generator matrix of the convolutional code. Hence, for a code rate R of $1/2$, let us use a second generator vector $\boldsymbol{g}^{(1)}$ of the convolutional code, and hence a second generator matrix $\boldsymbol{G}^{(1)}$, the code rate R of $1/3$ requires a third generator vector $\boldsymbol{g}^{(2)}$ of the convolutional code, and hence a third generator matrix $\boldsymbol{G}^{(2)}$ and so on. The nonzero overall *generator matrix of the convolutional code* is thus given by

$$\boldsymbol{G} = (\boldsymbol{G}^{(0)}, \boldsymbol{G}^{(1)} \cdots \boldsymbol{G}^{(K_c-1)}) \quad R = \frac{1}{K_c}, \quad K_c \in \mathbb{N}^*, \tag{1.1466}$$

$$= \Bigg(\overbrace{\begin{matrix} g_0^{(0)} & \cdots & g_{\nu-1}^{(0)} & 0 & \cdots \\ 0 & g_0^{(0)} & \cdots & g_{\nu-1}^{(0)} & \cdots \\ \vdots & \ddots & \ddots & \ddots & \ddots \end{matrix}}^{\boldsymbol{G}^{(0)}} \begin{matrix} \cdots \\ \cdots \\ \cdots \end{matrix} \overbrace{\begin{matrix} g_0^{(K_c-1)} & \cdots & g_{\nu-1}^{(K_c-1)} & 0 & \cdots \\ 0 & g_0^{(K_c-1)} & \cdots & g_{\nu-1}^{(K_c-1)} & \cdots \\ \vdots & \ddots & \ddots & \ddots & \ddots \end{matrix}}^{\boldsymbol{G}^{(K_c-1)}} \Bigg). \tag{1.1467}$$

Example 1.89. Let us consider K_c equal to 2 and

$$\boldsymbol{g}^{(0)} = (1,0,1,1), \quad \boldsymbol{g}^{(1)} = (1,1,1,1). \tag{1.1468}$$

Then we obtain

$$\boldsymbol{G} = \left(\overbrace{\begin{matrix} 1 & 0 & 1 & 1 & 0 & 0 & \cdots \\ 0 & 1 & 0 & 1 & 1 & 0 & \cdots \\ 0 & 0 & 1 & 0 & 1 & 1 & \cdots \\ \vdots & \ddots & \ddots & \ddots & \ddots & \ddots & \ddots \end{matrix}}^{\boldsymbol{G}^{(0)}} \; \overbrace{\begin{matrix} 1 & 1 & 1 & 1 & 0 & 0 & \cdots \\ 0 & 1 & 1 & 1 & 1 & 0 & \cdots \\ 0 & 0 & 1 & 1 & 1 & 1 & \cdots \\ \vdots & \ddots & \ddots & \ddots & \ddots & \ddots & \ddots \end{matrix}}^{\boldsymbol{G}^{(1)}} \right). \tag{1.1469}$$

Obviously, the binary convolutional code with the generator matrix $\boldsymbol{G}$ of (1.1466) generates $K_c \in \mathbb{N}^*$ code bits per message bit. Instead of using the form of (1.1466), we may consider the equivalent binary convolutional code with [15, equation (11.8), p. 456], [23, p. 235]

$$\boldsymbol{G} = \begin{pmatrix} g_0^{(0)} & \cdots & g_0^{(K_c-1)} & g_1^{(0)} & \cdots & g_1^{(K_c-1)} & \cdots & g_{v-1}^{(0)} & \cdots & g_{v-1}^{(K_c-1)} & 0 & \cdots \\ 0 & \cdots & 0 & g_0^{(0)} & \cdots & g_0^{(K_c-1)} & \cdots & g_{v-2}^{(0)} & \cdots & g_{v-2}^{(K_c-1)} & \cdots & \cdots \\ 0 & \cdots & 0 & 0 & \cdots & 0 & \cdots & g_{v-3}^{(0)} & \cdots & g_{v-3}^{(K_c-1)} & \cdots & \cdots \\ \vdots & \cdots & \vdots & \vdots & \cdots & \vdots & \ddots & \vdots & \cdots & \vdots & \vdots & \ddots \end{pmatrix},$$
$$K_c \in \mathbb{N}^*. \tag{1.1470}$$

Example 1.90. Setting out from Example 1.89, we obtain

$$\boldsymbol{G} = \begin{pmatrix} 1 & 1 & 0 & 1 & 1 & 1 & 1 & 1 & 0 & 0 & 0 & 0 & \cdots \\ 0 & 0 & 1 & 1 & 0 & 1 & 1 & 1 & 1 & 1 & 0 & 0 & \cdots \\ 0 & 0 & 0 & 0 & 1 & 1 & 0 & 1 & 1 & 1 & 1 & 1 & \cdots \\ 0 & 0 & 0 & 0 & 0 & 0 & 1 & 1 & 0 & 1 & 1 & 1 & \cdots \\ \vdots & \vdots & \vdots & \vdots & \vdots & \vdots & \vdots & \vdots & \vdots & \vdots & \vdots & \vdots & \ddots \end{pmatrix}. \tag{1.1471}$$

Figure 1.27 shows the block diagram of the encoder of a binary convolutional code with code rate R equal to 1/2 and constraint length v equal to 4 using $\boldsymbol{G}$ of (1.1471) [15, Figure 11.1, p. 455].

The polynomial $g^{(0)}(x)$ is equal to the irreducible, minimal polynomial $M^{(3)}(x)$ of $\mathbb{F}_8$ and is also represented by 15_8 in octal form [32, Table 8.2-1, p. 492], whereas the polynomial $g^{(1)}(x)$ is equal to $(M^{(0)}(x))^3$, i. e., $(x \oplus 1)^3$, also being represented by 17_8 in octal form [32, Table 8.2-1, p. 492]. The polynomial $g^{(1)}(x)$ is thus not irreducible in $\mathbb{F}_8$. However, $M^{(0)}(x)$ is irreducible in $\mathbb{F}_2$ and is the minimal polynomial of the element 1 in $\mathbb{F}_8$.

Defining the nonzero vectors [23, p. 234]

$$\boldsymbol{G}_m = \begin{pmatrix} g_m^{(0)} & \cdots & g_m^{(K_c-1)} \end{pmatrix}, \quad m \in \{0,1 \cdots (v-1)\}, \quad v \in \mathbb{N}^*, \tag{1.1472}$$

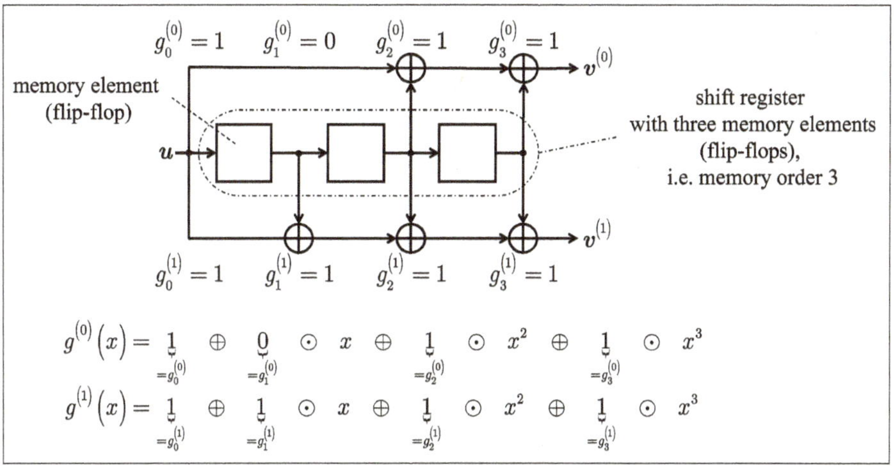

Figure 1.27: Block diagram of the encoder of a binary convolutional code with code rate R equal to 1/2 and constraint length ν equal to 4 using $\boldsymbol{G}$ of (1.1471), adapted according to [15, Figure 11.1, p. 455].

(1.1470) becomes [23, p. 234]

$$\boldsymbol{G} = \begin{pmatrix} \boldsymbol{G}_0 & \boldsymbol{G}_1 & \boldsymbol{G}_2 & \cdots & \boldsymbol{G}_{\nu-1} & \boldsymbol{0}_{1\times\nu} & \boldsymbol{0}_{1\times\nu} & \cdots & \boldsymbol{0}_{1\times\nu} & \cdots \\ \boldsymbol{0}_{1\times\nu} & \boldsymbol{G}_0 & \boldsymbol{G}_1 & \cdots & \boldsymbol{G}_{\nu-2} & \boldsymbol{G}_{\nu-1} & \boldsymbol{0}_{1\times\nu} & \cdots & \boldsymbol{0}_{1\times\nu} & \cdots \\ \boldsymbol{0}_{1\times\nu} & \boldsymbol{0}_{1\times\nu} & \boldsymbol{G}_0 & \cdots & \boldsymbol{G}_{\nu-3} & \boldsymbol{G}_{\nu-2} & \boldsymbol{G}_{\nu-1} & \cdots & \boldsymbol{0}_{1\times\nu} & \cdots \\ \vdots & \vdots & \ddots & \ddots & \vdots & \vdots & \ddots & \ddots & \vdots & \ddots \end{pmatrix}, \quad \nu \in \mathbb{N}^*. \tag{1.1473}$$

The most general version of $\boldsymbol{G}$ given by (1.1473) is obtained, when letting $\boldsymbol{G}_m$ and, consequently, also $\boldsymbol{0}_{1\times\nu}$, be matrices [23, p. 234]. We will however not consider this aspect further.

Since binary convolutional codes are obviously cyclic codes, we may associate the $K_c \in \mathbb{N}^*$ nonzero polynomials [5, equation (8.2.1), p. 249]

$$g^{(k_c)}(x) = \bigoplus_{m=0}^{\nu-1} g_m^{(k_c)} \odot x^m = g_0^{(k_c)} \oplus g_1^{(k_c)} \odot x \oplus \cdots \oplus g_{\nu-1}^{(k_c)} \odot x^{\nu-1}, \tag{1.1474}$$
$$k_c \in \{0, 1 \cdots (K_c - 1)\}, \quad K_c \in \mathbb{N}^*,$$

with the generator vectors $\boldsymbol{g}^{(k_c)}$, $k_c \in \{0, 1 \cdots (K_c - 1)\}$, $K_c \in \mathbb{N}^*$ (cf. (1.1464)) of the convolutional code. With the message polynomial $u(x)$, the code polynomials become [5, equation (8.2.5), p. 249]

$$v^{(k_c)}(x) = u(x) \odot g^{(k_c)}(x), \quad k_c \in \{0, 1 \cdots (K_c - 1)\}, \quad K_c \in \mathbb{N}^*. \tag{1.1475}$$

With the *generator polynomials matrix of the convolutional code*, which is often called the *generator matrix of the convolutional code* [5, equation (8.2.4), p. 249]

$$\boldsymbol{G}(x) = \left(g^{(0)}(x), g^{(1)}(x) \cdots g^{(K_c-1)}(x)\right), \quad K_c \in \mathbb{N}^*, \tag{1.1476}$$

and the *code polynomials vector* [5, equation (8.2.4), p. 249]

$$\boldsymbol{v}(x) = \left(v^{(0)}(x), v^{(1)}(x) \cdots v^{(K_c-1)}(x)\right), \quad K_c \in \mathbb{N}^*, \tag{1.1477}$$

we yield [5, equation (8.2.4), p. 249]

$$\boldsymbol{v}(x) = u(x) \odot \boldsymbol{G}(x) \quad (= u(x)\boldsymbol{G}(x)), \quad k_c \in \{0, 1 \cdots (K_c - 1)\}, \quad K_c \in \mathbb{N}^*. \tag{1.1478}$$

The code can hence by defined as follows [5, equation (8.2.6), p. 249]:

$$\mathbb{V} = \left\{u(x) \odot \boldsymbol{G}(x) \mid u(x) \in \mathbb{F}_2[x]\right\}. \tag{1.1479}$$

Using the generator matrix of $\boldsymbol{G}$ of (1.1466), respectively, $\boldsymbol{G}(x)$ of (1.1476), the encoder of the binary convolutional code has a *feedforward* structure [15, p. 454]; cf., e. g., Figure 1.27. These are also termed *finite impulse response (FIR)* encoders [23, p. 232] because of the finite constraint length $\nu \in \mathbb{N}^*$.

Do (1.1466) and (1.1476) ring a bell?

Looking at [1, Section 3.9], [67, equation (5.43), p. 209] and Remark 1.31, we find that $\boldsymbol{G}$ of (1.1466) has the same structure as $\underline{\boldsymbol{A}}^{\mathrm{T}}$ of [67, equation (5.43), p. 209]. Obviously, the binary convolutional code that we created by using $\boldsymbol{G}$ of (1.1466) resembles *coherent receiver antenna diversity (CRAD)* (cf., e. g., [68, Figure 1, pp. 76, 78]) with a transmitter having a single transmit antenna and a receiver having $K_c \in \mathbb{N}^*$ receive antennas; cf. [67, Bild 4.24, p. 184]. Furthermore, in the case of the considered binary convolutional code, each $\boldsymbol{G}^{(k_c)}$, $k_c \in \{0, 1 \cdots (K_c - 1)\}$, $K_c \in \mathbb{N}^*$ represents a distinct "diversity branch," i. e., a distinct link between the single transmit antenna and the k_cth, $k_c \in \{0, 1 \cdots (K_c-1)\}$, $K_c \in \mathbb{N}^*$ receive antenna.

An important difference between the binary convolutional code and CRAD lies in the fact

- that in the case of the binary convolutional code, the said "diversity branches" have to be taken care of completely by the transmitter implementation, which yields the reduction of the code rate R and, consequently, of the transmission rate when increasing $K_c \in \mathbb{N}^*$,
- whereas in the case of CRAD the realization of the "diversity branches" is mainly governed by physics, i. e., by the spatial character of the physical mobile radio channel, which avoids the reduction of the transmission rate when increasing the number of receive antennas $K_R \in \mathbb{N}^*$.

"*That's Life*" ("Blue Eyes Himself" Frank Sinatra, 1966), "You're riding high in April, shot down in May, but I know I'm gonna change that tune when I'm back on top, back on top in June! My, my."

Diversity schemes provide the largest gains in the case that the distinct "diversity branches" are mutually independent of each other [32, p. 828f.]. Other expressions one may encounter in this context are "uncorrelated" or "orthogonal."

How does this translate into "the speak of binary convolutional codes"? Obviously, we must require that the generator polynomials $g^{(k_c)}(x)$, $k_c \in \{0, 1 \cdots (K_c - 1)\}$, $K_c \in \mathbb{N}^*$ "mutually differ as much as possible." In the "speak of polynomials" this requirement means that the nonzero generator polynomials $g^{(k_c)}(x)$, $k_c \in \{0, 1 \cdots (K_c - 1)\}$, $K_c \in \mathbb{N}^*$ must be coprime, i. e., relatively prime. Hence, we require

$$\gcd\{g^{(0)}(x), g^{(1)}(x) \cdots g^{(K_c-1)}(x)\} = 1, \quad K_c \in \mathbb{N}^*. \tag{1.1480}$$

Otherwise, the decoder will occasionally produce an infinite number of errors although the message polynomial $u(x)$ has a finite degree; cf., e. g., [5, p. 254], [25, p. 284]. Such a behavior of the decoding, which has been enabled by an improper choice of the encoding process, is truly catastrophic [5, p. 254], [25, p. 284]. Obviously, the requirement (1.1480) relates to the encoding and decoding process, but not to the code itself [25, p. 284].

Definition 1.69 (Noncatastrophic Generator Matrix). The generator polynomials matrix of the convolutional code, i. e., the generator matrix of the convolutional code, $\boldsymbol{G}(x)$ of (1.1476) is termed *noncatastrophic generator matrix* if (1.1480) is fulfilled [5, Satz 8.1, p. 254f.], [25, Definition 9.4.3, p. 284].

Remark 1.43 (On the Noncatastrophic Generator Matrix). We require the generator polynomials $g^{(k_c)}(x)$, $k_c \in \{0, 1 \cdots (K_c - 1)\}$, $K_c \in \mathbb{N}^*$ to be nonzero. Then, in the case of

$$\gcd\{g^{(0)}(x), g^{(1)}(x) \cdots g^{(K_c-1)}(x)\} = p(x), \quad p(x) \neq 0, \quad K_c \in \mathbb{N}^*, \tag{1.1481}$$

we establish the noncatastrophic generator matrix as follows:

$$\boldsymbol{G}(x) = \left(\frac{g^{(0)}(x)}{p(x)}, \frac{g^{(1)}(x)}{p(x)} \cdots \frac{g^{(K_c-1)}(x)}{p(x)}\right), \quad K_c \in \mathbb{N}^*. \tag{1.1482}$$

Clearly, all the polynomials $g^{(k_c)}(x)/p(x)$, $k_c \in \{0, 1 \cdots (K_c - 1)\}$, $K_c \in \mathbb{N}^*$ have a finite degree, and hence provide a feedforward encoder.

In what follows, we will assume that (1.1480) is fulfilled and, consequently, $\boldsymbol{G}$ is noncatastrophic.

Example 1.91. Let us continue with Example 1.89 and Example 1.90. We have

$$g^{(0)}(x) = x^3 \oplus x^2 \oplus 1 \tag{1.1483}$$

and

$$g^{(1)}(x) = x^3 \oplus x^2 \oplus x \oplus 1; \tag{1.1484}$$

cf. Figure 1.27.

We immediately find

$$\gcd\{(x^3 \oplus x^2 \oplus x \oplus 1), (x^3 \oplus x^2 \oplus 1)\} = 1. \tag{1.1485}$$

Hence, $g^{(0)}(x)$ and $g^{(1)}(x)$ are relatively prime.

Comparing the structure of the encoder of a binary convolutional code with the multipath transmission channel occurring in mobile communications, termed the mobile radio channel, we immediately come to the conclusion that the encoding as well as the decoding of the said binary convolutional code can be described by a *trellis*, also termed *trellis diagram*; cf., e. g., [1]. Since the messages transmitted in advanced mobile communications are of finite length, consisting of, e. g., n, $n \in \mathbb{N}^*$ message bits, the determination of the initial state with the initial state vector $\boldsymbol{s}_0$ as well as the determination of the final or terminal state are most important. In the case of the mobile radio channel, the message is therefore preceded and succeeded by tail symbols, i. e., tail bits in the case of messages composed of bits.

Supposing binary transmission, the mobile radio channel being represented by a *discrete-time transversal filter*, i. e., a *finite impulse response (FIR) filter*, which models $W \in \mathbb{N}^*$ paths, corresponds to a *finite state machine (FSM)* having 2^{W-1}, $W \in \mathbb{N}^*$ distinct states at each time instant, assuming the *Mealy modeling* [69, Section 4.3.1, pp. 116–118].

Since the mentioned modeling is only virtual, which means that there is no real hardware making the mobile radio channel a finite state machine (FSM), there is also no reset button to determine the initial state with the initial state vector $\boldsymbol{s}_0$. Presetting the initial state with the initial state vector $\boldsymbol{s}_0$ hence requires the transmission of $(W-1)$ tail bits preceding the message, called preceding tail bits in what follows. Moreover, the final state must be set by transmitting $(W-1)$ tail bits succeeding the message, called succeeding tail bits in what follows. Without loss of generality, the preceding and the succeeding tail bits can be set to 0. In this case, both the initial state and the terminal state are identical, having the state vectors

$$\boldsymbol{s}_0 = \underbrace{(0, 0 \cdots 0)}_{W-1 \text{ zeros}} = \boldsymbol{0}_{W-1} = \boldsymbol{s}_{n+W-1}, \quad W, n \in \mathbb{N}^*. \tag{1.1486}$$

This looks like the terminal state was copied to the initial state in a manner similar to the cyclic prefix known from *orthogonal frequency division multiplexing (OFDM)*; cf., e. g., [1]. This situation is depicted in Figure 1.28. The "code rate" associated with the discrete-time transmit signal is hence given by

$$\frac{n}{n + 2 \cdot (W-1)} \leq 1, \quad W, n \in \mathbb{N}^*, \tag{1.1487}$$

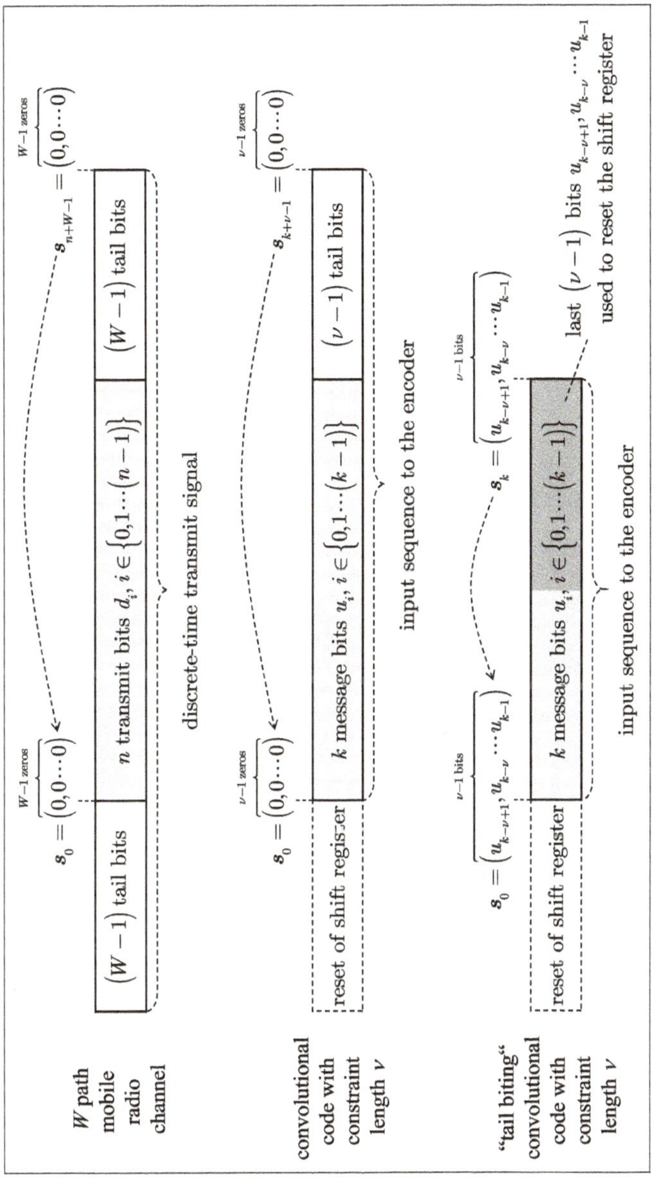

Figure 1.28: On the similarity of the mobile radio channel and the encoding of a binary convolutional code.

i. e., there is a small rate loss. However, the “code rate” approaches 1 for large $n \in \mathbb{N}^*$ and small $W \in \mathbb{N}^*$, which means that the rate loss diminishes and yields 0 in the limit.

In the case of the encoder of a binary convolutional code, the constraint length ν takes on the role of the number W of paths of the discrete-time mobile radio channel. Also, we will consider the message $\boldsymbol{u}$ to be composed of $k \in \mathbb{N}^*$ message bits. Now, there is no need for any preceding tail bits because we do have a reset button on the shift register contained in the encoder of the binary convolutional code, however, the $(\nu - 1)$ succeeding tail bits are still required; cf. Figure 1.28. Again, without loss of generality, the succeeding tail bits can be set to 0 and the initial state is reset to $\mathbf{0}_{\nu-1}$, i. e.,

$$\boldsymbol{s}_0 = \underbrace{(0, 0 \cdots 0)}_{\nu-1 \text{ zeros}} = \mathbf{0}_{\nu-1} = \boldsymbol{s}_{k+\nu-1}, \quad k, \nu \in \mathbb{N}^*. \tag{1.1488}$$

This situation also feels like the terminal state was copied to the initial state in a manner similar to the cyclic prefix known from orthogonal frequency division multiplexing (OFDM); cf., e. g., [1]. The associated code rate is thus given by

$$\frac{1}{K_\mathrm{c}} \cdot \frac{k}{k + \nu - 1} \le \frac{1}{K_\mathrm{c}}, \quad K_\mathrm{c}, k, \nu \in \mathbb{N}^*, \tag{1.1489}$$

K_c being the number of "diversity branches" of the binary convolutional code Of course, this code rate approaches $1/K_\mathrm{c}$ for large $k \in \mathbb{N}^*$ and small $\nu \in \mathbb{N}^*$.

We may further improve the code rate to become

$$\frac{1}{K_\mathrm{c}}, \quad K_\mathrm{c} \in \mathbb{N}^*, \tag{1.1490}$$

in any case. A first option is the renunciation on the $(\nu - 1)$ succeeding tail bits with the effect of performance degradation [70, p. 104]. A second option is the application of the concept of "*tail biting*" as in, e. g., the fourth generation mobile communications system "*Long Term Evolution (LTE)*," also abbreviated by 4G/LTE [71, Section 5.1.3, pp. 12–14]. Tail biting was first proposed in [70]. In brief and without specifics, using tail biting, we take the final $(\nu - 1)$ message bits to preset the initial state $\boldsymbol{s}_0$ [70, p. 104]; cf. Figure 1.28. Since the initial and the terminal states are however not known to the decoder, they must be estimated, which yields a little performance loss compared to the case of using a known terminal state [70, Table I, p. 107].

In general, the binary convolutional code generated by using the generator matrix $\boldsymbol{G}(x)$ of the convolutional code given by (1.1476) is nonsystematic [15, Figure 11.1, p. 454f.]. However, since all the generator polynomials $g^{(k_\mathrm{c})}(x)$, $k_\mathrm{c} \in \{0, 1 \cdots (K_\mathrm{c} - 1)\}$, $K_\mathrm{c} \in \mathbb{N}^*$ are nonzero, we immediately find

$$\boldsymbol{G}(x) = g^{(0)}(x)\left(1, \frac{g^{(1)}(x)}{g^{(0)}(x)} \cdots \frac{g^{(K_\mathrm{c}-1)}(x)}{g^{(0)}(x)}\right), \quad K_\mathrm{c} \in \mathbb{N}^*. \tag{1.1491}$$

Hence, an equivalent *systematic* binary convolutional code can, e. g., be encoded by using the generator matrix of the convolutional code in the following form [15, equation (11.44), p. 465; equation (11.64), p. 471], [23, equation (8.19), p. 254]:

$$\boldsymbol{G}_\mathrm{sys}(x) = \left(1, \frac{g^{(1)}(x)}{g^{(0)}(x)} \cdots \frac{g^{(K_\mathrm{c}-1)}(x)}{g^{(0)}(x)}\right), \quad K_\mathrm{c} \in \mathbb{N}^*. \tag{1.1492}$$

Note that since (1.1480) is fulfilled, all the polynomials $g^{(k_c)}(x)/g^{(0)}(x)$, $k_c \in \{1, 2 \cdots (K_c - 1)\}$, $K_c \in \mathbb{N}^*$ have an infinite degree, which means that the encoder of the binary convolutional code is an *infinite impulse response (IIR)* system [23, p. 233], also termed a *recursive system* [23, p. 245]. The binary convolutional code encoded by $\boldsymbol{G}_{\text{sys}}(x)$ of (1.1492) is also termed *recursive systematic convolutional (RSC) code* [72, p. 1064]. We will need these RSC codes when illustrating turbo codes.

Example 1.92. Let us continue with Example 1.89, Example 1.90 and Example 1.91. One possible recursive systematic convolutional (RSC) code corresponding to Figure 1.27 can be realized by using

$$\boldsymbol{G}_{\text{sys}}(x) = \left(1 \quad \frac{1 \oplus x \oplus x^2 \oplus x^3}{1 \oplus x^2 \oplus x^3}\right) = \left(1 \quad \frac{x^3 \oplus x^2 \oplus x \oplus 1}{x^3 \oplus x^2 \oplus 1}\right) = \left(1 \quad \frac{(M^{(0)}(x))^3}{M^{(3)}(x)}\right). \tag{1.1493}$$

The feedback polynomial $g^{(1)}(x)$ is the irreducible, minimal polynomial

$$g^{(1)}(x) = M^{(3)}(x) = x^3 \oplus x^2 \oplus 1 \tag{1.1494}$$

of $\mathbb{F}_8$ defined by $(a^3 \oplus a \oplus 1) = 0$. The feedforward polynomial $g^{(0)}(x)$ is given by the reducible polynomial

$$g^{(0)}(x) = \left(M^{(0)}(x)\right)^3 = (x \oplus 1)^3 = x^3 \oplus x^2 \oplus x \oplus 1, \tag{1.1495}$$

$M^{(0)}(x)$ being the minimal polynomial of the element 1 of $\mathbb{F}_8$. $M^{(3)}(x)$ of $\mathbb{F}_8$ is also represented by 15_8 in octal form [32, Table 8.2-1, p. 492], and $(M^{(0)}(x))^3$ can also be given by 17_8 in octal form [32, Table 8.2-1, p. 492].

Figure 1.29 illustrates the block diagram of the encoder of the recursive systematic convolutional (RSC) code with code rate R equal to 1/2, constraint length ν equal to 4 and $\boldsymbol{G}_{\text{sys}}(x)$ of (1.1493). The division by $M^{(3)}(x)$ can be realized by a feedback circuit; cf., e. g., [4, Figure 7.6, p. 210]. Consequently, we may design the encoder of a recursive systematic convolutional (RSC) code using a feedback circuit.

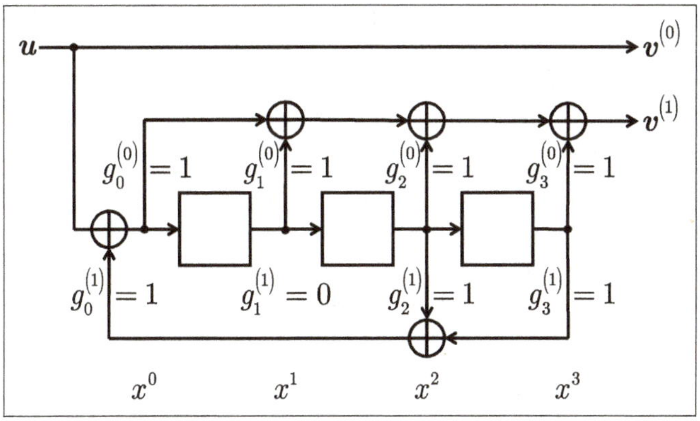

Figure 1.29: Block diagram of the encoder of the recursive systematic convolutional (RSC) code with code rate R equal to 1/2, constraint length ν equal to 4 and $\boldsymbol{G}_{\text{sys}}(x)$ of (1.1493).

Keeping in mind the *random coding argument* of [6, Theorem 11, pp. 22–24] and remembering that the linear feedback shift register (LFSR) based on the irreducible polynomial $M^{(3)}(x)$ produces an m-sequence, which looks random, choosing a primitive or irreducible polynomial as the feedback polynomial in a recursive systematic convolutional (RSC) code seems to be wise.

Remark 1.44 (On the Control of the Code Rate R). The above presented control of the code rate R seems quite coarse, doesn't it? This obvious disadvantage can be overcome by puncturing of the codewords; cf., e. g., Section 1.10 as well as [15, pp. 554, 582–598], [5, pp. 251–253], [23, pp. 255–259] and [14, pp. 217–219].

In 1967, Andrew J. *Viterbi* proposed a *maximum likelihood (ML)* sequence decoding algorithm termed the *Viterbi algorithm*, which was fairly simple to implement and helped to promote the deployment of binary convolutional codes in many applications, e. g., in deep-space and satellite communication systems in the 1970s [15, p. 453]. The *Viterbi algorithm* makes the generation of symbol-based soft decision outputs difficult.

In 1974, *Bahl, Cocke, Jelinek* and *Raviv* introduced a *maximum a posteriori probability (MAP)* symbol-by-symbol decoding algorithm, also called the *BCJR* algorithm [15, p. 453], which overcomes the problematic generation of symbol-based soft decision outputs encountered with the Viterbi algorithm and lends itself for the decoding of sophisticated coding schemes like turbo codes.

The reader is referred to [1] for more details on the said algorithms, which can be deployed for the decoding of binary convolutional codes.

Remark 1.45 (Quantum Communications System With Nonquantum Channel Coding—Convolutional Codes). Setting out from the two Remarks included in [1] on the simulation of coherent states and of the background thermal noise radiation field in quantum communications and on the quantum detection, we will now consider the wireless laser based binary on–off keying (OOK) quantum communications system, illustrated in the said two Remarks of [1], and combine it with classical, i. e., nonquantum, channel coding. Hence, in the Remarks 1.45, 2.1 and 3.2, including the Figures 1.30, 1.31, 1.32, 2.10 and 3.11, the attribute "*classical*" means "*nonquantum.*"

The reader may ask "why not use quantum error correction?" First of all, we should look at what the expression "quantum error correction?" means. Quantum error correction sets out from the situation that the information to be protected prevails in the form of at least one quantum bit, termed "*qubit,*" [73, p. 425], [74, pp. 237, 241]. The said information is not classical information but quantum information and, therefore, quantum error correction considers the encoding of quantum states to make them robust against the effects of any disturbances, e. g., noise and interactions with other quantum mechanical systems leading to decoherence [73, p. 425], [74, p. 237]. Often, entanglement is exploited to accomplish quantum error correction [73, pp. 459, 461], [74, p. 241]. In advanced mobile communications systems, we are, however, concerned about the transmission of classical information in the first place. Therefore, using quantum error correction does not seem appropriate, at least at first glance. Indeed, the main area of application of quantum error correction is quantum computing [73, pp. 425, 464; Section 10.6], [74, p. 237].

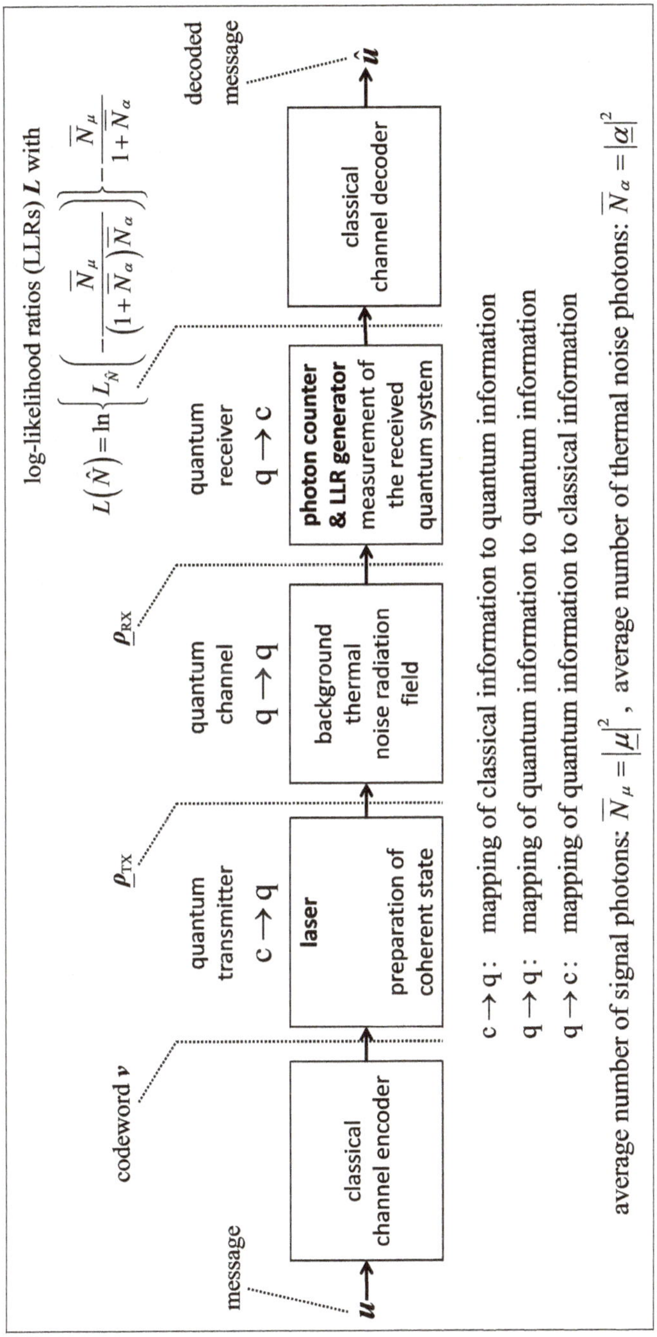

Figure 1.30: Wireless laser based binary on–off keying (OOK) quantum communications system with a classical channel encoder at its input and a classical channel decoder at its output (adapted according to [76, Bild 11, p. 115] and [77, Figure 5.1, p. 184; Figure 8.1, p. 382]).

Often, quantum error correction aims at the protection of one logical qubit by using a number n of physical qubits, thus setting the dimension k of the code equal to one [74, pp. 245–247, 251], [75, p. 31]. Hence, the length of the quantum error correction code is n. The corresponding quantum error correction codes usually have lengths of 9 or less, with n equal to 5 representing a lower bound [74, p. 251], [75, p. 32]. It seems that quantum error correction is more concerned about the code rate R equal to $1/n$ rather than about the dimension k and the length of the code n. Regrettably, the code rate $R \leq 1/5$ is nearly prohibitively low from the point of view of a mobile communications engineer.

In a nutshell, the question to be answered is *whether the radio front-ends in an advanced mobile communications system could just be replaced by, e. g., a wireless laser based binary on–off keying (OOK) quantum communications system without any quantum error correction, hence avoiding prohibitively low code rates, and without touching the remaining components of the advanced mobile communications system, thus still providing the transmission of classical information*. We will see that this is indeed possible. Of course, the inclusion of quantum error correction still remains an option.

Figure 1.30 depicts a simple test environment made up of the aforementioned wireless laser based binary on–off keying (OOK) quantum communications system with a classical channel encoder at its input and a classical channel decoder at its output (adapted according to [76, Bild 11, p. 115] and [77, Figure 5.1, p. 184; Figure 8.1, p. 382]). Referring to the above mentioned two Remarks of [1], the shown wireless laser based binary on–off keying (OOK) quantum communications system comprises of

- the classical channel encoder,
- the laser, preparing the coherent state, acting as the quantum transmitter, and thus as the transmitter front-end,
- the quantum channel, representing the background thermal noise radiation field,
- the photon counter and log-likelihood ratio (LLR) generator, carrying out the measurement of the received quantum mechanical systems, acting as the quantum receiver and thus as receiver front-end, and
- the classical channel decoder.

In Remark 1.45, the classical channel encoder implements the encoding of a convolutional code, and the classical channel decoder is a soft input maximum-likelihood (ML) decoder using the Viterbi algorithm [32, pp. 482–489], decoding the said convolutional code.

According to Figure 1.30, the *message* $\boldsymbol{u}$, comprising of the message bits u_i, $i \in \{0, 1 \cdots (k-1)\}$, $k \in \mathbb{N}^*$ is fed into the classical channel encoder, which generates the *codeword* $\mathbf{v}$, comprising of the code bits $v_j, j \in \{0, 1 \cdots (n-1)\}$, $k \leq n, k, n \in \mathbb{N}^*$. Then one code bit $v_j, j \in \{0, 1 \cdots (n-1)\}$, $k \leq n, k, n \in \mathbb{N}^*$ after the other is applied to the laser, which realizes the on–off keying (OOK) modulation for the code bit prevailing at its input in the following way [77, p. 318]:

- In the case of the present code bit $v_j, j \in \{0, 1 \cdots (n-1)\}$, $k \leq n, k, n \in \mathbb{N}^*$, having the value 0, the laser uses the coherent state $|0\rangle$, i. e., it emits no photons at all.
- In the case of the present code bit $v_j, j \in \{0, 1 \cdots (n-1)\}$, $k \leq n, k, n \in \mathbb{N}^*$, having the value 1, the laser uses the coherent state $|\underline{\mu}\rangle$, i. e., it emits on average $\bar{N}_\mu$ equal to $|\underline{\mu}|^2 > 0$ photons, which we will call *signal photons*.

When representing the mentioned pure states $|0\rangle$ and $|\underline{\mu}\rangle$ by the quantum mechanical density operator $\underline{\boldsymbol{\rho}}_{\mathrm{TX}}$, we yield

$$\underline{\boldsymbol{\rho}}_{\mathrm{TX}} = \begin{cases} |0\rangle\langle 0| & \text{if } v_j = 0, \\ |\underline{\mu}\rangle\langle\underline{\mu}| & \text{if } v_j = 1, \end{cases} \qquad j \in \left\{0, 1 \cdots (n-1)\right\}, k \leq n, k, n \in \mathbb{N}^*. \tag{1.1496}$$

Just to complete the picture, being in the state $|0\rangle$ means that the quantum mechanical harmonic oscillator, i. e., the laser does not emit any photons but it still has its zero point energy (ZPE), which it cannot emit and which consequently cannot be observed [78, p. 145f.], [79, second column, p. 749]. This means that the said zero point energy (ZPE) is *"not available for grilling steaks"* [79, second column, p. 749] but it is part of the quantum mechanical radiation field, which always comprises of oscillations and oscillators.

The quantum channel, which represents the background thermal noise radiation field, distorts the transmitted quantum mechanical radiation field with the quantum mechanical density operator $\underline{\boldsymbol{\rho}}_{\mathrm{TX}}$ generated by the laser by providing on average $\bar{N}_a$ equal to $|\underline{\alpha}|^2$ photons, which we will call *noise photons*. At the output of the quantum channel and, therefore, at the input of the quantum receiver, the quantum mechanical radiation field has the quantum mechanical density operator [77, equation (8.2), p. 363; equation (8.8), p. 365]

$$\underline{\boldsymbol{\rho}}_{\mathrm{RX}} = \frac{1}{\pi \bar{N}_a} \begin{cases} \int_{\mathbb{C}} \exp\{-|\underline{\alpha}|^2/\bar{N}_a\} |\underline{\alpha}\rangle\langle\underline{\alpha}| \, \mathrm{d}\underline{\alpha} & \text{if } v_j = 0, \\ \int_{\mathbb{C}} \exp\{-|\underline{\alpha} - \underline{\mu}|^2/\bar{N}_a\} |\underline{\alpha}\rangle\langle\underline{\alpha}| \, \mathrm{d}\underline{\alpha} & \text{if } v_j = 1, \end{cases} \tag{1.1497}$$

$$j \in \{0, 1 \cdots (n-1)\}, k \le n, k, n \in \mathbb{N}^*. \tag{1.1498}$$

The quantum receiver is a photon counter, which determines the number $\hat{N}$ of received photons and using $\hat{N}$ then generates the log-likelihood ratio (LLR)

$$L(\hat{N}) = \ln\left\{ L_{\hat{N}}\left(-\frac{\bar{N}_\mu}{(1+\bar{N}_a)\bar{N}_a} \right) \right\} - \frac{\bar{N}_\mu}{1+\bar{N}_a}, \tag{1.1499}$$

$L_{\hat{N}}(\cdot)$ denoting the regular *Laguerre polynomial*; cf., e. g., the Remark on the simulation of coherent states and of the background thermal noise radiation field in quantum communications [1]. The generation of the log-likelihood ratios (LLRs) is done consecutively for all the n code bits $v_j, j \in \{0, 1 \cdots (n-1)\}, k \le n, k, n \in \mathbb{N}^*$ forming the log-likelihood ratios (LLRs) vector $\boldsymbol{L}$.

Setting out from the log-likelihood ratios (LLRs) vector $\boldsymbol{L}$, the classical channel decoder determines the decoded message $\hat{\boldsymbol{u}}$.

Figure 1.31 illustrates the simulation results of the uncoded bit error ratios without any classical channel decoding and the coded bit error ratios with classical channel decoding of the convolutional code with code rate R equal to 1/2 and the generators in octal 5_8 and 7_8 [32, Table 8.2-1, p. 492]. The average number $\bar{N}_a \in \{0, 0.01, 0.03\}$ is the curve parameter. The code rate R equal to 1/2 is explicitly taken into account in the presented simulation results. This means that a message bit $u_i, i \in \{0, 1 \cdots (k-1)\}, k \in \mathbb{N}^*$ is represented by $1/R$ equal to 2, i. e., twice as many signal photons than a code bit $v_j, j \in \{0, 1 \cdots (n-1)\}$, $k = Rn = n/2, k, n \in \mathbb{N}^*$.

For instance, looking at $\bar{N}_a$ equal to 0, the uncoded bit error ratio of approximately $7 \cdot 10^{-2}$ is obtained at $\bar{N}_\mu$ equal to 2, whereas the coded bit error ratio at the corresponding $\bar{N}_\mu$ equal to 4 is approximately $5 \cdot 10^{-3}$. Looking at the simulation results of the uncoded bit error ratios discussed in the Remark on the quantum detection included in [1], the uncoded bit error ratio is approximately equal to $2 \cdot 10^{-4}$ at $\bar{N}_\mu$ equal to 8, whereas the coded bit error ratio of approximately $2 \cdot 10^{-4}$ requires only $\bar{N}_\mu$ approximately equal to 5.9 according to Figure 1.31. This is a saving of more than two signal photons on average.

Furthermore, Figure 1.32 shows the simulation results of the uncoded bit error ratios without any classical channel decoding and the coded bit error ratios with classical channel decoding of the convolutional code with code rate R equal to 1/2 and the generators in octal 561_8 and 753_8 [32, Table 8.2-1, p. 492]. The average number $\bar{N}_a \in \{0, 0.01, 0.03\}$ is the curve parameter. Again, the code rate R equal to 1/2 is explicitly taken into account in the presented simulation results. This means that a message bit u_i, $i \in \{0, 1 \cdots (k-1)\}, k \in \mathbb{N}^*$ is represented by $1/R$ equal to 2, i. e., twice as many signal photons than a code bit $v_j, j \in \{0, 1 \cdots (n-1)\}, k = Rn = n/2, k, n \in \mathbb{N}^*$.

For instance, looking at $\bar{N}_\alpha$ equal to 0, the uncoded bit error ratio of approximately $7 \cdot 10^{-2}$ is obtained at $\bar{N}_\mu$ equal to 2, whereas the coded bit error ratio at the corresponding $\bar{N}_\mu$ equal to 4 is less than $2 \cdot 10^{-4}$. Looking at the simulation results of the uncoded bit error ratios discussed in the Remark on the quantum detection included in [1], the uncoded bit error ratio is approximately equal to $2 \cdot 10^{-4}$ at $\bar{N}_\mu$ equal to 8, whereas the coded bit error ratio of approximately $2 \cdot 10^{-4}$ requires only $\bar{N}_\mu$ approximately equal to 4 according to Figure 1.32. This is a saving of approximately four signal photons on average.

Let us now discuss the simulated performance in the light of quantum error correction. Since we can regard each photon as the realization of a physical qubit used in quantum error correction (cf., e. g., [75, p. 30]), and knowing that the minimum number of physical qubits in well-known quantum error correction codes is 5 (see above), let us look at the achievable bit error ratio at $\bar{N}_\mu$ equal to 5. In the case of Figure 1.31, the corresponding coded bit error ratio will not drop well below 10^{-3} at $\bar{N}_\mu$ equal to 5, which means that quantum error correction shows a clear advantage. However, when the convolutional code with code rate R equal to 1/2 and the generators in octal 561_8 and 753_8 is considered, the coded bit error ratio may drop below 10^{-4} at $\bar{N}_\mu$ equal to 5, which does not look so unappealing compared to quantum error correction.

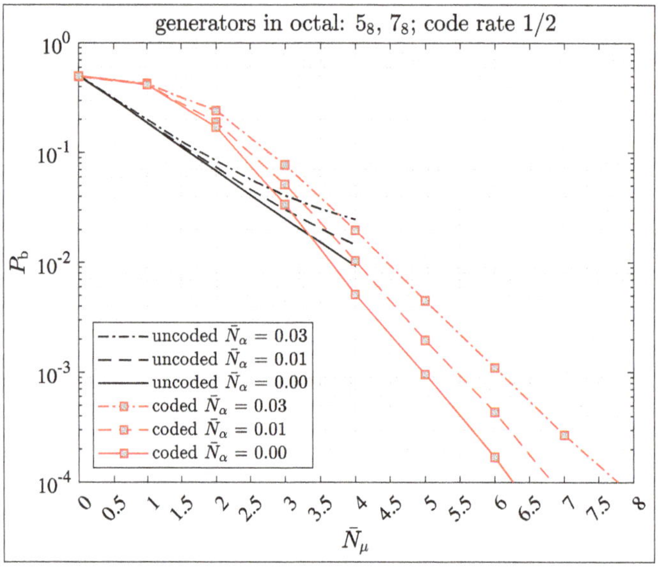

Figure 1.31: Simulation results of the uncoded bit error ratios without any classical channel decoding and the coded bit error ratios with decoding of the convolutional code with code rate R equal to 1/2 and the generators in octal 5_8 and 7_8. The average number $\bar{N}_\alpha \in \{0, 0.01, 0.03\}$ is the curve parameter.

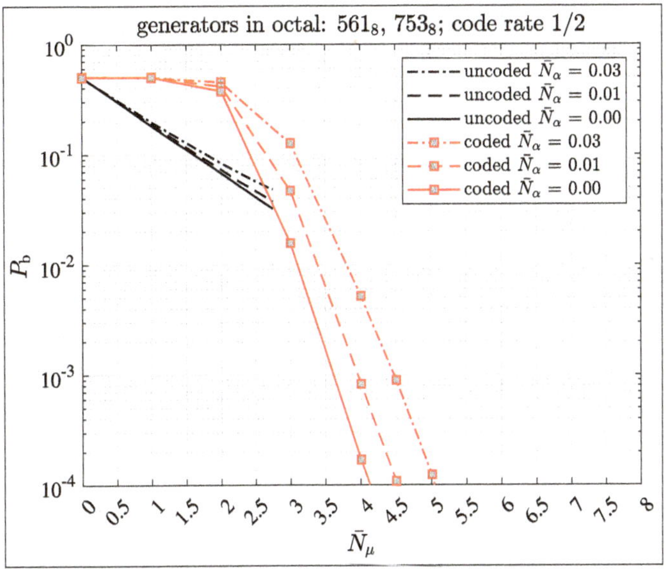

Figure 1.32: Simulation results of the uncoded bit error ratios without any classical channel decoding and the coded bit error ratios with classical channel decoding of the convolutional code with code rate R equal to 1/2 and the generators in octal 561_8 and 753_8. The average number $\bar{N}_\alpha \in \{0, 0.01, 0.03\}$ is the curve parameter.

2 Convolutional Turbo Codes

2.1 Rise and Fall

Turbo codes were first proposed by *Berrou*, *Glavieux* and *Thitimajshima* in 1993 during the International Conference on Communications (ICC'93), which was held in Geneva in May 1993 [72]. Indeed, Claude *Berrou*, Alain *Glavieux* and Punya *Thitimajshima* showed one single *turbo code*, yet suggesting that we were facing the birth of a new class of codes that had the potential to approach capacity limits as close as never before. Looking back at what happened about 30 years ago, it may sound quite unemotional, if not to say a bit "disdainful," having to say that, essentially, turbo codes just use what had been there before, however, not a single information theorist had seen the potential until then. And so it was a microelectronics engineer, Claude Berrou, who had to teach the world what you could do if you cleverly combined known pieces, namely

- two copies of a small recursive systematic convolutional (RSC) code, termed the two *constituent codes*, and
- parallel concatenation of these two constituent codes, deploying a properly chosen turbo code interleaver.

Since it deploys only convolutional codes as constituent codes, this type of turbo code is termed the *convolutional turbo code*. Only convolutional turbo codes have become relevant in advanced mobile communications systems. Therefore, we will neglect other types of turbo codes.

Figure 2.1 depicts an overview of the convolutional turbo code encoder structure based on [72, Figure 2, p. 1065]; see also [80, Figure 2, p. 136]. As already indicated above, the core of the convolutional turbo code encoder is composed of the turbo code interleaver and two recursive systematic convolutional (RSC) code encoders, which are usually chosen to be identical.

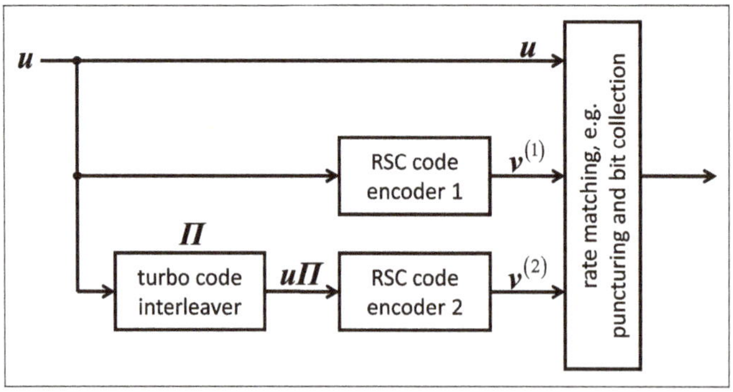

Figure 2.1: Convolutional turbo code encoder (adapted according to [72, Figure 2, p. 1065] and [80, Figure 2, p. 136]).

https://doi.org/10.1515/9783111239651-002

The basic concept of the convolutional turbo code provides the code rate R equal to 1/3. The *message vector* $\boldsymbol{u}$ is the input to the convolutional turbo code encoder, which provides the systematic information vector $\boldsymbol{u}$ together with two *parity check vectors* $\boldsymbol{v}^{(1)}$ and $\boldsymbol{v}^{(2)}$. At the output of the convolutional turbo code encoder, rate matching is accomplished by, e. g., puncturing and repetition as well as bit collection. Rate matching facilitates the appropriate variation of the code rate R. For instance, the alternating puncturing of the bits in the two parity check vectors $\boldsymbol{v}^{(1)}$ and $\boldsymbol{v}^{(2)}$ leaving only half of the bits of $\boldsymbol{v}^{(1)}$ and $\boldsymbol{v}^{(2)}$ without any puncturing of the message bits leads to the code rate R equal to 1/2; cf. [72, Figure 2, p. 1065].

Proposing a new code is not such a particular special thing, one might think—well, not with turbo codes. Turbo codes have at least two striking assets that everybody could understand immediately without doing any difficult rocket science, namely

- a splendid name, at least suggesting "formula one racing for communications"; there was no way out of thinking of Sammy Hagar's "*I Can't Drive 55*" released in 1984; this song just never becomes old; marketing par excellence, chapeau, Claude, that is genius;
- a breathtaking bit error ratio (BER) performance that no one had seen before for large message vector sizes, which have become increasingly important in advanced mobile communications.

Figure 2.2 shows the typical bit error ratio (BER) P_b performance of the turbo code with code rate 1/2 proposed by Berrou, Glavieux and Thitimajshima (cf., e. g., [72, Figure 5, p. 1069]) versus the average signal-to-noise ratio $10 \cdot \log_{10}(E_\text{b}/N_0)$ in dB when considering the transmission over an additive white Gaussian noise (AWGN) channel without any fading. As mentioned above, the code rate R equal to 1/2 is obtained by the alternating puncturing of the bits in the two parity check vectors $\boldsymbol{v}^{(1)}$ and $\boldsymbol{v}^{(2)}$, leaving only half of the bits of $\boldsymbol{v}^{(1)}$ and $\boldsymbol{v}^{(2)}$, while the message bits are left untouched; cf. [72, Figure 2, p. 1065].

According to Figure 2.2, the convolutional turbo code suggested by [72] requires only an average signal-to-noise ratio (SNR) $10 \log_{10}(E_\text{b}/N_0)$ of approximately 0.7 dB to achieve a BER of 10^{-5}, which is only about 0.5 dB more than the binary phase shift keying (BPSK) capacity limit [15, p. 771]. Convolutional turbo codes hence show the potential to considerably outperform classical channel codes, e. g., convolutional codes [15, p. 771].

Gérard *Battail* showed [35, Figure 5, p. 87f.] that the bit error ratio (BER) P_b performance shown in Figure 2.2 is typical for *weakly random-like codes*, hence identifying convolutional turbo codes as weakly random-like codes [35, pp. 88, 91f.]. The reason for calling such a code "*weakly* random-like" and not "*strongly* random-like" or just "random-like" is that only the average properties of the Hamming weight distribution of the codewords, in particular, the mean and the variance and the average shape, are like those of a strongly random-like code [35, p. 76]. In contrast, in the case of strongly random-like codes, each term of the Hamming weight distribution of the codewords is close to that of random coding [35, p. 76]. The flattening of the curve, which governs the

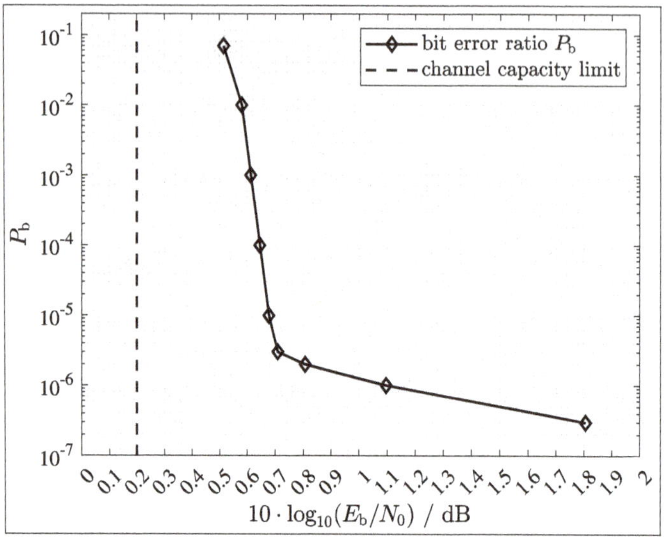

Figure 2.2: Bit error ratio (BER) P_b performance of the convolutional turbo code with code rate 1/2 proposed by Berrou, Glavieux and Thitimajshima, adapted according to [72, Figure 5, p. 1069] and [15, Figure 16.3, p. 771] versus the average signal-to-noise ratio (SNR) $10 \cdot \log_{10}(E_b/N_0)$ in dB, considering the transmission over an additive white Gaussian noise (AWGN) channel without any fading.

image at an average signal-to-noise ratio (SNR) above approx. 0.7 dB is due to the rather low "*free distance*," i. e., the minimum Hamming distance [35, Figure 5, p. 87f.].

A second "stab in the heart" of one or the other established information theorists was that it only took three mobile radio engineers, namely Markus *Naßhan*, Josef *Blanz* and the author [81, Figure 4, p. 774] to show to the world that convolutional turbo codes were the best choice for advanced mobile communications, beginning with the third generation mobile communications system "*Universal Mobile Telecommunications System (UMTS)*," also abbreviated by 3G/UMTS [82, p. 112f.]. In fact, to the best knowledge of the author [81] was the very first publication dealing with convolutional turbo codes in mobile radio systems.

And, "heartache number three," it had been the author and Markus *Naßhan* who were the first to discover that a most remarkable performance asset of turbo codes is the improvement of the *block error ratio (BLER)*, also called "*frame error ratio (FER)*" (cf., e. g., [81, Figure 4, p. 774]) at moderate average signal-to-noise ratios (SNRs), when compared with classical convolutional codes. The block error ratio (BLER) is the ratio of the number of erroneously decoded messages over the total number of transmitted messages. Although Battail calls the block error ratio (BLER) of weakly random-like codes "bad" at low average signal-to-noise ratios (SNRs) [35, pp. 76, 92], it shows that at moderate and at high average signal-to-noise ratios (SNRs) the block error ratio (BLER) performance is quite appealing. Since the early 2000s, the block error ratio (BLER) has become a most important performance criterion in 3G/UMTS and the fourth generation mobile communications system *"Long Term Evolution (LTE),"* also abbreviated by 4G/LTE, as well as more recent advanced mobile communications systems.

According to Figure 2.1, the convolutional turbo code proposed by Berrou, Glavieux and Thitimajshima in 1993 [72] is a systematic code, i. e., there is no puncturing of any message bits. To the best knowledge of the author, he, Jörg *Plechinger*, Markus *Doetsch* and Friedbert Manfred *Berens* were the first

- to discuss *rate compatible punctured turbo codes (RCPTC)* [80, Figure 2, pp. 136, 138f.], which are required for link adaptation [80, Figure 1, pp. 135–137], and
- to show that puncturing of some of the message bits provides performance benefits, although the resulting convolutional turbo code is no longer systematic. This kind of puncturing is termed "UKL puncturing" in [80, p. 140f.].

Link adaptation concepts already discussed in [80] have entered advanced mobile communications standards. Since advanced mobile communications systems such as 3G/UMTS and 4G/LTE deploy hybrid automatic repeat request (HARQ) schemes, retransmissions with partially punctured systematic information have become part of the standards [83, Figure 8.15; Figure 8.16; Figure 9.3; Figure 9.23; pp. 83, 85f., 139–142, 169, 172, 200].

After being "en vogue" for nearly three decades, being a celebrated asset of 3G/UMTS [82, p. 112f.] and 4G/LTE [84, pp. 61f., 145], [83, Figure 8.15; Figure 8.16; Figure 9.3; Figure 9.23; pp. 83, 85f., 139–142, 169, 172, 200], convolutional turbo codes began to somewhat resemble a worn out recording of a favorite song and, in the end, "fell out of favor" with standards delegates setting up the fifth generation mobile communications system *"5G,"* who replaced convolutional turbo codes by *"low density parity check (LDPC) codes"* [85, pp. 41, 102, 134–140, 402] [86, pp. 66, 156–160].

The reason for this decision that is commonly cited is "implementation complexity." The author has considerable reservations because he believes that the obtainable link and system performance should always be the guideline. If implementation complexity was such an issue, then one should reconsider the deployment of multiple antennas in the first place. Bearing in mind that most mobile terminals and probably all base stations are multistandard devices, two mutually incompatible sophisticated channel coding schemes, namely convolutional turbo codes and LDPC codes, now must be implemented at the same time both in mobile terminals and in base stations. In particular, the implementation complexity of modem chips, e. g., smart phones have to be observed carefully in this context. In contrast, the further evolution of the convolutional turbo code used in 3G/UMTS and 4G/LTE toward 5G applications would have provided the asset of backward compatibility with all implementation complexity benefits. Moreover, the implementation complexity of, e. g., mobile terminals is certainly not governed by the modem chip today but rather by units much closer to the human interface such as the applications that the user operates, as well as the display.

So, the given reason for dropping convolutional turbo codes does not appear to be the complete truth. Rather, the author is reminded of the 3G/UMTS standardization, which never seemed to be purely technology driven and in a couple of aspects also looked quite like "knitted with hot needle." In contrast, the author believes that the second generation mobile communications system 2G/GSM (global system for mobile communications) as well as 4G/LTE have both been marvelous pieces of engineering art. Regrettably, the author therefore just cannot push back the notion that the quality of mobile communications standards seems to be alternating like an antipodal binary sequence, swaying between "well done" and "well meant." Nevertheless, LDPC codes being discussed in Chapter 3 are certainly a good choice, just like convolutional turbo codes were when it was about 3G/UMTS and 4G/LTE. Since the importance of convolutional turbo codes is gradually diminishing, the author has decided to limit their discussion to this short chapter.

2.2 Parallel Concatenation of Recursive Systematic Convolutional (RSC) Codes

Figure 2.3 depicts the rate-1/3 turbo code encoder, which forms the basis of the turbo code encoder deployed by Berrou, Glavieux and Thitimajshima [72, Figure 2, p. 1065] with the *reducible polynomial* $(M^{(0)}(x))^4$ as the feedforward polynomial, $M^{(0)}(x)$ being the *minimal polynomial* of the element 1 in $\mathbb{F}_{16}$ and the minimal polynomial $M^{(3)}(x)$ of $\mathbb{F}_{16}$ as the *feedback polynomial*, $\mathbb{F}_{16}$ defined by $(\alpha^4 \oplus \alpha \oplus 1)$ equal to 0. $M^{(0)}(x)$ of $\mathbb{F}_{16}$ and $M^{(3)}(x)$ of $\mathbb{F}_{16}$ are relatively prime, i. e., their greatest common divisor (gcd) is equal to 1. Choosing the minimal polynomial $M^{(3)}(x)$ of $\mathbb{F}_{16}$ as the feedback polynomial is a

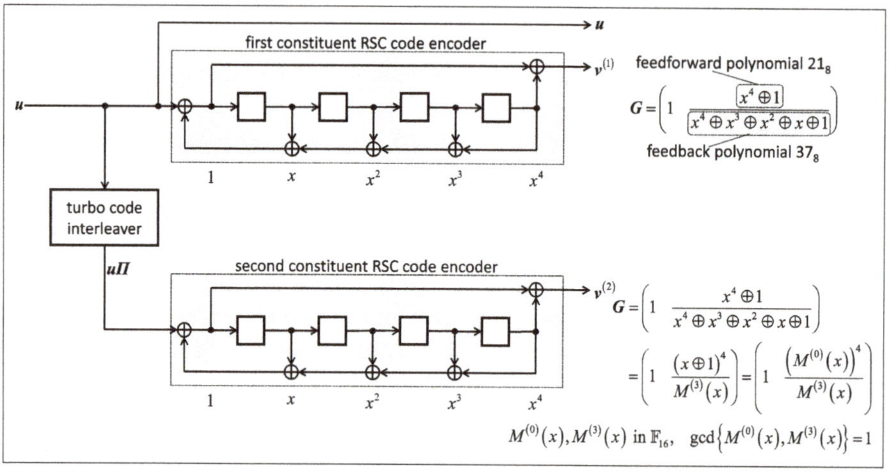

Figure 2.3: Turbo code encoder deployed by Berrou, Glavieux and Thitimajshima, adapted according to [72, Figure 2, p. 1065], using the reducible polynomial $(M^{(0)}(x))^4$ as the feedforward polynomial, $M^{(0)}(x)$ being the minimal polynomial of the element 1 in $\mathbb{F}_{16}$ and the minimal polynomial $M^{(3)}(x)$ of $\mathbb{F}_{16}$ as the feedback polynomial; the constraint length ν of the constituent RSC codes is equal to 5.

consequence of the *random coding argument* of [6, Theorem 11, pp. 22–24] as we already know.

Each constituent code represents the same convolutional turbo code. In the case of Figure 2.3, each constituent code has the constraint length ν equal to 5 and is therefore described by polynomials with a degree not greater than $(\nu-1)$, i. e., 4. This is the number of flip-flops required in each linear shift register contained in each of the constituent codes. Therefore, $(\nu - 1)$ is also called the *memory order* [15, Definition 11.2, p. 459].

Figure 2.4 depicts the rate-1/3 turbo code encoder deployed by Jung et al. (cf., e. g., [80, Figure 2, p. 136; Figure 4, p. 138] and [87, Figure 5.9, p. 139]) with the reducible polynomial $(M^{(0)}(x))^2$ as the feedforward polynomial, $M^{(0)}(x)$ being the minimal polynomial of the element 1 in $\mathbb{F}_4$ and the minimal polynomial $M^{(1)}(x)$ of $\mathbb{F}_4$ as the feedback polynomial, $\mathbb{F}_4$ being defined by $M^{(1)}(\alpha)$ equal to 0, $M^{(0)}(x)$ of $\mathbb{F}_4$ and $M^{(1)}(x)$ of $\mathbb{F}_4$ are relatively prime, i. e., their greatest common divisor (gcd) is equal to 1. Choosing the minimal polynomial $M^{(1)}(x)$ of $\mathbb{F}_4$ as the feedback polynomial is a consequence of the *random coding argument* of [6, Theorem 11, pp. 22–24] as we already know.

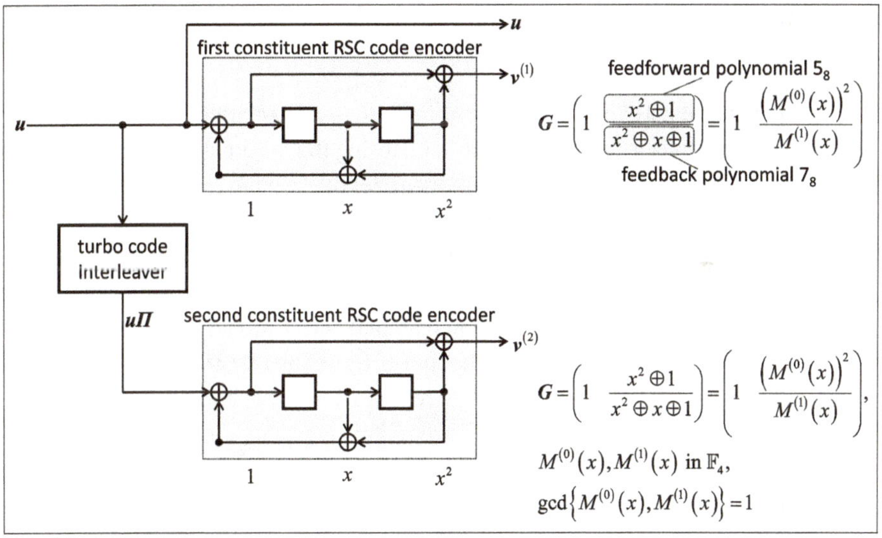

Figure 2.4: Turbo code encoder deployed by Jung et al., adapted according to [80, Figure 2, p. 136; Figure 4, p. 138], using the reducible polynomial $(M^{(0)}(x))^2$ as the feedforward polynomial, $M^{(0)}(x)$ being the minimal polynomial of the element 1 in $\mathbb{F}_4$ and the minimal polynomial $M^{(1)}(x)$ of $\mathbb{F}_4$ as the feedback polynomial; the constraint length ν of the constituent RSC codes is equal to 3.

Figure 2.5 depicts a second rate-1/3 turbo code encoder deployed by the author [88, p. 86] using the reducible polynomial $(M^{(0)}(x))^3$ as the feedforward polynomial, $M^{(0)}(x)$ being the minimal polynomial of the element 1 in $\mathbb{F}_8$ and $\mathbb{F}_8$ being defined by $(\alpha^3 \oplus \alpha \oplus 1)$ is equal to 0, and with the irreducible minimal polynomial $M^{(3)}(x)$ of $\mathbb{F}_8$ as the

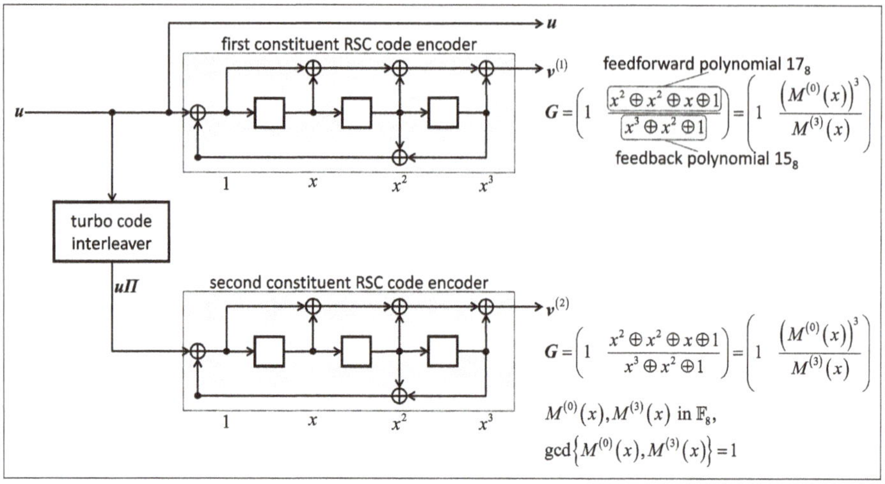

Figure 2.5: Turbo code encoder deployed by the author [88, p. 86], using the reducible polynomial $(M^{(0)}(x))^3$ as the feedforward polynomial, $M^{(0)}(x)$ being the minimal polynomial of the element 1 in $\mathbb{F}_8$ and the irreducible minimal polynomial $M^{(3)}(x)$ of $\mathbb{F}_8$ as the feedback polynomial; the constraint length ν of the constituent RSC codes is equal to 4.

feedback polynomial. $M^{(0)}(x)$ of $\mathbb{F}_8$ and $M^{(3)}(x)$ of $\mathbb{F}_8$ are relatively prime, i. e., their greatest common divisor (gcd) is equal to 1. Choosing the minimal polynomial $M^{(3)}(x)$ of $\mathbb{F}_4$ as the feedback polynomial is a consequence of the *random coding argument* of [6, Theorem 11, pp. 22–24] as we already know.

Figure 2.6 depicts the UMTS/LTE rate-1/3 turbo code encoder [56, Figure 4, p. 23], [71, Figure 5.2.3-2, p. 15], using the minimal polynomial $M^{(1)}(x)$ of $\mathbb{F}_8$ as the feedforward polynomial and $M^{(3)}(x)$ of $\mathbb{F}_8$ as the feedback polynomial. As we already know, the minimal polynomial $M^{(1)}(x)$ is the primitive polynomial of $\mathbb{F}_8$ defined by $M^{(1)}(\alpha)$ equal to 0 and the minimal polynomial $M^{(3)}(x)$ of $\mathbb{F}_8$ is the second irreducible polynomial of degree 3. Of course, $M^{(1)}(x)$ of $\mathbb{F}_8$ and $M^{(3)}(x)$ of $\mathbb{F}_8$ are relatively prime, i. e., their greatest common divisor (gcd) is equal to 1. Choosing the minimal polynomial $M^{(3)}(x)$ of $\mathbb{F}_4$ as the feedback polynomial is a consequence of the *random coding argument* of [6, Theorem 11, pp. 22–24] as we already know. Figure 2.6(a) illustrates the regular mode of operation, whereas Figure 2.6(b) depicts the termination mode of operation.

Looking at convolutional codes, decoding is usually accomplished by exploiting the trellis structure associated with the convolutional code. Furthermore, since advanced mobile communications systems use messages that are always finite in length, the initial and the final state of the encoding become of importance. This means that decoding performs best when both the initial and the final states are known to the decoder. In the case of the initial state, this is easy to accomplish by agreeing on a particular state in which the linear shift register shall set out from. Usually, this is the state 0. In the case of conventional convolutional codes with nonsystematic encoders using only feedforward

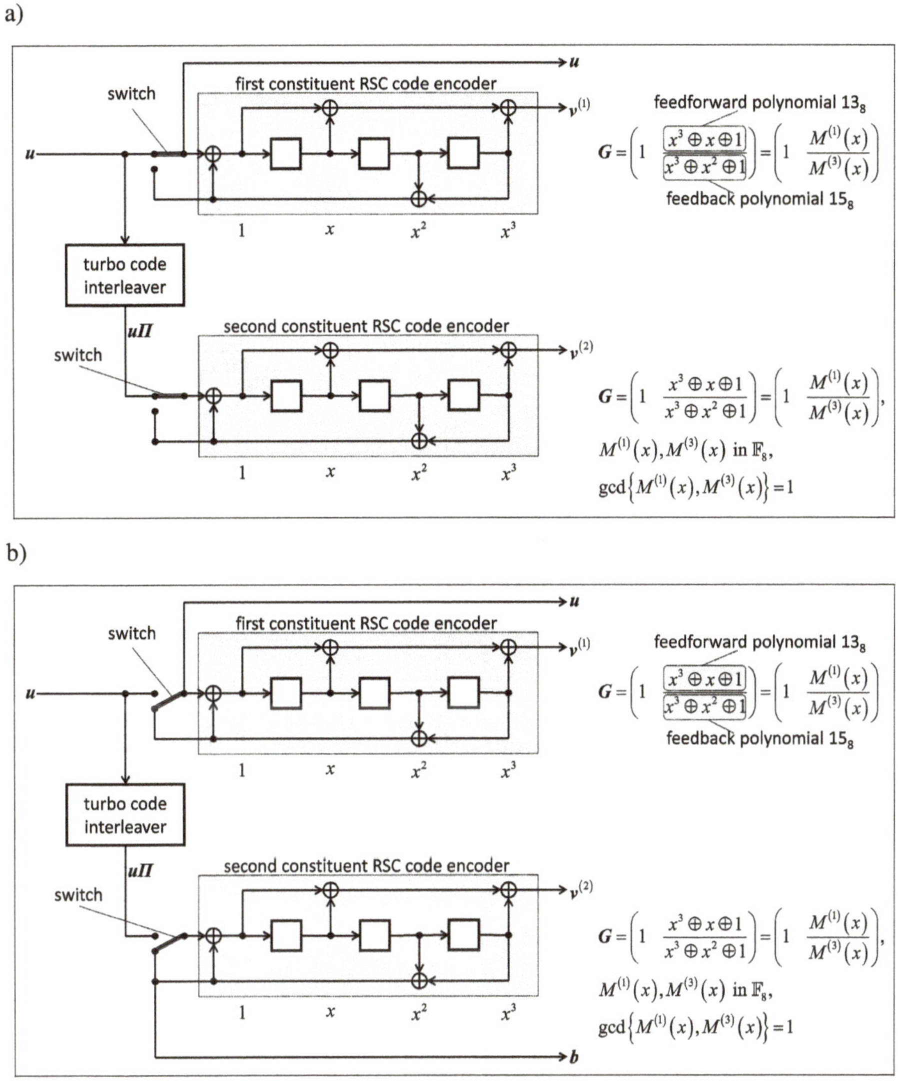

Figure 2.6: UMTS/LTE convolutional turbo code encoder, adapted according to [56, Figure 4, p. 23] and [71, Figure 5.2.3-2, p. 15], using the minimal polynomial $M^{(1)}(x)$ of $\mathbb{F}_8$ as the feedforward polynomial and $M^{(3)}(x)$ of $\mathbb{F}_8$ as the feedback polynomial; a) regular mode of operation; b) termination mode of operation; the constraint length v of the constituent RSC codes is equal to 4.

architectures, a desired final state can be easily accomplished by appending each message with $(v - 1)$ tail bits, each of which is usually chosen to be 0, yielding the final state 0. However, in the case of recursive systematic convolutional (RSC) codes, the choice of the tail bits required to obtain the final state 0 depends on the state that is assumed at the end of the message. In particular, in regard of the second constituent RSC code encoder, selecting these tailbits and placing them inside the message vector is quite a task. The 3GPP delegates, however, came up with a genius solution to this issue, which lead to the inclusion of two additional switches in the convolutional turbo code encoder; cf. Figure 2.6.

The encoding works in the following way. First, the states of both constituent RSC code encoders are initialized, allowing only 0's in each flip-flop, and the switches are in the "up" positions; cf. Figure 2.6(a). Then the message is encoded. After inputting the last message bit, the two switches are thrown in the "down" position and three further clock cycles are made using the configuration depicted in Figure 2.6(b). Figure 2.7 illustrates the termination mode of operation of the second constituent RSC code encoder in the UMTS/LTE convolutional turbo code encoder [56, Figure 4, p. 23], [71, Figure 5.2.3-2, p. 15].

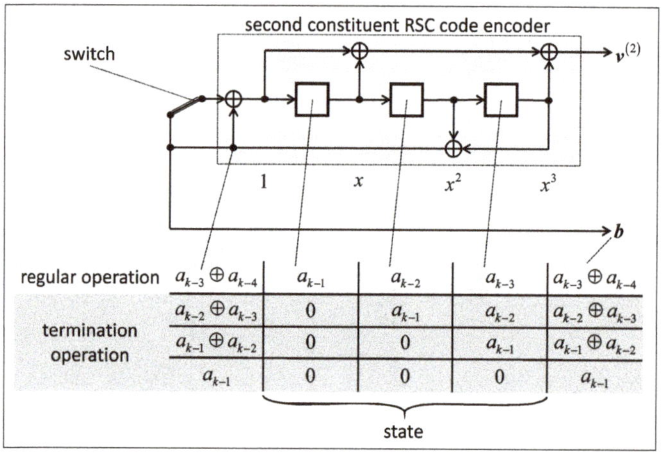

Figure 2.7: Termination mode of operation of the second constituent RSC code encoder in the UMTS/LTE convolutional turbo code encoder; cf. Figure 2.6.

The bit error ratio (BER) performance of convolutional turbo codes is considerably dependent on the choice of the turbo code interleaver, both regarding its size, which is identical with the dimension $k \in \mathbb{N}^*$ of the code and the structure of the interleaver [87, Chapter 7, pp. 193–229], [89, p. 682], [90, p. 1301f.; Figure 5, p. 1303]. The turbo code interleaver is an integral part of the overall turbo code encoder.

In particular, random interleavers show the potential to considerably improve the bit error ratio (BER) performance of convolutional turbo codes over regular interleaver structures; cf., e. g., [15, p. 778], [87, Chapter 7, pp. 193–229]. Essentially, it is due to the turbo code interleaver that very small and simple recursive systematic convolutional (RSC) codes can be made into a long and random-looking channel code.

Regarding the design of good convolutional turbo codes, it seems that choosing the constituent recursive systematic convolutional (RSC) codes is a more or less solved issue.

Thus, the interleaver design is probably the most creative part of the convolutional turbo code design; cf. [15, p. 778], [87, Chapter 7, pp. 193–229]. Choosing the right turbo code interleaver makes the convolutional turbo code design cumbersome and it seems to be hard to predict, whether one can find the best possible convolutional turbo code for the desired application and how long it will take to obtain it.

However, due to the very special Hamming weight distribution of the codewords of a convolutional turbo code, the bit error ratio (BER) P_b curve versus the average signal-to-noise ratio (SNR) $10 \cdot \log_{10}(E_\mathrm{b}/N_0)$, which is very steeply strictly monotonically decreasing at low average signal-to-noise ratio (SNR) $10 \cdot \log_{10}(E_\mathrm{b}/N_0)$ values, flattens at moderate and high average signal-to-noise ratio (SNR) $10 \cdot \log_{10}(E_\mathrm{b}/N_0)$ values; cf. Figure 2.2 [87, p. 171]. This flattening is sometimes called "error floor" [87, p. 171]. However, especially regarding the analysis by Battail [35, Figure 5, p. 87f.], this expression does not seem appropriate.

Regarding the design and choice of the constituent RSC codes, we could make the following observations:

- The feedforward polynomial and the feedback polynomial have the same degree $(\nu - 1)$, $\nu \in \mathbb{N}^* \setminus \{1\}$ being the constraint length.
- The feedforward polynomial and the feedback polynomial are relatively prime, i. e., their greatest common divisor (gcd) is equal to 1.
- The feedforward polynomial can be an irreducible polynomial of degree $(\nu - 1)$ of the finite field $\mathbb{F}_{2^{\nu-1}}$, $\nu \in \mathbb{N}^* \setminus \{1\}$ being the constraint length, e. g., the primitive minimal polynomial $M^{(1)}(x)$ of the finite field $\mathbb{F}_{2^{\nu-1}}$, but it can also be the reducible polynomial $(M^{(0)}(x))^{\nu-1}$ with $M^{(0)}(x)$ being the minimal polynomial of the element 1 of the field $\mathbb{F}_{2^{\nu-1}}$; see, e. g., [35, eq. (8), p. 90].
- The feedback polynomial is an irreducible polynomial or even the primitive polynomial of the finite field $\mathbb{F}_{2^{\nu-1}}$, $\nu \in \mathbb{N}^* \setminus \{1\}$ being the constraint length, e. g., the primitive minimal polynomial $M^{(1)}(x)$ or the irreducible minimal polynomial $M^{(3)}(x)$; see, e. g., [35, equation (8), p. 90].

Even in the absence of any message vector, the encoder of a constituent RSC code complying with the above list generates an m-sequence of period $2^{\nu-1}$, $\nu \in \mathbb{N}^* \setminus \{1\}$ being the constraint length, if the initial state is not 0. Since an m-sequence is also termed a *pseudo-noise (PN) sequence* [4, pp. 406–412] or a *pseudo-random sequence* [35, p. 90], Battail also calls the encoder of the said constituent RSC code a *pseudo-random encoder* [35, p. 90].

Like the Fibonacci numbers (cf. Example 1.29), turbo coding also underlines the importance of primes and irreducibles in nature. Nature just seems to love number theory.

It should be noted that the convolutional turbo codes considered in Section 2.2 refer to mobile radio applications. In the case of, e. g., deep space communications, convolutional turbo codes with code rates considerably lower than 1/3 have been investigated. These shall however not be discussed here because they are not regarded relevant for advanced mobile communications.

2.3 Decoding of Convolutional Turbo Codes

Figure 2.8 shows the generic block diagram of a receiver with receive antenna diversity using $K_\mathrm{R} \in \mathbb{N}^*$ receive antennas, K_R matched filters and maximal ratio combining (MRC) [1, Section 3.9], a soft output detector, being based on, e. g., the *soft output Viterbi algorithm (SOVA)* (cf., e. g., [1, Sections 3.9.12–3.9.14]), or the *maximum a posteriori probability symbol detection (MAPSD)* (cf., e. g., [1, Sections 3.9.15]), and a convolutional turbo code decoder.

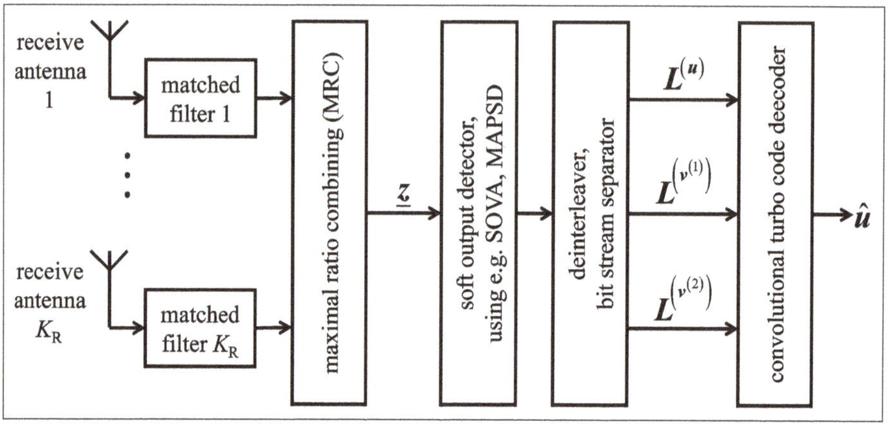

Figure 2.8: Generic block diagram of a receiver with receive antenna diversity [1, Section 3.9], soft output detector and convolutional turbo code decoder.

In order to mitigate the effect of error bursts and to benefit from time diversity, the transmitter uses an interleaver between the convolutional turbo code encoder and the digital modulator; cf., e. g., [67, Bild 4.24, pp. 184, 186]. This interleaving at the transmitter must be inverted at the receiver by an appropriate deinterleaver connected to the output of the soft output detector and the input of the convolutional turbo code decoder; cf., e. g., [67, p. 187].

The interleaver shown in Figure 2.8 generates three equally long vectors, namely

– the log-likelihood ratios (LLRs) vector

$$\boldsymbol{L}^{(u)} = \Big(L_0^{(u)}, L_1^{(u)} \cdots L_{k-1}^{(u)}, \underbrace{L_k^{(u)}, L_{k+1}^{(u)}, \cdots L_{k+v-2}^{(u)}}_{\text{created by the termination}}, \Big), \quad k, v \in \mathbb{N}^*, \tag{2.1}$$

$$\begin{aligned} L_i^{(u)} &= \ln\left\{\frac{\mathrm{P}\{u_i = 0 \mid \underline{\boldsymbol{z}}\}}{\mathrm{P}\{u_i = 1 \mid \underline{\boldsymbol{z}}\}}\right\} \\ &= \ln\left\{\frac{\mathrm{p}(\underline{\boldsymbol{z}} \mid u_i = 0)\mathrm{P}\{u_i = 0\}}{\mathrm{p}(\underline{\boldsymbol{z}})} \cdot \frac{\mathrm{p}(\underline{\boldsymbol{z}})}{\mathrm{p}(\underline{\boldsymbol{z}} \mid u_i = 1)\mathrm{P}\{u_i = 1\}}\right\} \end{aligned}$$

$$= \ln\left\{\frac{\mathrm{p}(\underline{\mathbf{z}} \mid u_i = 0)}{\mathrm{p}(\underline{\mathbf{z}} \mid u_i = 1)}\,\frac{\mathrm{P}\{u_i = 0\}}{\mathrm{P}\{u_i = 1\}}\right\}$$

$$= \ln\left\{\frac{\mathrm{p}(\underline{\mathbf{z}} \mid u_i = 0)}{\mathrm{p}(\underline{\mathbf{z}} \mid u_i = 1)}\right\} + \underbrace{\ln\left\{\frac{\mathrm{P}\{u_i = 0\}}{\mathrm{P}\{u_i = 1\}}\right\}}_{=0 \text{ for equally probable 0 and 1}}$$

$$= \ln\left\{\frac{\mathrm{p}(\underline{\mathbf{z}} \mid u_i = 0)}{\mathrm{p}(\underline{\mathbf{z}} \mid u_i = 1)}\right\}, \quad i \in \{0, 1 \cdots (k-1), k \cdots (k+\nu-2)\}, \quad k, \nu \in \mathbb{N}^*, \tag{2.2}$$

containing the log-likelihood ratios (LLRs) associated with the transmission and reception of the systematic output vector $\boldsymbol{u}$ of the convolutional turbo code encoder; cf. the Figures 2.1, 2.3, 2.4 and 2.6,

- the log-likelihood ratios (LLRs) vector

$$\boldsymbol{L}^{(\boldsymbol{v}^{(1)})} = \Big(L_0^{(\boldsymbol{v}^{(1)})}, L_1^{(\boldsymbol{v}^{(1)})} \cdots L_{k-1}^{(\boldsymbol{v}^{(1)})}, \underbrace{L_k^{(\boldsymbol{v}^{(1)})}, L_{k+1}^{(\boldsymbol{v}^{(1)})}, \cdots L_{k+\nu-2}^{(\boldsymbol{v}^{(1)})}}_{\text{created by the termination}}\Big), \quad k, \nu \in \mathbb{N}^*, \tag{2.3}$$

$$L_i^{(\boldsymbol{v}^{(1)})} = \ln\left\{\frac{\mathrm{P}\{v_i^{(1)} = 0 \mid \underline{\mathbf{z}}\}}{\mathrm{P}\{v_i^{(1)} = 1 \mid \underline{\mathbf{z}}\}}\right\}$$

$$= \ln\left\{\frac{\mathrm{p}(\underline{\mathbf{z}} \mid v_i^{(1)} = 0)}{\mathrm{p}(\underline{\mathbf{z}} \mid v_i^{(1)} = 1)}\right\} + \underbrace{\ln\left\{\frac{\mathrm{P}\{v_i^{(1)} = 0\}}{\mathrm{P}\{v_i^{(1)} = 1\}}\right\}}_{=0 \text{ for equally probable 0 and 1}}$$

$$= \ln\left\{\frac{\mathrm{p}(\underline{\mathbf{z}} \mid v_i^{(1)} = 0)}{\mathrm{p}(\underline{\mathbf{z}} \mid v_i^{(1)} = 1)}\right\}, \quad i \in \{0, 1 \cdots (k-1), k \cdots (k+\nu-2)\}, \quad k, \nu \in \mathbb{N}^*, \tag{2.4}$$

containing the log-likelihood ratios (LLRs) associated with the transmission and reception of the first parity check vector $\boldsymbol{v}^{(1)}$ of the convolutional turbo code encoder; cf. the Figures 2.1, 2.3, 2.4 and 2.6, and

- the log-likelihood ratios (LLRs) vector

$$\boldsymbol{L}^{(\boldsymbol{v}^{(2)})} = \Big(L_0^{(\boldsymbol{v}^{(2)})}, L_1^{(\boldsymbol{v}^{(2)})} \cdots L_{k-1}^{(\boldsymbol{v}^{(2)})}, \underbrace{L_k^{(\boldsymbol{v}^{(2)})}, L_{k+1}^{(\boldsymbol{v}^{(2)})}, \cdots L_{k+\nu-2}^{(\boldsymbol{v}^{(2)})}}_{\text{created by the termination}}\Big), \quad k, \nu \in \mathbb{N}^*, \tag{2.5}$$

$$L_i^{(\boldsymbol{v}^{(2)})} = \ln\left\{\frac{\mathrm{P}\{v_i^{(2)} = 0 \mid \underline{\mathbf{z}}\}}{\mathrm{P}\{v_i^{(2)} = 1 \mid \underline{\mathbf{z}}\}}\right\}$$

$$= \ln\left\{\frac{\mathrm{p}(\underline{\mathbf{z}} \mid v_i^{(2)} = 0)}{\mathrm{p}(\underline{\mathbf{z}} \mid v_i^{(2)} = 1)}\right\} + \underbrace{\ln\left\{\frac{\mathrm{P}\{v_i^{(2)} = 0\}}{\mathrm{P}\{v_i^{(2)} = 1\}}\right\}}_{=0 \text{ for equally probable 0 and 1}}$$

$$= \ln\left\{\frac{\mathrm{p}(\underline{\mathbf{z}} \mid v_i^{(2)} = 0)}{\mathrm{p}(\underline{\mathbf{z}} \mid v_i^{(2)} = 1)}\right\}, \quad i \in \{0, 1 \cdots (k-1), k \cdots (k+\nu-2)\}, \quad k, \nu \in \mathbb{N}^*, \tag{2.6}$$

containing the log-likelihood ratios (LLRs) associated with the transmission and reception of the second parity check vector $\boldsymbol{v}^{(2)}$ of the convolutional turbo code encoder; cf. the Figures 2.1, 2.3, 2.4 and 2.6.

We immediately find

$$\mathrm{p}(\underline{\mathbf{z}} \mid u_i = 0) = \frac{\exp\{L_i^{(\boldsymbol{u})}\}}{\exp\{L_i^{(\boldsymbol{u})}\} + 1}, \quad \mathrm{p}(\underline{\mathbf{z}} \mid u_i = 1) = \frac{1}{\exp\{L_i^{(\boldsymbol{u})}\} + 1}, \tag{2.7}$$

yielding

$$\mathrm{p}(\underline{\mathbf{z}} \mid u_i) = \frac{\exp\{(1 \oplus u_i) \cdot L_i^{(\boldsymbol{u})}\}}{\exp\{L_i^{(\boldsymbol{u})}\} + 1}, \quad i \in \{0, 1 \cdots (k-1), k \cdots (k+\nu-2)\}, \quad k, \nu \in \mathbb{N}^*, \tag{2.8}$$

$$\mathrm{p}(\underline{\mathbf{z}} \mid v_i^{(1)}) = \frac{\exp\{(1 \oplus v_i^{(1)}) \cdot L_i^{(\boldsymbol{v}^{(1)})}\}}{\exp\{L_i^{(\boldsymbol{v}^{(1)})}\} + 1}, \quad i \in \{0, 1 \cdots (k-1), k \cdots (k+\nu-2)\}, \quad k, \nu \in \mathbb{N}^*, \tag{2.9}$$

$$\mathrm{p}(\underline{\mathbf{z}} \mid v_i^{(2)}) = \frac{\exp\{(1 \oplus v_i^{(2)}) \cdot L_i^{(\boldsymbol{v}^{(2)})}\}}{\exp\{L_i^{(\boldsymbol{v}^{(2)})}\} + 1}, \quad i \in \{0, 1 \cdots (k-1), k \cdots (k+\nu-2)\}, \quad k, \nu \in \mathbb{N}^*. \tag{2.10}$$

Suboptimum iterative decoding, which employs individual soft input soft output (SISO) decoders for each one of the constituent RSC codes in an iterative manner, typically achieves favorable performance at reasonably low implementation complexity [87, p. 157]. The corresponding generic structure of a convolutional turbo code decoder is shown in Figure 2.9 [15, Figure 16.6, p. 780], [80, Figure 3, p. 137], [91, Figure 2, p. 532]. The convolutional turbo code decoder implements a suboptimum iterative decoding strategy, using

- a first constituent decoder for the decoding of the first constituent RSC code,
- a second constituent decoder for the decoding of the second constituent RSC code,
- two turbo code interleavers, each using the identical permutation matrix $\boldsymbol{\Pi}$,
- one turbo code deinterleaver, using the inverted permutation matrix $\boldsymbol{\Pi}^{-1}$, and
- two addition devices, each having a first input with positive weighting and a second and a third input each with negative weighting, computing the difference of the first and the further two inputs.

The three log-likelihood ratios (LLRs) vectors $\boldsymbol{L}^{(\boldsymbol{u})}$, $\boldsymbol{L}^{(\boldsymbol{v}^{(1)})}$ and $\boldsymbol{L}^{(\boldsymbol{v}^{(2)})}$ are the inputs to the convolutional turbo code decoder; cf. Figure 2.8.

The two log-likelihood ratios (LLRs) vectors $\boldsymbol{L}^{(\boldsymbol{u})}$ and $\boldsymbol{L}^{(\boldsymbol{v}^{(1)})}$, which are inputs to the convolutional turbo code decoder, are processed by the first constituent decoder for the decoding of the first constituent RSC code. $\boldsymbol{L}^{(\boldsymbol{u})}$ represents the systematic information

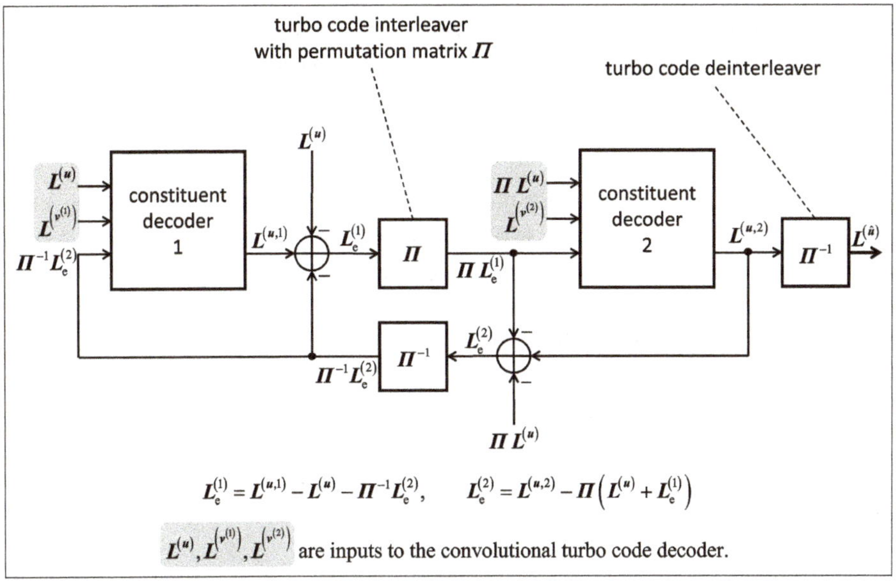

Figure 2.9: Convolutional turbo code decoder (adapted according to [15, Figure 16.6, p. 780], [80, Figure 3, p. 137] and [91, Figure 2, p. 532]).

and is therefore termed the *systematic log-likelihood ratios (LLRs) vector*. $\boldsymbol{L}^{(v^{(1)})}$ represents the parity check information, which is intrinsic to the coding using the first constituent RSC code and is therefore termed the *first intrinsic log-likelihood ratios (LLRs) vector*.

At the beginning of the iterative decoding process, the vector $\boldsymbol{\Pi}^{-1}\boldsymbol{L}_{\mathrm{e}}^{(2)}$ is initialized to be the zero vector $\mathbf{0}_{k+v-1}$, $k, v \in \mathbb{N}^*$. The first constituent decoder for the decoding of the first constituent RSC code generates the output vector $\boldsymbol{L}^{(\boldsymbol{u},1)}$, which is the positively weighted input to the left addition device. The vector $\boldsymbol{L}^{(\boldsymbol{u})}$ and the vector $\boldsymbol{\Pi}^{-1}\boldsymbol{L}_{\mathrm{e}}^{(2)}$ are the negatively weighted inputs to the same addition device, which computes

$$\boldsymbol{L}_{\mathrm{e}}^{(1)} = \boldsymbol{L}^{(\boldsymbol{u},1)} - \boldsymbol{L}^{(\boldsymbol{u})} - \boldsymbol{\Pi}^{-1}\boldsymbol{L}_{\mathrm{e}}^{(2)}. \tag{2.11}$$

After processing $\boldsymbol{L}_{\mathrm{e}}^{(1)}$ by the turbo code interleaver, the *interleaved systematic log-likelihood ratios (LLRs) vector* $\boldsymbol{\Pi}\boldsymbol{L}^{(\boldsymbol{u})}$, the interleaved vector $\boldsymbol{\Pi}\boldsymbol{L}_{\mathrm{e}}^{(1)}$ and the *second intrinsic log-likelihood ratios (LLRs) vector* $\boldsymbol{L}^{(v^{(2)})}$ are processed by the second constituent decoder for the decoding of the second constituent RSC code.

The interleaved vector $\boldsymbol{\Pi}\boldsymbol{L}_{\mathrm{e}}^{(1)}$ represents the interleaved version of additional information about the message vector $\boldsymbol{u}$, which was generated by the decoding process carried out by the first constituent decoder for the decoding of the first constituent RSC code. Since $\boldsymbol{\Pi}\boldsymbol{L}_{\mathrm{e}}^{(1)}$ does not relate to the encoding or decoding of the second constituent decoder for the decoding of the second constituent RSC code, it is regarded as the *interleaved first extrinsic log-likelihood ratios (LLRs) vector*. We have

$$\boldsymbol{\Pi L}_{\mathrm{e}}^{(1)} = ([\boldsymbol{\Pi L}_{\mathrm{e}}^{(1)}]_0, [\boldsymbol{\Pi L}_{\mathrm{e}}^{(1)}]_1 \cdots [\boldsymbol{\Pi L}_{\mathrm{e}}^{(1)}]_{k-1} \cdots [\boldsymbol{\Pi L}_{\mathrm{e}}^{(1)}]_{k+\nu-2}), \quad k, \nu \in \mathbb{N}^*. \tag{2.12}$$

The second constituent decoder for the decoding of the second constituent RSC code generates the output vector $\boldsymbol{L}^{(\boldsymbol{u},2)}$, which is the positively weighted input to the right addition device. The interleaved systematic log-likelihood ratios (LLRs) vector $\boldsymbol{\Pi L}^{(\boldsymbol{u})}$ and the interleaved first extrinsic log-likelihood ratios (LLRs) vector $\boldsymbol{\Pi L}_{\mathrm{e}}^{(1)}$ are the negatively weighted inputs to the same addition device, which computes

$$\boldsymbol{L}_{\mathrm{e}}^{(2)} = \boldsymbol{L}^{(\boldsymbol{u},2)} - \boldsymbol{\Pi L}^{(\boldsymbol{u})} - \boldsymbol{\Pi L}_{\mathrm{e}}^{(1)}. \tag{2.13}$$

After turbo code deinterleaving, we yield the *deinterleaved second extrinsic log-likelihood ratios (LLRs) vector*

$$\boldsymbol{\Pi}^{-1}\boldsymbol{L}_{\mathrm{e}}^{(2)} = ([\boldsymbol{\Pi}^{-1}\boldsymbol{L}_{\mathrm{e}}^{(2)}]_0, [\boldsymbol{\Pi}^{-1}\boldsymbol{L}_{\mathrm{e}}^{(2)}]_1 \cdots [\boldsymbol{\Pi}^{-1}\boldsymbol{L}_{\mathrm{e}}^{(2)}]_{k-1} \cdots [\boldsymbol{\Pi}^{-1}\boldsymbol{L}_{\mathrm{e}}^{(2)}]_{k+\nu-2}), \quad k, \nu \in \mathbb{N}^*, \tag{2.14}$$

with

$$\mathrm{p}(\mathrm{e}2 \mid u_i) = \frac{\exp\{(1 \oplus u_i) \cdot [\boldsymbol{\Pi}^{-1}\boldsymbol{L}_{\mathrm{e}}^{(2)}]_i\}}{\exp\{[\boldsymbol{\Pi}^{-1}\boldsymbol{L}_{\mathrm{e}}^{(2)}]_i\} + 1}, \quad i \in \{0, 1 \cdots (k-1), k \cdots (k+\nu-2)\}, \quad k, \nu \in \mathbb{N}^*. \tag{2.15}$$

After a certain number of decoding iterations, the convolutional turbo code decoder generates the soft output vector $\boldsymbol{L}^{(\hat{\boldsymbol{u}})}$ [87, p. 164f.], which is delivered to a successive receiver stage.

Usually, the constituent decoders deploy some variant of the *maximum a posteriori probability symbol detection (MAPSD)* (cf., e. g., [1, Sections 3.9.15]) scheme, which was first proposed in 1974 by *Bahl, Cocke, Jelinek* and *Raviv* [92], called the BCJR algorithm. The BCJR algorithm can be applied to any linear code, which makes it perfectly suited for the implementation in the constituent decoders of a turbo code [87, pp. 138–149, pp. 159–163]. Moreover, the BCJR algorithm can be performed in the log domain, also saving computational effort in a considerable way [87, p. 151f.]. Simplified versions of this decoding algorithm exist also, e. g.,
- the max*-log-MAP decoder and
- the max-log-MAP decoder, to the best knowledge of the author first proposed by him [88]; cf. also [87, pp. 149–151].

Considering the BCJR algorithm, the first constituent decoder for the decoding of the first constituent RSC code relies on the *transition metric increment* [1, Section 3.9.15]

$$\begin{aligned}\gamma(\boldsymbol{s}_i, \boldsymbol{s}_{i+1}) &= \mathrm{p}(\underline{\boldsymbol{z}} \mid u_i) \cdot \mathrm{p}(\underline{\boldsymbol{z}} \mid v_i^{(1)}) \cdot \mathrm{p}(\mathrm{e2} \mid u_i) \\ &= \frac{\exp\{(1 \oplus u_i) \cdot L_i^{(\boldsymbol{u})}\}}{\exp\{L_i^{(\boldsymbol{u})}\} + 1} \cdot \frac{\exp\{(1 \oplus v_i^{(1)}) \cdot L_i^{(\boldsymbol{v}^{(1)})}\}}{\exp\{L_i^{(\boldsymbol{v}^{(1)})}\} + 1} \cdot \frac{\exp\{(1 \oplus u_i) \cdot [\boldsymbol{\Pi}^{-1}\boldsymbol{L}_{\mathrm{e}}^{(2)}]_i\}}{\exp\{[\boldsymbol{\Pi}^{-1}\boldsymbol{L}_{\mathrm{e}}^{(2)}]_i\} + 1} \\ &= \frac{\exp\{(1 \oplus u_i) \cdot (L_i^{(\boldsymbol{u})} + [\boldsymbol{\Pi}^{-1}\boldsymbol{L}_{\mathrm{e}}^{(2)}]_i) + (1 \oplus v_i^{(1)}) \cdot L_i^{(\boldsymbol{v}^{(1)})}\}}{(\exp\{L_i^{(\boldsymbol{u})}\} + 1) \cdot (\exp\{[\boldsymbol{\Pi}^{-1}\boldsymbol{L}_{\mathrm{e}}^{(2)}]_i\} + 1) \cdot (\exp\{L_i^{(\boldsymbol{v}^{(1)})}\} + 1)}, \\ &\quad i \in \{0, 1 \cdots (k-1), k \cdots (k+v-2)\}, \quad k, v \in \mathbb{N}^*. \end{aligned} \tag{2.16}$$

Let

$$[\boldsymbol{\Pi}\boldsymbol{L}^{(\boldsymbol{u})}]_i, \quad i \in \{0, 1 \cdots (k-1), k \cdots (k+v-2)\}, \quad k, v \in \mathbb{N}^*, \tag{2.17}$$

be the ith, $i \in \{0, 1 \cdots (k-1), k \cdots (k+v-2)\}$, $k, v \in \mathbb{N}^*$, component of $\boldsymbol{\Pi}\boldsymbol{L}^{(\boldsymbol{u})}$, and let

$$[\boldsymbol{\Pi}\boldsymbol{L}_{\mathrm{e}}^{(1)}]_i, \quad i \in \{0, 1 \cdots (k-1), k \cdots (k+v-2)\}, \quad k, v \in \mathbb{N}^*, \tag{2.18}$$

be the ith, $i \in \{0, 1 \cdots (k-1), k \cdots (k+v-2)\}$, $k, v \in \mathbb{N}^*$, component of $\boldsymbol{\Pi}\boldsymbol{L}_{\mathrm{e}}^{(1)}$, and let

$$[\boldsymbol{\Pi}\boldsymbol{u}]_i, \quad i \in \{0, 1 \cdots (k-1), k \cdots (k+v-2)\}, \quad k, v \in \mathbb{N}^*, \tag{2.19}$$

be the ith, $i \in \{0, 1 \cdots (k-1), k \cdots (k+v-2)\}$, $k, v \in \mathbb{N}^*$, component of $\boldsymbol{\Pi}\boldsymbol{u}$, corresponding to (2.16). Furthermore, let us define

$$\mathrm{p}(\underline{\boldsymbol{z}} \mid [\boldsymbol{\Pi}\boldsymbol{u}]_i) = \frac{\exp\{(1 \oplus [\boldsymbol{\Pi}\boldsymbol{u}]_i)[\boldsymbol{\Pi}\boldsymbol{L}^{(\boldsymbol{u})}]_i\}}{\exp\{[\boldsymbol{\Pi}\boldsymbol{L}^{(\boldsymbol{u})}]_i\} + 1}, \quad i \in \{0, 1 \cdots (k-1), k \cdots (k+v-2)\}, \quad k, v \in \mathbb{N}^*, \tag{2.20}$$

$$\mathrm{p}(\mathrm{e1} \mid [\boldsymbol{\Pi}\boldsymbol{u}]_i) = \frac{\exp\{(1 \oplus [\boldsymbol{\Pi}\boldsymbol{u}]_i) \cdot [\boldsymbol{\Pi}\boldsymbol{L}_{\mathrm{e}}^{(1)}]_i\}}{\exp\{[\boldsymbol{\Pi}\boldsymbol{L}_{\mathrm{e}}^{(1)}]_i\} + 1}, \quad i \in \{0, 1 \cdots (k-1), k \cdots (k+v-2)\}, \quad k, v \in \mathbb{N}^*. \tag{2.21}$$

Then the second constituent decoder for the decoding of the second constituent RSC code relies on the transition metric increment [1, Section 3.9.15]

$$\begin{aligned}\gamma(\boldsymbol{s}_i, \boldsymbol{s}_{i+1}) &= \mathrm{p}(\underline{\boldsymbol{z}} \mid [\boldsymbol{\Pi}\boldsymbol{u}]_i) \cdot \mathrm{p}(\underline{\boldsymbol{z}} \mid v_i^{(2)}) \cdot \mathrm{p}(\mathrm{e1} \mid [\boldsymbol{\Pi}\boldsymbol{u}]_i) \\ &= \frac{\exp\{(1 \oplus [\boldsymbol{\Pi}\boldsymbol{u}]_i)[\boldsymbol{\Pi}\boldsymbol{L}^{(\boldsymbol{u})}]_i\}}{\exp\{[\boldsymbol{\Pi}\boldsymbol{L}^{(\boldsymbol{u})}]_i\} + 1} \\ &\quad \cdot \frac{\exp\{(1 \oplus v_i^{(2)}) \cdot L_i^{(\boldsymbol{v}^{(2)})}\}}{\exp\{L_i^{(\boldsymbol{v}^{(2)})}\} + 1}\end{aligned}$$

$$\cdot \frac{\exp\{(1 \oplus [\boldsymbol{\Pi u}]_i) \cdot [\boldsymbol{\Pi L}_{\mathrm{e}}^{(1)}]_i\}}{\exp\{[\boldsymbol{\Pi L}_{\mathrm{e}}^{(1)}]_i\} + 1} \tag{2.22}$$

$$= \frac{\exp\{(1 \oplus [\boldsymbol{\Pi u}]_i) \cdot ([\boldsymbol{\Pi L}^{(\boldsymbol{u})}]_i + [\boldsymbol{\Pi L}_{\mathrm{e}}^{(1)}]_i) + (1 \oplus v_i^{(2)}) \cdot L_i^{(\boldsymbol{v}^{(2)})}\}}{(\exp\{[\boldsymbol{\Pi L}^{(\boldsymbol{u})}]_i\} + 1) \cdot (\exp\{[\boldsymbol{\Pi L}_{\mathrm{e}}^{(1)}]_i\} + 1) \cdot (\exp\{L_i^{(\boldsymbol{v}^{(2)})}\} + 1)}, \tag{2.23}$$

$$i \in \{0, 1 \cdots (k-1), k \cdots (k+\nu-2)\}, \quad k, \nu \in \mathbb{N}^*.$$

Obviously, the transition metric increments $\gamma(\boldsymbol{s}_i, \boldsymbol{s}_{i+1})$ in the case of the first constituent decoder for the decoding of the first constituent RSC code and in the case of the second constituent decoder for the decoding of the second constituent RSC code have the form of a transition metric increment, which typically occurs in conventional diversity receivers with three diversity branches. However, in contrast to conventional diversity receivers, the terms $\mathrm{p}(\mathrm{e}2 \mid u_i)$ and $\mathrm{p}(\mathrm{e}1 \mid [\boldsymbol{\Pi u}]_i)$, which represent the third "diversity branch" in the respective constituent decoders, are created in the convolutional turbo code decoder, instead of being real existing diversity branches, which provides inputs to the convolutional turbo code decoder. Hence, $\mathrm{p}(\mathrm{e}2 \mid u_i)$ and $\mathrm{p}(\mathrm{e}1 \mid [\boldsymbol{\Pi u}]_i)$ are no inputs to the convolutional turbo code decoder. The circumstance of creating the third "diversity branch" during the decoding process facilitates the updating and improving of the contained information iteratively. The improvement can be directly seen from the iterative reduction of the bit error ratio (BER) obtained at a given preferably low average signal-to-noise ratio (SNR). However, the improvement obtained from one iteration to the next decreases with increasing iteration number. Typically, no relevant improvements are observed after the eighth iteration.

The choice of the transition metric increment $\mathrm{p}(\underline{\boldsymbol{z}} \mid u_i) \cdot \mathrm{p}(\underline{\boldsymbol{z}} \mid v_i^{(1)}) \cdot \mathrm{p}(\mathrm{e}2 \mid u_i)$ in the case of the first constituent decoder for decoding the first constituent RSC code shows that the respective components of the corresponding log-likelihood ratios (LLRs) vectors $\boldsymbol{L}^{(\boldsymbol{u})}$, $\boldsymbol{L}^{(\boldsymbol{v}^{(1)})}$ as well as $\boldsymbol{\Pi}^{-1}\boldsymbol{L}_{\mathrm{e}}^{(2)}$ are assumed to be independent. The same assumption of independence is made for the transition metric increment $\mathrm{p}(\underline{\boldsymbol{z}} \mid [\boldsymbol{\Pi u}]_i) \cdot \mathrm{p}(\underline{\boldsymbol{z}} \mid v_i^{(2)}) \cdot \mathrm{p}(\mathrm{e}1 \mid [\boldsymbol{\Pi u}]_i)$ in the case of the second constituent decoder for decoding the second constituent RSC code, regarding the respective components of the corresponding log-likelihood ratios (LLRs) vectors $\boldsymbol{\Pi L}^{(\boldsymbol{u})}$, $\boldsymbol{L}^{(\boldsymbol{v}^{(2)})}$ as well as $\boldsymbol{\Pi L}_{\mathrm{e}}^{(1)}$.

The assumption of independence is probably a little euphemistic [72, p. 1067]. It has been observed that the turbo code interleaver has a considerable impact on the dependence [72, p. 1067]. Both the size and the structure of the turbo code interleaver are relevant in this context. The larger the interleaver size, the lower is the said dependence. Beneficially, the achievable bit error ratio (BER) at a given low signal-to-noise ratio will typically reduce with increasing interleaver size [87, pp. 166–169]. Also, random interleaving provides a better reduction of the achievable bit error ratio (BER) at a given low signal-to-noise ratio than regularly structured interleavers [87, pp. 193–225].

Further discussions on turbo code decoding can, e. g., be found in [67, Appendix E, pp. 343–368], [1] and [87, Chapter 6, pp. 157–178].

Many, including the author, consider the processing of the log-likelihood ratios (LLRs) in the convolutional turbo code decoder to be the most enlightening part of the turbo codes. Indeed, there does not seem to exist a better testbed for the development of a solid "guts feeling" about reliability information and mutual information exchange. In particular, extrinsic information transfer (EXIT) charts, developed by Stephan *ten Brink* [93], have helped to shed some light on the "LLR game," which helped us all to become a bit more informed.

Several publications seem to implicitly suggest that convolutional turbo codes are a done deal. However, the author is not that much convinced of this notion. In particular, diving deeper into the sea of independence seems rewarding.

Remark 2.1 (Quantum Communications System With Nonquantum Channel Coding—Convolutional Turbo Code). Let us reuse the wireless laser based binary on–off keying (OOK) quantum communications system with a classical channel encoder at its input and a classical channel decoder at its output (adapted according to [76, Bild 11, p. 115] and [77, Figure 5.1, p. 184; Figure 8.1, p. 382]), introduced in Figure 1.30 and Remark 1.45. Recall that in Remarks 1.45, 2.1 and 3.2, including Figures 1.30, 1.31, 1.32, 2.10 and 3.11, the attribute "*classical*" means "*nonquantum*."

In Remark 2.1, the classical channel encoder implements the encoding of the convolutional turbo code used in the fourth generation mobile communications system "Long Term Evolution (LTE)," also abbreviated by 4G/LTE, with k equal to 6144 message bits and the code rate R equal to 1/3 [71, Figure 5.2.3-2, p. 15]. The classical channel decoder is the soft input turbo decoder using five decoding iterations and the max-log maximum a posteriori probability symbol detection ("max-log-MAP") algorithm in each constituent decoder. To his best knowledge, the "max-log-MAP" algorithm for the constituent decoding was first proposed by the author in January 1995 [88].

Figure 2.10 shows the simulation results of the uncoded bit error ratios without any classical channel decoding and the coded bit error ratios with classical channel decoding using the convolutional turbo code of 4G/LTE having the dimension k equal to 6144 and the code rate R equal to 1/3 [71, Figure 5.2.3-2, p. 15]. The average number $\bar{N}_a \in \{0, 0.01, 0.03\}$ is the curve parameter. As in Remark 1.45, the code rate R equal to 1/3 is explicitly taken into account in the presented simulation results. This means that a message bit $u_i, i \in \{0, 1 \cdots (k-1)\}, k \in \mathbb{N}^*$ is represented by $1/R$ equal to 3, i. e., three times as many signal photons than a code bit $v_j, j \in \{0, 1 \cdots (n-1)\}, k = Rn = n/3, k, n \in \mathbb{N}^*$.

For instance, looking at $\bar{N}_a$ equal to 0, the uncoded bit error ratio of approximately 0.2 is obtained at $\bar{N}_\mu$ equal to 1, whereas the coded bit error ratio at the corresponding $\bar{N}_\mu$ equal to 3 is less than 10^{-4}. Looking at the simulation results of the uncoded bit error ratios discussed in the Remark on the quantum detection included in [1], the uncoded bit error ratio is approximately equal to $2 \cdot 10^{-4}$ at $\bar{N}_\mu$ equal to 8, whereas the coded bit error ratio of approximately $2 \cdot 10^{-4}$ requires only $\bar{N}_\mu$ approximately equal to 2.9 according to Figure 2.10. This is a saving of approximately five signal photons on average.

Let us now discuss the simulated performance in the light of quantum error correction. Since we can regard each photon as the realization of a physical qubit used in quantum error correction (cf., e. g., [75, p. 30]), and knowing that the minimum number of physical qubits in well-known quantum error correction codes is 5 (see above), we can say that Figure 2.10 gives a clear indication of the fact that the deployed convolutional turbo code represents a tremendous competition to quantum error correction.

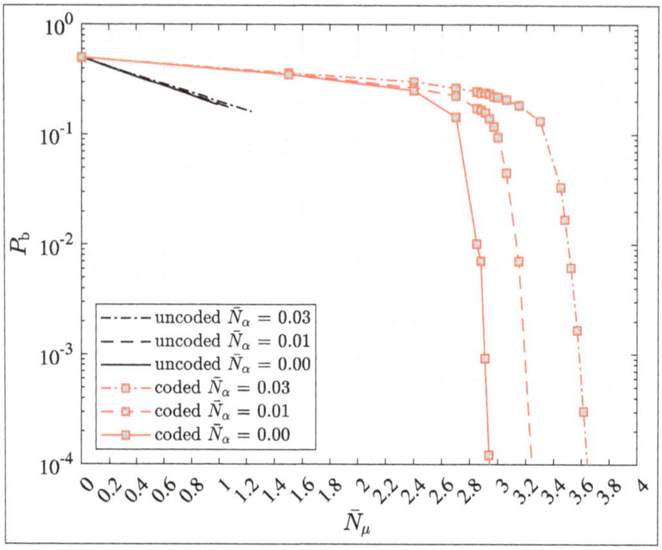

Figure 2.10: Simulation results of the uncoded bit error ratios without any classical channel decoding and the coded bit error ratios with classical channel decoding using the convolutional turbo code of the fourth generation mobile communications system "Long Term Evolution (LTE)," also abbreviated by 4G/LTE, having the dimension k equal to 6144 and the code rate R equal to 1/3 [71, Figure 5.2.3-2, p. 15]. The classical channel decoder is the soft input turbo decoder using five decoding iterations and the max-log maximum a posteriori probability symbol detection ("max-log-MAP") algorithm in each constituent decoder [88]. The average number $\bar{N}_\alpha \in \{0, 0.01, 0.03\}$ is the curve parameter.

3 Low Density Parity Check (LDPC) Codes

3.1 Definition of Low Density Parity Check (LDPC) Codes

In 1955, Peter *Elias* had proposed the convolutional codes [31]. Five years later, in 1960, one of his academic students, Robert Gray *Gallager*, submitted a doctoral (D.Sc.) dissertation [94], which was entitled

Low Density Parity Check Codes

and which introduced a new kind of (n, k, d_{min}) binary linear block codes of arbitrary length $n \in \mathbb{N}^*$, in which each component of a codeword is checked by a small but fixed number of parity check equations and each parity check set contains a small but fixed number of components of a codeword [94, p. 2]. In 1962, Gallager published the main concepts of his doctoral (D.Sc.) dissertation in a paper, which appeared in the IRE Transactions on Information Theory [95]. In 1963, Gallager provided us with a Research Monograph [96], which is an extended version of his doctoral (D.Sc.) dissertation [94].

After Gallager's publications of 1960 [94], 1962 [95] and 1963 [96], low density parity check codes were forgotten for 34 years, almost a "sleeping beauty's sleep." The next IEEE publication on low density parity check codes did not appear before 1997 [97]. In his 1997 contribution, [97] presented at the IEEE International Symposium on Information Theory held in Ulm, Germany. *MacKay* compared his results on low density parity check codes with the corresponding results found for turbo codes and showed that low density parity check codes performed almost as well as turbo codes. This clearly shows that the reason for the renewed interest in Gallager's invention after so many years of oblivion was the stunning performance of turbo codes introduced in 1993 by *Berrou*, *Glavieux* and *Thitimajshima* [72], which initiated quite an avalanche effect in coding research.

In 1997, low density parity check codes were still called *GL codes*. Meanwhile, and as the latest since the 1998 publication [98] by *McEliece*, *MacKay* and *Cheng*, Gallager's invention has become well known as *LDPC codes*; cf., e. g., [15, Chapter 17, pp. 851–945], [25, p. 354f.].

By the end of the past millennium, the standardization delegates were convinced that it should be turbo codes that had to become part of the third generation mobile communications system "*Universal Mobile Telecommunications System (UMTS)*," also abbreviated by 3G/UMTS [82, p. 112f.], and the fourth generation mobile communications system "*Long Term Evolution (LTE)*," also abbreviated by 4G/LTE [84, pp. 61f., 145], [83, Figure 8.15; Figure 8.16: Figure 9.3; Figure 9.23; pp. 83, 85f., 139–142, 169, 172, 200], However, in the fifth generation mobile communications system "*5G*," low density parity check codes won the race [85, pp. 41, 102, 134–140, 402] [86, pp. 66, 156–160].

The reason for this decision that is commonly cited is "implementation complexity." As already stated in Chapter 2, the author doubts the correctness of this decision because he believes that link and system performance should always be the guideline.

https://doi.org/10.1515/9783111239651-003

If implementation complexity was such an issue, then one should reconsider the deployment of multiple antennas in the first place. Bearing in mind that most mobile terminals and probably all base stations are multistandard devices, two mutually incompatible sophisticated channel coding schemes, namely convolutional turbo codes and LDPC codes, now must be implemented at the same time both in mobile terminals and in base stations. In particular, the implementation complexity of modem chips, e. g., smart phones have to be observed carefully in this context. In contrast, the further evolution of the convolutional turbo code used in 3G/UMTS and 4G/LTE toward 5G applications would have provided the asset of backward compatibility with all implementation complexity benefits. Moreover, the implementation complexity of, e. g., mobile terminals is certainly not governed by the modem chip today, but rather by units much closer to the human interface such as the applications that the user operates, as well as the display.

So, the given reason for dropping convolutional turbo codes does not appear to be the complete truth. Rather, the author is reminded of the 3G/UMTS standardization, which never seemed to be purely technology driven and in a couple of aspects also looked quite like "knitted with hot needle." In contrast, the author believes that the second generation mobile communications system 2G/GSM (global system for mobile communications) as well as 4G/LTE have both been marvelous pieces of engineering art. Regrettably, the author therefore just cannot push back the notion that the quality of mobile communications standards seems to be alternating like an antipodal binary sequence, swaying between "well done" and "well meant." Nevertheless, LDPC codes are certainly a good choice, just like convolutional turbo codes were when it was about 3G/UMTS and 4G/LTE.

So, why are LDPC codes so good? Gérard *Battail* showed [99, p. 202] that just like turbo codes LDPC codes are *weakly random-like codes*; cf. Sect. 2.1 and [35, pp. 88, 91f.]. The reason for calling an LDPC code "*weakly* random-like" and not "*strongly* random-like" or just "random-like" is that only the average properties of the Hamming weight distribution of the codewords, in particular, the mean and the variance and the average shape, are like those of a strongly random-like code [35, p. 76]. In contrast, in the case of strongly random-like codes, each term of the Hamming weight distribution of the codewords is close to that of random coding [35, p. 76].

Before illuminating this finding, we have to briefly illustrate what "low density parity check" means. First of all, we must remember that LDPC codes are $(n, k, d_{\min})$ binary linear block codes. Next, we must recognize that, when talking about parity checks, we are talking about the $(n-k) \times n$ parity check matrices $\boldsymbol{H}$. Clearly, "low density parity check" refers to the structure of the $(n-k) \times n$ parity check matrix $\boldsymbol{H}$, $k < n$, $k, n \in \mathbb{N}^*$ of an LDPC code. So, what does "low density" mean? Since a particular component v_j, $j \in \{0, 1 \cdots (n-1)\}$, $n \in \mathbb{N}^*$ of a codeword $\boldsymbol{v} \in \mathbb{F}_2^n$, $n \in \mathbb{N}^*$ is only "checked" if there is a 1 in the jth column of the considered row of the $(n-k) \times n$ parity check matrix $\boldsymbol{H}$, "low density" obviously refers to the number of 1s per column of the $(n-k) \times n$ parity check matrix $\boldsymbol{H}$ of an LDPC code; cf., e. g., [99, p. 202]. Also, we know that there are $(d_{\min} - 1)$ linearly independent columns of the parity check matrix $\boldsymbol{H}$.

So, let us set out from an $(n-k) \times n$ parity check matrix $\boldsymbol{H}$, which has n, $n \in \mathbb{N}^*$ columns, each of which is composed of $(n-k)$, $k < n$, $k, n \in \mathbb{N}^*$ components that can assume values taken from $\mathbb{F}_2$ equal to $\{0,1\}$. Each one of the n columns of the $(n-k) \times n$ parity check matrix $\boldsymbol{H}$ contains exactly y, $y \in \{2, 3 \cdots (n-k)\}$, $k < n$, $k, n \in \mathbb{N}^*$, 1s.

Gaussian elimination [11, pp. 918–920] allows us to transform $\boldsymbol{H}$ into the systematic form [15, equation (3.7), p. 70]

$$\boldsymbol{H}_{\text{sys}} = \begin{pmatrix} \boldsymbol{I}_{n-k} & \boldsymbol{P}^{\text{T}} \end{pmatrix}, \quad k < n, \quad k, n \in \mathbb{N}^*, \tag{3.1}$$

$\boldsymbol{P}^{\text{T}}$ being an $(n-k) \times k$ *transposed parity matrix* [15, equation (3.4), p. 69] by performing $(n-k)$ $k < n$, $k, n \in \mathbb{N}^*$, elimination steps [99, p. 202].

Prior to the Gaussian elimination [11, pp. 918–920], each column has the density

$$r_0 = \frac{y}{n-k}, \quad y \in \{2, 3 \cdots (n-k)\}, \quad k < n, \quad k, n \in \mathbb{N}^*, \tag{3.2}$$

of 1s [99, p. 202]. The average density r_i of 1s per column after the ith, $i \in \{1, 2 \cdots (n-k)\}$, $k < n$, $k, n \in \mathbb{N}^*$ elimination step is given by the recursion [99, p. 202]

$$\begin{aligned} r_i = r_{i-1} \cdot \left[1 - \frac{1}{n-k} + \left(1 + \frac{2}{n-k}\right) \cdot r_{i-1} - 2r_{i-1}^2\right], \\ i \in \{1, 2 \cdots (n-k)\}, \quad k < n, \quad k, n \in \mathbb{N}^*. \end{aligned} \tag{3.3}$$

Let us now assume that $(n-k)$, $k < n$, $k, n \in \mathbb{N}^*$ is large. Then we may determine the density after $(n-k) \to \infty$ elimination steps, yielding

$$r_\infty = r_\infty \cdot [1 + r_\infty - 2r_\infty^2], \tag{3.4}$$

which immediately becomes

$$r_\infty \cdot \left[r_\infty - \frac{1}{2}\right] = 0. \tag{3.5}$$

The polynomial $(r_\infty \cdot [r_\infty - 1/2])$ has the two roots 0 and 1/2. Since according to (3.3), r_i is increasing with increasing i, only r_∞ equal to 1/2 is sensible [99, p. 202]. This means that each of the components of the $\boldsymbol{P}^{\text{T}}$ being an $(n-k) \times k$ transposed parity matrix is chosen at random and independently of all other components from $\mathbb{F}_2$ with probability 1/2. Obviously, $\boldsymbol{P}^{\text{T}}$, and hence also $\boldsymbol{P}$ are random with density 1/2 [99, p. 202]. Since we constructed $\boldsymbol{H}_{\text{sys}}$ using the Gaussian elimination [11, pp. 918–920] and found that $\boldsymbol{P}^{\text{T}}$ is random, we may choose $\boldsymbol{H}$ and, therefore, also the generator matrix $\boldsymbol{G}$ at random [99, p. 202]. Similar to turbo codes, an LDPC code is a weakly random-like code; see above [99, p. 202].

Definition 3.1 (Low Density Parity Check (LDPC) Code). A low density parity check (LDPC) code is defined as the *null space of a parity check matrix* $\boldsymbol{H}$ with the following properties [15, Definition 17.1, p. 852]:

(1) Each row of $\boldsymbol{H}$ consists of ρ, $\rho \in \{2, 3 \cdots n\}$, $n \in \mathbb{N}^*$, 1s.

(2) Each column of $\boldsymbol{H}$ consists of γ, $\gamma \in \{2, 3 \cdots (n-k)\}$, $k, n \in \mathbb{N}^*$, 1s.

(3) The number of 1s in common between any two columns, denoted by λ $k, n \in \mathbb{N}^*$ is no greater than 1, i. e., $\lambda \in \{0, 1\}$.

(4) Both ρ and γ are small compared with the length n, $n \in \mathbb{N}^*$ of the code and the number J, $J \in \mathbb{N}^*$ of rows of $\boldsymbol{H}$.

According to the Definition 3.1, we let the number of parity check equations $J \in \mathbb{N}^*$ be more flexible instead of restricting it to $(n-k)$, $k < n$, $k, n \in \mathbb{N}^*$.

According to the properties (1) and (2), the parity check matrix $\boldsymbol{H}$ has constant row weight ρ and constant column weight γ [15, p. 852]. The property (3) states that no two rows of the parity check matrix $\boldsymbol{H}$ have more than a single 1 in common [15, p. 852].

Since according to property (4) both ρ and γ are small compared with the length n, $n \in \mathbb{N}^*$ of the code and the number of rows, $J \in \mathbb{N}^*$ in the parity check matrix $\boldsymbol{H}$, the parity check matrix $\boldsymbol{H}$ has a small density of 1s [15, p. 852].

And since there are J rows each with ρ 1's, the total number of 1s in the parity check matrix $\boldsymbol{H}$ is given by $J\rho$. Since the parity check matrix $\boldsymbol{H}$ has Jn components, the density of 1s, termed the *density of the parity check matrix* $\boldsymbol{H}$, is given by [15, p. 852]

$$r = \frac{J\rho}{Jn} = \frac{\rho}{n}, \quad \rho \in \{2, 3 \cdots n\}, \quad J, n \in \mathbb{N}^*. \tag{3.6}$$

Let us now look at the columns of the parity check matrix $\boldsymbol{H}$. There are n columns each containing γ 1s. Therefore, we also find [15, p. 852]

$$r = \frac{\gamma n}{Jn} = \frac{\gamma}{J}, \quad \gamma \in \{2, 3 \cdots (n-k)\}, \quad k < n, \quad J, k, n \in \mathbb{N}^*, \tag{3.7}$$

which obviously yields [15, p. 852]

$$r = \frac{\rho}{n} = \frac{\gamma}{J}, \quad \gamma \in \{2, 3 \cdots (n-k)\}, \quad \rho \in \{2, 3 \cdots n\}, \quad k < n, \quad J, k, n \in \mathbb{N}^*. \tag{3.8}$$

A low density r given by (3.8) implies that $\boldsymbol{H}$ is a sparse matrix [15, p. 852]. Since ρ and γ are constant, the LDPC code is called a *(γ, ρ)-regular LDPC code* [15, p. 852].

Remark 3.1 (Irregular Low Density Parity Check (LDPC) Code). If the weights of the columns as well as the weights of the rows of the parity check matrix $\boldsymbol{H}$ vary, the LDPC code is said to be *irregular* [15, p. 852].

Consider an (n, k, d_{min}) binary linear LDPC code $\mathbb{L}$ specified by a $J \times n$ parity check matrix $\boldsymbol{H}$ [15, p. 854]. Let

$$\boldsymbol{h}^{(0)}, \boldsymbol{h}^{(1)} \cdots \boldsymbol{h}^{(J-1)}, \quad J \in \mathbb{N}^*, \tag{3.9}$$

denote the J rows of the $J \times n$ parity check matrix $\boldsymbol{H}$ [15, p. 854]. Each row vector $\boldsymbol{h}^{(j)}$, $j \in \{0,1\cdots(J-1)\}, J \in \mathbb{N}^*$ has n components taken from $\mathbb{F}_2$ [15, p. 854]

$$\boldsymbol{h}^{(j)} = (h_0^{(j)}, h_1^{(j)} \cdots h_{n-1}^{(j)}), \quad h_i^{(j)} \in \mathbb{F}_2, \quad j \in \{0,1\cdots(J-1)\}, \quad i \in \{0,1\cdots(n-1)\}, \quad J, n \in \mathbb{N}^*. \tag{3.10}$$

The row vectors $\boldsymbol{h}^{(j)}, j \in \{0,1\cdots(J-1)\}, J \in \mathbb{N}^*$ of the parity check matrix $\boldsymbol{H}$ are not necessarily all linearly independent over $\mathbb{F}_2$. This means that the parity check matrix $\boldsymbol{H}$ does not necessarily have full rank. The rank rank$\{\boldsymbol{H}\}$ of the parity check matrix $\boldsymbol{H}$ is given by the number of linearly independent row vectors and relates to the length n and the dimension k of the code in the following way:

$$\text{rank}\{\boldsymbol{H}\} = n - k, \quad k < n, \quad k, n \in \mathbb{N}^*. \tag{3.11}$$

Therefore, in the case of

$$J > n - k, \quad k < n, \quad k, n \in \mathbb{N}^*, \tag{3.12}$$

there are $(J - n + k)$ row vectors, which linearly depend on the remaining $(n - k)$ row vectors. In order to determine the dimension k of the code, we must compute rank$\{\boldsymbol{H}\}$.

Example 3.1. Consider the cyclic (7, 3) binary linear block code with the parity check matrix [15, p. 858]:

$$\boldsymbol{H} = \begin{pmatrix} 1 & 1 & 0 & 1 & 0 & 0 & 0 \\ 0 & 1 & 1 & 0 & 1 & 0 & 0 \\ 0 & 0 & 1 & 1 & 0 & 1 & 0 \\ 0 & 0 & 0 & 1 & 1 & 0 & 1 \\ 1 & 0 & 0 & 0 & 1 & 1 & 0 \\ 0 & 1 & 0 & 0 & 0 & 1 & 1 \\ 1 & 0 & 1 & 0 & 0 & 0 & 1 \end{pmatrix}. \tag{3.13}$$

This cyclic (7, 3) binary linear block code has the following properties [15]:

- Dimension of the code: $k = 3$;
- Length of the code: $n = 7$;
- Code rate: $R = 3/7$;
- Number of 1s per row: $\rho = 3 \quad (< n = 7)$;
- Number of 1s per column: $\gamma = 3 \quad (< n = 7)$;
- Number of 1s is common between any two columns: $\lambda = 1$.

Therefore, it can be considered an LDPC code. The $J = 7$ row vectors of $\boldsymbol{H}$ given in (3.13) are

$$\begin{aligned} \boldsymbol{h}^{(0)} &= (1,1,0,1,0,0,0), \\ \boldsymbol{h}^{(1)} &= (0,1,1,0,1,0,0), \\ \boldsymbol{h}^{(2)} &= (0,0,1,1,0,1,0), \\ \boldsymbol{h}^{(3)} &= (0,0,0,1,1,0,1), \\ \boldsymbol{h}^{(4)} &= (1,0,0,0,1,1,0), \\ \boldsymbol{h}^{(5)} &= (0,1,0,0,0,1,1), \\ \boldsymbol{h}^{(6)} &= (1,0,1,0,0,0,1). \end{aligned} \tag{3.14}$$

Let us now compute all

$$\binom{4}{2}+\binom{4}{3}+\binom{4}{4}=6+4+1=11 \tag{3.15}$$

linear combinations of the first four row vectors $\boldsymbol{h}^{(0)}$, $\boldsymbol{h}^{(1)}$, $\boldsymbol{h}^{(2)}$ and $\boldsymbol{h}^{(3)}$,

$$\begin{aligned}
\boldsymbol{h}^{(0)} \oplus \boldsymbol{h}^{(1)} &= (1,0,1,1,1,0,0),\\
\boldsymbol{h}^{(0)} \oplus \boldsymbol{h}^{(2)} &= (1,1,1,0,0,1,0),\\
\boldsymbol{h}^{(0)} \oplus \boldsymbol{h}^{(3)} &= (1,1,0,0,1,0,1),\\
\boldsymbol{h}^{(1)} \oplus \boldsymbol{h}^{(2)} &= (0,1,0,1,1,1,0),\\
\boldsymbol{h}^{(1)} \oplus \boldsymbol{h}^{(3)} &= (0,1,1,1,0,0,1),\\
\boldsymbol{h}^{(2)} \oplus \boldsymbol{h}^{(3)} &= (0,0,1,0,1,1,1),\\
\boldsymbol{h}^{(0)} \oplus \boldsymbol{h}^{(1)} \oplus \boldsymbol{h}^{(2)} &= (1,0,0,0,1,1,0) = \boldsymbol{h}^{(4)},\\
\boldsymbol{h}^{(0)} \oplus \boldsymbol{h}^{(1)} \oplus \boldsymbol{h}^{(3)} &= (1,0,1,0,0,0,1) = \boldsymbol{h}^{(6)},\\
\boldsymbol{h}^{(0)} \oplus \boldsymbol{h}^{(2)} \oplus \boldsymbol{h}^{(3)} &= (1,1,1,1,1,1,1),\\
\boldsymbol{h}^{(1)} \oplus \boldsymbol{h}^{(2)} \oplus \boldsymbol{h}^{(3)} &= (0,1,0,0,0,1,1) = \boldsymbol{h}^{(5)},\\
\boldsymbol{h}^{(0)} \oplus \boldsymbol{h}^{(1)} \oplus \boldsymbol{h}^{(2)} \oplus \boldsymbol{h}^{(3)} &= (1,0,0,1,0,1,1).
\end{aligned} \tag{3.16}$$

According to (3.16), $\boldsymbol{h}^{(0)}$, $\boldsymbol{h}^{(1)}$, $\boldsymbol{h}^{(2)}$ and $\boldsymbol{h}^{(3)}$ are linearly independent because they cannot be obtained by any of the above linear combinations whereas $\boldsymbol{h}^{(4)}$, $\boldsymbol{h}^{(5)}$ and $\boldsymbol{h}^{(6)}$ are linearly dependent on $\boldsymbol{h}^{(0)}$, $\boldsymbol{h}^{(1)}$, $\boldsymbol{h}^{(2)}$ and $\boldsymbol{h}^{(3)}$. Therefore, the rank of the parity check matrix $\boldsymbol{H}$ is 4.

Let us now consider the standard procedure to determine the rank of the parity check matrix $\boldsymbol{H}$. Let us transform $\boldsymbol{H}$ into the reduced row echelon form by the *Gauss–Jordan algorithm*, i. e., by the *Gaussian elimination* [11, pp. 918–920]. We identify the first row as the starting row. Since the first four rows are in row echelon form, we add the first row to the rows 5 and 7 in the next two steps. Adding row 1 to row 5 yields

$$\boldsymbol{H}^{(1)} = \begin{pmatrix}
1 & 1 & 0 & 1 & 0 & 0 & 0\\
0 & 1 & 1 & 0 & 1 & 0 & 0\\
0 & 0 & 1 & 1 & 0 & 1 & 0\\
0 & 0 & 0 & 1 & 1 & 0 & 1\\
0 & 1 & 0 & 1 & 1 & 1 & 0\\
0 & 1 & 0 & 0 & 0 & 1 & 1\\
1 & 0 & 1 & 0 & 0 & 0 & 1
\end{pmatrix}. \tag{3.17}$$

Adding row 1 to row 7 yields

$$\boldsymbol{H}^{(2)} = \begin{pmatrix}
1 & 1 & 0 & 1 & 0 & 0 & 0\\
0 & 1 & 1 & 0 & 1 & 0 & 0\\
0 & 0 & 1 & 1 & 0 & 1 & 0\\
0 & 0 & 0 & 1 & 1 & 0 & 1\\
0 & 1 & 0 & 1 & 1 & 1 & 0\\
0 & 1 & 0 & 0 & 0 & 1 & 1\\
0 & 1 & 1 & 1 & 0 & 0 & 1
\end{pmatrix}. \tag{3.18}$$

We cannot improve the first column any more, which means that adding row 1 to any further row of $\boldsymbol{H}$ will not help any more. So, let us continue with the row 2. Adding row 2 to row 5 yields

$$\boldsymbol{H}^{(3)} = \begin{pmatrix} 1 & 1 & 0 & 1 & 0 & 0 & 0 \\ 0 & 1 & 1 & 0 & 1 & 0 & 0 \\ 0 & 0 & 1 & 1 & 0 & 1 & 0 \\ 0 & 0 & 0 & 1 & 1 & 0 & 1 \\ 0 & 0 & 1 & 1 & 0 & 1 & 0 \\ 0 & 1 & 0 & 0 & 0 & 1 & 1 \\ 0 & 1 & 1 & 1 & 0 & 0 & 1 \end{pmatrix}. \tag{3.19}$$

Next, adding row 2 to row 6 yields

$$\boldsymbol{H}^{(4)} = \begin{pmatrix} 1 & 1 & 0 & 1 & 0 & 0 & 0 \\ 0 & 1 & 1 & 0 & 1 & 0 & 0 \\ 0 & 0 & 1 & 1 & 0 & 1 & 0 \\ 0 & 0 & 0 & 1 & 1 & 0 & 1 \\ 0 & 0 & 1 & 1 & 0 & 1 & 0 \\ 0 & 0 & 1 & 0 & 1 & 1 & 1 \\ 0 & 1 & 1 & 1 & 0 & 0 & 1 \end{pmatrix}. \tag{3.20}$$

Next, adding row 2 to row 7 yields

$$\boldsymbol{H}^{(5)} = \begin{pmatrix} 1 & 1 & 0 & 1 & 0 & 0 & 0 \\ 0 & 1 & 1 & 0 & 1 & 0 & 0 \\ 0 & 0 & 1 & 1 & 0 & 1 & 0 \\ 0 & 0 & 0 & 1 & 1 & 0 & 1 \\ 0 & 0 & 1 & 1 & 0 & 1 & 0 \\ 0 & 0 & 1 & 0 & 1 & 1 & 1 \\ 0 & 0 & 0 & 1 & 1 & 0 & 1 \end{pmatrix}. \tag{3.21}$$

We cannot improve the second column any more, which means that adding row 2 to any further row of $\boldsymbol{H}$ will not help any more. So, let us continue with the row 3. Adding row 3 to row 5 yields

$$\boldsymbol{H}^{(6)} = \begin{pmatrix} 1 & 1 & 0 & 1 & 0 & 0 & 0 \\ 0 & 1 & 1 & 0 & 1 & 0 & 0 \\ 0 & 0 & 1 & 1 & 0 & 1 & 0 \\ 0 & 0 & 0 & 1 & 1 & 0 & 1 \\ 0 & 0 & 0 & 0 & 0 & 0 & 0 \\ 0 & 0 & 1 & 0 & 1 & 1 & 1 \\ 0 & 0 & 0 & 1 & 1 & 0 & 1 \end{pmatrix}. \tag{3.22}$$

Adding row 3 to row 6 yields

$$\boldsymbol{H}^{(7)} = \begin{pmatrix} 1 & 1 & 0 & 1 & 0 & 0 & 0 \\ 0 & 1 & 1 & 0 & 1 & 0 & 0 \\ 0 & 0 & 1 & 1 & 0 & 1 & 0 \\ 0 & 0 & 0 & 1 & 1 & 0 & 1 \\ 0 & 0 & 0 & 0 & 0 & 0 & 0 \\ 0 & 0 & 0 & 1 & 1 & 0 & 1 \\ 0 & 0 & 0 & 1 & 1 & 0 & 1 \end{pmatrix}. \tag{3.23}$$

We cannot improve the third column any more, which means that adding row 3 to any further row of $\boldsymbol{H}$ will not help any more. So, let us continue with the row 4. Adding row 4 to row 6 yields

$$\boldsymbol{H}^{(8)} = \begin{pmatrix} 1 & 1 & 0 & 1 & 0 & 0 & 0 \\ 0 & 1 & 1 & 0 & 1 & 0 & 0 \\ 0 & 0 & 1 & 1 & 0 & 1 & 0 \\ 0 & 0 & 0 & 1 & 1 & 0 & 1 \\ 0 & 0 & 0 & 0 & 0 & 0 & 0 \\ 0 & 0 & 0 & 0 & 0 & 0 & 0 \\ 0 & 0 & 0 & 1 & 1 & 0 & 1 \end{pmatrix}. \quad (3.24)$$

Finally, adding row 4 to row 7 yields

$$\boldsymbol{H}^{(9)} = \begin{pmatrix} 1 & 1 & 0 & 1 & 0 & 0 & 0 \\ 0 & 1 & 1 & 0 & 1 & 0 & 0 \\ 0 & 0 & 1 & 1 & 0 & 1 & 0 \\ 0 & 0 & 0 & 1 & 1 & 0 & 1 \\ 0 & 0 & 0 & 0 & 0 & 0 & 0 \\ 0 & 0 & 0 & 0 & 0 & 0 & 0 \\ 0 & 0 & 0 & 0 & 0 & 0 & 0 \end{pmatrix}. \quad (3.25)$$

$\boldsymbol{H}^{(9)}$ is in row echelon form. The rank of $\boldsymbol{H}^{(9)}$ is the number of nonzero rows, i. e., rank$\{\boldsymbol{H}^{(9)}\}$ is equal to 4.

Remember that we only used elementary row operations to transform $\boldsymbol{H}$ into $\boldsymbol{H}^{(9)}$, which means that the resulting parity check matrix $\boldsymbol{H}^{(9)}$ checks the same code. Therefore, rank$\{\boldsymbol{H}\}$ of the parity check matrix $\boldsymbol{H}$ is also equal to 4.

Since the length of the code is n equal to 7 and rank$\{\boldsymbol{H}\}$ of the parity check matrix $\boldsymbol{H}$ is 4, and using (3.11), the dimension of the code is given by

$$k = n - \text{rank}\{\boldsymbol{H}\} = 7 - 4 = 3. \quad (3.26)$$

With the codeword

$$\boldsymbol{v} = (v_0, v_1 \cdots v_{n-1}), \quad v_i \in \mathbb{F}_2, \quad i \in \{0, 1 \cdots (n-1)\}, \quad n \in \mathbb{N}^*, \quad (3.27)$$

of the $(n, k, d_{\min})$ binary linear LDPC code $\mathbb{L}$, the inner product [15, equation (17.2), p. 854]

$$s_j = \boldsymbol{v}[\boldsymbol{h}^{(j)}]^{\mathrm{T}} = \bigoplus_{i=0}^{n-1} v_i \odot h_i^{(j)} = 0, \quad j \in \{0, 1 \cdots (J-1)\}, \quad J \in \mathbb{N}^*, \quad (3.28)$$

gives the jth of J *parity check sums*, also termed the *parity check equations*, specified by the J rows of the parity check matrix $\boldsymbol{H}$ [15, p. 854]. A code bit $v_i \in \mathbb{F}_2$, $i \in \{0, 1 \cdots (n-1)\}$, $n \in \mathbb{N}^*$ is said to be *checked by the sum* s_j, $j \in \{0, 1 \cdots (J-1)\}$, $J \in \mathbb{N}^*$, according to (3.28) or by the jth row $\boldsymbol{h}^{(j)}$, $j \in \{0, 1 \cdots (J-1)\}$, $J \in \mathbb{N}^*$ of the parity check matrix $\boldsymbol{H}$ if and only if [15, p. 854f.]

$$h_i^{(j)} = 1, \quad i \in \{0, 1 \cdots (n-1)\}, \quad n \in \mathbb{N}^*. \quad (3.29)$$

The column weight $\gamma < J$ of the parity check matrix $\boldsymbol{H}$ is equal to the number of row vectors, which fulfill (3.29).

Let us denote these γ row vectors of the parity check matrix $\boldsymbol{H}$, which check on the code bit $v_i \in \mathbb{F}_2$ by

$$\boldsymbol{h}^{(j_0)}, \boldsymbol{h}^{(j_1)} \cdots \boldsymbol{h}^{(j_{\gamma-1})}, \quad h_i^{(j)} = 1 \, \forall j \in \{j_0, j_1 \cdots j_{\gamma-1}\}, \tag{3.30}$$

with

$$\boldsymbol{h}^{(j)} = \Big(h_0^{(j)}, h_1^{(j)} \cdots \underbrace{1}_{i\text{th position}} \cdots h_{n-1}^{(j)}\Big) \quad j \in \{j_0, j_1 \cdots j_{\gamma-1}\}. \tag{3.31}$$

These γ row vectors form the set

$$\mathbb{A}^{(i)} = \{\boldsymbol{h}^{(j_0)}, \boldsymbol{h}^{(j_1)} \cdots \boldsymbol{h}^{(j_{\gamma-1})}\}, \quad h_i^{(j)} = 1 \, \forall j \in \{j_0, j_1 \cdots j_{\gamma-1}\}, \quad i \in \{0, 1 \cdots (n-1)\}, \quad n \in \mathbb{N}^*. \tag{3.32}$$

It follows from property (3) in Definition 3.1 that any code bit other than v_i is *checked by at most one row vector in* $\mathbb{A}^{(i)}$ of (3.32) [15, p. 872]. Therefore, the γ row vectors in $\mathbb{A}^{(i)}$ of (3.32) are *orthogonal* on the code bit v_i [15, p. 872].

Any error pattern with $\lfloor \gamma/2 \rfloor$ or fewer errors can be corrected by the LDPC code of Definition 3.1 [15, p. 872]. Consequently, the minimum distance $d_{\min}$ is lower bounded by $(\gamma + 1)$, i. e., [15, p. 872]

$$d_{\min} \geq \gamma + 1. \tag{3.33}$$

Regrettably, there is no "cooking recipe" for the design of a good $(n, k, d_{\min})$ binary linear LDPC code $\mathbb{L}$; cf., e. g., [15, p. 853f.]. Often computer searches are hence employed to find a long $(n, k, d_{\min})$ binary linear LDPC code $\mathbb{L}$. This seems to be the downside of the design of low density parity check (LDPC) codes. It is hence hard to tell whether a search will be successful in a given design time window or not.

3.2 Decoding of Low Density Parity Check (LDPC) Codes

3.2.1 Tanner Graphs

In what follows, we will consider the decoding of LDPC codes based on a special kind of *graphs* [100, p. 387], [24, Definition B.1, p. 244], which are called *Tanner graphs* [15, Figure 17.6, pp. 355–358], first introduced in 1981 by Robert Michael *Tanner* [101]. A graph [15, p. 855], [100, p. 387], [24, Definition B.1, p. 244]

$$\mathcal{G} = (\mathbb{V}, \mathbb{E}) \tag{3.34}$$

consists of a set of *vertices (nodes)*, denoted by

$$\mathbb{V} = \{v_1, v_2, \cdots\} \tag{3.35}$$

and a set of *edges*, denoted by

$$\mathbb{E} = \{\varepsilon_1, \varepsilon_2 \cdots\} \tag{3.36}$$

such that each edge ε_k, $k \in \mathbb{N}^*$ is identified with an unordered pair (v_i, v_j) of vertices [15, p. 855]. The vertices v_i and v_j associated with the edge ε_k are called the *end vertices* of ε_k [15, p. 855]. The vertices v_i and v_j are *connected by the edge* ε_k [15, p. 855]. The vertices v_i and v_j are *adjacent* [15, p. 856]. The edge ε_k is said to be *incident with (incident on)* its vertices v_i and v_j [15, p. 855]. The number of edges that are incident on a vertex v_i is called the *degree of the vertex* v_i, denoted by $d(v_i)$ [15, p. 855]. Two edges that are incident on a common vertex are said to be *adjacent* or *connected* [15, p. 856].

A graph $\mathcal{G}$ with a finite number of vertices and edges is termed a *finite graph* [15, p. 856]. A graph $\mathcal{G}$ is most commonly represented by a diagram, in which the vertices are identified by points and each edge by a line, joining its end vertices [15, p. 855]; cf. Figure 3.1.

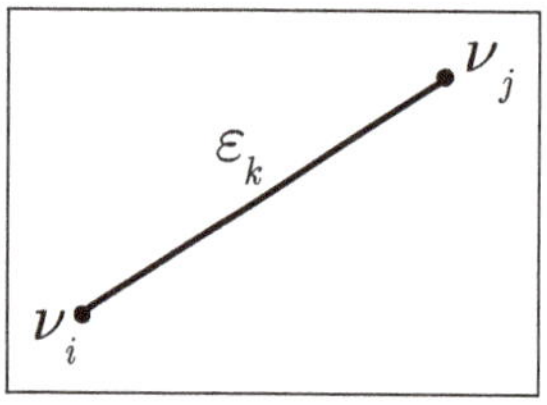

Figure 3.1: Vertices v_i and v_j connected by the edge ε_k.

Example 3.2. The graph $\mathcal{G}$ consists of six vertices and ten edges; cf. Figure 3.2 and [15, Figure 17.3, p. 855]. For instance, we have the following properties:

$$\begin{aligned} d(v_4) &= 4, \\ d(v_6) &= 2. \end{aligned} \tag{3.37}$$

The vertices v_2 and v_4 are connected by ε_3. The vertices v_3 and v are connected by ε_9. The edges ε_1 and ε_2 are adjacent. The edges ε_4 and ε_6 are adjacent.

A *path* in a graph $\mathcal{G}$ is a finite alternating sequence of vertices and edges, beginning and ending with vertices [15, p. 856]. Each edge is incident with vertices preceding and following it, and no vertex appears more than once [15, p. 856]. The number of edges on a path is termed the *length* of the path [15, p. 856].

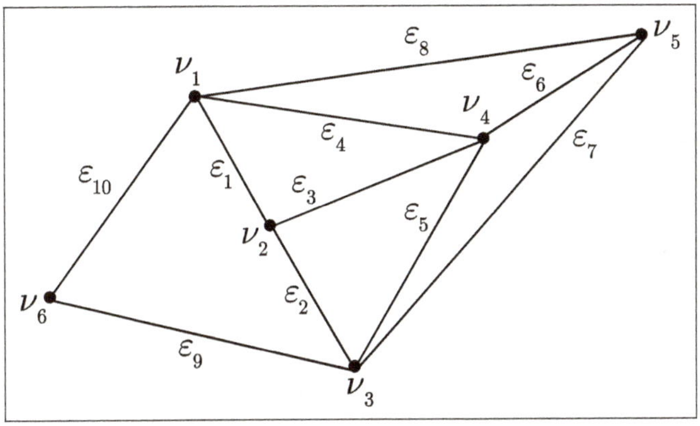

Figure 3.2: Graph $\mathcal{G}$ of Example 3.2 (adapted according to [15, Figure 17.3, p. 855]).

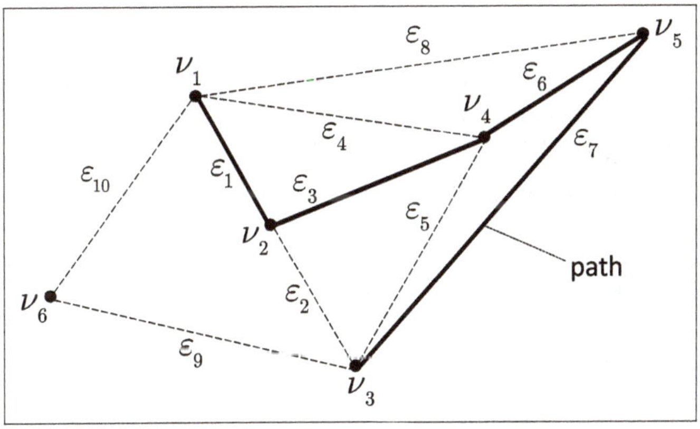

Figure 3.3: Exemplary path in the graph $\mathcal{G}$ of the Examples 3.2 and 3.3; cf. Figure 3.2.

Example 3.3. An example path shown in Figure 3.3 is the following:

$$\nu_1,\ \varepsilon_1,\ \nu_2,\ \varepsilon_3,\ \nu_4,\ \varepsilon_6,\ \nu_5,\ \varepsilon_7,\ \nu_3; \tag{3.38}$$

see Figure 3.3 [15, p. 856]. The path length is 4 [15, p. 856].

It is possible for a path to begin and end at the same vertex [15, p. 856]. Such a closed path is called a *cycle* [15, p. 856]. No vertex on a cycle except the initial and the final vertex appears more than once [15, p. 856].

Example 3.4. An exemplary cycle in the graph $\mathcal{G}$ of the Examples 3.2 and 3.3; cf. [15, Figure 17.3, p. 855], is shown in Figure 3.4,

$$\nu_1,\ \varepsilon_1,\ \nu_2,\ \varepsilon_3,\ \nu_4,\ \varepsilon_4,\ \nu_1; \tag{3.39}$$

see Figure 3.4. This cycle has the path length 3.

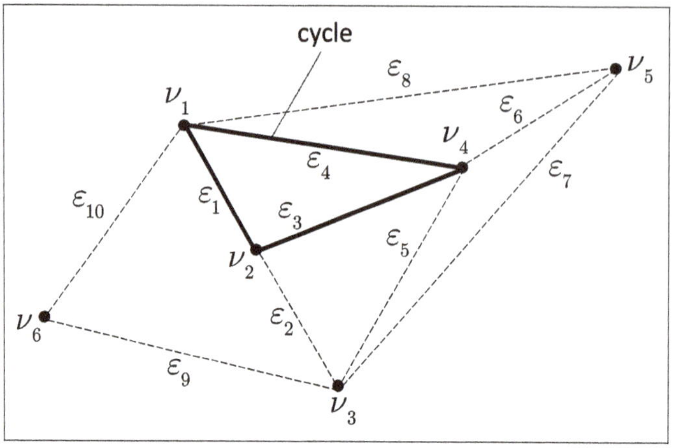

Figure 3.4: Exemplary cycle in the graph $\mathcal{G}$ of the Examples 3.2 and 3.3; cf. Figure 3.2.

An edge, for which the initial and the final vertices are the same, is a cycle of length 1 [15, p. 856]. This cycle is called a *self-loop* [15, p. 856]. The length of the shortest cycle in a graph is called the *girth* of the graph [15, p. 856].

A graph without any cycles is called *acyclic* [15, p. 856]. An acyclic graph is a *tree* [15, p. 856].

Example 3.5. The girth of the graph $\mathcal{G}$ shown in Figure 3.2 is 3; see also Figure 3.4.

A graph is said to be *connected* if there is at least one path between every pair of vertices in $\mathcal{G}$ [15, p. 856].

Example 3.6. The graph $\mathcal{G}$ shown in Figure 3.2 is connected.

A graph $\mathcal{G}$ is *bipartite* if its set of vertices $\mathbb{V}$ can be *partitioned into two disjoint subsets* $\mathbb{V}^{(1)}$ and $\mathbb{V}^{(2)}$, i. e.,

$$\mathbb{V}^{(1)} \cap \mathbb{V}^{(2)} = \mathbb{V}^{(1)}\mathbb{V}^{(2)} = \emptyset, \tag{3.40}$$

such that every edge in $\mathbb{E}$ joins a vertex in $\mathbb{V}^{(1)}$ with a vertex in $\mathbb{V}^{(2)}$ and no vertices in either set $\mathbb{V}^{(1)}$ and $\mathbb{V}^{(2)}$ are connected [15, p. 856]. It is obvious that *a bipartite graph has no self-loops* [15, p. 856].

Example 3.7. Figure 3.5 shows an exemplary bipartite graph $\mathcal{G}$ [15, Figure 17.5, p. 857]. We have

$$\begin{aligned}\mathbb{V}^{(1)} &= \{\nu_1, \nu_2, \nu_3\},\\ \mathbb{V}^{(2)} &= \{\nu_4, \nu_5, \nu_6, \nu_7, \nu_8\}.\end{aligned} \tag{3.41}$$

The girth of the graph is 6.

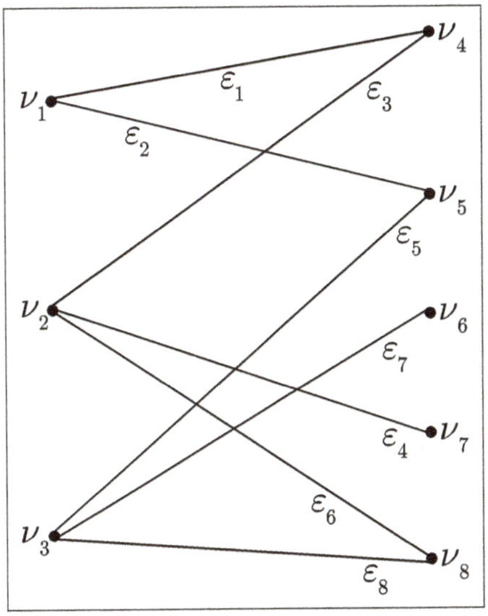

Figure 3.5: Exemplary bipartite graph (adapted according to [15, Figure 17.5, p. 857]).

If a bipartite graph contains cycles, then all cycles have even lengths because leaving a vertex in $\mathbb{V}^{(1)}$ and returning to another vertex in $\mathbb{V}^{(1)}$ requires two edges each time [15, p. 856].

For an $(n, k, d_{\min})$ binary linear LDPC code $\mathbb{L}$ of length n specified by a parity check matrix $\boldsymbol{H}$ with J row vectors $\boldsymbol{h}^{(j)}, j \in \{0, 1 \cdots (J-1)\}, J \in \mathbb{N}^*$, we create a bipartite graph $\mathcal{G}_{\mathrm{T}}$, which consists of two sets of vertices $\mathbb{V}^{(1)}$ and $\mathbb{V}^{(2)}$ [15, p. 857]. The graph $\mathcal{G}_{\mathrm{T}}$ is called Tanner graph [15, p. 857].

The first set $\mathbb{V}^{(1)}$ consists of n vertices, which represent the n code bits $v_0, v_1 \cdots v_{n-1}$, $n \in \mathbb{N}^*$ [15, p. 857]. These vertices are called the *code bit vertices* or *variable nodes* [15]. In what follows, we will hence identify the variable nodes $v_i^{(1)} \in \mathbb{V}^{(1)}$ by the n code bits $v_i, i \in \{0, 1 \cdots (n-1)\}, n \in \mathbb{N}^*$. The second set $\mathbb{V}^{(2)}$ consists of J vertices, which represent the J parity check sums or parity check equations $s_0, s_1 \cdots s_{J-1}, J \in \mathbb{N}^*$ given by (3.28) [15, p. 857]. These vertices are called the *check sum vertices* or *check nodes* [15, p. 857]. In what follows, we will hence identify the check nodes $v_j^{(2)} \in \mathbb{V}^{(2)}$ by the n parity check sums $s_j, j \in \{0, 1 \cdots (J-1)\}, J \in \mathbb{N}^*$.

A variable node v_i, $i \in \{0, 1 \cdots (n-1)\}$, $n \in \mathbb{N}^*$ is connected to a check node s_j, $j \in \{0, 1 \cdots (J-1)\}, J \in \mathbb{N}^*$, by an edge, denoted by (v_i, s_j), if and only if the code bit v_i, $i \in \{0, 1 \cdots (n-1)\}$, $n \in \mathbb{N}^*$ is checked by the parity check sum $s_j, j \in \{0, 1 \cdots (J-1)\}$, $J \in \mathbb{N}^*$ [15, p. 857]. As required, no two variable nodes are connected and no two check nodes are connected [15, p. 857].

The degree of a variable node $v_i, i \in \{0, 1 \cdots (n-1)\}, n \in \mathbb{N}^*$ is equal to the number of parity check sums, which check on $v_i, i \in \{0, 1 \cdots (n-1)\}, n \in \mathbb{N}^*$ [15, p. 857]. Thus, the degree of a check node $s_j, j \in \{0, 1 \cdots (J-1)\}, J \in \mathbb{N}^*$ is the number of code bits checked by $s_j, j \in \{0, 1 \cdots (J-1)\}, J \in \mathbb{N}^*$ [15, p. 857].

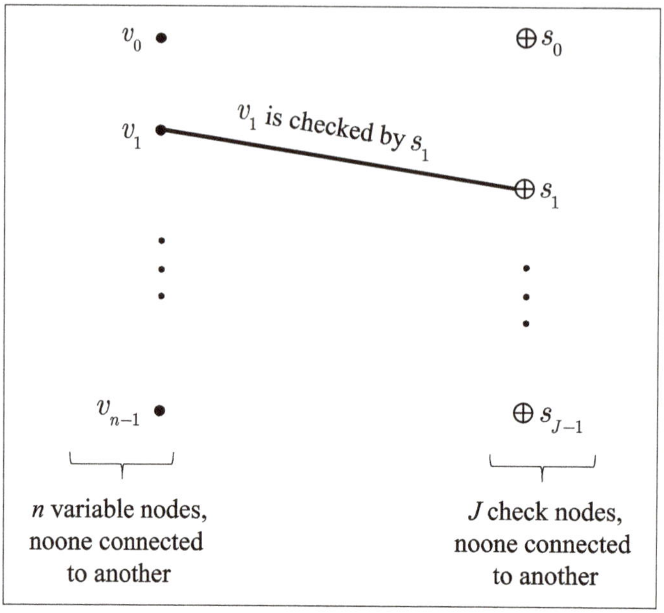

Figure 3.6: Construction of a Tanner graph $\mathcal{G}_T$; cf. [15, p. 857f.].

The construction of a Tanner graph $\mathcal{G}_T$ is illustrated in Figure 3.6; cf. [15, p. 857f.].

Example 3.8. The $(n, k, d_{\min})$ binary linear LDPC code $\mathbb{L}$ with the parity check matrix

$$\boldsymbol{H} = \begin{pmatrix} 1 & 1 & 0 & 1 & 0 & 0 & 0 \\ 0 & 1 & 1 & 0 & 1 & 0 & 0 \\ 0 & 0 & 1 & 1 & 0 & 1 & 0 \\ 0 & 0 & 0 & 1 & 1 & 0 & 1 \\ 1 & 0 & 0 & 0 & 1 & 1 & 0 \\ 0 & 1 & 0 & 0 & 0 & 1 & 1 \\ 1 & 0 & 1 & 0 & 0 & 0 & 1 \end{pmatrix} \tag{3.42}$$

of (3.13) has the Tanner graph $\mathcal{G}_T$ shown in Figure 3.7; cf. [15, p. 858]. This LDPC code has the following parity check sums:

$$\begin{aligned}
s_0 &= v_0 \oplus v_1 \oplus v_3, \\
s_1 &= v_1 \oplus v_2 \oplus v_4, \\
s_2 &= v_2 \oplus v_3 \oplus v_5, \\
s_3 &= v_3 \oplus v_4 \oplus v_6, \\
s_4 &= v_0 \oplus v_4 \oplus v_5, \\
s_5 &= v_1 \oplus v_5 \oplus v_6, \\
s_6 &= v_0 \oplus v_2 \oplus v_6.
\end{aligned} \tag{3.43}$$

The girth of this Tanner graph $\mathcal{G}_T$ is 6.

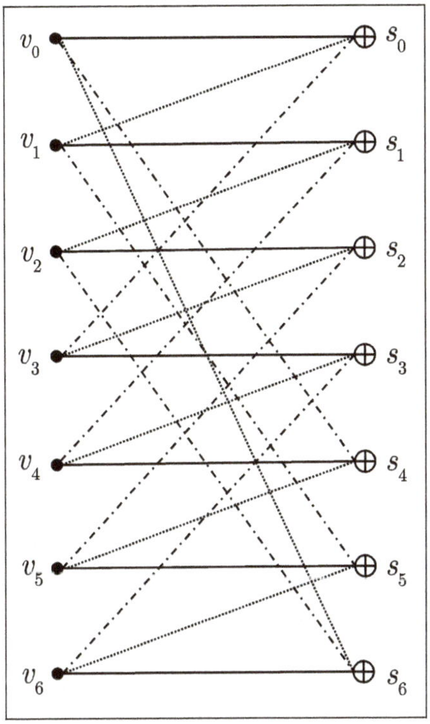

Figure 3.7: Tanner graph $\mathcal{G}_\mathrm{T}$ for $\boldsymbol{H}$ (adapted according to [15, Figure 17.6, p. 858]).

For a (γ, ρ)-regular $(n, k, d_{\min})$ binary linear LDPC code $\mathbb{L}$, the degrees of all variable nodes in the Tanner graph $\mathcal{G}_\mathrm{T}$ are the same and equal to γ, i. e., they are equal to the column weight of the parity check matrix $\boldsymbol{H}$ [15, p. 857]. Furthermore, the degrees of all the check nodes are the same and equal to ρ, i. e., the row weight of the parity check matrix $\boldsymbol{H}$ [15, p. 857]. Consequently, the corresponding Tanner graph $\mathcal{G}_\mathrm{T}$ is called *regular* [15, p. 857].

Furthermore, it follows from Definition 3.1 that no two code bits of an $(n, k, d_{\min})$ binary linear LDPC code $\mathbb{L}$ are checked by two different parity check sums [15, p. 857]. This implies that the Tanner graph of an $(n, k, d_{\min})$ binary linear LDPC code $\mathbb{L}$ *does not contain cycles of length* 4 [15, p. 857].

3.2.2 Log-Likelihood Ratio Processing in Parity Check Sums

In addition to being weakly random-like codes, low density parity check (LDPC) codes and convolutional turbo codes also have the enlightening about the processing of log-likelihood ratios (LLRs) in common. In this regard, the low density parity check (LDPC) codes extend our knowledge to the processing of log-likelihood ratios (LLRs) in parity check sums. We will illustrate this aspect.

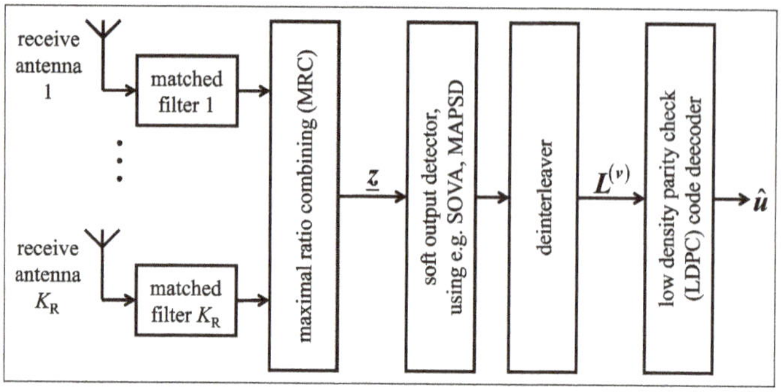

Figure 3.8: Generic block diagram of a receiver with receive antenna diversity [1, Section 3.9], soft output detector and $(n, k, d_{\min})$ binary linear LDPC code $\mathbb{L}$ decoder.

Figure 3.8 shows the generic block diagram of a receiver with receive antenna diversity using $K_R \in \mathbb{N}^*$ receive antennas, K_R matched filters and maximal ratio combining (MRC) [1, Section 3.9], a soft output detector, being based on, e. g., the *soft output Viterbi algorithm (SOVA)* (cf., e. g., [1, Sections 3.9.12–3.9.14]) or the *maximum a posteriori probability symbol detection (MAPSD)* (cf., e. g., [1, Sections 3.9.15]) and an $(n, k, d_{\min})$ binary linear LDPC code $\mathbb{L}$ decoder. The shown block diagram is very similar to Figure 2.8. The difference mainly lies in the fact that the deinterleaver only generates a single output vector, namely the log-likelihood ratios (LLRs) vector

$$\boldsymbol{L}^{(v)} = (L_0^{(v)}, L_1^{(v)} \cdots L_{n-1}^{(v)}), \quad n \in \mathbb{N}^*, \tag{3.44}$$

$$\begin{aligned} L_i^{(v)} &= \ln\left\{\frac{\mathrm{P}\{v_i = 0 \,|\, \underline{\mathbf{z}}\}}{\mathrm{P}\{v_i = 1 \,|\, \underline{\mathbf{z}}\}}\right\} \\ &= \ln\left\{\frac{\mathrm{p}(\underline{\mathbf{z}} \,|\, v_i = 0)}{\mathrm{p}(\underline{\mathbf{z}} \,|\, v_i = 1)}\right\} + \underbrace{\ln\left\{\frac{\mathrm{P}\{v_i = 0\}}{\mathrm{P}\{v_i = 1\}}\right\}}_{=0 \text{ for equally probable 0 and 1}} \\ &= \ln\left\{\frac{\mathrm{p}(\underline{\mathbf{z}} \,|\, v_i = 0)}{\mathrm{p}(\underline{\mathbf{z}} \,|\, v_i = 1)}\right\}, \quad i \in \{0, 1 \cdots (n-1)\}, \quad n \in \mathbb{N}^*, \end{aligned} \tag{3.45}$$

containing the log-likelihood ratios (LLRs) associated with the transmission and reception of the code bits $v_i \in \mathbb{F}_2$, $i \in \{0, 1 \cdots (n-1)\}$, $n \in \mathbb{N}^*$. Like in the case of the convolutional turbo codes, we find

$$\mathrm{p}(\underline{\mathbf{z}} \,|\, v_i) = \frac{\exp\{(1 \oplus v_i) \cdot L_i^{(v)}\}}{\exp\{L_i^{(v)}\} + 1}, \quad i \in \{0, 1 \cdots (n-1)\}, \quad n \in \mathbb{N}^*. \tag{3.46}$$

The new aspect in the decoding is the processing of the log-likelihood ratios (LLRs) in the check nodes. We will therefore consider this aspect in detail.

Let us begin with the simplest case, the processing of the log-likelihood ratios (LLRs) associated with a check sum, which consists only of two terms, i. e., checks on only two code bits, v_p and v_q, say. Then the conditional LLR associated with v_1 is denoted by

$$L_p^{(v)} = \ln\left\{\frac{\mathrm{p}(\underline{\mathbf{z}}\,|\,v_p = 0)}{\mathrm{p}(\underline{\mathbf{z}}\,|\,v_p = 1)}\right\}, \quad L_q^{(v)} = \ln\left\{\frac{\mathrm{p}(\underline{\mathbf{z}}\,|\,v_q = 0)}{\mathrm{p}(\underline{\mathbf{z}}\,|\,v_q = 1)}\right\}, \tag{3.47}$$
$$p \neq q, \quad p, q \in \{0, 1 \cdots (n-1)\}, \quad n \in \mathbb{N}^*.$$

Let us consider the parity check sum

$$s_j = v_p \oplus v_q \quad p \neq q, \quad p, q \in \{0, 1 \cdots (n-1)\}, \quad j \in \{0, 1 \cdots (J-1)\}, \quad J, n \in \mathbb{N}^*, \tag{3.48}$$

which is the jth component of the *parity check sum vector*

$$\mathbf{s} = (s_0, s_1 \cdots s_{J-1}), \quad J \in \mathbb{N}^*. \tag{3.49}$$

We would like to compute

$$L_j^{(s)} = \ln\left\{\frac{\mathrm{P}\{s_j = 0\,|\,\underline{\mathbf{z}}\}}{\mathrm{P}\{s_j = 1\,|\,\underline{\mathbf{z}}\}}\right\}, \quad j \in \{0, 1 \cdots (J-1)\}, \quad J, n \in \mathbb{N}^*, \tag{3.50}$$

as a function of $L_p^{(v)}$ and $L_q^{(v)}$.

To arrive at the desired representation, let us distinguish between the cases

$$s_j = v_p \oplus v_q = 0 \tag{3.51}$$

and

$$s_j = v_p \oplus v_q = 1. \tag{3.52}$$

Let us consider the first of these two cases. Clearly, s_j can only assume the value 0:

- in the event $\mathbb{E}_1$ equal to $\{(v_p = 0)(v_q = 0)\}$ that v_p and v_q are jointly equal to 0 or
- in the event $\mathbb{E}_2$ equal to $\{(v_p = 1)(v_q = 1)\}$ that v_p and v_q are jointly equal to 1.

Since the two events $\mathbb{E}_1$ and $\mathbb{E}_2$ are disjoint, i. e., the cannot occur at the same time, and hence

$$\mathbb{E}_1 \cap \mathbb{E}_2 = \mathbb{E}_1\mathbb{E}_2 = \emptyset, \tag{3.53}$$

we yield [102, equation (10), p. 7]

$$\begin{aligned}\mathrm{P}\{s_j = 0\,|\,\underline{\mathbf{z}}\} &= \mathrm{P}\{\mathbb{E}_1 + \mathbb{E}_2\,|\,\underline{\mathbf{z}}\}\\ &= \mathrm{P}\{\mathbb{E}_1\,|\,\underline{\mathbf{z}}\} + \mathrm{P}\{\mathbb{E}_2\,|\,\underline{\mathbf{z}}\}\end{aligned}$$

$$= \mathsf{P}\{(v_p = 0)(v_q = 0) \mid \underline{\boldsymbol{z}}\} + \mathsf{P}\{(v_p = 1)(v_q = 1) \mid \underline{\boldsymbol{z}}\}. \tag{3.54}$$

Considering the events $\{v_p\}$ and $\{v_q\}$ independent, it follows [102, equation (2), p. 9]

$$\mathsf{P}\{s_j = 0 \mid \underline{\boldsymbol{z}}\} = \mathsf{P}\{v_p = 0 \mid \underline{\boldsymbol{z}}\} \cdot \mathsf{P}\{v_q = 0 \mid \underline{\boldsymbol{z}}\} + \mathsf{P}\{v_p = 1 \mid \underline{\boldsymbol{z}}\} \cdot \mathsf{P}\{v_q = 1 \mid \underline{\boldsymbol{z}}\}. \tag{3.55}$$

In the second case s_j equal to 1, we obtain

$$\mathsf{P}\{s_j = 1 \mid \underline{\boldsymbol{z}}\} = \mathsf{P}\{v_p = 0 \mid \underline{\boldsymbol{z}}\} \cdot \mathsf{P}\{v_q = 1 \mid \underline{\boldsymbol{z}}\} + \mathsf{P}\{v_p = 1 \mid \underline{\boldsymbol{z}}\} \cdot \mathsf{P}\{v_q = 0 \mid \underline{\boldsymbol{z}}\} \tag{3.56}$$

in the same way.

Therefore, we yield (cf., e. g., [103, equation (1), p. 1033], [104, equation (5), p. 306])

$$\begin{aligned}
L_j^{(s)} &= \ln\left\{\frac{\mathsf{P}\{v_p = 0 \mid \underline{\boldsymbol{z}}\} \cdot \mathsf{P}\{v_q = 0 \mid \underline{\boldsymbol{z}}\} + \mathsf{P}\{v_p = 1 \mid \underline{\boldsymbol{z}}\} \cdot \mathsf{P}\{v_q = 1 \mid \underline{\boldsymbol{z}}\}}{\mathsf{P}\{v_p = 0 \mid \underline{\boldsymbol{z}}\} \cdot \mathsf{P}\{v_q = 1 \mid \underline{\boldsymbol{z}}\} + \mathsf{P}\{v_p = 1 \mid \underline{\boldsymbol{z}}\} \cdot \mathsf{P}\{v_q = 0 \mid \underline{\boldsymbol{z}}\}}\right\} \\
&= \ln\left\{\frac{1 + \frac{\mathsf{P}\{v_p=0 \mid \underline{\boldsymbol{z}}\}}{\mathsf{P}\{v_p=1 \mid \underline{\boldsymbol{z}}\}} \cdot \frac{\mathsf{P}\{v_q=0 \mid \underline{\boldsymbol{z}}\}}{\mathsf{P}\{v_q=1 \mid \underline{\boldsymbol{z}}\}}}{\frac{\mathsf{P}\{v_p=0 \mid \underline{\boldsymbol{z}}\} \cdot \mathsf{P}\{v_q=1 \mid \underline{\boldsymbol{z}}\}}{\mathsf{P}\{v_p=1 \mid \underline{\boldsymbol{z}}\} \cdot \mathsf{P}\{v_q=1 \mid \underline{\boldsymbol{z}}\}} + \frac{\mathsf{P}\{v_p=1 \mid \underline{\boldsymbol{z}}\} \cdot \mathsf{P}\{v_q=0 \mid \underline{\boldsymbol{z}}\}}{\mathsf{P}\{v_p=1 \mid \underline{\boldsymbol{z}}\} \cdot \mathsf{P}\{v_q=1 \mid \underline{\boldsymbol{z}}\}}}\right\} \\
&= \ln\left\{\frac{1 + \frac{\mathsf{P}\{v_p=0 \mid \underline{\boldsymbol{z}}\}}{\mathsf{P}\{v_p=1 \mid \underline{\boldsymbol{z}}\}} \cdot \frac{\mathsf{P}\{v_q=0 \mid \underline{\boldsymbol{z}}\}}{\mathsf{P}\{v_q=1 \mid \underline{\boldsymbol{z}}\}}}{\frac{\mathsf{P}\{v_p=0 \mid \underline{\boldsymbol{z}}\}}{\mathsf{P}\{v_p=1 \mid \underline{\boldsymbol{z}}\}} + \frac{\mathsf{P}\{v_q=0 \mid \underline{\boldsymbol{z}}\}}{\mathsf{P}\{v_q=1 \mid \underline{\boldsymbol{z}}\}}}\right\} \\
&= \ln\left\{\frac{1 + \exp\{L_p^{(v)}\} \cdot \exp\{L_q^{(v)}\}}{\exp\{L_p^{(v)}\} + \exp\{L_q^{(v)}\}}\right\} \\
&= \ln\left\{\frac{1 + \exp\{L_p^{(v)} + L_q^{(v)}\}}{\exp\{L_p^{(v)}\} + \exp\{L_q^{(v)}\}}\right\}, \quad j \in \{0, 1 \cdots (J-1)\}, \quad J \in \mathbb{N}^*.
\end{aligned} \tag{3.57}$$

Alternatively, we can write

$$\begin{aligned}
L_j^{(s)} &= \ln\left\{\frac{(\exp\{L_p^{(v)}\} + 1)(\exp\{L_q^{(v)}\} + 1) + (\exp\{L_p^{(v)}\} - 1)(\exp\{L_q^{(v)}\} - 1)}{(\exp\{L_p^{(v)}\} + 1)(\exp\{L_q^{(v)}\} + 1) - (\exp\{L_p^{(v)}\} - 1)(\exp\{L_q^{(v)}\} - 1)}\right\} \\
&= \ln\left\{\frac{\prod_{i \in \{p,q\}}(\exp\{L_i^{(v)}\} + 1) + \prod_{i \in \{p,q\}}(\exp\{L_i^{(v)}\} - 1)}{\prod_{i \in \{p,q\}}(\exp\{L_i^{(v)}\} + 1) - \prod_{i \in \{p,q\}}(\exp\{L_i^{(v)}\} - 1)}\right\} \\
&= \ln\left\{\frac{1 + \prod_{i \in \{p,q\}} \frac{(\exp\{L_i^{(v)}\}-1)}{(\exp\{L_i^{(v)}\}+1)}}{1 - \prod_{i \in \{p,q\}} \frac{(\exp\{L_i^{(v)}\}-1)}{(\exp\{L_i^{(v)}\}+1)}}\right\} = \ln\left\{\frac{1 + \prod_{i \in \{p,q\}} \frac{\exp\{L_i^{(v)}/2\} - \exp\{-L_i^{(v)}/2\}}{\exp\{L_i^{(v)}/2\} + \exp\{-L_i^{(v)}/2\}}}{1 - \prod_{i \in \{p,q\}} \frac{\exp\{L_i^{(v)}/2\} - \exp\{-L_i^{(v)}/2\}}{\exp\{L_i^{(v)}/2\} + \exp\{-L_i^{(v)}/2\}}}\right\},
\end{aligned} \tag{3.58}$$

which yields [11, equation (3.13c), p. 134]

$$L_j^{(s)} = \ln\left\{\frac{1 + \prod_{i \in \{p,q\}} \tanh\{L_i^{(v)}/2\}}{1 - \prod_{i \in \{p,q\}} \tanh\{L_i^{(v)}/2\}}\right\}. \tag{3.59}$$

Let us now consider the parity check sum

$$s_j = \bigoplus_{i \in \{i_0, i_1 \cdots i_{p-1}\}} v_i, \quad j \in \{0, 1 \cdots (J-1)\}, \quad J \in \mathbb{N}^*, \tag{3.60}$$

in which $\{i_0, i_1 \cdots i_{p-1}\}$ is the set of indices of those code bits, which are checked by s_j of (3.60). We will prove by induction that [105, equation (48), p. 240]

$$L_j^{(s)} = \ln\left\{\frac{\prod_{i \in \{i_0, i_1 \cdots i_{p-1}\}}(\exp\{L_i^{(v)}\} + 1) + \prod_{i \in \{i_0, i_1 \cdots i_{p-1}\}}(\exp\{L_i^{(v)}\} - 1)}{\prod_{i \in \{i_0, i_1 \cdots i_{p-1}\}}(\exp\{L_i^{(v)}\} + 1) - \prod_{i \in \{i_0, i_1 \cdots i_{p-1}\}}(\exp\{L_i^{(v)}\} - 1)}\right\}, \tag{3.61}$$
$$j \in \{0, 1 \cdots (J-1)\}, \quad J \in \mathbb{N}^*.$$

Proof 3.1 (Proof of (3.61)).

Base Case

In the case of y equal to 1, we have

$$L_j^{(s)} = \ln\left\{\frac{(\exp\{L_{i_0}^{(v)}\} + 1) + (\exp\{L_{i_0}^{(v)}\} - 1)}{(\exp\{L_{i_0}^{(v)}\} + 1) - (\exp\{L_{i_0}^{(v)}\} - 1)}\right\} = \ln\left\{\frac{2 \cdot \exp\{L_{i_0}^{(v)}\}}{2}\right\}$$
$$= \ln\left\{\exp\left\{L_{i_0}^{(v)}\right\}\right\} = L_{i_0}^{(v)}, \tag{3.62}$$

which is true.

In the case of y equal to 2, (3.58) holds.

Induction Step

We set out from (3.61). Let us prove

$$\tilde{L}_j^{(s)} = \ln\left\{\frac{\prod_{i \in \{i_0, i_1 \cdots i_p\}}(\exp\{L_i^{(v)}\} + 1) + \prod_{i \in \{i_0, i_1 \cdots i_p\}}(\exp\{L_i^{(v)}\} - 1)}{\prod_{i \in \{i_0, i_1 \cdots i_p\}}(\exp\{L_i^{(v)}\} + 1) - \prod_{i \in \{i_0, i_1 \cdots i_p\}}(\exp\{L_i^{(v)}\} - 1)}\right\} \tag{3.63}$$

in the case of the set $\{i_0, i_1 \cdots i_p\}$.

It follows from (3.61) that

$$e^{L_j^{(s)}} + 1 = \frac{2 \cdot \prod_{i \in \{i_0, i_1 \cdots i_{p-1}\}}(e^{L_i^{(v)}} + 1)}{\prod_{i \in \{i_0, i_1 \cdots i_{p-1}\}}(e^{L_i^{(v)}} + 1) - \prod_{i \in \{i_0, i_1 \cdots i_{p-1}\}}(e^{L_i^{(v)}} - 1)}, \tag{3.64}$$

$$e^{L_j^{(s)}} - 1 = \frac{2 \cdot \prod_{i \in \{i_0, i_1 \cdots i_{p-1}\}}(e^{L_i^{(v)}} - 1)}{\prod_{i \in \{i_0, i_1 \cdots i_{p-1}\}}(e^{L_i^{(v)}} + 1) - \prod_{i \in \{i_0, i_1 \cdots i_{p-1}\}}(e^{L_i^{(v)}} - 1)}. \tag{3.65}$$

In addition, we have

$$\left(e^{L_j^{(s)}} + 1\right)\left(e^{L_{i_p}^{(v)}} + 1\right) = \frac{2 \cdot \prod_{i \in \{i_0, i_1 \cdots i_p\}}(e^{L_i^{(v)}} + 1)}{\prod_{i \in \{i_0, i_1 \cdots i_{p-1}\}}(e^{L_i^{(v)}} + 1) - \prod_{i \in \{i_0, i_1 \cdots i_{p-1}\}}(e^{L_i^{(v)}} - 1)}, \tag{3.66}$$

$$\left(e^{L_j^{(s)}} - 1\right)\left(e^{L_{i_p}^{(v)}} - 1\right) = \frac{2 \cdot \prod_{i \in \{i_0, i_1 \cdots i_p\}}(e^{L_i^{(v)}} - 1)}{\prod_{i \in \{i_0, i_1 \cdots i_{p-1}\}}(e^{L_i^{(v)}} + 1) - \prod_{i \in \{i_0, i_1 \cdots i_{p-1}\}}(e^{L_i^{(v)}} - 1)}. \tag{3.67}$$

Therefore, we obtain

$$\tilde{L}_j^{(s)} = \frac{(e^{L_j^{(s)}}+1)(e^{L_{i_p}^{(v)}}+1)+(e^{L_j^{(s)}}-1)(e^{L_{i_p}^{(v)}}-1)}{(e^{L_j^{(s)}}+1)(e^{L_{i_p}^{(v)}}+1)-(e^{L_j^{(s)}}-1)(e^{L_{i_p}^{(v)}}-1)} = \frac{\prod_{i\in\{i_0,i_1\cdots i_p\}}(e^{L_i^{(v)}}+1)+\prod_{i\in\{i_0,i_1\cdots i_p\}}(e^{L_i^{(v)}}-1)}{\prod_{i\in\{i_0,i_1\cdots i_p\}}(e^{L_i^{(v)}}+1)-\prod_{i\in\{i_0,i_1\cdots i_p\}}(e^{L_i^{(v)}}-1)}, \tag{3.68}$$

which completes the proof. □

Furthermore, we find [15, equation (17.61), p. 879], [105, equation (48), p. 240], [11, equation (3.13c), p. 134],

$$L_j^{(s)} = \ln\left\{\frac{1+\prod_{i\in\{i_0,i_1\cdots i_{p-1}\}}\frac{\exp\{L_i^{(v)}\}-1}{\exp\{L_i^{(v)}\}+1}}{1-\prod_{i\in\{i_0,i_1\cdots i_{p-1}\}}\frac{\exp\{L_i^{(v)}\}-1}{\exp\{L_i^{(v)}\}+1}}\right\} = \ln\left\{\frac{1+\prod_{i\in\{i_0,i_1\cdots i_{p-1}\}}\tanh\{L_i^{(v)}/2\}}{1-\prod_{i\in\{i_0,i_1\cdots i_{p-1}\}}\tanh\{L_i^{(v)}/2\}}\right\}, \tag{3.69}$$
$$j\in\{0,1\cdots(J-1)\},\quad J\in\mathbb{N}^*.$$

3.2.3 Sum Product Algorithm

In this Section 3.2.3, we will illustrate the iterative decoding scheme termed *sum product algorithm (SPA)* [15, pp. 875–880]. The goal of this form of decoding is to iteratively update the log-likelihood ratios (LLRs) $L_i^{(v)}$, $i\in\{0,1\cdots(n-1)\}$, $n\in\mathbb{N}^*$ before hard decision decoding by exploiting the code structure, i. e., the Tanner graph $\mathcal{G}_\mathrm{T}$; cf., e. g., [15, p. 875].

Let us set out from the Tanner graph $\mathcal{G}_\mathrm{T}$ of a low density parity check (LDPC) code. Let $\mathbb{M}(i)$ be the set of check nodes connected to the variable node, which corresponds to the code bit v_i and the log-likelihood ratio (LLR) $L_i^{(v)}$; cf. Figure 3.9 [15, p. 857]. Fur-

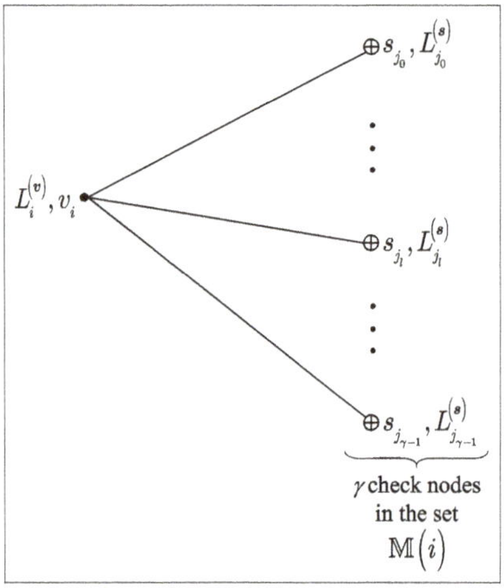

Figure 3.9: Check nodes in the set $\mathbb{M}(i)$ [15, p. 857].

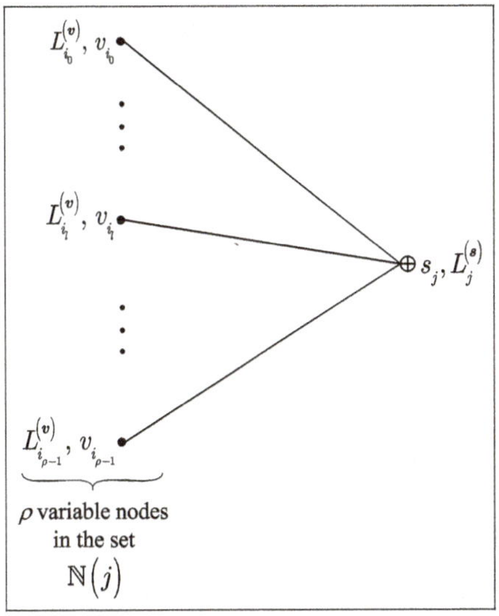

Figure 3.10: Variable nodes in the set $\mathbb{N}(j)$ [15, p. 857].

thermore, let $\mathbb{N}(j)$ be the set of variable nodes connected to the check node s_j, which corresponds to the jth parity check sum; cf. Figure 3.10 [15, p. 857].

The sum product algorithm (SPA) comprises an initialization step and three further decoding steps [15, pp. 875–880]. Let the message sent from variable node v_i to check node s_j be denoted by $\lambda_{i\to j}$ [15, pp. 875–880]. Let the message sent back from check node s_j to variable node v_i be $\Lambda_{j\mapsto i}$ [15, pp. 875–880].

The SPA algorithm works as follows [15, pp. 875–880].

Definition 3.2 (Sum Product Algorithm (SPA)).
1) Initialization Step
Let

$$\lambda_{i\mapsto j} = L_i^{(v)}, \quad i \in \{0, 1 \cdots (n-1)\}, \quad j \in \{0, 1 \cdots (J-1)\}, \quad J, n \in \mathbb{N}^*, \tag{3.70}$$

$$\Lambda_{j\mapsto i} = 0, \quad i \in \{0, 1 \cdots (n-1)\}, \quad j \in \{0, 1 \cdots (J-1)\}, \quad J, n \in \mathbb{N}^*. \tag{3.71}$$

2) Check Node Update Step
For each check node $s_j, j \in \{0, 1 \cdots (J-1)\}, J \in \mathbb{N}^*$, and for each $v_{i_l} \in \mathbb{N}(j) \setminus \{v_i\}$ compute the *extrinsic information* associated with $\lambda_{i_l\mapsto j}$ as the new value for $\Lambda_{j\mapsto i}$, as illustrated in Section 3.2.2. This extrinsic information is sent back to the variable node v_i for further processing.

3) Variable Node Update Step
For each variable node v_i in the set of variable nodes and for each $s_{j_l} \in \mathbb{M}(i) \setminus \{s_j\}$ compute

$$\lambda_{i\mapsto j} = L_i^{(v)} + \sum_{s_{j_l} \in \mathbb{M}(i)\setminus\{s_j\}} \Lambda_{j_l\mapsto i}, \quad i \in \{0, 1 \cdots (n-1)\}, \quad n \in \mathbb{N}^*. \tag{3.72}$$

The Check Node Update Step and the Variable Node Update Step are the two steps of the iteration loop. Typically around one hundred (100) iterations are performed prior to the Decision Step.

4) Decision Step

In the Decision Step, which replaces the Variable Node Update Step in the last iteration, compute

$$\lambda_i = L_i^{(v)} + \sum_{s_j \in \mathbb{M}(i)} \Lambda_{j\mapsto i}, \quad i \in \{0, 1 \cdots (n-1)\}, \quad n \in \mathbb{N}^*. \tag{3.73}$$

Then quantize λ_i, $i \in \{0, 1 \cdots (n-1)\}$, $n \in \mathbb{N}^*$, using the following rule:

$$\lambda_i^{(q)} = \begin{cases} 0 & \text{if } \lambda_i \geq 0, \\ 1 & \text{if } \lambda_i < 0, \end{cases} \quad i \in \{0, 1 \cdots (n-1)\}, \quad n \in \mathbb{N}^*. \tag{3.74}$$

Next, compute

$$\left(\lambda_0^{(q)}, \lambda_1^{(q)} \cdots \lambda_{n-1}^{(q)}\right) \boldsymbol{H}^{\mathrm{T}}. \tag{3.75}$$

If these check sums are all equal to zero, then halt; otherwise resume the iteration, i. e., the Check Node Update Step and the Variable Node Update Step. If the SPA does not halt within some maximum number of iterations, then declare decoding failure.

Remark 3.2 (Quantum Communications System With Nonquantum Channel Coding—LDPC Code). Let us reuse the wireless laser based binary on-off keying (OOK) quantum communications system with a classical channel encoder at its input and a classical channel decoder at its output (adapted according to [76, Bild 11, p. 115] and [77, Figure 5.1, p. 184; Figure 8.1, p. 382]), introduced in Figure 1.30 and Remark 1.45. Again, recall that in the Remarks 1.45, 2.1 and 3.2, including the Figures 1.30, 1.31, 1.32, 2.10 and 3.11, the attribute "*classical*" means "*nonquantum*."

In Remark 3.2, the classical channel encoder implements the encoding of the LDPC code of the wireless local area network (WLAN) IEEE 802.11 standard [106, Table F-1, p. 4130] with k equal to 540 message bits, with n equal to 648 code bits and with the code rate R equal to $5/6$. The classical channel decoder is the soft input LDPC decoder using the sum product algorithm (SPA) and ten decoding iterations.

Figure 3.11 shows the corresponding simulation results of the uncoded bit error ratios without any classical channel decoding and the coded bit error ratios with classical channel decoding using the LDPC code of the wireless local area network (WLAN) IEEE 802.11 standard [106, Table F-1, p. 4130] with k equal to 540 message bits, with n equal to 648 code bits and with the code rate R equal to $5/6$. The average number $\bar{N}_\alpha \in \{0, 0.01, 0.03\}$ is the curve parameter. As in Remarks 1.45 and 2.1, the code rate R equal to $5/6$ is explicitly taken into account in the presented simulation results.

For instance, looking at $\bar{N}_\alpha$ equal to 0 and the simulation results of the uncoded bit error ratios discussed in the Remark on the quantum detection included in [1], the uncoded bit error ratio is approximately equal to $2 \cdot 10^{-4}$ at $\bar{N}_\mu$ equal to 8, whereas the coded bit error ratio of approximately $2 \cdot 10^{-4}$ requires only $\bar{N}_\mu$ approximately equal to 4.9 according to Figure 2.10. This is a saving of approximately three signal photons on average.

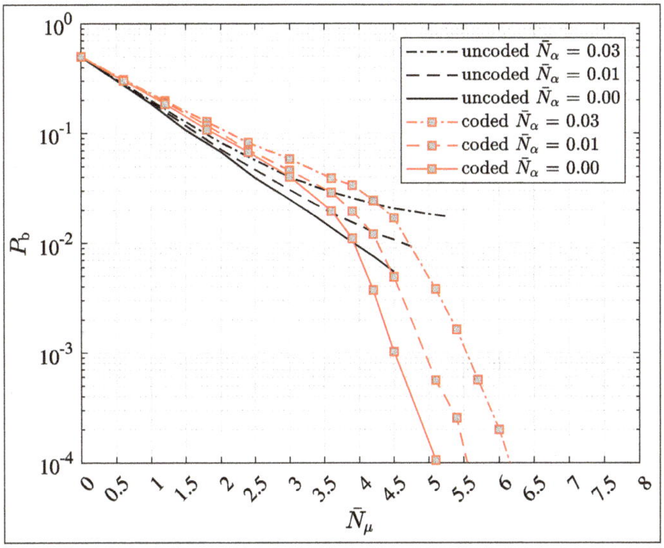

Figure 3.11: Simulation results of the uncoded bit error ratios without any classical channel decoding and the coded bit error ratios with classical channel decoding using the LDPC code of the wireless local area network (WLAN) IEEE 802.11 standard [106, Table F-1, p. 4130] with k equal to 540 message bits, with n equal to 648 code bits and with the code rate R equal to 5/6. The classical channel decoder is the soft input LDPC decoder using the sum product algorithm (SPA) and ten decoding iterations. The average number $\bar{N}_\alpha \in \{0, 0.01, 0.03\}$ is the curve parameter.

4 Polar Codes

4.1 Channel Polarization Paradigm

To the best knowledge of the author, *polar codes* were first discussed in 2009; cf., e. g., [16]. Essentially, polar codes can be regarded as an extension of the Reed–Muller (RM) codes, which have been well known for decades [107, Section III]. This fact has already been pointed out by Erdal Arikan, who is said to be the inventor of polar codes [16, pp. 3055f., 3068f.]. Arikan states [16, pp. 3055f., 3069]

> *"As the present work progressed, it became clear that polar coding had much in common with Reed–Muller (RM) coding (...). Indeed, recursive code construction and SC* (remark by the author: successive cancellation) *decoding, which are two essential ingredients of polar coding, appear to have been introduced into coding theory by RM codes.*
> (...)
> *Unlike polar coding, RM coding selects the information set in a channel-independent manner; it is not as fine-tuned to the channel polarization phenomenon as polar coding is.*
> (...)
> *So, polar coding goes beyond RM coding* (...)
> (...)
> *It is interesting that the possibility of RM codes being capacity-achieving codes under ML* (remark by the author: maximum-likelihood) *decoding seems to have received no attention in the literature."*

The most appealing aspects of polar codes are
- the establishment of the generator matrices by applying the well-known Kronecker product calculus, already discussed in Section 1.17.4 and
- the exploitation of the also well-known *chain rule of information* [7, equation (2.2.29), p. 22], which is the essence of what some term the "channel polarization" method [107, Section III].

In [107, Section III], predecessors of polar codes are mentioned. In this context, the doctoral dissertation by Korada of 2009 [108] should be noted. Although it seems debatable to call polar codes new or "nontrivial," the eminent exploitation of information theory in the code design is most remarkable and, in the opinion of the author, earns polar codes a place among the *sophisticated channel codes*.

So, why have polar codes become important in the context of advanced mobile communications? Let us see. We already know that advanced mobile communications provide a multitude of communications services to the users who may want to apply such services even simultaneously. This paradigm led to the introduction of the *transport format combination indicator (TFCI)* already in the third generation mobile communications system "*Universal Mobile Telecommunications System (UMTS)*," also abbreviated 3G/UMTS [82, Figure 6.19, p. 113], [71, Section 4.3.3, p. 55]. In addition, the fourth generation mobile communications system "*Long Term Evolution (LTE)*," also abbreviated by 4G/LTE, relies on Reed–Muller (RM) codes for control information; cf., e. g., [71, Sec-

https://doi.org/10.1515/9783111239651-004

tion 5.2.2.64, p. 120f.]. It is hence the close relationship between Reed–Muller (RM) codes and polar codes, which makes polar codes a perfect choice for the inclusion in 5G to further evolve the encoding of the control channels.

Let us plunge into the engineering details of polar codes now, and first revisit the information theoretical aspects, which will be helpful in understanding polar codes. Suppose the transmission of a multitude of N transmit signals, comprising of the N binary input symbols $X_0, X_1 \cdots X_{N-1}, N \in \mathbb{N}^*$, of the transmission channel, gathered in the transmit vector

$$\boldsymbol{X} = (X_0, X_1 \cdots X_{N-1}), \quad X_i \in \mathbb{F}_2, \quad i \in \{0, 1 \cdots (N-1)\}, \quad N \in \mathbb{N}^*, \tag{4.1}$$

and the reception of a multitude of N receive signals, comprised of the N output symbols $Y_0, Y_1 \cdots Y_{N-1}, N \in \mathbb{N}^*$, of the transmission channel, combined in the receive vector

$$\boldsymbol{Y} = (Y_0, Y_1 \cdots Y_{N-1}), \quad N \in \mathbb{N}^*. \tag{4.2}$$

The input symbols $X_i, i \in \{0, 1 \cdots (N-1)\}, N \in \mathbb{N}^*$ associated with the elementary events $\{X_i\}, i \in \{0, 1 \cdots (N-1)\}, N \in \mathbb{N}^*$ are the outcomes of N independent random experiments with identical a priori probabilities

$$\mathsf{P}\{\{X_i\}\} = \mathsf{P}\{\{X\}\}, \quad i \in \{0, 1 \cdots (N-1)\}, \quad N \in \mathbb{N}^*, \tag{4.3}$$

of the occurrence of the events $\{X_i\}, i \in \{0, 1 \cdots (N-1)\}, N \in \mathbb{N}^*$. In this case, the channel capacity of the transmission channel with a single input symbol X_i and a single output symbol Y_i is given by

$$C_1 = \max_{\mathsf{P}\{\{X\}\}} I\{X_i; Y_i\}, \quad i \in \{0, 1 \cdots (N-1)\}, \quad N \in \mathbb{N}^*. \tag{4.4}$$

Let us set out from the task to design a $(n, k, d_{\min})$ binary linear block code $\mathbb{V}$. This inevitably leads to the fact that out of the set of the two operations {additive operation $\oplus$, multiplicative operation $\odot$} of the finite field $(\mathbb{F}_2, \oplus, \odot)$ only the additive operation $\oplus$ can be used for the encoding, just like in the case of the classical codes discussed in Chapter 1. The task of the polar codes design to be accomplished is hence the following:

> *"Find a method, relying only on the additive operation "$\oplus$" of $\mathbb{F}_2$, in which each channel output Y_i, $i \in \{0, 1 \cdots (N-1)\}, N \in \mathbb{N}^*$ does not only rely on the channel input X_i, $i \in \{0, 1 \cdots (N-1)\}, N \in \mathbb{N}^*$, but also on at least some of the other inputs in most cases, as well, preserving the equality in $I\{\boldsymbol{X}; \boldsymbol{Y}\} \leq N \cdot C$ taught by [7, Theorem 4.2.1, p. 75],"*

which means that every single transmission shall achieve the channel capacity C.

Furthermore, bearing in mind the well-known analysis made, e. g., by Gérard *Battail* [35, Sections 3.2, 3.3 and 4.1, pp. 85–88], the minimum distance $d_{\min}$ must be chosen as large as possible to provide a best possible error performance of the new code. Looking at an $(n, k, d_{\min})$ binary linear LDPC code $\mathbb{L}$, we already learned from Definition 3.1

that the parity checks provide the best possible performance of the code in the case of “mutually orthogonal” columns and rows; cf. requirement item (3) of Definition 3.1.

Starting from a conceptually smallest possible $(n, k, d_{\min})$ binary linear block code $\mathbb{V}$, which must have the dimension k equal to 2, because otherwise the expression “coding” might rather cause exhilaration than anything else and, consequently, the code length n equal to 2 because for the code length 1 flipping a coin will be a probate means of decoding rather than a reception driven concept. We might ask ourselves how could a 2×2 matrix taken from $\mathbb{F}_2^{2\times 2}$ look like, which fulfills the requirement of at most a single 1 in common between two distinct rows or two distinct columns? Admittedly, the reader might think that in classical coding the parity check matrix must be void because it has $(n-k)$ equal to 0 rows. But this is not quite right; just look at an $(n, k, d_{\min})$ binary linear LDPC code $\mathbb{L}$ and its parity check matrix, which has a number of rows that is not necessarily related to the result of $(n - k)$. So, asking for a quadratic matrix does not seem that futile any more, and hence trying out a 2×2 matrix is not such a bad idea.

So, let us get back to the question above. When creating a 2×2 matrix with elements taken from $\mathbb{F}_2$, we can only rely on four different possible row vectors, namely $(0, 0)$, $(1, 0)$, $(0, 1)$ and $(1, 1)$. Since our matrix requires full rank, the zero vector $(0, 0)$ is out of the game and sent to the cabin before the match begins. Hence, there are only three combinations to go. Choosing $(1, 0)$ and $(0, 1)$ seem perfect at first glance. Regrettably, the transmission of two consecutive bits would then be mutually independent, and hence the above requirement “*... but also on at least some of the other inputs as well in most cases ...*” would never be fulfilled. So, only two combinations are left, namely

- $(1, 0)$ and $(1, 1)$ as well as
- $(0, 1)$ and $(1, 1)$.

So, we have the following choices:

$$\begin{pmatrix} 1 & 0 \\ 1 & 1 \end{pmatrix}, \begin{pmatrix} 1 & 1 \\ 1 & 0 \end{pmatrix}, \begin{pmatrix} 0 & 1 \\ 1 & 1 \end{pmatrix}, \begin{pmatrix} 1 & 1 \\ 0 & 1 \end{pmatrix}. \tag{4.5}$$

Since permuting the rows yield the same code, only

$$\begin{pmatrix} 1 & 0 \\ 1 & 1 \end{pmatrix}, \begin{pmatrix} 0 & 1 \\ 1 & 1 \end{pmatrix} \tag{4.6}$$

remain, which only differ in the sequence of the columns. Changing this sequence yields equivalent codes. Hence, only one choice remains, e. g.,

$$\begin{pmatrix} 1 & 0 \\ 1 & 1 \end{pmatrix}. \tag{4.7}$$

In fact, it also does not really matter much whether we consider (4.7) the parity check matrix of the generator matrix because swapping yields dual codes.

So, let us agree on the generator matrix

$$\boldsymbol{G}_2 = \begin{pmatrix} 1 & 0 \\ 1 & 1 \end{pmatrix}. \tag{4.8}$$

Now, bearing in mind Reed–Muller (RM) codes and looking at (1.1425), we immediately see that $\boldsymbol{G}_2$ of (4.8) and $\boldsymbol{H}_2$ of (1.1425) generate equivalent codes. Longer codes could hence be created in the same way as described in, e. g., Section 1.17.4.

Now, we could try to bring some order in the selected *"other inputs in most cases."* For instance, we could require that the average mutual information $I\{X_i; \boldsymbol{Y}\}$ $i \in \{0, 1 \cdots (N-1)\}, N \in \mathbb{N}^*$ is conditioned on the preceding channel inputs $X_0, X_1 \cdots X_{i-1}$, $i \in \{0, 1 \cdots (N-1)\}, N \in \mathbb{N}^*$. We hence replace $I\{X_i; \boldsymbol{Y}\}$ by the average conditional mutual information

$$\begin{aligned} I\{X_i; \boldsymbol{Y} \,|\, X_0, X_1 \cdots X_{i-1}\} &= \mathrm{E}\left\{\log_2\left\{\frac{\mathrm{P}\{\{X_i\}\{\boldsymbol{Y}\} \,|\, \{X_0\}\{X_1\} \cdots \{X_{i-1}\}\}}{\mathrm{P}\{\{X_i\} \,|\, \{X_0\} \cdots \{X_{i-1}\}\}\mathrm{P}\{\{\boldsymbol{Y}\} \,|\, \{X_0\} \cdots \{X_{i-1}\}\}}\right\}\right\} \\ &= \sum_{\boldsymbol{Y}, X_0, X_1 \cdots X_i} \mathrm{P}\{\{X_i\}\{\boldsymbol{Y}\}\{X_0\}\{X_1\} \cdots \{X_{i-1}\}\} \\ &\quad \log_2\left\{\frac{\mathrm{P}\{\{X_i\}\{\boldsymbol{Y}\} \,|\, \{X_0\}\{X_1\} \cdots \{X_{i-1}\}\}}{\mathrm{P}\{\{X_i\} \,|\, \{X_0\} \cdots \{X_{i-1}\}\}\mathrm{P}\{\{\boldsymbol{Y}\} \,|\, \{X_0\} \cdots \{X_{i-1}\}\}}\right\}, \\ &\quad i \in \{0, 1 \cdots (N-1)\}, \quad N \in \mathbb{N}^*. \end{aligned} \tag{4.9}$$

Now, the maximum average conditional mutual information $\max_{\mathrm{P}\{\{X\}\}} I\{X_i; \boldsymbol{Y} \,|\, X_0, X_1 \cdots X_{i-1}\}$ will not be equal for all $i \in \{0, 1 \cdots (N-1)\}, N \in \mathbb{N}^*$. So, it is likely that there are some transmissions with a low capacity and some transmissions achieving a high capacity.

Does all this sound inventive? The author believes that a ballot among the readers and coding experts might not result in a definite yes.

Applying the *channel polarization* operation, one creates out of independent copies of a given binary input discrete memoryless channel (DMC) a second set of channels, which exhibit a *"polarization effect"* in the sense that, as the number N of such channels approaches infinity, the capacity terms tend toward 1 or 0 for all but a vanishing fraction of indices $i \in \{0, 1 \cdots (N-1)\}, N \in \mathbb{N}^*$ [16, Section I.B, p. 3052]. This is achieved by creating virtual channels that operate according to the maximum average conditional mutual information $\max_{\mathrm{P}\{\{X\}\}} I\{X_i; \boldsymbol{Y} \,|\, X_0, X_1 \cdots X_{i-1}\}$. The channel polarization operation will be illustrated using the binary erasure channel (BEC). The said channel polarization operation consists of a *channel combining phase* and a *channel splitting phase* [16, Section I.B, p. 3052]. Indeed, conceptually, this approach looks novel.

Let us consider the channel combining first.

We set out from the uncoded transmission of N mutually independent outcomes, i. e., the transmit symbols $X_0, X_1 \cdots X_{N-1}$ of the corresponding events, all gathered in the transmit vector $\boldsymbol{X}$ of (4.1). Each transmit symbol is transmitted via its own copy of the aforementioned binary input discrete memoryless channel (DMC), which shall be denoted by W in what follows. All the N binary input discrete memoryless channels (DMCs) W are assumed to be identical. Consequently, the probability $\mathsf{P}\{\boldsymbol{X}\}$ is given by

$$\begin{aligned}\mathsf{P}\{\{\boldsymbol{X}\}\} &= \mathsf{P}\{\{X_0\}\{X_1\}\cdots\{X_{N-1}\}\} \\ &= \mathsf{P}\{\{X_0\}\}\cdot\mathsf{P}\{\{X_1\}\}\cdot\cdots\cdot\mathsf{P}\{\{X_{N-1}\}\}, \quad N\in\mathbb{N}^*. \end{aligned} \tag{4.10}$$

The input of the first binary input discrete memoryless channel (DMC) W is $X_0 \in \mathbb{F}_2$, the input of the second binary input discrete memoryless channel (DMC) W is $X_1 \in \mathbb{F}_2$ and so on. The associated random experiments shall be independent. At the output of the first binary input discrete memoryless channel (DMC) W, the random variable, i. e., the receive symbol, Y_0 is measured at the output of the second binary input discrete memoryless channel (DMC) W, the random variable, i. e., the receive symbol, Y_1 is measured, and so on. The N random variables, i. e., the receive symbols, $Y_0, Y_1\cdots Y_{N-1}$, $N \in \mathbb{N}^*$ form the receive vector $\boldsymbol{Y}$ of (4.2).

Owing to the assumed independence, the average mutual information is given by

$$I\{\boldsymbol{X};\boldsymbol{Y}\} = \sum_{i=0}^{N-1} I\{X_i;\boldsymbol{Y}\} = \sum_{i=0}^{N-1} I\{X_i;Y_i\}, \quad N\in\mathbb{N}^*. \tag{4.11}$$

Consequently, the channel capacity of the combination of the N binary input discrete memoryless channels (DMCs) W is given by

$$C_N = \max_{\mathsf{P}\{\{\boldsymbol{X}\}\}} I\{\boldsymbol{X};\boldsymbol{Y}\}, \quad N\in\mathbb{N}^*. \tag{4.12}$$

The N binary input discrete memoryless channels (DMCs) W form the *combined discrete memoryless channel (DMC)*

$$W_N : \boldsymbol{X} \mapsto \boldsymbol{Y}, \quad \text{i. e.,} \quad W_N : (X_0, X_1\cdots X_{N-1}) \mapsto (Y_0, Y_1\cdots Y_{N-1}), \quad N\in\mathbb{N}^*, \tag{4.13}$$

with N independent random variables $X_0, X_1\cdots X_{N-1}$ at its input and N independent random variables $Y_0, Y_1\cdots Y_{N-1}$ at its output. W_N is also simply called the *combined channel* [33, Figure 4.4, p. 92f.].

Using (4.4), the channel capacity (4.12) of the combined discrete memoryless channel (DMC) W_N becomes

$$C_N = \max_{\mathsf{P}\{\{\boldsymbol{X}\}\}} \sum_{i=0}^{N-1} I\{X_i;Y_i\} = \sum_{i=0}^{N-1} \max_{\mathsf{P}\{\{X_i\}\}} I\{X_i;Y_i\} = NC_1, \quad N\in\mathbb{N}^*. \tag{4.14}$$

The situation described above is depicted in Figure 4.1; cf., e. g., [33, pp. 91f.].

Next, let us look at the channel splitting.

Assert that the maximum conditional average mutual information $\max_{\mathsf{P}\{\{X\}\}} I\{X_i;\boldsymbol{Y}\,|\,X_0,X_1\cdots X_{i-1}\}$ is not equal for all $i\in\{0,1\cdots(N-1)\}$, $N\in\mathbb{N}^*$ again. We call the

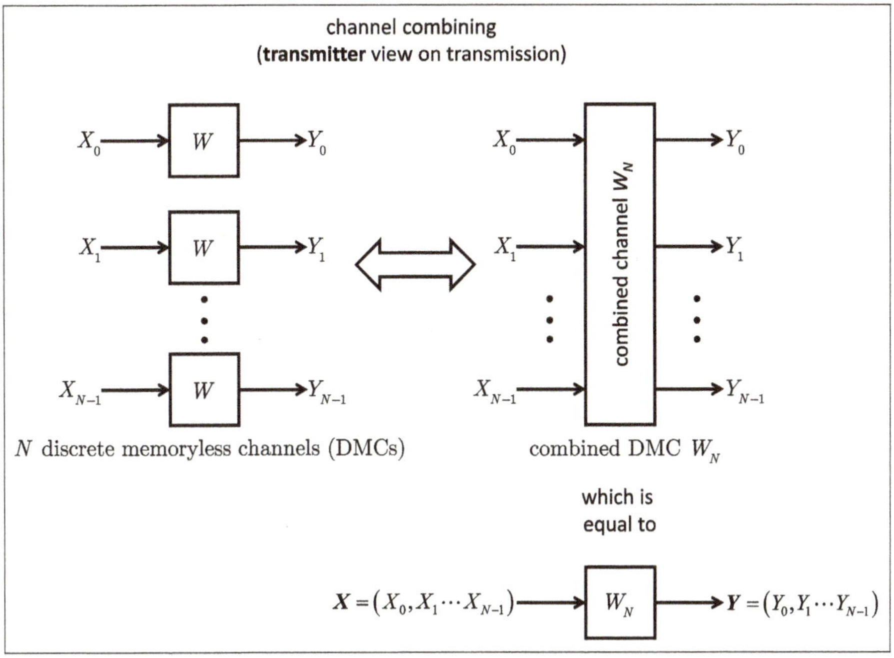

Figure 4.1: Channel combining, i. e., transmission over N discrete memoryless channels (DMCs), forming the combined discrete memoryless channel (DMC) W_N; cf., e. g., [33, p. 91f.].

associated N channels *polarized*. These associated polarized channels are also termed *split virtual channels*. In addition, the split virtual channels are called *split bit channels* or just *split channels* [33, Figure 4.9, pp. 104, 106]. We will now create split virtual channels.

The expression $I\{X_i; \boldsymbol{Y} \mid X_0, X_1 \cdots X_{i-1}\}$, $i \in \{0, 1 \cdots (N-1)\}$, $N \in \mathbb{N}^*$ already tells us the shape of the split virtual channels.

- The first split virtual channel W_N^0 has the input $X_0 \in \mathbb{F}_2$, and the output is given by the receive vector $\boldsymbol{Y}$ of (4.2) prevailing at the output of the combined channel W_N according to Figure 4.1. We yield

$$W_N^0 : X_0 \mapsto \boldsymbol{Y}. \tag{4.15}$$

- The second split virtual channel W_N^1 has the input $X_1 \in \mathbb{F}_2$, and the output is given by the receive vector $\boldsymbol{Y}$ of (4.2) prevailing at the output of the combined channel W_N according to Figure 4.1 and the previously detected transmit symbol $\hat{X}_0$, which we assume to be identical with the actual transmit symbol X_0. We thus have

$$W_N^1 : X_1 \mapsto \boldsymbol{Y}, X_0. \tag{4.16}$$

- The third split virtual channel W_N^2 has the input $X_2 \in \mathbb{F}_2$, and the output is given by the receive vector $\boldsymbol{Y}$ of (4.2) prevailing at the output of the combined channel

W_N according to Figure 4.1 and the previously detected transmit symbols $\hat{X}_0$ and $\hat{X}_1$, which we assume to be identical with the actual transmit symbols X_0 and X_1. We yield

$$W_N^2 : X_2 \mapsto \boldsymbol{Y}, X_0, X_1. \tag{4.17}$$

– ...
– The ith split virtual channel W_N^i, $i \in \{0, 1 \cdots (N-1)\}$, $N \in \mathbb{N}^*$ has the input X_i, $i \in \{0, 1 \cdots (N-1)\}$, $N \in \mathbb{N}^*$, and the output is given by the receive vector $\boldsymbol{Y}$ of (4.2) prevailing at the output of the combined channel W_N according to Figure 4.1 and the previously detected transmit symbols $\hat{X}_0, \hat{X}_1 \cdots \hat{X}_{i-1}$, $i \in \{0, 1 \cdots (N-1)\}$, $N \in \mathbb{N}^*$, which we assume to be identical with the actual transmit symbols $X_0, X_1 \cdots X_{i-1}$, $i \in \{0, 1 \cdots (N-1)\}$, $N \in \mathbb{N}^*$. We thus have

$$W_N^i : X_i \mapsto \boldsymbol{Y}, X_0, X_1 \cdots X_{i-1}, \quad i \in \{0, 1 \cdots (N-1)\}, \quad N \in \mathbb{N}^*. \tag{4.18}$$

– ...
– The Nth split virtual channel W_N^{N-1}, $N \in \mathbb{N}^*$ has the input X_{N-1}, $N \in \mathbb{N}^*$, and the output is given by the receive vector $\boldsymbol{Y}$ of (4.2) prevailing at the output of the combined channel W_N according to Figure 4.1 and the previously detected transmit symbols $\hat{X}_0, \hat{X}_1 \cdots \hat{X}_{N-2}$, $N \in \mathbb{N}^*$, which we assume to be identical with the actual transmit symbols $X_0, X_1 \cdots X_{N-2}$, $N \in \mathbb{N}^*$. We hence obtain

$$W_N^{N-1} : X_{N-1} \mapsto \boldsymbol{Y}, X_0, X_1 \cdots X_{N-2}. \tag{4.19}$$

The split channels described above are illustrated in Figure 4.2; cf., e. g., [33, pp. 103–106].

Clearly, none of the detected transmit symbols $\hat{X}_0, \hat{X}_1 \cdots \hat{X}_{N-2}$, $N \in \mathbb{N}^*$ at the output of any split virtual channel can be considered a “real” channel output. Hence, the approach of channel splitting cannot be considered meaningful without any receiver. It must be seen as the a posteriori point of view on the transmission, i. e., the look onto the transmission from the receiver toward the transmitter rather than from the transmitter to the receiver, the latter view being based on the combined channel discussed above and illustrated in Figure 4.1.

So far, the assertion that the conditional average mutual information values, which are denoted by $I\{X_i; \boldsymbol{Y} \mid X_0, X_1 \cdots X_{i-1}\}$, $i \in \{0, 1 \cdots (N-1)\}$, $N \in \mathbb{N}^*$, vary with $i \in \{0, 1 \cdots (N-1)\}$, $N \in \mathbb{N}^*$ sounds appealing, but can we prove it? Let us see.

However, we first must establish a solid common understanding about the channel capacity of a binary input discrete memoryless channel (DMC). Of course, we know that the channel capacity of any channel is the maximum of the average mutual information and the maximization is done with respect to the probability of the binary transmit signals, i. e., the a priori probability of the source.

Also, we know that in the case of a source generating binary transmit signals, i. e., binary symbols or bits, the source entropy is given by the binary entropy function $\mathrm{H}_2\{q\}$

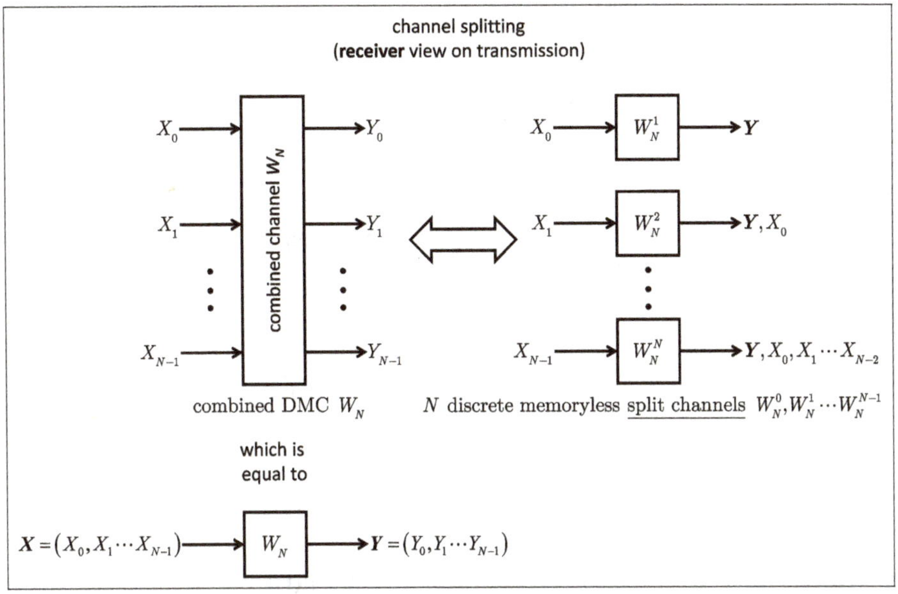

Figure 4.2: Channel splitting; cf., e. g., [33, pp. 103–106].

of (1.216) (cf. Definition 1.30), which assumes its maximum for the a priori probability q of the symbol 0 being equal to 1/2, and hence being equal to the a priori probability $(1-q)$ of the symbol 1; cf. Remark 1.18. In this case, we obtain the channel capacity; cf., e. g., Remark 1.19 in the case of the binary symmetric (noisy) channel (BSC) and Remark 1.20 in the case of the binary erasure channel (BEC). Let us set out from the average mutual information of a binary input discrete memoryless channel (DMC) [7, equation (2.2.7), p. 18] with the input symbol X taken from $\mathbb{F}_2$ and the output symbol Y taken from $\mathbb{Y}$, e. g., $\mathbb{F}_2$ in the case of the binary symmetric (noisy) channel (BSC) or $(\mathbb{F}_2 + \{\mathrm{e}\})$ equal to $\{0, 1, \mathrm{e}\}$ in the case of the binary erasure channel (BEC)

$$
\begin{aligned}
I\{X;Y\} &= \sum_{X \in \mathbb{F}_2} \sum_{Y \in \mathbb{Y}} \mathsf{P}\{\{X\}\{Y\}\} \cdot \log_2 \left\{ \frac{\mathsf{P}\{\{X\}\{Y\}\}}{\mathsf{P}\{\{X\}\} \cdot \mathsf{P}\{\{Y\}\}} \right\} \\
&= \sum_{X \in \mathbb{F}_2} \sum_{Y \in \mathbb{Y}} \mathsf{P}\{\{Y\} \mid \{X\}\} \cdot \mathsf{P}\{\{X\}\} \cdot \log_2 \left\{ \frac{\mathsf{P}\{\{Y\} \mid \{X\}\} \cdot \mathsf{P}\{\{X\}\}}{\mathsf{P}\{\{X\}\} \cdot \mathsf{P}\{\{Y\}\}} \right\} \\
&= \sum_{X \in \mathbb{F}_2} \sum_{Y \in \mathbb{Y}} \mathsf{P}\{\{Y\} \mid \{X\}\} \cdot \mathsf{P}\{\{X\}\} \cdot \log_2 \left\{ \frac{\mathsf{P}\{\{Y\} \mid \{X\}\}}{\sum_{\tilde{X} \in \mathbb{F}_2} \mathsf{P}\{\{Y\}\{\tilde{X}\}\}} \right\} \\
&= \sum_{X \in \mathbb{F}_2} \sum_{Y \in \mathbb{Y}} \mathsf{P}\{\{Y\} \mid \{X\}\} \cdot \mathsf{P}\{\{X\}\} \cdot \log_2 \left\{ \frac{\mathsf{P}\{\{Y\} \mid \{X\}\}}{\sum_{\tilde{X} \in \mathbb{F}_2} \mathsf{P}\{\{Y\} \mid \{\tilde{X}\}\} \cdot \mathsf{P}\{\{\tilde{X}\}\}} \right\} \\
&= \sum_{X \in \mathbb{F}_2} \sum_{Y \in \mathbb{Y}} \mathsf{P}\{\{Y\} \mid \{X\}\} \cdot \mathsf{P}\{\{X\}\}
\end{aligned}
$$

$$\cdot \log_2 \left\{ \frac{\mathsf{P}\{\{Y\} \mid \{X\}\}}{\mathsf{P}\{\{Y\} \mid \{0\}\} \cdot \mathsf{P}\{\{0\}\} + \mathsf{P}\{\{Y\} \mid \{1\}\} \cdot \mathsf{P}\{\{1\}\}} \right\}, \tag{4.20}$$

which assumes its maximum for

$$\mathsf{P}\{\{X\}\} = \mathsf{P}\{\{0\}\} = \mathsf{P}\{\{1\}\} = \frac{1}{2}. \tag{4.21}$$

Denoting the binary input discrete memoryless channel (DMC) by W, we thus obtain its channel capacity [16, p. 3051]

$$\begin{aligned} C\{W\} &= \max_{\mathsf{P}\{\{X\}\}} I\{X; Y\} \\ &= \sum_{X \in \mathbb{F}_2} \sum_{Y \in \mathbb{Y}} \frac{1}{2} \mathsf{P}\{\{Y\} \mid \{X\}\} \cdot \log_2 \left\{ \frac{\mathsf{P}\{\{Y\} \mid \{X\}\}}{\frac{1}{2}\mathsf{P}\{\{Y\} \mid \{0\}\} + \frac{1}{2}\mathsf{P}\{\{Y\} \mid \{1\}\}} \right\} \\ &= \frac{1}{2} \sum_{X \in \mathbb{F}_2} \sum_{Y \in \mathbb{Y}} \mathsf{P}\{\{Y\} \mid \{X\}\} \cdot \log_2 \left\{ \frac{2 \cdot \mathsf{P}\{\{Y\} \mid \{X\}\}}{\mathsf{P}\{\{Y\} \mid \{0\}\} + \mathsf{P}\{\{Y\} \mid \{1\}\}} \right\} \end{aligned} \tag{4.22}$$

also termed *symmetric capacity* [16, p. 3051]. Now, we are prepared for the evaluation of our above assertion; cf., e. g., [16, Proposition 4, pp. 3057, 3071f.].

Theorem 4.1 (Channel Capacity of a Pair of Split Virtual DMCs). *Suppose that two original binary input discrete memoryless channels (DMCs) W were transformed into a pair of binary split virtual discrete memoryless channels (DMCs) W' and W''* [16, *Proposition 4, pp. 3057, 3071f.*]. *Then we have* [16, *Proposition 4, pp. 3057, 3071f.*]

$$C\{W'\} + C\{W''\} = 2C\{W\} \tag{4.23}$$

and

$$C\{W'\} \leq C\{W''\} \tag{4.24}$$

with equality if and only if $C\{W\}$ is equal to 0 *or* 1.

Proof. We will only consider the case $\mathsf{P}\{\{X\}\}$ being equal to 1/2. Since we have two channels W, we have two input symbols, X_0 and X_1 say, and two output symbols, Y_0 and Y_1 say. Then we can write

$$\begin{aligned} C\{W'\} &= I\{X_0; Y_0, Y_1\} \\ &= \sum_{X_0, Y_0, Y_1} \mathsf{P}\{\{X_0\}\{Y_0\}\{Y_1\}\} \log_2 \left\{ \frac{\mathsf{P}\{\{X_0\}\{Y_0\}\{Y_1\}\}}{\mathsf{P}\{\{X_0\}\} \cdot \mathsf{P}\{\{Y_0\}\{Y_1\}\}} \right\} \end{aligned} \tag{4.25}$$

and

$$\begin{aligned} C\{W''\} &= I\{X_1; X_0, Y_0, Y_1\} \\ &= \sum_{X_0, X_1, Y_0, Y_1} \mathsf{P}\{\{X_0\}\{X_1\}\{Y_0\}\{Y_1\}\} \log_2 \left\{ \frac{\mathsf{P}\{\{X_0\}\{X_1\}\{Y_0\}\{Y_1\}\}}{\mathsf{P}\{\{X_0\}\} \cdot \mathsf{P}\{\{X_0\}\{Y_0\}\{Y_1\}\}} \right\}. \end{aligned} \tag{4.26}$$

Furthermore, since X_0 and X_1 are independent, and hence

$$\mathrm{P}\{\{X_1\}\mid\{X_2\}\} = \mathrm{P}\{\{X_1\}\} \tag{4.27}$$

holds, we find

$$\begin{aligned}
I\{X_1;Y_0,Y_1\mid X_0\} &= \sum_{X_0,X_1,Y_0,Y_1} \mathrm{P}\{\{X_0\}\{X_1\}\{Y_0\}\{Y_1\}\} \\
&\quad \cdot \log_2\left\{\frac{\mathrm{P}\{\{X_1\}\{Y_0\}\{Y_1\}\mid\{X_0\}\}}{\mathrm{P}\{\{X_1\}\mid\{X_0\}\}\mathrm{P}\{\{Y_0\}\{Y_1\}\mid\{X_0\}\}}\right\} \\
&= \sum_{X_0,X_1,Y_0,Y_1} \mathrm{P}\{\{X_0\}\{X_1\}\{Y_0\}\{Y_1\}\} \\
&\quad \cdot \log_2\left\{\frac{\mathrm{P}\{\{X_1\}\{Y_0\}\{Y_1\}\mid\{X_0\}\}\mathrm{P}\{\{X_0\}\}}{\mathrm{P}\{\{X_1\}\}\mathrm{P}\{\{Y_0\}\{Y_1\}\mid\{X_0\}\}\mathrm{P}\{\{X_0\}\}}\right\} \\
&= \sum_{X_0,X_1,Y_0,Y_1} \mathrm{P}\{\{X_0\}\{X_1\}\{Y_0\}\{Y_1\}\} \\
&\quad \cdot \log_2\left\{\frac{\mathrm{P}\{\{X_0\}\{X_1\}\{Y_0\}\{Y_1\}\}}{\mathrm{P}\{\{X_1\}\}\mathrm{P}\{\{X_0\}\{Y_0\}\{Y_1\}\}}\right\} \\
&= I\{X_1;X_0,Y_0,Y_1\} \\
&= C\{W''\}.
\end{aligned} \tag{4.28}$$

The chain rule of information (cf., e. g., [7, equations (2.2.28) and (2.2.29), p. 22]) yields

$$I\{X_0,X_1;Y_0,Y_1\} = I\{X_0;Y_0,Y_1\} + I\{X_1;Y_0,Y_1\mid X_0\} = C\{W'\} + C\{W''\}. \tag{4.29}$$

Using (4.11) and (4.14), we also have

$$I\{X_0,X_1;Y_0,Y_1\} = 2\cdot C\{W\}. \tag{4.30}$$

Thus, (4.29) takes the form

$$C\{W'\} + C\{W''\} = 2\cdot C\{W\}. \tag{4.31}$$

With

$$C\{W\} = I\{X_1;Y_1\} = \sum_{X_0,X_1,Y_0,Y_1} \mathrm{P}\{\{X_0\}\{X_1\}\{Y_0\}\{Y_1\}\} \log_2\left\{\frac{\mathrm{P}\{\{X_1\}\{Y_1\}\}}{\mathrm{P}\{\{X_1\}\}\cdot\mathrm{P}\{\{Y_1\}\}}\right\} \tag{4.32}$$

and

$$\begin{aligned}
I\{X_1;X_0,Y_0\mid Y_1\} &= \sum_{X_0,X_1,Y_0,Y_1} \mathrm{P}\{\{X_0\}\{X_1\}\{Y_0\}\{Y_1\}\} \\
&\quad \cdot \log_2\left\{\frac{\mathrm{P}\{\{X_0\}\{X_1\}\{Y_0\}\mid\{Y_1\}\}}{\mathrm{P}\{\{X_1\}\mid\{Y_1\}\}\cdot\mathrm{P}\{\{X_0\}\{Y_0\}\mid\{Y_1\}\}}\right\},
\end{aligned} \tag{4.33}$$

we yield

$$\begin{aligned}
C\{W\} + I\{X_1;X_0,Y_0\mid Y_1\} &= \sum_{X_0,X_1,Y_0,Y_1} \mathrm{P}\{\{X_0\}\{X_1\}\{Y_0\}\{Y_1\}\} \\
&\quad \cdot \left\{\log_2\left\{\frac{\mathrm{P}\{\{X_1\}\{Y_1\}\}}{\mathrm{P}\{\{X_1\}\}\cdot\mathrm{P}\{\{Y_1\}\}}\right\}\right.
\end{aligned}$$

$$
\begin{aligned}
&+\log_2\left\{\frac{\mathrm{P}\{\{X_0\}\{X_1\}\{Y_0\}\mid\{Y_1\}\}}{\mathrm{P}\{\{X_1\}\mid\{Y_1\}\}\cdot\mathrm{P}\{\{X_0\}\{Y_0\}\mid\{Y_1\}\}}\right\}\\
&=\sum_{X_0,X_1,Y_0,Y_1}\mathrm{P}\big\{\{X_0\}\{X_1\}\{Y_0\}\{Y_1\}\big\}\\
&\cdot\log_2\left\{\frac{\mathrm{P}\{\{X_1\}\{Y_1\}\}\mathrm{P}\{\{X_0\}\{X_1\}\{Y_0\}\mid\{Y_1\}\}}{\mathrm{P}\{\{X_1\}\}\cdot\mathrm{P}\{\{X_1\}\{Y_1\}\}\cdot\mathrm{P}\{\{X_0\}\{Y_0\}\mid\{Y_1\}\}}\right\}\\
&=\sum_{X_0,X_1,Y_0,Y_1}\mathrm{P}\big\{\{X_0\}\{X_1\}\{Y_0\}\{Y_1\}\big\}\\
&\cdot\log_2\left\{\frac{\mathrm{P}\{\{X_0\}\{X_1\}\{Y_0\}\{Y_1\}\}}{\mathrm{P}\{\{X_1\}\}\cdot\mathrm{P}\{\{X_0\}\{Y_0\}\{Y_1\}\}}\right\}\\
&=I\{X_1;Y_0,Y_1,X_0\}\\
&=C\big\{W''\big\}.
\end{aligned}
\tag{4.34}
$$

Therefore,

$$C\big\{W''\big\}\geq C\{W\}. \tag{4.35}$$

With (4.31) and (4.35), we obtain

$$2\cdot C\{W\}=C\big\{W''\big\}+C\big\{W'\big\}\geq C\{W\}+C\big\{W'\big\} \tag{4.36}$$

and, therefore,

$$C\{W\}\geq C\big\{W'\big\}. \tag{4.37}$$

Combining (4.35) and (4.37) yield

$$C\big\{W''\big\}\geq C\{W\}\geq C\big\{W'\big\}. \tag{4.38}$$

Equality holds if $I\{X_1;X_0,Y_0\mid Y_1\}$ is equal to 0. According to (4.33), this requires

$$\mathrm{P}\big\{\{X_0\}\{X_1\}\{Y_0\}\mid\{Y_1\}\big\}=\mathrm{P}\big\{\{X_1\}\mid\{Y_1\}\big\}\cdot\mathrm{P}\big\{\{X_0\}\{Y_0\}\mid\{Y_1\}\big\} \tag{4.39}$$

or, equivalently,

$$\mathrm{P}\big\{\{X_0\}\{X_1\}\{Y_0\}\{Y_1\}\big\}\mathrm{P}\big\{\{Y_1\}\big\}=\mathrm{P}\big\{\{X_1\}\{Y_1\}\big\}\cdot\mathrm{P}\big\{\{X_0\}\{Y_0\}\{Y_1\}\big\}, \tag{4.40}$$

i. e.,

$$
\begin{aligned}
\mathrm{P}\big\{\{Y_0\}\{Y_1\}\mid\{X_0\}\{X_1\}\big\}\mathrm{P}\big\{\{Y_1\}\big\}&=\frac{\mathrm{P}\{\{X_0\}\{Y_0\}\{Y_1\}\}}{\mathrm{P}\{\{X_0\}\}}\,\frac{\mathrm{P}\{\{X_1\}\{Y_1\}\}}{\mathrm{P}\{\{X_1\}\}}\\
&=\mathrm{P}\big\{\{Y_0\}\{Y_1\}\mid\{X_0\}\big\}\mathrm{P}\big\{\{Y_1\}\mid\{X_1\}\big\}.
\end{aligned}
\tag{4.41}
$$

Since the elementary events $\{X_0\}$ and $\{X_1\}$ as well as the corresponding elementary events $\{Y_0\}$ and $\{Y_1\}$ are independent, using the generator matrix $\boldsymbol{G}_2$ of (4.8), which yields

$$(X_0,X_1)=(u_0,u_1)\boldsymbol{G}_2=(u_0,u_1)\begin{pmatrix}1&0\\1&1\end{pmatrix}=(u_0\oplus u_1,u_1), \tag{4.42}$$

we obtain

$$\mathrm{P}\big\{\{Y_0\}\{Y_1\}\mid\{X_0\}\{X_1\}\big\}=\mathrm{P}\big\{\{Y_0\}\mid\{u_0\oplus u_1\}\big\}\mathrm{P}\big\{\{Y_1\}\mid\{u_1\}\big\}. \tag{4.43}$$

Equation (4.41) becomes

$$\mathrm{P}\{\{Y_1\} \mid \{u_1\}\}\mathrm{P}\{\{Y_0\} \mid \{u_0 \oplus u_1\}\}\mathrm{P}\{\{Y_1\}\} - \mathrm{P}\{\{Y_0\}\{Y_1\} \mid \{u_0\}\} = 0. \tag{4.44}$$

Owing to

$$\mathrm{P}\{\{Y_1\}\} = \frac{1}{2}\mathrm{P}\{\{Y_1\} \mid \{u_1\}\} + \frac{1}{2}\mathrm{P}\{\{Y_1\} \mid \{u_1 \oplus 1\}\} \tag{4.45}$$

and

$$\begin{aligned}\mathrm{P}\{\{Y_0\}\{Y_1\} \mid \{u_0\}\} &= \frac{1}{2}\mathrm{P}\{\{Y_0\} \mid \{u_0 \oplus u_1\}\}\mathrm{P}\{\{Y_1\} \mid \{u_1\}\} \\ &+ \frac{1}{2}\mathrm{P}\{\{Y_0\} \mid \{u_0 \oplus u_1 \oplus 1\}\}\mathrm{P}\{\{Y_1\} \mid \{u_1 \oplus 1\}\}\end{aligned} \tag{4.46}$$

in the case of equiprobable information bits, we yield

$$\mathrm{P}\{\{Y_1\} \mid \{u_1\}\}\mathrm{P}\{\{Y_1\} \mid \{u_1 \oplus 1\}\}\{\mathrm{P}\{\{Y_0\} \mid \{u_0 \oplus u_1\}\} - \mathrm{P}\{\{Y_0\} \mid \{u_0 \oplus u_1 \oplus 1\}\}\} = 0, \tag{4.47}$$

i. e.,

$$\begin{aligned}
u_0 = 0,\ u_1 = 0, &\Rightarrow \mathrm{P}\{\{Y_1\} \mid \{0\}\}\mathrm{P}\{\{Y_1\} \mid \{1\}\}\{\mathrm{P}\{\{Y_0\} \mid \{0\}\} - \mathrm{P}\{\{Y_0\} \mid \{1\}\}\} = 0, \\
u_0 = 0,\ u_1 = 1, &\Rightarrow \mathrm{P}\{\{Y_1\} \mid \{1\}\}\mathrm{P}\{\{Y_1\} \mid \{0\}\}\{\mathrm{P}\{\{Y_0\} \mid \{1\}\} - \mathrm{P}\{\{Y_0\} \mid \{0\}\}\} = 0, \\
u_0 = 1,\ u_1 = 0, &\Rightarrow \mathrm{P}\{\{Y_1\} \mid \{0\}\}\mathrm{P}\{\{Y_1\} \mid \{1\}\}\{\mathrm{P}\{\{Y_0\} \mid \{1\}\} - \mathrm{P}\{\{Y_0\} \mid \{0\}\}\} = 0, \\
u_0 = 1,\ u_1 = 1, &\Rightarrow \mathrm{P}\{\{Y_1\} \mid \{1\}\}\mathrm{P}\{\{Y_1\} \mid \{0\}\}\{\mathrm{P}\{\{Y_0\} \mid \{0\}\} - \mathrm{P}\{\{Y_0\} \mid \{1\}\}\} = 0,
\end{aligned} \tag{4.48}$$

thus

$$\mathrm{P}\{\{Y_1\} \mid \{0\}\}\mathrm{P}\{\{Y_1\} \mid \{1\}\}\{\mathrm{P}\{\{Y_0\} \mid \{0\}\} - \mathrm{P}\{\{Y_0\} \mid \{1\}\}\} = 0. \tag{4.49}$$

Either $\mathrm{P}\{\{Y_1\} \mid \{0\}\}\mathrm{P}\{\{Y_1\} \mid \{1\}\}$ must vanish, which means that $C\{W\}$ has to be 1 because there are no erasures or errors at all, or $\mathrm{P}\{\{Y_0\} \mid \{0\}\}$ must be equal to $\mathrm{P}\{\{Y_0\} \mid \{1\}\}$, which means that the channel has no capacity at all and $C\{W\}$ has to be 0. □

Theorem 4.1 is a general result, which applies to all discrete memoryless channels (DMCs). Theorem 4.1 allows us to arrange the split virtual channels in an order from very bad to very good. The sequence of the split virtual channels is termed the *reliability sequence* [109, p. 1241].

The argument at the end of the proof of Theorem 4.1 illustrates the *polarization paradigm*. In the context of polar codes, polarization means that there are some split virtual channels with the channel capacity that tends to be equal to 1, and some split virtual channels with the channel capacity that tends to be equal to 0.

Now, let us look at the Bhattachacharyya parameter.

Theorem 4.2 (Bhattachacharyya Parameters of Split Virtual DMCs). *Suppose that two original binary input discrete memoryless channels (DMCs) W were transformed into a pair of binary split virtual discrete memoryless channel (DMC)s W′ and W″. Then the Bhattachacharyya parameters of the split virtual discrete memoryless channel (DMC)s are given by* [16, *p. 3072*]

$$Z(W'') = Z(W)^2 \tag{4.50}$$

and

$$Z(W') \leq 2Z(W) - Z(W)^2 \tag{4.51}$$

with equality in the case W is a binary erasure channel (BEC).

Proof. In general, the output of W'' is a function $f(\boldsymbol{Y}_0^1)$ of $\boldsymbol{Y}_0^1$. Using

$$\mathrm{P}\big\{\{f(\boldsymbol{Y}_0^1)\}\{u_0\} \mid \{u_1\}\big\} = \frac{1}{2} \cdot \mathrm{P}\big\{\{Y_0\} \mid \{u_0 \oplus u_1\}\big\} \cdot \mathrm{P}\big\{, \mid \{Y_1\}\big\}\{u_1\} \tag{4.52}$$

and hence

$$\begin{aligned} \mathrm{P}\big\{\{f(\boldsymbol{Y}_0^1)\}\{u_0\} \mid \{0\}\big\} &= \frac{1}{2} \cdot \mathrm{P}\big\{\{Y_0\} \mid \{u_0\}\big\} \cdot \mathrm{P}\big\{\{Y_1\} \mid \{0\}\big\}, \\ \mathrm{P}\big\{\{f(\boldsymbol{Y}_0^1)\}\{u_0\} \mid \{1\}\big\} &= \frac{1}{2} \cdot \mathrm{P}\big\{\{Y_0\} \mid \{u_0 \oplus 1\}\big\} \cdot \mathrm{P}\big\{\{Y_1\} \mid \{1\}\big\}, \end{aligned} \tag{4.53}$$

the Bhattacharyya parameter $Z(W'')$ is given by

$$\begin{aligned} Z(W'') &= \sum_{\boldsymbol{Y}_0^1, u_0} \sqrt{\mathrm{P}\big\{\{f(\boldsymbol{Y}_0^1)\}\{u_0\} \mid \{0\}\big\}\mathrm{P}\big\{\{f(\boldsymbol{Y}_0^1)\}\{u_0\} \mid \{1\}\big\}} \\ &= \sum_{\boldsymbol{Y}_0^1, u_0} \frac{1}{2} \sqrt{\mathrm{P}\big\{\{Y_0\} \mid \{u_0\}\big\} \cdot \mathrm{P}\big\{\{Y_1\} \mid \{0\}\big\} \cdot \mathrm{P}\big\{\{Y_0\} \mid \{u_0 \oplus 1\}\big\} \cdot \mathrm{P}\big\{\{Y_1\} \mid \{1\}\big\}} \\ &= \underbrace{\sum_{Y_1} \sqrt{\mathrm{P}\big\{\{Y_1\} \mid \{0\}\big\}\mathrm{P}\big\{\{Y_1\} \mid \{1\}\big\}}}_{=Z(W)} \\ &\quad \cdot \frac{1}{2} \sum_{Y_0} \sum_{u_0} \sqrt{\mathrm{P}\big\{\{Y_0\} \mid \{u_0\}\big\}\mathrm{P}\big\{\{Y_0\} \mid \{u_0 \oplus 1\}\big\}} \\ &= Z(W) \cdot \underbrace{\frac{1}{2} \sum_{Y_0} 2 \cdot \sqrt{\mathrm{P}\big\{\{Y_0\} \mid \{0\}\big\}\mathrm{P}\big\{\{Y_0\} \mid \{1\}\big\}}}_{=Z(W)} \\ &= Z(W)^2 = \epsilon^2. \end{aligned} \tag{4.54}$$

Using (4.107), i. e.,

$$\mathrm{P}\big\{\{f(\boldsymbol{Y}_0^1)\} \mid \{u_0\}\big\} = \frac{1}{2} \sum_{u_1} \mathrm{P}\big\{\{Y_0\} \mid \{u_0 \oplus u_1\}\big\}\mathrm{P}\big\{\{Y_1\} \mid \{u_1\}\big\}, \tag{4.55}$$

and hence

$$\begin{aligned} \mathrm{P}\big\{\{f(\boldsymbol{Y}_0^1)\} \mid \{0\}\big\} &= \frac{1}{2}\Big(\mathrm{P}\big\{\{Y_0\} \mid \{0\}\big\}\mathrm{P}\big\{\{Y_1\} \mid \{0\}\big\} + \mathrm{P}\big\{\{Y_0\} \mid \{1\}\big\}\mathrm{P}\big\{\{Y_1\} \mid \{1\}\big\}\Big), \\ \mathrm{P}\big\{\{f(\boldsymbol{Y}_0^1)\} \mid \{1\}\big\} &= \frac{1}{2}\Big(\mathrm{P}\big\{\{Y_0\} \mid \{1\}\big\}\mathrm{P}\big\{\{Y_1\} \mid \{0\}\big\} + \mathrm{P}\big\{\{Y_0\} \mid \{0\}\big\}\mathrm{P}\big\{\{Y_1\} \mid \{1\}\big\}\Big). \end{aligned} \tag{4.56}$$

The Bhattacharyya parameter $Z(W')$ is given by

$$Z(W') = \sum_{\boldsymbol{Y}_0^1} \sqrt{\mathrm{P}\big\{\{f(\boldsymbol{Y}_0^1)\} \mid \{0\}\big\}\mathrm{P}\big\{\{f(\boldsymbol{Y}_0^1)\} \mid \{1\}\big\}}$$

$$= \frac{1}{2} \cdot \sum_{Y_0^1} \sqrt{P\{\{Y_0\} \mid \{0\}\} P\{\{Y_1\} \mid \{0\}\} + P\{\{Y_0\} \mid \{1\}\} P\{\{Y_1\} \mid \{1\}\}} \cdot \sqrt{P\{\{Y_0\} \mid \{1\}\} P\{\{Y_1\} \mid \{0\}\} + P\{\{Y_0\} \mid \{0\}\} P\{\{Y_1\} \mid \{1\}\}}, \tag{4.57}$$

which can be upper bounded by [16, p. 3072]

$$\begin{aligned} Z(W') \leq \frac{1}{2} \Bigg[& \sum_{Y_0^1} \left(\sqrt{P\{\{Y_0\} \mid \{0\}\} P\{\{Y_1\} \mid \{0\}\}} + \sqrt{P\{\{Y_0\} \mid \{1\}\} P\{\{Y_1\} \mid \{1\}\}} \right) \\ & \cdot \left(\sqrt{P\{\{Y_0\} \mid \{0\}\} P\{\{Y_1\} \mid \{1\}\}} + \sqrt{P\{\{Y_0\} \mid \{1\}\} P\{\{Y_1\} \mid \{0\}\}} \right) \Bigg] \\ & - \underbrace{\sum_{Y_0^1} \sqrt{P\{\{Y_0\} \mid \{0\}\} P\{\{Y_1\} \mid \{0\}\} P\{\{Y_0\} \mid \{1\}\} P\{\{Y_1\} \mid \{1\}\}}}_{=Z(W)^2}. \end{aligned} \tag{4.58}$$

We hence yield

$$\begin{aligned} Z(W') \leq \frac{1}{2} \sum_{Y_0^1} \left(\begin{array}{c} (P\{\{Y_0\} \mid \{0\}\} + P\{\{Y_0\} \mid \{1\}\}) \cdot \sqrt{P\{\{Y_1\} \mid \{0\}\} \cdot P\{\{Y_1\} \mid \{1\}\}} \\ + \\ (P\{\{Y_1\} \mid \{0\}\} + P\{\{Y_1\} \mid \{1\}\}) \cdot \sqrt{P\{\{Y_0\} \mid \{0\}\} \cdot P\{\{Y_0\} \mid \{1\}\}} \end{array} \right) \\ - Z(W)^2, \end{aligned} \tag{4.59}$$

i. e.,

$$\begin{aligned} Z(W') \leq \frac{1}{2} \left(\begin{array}{c} 2 \cdot \underbrace{\sum_{Y_1} \sqrt{P\{\{Y_1\} \mid \{0\}\} \cdot P\{\{Y_1\} \mid \{1\}\}}}_{=Z(W)} \\ + \\ 2 \cdot \underbrace{\sum_{Y_0} \sqrt{P\{\{Y_0\} \mid \{0\}\} \cdot P\{\{Y_0\} \mid \{1\}\}}}_{=Z(W)} \end{array} \right) \\ - Z(W)^2. \end{aligned} \tag{4.60}$$

Equation (4.60) thus becomes

$$Z(W') \leq 2Z(W) - Z(W)^2. \tag{4.61}$$

In the case of the binary erasure channel (BEC), we have

$$\begin{aligned} Z(W') = & \frac{1}{2} \sqrt{P\{\{0\} \mid \{0\}\} \cdot P\{\{0\} \mid \{0\}\} + P\{\{0\} \mid \{1\}\} \cdot P\{\{0\} \mid \{1\}\}} \\ & \cdot \sqrt{P\{\{0\} \mid \{1\}\} \cdot P\{\{0\} \mid \{0\}\} + P\{\{0\} \mid \{0\}\} \cdot P\{\{0\} \mid \{1\}\}} \\ & + \frac{1}{2} \sqrt{P\{\{0\} \mid \{0\}\} \cdot P\{\{1\} \mid \{0\}\} + P\{\{0\} \mid \{1\}\} \cdot P\{\{1\} \mid \{1\}\}} \\ & \cdot \sqrt{P\{\{0\} \mid \{1\}\} \cdot P\{\{1\} \mid \{0\}\} + P\{\{0\} \mid \{0\}\} \cdot P\{\{1\} \mid \{1\}\}} \\ & + \frac{1}{2} \sqrt{P\{\{0\} \mid \{0\}\} \cdot P\{\{e\} \mid \{0\}\} + P\{\{0\} \mid \{1\}\} \cdot P\{\{e\} \mid \{1\}\}} \\ & \cdot \sqrt{P\{\{0\} \mid \{1\}\} \cdot P\{\{e\} \mid \{0\}\} + P\{\{0\} \mid \{0\}\} \cdot P\{\{e\} \mid \{1\}\}} \end{aligned}$$

$$
\begin{aligned}
&+\frac{1}{2}\sqrt{P\{\{1\}\mid\{0\}\}\cdot P\{\{0\}\mid\{0\}\}+P\{\{1\}\mid\{1\}\}\cdot P\{\{0\}\mid\{1\}\}}\\
&\cdot\sqrt{P\{\{1\}\mid\{1\}\}\cdot P\{\{0\}\mid\{0\}\}+P\{\{1\}\mid\{0\}\}\cdot P\{\{0\}\mid\{1\}\}}\\
&+\frac{1}{2}\sqrt{P\{\{1\}\mid\{0\}\}\cdot P\{\{1\}\mid\{0\}\}+P\{\{1\}\mid\{1\}\}\cdot P\{\{1\}\mid\{1\}\}}\\
&\cdot\sqrt{P\{\{1\}\mid\{1\}\}\cdot P\{\{1\}\mid\{0\}\}+P\{\{1\}\mid\{0\}\}\cdot P\{\{1\}\mid\{1\}\}}\\
&+\frac{1}{2}\sqrt{P\{\{1\}\mid\{0\}\}\cdot P\{\{e\}\mid\{0\}\}+P\{\{1\}\mid\{1\}\}\cdot P\{\{e\}\mid\{1\}\}}\\
&\cdot\sqrt{P\{\{1\}\mid\{1\}\}\cdot P\{\{e\}\mid\{0\}\}+P\{\{1\}\mid\{0\}\}\cdot P\{\{e\}\mid\{1\}\}}\\
&+\frac{1}{2}\sqrt{P\{\{e\}\mid\{0\}\}\cdot P\{\{0\}\mid\{0\}\}+P\{\{e\}\mid\{1\}\}\cdot P\{\{0\}\mid\{1\}\}}\\
&\cdot\sqrt{P\{\{e\}\mid\{1\}\}\cdot P\{\{0\}\mid\{0\}\}+P\{\{e\}\mid\{0\}\}\cdot P\{\{0\}\mid\{1\}\}}\\
&+\frac{1}{2}\sqrt{P\{\{e\}\mid\{0\}\}\cdot P\{\{1\}\mid\{0\}\}+P\{\{e\}\mid\{1\}\}\cdot P\{\{1\}\mid\{1\}\}}\\
&\cdot\sqrt{P\{\{e\}\mid\{1\}\}\cdot P\{\{1\}\mid\{0\}\}+P\{\{e\}\mid\{0\}\}\cdot P\{\{1\}\mid\{1\}\}}\\
&+\frac{1}{2}\sqrt{P\{\{e\}\mid\{0\}\}\cdot P\{\{e\}\mid\{0\}\}+P\{\{e\}\mid\{1\}\}\cdot P\{\{e\}\mid\{1\}\}}\\
&\cdot\sqrt{P\{\{e\}\mid\{1\}\}\cdot P\{\{e\}\mid\{0\}\}+P\{\{e\}\mid\{0\}\}\cdot P\{\{e\}\mid\{1\}\}},
\end{aligned}
\tag{4.62}
$$

which becomes

$$
\begin{aligned}
Z(W') &= \frac{1}{2}(1-\epsilon)\epsilon+\frac{1}{2}(1-\epsilon)\epsilon+\frac{1}{2}(1-\epsilon)\epsilon+\frac{1}{2}(1-\epsilon)\epsilon+\frac{1}{2}\epsilon\sqrt{2}\cdot\epsilon\sqrt{2}\\
&= 2(1-\epsilon)\epsilon+\epsilon^2\\
&= 2\epsilon-\epsilon^2\\
&= 2Z(W)-Z(W)^2.
\end{aligned}
\tag{4.63}
$$

□

Theorem 4.2 is a general result, which applies to all discrete memoryless channels (DMCs). However, there is a special case, namely the binary erasure channel (BEC), which exhibits multiplicative noise and is hence a particularly important kind of fading channel, and thus remarkable for advanced mobile communications. In the case of the binary erasure channel (BEC), we find (cf., e. g., [33, equation (2.65), pp. 58, 88])

$$C\{W\}=1-Z(W), \tag{4.64}$$

i. e., the channel capacity $C\{W\}$ and the Bhattachacharyya parameter $Z(W)$ are oppositely related to each other [33, p. 88]; hence we find

$$
\begin{aligned}
C\{W''\} &= 1-Z(W') = 1-2\epsilon+\epsilon^2=(1-\epsilon)^2=C\{W\}^2,\\
C\{W'\} &= 1-Z(W'') = 1-\epsilon^2 = 2C\{W\}-C\{W\}^2.
\end{aligned}
\tag{4.65}
$$

4.2 Recursive Definition of Generator Matrices

We will set out from the 2×2 matrix

$$\boldsymbol{G}_2 = \begin{pmatrix} 1 & 0 \\ 1 & 1 \end{pmatrix} \in \mathbb{F}_2^{2\times 2}; \tag{4.66}$$

see also (4.8). Referring to Section 1.16, we obtain

$$\begin{aligned} \boldsymbol{G}_2^{\otimes\mu} &= \boldsymbol{G}_2^{\otimes(\mu-1)} \otimes \boldsymbol{G}_2 = \boldsymbol{\Pi}_N(\boldsymbol{G}_2 \otimes \boldsymbol{G}_2^{\otimes(\mu-1)})\boldsymbol{\Pi}_N^{\mathrm{T}} = \boldsymbol{\Pi}_N \boldsymbol{G}_2^{\otimes\mu}\boldsymbol{\Pi}_N^{\mathrm{T}}, \\ &\mu = \log_2\{N\}, \quad N \in \{2, 4, 8, 16 \cdots 2^\mu \cdots\}, \quad \mu \in \mathbb{N}^*. \end{aligned} \tag{4.67}$$

Let us now define the $N \times N$ polar code generator matrix

$$\begin{aligned} \boldsymbol{G}_N &= \boldsymbol{B}_N \boldsymbol{G}_2^{\otimes\mu} = \boldsymbol{\Pi}_N(\boldsymbol{I}_2 \otimes \boldsymbol{B}_{N/2})\boldsymbol{G}_2^{\otimes\mu}, \\ &\mu = \log_2\{N\}, \quad N \in \{2, 4, 8, 16 \cdots 2^\mu \cdots\}, \quad \mu \in \mathbb{N}^*. \end{aligned} \tag{4.68}$$

Furthermore, we can write

$$\begin{aligned} \boldsymbol{G}_N &= \boldsymbol{\Pi}_N(\boldsymbol{I}_2 \otimes \boldsymbol{B}_{N/2})\boldsymbol{G}_2^{\otimes\mu} \\ &= \boldsymbol{\Pi}_N(\boldsymbol{I}_2 \otimes \boldsymbol{B}_{N/2})(\boldsymbol{G}_2 \otimes \boldsymbol{G}_2^{\otimes(\mu-1)}) \\ &= \boldsymbol{\Pi}_N(\boldsymbol{G}_2 \otimes \boldsymbol{B}_{N/2}\boldsymbol{G}_2^{\otimes(\mu-1)}) \\ &= \boldsymbol{\Pi}_N(\boldsymbol{G}_2 \otimes \boldsymbol{G}_{N/2}) \\ &= \boldsymbol{\Pi}_N(\boldsymbol{G}_2\boldsymbol{I}_2 \otimes \boldsymbol{I}_{N/2}\boldsymbol{G}_{N/2}), \\ &\mu = \log_2\{N\}, \quad N \in \{2, 4, 8, 16 \cdots 2^\mu \cdots\}, \quad \mu \in \mathbb{N}^*, \end{aligned} \tag{4.69}$$

and thus

$$\begin{aligned} \boldsymbol{G}_N &= \boldsymbol{\Pi}_N(\boldsymbol{G}_2 \otimes \boldsymbol{I}_{N/2})(\boldsymbol{I}_2 \otimes \boldsymbol{G}_{N/2}), \\ &N \in \{2, 4, 8, 16 \cdots 2^\mu \cdots\}, \quad \mu \in \mathbb{N}^*. \end{aligned} \tag{4.70}$$

Again, referring to Section 1.16, we obtain

$$\boldsymbol{I}_{N/2} \otimes \boldsymbol{G}_2 = \boldsymbol{\Pi}_N(\boldsymbol{G}_2 \otimes \boldsymbol{I}_{N/2})\boldsymbol{\Pi}_N^{\mathrm{T}}, \tag{4.71}$$

which immediately becomes

$$\boldsymbol{\Pi}_N(\boldsymbol{G}_2 \otimes \boldsymbol{I}_{N/2}) = (\boldsymbol{I}_{N/2} \otimes \boldsymbol{G}_2)\boldsymbol{\Pi}_N, \tag{4.72}$$

leading to

$$\boldsymbol{G}_N = (\boldsymbol{I}_{N/2} \otimes \boldsymbol{G}_2)\boldsymbol{\Pi}_N(\boldsymbol{I}_2 \otimes \boldsymbol{G}_{N/2}). \tag{4.73}$$

Remark 4.1 (Relationship With the Fourier Matrices). In the context of (4.73), let us remember the discussion in [41, equations (2.5.15)–(2.5.22), Section 2.5, p. 36f.] of the *Fourier matrices* and in [41, Section 2.6, problem 3, p. 39] of the *Hadamard matrices*.

Combining the findings of [41, equations (2.5.15)–(2.5.22), Section 2.5, p. 36f.] with *Rademacher functions* [54, Chapter 3.1, pp. 46–49], [60] (cf., e. g., Definition 1.67) introduced in the context of Reed–Muller (RM) codes and some of their properties discussed in Section 1.17.2 clearly, it cannot be denied that polar codes are descendants of Reed–Muller (RM) codes.

Also, the mathematical treatment of the recursive definition of the generator matrices of polar codes illustrated in Section 4.2 can be anticipated in, e. g., [41, equations (2.5.15)–(2.5.22), Section 2.5, p. 36f.] and [41, Section 2.6, problem 3, p. 39].

Using the message vector,

$$\boldsymbol{u} = (u_0, u_1 \cdots u_{N-1}), \quad N \in \mathbb{N}^*, \tag{4.74}$$

computing

$$\boldsymbol{u}\boldsymbol{G}_N = \boldsymbol{u}(\boldsymbol{I}_{N/2} \otimes \boldsymbol{G}_2)\boldsymbol{\Pi}_N(\boldsymbol{I}_2 \otimes \boldsymbol{G}_{N/2}) \tag{4.75}$$

means that first we are using $N/2$ local polar transformations, represented by $\boldsymbol{G}_2$, followed by a reverse shuffle $\boldsymbol{\Pi}_N$ equal to $\boldsymbol{\Sigma}^{\mathrm{T}}_{2,\frac{N}{2}}$ and finally two parallel $\boldsymbol{G}_{N/2}$ polar transformations.

We will now show by induction that

$$\begin{aligned} \boldsymbol{G}_N &= \boldsymbol{B}_N \boldsymbol{G}_2^{\otimes\mu} = \boldsymbol{G}_2^{\otimes\mu}\boldsymbol{B}_N, \\ &\mu = \log_2\{N\}, \quad N \in \{2, 4, 8, 16 \cdots 2^\mu \cdots\}, \quad \mu \in \mathbb{N}^*, \end{aligned} \tag{4.76}$$

holds, i. e., $\boldsymbol{B}_N$ commutes with $\boldsymbol{G}_2^{\otimes\mu}$.

Let us begin with the base case. For N equal to 2, we already know that $\boldsymbol{B}_2$ is equal to $\boldsymbol{B}_2^{\mathrm{T}}$, which is the 2×2 identity matrix $\boldsymbol{I}_2$. Of course, $\boldsymbol{B}_2$ commutes with $\boldsymbol{G}_2$.

In the induction step, we set out from

$$\begin{aligned} \boldsymbol{G}_{N/2}\boldsymbol{B}_{N/2} &= \boldsymbol{B}_{N/2}\boldsymbol{G}_2^{\otimes(\mu-1)}\boldsymbol{B}_{N/2} = \boldsymbol{G}_2^{\otimes(\mu-1)}\boldsymbol{B}_{N/2}\boldsymbol{B}_{N/2} = \boldsymbol{G}_2^{\otimes(\mu-1)}, \\ &\mu = \log_2\{N\}, \quad N \in \{2, 4, 8, 16 \cdots 2^\mu \cdots\}, \quad \mu \in \mathbb{N}^*. \end{aligned} \tag{4.77}$$

Let us now determine

$$\begin{aligned} \boldsymbol{G}_N\boldsymbol{B}_N &= \boldsymbol{B}_N\boldsymbol{G}_2^{\otimes\mu}\boldsymbol{B}_N \\ &= \boldsymbol{\Pi}_N(\boldsymbol{I}_2 \otimes \boldsymbol{B}_{N/2})\boldsymbol{G}_2^{\otimes\mu}(\boldsymbol{I}_2 \otimes \boldsymbol{B}_{N/2})\boldsymbol{\Pi}_N^{\mathrm{T}} \\ &= (\boldsymbol{B}_{N/2} \otimes \boldsymbol{I}_2)\boldsymbol{\Pi}_N\boldsymbol{G}_2^{\otimes\mu}\boldsymbol{\Pi}_N^{\mathrm{T}}(\boldsymbol{B}_{N/2} \otimes \boldsymbol{I}_2) \\ &= (\boldsymbol{B}_{N/2} \otimes \boldsymbol{I}_2)\boldsymbol{G}_2^{\otimes\mu}(\boldsymbol{B}_{N/2} \otimes \boldsymbol{I}_2) \\ &= (\boldsymbol{B}_{N/2} \otimes \boldsymbol{I}_2)(\boldsymbol{G}_2^{\otimes(\mu-1)} \otimes \boldsymbol{G}_2)(\boldsymbol{B}_{N/2} \otimes \boldsymbol{I}_2) \\ &= (\boldsymbol{B}_{N/2}\boldsymbol{G}_2^{\otimes(\mu-1)} \otimes \boldsymbol{G}_2)(\boldsymbol{B}_{N/2} \otimes \boldsymbol{I}_2) \end{aligned}$$

$$
\begin{aligned}
&= \boldsymbol{B}_{N/2}\boldsymbol{G}_2^{\otimes(\mu-1)}\boldsymbol{B}_{N/2} \otimes \boldsymbol{G}_2 \\
&= \boldsymbol{G}_2^{\otimes(\mu-1)} \otimes \boldsymbol{G}_2 \\
&= \boldsymbol{G}_2^{\otimes\mu}, \quad \mu = \log_2\{N\}, \quad N \in \{2,4,8,16\cdots 2^{\mu}\cdots\}, \quad \mu \in \mathbb{N}^*,
\end{aligned} \tag{4.78}
$$

which completes our proof.

Now, using (4.68), i. e., $\boldsymbol{G}_N = \boldsymbol{B}_N\boldsymbol{G}_2^{\otimes\mu}$, and (1.1232), i. e., $\boldsymbol{B}_N = (\boldsymbol{I}_2 \otimes \boldsymbol{B}_{N/2})\Pi_N^{\mathrm{T}}$, we can write

$$
\begin{aligned}
\boldsymbol{G}_N &= \boldsymbol{G}_2^{\otimes\mu}\boldsymbol{B}_N \\
&= \boldsymbol{G}_2^{\otimes\mu}(\boldsymbol{I}_2 \otimes \boldsymbol{B}_{N/2})\boldsymbol{\Pi}_N^{\mathrm{T}} \\
&= (\boldsymbol{G}_2 \otimes \boldsymbol{G}_2^{\otimes(\mu-1)})(\boldsymbol{I}_2 \otimes \boldsymbol{B}_{N/2})\boldsymbol{\Pi}_N^{\mathrm{T}} \\
&= (\boldsymbol{G}_2 \otimes \boldsymbol{G}_2^{\otimes(\mu-1)}\boldsymbol{B}_{N/2})\boldsymbol{\Pi}_N^{\mathrm{T}} \\
&= (\boldsymbol{G}_2 \otimes \boldsymbol{G}_{N/2})\boldsymbol{\Pi}_N^{\mathrm{T}} \\
&= (\boldsymbol{I}_2\boldsymbol{G}_2 \otimes \boldsymbol{G}_{N/2}\boldsymbol{I}_{N/2})\boldsymbol{\Pi}_N^{\mathrm{T}} \\
&= (\boldsymbol{I}_2 \otimes \boldsymbol{G}_{N/2})(\boldsymbol{G}_2 \otimes \boldsymbol{I}_{N/2})\boldsymbol{\Pi}_N^{\mathrm{T}}, \\
&\quad \mu = \log_2\{N\}, \quad N \in \{2,4,8,16\cdots 2^{\mu}\cdots\}, \quad \mu \in \mathbb{N}^*.
\end{aligned} \tag{4.79}
$$

Once more referring to Section 1.16, we yield

$$
(\boldsymbol{G}_2 \otimes \boldsymbol{I}_{N/2})\boldsymbol{\Pi}_N^{\mathrm{T}} = \boldsymbol{\Pi}_N^{\mathrm{T}}(\boldsymbol{I}_{N/2} \otimes \boldsymbol{G}_2). \tag{4.80}
$$

Hence, (4.79) becomes

$$
\boldsymbol{G}_N = (\boldsymbol{I}_2 \otimes \boldsymbol{G}_{N/2})\boldsymbol{\Pi}_N^{\mathrm{T}}(\boldsymbol{I}_{N/2} \otimes \boldsymbol{G}_2). \tag{4.81}
$$

With the message vector $\boldsymbol{u}$ of (4.74), computing

$$
\boldsymbol{u}\boldsymbol{G}_N = \boldsymbol{u}(\boldsymbol{I}_2 \otimes \boldsymbol{G}_{N/2})\boldsymbol{\Pi}_N^{\mathrm{T}}(\boldsymbol{I}_{N/2} \otimes \boldsymbol{G}_2) \tag{4.82}
$$

means that first we carry out two parallel polar transformations, represented by $\boldsymbol{G}_{N/2}$, then a shuffle $\boldsymbol{\Pi}_N^{\mathrm{T}}$ equal to $\boldsymbol{\Sigma}_{2,\frac{N}{2}}$ and finally $N/2$ local polar transformations, given by $\boldsymbol{G}_2$.

In summary, the polar code generator matrix can be represented by

$$
\begin{aligned}
\boldsymbol{G}_N &= \boldsymbol{B}_N\boldsymbol{G}_2^{\otimes\mu} && (4.83)\\
&= \boldsymbol{\Pi}_N(\boldsymbol{I}_2 \otimes \boldsymbol{B}_{N/2})\boldsymbol{G}_2^{\otimes\mu} && (4.84)\\
&= (\boldsymbol{I}_2 \otimes \boldsymbol{G}_{N/2})\boldsymbol{\Pi}_N^{\mathrm{T}}(\boldsymbol{I}_{N/2} \otimes \boldsymbol{G}_2) && (4.85)\\
&= (\boldsymbol{I}_{N/2} \otimes \boldsymbol{G}_2)\boldsymbol{\Pi}_N(\boldsymbol{I}_2 \otimes \boldsymbol{G}_{N/2}), && (4.86)\\
&\quad \mu = \log_2\{N\}, \quad N \in \{2,4,8,16\cdots 2^{\mu}\cdots\}, \quad \mu \in \mathbb{N}^*.
\end{aligned}
$$

The corresponding polar code based on (4.86) is illustrated in Figure 4.3.

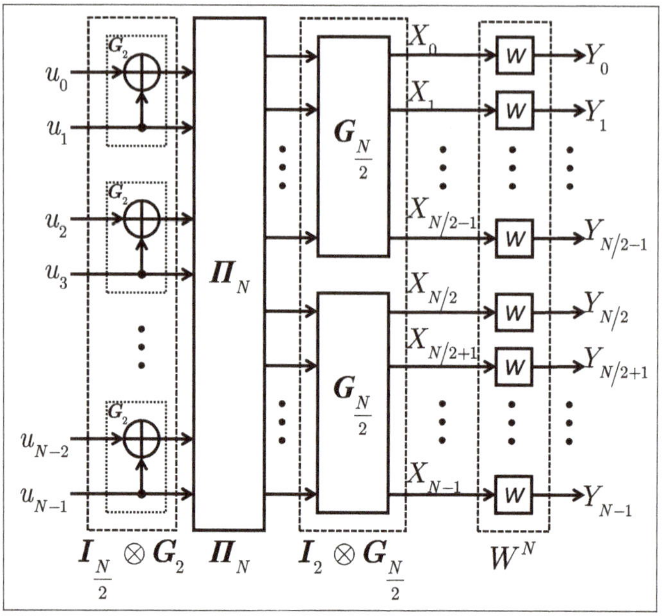

Figure 4.3: Polar code based on (4.86) for $N = 2^{\mu}$, $\mu \in \mathbb{N}^*$ (adapted according to [33, Figure 4.7, p. 98]).

4.3 Channel Combining Revisited

Since encoding will now enter the coding "playing field," we first need to introduce the concept of the raw combined discrete memoryless channel (DMC)

$$W^N : \boldsymbol{X} \mapsto \boldsymbol{Y}, \quad \text{i. e.,} \quad W^N : (X_0, X_1 \cdots X_{N-1}) \mapsto (Y_0, Y_1 \cdots Y_{N-1}), \qquad (4.87)$$
$$\mu = \log_2\{N\}, \quad N \in \{2, 4, 8, 16 \cdots 2^{\mu} \cdots\}, \quad \mu \in \mathbb{N}^*,$$

which consists of N binary input discrete memoryless channels (DMCs) being combined to form W^N. The raw combined discrete memoryless channel (DMC) W^N of (4.87) has the input vector $\boldsymbol{X}$ given by (4.1) and the output vector $\boldsymbol{Y}$ introduced by (4.2). The combined discrete memoryless channel (DMC)

$$W_N : \boldsymbol{u} \mapsto \boldsymbol{Y}, \qquad (4.88)$$

has the message vector $\boldsymbol{u}$ given by (4.74) as its input, and the output vector is $\boldsymbol{Y}$ like in the case of the raw combined discrete memoryless channel (DMC) W^N. The message vector $\boldsymbol{u}$ consists of the elementary events $\{u_i\}$, $i \in \{0, 1 \cdots (N-1)\}$, $N \in \{2, 4, 8, 16 \cdots 2^{\mu} \cdots\}$, $\mu \in \mathbb{N}^*$ associated with the message bits $u_i \in \mathbb{F}_2$, $i \in \{0, 1 \cdots (N-1)\}$, $N \in \{2, 4, 8, 16 \cdots 2^{\mu} \cdots\}$, $\mu \in \mathbb{N}^*$, which are the components of the message vector $\boldsymbol{u}$. Hence, the code length n is equal to N, which is a power of 2 [16, p. 3052].

The relation between the raw combined discrete memoryless channel (DMC) W^N and the combined discrete memoryless channel (DMC) W_N is illustrated in Figure 4.4.

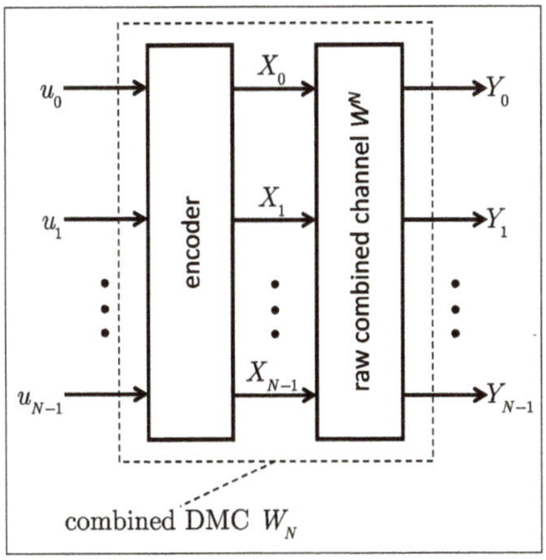

Figure 4.4: Combined discrete memoryless channel (DMC) W_N and raw combined discrete memoryless channel (DMC) W^N.

4.4 Channel Splitting Revisited

Setting out from the raw combined discrete memoryless channel (DMC) W^N, i. e., hence having synthesized the combined discrete memoryless channel (DMC) W_N, the next step of the channel polarization is to split back into a set of binary input coordinate channels

$$\begin{aligned}
W_N^0 &: u_0 \mapsto \boldsymbol{Y}, \\
W_N^1 &: u_1 \mapsto \boldsymbol{Y}, u_0, \\
W_N^2 &: u_2 \mapsto \boldsymbol{Y}, u_0, u_1, \\
&\vdots \\
W_N^i &: u_i \mapsto \boldsymbol{Y}, u_0, u_1 \cdots u_{i-1}, \quad 2 < i < (N-1), \\
&\vdots \\
W_N^{N-1} &: u_{N-1} \mapsto \boldsymbol{Y}, u_0, u_1 \cdots u_{N-2}.
\end{aligned} \tag{4.89}$$

Now, we have to consider the transition probabilities of these channels in order to evaluate the performance.

Theorem 4.3 (Transition Probability of a Split Virtual Channel). *In the case of u_i equal to $\{0\}$ or equal to $\{1\}$ being equally probable, the split virtual channel W_N^i has the transition probability*

$$\mathrm{P}_N^i\left\{\{\boldsymbol{Y}\}\{u_0\}\cdots\{u_{i-1}\} \mid \{u_i\}\right\} = \frac{1}{2^{N-1}} \sum_{u_{i+1}\cdots u_{N-1}} \mathrm{P}_N\left\{\{\boldsymbol{Y}\} \mid \{\boldsymbol{u}\}\right\}, \tag{4.90}$$

$$i \in \left\{0, 1 \cdots (N-1)\right\}, \quad N \in \left\{2, 4, 8, 16 \cdots 2^{\mu} \cdots\right\}, \quad \mu \in \mathbb{N}^*.$$

Proof. The joint probability $\mathrm{P}\{\{\boldsymbol{Y}\}\{u_0\}\cdots\{u_{i-1}\}\{u_i\}\}$ is given by

$$\mathrm{P}\Big\{\{\boldsymbol{Y}\}\{u_0\}\cdots\{u_{i-1}\}\{u_i\}\Big\} = \sum_{u_{i+1}\cdots u_{N-1}} \mathrm{P}_N\Big\{\{\boldsymbol{Y}\}\{\boldsymbol{u}\}\Big\}, \tag{4.91}$$
$$i \in \Big\{0, 1 \cdots (N-1)\Big\}, \quad N \in \Big\{2, 4, 8, 16 \cdots 2^{\mu} \cdots\Big\}, \quad \mu \in \mathbb{N}^*.$$

Since

$$\mathrm{P}\Big\{\{\boldsymbol{Y}\}\{u_0\}\cdots\{u_{i-1}\}\{u_i\}\Big\} = \mathrm{P}_N^i\Big\{\{\boldsymbol{Y}\}\{u_0\}\cdots\{u_{i-1}\} \mid \{u_i\}\Big\} \cdot \mathrm{P}\Big\{\{u_i\}\Big\} \tag{4.92}$$
$$i \in \Big\{0, 1 \cdots (N-1)\Big\}, \quad N \in \Big\{2, 4, 8, 16 \cdots 2^{\mu} \cdots\Big\}, \quad \mu \in \mathbb{N}^*$$

holds, (4.91) leads to

$$\begin{aligned}\mathrm{P}_N^i\Big\{\{\boldsymbol{Y}\}\{u_0\}\cdots\{u_{i-1}\} \mid \{u_i\}\Big\} &= \frac{\mathrm{P}\{\{\boldsymbol{Y}\}\{u_0\}\cdots\{u_{i-1}\}\{u_i\}\}}{\mathrm{P}\{\{u_i\}\}} \\ &= \frac{\sum_{u_{i+1}\cdots u_{N-1}} \mathrm{P}_N\{\{\boldsymbol{Y}\}\{\boldsymbol{u}\}\}}{\mathrm{P}\{\{u_i\}\}},\end{aligned} \tag{4.93}$$
$$i \in \Big\{0, 1 \cdots (N-1)\Big\}, \quad N \in \Big\{2, 4, 8, 16 \cdots 2^{\mu} \cdots\Big\}, \quad \mu \in \mathbb{N}^*.$$

Using

$$\mathrm{P}_N\Big\{\{\boldsymbol{Y}\}\{\boldsymbol{u}\}\Big\} = \mathrm{P}_N\Big\{\{\boldsymbol{Y}\} \mid \{\boldsymbol{u}\}\Big\} \cdot \mathrm{P}\Big\{\{\boldsymbol{u}\}\Big\}, \tag{4.94}$$

(4.93) leads to

$$\mathrm{P}_N^i\Big\{\{\boldsymbol{Y}\}\{u_0\}\cdots\{u_{i-1}\} \mid \{u_i\}\Big\} = \frac{\sum_{u_{i+1}\cdots u_{N-1}} \mathrm{P}_N\{\{\boldsymbol{Y}\} \mid \{\boldsymbol{u}\}\} \cdot \mathrm{P}\{\{\boldsymbol{u}\}\}}{\mathrm{P}\{\{u_i\}\}}, \tag{4.95}$$
$$i \in \Big\{0, 1 \cdots (N-1)\Big\}, \quad N \in \Big\{2, 4, 8, 16 \cdots 2^{\mu} \cdots\Big\}, \quad \mu \in \mathbb{N}^*.$$

Since all elementary events $\{u_i\}$, $i \in \{0, 1 \cdots (N-1)\}$, $N \in \{2, 4, 8, 16 \cdots 2^{\mu} \cdots\}$, $\mu \in \mathbb{N}^*$ are considered to be independent, and assuming that u_i equal to 0 and u_i equal to 1 are equally likely, i. e.,

$$\mathrm{P}\Big\{\{u_i = 0\}\Big\} = \mathrm{P}\Big\{\{u_i = 1\}\Big\} = \frac{1}{2}, \tag{4.96}$$
$$i \in \Big\{0, 1 \cdots (N-1)\Big\}, \quad N \in \Big\{2, 4, 8, 16 \cdots 2^{\mu} \cdots\Big\}, \quad \mu \in \mathbb{N}^*,$$

we find

$$\mathrm{P}\Big\{\{\boldsymbol{u}\}\Big\} = \prod_{i=0}^{N-1} \mathrm{P}\Big\{\{u_i\}\Big\} = \frac{1}{2^N}, \quad N \in \Big\{2, 4, 8, 16 \cdots 2^{\mu} \cdots\Big\}, \quad \mu \in \mathbb{N}^*. \tag{4.97}$$

Equation (4.95) becomes

$$\begin{aligned}\mathrm{P}_N^i\Big\{\{\boldsymbol{Y}\}\{u_0\}\cdots\{u_{i-1}\} \mid \{u_i\}\Big\} &= \frac{\sum_{u_{i+1}\cdots u_{N-1}} \mathrm{P}_N\{\{\boldsymbol{Y}\} \mid \{\boldsymbol{u}\}\} \cdot \frac{1}{2^N}}{\frac{1}{2}} \\ &= \frac{1}{2^{N-1}} \sum_{u_{i+1}\cdots u_{N-1}} \mathrm{P}_N\Big\{\{\boldsymbol{Y}\} \mid \{\boldsymbol{u}\}\Big\},\end{aligned} \tag{4.98}$$
$$i \in \Big\{0, 1 \cdots (N-1)\Big\}, \quad N \in \Big\{2, 4, 8, 16 \cdots 2^{\mu} \cdots\Big\}, \quad \mu \in \mathbb{N}^*. \qquad \square$$

Therefore, we have

$$\mathsf{P}_N^0\{\{\boldsymbol{Y}\}\mid\{u_0\}\} = \frac{1}{2^{N-1}}\sum_{u_1,u_2\cdots u_{N-1}} \mathsf{P}_N\{\{\boldsymbol{Y}\}\mid\{\boldsymbol{u}\}\},$$

$$\mathsf{P}_N^1\{\{\boldsymbol{Y}\}\{u_0\}\mid\{u_1\}\} = \frac{1}{2^{N-1}}\sum_{u_2,u_3\cdots u_{N-1}} \mathsf{P}_N\{\{\boldsymbol{Y}\}\mid\{\boldsymbol{u}\}\},$$

$$\mathsf{P}_N^2\{\{\boldsymbol{Y}\}\{u_0\}\{u_1\}\mid\{u_2\}\} = \frac{1}{2^{N-1}}\sum_{u_3,u_4\cdots u_{N-1}} \mathsf{P}_N\{\{\boldsymbol{Y}\}\mid\{\boldsymbol{u}\}\},$$

$$\vdots$$

$$\mathsf{P}_N^i\{\{\boldsymbol{Y}\}\{u_0\}\{u_1\}\cdots\{u_{i-1}\}\mid\{u_i\}\} = \frac{1}{2^{N-1}}\sum_{u_{i+1},u_{i+2}\cdots u_{N-1}} \mathsf{P}_N\{\{\boldsymbol{Y}\}\mid\{\boldsymbol{u}\}\}, \quad (4.99)$$

$$3 < n < N-1,$$

$$\vdots$$

$$\mathsf{P}_N^{N-2}\{\{\boldsymbol{Y}\}\{u_0\}\{u_1\}\cdots\{u_{N-3}\}\mid\{u_{N-2}\}\} = \frac{1}{2^{N-1}}\sum_{u_{N-1}} \mathsf{P}_N\{\{\boldsymbol{Y}\}\mid\{\boldsymbol{u}\}\},$$

$$\mathsf{P}_N^{N-1}\{\{\boldsymbol{Y}\}\{u_0\}\{u_1\}\cdots\{u_{N-2}\}\mid\{u_{N-1}\}\} = \frac{1}{2^{N-1}}\mathsf{P}_N\{\{\boldsymbol{Y}\}\mid\{\boldsymbol{u}\}\},$$

where $(\boldsymbol{Y}, u_0, u_1\cdots u_{i-1})$ denotes the output of W_N^i and u_i denotes its input [16, p. 3053]. Using

$$\boldsymbol{u}_a^b = (u_a, u_{a+1}\cdots u_b),\quad \boldsymbol{X}_a^b = (X_a, X_{a+1}\cdots X_b),\quad \boldsymbol{Y}_a^b = (Y_a, Y_{a+1}\cdots Y_b), \quad (4.100)$$
$$a \le b,\quad a,b\in\{0,1\cdots(N-1)\},\quad N\in\mathbb{N}^*,$$

replacing $\mathsf{P}_N\{\{\boldsymbol{Y}\}\mid\{\boldsymbol{u}\}\}$ by $\mathsf{P}_N\{\{\boldsymbol{Y}_0^{N-1}\}\mid\{\boldsymbol{u}_0^{N-1}\}\}$, we can write [16, equation (4), p. 3052]

$$\mathsf{P}_N\{\{\boldsymbol{Y}_0^{N-1}\}\mid\{\boldsymbol{u}_0^{N-1}\}\} = \mathsf{P}^N\{\{\boldsymbol{Y}_0^{N-1}\}\mid\{\boldsymbol{X}_0^{N-1}\}\} = \mathsf{P}^N\{\{\boldsymbol{Y}_0^{N-1}\}\mid\{\boldsymbol{u}_0^{N-1}\boldsymbol{G}_N\}\}. \quad (4.101)$$

Therefore, (4.99) can be written as

$$\mathsf{P}_N^0\{\{\boldsymbol{Y}_0^{N-1}\}\mid\{u_0\}\} = \frac{1}{2^{N-1}}\sum_{\boldsymbol{u}_1^{N-1}} \mathsf{P}^N\{\{\boldsymbol{Y}_0^{N-1}\}\mid\{\boldsymbol{u}_0^{N-1}\boldsymbol{G}_N\}\},$$

$$\mathsf{P}_N^1\{\{\boldsymbol{Y}_0^{N-1}\}\{u_0\}\mid\{u_1\}\} = \frac{1}{2^{N-1}}\sum_{\boldsymbol{u}_2^{N-1}} \mathsf{P}^N\{\{\boldsymbol{Y}_0^{N-1}\}\mid\{\boldsymbol{u}_0^{N-1}\boldsymbol{G}_N\}\},$$

$$\mathsf{P}_N^2\{\{\boldsymbol{Y}_0^{N-1}\}\{\boldsymbol{u}_0^1\}\mid\{u_2\}\} = \frac{1}{2^{N-1}}\sum_{\boldsymbol{u}_3^{N-1}} \mathsf{P}^N\{\{\boldsymbol{Y}_0^{N-1}\}\mid\{\boldsymbol{u}_0^{N-1}\boldsymbol{G}_N\}\},$$

$$\vdots$$

$$\mathsf{P}_N^i\{\{\boldsymbol{Y}_0^{N-1}\}\{\boldsymbol{u}_0^{i-1}\}\mid\{u_i\}\} = \frac{1}{2^{N-1}}\sum_{\boldsymbol{u}_{i+1}^{N-1}} \mathsf{P}^N\{\{\boldsymbol{Y}_0^{N-1}\}\mid\{\boldsymbol{u}_0^{N-1}\boldsymbol{G}_N\}\}, \quad (4.102)$$

$$3 < n < N-1,$$

$$\vdots$$

$$\mathsf{P}_N^{N-2}\{\{\boldsymbol{Y}_0^{N-1}\}\{\boldsymbol{u}_0^{N-3}\} \mid \{u_{N-2}\}\} = \frac{1}{2^{N-1}} \sum_{u_{N-1}} \mathsf{P}^N\{\{\boldsymbol{Y}_0^{N-1}\} \mid \{\boldsymbol{u}_0^{N-1}\boldsymbol{G}_N\}\},$$

$$\mathsf{P}_N^{N-1}\{\{\boldsymbol{Y}_0^{N-1}\}\{\boldsymbol{u}_0^{N-2}\} \mid \{u_{N-1}\}\} = \frac{1}{2^{N-1}} \mathsf{P}^N\{\{\boldsymbol{Y}_0^{N-1}\} \mid \{\boldsymbol{u}_0^{N-1}\boldsymbol{G}_N\}\}.$$

Assuming equiprobable information bits, the ith split virtual discrete memoryless channel (DMC) therefore has the transition probability [16, equation (5), p. 3053]

$$\mathsf{P}_N^i\{\{\boldsymbol{Y}_0^{N-1}\}\{\boldsymbol{u}_0^{i-1}\} \mid \{u_i\}\} = \frac{1}{2^{N-1}} \sum_{\boldsymbol{u}_{i+1}^{N-1}} \mathsf{P}_N\{\{\boldsymbol{Y}_0^{N-1}\} \mid \{\boldsymbol{u}_0^{N-1}\}\} \tag{4.103}$$

$$= \frac{1}{2^{N-1}} \sum_{\boldsymbol{u}_{i+1}^{N-1}} \mathsf{P}^N\{\{\boldsymbol{Y}_0^{N-1}\} \mid \{\boldsymbol{u}_0^{N-1}\boldsymbol{G}_N\}\}. \tag{4.104}$$

Let us denote the ith component of $\boldsymbol{u}_0^{N-1}\boldsymbol{G}_N$ by $[\boldsymbol{u}_0^{N-1}\boldsymbol{G}_N]_i$. Since the transitions in the raw combined discrete memoryless channel (DMC) $W^N : \boldsymbol{X} \mapsto \boldsymbol{Y}$ are mutually independent, we can write (cf. [16, equations (17)–(20), p. 3056])

$$\mathsf{P}^N\{\{\boldsymbol{Y}_0^{N-1}\} \mid \{\boldsymbol{u}_0^{N-1}\boldsymbol{G}_N\}\} = \prod_{i=0}^{N-1} \mathsf{P}\{\{y_i\} \mid \{[\boldsymbol{u}_0^{N-1}\boldsymbol{G}_N]_i\}\} \tag{4.105}$$

$$= \mathsf{P}\{\{y_0\} \mid \{[\boldsymbol{u}_0^{N-1}\boldsymbol{G}_N]_0\}\} \cdot \cdots \cdot \mathsf{P}\{\{y_{N-1}\} \mid \{[\boldsymbol{u}_0^{N-1}\boldsymbol{G}_N]_{N-1}\}\}.$$

Example 4.1. In the case of N equal to 2, using (4.105), we yield (cf. [16, equations (17)–(20), p. 3056])

$$\begin{aligned}\mathsf{P}_2\{\{\boldsymbol{Y}_0^1\} \mid \{\boldsymbol{u}_0^1\}\} &= \mathsf{P}^2\{\{\boldsymbol{Y}_0^1\} \mid \{\boldsymbol{u}_0^1\boldsymbol{G}_N\}\}\\ &= \mathsf{P}^2\left\{\{(Y_0, Y_1)\} \mid \left\{(u_0, u_1)\begin{pmatrix} 1 & 0 \\ 1 & 1 \end{pmatrix}\right\}\right\}\\ &= \mathsf{P}^2\{\{(Y_0, Y_1)\} \mid \{(u_0 \oplus u_1, u_1)\}\}\\ &= \mathsf{P}\{\{Y_0\} \mid \{u_0 \oplus u_1\}\} \cdot \mathsf{P}\{\{Y_1\} \mid \{u_1\}\}\end{aligned} \tag{4.106}$$

and [16, equations (17)–(20), p. 3056]

$$\begin{aligned}\mathsf{P}_2^0\{\{\boldsymbol{Y}_0^1\} \mid \{u_0\}\} &= \frac{1}{2}\sum_{u_1} \mathsf{P}^2\{\{\boldsymbol{Y}_0^1\} \mid \{\boldsymbol{u}_0^1\boldsymbol{G}_N\}\}\\ &= \frac{1}{2}\sum_{u_1} \mathsf{P}\{\{Y_0\} \mid \{u_0 \oplus u_1\}\} \cdot \mathsf{P}\{\{Y_1\} \mid \{u_1\}\},\end{aligned} \tag{4.107}$$

$$\begin{aligned}\mathsf{P}_2^1\{\{\boldsymbol{Y}_0^1\}\{u_0\} \mid \{u_1\}\} &= \frac{1}{2}\mathsf{P}^2\{\{\boldsymbol{Y}_0^1\} \mid \{\boldsymbol{u}_0^1\boldsymbol{G}_N\}\}\\ &= \frac{1}{2}\mathsf{P}\{\{Y_0\} \mid \{u_0 \oplus u_1\}\} \cdot \mathsf{P}\{\{Y_1\} \mid \{u_1\}\}.\end{aligned} \tag{4.108}$$

Let

$$\boldsymbol{u}_{0,0}^{N-1} = (u_1, u_3, u_5, u_7 \cdots u_{N-1}) \quad \text{(since } N \text{ is even)} \tag{4.109}$$

denote the information bit vector, which contains only the $N/2$ message bits with *odd* index, and let

$$\boldsymbol{u}_{\mathrm{e},0}^{N-1} = (u_0, u_2, u_4, u_6 \cdots u_{N-2}) \quad \text{(since } N \text{ is even)} \tag{4.110}$$

denote the information bit vector which contains only the $N/2$ message bits with *even* index. Also, let us keep in mind that

$$\boldsymbol{u}_{\mathrm{o},0}^{N-1} \oplus \boldsymbol{u}_{\mathrm{e},0}^{N-1} = (u_0 \oplus u_1, u_2 \oplus u_3, u_4 \oplus u_5, u_6 \oplus u_7 \cdots u_{N-2} \oplus u_{N-1}) \tag{4.111}$$

holds.

Theorem 4.4 (Recursive Computation of Transition Probabilities). *Let $i \in \{1, 2 \cdots N\}$, $N \in \{2, 4, 8, 16 \cdots 2^\mu \cdots\}$, $\mu \in \mathbb{N}^*$. For the split virtual discrete memoryless channel (DMC) $W_{2N}^{2i-2} : u_{2i-2} \mapsto \boldsymbol{Y}, u_0 \cdots u_{2i-3}$, the transition probability is given by*

$$\begin{aligned}
&\mathrm{P}_{2N}^{2i-2}\left\{\left\{\boldsymbol{Y}_0^{2N-1}\right\}\left\{\boldsymbol{u}_0^{2i-3}\right\} \mid \{u_{2i-2}\}\right\} \\
&= \frac{1}{2}\sum_{u_{2i-1}} \mathrm{P}_N^{i-1}\left\{\left\{\boldsymbol{Y}_0^{N-1}\right\}\left\{\boldsymbol{u}_{\mathrm{e},0}^{2i-3} \oplus \boldsymbol{u}_{\mathrm{o},0}^{2i-3}\right\} \mid \{u_{2i-2} \oplus u_{2i-1}\}\right\} \\
&\cdot \mathrm{P}_N^{i-1}\left\{\left\{\boldsymbol{Y}_N^{2N-1}\right\}\left\{\boldsymbol{u}_{\mathrm{o},0}^{2i-3}\right\} \mid \{u_{2i-1}\}\right\},
\end{aligned} \tag{4.112}$$

and for the split virtual discrete memoryless channel (DMC) $W_{2N}^{2i-1} : u_{2i-1} \mapsto \boldsymbol{Y}, u_0 \cdots u_{2i-2}$, the transition probability is given by

$$\begin{aligned}
&\mathrm{P}_{2N}^{2i-1}\left\{\left\{\boldsymbol{Y}_0^{2N-1}\right\}\left\{\boldsymbol{u}_0^{2i-2}\right\} \mid \{u_{2i-1}\}\right\} \\
&= \frac{1}{2}\mathrm{P}_N^{i-1}\left\{\left\{\boldsymbol{Y}_0^{N-1}\right\}\left\{\boldsymbol{u}_{\mathrm{e},0}^{2i-3} \oplus \boldsymbol{u}_{\mathrm{o},0}^{2i-3}\right\} \mid \{u_{2i-2} \oplus u_{2i-1}\}\right\} \\
&\cdot \mathrm{P}_N^{i-1}\left\{\left\{\boldsymbol{Y}_N^{2N-1}\right\}\left\{\boldsymbol{u}_{\mathrm{o},0}^{2i-3}\right\} \mid \{u_{2i-1}\}\right\}.
\end{aligned} \tag{4.113}$$

Proof. We already know that in the case of N equal to 2, using (4.105), we yield

$$\begin{aligned}
\mathrm{P}_2\left\{\left\{\boldsymbol{Y}_0^1\right\} \mid \left\{\boldsymbol{u}_0^1\right\}\right\} &= \mathrm{P}^2\left\{\left\{\boldsymbol{Y}_0^1\right\} \mid \left\{\boldsymbol{u}_0^1 \boldsymbol{G}\right\}\right\} \\
&= \mathrm{P}^2\left\{\left\{(Y_0, Y_1)\right\} \mid \left\{(u_0 \oplus u_1, u_1)\right\}\right\} \\
&= \mathrm{P}\left\{\{Y_0\} \mid \{u_0 \oplus u_1\}\right\}\mathrm{P}\left\{\{Y_1\} \mid \{u_1\}\right\},
\end{aligned} \tag{4.114}$$

because u_0 and u_1 are independent of each other, and

$$\begin{aligned}
\mathrm{P}_2^0\left\{\left\{\boldsymbol{Y}_0^1\right\} \mid \{u_0\}\right\} &= \frac{1}{2}\sum_{u_1} \mathrm{P}_2\left\{\left\{\boldsymbol{Y}_0^1\right\} \mid \left\{\boldsymbol{u}_0^1\right\}\right\} \\
&= \frac{1}{2}\sum_{u_1} \mathrm{P}\left\{\{Y_0\} \mid \{u_0 \oplus u_1\}\right\}\mathrm{P}\left\{\{Y_1\} \mid \{u_1\}\right\}
\end{aligned} \tag{4.115}$$

as well as

$$\begin{aligned}
\mathrm{P}_2^1\left\{\left\{\boldsymbol{Y}_0^1\right\}\{u_0\} \mid \{u_1\}\right\} &= \frac{1}{2}\mathrm{P}_2\left\{\left\{\boldsymbol{Y}_0^1\right\} \mid \left\{\boldsymbol{u}_0^1\right\}\right\} \\
&= \frac{1}{2}\mathrm{P}\left\{\{Y_0\} \mid \{u_0 \oplus u_1\}\right\}\mathrm{P}\left\{\{Y_1\} \mid \{u_1\}\right\};
\end{aligned} \tag{4.116}$$

cf. [16, equations (17)–(20), p. 3056]. We will generalize these results in what follows.

We will set out from

$$\begin{aligned}\mathrm{P}_{2N}^{2i-2}\{\{\boldsymbol{Y}_0^{2N-1}\}\{\boldsymbol{u}_0^{2i-3}\}\mid\{u_{2i-2}\}\} &= \frac{\mathrm{P}\{\{\boldsymbol{Y}_0^{2N-1}\}\{\boldsymbol{u}_0^{2i-3}\}\{u_{2i-2}\}\}}{\mathrm{P}\{\{u_{2i-2}\}\}}\\ &= \frac{\mathrm{P}\{\{\boldsymbol{Y}_0^{2N-1}\}\{\boldsymbol{u}_0^{2i-2}\}\}}{\frac{1}{2}}\\ &= 2\cdot\sum_{\boldsymbol{u}_{2i-1}^{2N-1}}\mathrm{P}\{\{\boldsymbol{Y}_0^{2N-1}\}\{\boldsymbol{u}_0^{2N-1}\}\}\\ &= 2\cdot\sum_{\boldsymbol{u}_{2i-1}^{2N-1}}\mathrm{P}\{\{\boldsymbol{Y}_0^{2N-1}\}\mid\{\boldsymbol{u}_0^{2N-1}\}\}\underbrace{\mathrm{P}\{\{\boldsymbol{u}_0^{2N-1}\}\}}_{=1/2^{2N}}\\ &= \frac{1}{2^{2N-1}}\cdot\sum_{\boldsymbol{u}_{2i-1}^{2N-1}}\mathrm{P}\{\{\boldsymbol{Y}_0^{2N-1}\}\mid\{\boldsymbol{u}_0^{2N-1}\}\}.\end{aligned}\tag{4.117}$$

Since

$$\mathrm{P}\{\{\boldsymbol{Y}_0^{2N-1}\}\mid\{\boldsymbol{u}_0^{2N-1}\}\} = \mathrm{P}_N\{\{\boldsymbol{Y}_0^{N-1}\}\mid\{\boldsymbol{u}_{\mathrm{e},0}^{2N-1}\oplus\boldsymbol{u}_{\mathrm{o},0}^{2N-1}\}\}\cdot\mathrm{P}_N\{\{\boldsymbol{Y}_N^{2N-1}\}\mid\{\boldsymbol{u}_{\mathrm{o},0}^{2N-1}\}\}\tag{4.118}$$

holds, which means that the first N samples in $\boldsymbol{Y}_0^{2N-1}$ depend always on $\boldsymbol{u}_{\mathrm{e},0}^{2N-1}\oplus\boldsymbol{u}_{\mathrm{o},0}^{2N-1}$, (cf. (4.111)) and the last N samples in $\boldsymbol{Y}_0^{2N-1}$ depend only on $\boldsymbol{u}_{\mathrm{o},0}^{2N-1}$, (4.117) becomes

$$\begin{aligned}&\mathrm{P}_{2N}^{2i-2}\{\{\boldsymbol{Y}_0^{2N-1}\}\{\boldsymbol{u}_0^{2i-3}\}\mid\{u_{2i-2}\}\}\\ &\quad= \frac{1}{2^{2N-1}}\cdot\sum_{\boldsymbol{u}_{2i-1}^{2N-1}}\mathrm{P}_N\{\{\boldsymbol{Y}_0^{N-1}\}\mid\{\boldsymbol{u}_{\mathrm{e},0}^{2N-1}\oplus\boldsymbol{u}_{\mathrm{o},0}^{2N-1}\}\}\cdot\mathrm{P}_N\{\{\boldsymbol{Y}_N^{2N-1}\}\mid\{\boldsymbol{u}_{\mathrm{o},0}^{2N-1}\}\},\end{aligned}\tag{4.119}$$

which leads to

$$\begin{aligned}&\mathrm{P}_{2N}^{2i-2}\{\{\boldsymbol{Y}_0^{2N-1}\}\{\boldsymbol{u}_0^{2i-3}\}\mid\{u_{2i-2}\}\}\\ &\quad= \frac{1}{2}\frac{1}{2^{2N-2}}\cdot\sum_{\boldsymbol{u}_{2i-1}^{2N-1}}\mathrm{P}_N\{\{\boldsymbol{Y}_0^{N-1}\}\mid\{\boldsymbol{u}_{\mathrm{e},0}^{2N-1}\oplus\boldsymbol{u}_{\mathrm{o},0}^{2N-1}\}\}\cdot\mathrm{P}_N\{\{\boldsymbol{Y}_N^{2N-1}\}\mid\{\boldsymbol{u}_{\mathrm{o},0}^{2N-1}\}\}\\ &\quad= \frac{1}{2}\sum_{\boldsymbol{u}_{\mathrm{o},2i-1}^{2N-1}}\frac{1}{2^{N-1}}\cdot\mathrm{P}_N\{\{\boldsymbol{Y}_N^{2N-1}\}\mid\{\boldsymbol{u}_{\mathrm{o},0}^{2N-1}\}\}\\ &\quad\cdot\sum_{\boldsymbol{u}_{\mathrm{e},2i-1}^{2N-1}}\frac{1}{2^{N-1}}\cdot\mathrm{P}_N\{\{\boldsymbol{Y}_0^{N-1}\}\mid\{\boldsymbol{u}_{\mathrm{e},0}^{2N-1}\oplus\boldsymbol{u}_{\mathrm{o},0}^{2N-1}\}\}\\ &\quad= \sum_{u_{2i-1}}\frac{1}{2}\cdot\left(\frac{1}{2^{N-1}}\cdot\sum_{\boldsymbol{u}_{\mathrm{o},2i}^{2N-1}}\mathrm{P}_N\{\{\boldsymbol{Y}_N^{2N-1}\}\mid\{\boldsymbol{u}_{\mathrm{o},0}^{2N-1}\}\}\right)\\ &\quad\cdot\left(\frac{1}{2^{N-1}}\cdot\sum_{\boldsymbol{u}_{\mathrm{e},2i}^{2N-1}}\mathrm{P}_N\{\{\boldsymbol{Y}_0^{N-1}\}\mid\{\boldsymbol{u}_{\mathrm{e},0}^{2N-1}\oplus\boldsymbol{u}_{\mathrm{o},0}^{2N-1}\}\}\right).\end{aligned}\tag{4.120}$$

Since

$$\mathrm{P}_N^{i-1}\{\{\boldsymbol{Y}_0^{N-1}\}\{\boldsymbol{u}_0^{i-2}\}\mid\{u_{i-1}\}\} = \frac{1}{2^{N-1}}\sum_{\boldsymbol{u}_i^{N-1}}\mathrm{P}_N\{\{\boldsymbol{Y}_0^{N-1}\}\mid\{\boldsymbol{u}_0^{N-1}\}\}\tag{4.121}$$

holds according to (4.104), we find

$$\mathsf{P}_N^{i-1}\{\{\boldsymbol{Y}_0^{N-1}\}\{\boldsymbol{u}_{e,0}^{2i-3}\oplus\boldsymbol{u}_{o,0}^{2i-3}\}\mid\{u_{2i-2}\oplus u_{2i-1}\}\} = \frac{1}{2^{N-1}}\cdot\sum_{u_{e,2i}^{2N-1}}\mathsf{P}_N\{\{\boldsymbol{Y}_0^{N-1}\}\mid\{\boldsymbol{u}_{e,0}^{2N-1}\oplus\boldsymbol{u}_{o,0}^{2N-1}\}\} \tag{4.122}$$

and

$$\mathsf{P}_N^{i-1}\{\{\boldsymbol{Y}_N^{2N-1}\}\{\boldsymbol{u}_{o,0}^{2i-3}\}\mid\{u_{2i-1}\}\} = \frac{1}{2^{N-1}}\cdot\sum_{u_{o,2i}^{2N-1}}\mathsf{P}_N\{\{\boldsymbol{Y}_N^{2N-1}\}\mid\{\boldsymbol{u}_{o,0}^{2N-1}\}\}. \tag{4.123}$$

Hence, we yield

$$\begin{aligned}\mathsf{P}_{2N}^{2i-2}\{\{\boldsymbol{Y}_0^{2N-1}\}\{\boldsymbol{u}_0^{2i-3}\}\mid\{u_{2i-2}\}\} &= \frac{1}{2}\sum_{u_{2i-1}}\mathsf{P}_N^{i-1}\{\{\boldsymbol{Y}_N^{2N-1}\}\{\boldsymbol{u}_{o,0}^{2i-3}\}\mid\{u_{2i-1}\}\}\\ &\cdot\mathsf{P}_N^{i-1}\{\{\boldsymbol{Y}_0^{N-1}\}\{\boldsymbol{u}_{e,0}^{2i-3}\oplus\boldsymbol{u}_{o,0}^{2i-3}\}\mid\{u_{2i-2}\oplus u_{2i-1}\}\}.\end{aligned} \tag{4.124}$$

Now, we will set out from

$$\begin{aligned}\mathsf{P}_{2N}^{2i-1}\{\{\boldsymbol{Y}_0^{2N-1}\}\{\boldsymbol{u}_0^{2i-2}\}\mid\{u_{2i-1}\}\} &= 2\sum_{\boldsymbol{u}_{2i}^{2N-1}}\mathsf{P}\{\{\boldsymbol{Y}_0^{2N-1}\}\{\boldsymbol{u}_0^{2N-1}\}\}\\ &= \frac{1}{2^{2N-1}}\sum_{\boldsymbol{u}_{2i}^{2N-1}}\mathsf{P}\{\{\boldsymbol{Y}_0^{2N-1}\}\mid\{\boldsymbol{u}_0^{2N-1}\}\}\\ &= \frac{1}{2}\cdot\sum_{\boldsymbol{u}_{2i}^{2N-1}}\frac{1}{2^{N-1}}\cdot\mathsf{P}_N\{\{\boldsymbol{Y}_0^{N-1}\}\mid\{\boldsymbol{u}_{e,0}^{2N-1}\oplus\boldsymbol{u}_{o,0}^{2N-1}\}\}\\ &\quad\frac{1}{2^{N-1}}\cdot\mathsf{P}_N\{\{\boldsymbol{Y}_N^{2N-1}\}\mid\{\boldsymbol{u}_{o,0}^{2N-1}\}\}\\ &= \frac{1}{2}\cdot\left(\frac{1}{2^{N-1}}\cdot\sum_{\boldsymbol{u}_{o,2i}^{2N-1}}\mathsf{P}_N\{\{\boldsymbol{Y}_N^{2N-1}\}\mid\{\boldsymbol{u}_{o,0}^{2N-1}\}\}\right)\\ &\quad\cdot\left(\frac{1}{2^{N-1}}\cdot\sum_{\boldsymbol{u}_{e,2i}^{2N-1}}\mathsf{P}_N\{\{\boldsymbol{Y}_0^{N-1}\}\mid\{\boldsymbol{u}_{e,0}^{2N-1}\oplus\boldsymbol{u}_{o,0}^{2N-1}\}\}\right)\end{aligned} \tag{4.125}$$

and yield

$$\begin{aligned}\mathsf{P}_{2N}^{2i-1}\{\{\boldsymbol{Y}_0^{2N-1}\}\{\boldsymbol{u}_0^{2i-2}\}\mid\{u_{2i-1}\}\} &= \frac{1}{2}\cdot\mathsf{P}_N^{i-1}\{\{\boldsymbol{Y}_0^{N-1}\}\{\boldsymbol{u}_{e,0}^{2i-3}\oplus\boldsymbol{u}_{o,0}^{2i-3}\}\mid\{u_{2i-2}\oplus u_{2i-1}\}\}\\ &\cdot\mathsf{P}_N^{i-1}\{\{\boldsymbol{Y}_N^{2N-1}\}\{\boldsymbol{u}_{o,0}^{2i-3}\}\mid\{u_{2i-1}\}\}.\end{aligned} \tag{4.126}$$

□

Hence, $\mathsf{P}_{2N}^{2i-2}\{\{\boldsymbol{Y}_0^{2N-1}\}\{\boldsymbol{u}_0^{2i-3}\}\mid\{u_{2i-2}\}\}$ and $\mathsf{P}_{2N}^{2i-1}\{\{\boldsymbol{Y}_0^{2N-1}\}\{\boldsymbol{u}_0^{2i-2}\}\mid\{u_{2i-1}\}\}$ can be computed recursively.

Example 4.2 (Simplest Polar Code of Length 2). We will restrict ourselves to the binary erasure channel (BEC). In the case of N equal to 2, we now have the combined discrete memoryless channel (DMC)

$$W_2 : \boldsymbol{u} \mapsto \boldsymbol{Y}$$
$$\mathrm{P}_N\{\{\boldsymbol{Y}_0^1\} \mid \{\boldsymbol{u}_0^1\}\} = \mathrm{P}\{\{Y_0\} \mid \{u_0 \oplus u_1\}\} \cdot \mathrm{P}\{\{Y_1\} \mid \{u_1\}\}. \tag{4.127}$$

Since X_0 is equal to $(u_0 \oplus u_1)$ and X_1 is equal to u_1, the encoding is invertible. Applying the chain rule of the average mutual information

$$I\{\boldsymbol{u}; \boldsymbol{Y}\} = I\{u_0; \boldsymbol{Y}\} + I\{u_1; \boldsymbol{Y}\} \tag{4.128}$$

leading to

$$C_2 = 2C_1, \tag{4.129}$$

i. e.,

$$C_2 = 2(1 - \epsilon). \tag{4.130}$$

Thus, the encoding (X_0, X_1) equal to $\boldsymbol{u}\boldsymbol{G}_2$ using $\boldsymbol{G}_2$ of (4.66), i. e., $(u_0 \oplus u_1, u_1)$, preserves the sum capacity.

Figure 4.5 shows the polar code for N equal to 2 with
- the two binary input discrete memoryless channels (DMCs) W, forming the raw combined discrete memoryless channel (DMC) W^2,
- the generator matrix $\boldsymbol{G}_2$ and
- the combined discrete memoryless channel (DMC) W_2, which is composed of the generator matrix $\boldsymbol{G}_2$ and the raw combined discrete memoryless channel (DMC) W^2.

In the case of N equal to 2, we have two split virtual channels

$$W_2^0 : u_0 \mapsto \boldsymbol{Y}, \quad W_2^1 : u_1 \mapsto \boldsymbol{Y}, u_0. \tag{4.131}$$

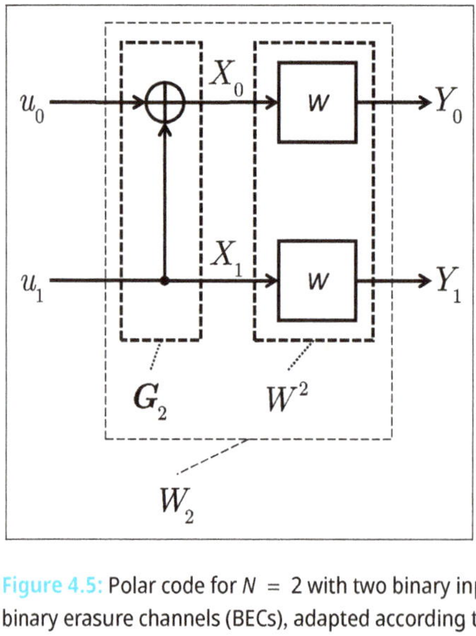

Figure 4.5: Polar code for $N = 2$ with two binary input discrete memoryless channels (DMCs), e. g., two binary erasure channels (BECs), adapted according to [33, Figure 3.2, p. 76; Figure 4.6, p. 96].

We have

$$I\{\boldsymbol{u}; \boldsymbol{Y}\} = I\{u_0; \boldsymbol{Y}\} + I\{u_1; \boldsymbol{Y}, u_0\} \tag{4.132}$$

leading to C_2 equal to $2C_1$.

Let us recall that

- $\max I\{u_0; \boldsymbol{Y}\}$ is the maximum average mutual information about the event $\{u_0\}$ by the occurrence of the events from the joint ensemble $\boldsymbol{Y}$ and
- $\max I\{u_1; \boldsymbol{Y}, u_0\}$ is the maximum average mutual information about the event $\{u_1\}$ by the occurrence of the events from the joint ensemble $\boldsymbol{Y}$ given the event $\{u_0\}$.

Figure 4.5 implies that

- $\max I\{u_0; \boldsymbol{Y}\}$ refers to a *virtual channel* W_2^-, i. e., W_2^0, which relates the output symbols Y_0 and Y_1 with the input bit u_0 being equal to $(X_0 \oplus X_1)$ and
- $\max I\{u_1; \boldsymbol{Y}, u_0\}$ refers to a *virtual channel* W_2^+, i. e., W_2^1, which relates the output symbols Y_0 and Y_1 with the input bit u_1 equal to X_1 or $(v_1' \oplus u_1)$, given the input bit u_0.

Since these two split virtual channels W_2^-, i. e., W_2^0, and W_2^+, i. e., W_2^1 do not relate in a symmetric way, their channel capacities

$$C_2^- = C_2^0 = \max I(u_1; \mathbb{Y}) \tag{4.133}$$

and

$$C_2^+ = C_2^1 = \max I(u_2; \mathbb{Y} \mid u_1) \tag{4.134}$$

are not equal.

Let us now only consider the binary erasure channel (BEC). Regarding W_2^-, i. e., W_2^0, which is the first of these two split virtual channels, u_0, which equals $(X_0 \oplus X_1)$ is erased if either one of X_0 or X_1 is erased, that is with erasure probability,

$$\epsilon_2^{(0)} = \underbrace{\epsilon(1-\epsilon)}_{\substack{X_0 \text{ is erased,} \\ X_1 \text{ is not erased}}} + \underbrace{(1-\epsilon)\epsilon}_{\substack{X_1 \text{ is erased,} \\ X_0 \text{ is not erased}}} + \underbrace{\epsilon^2}_{\substack{X_0 \text{ and } X_1 \\ \text{are erased}}} = \epsilon - \epsilon^2 + \epsilon - \epsilon^2 + \epsilon^2 = 2\epsilon - \epsilon^2. \tag{4.135}$$

The channel capacity of this first channel is given by

$$C_2^- = C_2^0 = \max I(u_1; \mathbb{Y}) = 1 - \epsilon_2^{(0)} = 1 - 2\epsilon + \epsilon^2 = (1-\epsilon)^2 = C_1^2. \tag{4.136}$$

Regarding the second of these two channels, W_2^+, i. e., W_2^1, u_1, which equals $(X_0 \oplus u_0)$ or X_1, is erased only if both X_0 and X_1 are jointly erased, that is with erasure probability

$$\epsilon_2^{(1)} = \epsilon^2. \tag{4.137}$$

The channel capacity of this second channel is given by

$$C_2^+ = C_2^1 = \max I(u_2; \mathbb{Y} \mid u_1) = 1 - \epsilon_2^{(1)} = 1 - \epsilon^2 = 2C_1 - C_1^2. \tag{4.138}$$

We yield

$$C_2^+ \geq C_2^- \tag{4.139}$$

with equality only for ϵ equal to 0 (best case, no erasures at all) and 1 (worst case, everything is erased) and

$$C_2^+ + C_2^- = 2C_1, \tag{4.140}$$

as expected.

The channel capacities C_1 of the binary erasure channel (BEC) as well as $C_2^-(C_2^0)$ and $C_2^+(C_2^1)$ of the two split virtual channels $W_2^-(W_2^0)$ and $W_2^+(W_2^1)$ are depicted in Figure 4.6. C_2^- of (4.136) and C_2^+ of (4.138) were also given in [16]. Although that the argument about and the derivation of C_2^- of (4.136) and C_2^+ of (4.138) are obvious results of information theory, they were first stated by Erdal *Arikan* [16].

The generation of the codeword can be conveniently illustrated using the binary tree diagram of height 1; cf. Figure 4.7. The height of the binary tree is the maximum number of links, i. e., branches in a path.

Also, the elements $\rho_i^{(N)}$, $i \in \{0, 1 \cdots (N-1)\}$, $N \in \mathbb{N}^*$ of the *reliability sequence*, which is represented by an ordered N-tuple termed *reliability vector*

$$\boldsymbol{\rho}^{(N)} = \left(\rho_0^{(N)}, \rho_1^{(N)} \cdots \rho_{N-1}^{(N)}\right) \tag{4.141}$$

are given in Figure 4.7. As already indicated above, the reliability sequence, and hence the reliability vector represent the reliability of the message bits, and hence of the split virtual channels in an order from very bad to very good. The components $\rho_i^{(N)}$, $i \in \{0, 1 \cdots (N-1)\}$, $N \in \mathbb{N}^*$ of $\boldsymbol{\rho}^{(N)}$ given by (4.141) can assume integer values between 0 and $(N-1)$ and must be distinct. In the case of N equal to 2, the first component $\rho_0^{(2)}$ is equal to 0 and represents the reliability of the message bit u_0, and the second component $\rho_1^{(2)}$ is equal to 1 and represents the reliability of the message bit u_1. The greater the index of the component $\rho_i^{(N)}$, $i \in \{0, 1 \cdots (N-1)\}$, $N \in \mathbb{N}^*$ the more reliable is the message bit u_i, $i \in \{0, 1 \cdots (N-1)\}$, $N \in \mathbb{N}^*$.

Owing to (4.139), the split virtual channel W_2^+, i. e., W_2^1 is considered good and the split virtual channel W_2^-, i. e., W_2^0 is considered bad. The reliability of these two virtual channels in ascending order is represented by the reliability vector

$$\boldsymbol{\rho}^{(2)} = (0, 1). \tag{4.142}$$

A clever transmission hence follows the following concept:

1) Do not assign any information to u_0, rather freeze it, i. e., fix its value, e. g., use u_0 equal to 0.
2) Use only u_1 to transmit information.

Therefore, the message vector is given by

$$\boldsymbol{u} = (0, u_1). \tag{4.143}$$

By doing so, we arrive at the polar code

$$\{(0,0), (1,1)\} \tag{4.144}$$

where

$$\boldsymbol{X} = (0, u_1)\begin{pmatrix} 1 & 0 \\ 1 & 1 \end{pmatrix} = (u_1, u_1), \tag{4.145}$$

which is a repetition code like, e. g., the zeroth-order Reed–Muller (RM) code $\mathcal{RM}(0,1)$ with dimension K equal to 1, length N equal to 2 and minimum distance $d_{\min}$ equal to 2, having the code rate

$$R = \frac{K}{N} = \frac{1}{2}. \tag{4.146}$$

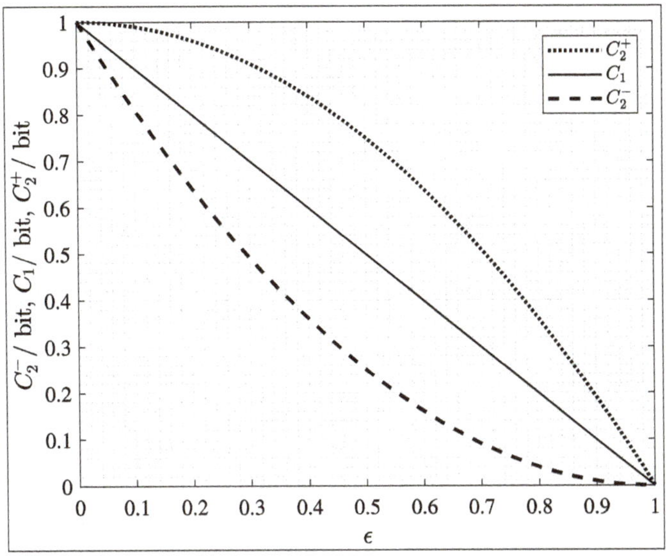

Figure 4.6: Channel capacities C_1 of the binary erasure channel (BEC) as well as $C_2^-(C_2^0)$ and $C_2^+(C_2^1)$ of the two split virtual channels $W_2^-(W_2^0)$ and $W_2^+(W_2^1)$.

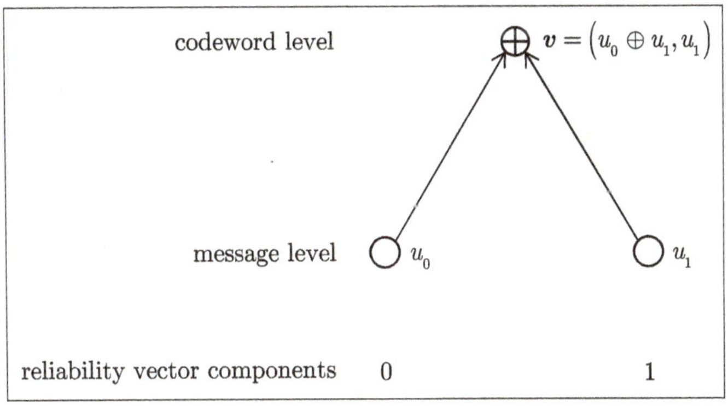

Figure 4.7: Binary tree diagram of the polar code encoding with code length $N = 2$.

Example 4.3 (Polar Code of Length 4). We will restrict ourselves to the binary erasure channel (BEC). Let us look at the case of N equal to 4. The *two-fold Kronecker product* of $\boldsymbol{G}_2$ given (4.66) yields

$$\boldsymbol{G}_2^{\otimes 2} = \boldsymbol{G}_2 \otimes \boldsymbol{G}_2 = \begin{pmatrix} \boldsymbol{G}_2 & \boldsymbol{0}_{2\times 2} \\ \boldsymbol{G}_2 & \boldsymbol{G}_2 \end{pmatrix} = \begin{pmatrix} 1 & 0 & 0 & 0 \\ 1 & 1 & 0 & 0 \\ 1 & 0 & 1 & 0 \\ 1 & 1 & 1 & 1 \end{pmatrix}. \tag{4.147}$$

Furthermore, we have

$$\boldsymbol{\Pi}_4 = \boldsymbol{\Sigma}_{2,2}^{\mathrm{T}} = \begin{pmatrix} 1 & 0 & 0 & 0 \\ 0 & 0 & 1 & 0 \\ 0 & 1 & 0 & 0 \\ 0 & 0 & 0 & 1 \end{pmatrix} \tag{4.148}$$

and

$$\boldsymbol{I}_2 \otimes \boldsymbol{B}_2 = \boldsymbol{I}_2 \otimes \boldsymbol{I}_2 = \boldsymbol{I}_4. \tag{4.149}$$

Hence, for N equal to 4, (4.83) becomes

$$\boldsymbol{G}_4 = \boldsymbol{B}_4 \boldsymbol{G}_2^{\otimes 2} = \boldsymbol{\Pi}_4 (\boldsymbol{I}_2 \otimes \boldsymbol{B}_2) \boldsymbol{G}_2^{\otimes 2} = \boldsymbol{\Pi}_4 \boldsymbol{I}_4 \boldsymbol{G}_2^{\otimes 2} = \boldsymbol{\Pi}_4 \boldsymbol{G}_2^{\otimes 2}, \tag{4.150}$$

yielding

$$\boldsymbol{G}_4 = \begin{pmatrix} 1 & 0 & 0 & 0 \\ 0 & 0 & 1 & 0 \\ 0 & 1 & 0 & 0 \\ 0 & 0 & 0 & 1 \end{pmatrix} \begin{pmatrix} 1 & 0 & 0 & 0 \\ 1 & 1 & 0 & 0 \\ 1 & 0 & 1 & 0 \\ 1 & 1 & 1 & 1 \end{pmatrix} = \begin{pmatrix} 1 & 0 & 0 & 0 \\ 1 & 0 & 1 & 0 \\ 1 & 1 & 0 & 0 \\ 1 & 1 & 1 & 1 \end{pmatrix}. \tag{4.151}$$

Obviously, we permuted the second and the third rows and obtained the standard form of the generator matrix of the polar code with N equal to 4. The corresponding polar code with length N equal to 4 is shown in Figure 4.8.

Using

$$\boldsymbol{G}_N = (\boldsymbol{I}_{N/2} \otimes \boldsymbol{G}_2) \boldsymbol{\Pi}_N (\boldsymbol{I}_2 \otimes \boldsymbol{G}_{N/2}) \tag{4.152}$$

leads to an alternative encoding structure given by

$$\boldsymbol{G}_4 = (\boldsymbol{I}_2 \otimes \boldsymbol{G}_2) \boldsymbol{\Pi}_4 (\boldsymbol{I}_2 \otimes \boldsymbol{G}_2); \tag{4.153}$$

cf. Figure 4.9.

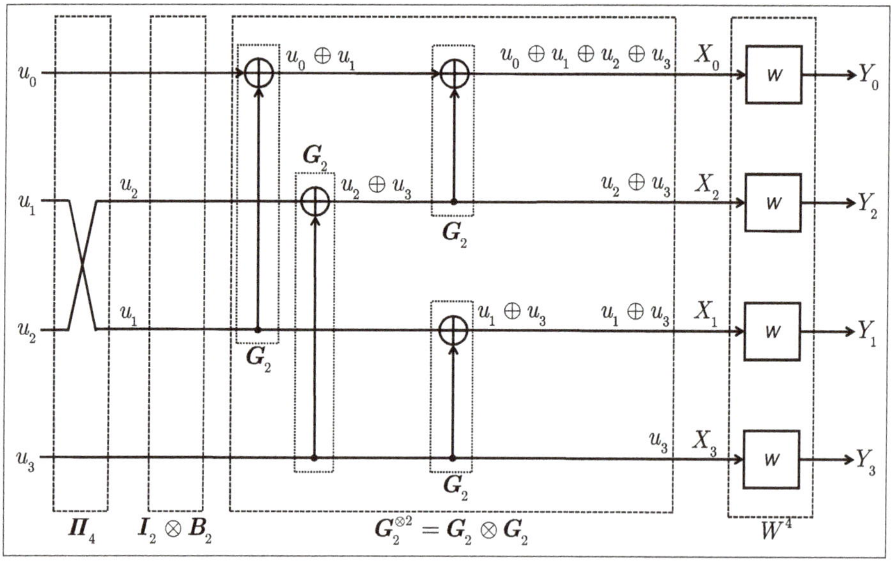

Figure 4.8: Polar code for N equal to 4 with two binary erasure channels (BECs).

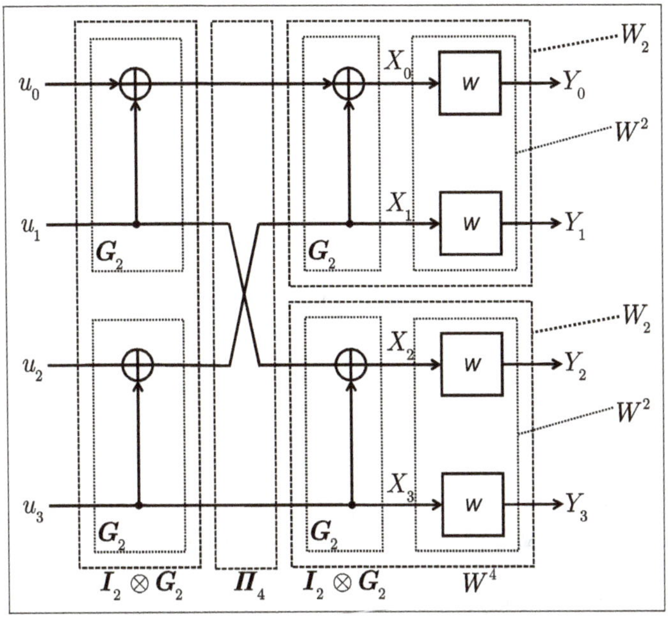

Figure 4.9: Alternative representation of the polar code for N equal to 4 with two binary erasure channels (BECs), adapted according to [33, Figure 4.8, p. 101].

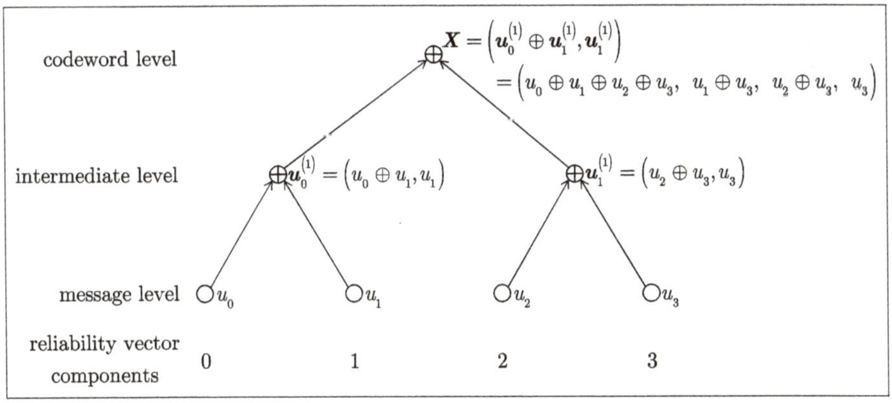

Figure 4.10: Binary tree diagram of the polar code encoding with code length $N = 4$; cf., e. g., [33, Figure 2.11, p. 33] for the decoding.

The generation of the codeword can be conveniently illustrated using the binary tree diagram of height 2; cf. Figure 4.10 [33, Figure 2.11, p. 33]. Also, the components $\rho_i^{(4)}$, $i \in \{0,1,2,3\}$ of the reliability vector $\boldsymbol{\rho}^{(4)}$ are shown in Figure 4.10.

The first component $\rho_0^{(4)}$ equal to 0 represents the reliability of the message bit u_0, the second component $\rho_1^{(4)}$ equal to 1 represents the reliability of the message bit u_1 and so on.

With

$$\boldsymbol{u} = (u_0, u_1, u_2, u_3) \tag{4.154}$$

we find

$$\boldsymbol{X} = (u_0, u_1, u_2, u_3)\begin{pmatrix} 1 & 0 & 0 & 0 \\ 1 & 0 & 1 & 0 \\ 1 & 1 & 0 & 0 \\ 1 & 1 & 1 & 1 \end{pmatrix} = \big([u_0 \oplus u_1 \oplus u_2 \oplus u_3], [u_2 \oplus u_3], [u_1 \oplus u_3], u_3\big). \tag{4.155}$$

The corresponding polar code with length N equal to 4 is shown in Figure 4.8 and in Figure 4.9.

Using the chain rule of the average mutual information, we yield

$$I\{\boldsymbol{u}; \boldsymbol{Y}\} = I\{u_0; \boldsymbol{Y}\} + I\{u_1; \boldsymbol{Y} \mid u_0\} + I\{u_2; \boldsymbol{Y} \mid \boldsymbol{u}_0^1\} + I\{u_3; \boldsymbol{Y} \mid \boldsymbol{u}_0^2\}. \tag{4.156}$$

Hence,

- $I\{u_0; \boldsymbol{Y}\}$ refers to a *virtual channel* W_4^{--}, i. e., W_4^0, which relates the output symbols y_0, y_1, y_2, y_3 with the input bit u_0, which is equal to $(X_0 \oplus X_1 \oplus X_2 \oplus X_3)$,
- $I\{u_1; \boldsymbol{Y} \mid u_0\}$ refers to a *virtual channel* W_4^{-+}, i. e., W_4^1, which relates the output symbols y_0, y_1, y_2, y_3 with the input bit u_1 equal to $(X_2 \oplus X_3)$ or $(X_0 \oplus X_1 \oplus u_0)$, given the input bit u_0,
- $I\{u_2; \boldsymbol{Y} \mid \boldsymbol{u}_0^1\}$ refers to a *virtual channel* W_4^{+-}, i. e., W_4^2, which relates the output symbols y_0, y_1, y_2, y_3 with the input bit u_2, being equal to $(X_0 \oplus X_2 \oplus u_0)$ or $(X_1 \oplus X_2 \oplus u_1)$ or $(X_0 \oplus X_3 \oplus u_0 \oplus u_1)$ or $(X_1 \oplus X_3)$, given the input bits u_0 and u_1,
- $I\{u_3; \boldsymbol{Y} \mid \boldsymbol{u}_0^2\}$ refers to a *virtual channel* W_4^{++}, i. e., W_4^3, which relates the output symbols y_0, y_1, y_2, y_3 with the input bit u_3 equal to $(X_0 \oplus u_0 \oplus u_1 \oplus u_2)$ or $(X_1 \oplus u_2)$ or $(X_2 \oplus u_1)$ or X_3 or $(X_0 \oplus X_1 \oplus X_2 \oplus u_0)$ or $(X_0 \oplus X_1 \oplus X_3 \oplus u_0 \oplus u_1)$ or $(X_0 \oplus X_2 \oplus X_3 \oplus u_0 \oplus u_2)$ or $(X_1 \oplus X_2 \oplus X_3 \oplus u_0 \oplus u_1 \oplus u_2)$.

Regarding W_4^{--}, i. e., W_4^0, the first of these four split virtual channels, u_0 is erased if either one of X_0, X_1, X_2 or X_3 or combinations thereof are erased, that is with the erasure probability

$$\epsilon_4^{(0)} = \begin{pmatrix} 4 \\ 1 \end{pmatrix} \epsilon(1-\epsilon)^3 + \begin{pmatrix} 4 \\ 2 \end{pmatrix} \epsilon^2(1-\epsilon)^2 + \begin{pmatrix} 4 \\ 4 \end{pmatrix} \epsilon^3(1-\epsilon) + \begin{pmatrix} 4 \\ 4 \end{pmatrix} \epsilon^4, \tag{4.157}$$

yielding

$$\epsilon_4^{(0)} = 4\epsilon - 6\epsilon^2 + 4\epsilon^3 - \epsilon^4. \tag{4.158}$$

Consequently,

$$C_4^{--} = C_4^0 = \max I\{u_0; \boldsymbol{Y}\} = 1 - \epsilon_4^{(0)} = 1 - 4\epsilon + 6\epsilon^2 - 4\epsilon^3 + \epsilon^4 = (1-\epsilon)^4, \tag{4.159}$$

i. e.,

$$C_4^{--} = C_4^0 = \left(C_2^{-}\right)^2 = \left(C_2^0\right)^2. \tag{4.160}$$

Regarding W_4^{-+}, i. e., W_4^1, the second of these four split virtual channels, u_1, which equals $(X_2 \oplus X_3)$ or $(X_0 \oplus X_1 \oplus u_0)$ is erased only if both $(X_2 \oplus X_3)$ and $(X_0 \oplus X_1)$ are erased. Since $(X_2 \oplus X_3)$ and $(X_0 \oplus X_1)$ are independent, that is with probability

$$\epsilon_4^{(1)} = \Big[\underbrace{\epsilon(1-\epsilon)}_{\substack{X_0 \text{ is erased,} \\ X_1 \text{ is not erased}}} + \underbrace{\epsilon(1-\epsilon)}_{\substack{X_1 \text{ is erased,} \\ X_0 \text{ is not erased}}} + \underbrace{\epsilon^2}_{\substack{X_0 \text{ is erased and} \\ X_1 \text{ is erased}}}\Big]$$

$$\cdot\Bigg[\underbrace{\epsilon(1-\epsilon)}_{\substack{X_2\text{ is erased,}\\X_3\text{ is not erased}}}+\underbrace{\epsilon(1-\epsilon)}_{\substack{X_3\text{ is erased,}\\X_2\text{ is not erased}}}+\underbrace{\epsilon^2}_{\substack{X_2\text{ is erased and}\\X_3\text{ is erased}}}\Bigg] \tag{4.161}$$
$$=\left[\binom{2}{1}\epsilon(1-\epsilon)+\binom{2}{2}\epsilon^2\right]^2$$
$$=\left[2\epsilon(1-\epsilon)+\epsilon^2\right]^2$$
$$=\left[2\epsilon-\epsilon^2\right]^2,$$

yielding

$$\epsilon_4^{(1)}=4\epsilon^2-4\epsilon^3+\epsilon^4. \tag{4.162}$$

Consequently,

$$\begin{aligned}C_4^{-+}&=C_4^1\\&=I\{u_1;\boldsymbol{Y}\mid u_0\}\\&=1-\epsilon_4^{(1)}\\&=1-4\epsilon^2+4\epsilon^3-\epsilon^4\\&=2-4\epsilon+2\epsilon^2-\left[1-4\epsilon+6\epsilon^2-4\epsilon^3+\epsilon^4\right]\\&=2(1-\epsilon)^2-(1-\epsilon)^4,\end{aligned} \tag{4.163}$$

yielding

$$C_4^{-+}=C_4^1=2C_2^- -\left(C_2^-\right)^2=2C_2^0-\left(C_2^0\right)^2. \tag{4.164}$$

Regarding W_4^{+-}, i. e., W_4^2, the third of these four split virtual channels, u_2, which equals $(X_0\oplus X_2\oplus u_0)$ or $(X_1\oplus X_2\oplus u_1)$ or $(X_0\oplus X_3\oplus u_0\oplus u_1)$ or $(X_1\oplus X_3)$ is erased only if $(X_0\oplus X_2)$, $(X_1\oplus X_2)$, $(X_0\oplus X_3)$ and $(X_1\oplus X_3)$ are jointly erased, that is with probability

$$\begin{aligned}\epsilon_4^{(2)}=&\underbrace{\epsilon}_{X_0\text{ is erased}}\cdot\underbrace{\epsilon}_{X_1\text{ is erased}}\cdot\underbrace{(1-\epsilon)}_{X_2\text{ is not erased}}\cdot\underbrace{(1-\epsilon)}_{X_3\text{ is not erased}}\\&+\underbrace{\epsilon}_{X_2\text{ is erased}}\cdot\underbrace{\epsilon}_{X_3\text{ is erased}}\cdot\underbrace{(1-\epsilon)}_{X_0\text{ is not erased}}\cdot\underbrace{(1-\epsilon)}_{X_1\text{ is not erased}}\\&+\underbrace{\epsilon}_{X_0\text{ is erased}}\cdot\underbrace{\epsilon}_{X_1\text{ is erased}}\cdot\underbrace{\epsilon}_{X_2\text{ is erased}}\cdot\underbrace{(1-\epsilon)}_{X_3\text{ is not erased}}\\&+\underbrace{\epsilon}_{X_0\text{ is erased}}\cdot\underbrace{\epsilon}_{X_1\text{ is erased}}\cdot\underbrace{\epsilon}_{X_3\text{ is erased}}\cdot\underbrace{(1-\epsilon)}_{X_2\text{ is not erased}}\\&+\underbrace{\epsilon}_{X_0\text{ is erased}}\cdot\underbrace{\epsilon}_{X_2\text{ is erased}}\cdot\underbrace{\epsilon}_{X_3\text{ is erased}}\cdot\underbrace{(1-\epsilon)}_{X_1\text{ is not erased}}\\&+\underbrace{\epsilon}_{X_1\text{ is erased}}\cdot\underbrace{\epsilon}_{X_2\text{ is erased}}\cdot\underbrace{\epsilon}_{X_3\text{ is erased}}\cdot\underbrace{(1-\epsilon)}_{X_0\text{ is not erased}}\\&+\underbrace{\epsilon}_{X_0\text{ is erased}}\cdot\underbrace{\epsilon}_{X_1\text{ is erased}}\cdot\underbrace{\epsilon}_{X_2\text{ is erased}}\cdot\underbrace{(1-\epsilon)}_{X_3\text{ is erased}}\\&=2\epsilon^2(1-\epsilon)^2+4\epsilon^3(1-\epsilon)+\epsilon^4\\&=2\epsilon^2\left(1-2\epsilon+\epsilon^2\right)+4\epsilon^3-3\epsilon^4\\&=2\epsilon^2-4\epsilon^3+2\epsilon^4+4\epsilon^3-3\epsilon^4,\end{aligned} \tag{4.165}$$

i. e.,

$$\epsilon_4^{(2)} = 2\epsilon^2 - \epsilon^4. \tag{4.166}$$

We observe the following:

- In order to jointly erase $(X_0 \oplus X_2)$, $(X_1 \oplus X_2)$, $(X_0 \oplus X_3)$ and $(X_1 \oplus X_3)$, the erasing of a single bit X_0, X_1, X_2 or X_3 is not sufficient; consequently, there is no contribution of the form $\epsilon(1-\epsilon)^3$ to $\epsilon_4^{(2)}$.
- If X_0, X_2 would be erased and X_1, X_3 would not, $(X_1 \oplus X_3)$ would not be erased.
- If X_0, X_3 would be erased and X_1, X_2 would not, $(X_1 \oplus X_2)$ would not be erased.
- If X_1, X_2 would be erased and X_0, X_3 would not, $(X_0 \oplus X_3)$ would not be erased.
- If X_1, X_3 would be erased and X_0, X_2 would not, $(X_0 \oplus X_2)$ would not be erased.

Hence, the last four cases do not contribute to $\epsilon_4^{(2)}$.

Consequently,

$$C_4^{+-} = C_4^2 = \max I\{u_2; \mathbf{Y} \mid \mathbf{u}_0^1\} = 1 - \epsilon_4^{(2)} = 1 - 2\epsilon^2 + \epsilon^4 = \left(1-\epsilon^2\right)^2, \tag{4.167}$$

yielding

$$C_4^{+-} = C_4^2 = \left(C_2^+\right)^2 = \left(C_2^1\right)^2. \tag{4.168}$$

Regarding W_4^{++}, i. e., W_4^3, the fourth of these four split virtual channels, u_3, which equals $(X_0 \oplus u_0 \oplus u_1 \oplus u_2)$ or $(X_1 \oplus u_2)$ or $(X_2 \oplus u_1)$ or X_3 or $(X_0 \oplus X_1 \oplus X_2 \oplus u_0)$ or $(X_0 \oplus X_1 \oplus X_3 \oplus u_0 \oplus u_1)$ or $(X_0 \oplus X_2 \oplus X_3 \oplus u_0 \oplus u_2)$ or $(X_1 \oplus X_2 \oplus X_3 \oplus u_0 \oplus u_1 \oplus u_2)$ is erased only if X_0, X_1, X_2 and X_3 are jointly erased, that is with probability

$$\epsilon_4^{(3)} = \epsilon^4. \tag{4.169}$$

Consequently,

$$\begin{aligned} C_4^{++} &= C_4^3 \\ &= \max I\{u_3; \mathbf{Y} \mid \mathbf{u}_0^2\} \\ &= 1 - \epsilon_4^{(3)} \\ &= 1 - \epsilon^4 \\ &= 2 - 2\epsilon^2 - 1 + 2\epsilon^2 - \epsilon^4 \\ &= 2\left(1-\epsilon^2\right) - \left(1-\epsilon^2\right)^2 \end{aligned} \tag{4.170}$$

yielding

$$C_4^{++} = C_4^3 = 2C_2^+ - \left(C_2^+\right)^2 = 2C_2^1 - \left(C_2^1\right)^2. \tag{4.171}$$

The channel capacities C_1, C_4^{--}, C_4^{-+}, C_4^{+-} and C_4^{++} are depicted in Figure 4.11.

As expected, we find

$$C_4^{--} \le C_4^{-+} \le C_4^{+-} \le C_4^{++} \quad \left(C_4^0 \le C_4^1 \le C_4^2 \le C_4^3\right) \tag{4.172}$$

with equality only for ϵ equal to 1 (noise only) or 0 (perfect channel) and

$$\begin{aligned} C_4 &= C_4^{--} + C_4^{-+} + C_4^{+-} + C_4^{++} \\ &= \left(C_2^-\right)^2 + 2C_2^- - \left(C_2^-\right)^2 + \left(C_2^+\right)^2 + 2C_2^+ - \left(C_2^+\right)^2 \\ &= 2\left(C_2^- + C_2^+\right) \end{aligned} \tag{4.173}$$

$$= 2C_2$$
$$= 4C_1.$$

The reliability vector is given by

$$\boldsymbol{\rho}^{(4)} = (0, 1, 2, 3). \tag{4.174}$$

A clever transmission hence follows the following concept:
1) Do not assign any information to u_0, rather freeze it, i. e., assign a fixed value to it, 0 say.
2) Use only u_1, u_2 and u_3 or preferably only u_2 and u_3 with u_1 frozen to, e. g., 0, as well for information transmission.

Let us consider the latter case, i. e.,

$$\boldsymbol{u} = (0, 0, u_2, u_3). \tag{4.175}$$

With (4.175), we have

$$\boldsymbol{X} = (0, 0, u_2, u_3) \begin{pmatrix} 1 & 0 & 0 & 0 \\ 1 & 0 & 1 & 0 \\ 1 & 1 & 0 & 0 \\ 1 & 1 & 1 & 1 \end{pmatrix} = \big([u_2 \oplus u_3], [u_2 \oplus u_3], u_3, u_3\big), \tag{4.176}$$

which yields the polar code

$$\big\{(0, 0, 0, 0), (1, 1, 0, 0), (1, 1, 1, 1), (0, 0, 1, 1)\big\}, \tag{4.177}$$

with length N equal to 4, dimension K equal to 2 and minimum distance $d_{\min}$ equal to 2, having the code rate

$$R = \frac{K}{N} = \frac{1}{2}. \tag{4.178}$$

This is a subcode of the first-order Reed–Muller (RM) code $\mathcal{RM}(1, 2)$. The following codewords of the first-order Reed–Muller code $\mathcal{RM}(1, 2)$ were left out:

$$\big\{(0, 1, 0, 1), (1, 0, 1, 0), (0, 1, 1, 0), (1, 0, 0, 1)\big\}. \tag{4.179}$$

Now, let us use

$$\boldsymbol{u} = (0, u_1, u_2, u_3), \tag{4.180}$$

yielding

$$\boldsymbol{X} = (0, u_1, u_2, u_3) \begin{pmatrix} 1 & 0 & 0 & 0 \\ 1 & 0 & 1 & 0 \\ 1 & 1 & 0 & 0 \\ 1 & 1 & 1 & 1 \end{pmatrix} = \big([u_1 \oplus u_2 \oplus u_3], [u_2 \oplus u_3], [u_1 \oplus u_3], u_3\big) \tag{4.181}$$

and the polar code

$$\left\{ \begin{matrix} (0, 0, 0, 0), (1, 1, 1, 1), \\ (0, 1, 0, 1), (1, 0, 1, 0), \\ (0, 0, 1, 1), (1, 1, 0, 0), \\ (0, 1, 1, 0), (1, 0, 0, 1) \end{matrix} \right\}, \tag{4.182}$$

which is the first-order Reed–Muller (RM) code $\mathcal{RM}(1, 2)$.

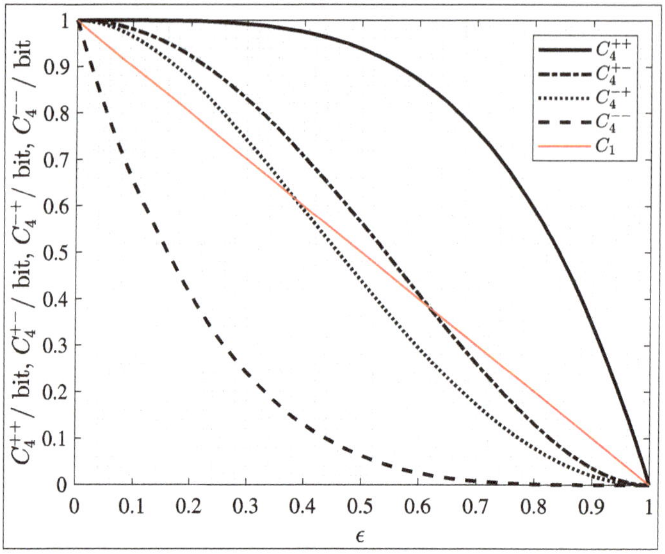

Figure 4.11: Channel capacities C_1 of the binary erasure channel (BEC) as well as C_4^{--}, C_4^{-+}, C_4^{+-} and C_4^{++} of the four split virtual channels W_4^{--}, W_4^{-+}, W_4^{+-} and W_4^{++}.

Example 4.4 (Polar Code of Length 8). The *three-fold Kronecker product* of $\boldsymbol{G}_2$ of (4.66) yields

$$\boldsymbol{G}_2^{\otimes 3} = \boldsymbol{G}_2 \otimes \boldsymbol{G}_2 \otimes \boldsymbol{G}_2 = \begin{pmatrix} 1 & 0 & 0 & 0 & 0 & 0 & 0 & 0 \\ 1 & 1 & 0 & 0 & 0 & 0 & 0 & 0 \\ 1 & 0 & 1 & 0 & 0 & 0 & 0 & 0 \\ 1 & 1 & 1 & 1 & 0 & 0 & 0 & 0 \\ 1 & 0 & 0 & 0 & 1 & 0 & 0 & 0 \\ 1 & 1 & 0 & 0 & 1 & 1 & 0 & 0 \\ 1 & 0 & 1 & 0 & 1 & 0 & 1 & 0 \\ 1 & 1 & 1 & 1 & 1 & 1 & 1 & 1 \end{pmatrix}, \tag{4.183}$$

which can be transformed into the standard form of the generator matrix of the polar code with N equal to 8 by swapping rows 2 and 5 as well as rows 4 and 7, while the rows 1, 3, 6 and 8 remain in place. We yield

$$\boldsymbol{G}_8 = \boldsymbol{B}_8 \boldsymbol{G}_2^{\otimes 3} = \begin{pmatrix} 1 & 0 & 0 & 0 & 0 & 0 & 0 & 0 \\ 0 & 0 & 0 & 0 & 1 & 0 & 0 & 0 \\ 0 & 0 & 1 & 0 & 0 & 0 & 0 & 0 \\ 0 & 0 & 0 & 0 & 0 & 0 & 1 & 0 \\ 0 & 1 & 0 & 0 & 0 & 0 & 0 & 0 \\ 0 & 0 & 0 & 0 & 0 & 1 & 0 & 0 \\ 0 & 0 & 0 & 1 & 0 & 0 & 0 & 0 \\ 0 & 0 & 0 & 0 & 0 & 0 & 0 & 1 \end{pmatrix} \begin{pmatrix} 1 & 0 & 0 & 0 & 0 & 0 & 0 & 0 \\ 1 & 1 & 0 & 0 & 0 & 0 & 0 & 0 \\ 1 & 0 & 1 & 0 & 0 & 0 & 0 & 0 \\ 1 & 1 & 1 & 1 & 0 & 0 & 0 & 0 \\ 1 & 0 & 0 & 0 & 1 & 0 & 0 & 0 \\ 1 & 1 & 0 & 0 & 1 & 1 & 0 & 0 \\ 1 & 0 & 1 & 0 & 1 & 0 & 1 & 0 \\ 1 & 1 & 1 & 1 & 1 & 1 & 1 & 1 \end{pmatrix}$$

$$
= \begin{pmatrix}
1 & 0 & 0 & 0 & 0 & 0 & 0 & 0 \\
1 & 0 & 0 & 0 & 1 & 0 & 0 & 0 \\
1 & 0 & 1 & 0 & 0 & 0 & 0 & 0 \\
1 & 0 & 1 & 0 & 1 & 0 & 1 & 0 \\
1 & 1 & 0 & 0 & 0 & 0 & 0 & 0 \\
1 & 1 & 0 & 0 & 1 & 1 & 0 & 0 \\
1 & 1 & 1 & 1 & 0 & 0 & 0 & 0 \\
1 & 1 & 1 & 1 & 1 & 1 & 1 & 1
\end{pmatrix}. \tag{4.184}
$$

Using the chain rule of the average mutual information, we obtain

$$
\begin{aligned}
I\{\boldsymbol{u};\boldsymbol{Y}\} = I\{u_0;\boldsymbol{Y}\} &+ I\{u_1;\boldsymbol{Y} \mid u_0\} \\
&+ I\{u_2;\boldsymbol{Y} \mid \boldsymbol{u}_0^1\} + I\{u_3;\boldsymbol{Y} \mid \boldsymbol{u}_0^2\} \\
&+ I\{u_4;\boldsymbol{Y} \mid \boldsymbol{u}_0^3\} + I\{u_5;\boldsymbol{Y} \mid \boldsymbol{u}_0^4\} \\
&+ I\{u_6;\boldsymbol{Y} \mid \boldsymbol{u}_0^5\} + I\{u_7;\boldsymbol{Y} \mid \boldsymbol{u}_0^6\}.
\end{aligned} \tag{4.185}
$$

We find the following capacities of the eight split virtual channels W_8^{---}, i. e., W_8^0, to W_8^{+++}, i. e., W_8^7:

$$
\begin{aligned}
C_8^{---} &= C_8^0 = \max I\{u_0;\boldsymbol{Y}\} = \left(C_4^{--}\right)^2 = \left(C_4^0\right)^2, \\
C_8^{--+} &= C_8^1 = \max I\{u_1;\boldsymbol{Y} \mid u_0\} = 2C_4^{--} - \left(C_4^{--}\right)^2 = 2C_4^0 - \left(C_4^0\right)^2, \\
C_8^{-+-} &= C_8^2 = \max I\{u_2;\boldsymbol{Y} \mid \boldsymbol{u}_0^1\} = \left(C_4^{-+}\right)^2 = \left(C_4^1\right)^2, \\
C_8^{-++} &= C_8^3 = \max I\{u_3;\boldsymbol{Y} \mid \boldsymbol{u}_0^2\} = 2C_4^{-+} - \left(C_4^{-+}\right)^2 = 2C_4^1 - \left(C_4^1\right)^2, \\
C_8^{+--} &= C_8^4 = \max I\{u_4;\boldsymbol{Y} \mid \boldsymbol{u}_0^3\} = \left(C_4^{+-}\right)^2 = \left(C_4^2\right)^2, \\
C_8^{+-+} &= C_8^5 = \max I\{u_5;\boldsymbol{Y} \mid \boldsymbol{u}_0^4\} = 2C_4^{+-} - \left(C_4^{+-}\right)^2 = 2C_4^2 - \left(C_4^2\right)^2, \\
C_8^{++-} &= C_8^6 = \max I\{u_6;\boldsymbol{Y} \mid \boldsymbol{u}_0^5\} = \left(C_4^{++}\right)^2 = \left(C_4^3\right)^2, \\
C_8^{+++} &= C_8^7 = \max I\{u_7;\boldsymbol{Y} \mid \boldsymbol{u}_0^6\} = 2C_4^{++} - \left(C_4^{++}\right)^2 = 2C_4^3 - \left(C_4^3\right)^2.
\end{aligned} \tag{4.186}
$$

The channel capacities C_8^{---} to C_8^{+++} of the eight split virtual channels W_8^{---} to W_8^{+++} are shown in Figure 4.12. We find that

$$
C_8^{---} \le C_8^{--+} \le C_8^{-+-} \le C_8^{+--} \le C_8^{-++} \le C_8^{+-+} \le C_8^{++-} \le C_8^{+++}, \tag{4.187}
$$

i. e.,

$$
C_8^0 \le C_8^1 \le C_8^2 \le C_8^4 \le C_8^3 \le C_8^5 \le C_8^6 \le C_8^7. \tag{4.188}
$$

Clearly,

$$
\begin{aligned}
C_8 &= C_8^{---} + C_8^{--+} + C_8^{-+-} + C_8^{+--} + C_8^{-++} + C_8^{+-+} + C_8^{++-} + C_8^{+++} \\
&= 2C_4 \\
&= 4C_2 \\
&= 8C_1.
\end{aligned} \tag{4.189}
$$

The reliability vector is given by

$$
\boldsymbol{\rho}^{(8)} = (0, 1, 2, 4, 3, 5, 6, 7). \tag{4.190}
$$

Therefore, it is advisable to use only the bits u_4, u_6, u_7, u_8 for data transmission, i. e.,

$$\boldsymbol{u} = (0, 0, 0, u_3, 0, u_5, u_6, u_7) \tag{4.191}$$

yielding

$$\begin{aligned}\boldsymbol{X} = (0, 0, 0, u_3, 0, u_5, u_6, u_7) &\begin{pmatrix} 1 & 0 & 0 & 0 & 0 & 0 & 0 & 0 \\ 1 & 0 & 0 & 0 & 1 & 0 & 0 & 0 \\ 1 & 0 & 1 & 0 & 0 & 0 & 0 & 0 \\ 1 & 0 & 1 & 0 & 1 & 0 & 1 & 0 \\ 1 & 1 & 0 & 0 & 0 & 0 & 0 & 0 \\ 1 & 1 & 0 & 0 & 1 & 1 & 0 & 0 \\ 1 & 1 & 1 & 1 & 0 & 0 & 0 & 0 \\ 1 & 1 & 1 & 1 & 1 & 1 & 1 & 1 \end{pmatrix} \\ = ([u_3 \oplus u_5 \oplus u_6 \oplus u_7], &[u_5 \oplus u_6 \oplus u_7], [u_3 \oplus u_6 \oplus u_7], \\ [u_6 \oplus u_7], [u_3 \oplus u_5 \oplus u_7], &[u_5 \oplus u_7], [u_3 \oplus u_7], u_7),\end{aligned} \tag{4.192}$$

which results in the following polar code:

$$\left\{\begin{array}{l}(0,0,0,0,0,0,0,0), (1,1,1,1,1,1,1,1), (0,0,0,0,1,1,1,1), (1,1,1,1,0,0,0,0), \\ (0,0,1,1,1,1,0,0), (1,1,0,0,0,0,1,1), (0,0,1,1,0,0,1,1), (1,1,0,0,1,1,0,0), \\ (0,1,1,0,1,0,0,1), (1,0,0,1,0,1,1,0), (0,1,1,0,0,1,1,0), (1,0,0,1,1,0,0,1), \\ (0,1,0,1,1,0,1,0), (1,0,1,0,0,1,0,1), (0,1,0,1,0,1,0,1), (1,0,1,0,1,0,1,0)\end{array}\right\} \tag{4.193}$$

with dimension K equal to 4, length N equal to 8 and minimum distance $d_{\min}$ equal to 4, having the code rate

$$R = \frac{K}{N} = \frac{1}{2}. \tag{4.194}$$

This polar code is identical with the first-order Reed–Muller (RM) code $\mathcal{RM}(1, 3)$.

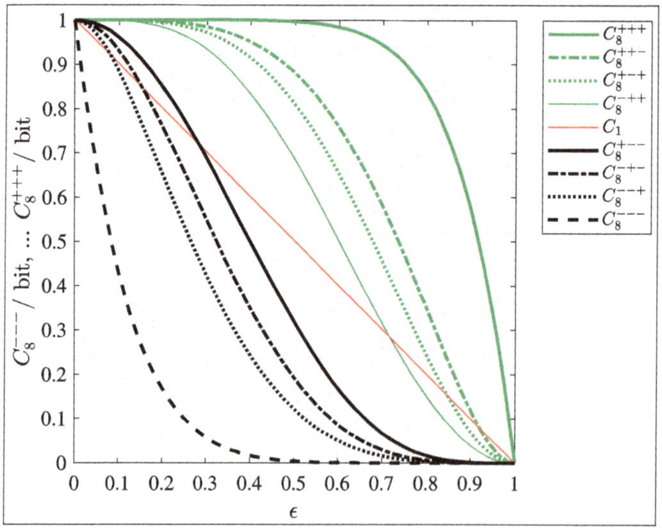

Figure 4.12: Channel capacities C_8^{---} to C_8^{+++} of the eight split virtual channels W_8^{---} to W_8^{+++}.

Let us now generalize the above findings. First, we introduce the binary representation $\beta_\nu\beta_{\nu-1}\cdots\beta_2\beta_1$ of a number $i \in \{0,1\cdots(N-1)\}$, $N = 2^\nu$, $\nu \in \mathbb{N}^*$,

$$i = \beta_\nu 2^{\nu-1} + \beta_{\nu-1}2^{\nu-2} + \cdots + \beta_1 2^0, \quad \beta_\nu, \beta_{\nu-1}\cdots\beta_\nu, \beta_1 \in \mathbb{F}_2. \tag{4.195}$$

Then, using

$$b_k = \begin{cases} + & \text{if } \beta_k = 1, \\ - & \text{if } \beta_k = 0, \end{cases} \quad k \in \{1,2\cdots\nu\}, \tag{4.196}$$

the capacity of the N virtual channels formed by setting out from the binary erasure channel (BEC) is given by [33, equation (3.32), p. 88]

$$C_N^{b_\nu\cdots b_2 b_1} = \begin{cases} (C_{N/2}^{b_\nu\cdots b_3 b_2})^2 & \text{if } b_1 = - \quad (\beta_1 = 0), \\ 2C_{N/2}^{b_\nu\cdots b_3 b_2} - (C_{N/2}^{b_\nu\cdots b_3 b_2})^2 & \text{if } b_1 = + \quad (\beta_1 = 1), \end{cases} \tag{4.197}$$

and the corresponding information bit is u_i with i according to (4.195).

Remark 4.2 (Doob Martingale). Let us consider all the values $C_N^{i_N}$, $i_N \in \mathbb{N}$, as random variables. Then the sequence

$$\{C_1, C_2^{i_2}, C_4^{i_4}\cdots C_{N/2}^{i_{N/2}}, C_N^{i_N}\cdots\}, \quad N = 2^\nu, \quad \nu \in \mathbb{N}^*, \quad i_2, i_4\cdots i_{N/2}, i_N\cdots \in \mathbb{N}, \tag{4.198}$$

is the realization of a discrete-time stochastic process. Furthermore, we know that $0 \le C_N^{i_N} \le 1$, $i_N \in \mathbb{N}$, which yields the mathematical expectation [16, equation (40), p. 3059]

$$\mathrm{E}\{|C_N^{i_N}|\} < \infty, \quad N = 2^\nu, \quad \nu \in \mathbb{N}^*, \quad i_2, i_4\cdots i_{N/2}, i_N\cdots \in \mathbb{N}. \tag{4.199}$$

According to (4.197), each $C_N^{i_N}$, $i_N \in \mathbb{N}$, $N = 2^\nu$, $\nu \in \mathbb{N}^*$, can only assume two values, namely $(C_{N/2}^{i_{N/2}})^2$ and $[2C_{N/2}^{i_{N/2}} - (C_{N/2}^{i_{N/2}})^2]$. Each one of these two values will be assumed with equal probability 1/2. Therefore, given $C_{N/2}^{i_{N/2}}$, the conditional mathematical expectation $\mathrm{E}\{C_N^{i_N} \mid C_{N/2}^{i_{N/2}}\}$ is given by [16, equation (42), p. 3059]

$$\mathrm{E}\{C_N^{i_N} \mid C_{N/2}^{i_{N/2}}\} = \frac{1}{2}\cdot\left(C_{N/2}^{i_{N/2}}\right)^2 + \frac{1}{2}\cdot\left[2C_{N/2}^{i_{N/2}} - \left(C_{N/2}^{i_{N/2}}\right)^2\right] = C_{N/2}^{i_{N/2}}. \tag{4.200}$$

The said stochastic process with realizations of the form given by (4.198), which obeys (4.199) and (4.200) is called *Doob martingale* [110, p. 91], named after the American mathematician Joseph Leo *Doob*.

Most importantly, $C_N^{i_N}$, $i_N \in \mathbb{N}$, $N = 2^\nu$, constituting the realization given by (4.198) converge to a finite limit for $N \to \infty$ [110, Theorem 4.1, p. 319].

Although there are multiple origins [111] in the context of probability theory the word martingale can be considered to be “borrowed directly from the vocabulary of gamblers” calling their winning strategies “martingales” [111, p. 3]. “To play the martingale is to always bet all that was lost” [111, p. 3].

Since the channel capacity and the Bhattacharyya parameter are oppositely related [33, p. 88], we yield the Bhattacharyya parameter [33, equation (3.31), p. 88]

$$Z_N^{b_\nu \cdots b_2 b_1} = \begin{cases} 2Z_{N/2}^{b_\nu \cdots b_3 b_2} - (Z_{N/2}^{b_\nu \cdots b_3 b_2})^2 & \text{if } b_1 = - \quad (\beta_1 = 0), \\ (Z_{N/2}^{b_\nu \cdots b_3 b_2})^2 & \text{if } b_1 = + \quad (\beta_1 = 1), \end{cases} \tag{4.201}$$

when setting out from the binary erasure channel (BEC).

Now, we can determine the code rate

$$R = \frac{K}{N}, \quad K \le N, \quad K, N \in \mathbb{N}^*. \tag{4.202}$$

Choosing those message bits with the K greatest capacity values determined by (4.197) and setting the remaining $(N - K)$ bits to a fixed value, 0 say, making these fixed bits "dummy bits" or "*frozen bits*," is most advisable for reliable communications.

Example 4.5. Let us extend Example 4.2. Again, we set out from the binary erasure channel (BEC), having the channel capacity

$$C_1 = 1 - \epsilon \tag{4.203}$$

and the Bhattacharyya parameter

$$Z_1 = \epsilon. \tag{4.204}$$

Let N equal to 2. Then we have

$$C_2^i\big|_{i=0} = C_2^0 = C_2^- = (C_1)^2 = (1-\epsilon)^2 = 1 - 2\epsilon + \epsilon^2, \tag{4.205}$$

$$\begin{aligned} C_2^i\big|_{i=1} &= C_2^1 = C_2^+ = 2C_1 - (C_1)^2 = 2(1-\epsilon) - (1-\epsilon)^2 = (1-\epsilon)(2-1+\epsilon) \\ &= (1+\epsilon)(1-\epsilon) = 1 - \epsilon^2 \ge C_2^-. \end{aligned} \tag{4.206}$$

The message bit corresponding to C_2^- is u_0. The message bit corresponding to C_2^+ is u_1. Since C_2^+ is greater than or equal to C_2^-, we choose u_0 to be a "dummy bit," also called "*frozen bit*" [16, p. 3054], freezing its value to 0, and use only u_1 for information transfer.

The Bhattacharyya parameter is given by

$$Z_2^i\big|_{i=0} = Z_2^0 = Z_2^- = 2Z_1 - (Z_1)^2 = 2\epsilon - \epsilon^2, \tag{4.207}$$

$$Z_2^i\big|_{i=1} = Z_2^1 = Z_2^+ = (Z_1)^2 = \epsilon^2 \le Z_2^-. \tag{4.208}$$

Example 4.6. Let us extend Example 4.3. Again, we set out from the binary erasure channel (BEC), having the channel capacity

$$C_1 = 1 - \epsilon. \tag{4.209}$$

Let N equal to 4. Then we have

$$C_4^{b_2 b_1} = \begin{cases} (C_2^{b_2})^2 & \text{if } b_1 = -, \\ 2C_2^{b_2} - (C_2^{b_2})^2 & \text{if } b_1 = +, \end{cases} \tag{4.210}$$

i. e.,

$$\begin{aligned}
C_4^i\big|_{i=0} &= C_4^0 = C_4^{--} = \left(C_2^-\right)^2, && \text{corresponding message bit } u_0,\\
C_4^i\big|_{i=1} &= C_4^1 = C_4^{-+} = 2C_2^- - \left(C_2^-\right)^2, && \text{corresponding message bit } u_1,\\
C_4^i\big|_{i=2} &= C_4^2 = C_4^{+-} = \left(C_2^+\right)^2, && \text{corresponding message bit } u_2,\\
C_4^i\big|_{i=3} &= C_4^3 = C_4^{++} = 2C_2^+ - \left(C_2^+\right)^2, && \text{corresponding message bit } u_3.
\end{aligned} \tag{4.211}$$

Since

$$C_4^{--} \le C_4^{-+} \le C_4^{+-} \le C_4^{++}, \tag{4.212}$$

the selection of "dummy bits," i. e., "frozen bits," should be according to the following scheme:

- First, let u_0 be the "dummy bit," yielding the code rate R equal to 3/4.
- Next, let u_0 and u_1 be the "dummy bits," yielding the code rate R equal to 1/2.
- Finally, let u_0, u_1 and u_2 be the "dummy bits," yielding the code rate R equal to 1/4.

The Bhattacharyya parameter is given by

$$\begin{aligned}
Z_4^i\big|_{i=0} &= Z_4^0 = Z_4^{--} = 2Z_2^- - \left(Z_2^-\right)^2, && \text{corresponding message bit } u_0,\\
Z_4^i\big|_{i=1} &= Z_4^1 = Z_4^{-+} = \left(Z_2^-\right)^2, && \text{corresponding message bit } u_1,\\
Z_4^i\big|_{i=2} &= Z_4^2 = Z_4^{+-} = 2Z_2^+ - \left(Z_2^+\right)^2, && \text{corresponding message bit } u_2,\\
Z_4^i\big|_{i=3} &= Z_4^3 = Z_4^{++} = \left(Z_2^+\right)^2, && \text{corresponding message bit } u_3.
\end{aligned} \tag{4.213}$$

Example 4.7. Let us extend Example 4.4. Again, we set out from the binary erasure channel (BEC), having the channel capacity

$$C_1 = 1 - \epsilon. \tag{4.214}$$

Let N equal to 8. Then we have

$$C_8^{b_3b_2b_1} = \begin{cases} (C_4^{b_3b_2})^2 & \text{if } b_1 = -,\\ 2C_4^{b_3b_2} - (C_4^{b_3b_2})^2 & \text{if } b_1 = +, \end{cases} \tag{4.215}$$

i. e.,

$$\begin{aligned}
C_8^i\big|_{i=0} &= C_8^0 = C_8^{---} = \left(C_4^{--}\right)^2, && \text{corresponding message bit } u_0,\\
C_8^i\big|_{i=1} &= C_8^1 = C_8^{--+} = 2C_4^{--} - \left(C_4^{--}\right)^2, && \text{corresponding message bit } u_1,\\
C_8^i\big|_{i=2} &= C_8^2 = C_8^{-+-} = \left(C_4^{-+}\right)^2, && \text{corresponding message bit } u_2,\\
C_8^i\big|_{i=3} &= C_8^3 = C_8^{-++} = 2C_4^{-+} - \left(C_4^{-+}\right)^2, && \text{corresponding message bit } u_3,\\
C_8^i\big|_{i=4} &= C_8^4 = C_8^{+--} = \left(C_4^{+-}\right)^2, && \text{corresponding message bit } u_4,\\
C_8^i\big|_{i=5} &= C_8^5 = C_8^{+-+} = 2C_4^{+-} - \left(C_4^{+-}\right)^2, && \text{corresponding message bit } u_5,\\
C_8^i\big|_{i=6} &= C_8^6 = C_8^{++-} = \left(C_4^{++}\right)^2, && \text{corresponding message bit } u_6,\\
C_8^i\big|_{i=7} &= C_8^7 = C_8^{+++} = 2C_4^{++} - \left(C_4^{++}\right)^2, && \text{corresponding message bit } u_7.
\end{aligned} \tag{4.216}$$

Since

$$C_8^{---} \le C_8^{--+} \le C_8^{-+-} \le C_8^{+--} \le C_8^{-++} \le C_8^{+-+} \le C_8^{++-} \le C_8^{+++} \tag{4.217}$$

the selection of "dummy bits," i. e., "frozen bits," should be according to the following scheme:

- First, let u_0 be the "dummy bit," yielding the code rate R equal to 7/8.
- Next, let u_0 and u_1 be the "dummy bits," yielding the code rate R equal to 3/4.
- Next, let u_0, u_1 and u_2 be the "dummy bits," yielding the code rate R equal to 5/8.
- Next, let u_0, u_1, u_2 and u_4 be the "dummy bits," yielding the code rate R equal to 1/2.
- Next, let u_0, u_1, u_2, u_4 and u_3 be the "dummy bits," yielding the code rate R equal to 3/8.
- Next, let u_0, u_1, u_2, u_4, u_3 and u_5 be the "dummy bits," yielding the code rate R equal to 1/4.
- Finally, let u_0, u_1, u_2, u_4, u_3, u_5 and u_6 be the "dummy bits," yielding the code rate R equal to 1/8.

The Bhattacharyya parameter is given by

$$\begin{aligned}
Z_8^i\big|_{i=0} &= Z_8^0 = Z_8^{---} = 2Z_4^{--} - \left(Z_4^{--}\right)^2, && \text{corresponding message bit } u_0,\\
Z_8^i\big|_{i=1} &= Z_8^1 = Z_8^{--+} = \left(Z_4^{--}\right)^2, && \text{corresponding message bit } u_1,\\
Z_8^i\big|_{i=2} &= Z_8^2 = Z_8^{-+-} = 2Z_4^{-+} - \left(Z_4^{-+}\right)^2, && \text{corresponding message bit } u_2,\\
Z_8^i\big|_{i=3} &= Z_8^3 = Z_8^{-++} = \left(Z_4^{-+}\right)^2, && \text{corresponding message bit } u_3,\\
Z_8^i\big|_{i=4} &= Z_8^4 = Z_8^{+--} = 2Z_4^{+-} - \left(Z_4^{+-}\right)^2, && \text{corresponding message bit } u_4,\\
Z_8^i\big|_{i=5} &= Z_8^5 = Z_8^{+-+} = \left(Z_4^{+-}\right)^2, && \text{corresponding message bit } u_5,\\
Z_8^i\big|_{i=6} &= Z_8^6 = Z_8^{++-} = 2Z_4^{++} - \left(Z_4^{++}\right)^2, && \text{corresponding message bit } u_6,\\
Z_8^i\big|_{i=7} &= Z_8^7 = Z_8^{+++} = \left(Z_4^{++}\right)^2, && \text{corresponding message bit } u_7.
\end{aligned} \tag{4.218}$$

Figure 4.13 depicts the channel capacities C_{256}^0 to C_{256}^{255} of the 256 split virtual channels W_{256}^0 to W_{256}^{255} and Figure 4.14 depicts the channel capacities C_{1024}^0 to C_{1024}^{1023} of the 1024 split virtual channels W_{1024}^0 to W_{1024}^{1023}. It is quite remarkable to see that even in bad channel conditions, split virtual channels approaching the channel capacity of 1 bit exist.

According to the above discussion, facilitating a best possible robust transmission requires the selection of those split virtual channels with the best performance. This selection is based on the above mentioned reliability sequence, i. e., the reliability vector, and completely defines the polar code. It is this reliability sequence, which determines the polar code and, therefore, the establishing of the reliability sequence can be regarded the essence of the *polar code construction*.

Indeed, polar codes or not universal, which means that the reliability sequence will not only depend on the transmission channel [16, p. 3055], but also on the channel state, e. g., the signal-to-noise ratio (SNR) at the receiver input, which is a time variant in the case of advanced wireless communications.

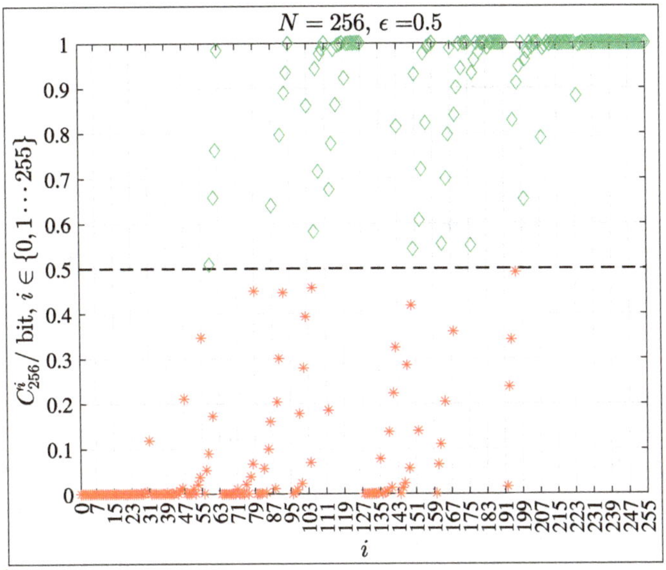

Figure 4.13: Channel capacities C^0_{256} to C^{255}_{256} of the 256 split virtual channels W^0_{256} to W^{255}_{256}.

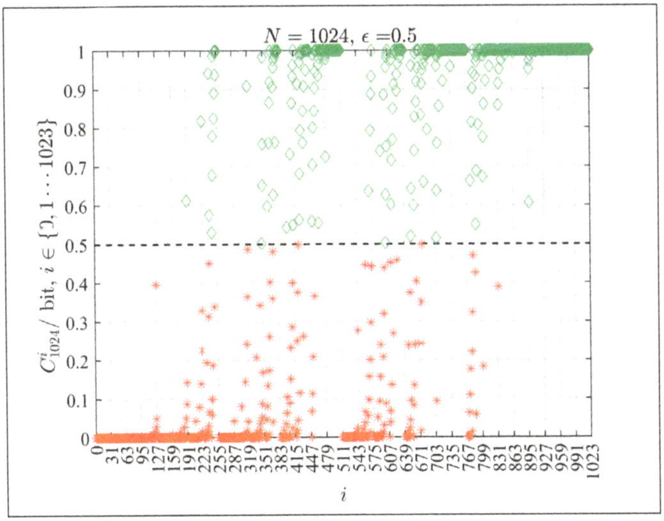

Figure 4.14: Channel capacities C^0_{1024} to C^{1023}_{1024} of the 1024 split virtual channels W^0_{1024} to W^{1023}_{1024}; cf., e. g., [33, Figure 3.8, p. 89] for a similar result.

Example 4.8. Let us consider the binary erasure channel (BEC). In the case of N equal to 32 and the erasure probability ϵ equal to 0.1, the reliability vector will be

$$\begin{aligned}\boldsymbol{\rho}^{(32)} = (&0, 1, 2, 4, 8, 16, 3, 5, 6, 9, 10, 17, 12, 18, 20, 24,\\ &7, 11, 13, 14, 19, 21, 22, 25, 26, 28, 15, 23, 27, 29, 30, 31)\,, \end{aligned} \tag{4.219}$$

whereas for the erasure probability ϵ equal to 0.5, the reliability vector will be

$$\boldsymbol{\rho}^{(32)} = (0,1,2,4,8,16,3,5,6,9,10,17,12,18,7,20,\\ 11,24,13,19,14,21,22,25,26,28,15,23,27,29,30,31)\,. \tag{4.220}$$

Regrettably, there is no unique approach to the establishment of the reliability sequence. In [16], two methods for the establishment of the reliability sequence are given, namely
- a first method based on the Bhattacharyya parameter, alternatively on the channel capacity [16, no. 4), p. 3055],
- a second method based on Monte Carlo simulations; cf., e. g., [16, p. 3055].

An analytical solution to the establishment of the reliability sequence seems to be known only for the binary erasure channel (BEC). In all other cases, especially in the case of the mobile radio channel, lengthy simulations are required, which is a considerable obstacle regarding the deployment of polar codes. For instance, there were some variations of the reliability vector until the final version reported in [112, Table 5.3.1.2-1, pp. 17–19] was agreed on.

Now, having studied some examples of polar codes and evaluated the channel capacities of the split virtual channels, the time has come to discuss what the expression "*capacity-achieving*" used in, e. g., the title of Arikan's publication [16] might mean. We will set out from the length N equal to 2^ν, $\nu \in \mathbb{N}^*$. Furthermore, we will abbreviate the sequence $\{b_\nu \cdots b_2 b_1\}$ used in the superscript of $C_N^{b_\nu \cdots b_2 b_1}$ introduced in (4.197) by the vector

$$\boldsymbol{b}_\nu = (b_\nu \cdots b_2, b_1), \quad \boldsymbol{b} \in \{+,-\}^\nu, \quad \nu \in \mathbb{N}^*. \tag{4.221}$$

Let us furthermore introduce the two real bounds $\delta_\mathrm{L} \in (0,1)$ and $\delta_\mathrm{U} \in (0,1)$, which comply with

$$0 < \delta_\mathrm{L} < \delta_\mathrm{U} < 1. \tag{4.222}$$

In addition, let us denote the number of split virtual channels having the channel capacity $C_N^{b_\nu \cdots b_2 b_1} \in (\delta_\mathrm{U}, 1]$ close to 1 by $n_{\mathrm{U},\nu}$. Then we have

$$0 \le n_{\mathrm{U},\nu} \le N = 2^\nu, \quad \alpha_{\mathrm{U},\nu} = \frac{n_{\mathrm{U},\nu}}{N} = \frac{n_{\mathrm{U},\nu}}{2^\nu} \in [0,1], \quad \nu \in \mathbb{N}^*. \tag{4.223}$$

Also, let us denote the number of split virtual channels having the channel capacity $C_N^{b_\nu \cdots b_2 b_1} \in [\delta_\mathrm{L}, \delta_\mathrm{U}]$ by n_I. Of course, we obtain

$$0 \le n_{\mathrm{I},\nu} \le N = 2^\nu, \quad \alpha_{\mathrm{I},\nu} = \frac{n_{\mathrm{I},\nu}}{N} = \frac{n_{\mathrm{I},\nu}}{2^\nu} \in [0,1], \quad \nu \in \mathbb{N}^*. \tag{4.224}$$

Finally, let us denote the number of split virtual channels having the channel capacity $C_N^{b_\nu \cdots b_2 b_1} \in [0, \delta_\mathrm{L})$ close to 0 by n_L. Analogously to (4.223), we yield

$$0 \le n_{\mathrm{L},\nu} \le N = 2^{\nu}, \quad \alpha_{\mathrm{L},\nu} = \frac{n_{\mathrm{L},\nu}}{N} = \frac{n_{\mathrm{L},\nu}}{2^{\nu}} \in [0,1], \quad \nu \in \mathbb{N}^{*}. \tag{4.225}$$

Now, let us look at channel polarization for a last time. The proof given below follows a similar path as presented in [113, Theorem 9.1, pp. 180–184].

Theorem 4.5 (Channel Polarization). *For any binary input discrete memoryless channel (DMC) having the channel capacity C_1, we obtain*

$$\lim_{N\to\infty} \alpha_{\mathrm{L},\nu} = 1 - C_1, \tag{4.226}$$

$$\lim_{N\to\infty} \alpha_{\mathrm{I},\nu} = 0, \tag{4.227}$$

$$\lim_{N\to\infty} \alpha_{\mathrm{U},\nu} = C_1, \tag{4.228}$$

i. e., the split virtual channels polarize.

Proof. We have

$$\frac{1}{2^{\nu+1}} \sum_{\boldsymbol{b}_{\nu+1}\in\{+,-\}^{\nu+1}} C_N^{\boldsymbol{b}_{\nu+1}} = \frac{1}{2^{\nu}} \sum_{\boldsymbol{b}_{\nu}\in\{+,-\}^{\nu}} \frac{1}{2}\left(C_N^{\boldsymbol{b}_{\nu}+} + C_N^{\boldsymbol{b}_{\nu}-}\right). \tag{4.229}$$

Using (4.197), we yield

$$\frac{1}{2}\left(C_N^{\boldsymbol{b}_{\nu}+} + C_N^{\boldsymbol{b}_{\nu}-}\right) = C_{N/2}^{\boldsymbol{b}_{\nu}}, \tag{4.230}$$

and thus we obtain

$$\frac{1}{2^{\nu+1}} \sum_{\boldsymbol{b}_{\nu+1}\in\{+,-\}^{\nu+1}} C_N^{\boldsymbol{b}_{\nu+1}} = \frac{1}{2^{\nu}} \sum_{\boldsymbol{b}_{\nu}\in\{+,-\}^{\nu}} C_{N/2}^{\boldsymbol{b}_{\nu}}. \tag{4.231}$$

Obviously, (4.231) is equal to C_1, which follows from induction; see below.

Proof 4.1 (Proof by Induction).
<u>Base Case</u>
Use (4.230) for N equal to 2. We obtain

$$\frac{1}{2}\left(C_2^{+} + C_2^{-}\right) = C_1, \tag{4.232}$$

which is identical to (4.140).

<u>Induction Step</u>
We set out from

$$\frac{1}{2^{\nu}} \sum_{\boldsymbol{b}_{\nu}\in\{+,-\}^{\nu}} C_{N/2}^{\boldsymbol{b}_{\nu}} = C_1. \tag{4.233}$$

Combining (4.231) and (4.233) yield

$$\frac{1}{2^{\nu+1}} \sum_{\boldsymbol{b}_{\nu+1}\in\{+,-\}^{\nu+1}} C_N^{\boldsymbol{b}_{\nu+1}} = C_1. \tag{4.234}$$

□

Now, we can proceed to the next aspect. Setting out from (4.197), we also find

$$\frac{1}{2}\left(C_N^{\boldsymbol{b}_\nu +} - C_N^{\boldsymbol{b}_\nu -}\right) = C_{N/2}^{\boldsymbol{b}_\nu} - \left(C_{N/2}^{\boldsymbol{b}_\nu}\right)^2. \tag{4.235}$$

Furthermore, with (4.230) and (4.235), we find the following identity:

$$\begin{aligned}
\frac{(C_N^{\boldsymbol{b}_\nu +})^2 + (C_N^{\boldsymbol{b}_\nu -})^2}{2} &= \frac{(C_N^{\boldsymbol{b}_\nu +})^2 + 2C_N^{\boldsymbol{b}_\nu +}C_N^{\boldsymbol{b}_\nu -} + (C_N^{\boldsymbol{b}_\nu -})^2}{4} + \frac{(C_N^{\boldsymbol{b}_\nu +})^2 - 2C_N^{\boldsymbol{b}_\nu +}C_N^{\boldsymbol{b}_\nu -} + (C_N^{\boldsymbol{b}_\nu -})^2}{4} \\
&= \frac{(C_N^{\boldsymbol{b}_\nu +} + C_N^{\boldsymbol{b}_\nu -})^2}{4} + \frac{(C_N^{\boldsymbol{b}_\nu +} - C_N^{\boldsymbol{b}_\nu -})^2}{4} \\
&= \left(\frac{C_N^{\boldsymbol{b}_\nu +} + C_N^{\boldsymbol{b}_\nu -}}{2}\right)^2 + \left(\frac{C_N^{\boldsymbol{b}_\nu +} - C_N^{\boldsymbol{b}_\nu -}}{2}\right)^2 && (4.236) \\
&= \left(C_{N/2}^{\boldsymbol{b}_\nu}\right)^2 + \left(C_{N/2}^{\boldsymbol{b}_\nu} - \left(C_{N/2}^{\boldsymbol{b}_\nu}\right)^2\right)^2 && (4.237) \\
&= \left(C_{N/2}^{\boldsymbol{b}_\nu}\right)^2\left\{1 + \left(1 - C_{N/2}^{\boldsymbol{b}_\nu}\right)^2\right\} \geq \left(C_{N/2}^{\boldsymbol{b}_\nu}\right)^2.
\end{aligned}$$

Setting out from (4.237), we yield

$$\begin{aligned}
\frac{1}{2^{\nu+1}} \sum_{\boldsymbol{b}_{\nu+1} \in \{+,-\}^{\nu+1}} \left(C_N^{\boldsymbol{b}_{\nu+1}}\right)^2 &= \frac{1}{2^\nu} \sum_{\boldsymbol{b}_\nu \in \{+,-\}^\nu} \frac{(C_N^{\boldsymbol{b}_\nu +})^2 + (C_N^{\boldsymbol{b}_\nu -})^2}{2} \\
&= \frac{1}{2^\nu} \sum_{\boldsymbol{b}_\nu \in \{+,-\}^\nu} \left(C_{N/2}^{\boldsymbol{b}_\nu}\right)^2 + \frac{1}{2^\nu} \sum_{\boldsymbol{b}_\nu \in \{+,-\}^\nu} \left(C_{N/2}^{\boldsymbol{b}_\nu} - \left(C_{N/2}^{\boldsymbol{b}_\nu}\right)^2\right)^2. && (4.238)
\end{aligned}$$

Using (4.238), we immediately find the following relationship:

$$\frac{1}{2^{\nu+1}} \sum_{\boldsymbol{b}_{\nu+1} \in \{+,-\}^{\nu+1}} \left(C_N^{\boldsymbol{b}_{\nu+1}}\right)^2 \geq \frac{1}{2^\nu} \sum_{\boldsymbol{b}_\nu \in \{+,-\}^\nu} \left(C_{N/2}^{\boldsymbol{b}_\nu}\right)^2. \tag{4.239}$$

Clearly,

$$1 \geq \frac{1}{2^{\nu+1}} \sum_{\boldsymbol{b}_{\nu+1} \in \{+,-\}^{\nu+1}} \left(C_N^{\boldsymbol{b}_{\nu+1}}\right)^2 \geq \frac{1}{2^\nu} \sum_{\boldsymbol{b}_\nu \in \{+,-\}^\nu} \left(C_{N/2}^{\boldsymbol{b}_\nu}\right)^2 \geq 0. \tag{4.240}$$

The first term on the right-hand side of (4.238) relates to those split virtual channels with channel capacity taken from $(\delta_\mathrm{U}, 1]$ and $[0, \delta_\mathrm{L})$, whereas the second term on the right-hand side of (4.238) relates to those split virtual channels with channel capacity taken from $[\delta_\mathrm{L}, \delta_\mathrm{U}]$. Therefore, we may bound the second term on the right-hand side of (4.238) by

$$\frac{1}{2^\nu} \sum_{\boldsymbol{b}_\nu \in \{+,-\}^\nu} \left(C_{N/2}^{\boldsymbol{b}_\nu} - \left(C_{N/2}^{\boldsymbol{b}_\nu}\right)^2\right)^2 \geq a_{\mathrm{I},\nu} \cdot f(\delta_\mathrm{L}, \delta_\mathrm{U}) \geq 0 \tag{4.241}$$

using some nonnegative function $f(\delta_\mathrm{L}, \delta_\mathrm{U})$. Using the binary entropy function $\mathrm{H}_2\{p\}$ given by (1.216) as well as its inverse denoted by $\mathrm{H}_2^{-1}\{x\}$, we may choose [113, p. 182]

$$f(\delta_\mathrm{L}, \delta_\mathrm{U}) = \min_{0 < \mathrm{H}_2^{-1}\{1-\delta_\mathrm{U}\} \leq p \leq \mathrm{H}_2^{-1}\{1-\delta_\mathrm{L}\} < 1/2} \left(\mathrm{H}_2\{2p(1-p)\} - \mathrm{H}_2\{p\}\right)^2, \tag{4.242}$$

which can be proven by using *Mrs. Gerber's Lemma (MGL)*; cf., e. g., [113, p. 182] and [114, p. 19].

Proof 4.2 (Proof of (4.242)). According to [113, Lemma 9.2, p. 182], if

$$C_1 = 1 - \mathrm{H}_2\{p\}, \quad p \in \left(0, \frac{1}{2}\right), \tag{4.243}$$

then with the equivocation $\mathrm{H}\{u_0 \oplus u_1 \mid \boldsymbol{Y}_0^1\}$, the channel capacity C_2^- becomes [113, Lemma 9.2, p. 182]

$$C_2^- = 1 - \mathrm{H}\{u_0 \oplus u_1 \mid \boldsymbol{Y}_0^1\} \leq 1 - \mathrm{H}_2\{2p(1-p)\}. \tag{4.244}$$

The inequality is exactly what Mrs. Gerber's lemma states, i. e., the equivocation $\mathrm{H}\{u_0 \oplus u_1 \mid \boldsymbol{Y}_0^1\}$ is lower bounded by the binary entropy function $\mathrm{H}_2\{2p(1-p)\}$ taken at $2p(1-p)$ [114, p. 19]. Using (4.232), we immediately obtain [113, Lemma 9.2, p. 182]

$$\begin{aligned} C_2^+ &= 2C_1 - C_2^- \geq 2 - 2\mathrm{H}_2\{p\} - 1 + \mathrm{H}_2\{2p(1-p)\} \\ &\geq 1 - 2\mathrm{H}_2\{p\} + \mathrm{H}_2\{2p(1-p)\}, \end{aligned} \tag{4.245}$$

which leads us to

$$C_2^+ - C_2^- \geq -2\mathrm{H}_2\{p\} + 2\mathrm{H}_2\{2p(1-p)\}, \tag{4.246}$$

i. e., [113, Lemma 9.2, p. 182]

$$\frac{C_2^+ - C_2^-}{2} \geq \mathrm{H}_2\{2p(1-p)\} - \mathrm{H}_2\{p\}. \tag{4.247}$$

This result leads us to the choice of (4.242). □

Now, (4.241) can be written in the following form:

$$\frac{\{\sum_{\boldsymbol{b}_{\nu+1} \in \{+,-\}^{\nu+1}} (C_N^{\boldsymbol{b}_{\nu+1}})^2\}/2^{\nu+1} - \{\sum_{\boldsymbol{b}_\nu \in \{+,-\}^\nu} (C_{N/2}^{\boldsymbol{b}_\nu})^2\}/2^\nu}{f(\delta_\mathrm{L}, \delta_\mathrm{U})} \geq a_{\mathrm{I},\nu}. \tag{4.248}$$

Since the left-hand side of (4.248) approaches zero as ν approaches infinity, $a_{\mathrm{I},\nu}$ also approaches zero.

In a way similar to (4.241), we find

$$\begin{aligned} C_1 = \frac{1}{2^\nu} \sum_{\boldsymbol{b}_\nu \in \{+,-\}^\nu} C_{N/2}^{\boldsymbol{b}_\nu} &\leq (\delta_\mathrm{L} - 0)a_{\mathrm{L},\nu} + (\delta_\mathrm{U} - \delta_\mathrm{L})a_{\mathrm{I},\nu} + (1 - \delta_\mathrm{L})a_{\mathrm{U},\nu} \\ &\leq \delta_\mathrm{L} + (\delta_\mathrm{U} - \delta_\mathrm{L})a_{\mathrm{I},\nu} + (1 - \delta_\mathrm{L})a_{\mathrm{U},\nu}, \end{aligned} \tag{4.249}$$

which immediately becomes

$$C_1 \leq \delta_\mathrm{L} + (1 - \delta_\mathrm{L}) \lim_{\nu \to \infty} \min a_{\mathrm{U},\nu}. \tag{4.250}$$

Equation (4.250) can only be fulfilled in all cases if $a_{\mathrm{U},\nu}$ converges to C_1.

Finally, we find

$$\begin{aligned} 1 - C_1 = 1 - \frac{1}{2^\nu} \sum_{\boldsymbol{b}_\nu \in \{+,-\}^\nu} C_{N/2}^{\boldsymbol{b}_\nu} &\leq (\delta_\mathrm{U} - 0)a_{\mathrm{L},\nu} + (\delta_\mathrm{U} - \delta_\mathrm{L})a_{\mathrm{I},\nu} + (1 - \delta_\mathrm{U})a_{\mathrm{U},\nu} \\ &\leq \delta_\mathrm{U} a_{\mathrm{L},\nu} + (1 - \delta_\mathrm{U})C_1, \end{aligned} \tag{4.251}$$

which becomes

$$1 - C_1 \leq \delta_\mathrm{U} \lim_{\nu \to \infty} \max a_{\mathrm{L},\nu} + (1 - \delta_\mathrm{U})C_1. \tag{4.252}$$

Equation (4.252) can only be fulfilled in all cases if $a_{\mathrm{L},\nu}$ approaches $(1 - C_1)$ as ν approaches infinity. □

Theorem 4.5 shows that the number of split virtual channels that can be used for information transfer approaches NC_1 as N grows toward infinity. Therefore, the dimension k_∞ of the polar code, which is obtained in the case of $N \to \infty$, is given by

$$k_\infty = \lim_{N\to\infty} NC_1 \quad (\to \infty), \quad 0 < C_1 \leq 1. \tag{4.253}$$

Since N is the length of the polar code, the achievable transmission rate, i. e., the code rate, R_∞ of the polar code, which is obtained in the case of $N \to \infty$, is equal to

$$R_\infty = \lim_{N\to\infty} \frac{NC_1}{N} = C_1 < \infty, \quad 0 < C_1 \leq 1. \tag{4.254}$$

Definition 4.1 (Capacity-achieving Binary Channel Code). An (n, k) binary channel code $\mathbb{V}$ with the dimension k and the length n, and thus with the code rate R equal to k/n, having a known construction rule, is called *"capacity-achieving,"* if the maximum achievable transmission rate, i. e., the code rate, R, for which an error-free transmission is possible, is equal to the channel capacity C_1 of the binary input discrete memoryless channel (DMC), over which the transmission of codewords taken from $\mathbb{V}$ is carried out.

A somewhat different definition of the expression *"capacity-achieving"* can, e. g., be found in [115, Definition 10, p. 4303]. However, this definition of [115, Definition 10, p. 4303] does not even mention the expression *"channel capacity,"* which makes the definition somewhat curious.

Polar codes obviously fulfill Definition 4.1 in the case of $N \to \infty$, which is the reason why the term *"capacity-achieving"* is used in the context of polar codes. Note that achieving the channel capacity requires the length N of the polar codes to grow beyond any limits, which leads to NC_1 split virtual channels that assume the channel capacity 1, whereas the remaining $N(1 - C_1)$ split virtual channels have the channel capacity 0.

Since the dimension k_∞ of any polar code, which is obtained in the case of $N \to \infty$, is equal to NC_1, the transmission of codewords, taken from the said polar code, over a considered binary input discrete memoryless channel (DMC) suffers from the loss equal to $N(1-C_1)$. Therefore, the equivocation is given by $(1-C_1)$ and, consequently, the source entropy is equal to 1.

In contrast to other channel codes discussed in this textbook, the construction rule of the said polar codes provides us with the identity of those split virtual channels, which have the channel capacity 0. Hence, we know exactly which bits will be lost during the transmission over the considered binary input discrete memoryless channel (DMC) *prior* to transmitting.

Note that N has to diverge to provide the capacity-achieving performance. Let us furthermore bear in mind that the binary input of the split virtual channel W_N^i, $i \in \{0, 1 \cdots N-1\}$, $N = 2^\nu$, $\nu \in \mathbb{N}^*$ is the distinct message bit $u_i \in \mathbb{F}_2$, $i \in \{0, 1 \cdots N-1\}$, $N = 2^\nu$, $\nu \in \mathbb{N}^*$; cf., e. g., Section 4.4 and (4.89). What has just been discussed above means that there are only $k_\infty \leq N$, $N = 2^\nu$, $\nu \in \mathbb{N}^*$ of the said split virtual channels with channel capacity 1, and hence there are only $k_\infty \leq N$, $N = 2^\nu$, $\nu \in \mathbb{N}^*$ message bits that can be transmitted without any transmission errors. The remaining $(N - k_\infty)$ message bits will be lost on the transmission channel. Therefore, we may consider these remaining $(N - k_\infty)$ message bits as dummy bits and "freeze" them by, e. g., setting there values to 0.

Using N message bits instead of only $k_\infty \leq N$ message bits in polar coding seems to be a considerable difference to other channel codes discussed in this textbook. Of course, we may view the "frozen" message bits as part of the computation of the redundancy required to overcome the equivocation, which tends to close the "conceptual gap" between polar codes and other channel codes.

Remark 4.3 (Finding the Philosopher's Stone). All this sounds like it was Erdal Arikan but not Harry Potter to find the Philosopher's Stone. In a way, it looks like the *"Midas Touch"* (Midnight Star, 1986) applied to ordinary channel codes, in our case the Reed–Muller (RM) codes, which then turns into gold, and in our case, the polar codes.

However, the celebrated channel polarization of Theorem 4.5 is regrettably only available for N becoming infinitely large. Oh my, oh my, the truth in this remark might be the seed of somebody's mutation into a *"Desperado"* (Eagles, 1973).

As long as N is finite as it is required in any communications system, Theorem 4.5 does not provide any information about the actual circumstances. These circumstances must be calculated (cf., e. g., Figure 4.13 and Figure 4.14) or determined by simulations. Looking at, e. g., Figure 4.13 and Figure 4.14 not all split virtual channels polarize and, moreover, the channel capacity values 1 and 0 seem to be extremely rare. However, the many split virtual channels with the channel capacity lower than 1 do not provide any error-free transmission. These facts govern the published findings are discussed in the following Section 4.5.

The "exuberant celebration" of polar codes somewhat reminds the author a little of similar events around the turn of the millennium in connection with spatial multiplexing. Although there were some scientists at the time who foresaw that the promises to increase the channel capacity by integer multiples were euphemistic to say the least, they were wise enough to remain silent and not object, thus figuratively saving themselves from crucifixion. Again figuratively speaking, the author was one of those few "well poisoners." Today, the author is glad to say that eventually the architects of the 4G/LTE standard found a sensible way to deal with spatial multiplexing.

With regard to 5G, however, the author again prefers to remain silent in most cases. However, the reader may hopefully deign to accept the opinion of an "aging oddball" who has doubts about the choice of the channel coding schemes for 5G. To put it in a nutshell, when thinking about, e. g., the "amazing findings" of spatial multiplexing and polar coding, the author cannot help to say that there is no way to fool nature. One might end like Icarus did. So, folks, science and research is like a *"Long Train Running"* (The Doobie Brothers, 1973). You must *"Carry On Wayward Son"* (Kansas, 1976).

4.5 Performance of Polar Codes With Finite Length

The probably most well-known decoding scheme for polar codes termed *successive cancellation decoding* was illustrated in [16, pp. 3054, 3065–3067], e. g., see also [116, pp. 4–8]. Meanwhile, a multitude of decoding schemes for polar codes have been proposed; cf., e. g., [116].

Remark 4.4 (Successive Cancellation Decoding of Polar Codes). Let the decoded message vectors for the odd and the even indexes of the message symbols be

$$\hat{\boldsymbol{u}}_{o,0}^{N-1} = (\hat{u}_1, \hat{u}_3 \cdots \hat{u}_{N-1}), \quad \hat{\boldsymbol{u}}_{e,0}^{N-1} = (\hat{u}_0, \hat{u}_2 \cdots \hat{u}_{N-2}). \tag{4.255}$$

Letting $i \in \{1, 2 \cdots N/2\}$, we now define the following *likelihood ratios (LRs):*

$$\mathcal{L}_{N/2}^{i-1}\left\{\boldsymbol{Y}_0^{N/2-1}, \hat{\boldsymbol{u}}_{e,0}^{2i-3} \oplus \hat{\boldsymbol{u}}_{o,0}^{2i-3}\right\} = \frac{\mathrm{P}_{N/2}^{i-1}\{\{\boldsymbol{Y}_0^{N/2-1}\}\{\hat{\boldsymbol{u}}_{e,0}^{2i-3} \oplus \hat{\boldsymbol{u}}_{o,0}^{2i-3}\} \mid \{0\}\}}{\mathrm{P}_{N/2}^{i-1}\{\{\boldsymbol{Y}_0^{N/2-1}\}\{\hat{\boldsymbol{u}}_{e,0}^{2i-3} \oplus \hat{\boldsymbol{u}}_{o,0}^{2i-3}\} \mid \{1\}\}}, \tag{4.256}$$

$$\mathcal{L}_{N/2}^{i-1}\left\{\boldsymbol{Y}_{N/2}^{N-1}, \hat{\boldsymbol{u}}_{o,0}^{2i-3}\right\} = \frac{\mathrm{P}_{N/2}^{i-1}\{\{\boldsymbol{Y}_{N/2}^{N-1}\}\{\hat{\boldsymbol{u}}_{o,0}^{2i-3}\} \mid \{0\}\}}{\mathrm{P}_{N/2}^{i-1}\{\{\boldsymbol{Y}_{N/2}^{N-1}\}\{\hat{\boldsymbol{u}}_{o,0}^{2i-3}\} \mid \{1\}\}}, \tag{4.257}$$

as well as the corresponding *log-likelihood ratios (LLRs)*

$$L_{N/2}^{i-1}\left\{\boldsymbol{Y}_0^{N/2-1}, \hat{\boldsymbol{u}}_{e,0}^{2i-3} \oplus \hat{\boldsymbol{u}}_{o,0}^{2i-3}\right\} = \ln\left\{\frac{\mathrm{P}_{N/2}^{i-1}\{\{\boldsymbol{Y}_0^{N/2-1}\}\{\hat{\boldsymbol{u}}_{e,0}^{2i-3} \oplus \hat{\boldsymbol{u}}_{o,0}^{2i-3}\} \mid \{0\}\}}{\mathrm{P}_{N/2}^{i-1}\{\{\boldsymbol{Y}_0^{N/2-1}\}\{\hat{\boldsymbol{u}}_{e,0}^{2i-3} \oplus \hat{\boldsymbol{u}}_{o,0}^{2i-3}\} \mid \{1\}\}}\right\}, \tag{4.258}$$

$$L_{N/2}^{i-1}\left\{\boldsymbol{Y}_{N/2}^{N-1}, \hat{\boldsymbol{u}}_{o,0}^{2i-3}\right\} = \ln\left\{\frac{\mathrm{P}_{N/2}^{i-1}\{\{\boldsymbol{Y}_{N/2}^{N-1}\}\{\hat{\boldsymbol{u}}_{o,0}^{2i-3}\} \mid \{0\}\}}{\mathrm{P}_{N/2}^{i-1}\{\{\boldsymbol{Y}_{N/2}^{N-1}\}\{\hat{\boldsymbol{u}}_{o,0}^{2i-3}\} \mid \{1\}\}}\right\}. \tag{4.259}$$

Taking $i \in \{1, 2 \cdots N/2\}$, we thus obtain the recursion formulas (cf., e. g., [16, equations (75) and (76), p. 3065f.], [117, equation (9), p. 1668])

$$\begin{aligned} L_N^{2i-2}\left\{\boldsymbol{Y}_0^{N-1}, \hat{\boldsymbol{u}}_0^{2i-3}\right\} = 2\,\mathrm{arctanh}\Bigg\{&\tanh\left\{\frac{L_{N/2}^{i-1}\{\boldsymbol{Y}_0^{N/2-1}, \hat{\boldsymbol{u}}_{e,0}^{2i-3} \oplus \hat{\boldsymbol{u}}_{o,0}^{2i-3}\}}{2}\right\} \\ &\cdot \tanh\left\{\frac{L_{N/2}^{i-1}\{\boldsymbol{Y}_{N/2}^{N-1}, \hat{\boldsymbol{u}}_{o,0}^{2i-3}\}}{2}\right\}\Bigg\}, \end{aligned} \tag{4.260}$$

$$\begin{aligned} L_N^{2i-1}\left\{\boldsymbol{Y}_0^{N-1}, \hat{\boldsymbol{u}}_0^{2i-2}\right\} &= (1 - 2\hat{u}_{2i-2}) L_{N/2}^{i-1}\left\{\boldsymbol{Y}_0^{N/2-1}, \hat{\boldsymbol{u}}_{e,0}^{2i-3} \oplus \hat{\boldsymbol{u}}_{o,0}^{2i-3}\right\} + L_{N/2}^{i-1}\left\{\boldsymbol{Y}_{N/2}^{N-1}, \hat{\boldsymbol{u}}_{o,0}^{2i-3}\right\} \\ &= (-1)^{\hat{u}_{2i-2}} L_{N/2}^{i-1}\left\{\boldsymbol{Y}_0^{N/2-1}, \hat{\boldsymbol{u}}_{e,0}^{2i-3} \oplus \hat{\boldsymbol{u}}_{o,0}^{2i-3}\right\} + L_{N/2}^{i-1}\left\{\boldsymbol{Y}_{N/2}^{N-1}, \hat{\boldsymbol{u}}_{o,0}^{2i-3}\right\}. \end{aligned} \tag{4.261}$$

Therefore, the successive cancelation decoder can be established in a recursive manner. The successive cancellation decoder then makes the following decisions [117, equation (1), p. 1665]:

$$\begin{aligned} \hat{u}_{i-1} &= 0 \text{ if } L_N^{i-1}\left\{\boldsymbol{Y}_0^{N-1}, \hat{\boldsymbol{u}}_0^{i-2}\right\} \geq 0, \\ \hat{u}_{i-1} &= 1 \text{ if } L_N^{i-1}\left\{\boldsymbol{Y}_0^{N-1}, \hat{\boldsymbol{u}}_0^{i-2}\right\} < 0, \end{aligned} \quad i \in \{1, 2 \cdots N\}. \tag{4.262}$$

Example 4.9. Let us look at N equal to 2. Hence, the decoded message vectors for the odd and the even indexes of the message symbols are

$$\hat{\boldsymbol{u}}_{o,0}^{1} = (\hat{u}_1) = \hat{u}_1, \quad \hat{\boldsymbol{u}}_{e,0}^{1} = (\hat{u}_0) = \hat{u}_0. \tag{4.263}$$

Furthermore, we have $i \in \{1\}$. We thus obtain

$$L_2^0\left\{\boldsymbol{Y}_0^1\right\} = 2\,\mathrm{arctanh}\left\{\tanh\left\{\frac{L_1^0\{Y_0\}}{2}\right\} \cdot \tanh\left\{\frac{L_1^0\{Y_1\}}{2}\right\}\right\}, \tag{4.264}$$

$$L_2^1\left\{\boldsymbol{Y}_0^1, \hat{u}_0\right\} = (1 - 2\hat{u}_0) L_1^0\{Y_0\} + L_1^0\{Y_1\}. \tag{4.265}$$

$L_1^0\{Y_0\}$ is the log-likelihood ratio (LLR) obtained at the output of the transmission channel associated with the first receive sample Y_0 and $L_1^0\{Y_1\}$ is the log-likelihood ratio (LLR) obtained at the output of the transmission channel associated with the second receive sample Y_1.

Taking Example 4.5 into consideration and letting u_0 be a "frozen" message bit equal to 0, the successive cancellation decoder decodes u_1, yielding

$$L_2^1\left\{\mathbf{Y}_0^1, 0\right\} = L_1^0\{Y_0\} + L_1^0\{Y_1\} \tag{4.266}$$

and

$$\begin{aligned} \hat{u}_1 &= 0 \text{ if } L_2^1\left\{\mathbf{Y}_0^1, 0\right\} \geq 0, \\ \hat{u}_1 &= 1 \text{ if } L_2^1\left\{\mathbf{Y}_0^1, 0\right\} < 0. \end{aligned} \tag{4.267}$$

Example 4.10. Let us look at N equal to 4. Hence, the decoded message vectors for the odd and the even indexes of the message symbols are

$$\hat{\boldsymbol{u}}_{o,0}^3 = (\hat{u}_1, \hat{u}_3), \quad \hat{\boldsymbol{u}}_{e,0}^3 = (\hat{u}_0, \hat{u}_2). \tag{4.268}$$

Furthermore, we have $i \in \{1, 2\}$. We thus obtain

$$L_4^0\left\{\mathbf{Y}_0^3\right\} = 2\operatorname{arctanh}\left\{\tanh\left\{\frac{L_2^0\{\mathbf{Y}_0^1\}}{2}\right\} \cdot \tanh\left\{\frac{L_2^0\{\mathbf{Y}_2^3\}}{2}\right\}\right\}, \tag{4.269}$$

$$L_4^1\left\{\mathbf{Y}_0^3, \hat{u}_0\right\} = (1 - 2\hat{u}_0)L_2^0\left\{\mathbf{Y}_0^1\right\} + L_2^0\left\{\mathbf{Y}_2^3\right\}, \tag{4.270}$$

$$L_4^2\left\{\mathbf{Y}_0^3, \hat{\boldsymbol{u}}_0^1\right\} = 2\operatorname{arctanh}\left\{\tanh\left\{\frac{L_2^1\{\mathbf{Y}_0^1, \hat{u}_0 \oplus \hat{u}_1\}}{2}\right\} \cdot \tanh\left\{\frac{L_2^1\{\mathbf{Y}_2^3, \hat{u}_1\}}{2}\right\}\right\}, \tag{4.271}$$

$$L_4^3\left\{\mathbf{Y}_0^3, \hat{\boldsymbol{u}}_0^2\right\} = (1 - 2\hat{u}_2)L_2^1\left\{\mathbf{Y}_0^1, \hat{u}_0 \oplus \hat{u}_1\right\} + L_2^1\left\{\mathbf{Y}_2^3, \hat{u}_1\right\}. \tag{4.272}$$

Taking Example 4.6 into consideration and letting u_0 and u_1 be a "frozen" message bits equal to 0, the successive cancellation decoder decodes u_2 and u_3, yielding

$$L_4^2\left\{\mathbf{Y}_0^3, \mathbf{0}_2\right\} = 2\operatorname{arctanh}\left\{\tanh\left\{\frac{L_2^1\{\mathbf{Y}_0^1, 0\}}{2}\right\} \cdot \tanh\left\{\frac{L_2^1\{\mathbf{Y}_2^3, 0\}}{2}\right\}\right\}, \tag{4.273}$$

$$L_4^3\left\{\mathbf{Y}_0^3, (0, 0, \hat{u}_2)\right\} = (1 - 2\hat{u}_2)L_2^1\left\{\mathbf{Y}_0^1, 0\right\} + L_2^1\left\{\mathbf{Y}_2^3, 0\right\} \tag{4.274}$$

and

$$\begin{aligned} \hat{u}_2 &= 0 \text{ if } L_4^2\left\{\mathbf{Y}_0^3, \mathbf{0}_2\right\} \geq 0, \\ \hat{u}_2 &= 1 \text{ if } L_4^2\left\{\mathbf{Y}_0^3, \mathbf{0}_2\right\} < 0, \\ \hat{u}_3 &= 0 \text{ if } L_4^3\left\{\mathbf{Y}_0^3, (0, 0, \hat{u}_2)\right\} \geq 0, \\ \hat{u}_3 &= 1 \text{ if } L_4^3\left\{\mathbf{Y}_0^3, (0, 0, \hat{u}_2)\right\} < 0. \end{aligned} \tag{4.275}$$

Although polar codes have been celebrated, several authors showed that polar codes are outperformed by comparable

- low density parity check (LDPC) codes [116, Figure 3.6, p. 52], [118, Figures 9–14], [119, Figures 11–16, pp. 1954–1957], as well as

- turbo codes [118, Figures 9–14], [119, Figures 11–16, pp. 1954-1957], [120, Figures 6–8, p. 5].

It seems that still weakly random-like codes such as low density parity check (LDPC) codes and turbo codes perform better than anything else, and thus Shannon has been right all the time. It is therefore more than surprising that polar codes were included in the 5G standard.

Bibliography

[1] Jung, P.: Advanced mobile communications — inner physical layer transceiver. Berlin: W. de Gruyter, 2024.

[2] Hawking, S. W.: Eine kurze Geschichte der Zeit — Die Suche nach der Urkraft des Universums. Thirty-second edition, Reinbek: Rowohlt, 2022.

[3] Heisenberg, W.: Der Teil und das Ganze. Fifteenth edition, München: Piper, 2022.

[4] MacWilliams, F. J.; Sloane, N. J. A.: The theory of error-correcting code. Amsterdam: North-Holland, 1977.

[5] Friedrichs, B.: Kanalcodierung. Berlin: Springer, 1996.

[6] Shannon, C. E.: A mathematical theory of communication. The Bell System Technical Journal, vol. 27 (1948), pp. 379–423 and pp. 623–656.

[7] Gallager, R. G.: Information theory and reliable communication. New York: Wiley, 1968.

[8] Rafelski, J.: Spezielle Relativitästheorie heute — schlüssig erklärt mit Beispielen, Aufgaben und Diskussionen. Berlin: Springer, 2019.

[9] Goscinny, R.; Tabary, J.: Sternstunden des Großwesirs Isnogud. Vol. 5 of the Series *"Die Abenteuer des Kalifen Harun Al Pussah"*, Stuttgart: Delta, 1976.

[10] Einstein, A.: Über die spezielle und die allgemeine Relativitätstheorie. Twenty-fourth edition, Heidelberg: Springer, 2001/2013.

[11] Bronstein, I. N.; Semendjajew, K. A.; Musiol, G.; Mühlig, H.: Taschenbuch der Mathematik. Sixth edition, Frankfurt a.M.: Harri Deutsch, 2005.

[12] 3GPP TS 36.211 V17.2.0 (2022-06): Physical channels and modulation. June 2022.

[13] IEEE Std 100-1992: The new IEEE standard dictionary of electrical and electronics terms. Fifth edition, New York: IEEE, 1993.

[14] Anderson, J. B.; Mohan, S.: Source and channel coding — an algorithmic approach. Boston: Kluwer, 1991.

[15] Lin, S.; Costello, D. J.: Error control coding. Second edition, Upper Saddle River: Pearson Prentice Hall, 2004.

[16] Arikan, E.: Channel polarization: a method for constructing capacity-achieving codes for symmetric binary-input memoryless channels. IEEE Transactions on Information Theory, vol. 55 (2009) no. 7, pp. 3051–3073.

[17] McEliece, R. J.: Finite fields for computer scientists and engineers. Boston: Kluwer, 1987.

[18] Lidl, R.; Niederreiter, H.: Finite fields. Second edition, Cambridge: Cambridge University Press, 1997.

[19] Bronstein, I. N.; Semendjajew, K. A.; Grosche, G.; Ziegler, V.; Ziegler, D.; Zeidler, E.: Teubner-Taschenbuch der Mathematik — Teil I. Second edition, Stuttgart: Teubner, 2003.

[20] Tietze, U.; Schenk, C.; Gamm, E.: Halbleiter-Schaltungstechnik. Sixteenth edition, Berlin: Springer Vieweg, 2019.

[21] Kantorowitsch, L. W.; Akilow, G. P.: Funktionalanalysis in normierten Räumen. Ninth edition, Frankfurt a.M.: Harri Deutsch, 1978.

[22] Burg, K.; Haf, H.; Wille, F.: Höhere Mathematik für Ingenieure — Band II Lineare Algebra. Third edition, Stuttgart: Teubner, 1992.

[23] Bossert, M.: Kanalcodierung. Second edition, Stuttgart: Teubner, 1998.

[24] Bossert, M.; Breitbach, M.: Digitale Netze Stuttgart: Teubner, 1999.

[25] Blahut, R. E.: Algebraic codes for data transmission. Cambridge: Cambridge University Press, 2003.

[26] Biglieri, E.: Coding for wireless channels. New York: Springer, 2005.

[27] Tomlinson, M.; Tjhai, C. J.; Ambroze, M. A.; Ahmed, M.; Jibril, M.: Error-correction coding and decoding — bounds, codes, decoders, analysis and applications. Cham: Springer Nature, 2017.

[28] Ball, S.: A course in algebraic error-correcting codes. Cham: Birkhäuser (Springer Nature), 2020.

[29] Blahut, R. E.: Theory and practice of error control codes. Reading: Addison-Wesley, 1983.

[30] Hausdorff, F.: Mengenlehre. Second edition, Berlin: W. de Gruyter, 1927.

https://doi.org/10.1515/9783111239651-005

[31] Elias, P.: Coding for two noisy channels. In: Colin Cherry (ed.). Information Theory, pp. 61–74, San Diego: Academic Press, 1956.
[32] Proakis, J. G.: Digital communications. Fourth edition, Boston: McGraw-Hill, 2001.
[33] Gazi, O.: Polar codes — a non-trivial approach to channel coding. Singapore: Springer, 2019.
[34] Viterbi, A. J.; Omura, J. K.: Principles of digital communication and coding. Tokyo: McGraw-Hill, 1979.
[35] Battail, G.: On random-like codes. In: Chouinard, J. Y., Fortier, P., Gulliver, T. A. (eds.): Information Theory and Applications II, Canadian Workshop on Information Theory (CWIT) 1995, Lecture Notes in Computer Science, vol. 1133, pp. 76–94, Berlin: Springer, 1996.
[36] Papula, L.: Mathematische Formelsammlung. Ninth edition, Wiesbaden: Vieweg, 2006.
[37] Hougardy, S.; Vygen, J.: Algorithmische Mathematik. Berlin: Springer, 2016.
[38] Iverson, K. E.: A programming language. New York: Wiley, 1962.
[39] Pless, V. S.; Huffmann, W. C. (eds.): Handbook of coding theory — volume I. Amsterdam: Elsevier, 2010.
[40] Lee, J. M.: Introduction to topological manifolds. Second edition, New York: Springer, 2011.
[41] Davis, F. J.: Circulant matrices. New York: Wiley, 1979.
[42] Fine, B.; Gaglione, A.; Moldenhauer, A.; Rosenberger, G.; Spellman, D.: Algebra and number theory — a selection of highlights. Berlin: W. de Gruyter, 2017.
[43] Bartholomé, A.; Rung, J.; Kern, H.: Zahlentheorie für Einsteiger. Seventh edition, Wiesbaden: Vieweg + Teubner, 2010.
[44] Ziegenbalg, J.: Elementare Zahlentheorie. Second edition, Wiesbaden: Springer, 2015.
[45] Dunham, W.: Euler — the master of us all. Volume 22, New York: The Mathematical Association of America, 1999.
[46] Nahin, P. J.: An imaginary tale — the story of $\sqrt{-1}$. Princeton: Princeton University Press, 1998.
[47] Knauer, U.; Knauer, K.: Diskrete und algebraische Strukturen — kurz gefasst. Second edition, Berlin: Springer, 2015.
[48] Karpfinger, C.; Meyberg, K.: Algebra — Gruppen — Ringe — Körper. Heidelberg: Spektrum, 2009.
[49] Jost, J.: Algebraische Strukturen. Wiesbaden: Springer, 2019.
[50] Dickson, L. E.: History of the theory of numbers — volume 1. Washington: Carnegie Institute of Washington, 1919.
[51] Grassl, M.: Code tables — bounds on the parameters of various types of codes. http://www.codetables.de, ongoing updates.
[52] Hardy, Y.; Steeb, W.-H.: Matrix calculus, Kronecker product and tensor product. Third edition, Singapore: World Scientific, 2019.
[53] Sattolo, S.: An algorithm to generate a random cyclic permutation. Information Processing Letters, vol. 22 (1986) no. 6, pp. 315–317.
[54] Gauß, E.: Walsh-Funktionen für Ingenieure und Naturwissenschaftler. Stuttgart: Teubner, 1994.
[55] Graham, A.: Kronecker products — matrix calculus with applications. Mineola: Dover, 2018.
[56] 3GPP TS 25.212 V17.0.0 (2022-03): Multiplexing and channel coding (FDD). March 2022.
[57] 3GPP TS 25.212 V1.0.0 (1999-04): Multiplexing and channel coding (FDD). April 1999.
[58] Muller, D. E.: Application of boolean algebra to switching circuit design and to error detection. Transactions of the IRE Professional Group on Electronic Computers, vol. EC-3 (1954) no. 3, pp. 6–12.
[59] Reed, I. S.: A class of multiple-error-correcting codes and the decoding scheme. Transactions of the IRE Professional Group on Information Theory, vol. 4 (1954) no. 4, pp. 38–49.
[60] Rademacher, H.: Einige Sätze über Reihen von allgemeinen Orthogonalfunktionen. Mathematische Annalen, vol. 87 (1922) no. 1/2, pp. 112–138.
[61] Lüke, H. D.: Korrelationssignale. Berlin: Springer, 1992.
[62] Gold, R.: Maximal recursive sequences with 3-valued cross-correlation functions. IEEE Transactions on Information Theory, vol. 14 (1968) no. 1, pp. 154–156.
[63] Gold, R.: Optimal binary sequences for spread spectrum multiplexing. IEEE Transactions on Information Theory, vol. 13 (1967) no. 4, pp. 619–621.

[64] Ziemer, R. E.; Peterson, R. L.: Digital communications and spread spectrum systems. Second edition, New York: Macmillan, 1985.

[65] 3GPP TS 25.212 V15.0.0 (2017-09): Multiplexing and channel coding (FDD). September 2017.

[66] Glover, I. A.; Grant, P. M.: Digital communications. Hemel Hempstead: Prentice Hall, 1998.

[67] Jung, P.: Analyse und Entwurf digitaler Mobilfunksysteme. Teubner: Stuttgart, 1997.

[68] Jung, P.; Blanz, J.: Joint detection with coherent receiver antenna diversity in CDMA mobile radio systems. IEEE Transactions on Vehicular Technology, vol. 44 (1995) no. 1, pp. 76–88.

[69] Sauer-Greff, W.: Optimale und suboptimale Empfänger für verzerrte und gestörte Datensignale unter besonderer Berücksichtigung der aufwandsreduzierten Detektion mittel M-Algorithmus. Doctoral Dissertation, Fachbereich Elektrotechnik, Universität Kaiserslautern, D 386, 1989.

[70] Ma, H. H.; Wolf, J. K.: On tail biting convolutional codes. IEEE Transactions on Communications, vol. 34 (1986) no. 2, pp. 104–111.

[71] 3GPP TS 36.212 V17.1.0 (2022-04): Multiplexing and channel coding. April 2022.

[72] Berrou, C.; Glavieux, A.; Thitimajshima, P.: Near Shannon limit error-correcting coding and decoding: turbo-codes (1). Proceedings of the IEEE International Conference on Communications (ICC'93), Geneva, Switzerland, pp. 1064–1070, 1993.

[73] Nielsen, M. A.; Chuang, I. L.: Quantum computation and quantum information. Tenth anniversary edition, Cambridge: Cambridge University Press, 2010.

[74] Homeister, M.: Quantum Computing verstehen. Fifth edition, Wiesbaden: Springer-Vieweg, 2018.

[75] Biercuk, M. J.; Stace, T. M.: Quantum error correction at the threshold: if technologists don't get beyond it, quantum computers will never be big. IEEE Spectrum, vol. 59 (July 2022) no. 7, pp. 28–46.

[76] Jung, P.: Einführung in die Quantenkommunikation. Düren: Shaker, 2022.

[77] Cariolaro, G.: Quantum communications. Cham: Springer, 2015.

[78] Planck, M.: Eine neue Strahlungshypothese. Verhandlungen der Deutschen Physikalischen Gesellschaft, vol. 13 (3 February 1911), pp. 138–148.

[79] Grau, G. K.; Kleen, W. J.: Comments on zero-point energy, quantum noise and spontaneous emission noise. Solid-State Electronics, vol. 25 (1982) no. 8, pp. 749–751.

[80] Jung, P.; Plechinger, J.; Doetsch, M.; Berens F. M.: Advances on the application of turbo-codes to data services in third generation mobile networks. Proceedings of the International Symposium on Turbo Codes & Related Topics, Brest, France, pp. 135–142, 1997.

[81] Jung, P.; Naßhan, M.; Blanz, J.: Application of turbo-codes to a CDMA mobile radio system using joint detection and antenna diversity. Proceedings of IEEE Vehicular Technology Conference (VTC'94), Stockholm, Sweden, pp. 54–59, 1994.

[82] Holma, H.; Toskala, A. (eds.): WCDMA for UMTS — radio access for third generation mobile communications. Chichester: Wiley, 2002.

[83] Chapman, T.; Larsson, E.; von Wrycza, P.; Dahlman, E.; Parkvall, S.; Sköld, J.: HSPA evolution — The fundamentals for mobile broadband. London: Academic Press, 2015.

[84] Cox, C.: An introduction to LTE — Lte, LTE-Advanced, SAE, VoLTE and 4G Mobile Communications. Second edition, Chichester: Wiley, 2014.

[85] Holma, H.; Toskala, A.; Nakamura, T.: 5G technology — 3GPP new radio. Chichester: Wiley, 2020.

[86] Dahlman, E.; Parkvall, S.; Sköld, J.: 5G NR — The next generation wireless access technology. London: Academic Press, 2018.

[87] Vucetic, B.; Yuan, J.: Turbo codes — principles and applications. Boston: Kluwer, 2000.

[88] Jung, P.: Novel low complexity decoder for turbo-codes. Electronics Letters, vol. 31 (1995) no. 2, pp. 86–87.

[89] Jung, P., Naßhan, M.: Results on turbo-codes for speech transmission in a joint detection CDMA mobile radio system with coherent receiver antenna diversity. IEEE Transactions on Vehicular Technology, vol. 46 (1997) no. 4, pp. 862–870.

[90] Robertson, P.: Illuminating the structure of code and decoder of parallel concatenated recursive systematic (turbo) codes. Proceedings of the 1994 IEEE GLOBECOM, San Francisco, California, USA, pp. 1298–1303, 1994.

[91] Jung, P.: Comparison of turbo-code decoders applied to short frame transmission systems. IEEE Journal on Selected Areas in Communications, vol. 14 (1996) no. 3, pp. 530–537.
[92] Bahl, L. R.; Cocke, J.; Jelinek, F.; Raviv, J.: Optimal decoding of linear codes for minimizing symbol error rate. IEEE Transactions on Information Theory, vol. 20 (March 1974) no. 2, pp. 284–287.
[93] ten Brink, S.: Convergence behavior of iteratively decoded parallel concatenated codes. IEEE Transactions on Communications, vol. 49 (2001) no. 10, pp. 1727–1737.
[94] Gallager, R. G.: Low density parity check codes. Doctoral (D.Sc.) dissertation: Massachusetts Institute of Technology (MIT), Cambridge, USA, 1960.
[95] Gallager, R. G.: Low-density parity-check codes. IRE Transactions on Information Theory, vol. 8 (1962) no. 1, pp. 21–28.
[96] Gallager, R. G.: Low density parity check codes. Research Monograph: Massachusetts Institute of Technology (MIT), Cambridge, USA, 1963.
[97] MacKay, D. J. C.: Good error-correcting codes based on very sparse matrices. Proceedings of IEEE International Symposium on Information Theory (ISIT 1997), Ulm, Germany, p. 113, 1997.
[98] McEliece, R. J.; MacKay, D. J. C.; Cheng, J.-F.: Finite fields for computer scientists and engineers. IEEE Journal on Selected Areas in Communications, vol. 16 (1998) no. 2, pp. 140–152.
[99] Battail, G.: On Gallager's low-density parity-check codes. Proceedings of IEEE International Symposium on Information Theory (ISIT 2000), Sorrento, Italy, p. 202, 2000.
[100] Bertsekas, D.; Gallager, R. G.: Data networks. Second edition, Englewood Cliffs: Prentice-Hall International, 1992.
[101] Tanner, R. M.: A recursive approach to low complexity codes. IEEE Transactions on Information Theory, vol. 27 (1981) no. 5, pp. 533–547.
[102] Kolmogorov, A. N.: Foundations of the theory of probability. Second English edition, Mineola: Dover, 2018.
[103] Freeman, D. F.; Michelson, A. M.: Performance of a two-dimensional product code with soft-decision decoding. Proceedings of the IEEE Military Communications Conference (MILCOM'94), Fort Monmouth, USA, pp. 1032–1037, 1994.
[104] Schnabl, G.; Bossert, M.: Soft-decision decoding of Reed-Muller codes as generalized multiple concatenated codes. IEEE Transactions on Information Theory, vol. 41 (1995) no. 1, pp. 304–308.
[105] Battail, G.; Decouvelaere, M. C.; Godlewski, P.: Replication coding. IEEE Transactions on Information Theory, vol. IT-25 (1979) no. 3, pp. 332–345.
[106] IEEE Standard for Information Technology Std 802.11-2020 (Revision of IEEE Std 802.11-2016): Part 11 — Wireless LAN medium access control (MAC) and physical layer (PHY) specifications. 2020.
[107] Arikan, E.: A survey of Reed-Muller codes from polar coding perspective. Proceedings of the 2010 IEEE Information Theory Workshop (ITW 2010), Cairo, Egypt, 2010.
[108] Korada, S. B..: Polar codes for channel and source coding. Doctoral dissertation no. 4461, Ecole Polytechnique Fédérale de Lausanne (EPFL), Lausanne, Switzerland, 2009.
[109] Goel, C.; Sarkar, A.; Kumaravelu, V. B.; Gudla, V. V.: Polar codes for 5G new radio. Proceedings of the International Conference on Communication and Signal Processing, Melmaruvathur, India, pp. 1240–1244, 2020.
[110] Doob, J. L.: Stochastic processes. New York: Wiley, 1953.
[111] Mansuy, R.: The origins of the word "martingale". Electronic Journal for History of Probability and Statistics, vol. 5 (2009) no. 1, pp. 1–10.
[112] 3GPP TS 38.212 V17.5.0 (2023-03): Multiplexing and channel coding. March 2023.
[113] Alsan, M.: Re-proving channel polarization theorems: an extremality and robustness analysis. Doctoral dissertation 6403, EPFL, Switzerland, 2015.
[114] El Gamal, A.; Kim, Y.-H.: Network information theory. Cambridge: Cambridge University Press, 2011.
[115] Kudekar, S.; Kumar, S.; Modelli, M.; Pfister, H. D.; Sasoglu, E.; Urbanke, R. L.: Reed-Muller codes achieve capacity on erasure channels. IEEE Transactions on Information Theory, vol. 63 (2017) no. 7, pp. 4298–4316.

[116] Girard, P.; Thibeault, C.; Gross, W. J.: High-speed decoders for polar codes. Cham: Springer, 2017.

[117] Leroux, C.; Tal, I.; Vardy, A.; Gross, W. J.: Hardware architectures for successive cancellation decoding of polar codes. Proceedings of the 2011 IEEE International Conference on Acoustics, Speech and Signal Processing (ICASSP), Prague, Czech Republic, pp. 1665–1668, 2011.

[118] Tahir, B.; Schwarz, S.; Rupp, M.: BER comparison between convolutional, turbo, LDPC, and polar codes. Proceedings of the 2017 24th International Conference on Telecommunications (ICT), Limassol, Cyprus, 2017.

[119] Cuc, A.-M.; Morgos, F. L.; Grava, C.: Performance analysis of turbo codes, LDPC codes and polar codes over an AWGN channel in the presence of inter symbol interference. Sensors, vol. 23 (2023) no. 4, pp. 1942–1961.

[120] Rao, K. D.: Performance analysis of enhanced turbo and polar codes with list decoding for URLLC in 5G systems. Proceedings of the Fifth International Conference for Convergence in Technology (I2CT), Pune, India, pp. 1–6, 2019.

Index

https://doi.org/10.1515/9783111239651-006

www.ingramcontent.com/pod-product-compliance
Lightning Source LLC
Chambersburg PA
CBHW080649220326
41598CB00033B/5149